The UK Pesticide Guide 2008

Editor: R. Whitehead BA, MSc

BCPC

www.cabi.org

CABI is one of the world's foremost publishers of databases, books and compendia in agriculture and applied life sciences. It has a worldwide reputation for producing high quality, value-added information, drawing on its links with the scientific community. CABI produces CAB Abstracts – the leading agricultural A&I database; together with the Crop Protection Compendium which provides solutions for identifying, preventing and solving crop health problems. CABI is a not-for-profit organization. For further information, please contact CABI, Nosworthy Way, Wallingford, Oxon OX10 8DE, UK.

Telephone: (01491) 832111
Fax: (01491) 833508
e-mail: enquiries@cabi.org
Web: www.cabi.org

BCPC (British Crop Production Council) is a self-supporting limited company with charity status, which was formed in 1968 to promote the knowledge and understanding of the science and practice of crop protection. The corporate members include government departments and research councils; advisory services; associations concerned with the farming industry; agrochemical manufacturers; agricultural engineering, agricultural contracting and distribution services; universities; scientific societies; organisations concerned with the environment; and some experienced independent members. Further details available from BCPC, 7 Omni Business Centre, Omega Park, Alton, Hampshire GU34 2QD, UK.

Telephone: (01420) 593200
Fax: (01420) 593209
e-mail: md@bcpc.org
Web: www.bcpc.org

ISBN 978-1-845934-16-3

Typeset and printed by Page Bros, Norwich, UK

Contents

Disclaimer

Every effort has been made to ensure that the information in this book is correct at the time of going to press, but the Editor and the publishers do not accept liability for any error or omission in the content, or for any loss, damage or other accident arising from the use of the products listed herein. Omission of a product does not necessarily mean that it is not approved or that it is unavailable.

It is essential to follow the instructions on the approved label when handling, storing or using any crop protection product. Approved 'off-label' extensions of use are undertaken entirely at the risk of the user.

Editor's Note

It is 21 years since the Control of Pesticides Regulations replaced the voluntary Agricultural Chemicals Approval Scheme, and official product approval before sale became mandatory in the UK. It is also 21 years since the publication of the first edition of this *Guide*. Over that period we have sought to provide a compact, yet comprehensive tool for farmers, advisors, students and all others engaged in the professional use of pesticides in agriculture, horticulture and amenity.

Our aim is to provide an independent practical *Guide* to the products available in the market, the purposes for which they may be used, and the controlling legislation that surrounds them. Although the number of products listed in the *Guide* has remained fairly constant over its 21 years, the identity and nature of the active ingredients they contain, and the structure of the industry that provides them, have changed significantly. Increasingly complex and demanding legislation has driven the need to reduce the environmental impact of pesticides, to deliver increased efficacy and operator safety, and to raise the professional competence of those who store, advise on, and use them. Access to up-to-date information has become increasingly important to meet the requirements of assurance schemes and cross-compliance in various government initiatives.

To meet this need, the *Guide* is fully revised every year. In this way we have ensured that it has remained topical and informative. Following cessation of publication of the printed form of the annual Ministry list of approved products (the 'Blue Book'), we added a section to this publication to show all registered products not requested for inclusion in the main body of the *Guide*. It thus became a unique printed reference for all approved products within its scope. More recently, we have included details of European pesticide legislation and the impact on users in UK of the Review Programme.

This 21st edition is no exception to our aim of continuous improvement. Over 250 product replacements or introductions have been made since the previous edition. Many of these reflect the industry-wide obligation to replace formulations containing nonylphenol or octylphenol ethoxylates; others include seven new active ingredients, listed for the first time. Product deletions include those containing active ingredients for which 'Essential Use' derogations were granted, and which finally expired in December 2007. The increasing problem of managing pesticide resistance has been addressed by the inclusion, for the first time, of mode of action codes, as published by the various resistance action committees. Progress with replacement of the Long Term Arrangements for Extension of Use, and the corresponding increased number of Specific Off-Label Approvals (SOLAs), are fully covered.

Over the years we have refined and improved the presentation of information in the *Guide*, seeking always to keep its size manageable and portable. In response to customer demand, we produced an electronic version in CD-ROM format in 1999. Improved technology and widespread access to broadband enables us to replace the CD with an online version in 2008.

Products are included in the main body of the *Guide* only if requested by the supplier, with an assurance that they are available on the market. We remain very grateful to those who provide the information, enabling the *Guide* to remain a valuable reference for all. As always, criticisms and suggestions are welcome and, in particular, notification of any errors or omissions.

R. Whitehead
Editor
ukpg@bcpc.org

The information in this publication has been selected from official sources and from suppliers' product labels and product manuals of pesticides approved for use in the UK under the Control of Pesticides Regulations 1986 or the Plant Protection Products Regulations 1995. The content is based on information received by the Editor up to October 2007.

For developments since publication of this edition, and new products notified to the Editor during the year, see: **www.ukpesticideguide.co.uk**

Changes Since 2007 Edition

Pesticides and adjuvants added or deleted since the 2007 edition are listed below. Every effort has been made to ensure the accuracy of this list, but late notifications to the Editor are not included.

Products Added

Products new to this edition are shown here. In addition, products that were listed in the previous edition whose PSD/HSE registration number and/or supplier have changed are included.

Product	Reg. No.	Supplier
Adjust	13148	Mandops
Agriguard Fluroxypyr	13335	AgriGuard
Agrovista Paclo	13424	Agrovista
Agrovista Penco	13425	Agrovista
Alpha Briotril	13580	Makhteshim
Alpha Briotril Plus 19/19	13579	Makhteshim
Alpha Bromolin 225 EC	13431	Makhteshim
Alpha Bromotril P	13326	Makhteshim
Alpha Chlorpyrifos 48 EC	13532	Makhteshim
Alpha Propachlor 50 SC	13585	Makhteshim
Alpha Trifluralin 48 EC	13433	Makhteshim
Amega Duo	13358	Nufarm UK
Amicron	13336	AgriGuard
Asulox	13175	United Phosphorus
Austral Plus	13314	Syngenta
Avail XT	13194	AgriGuard
Awac	13296	Globachem
Banner Maxx	13167	Syngenta
Barleyquat B	13187	Mandops
Barton WG	13284	Dow
Baythroid	13221	Makhteshim
Betosip Combi	13254	Sipcam
Bettaquat B	13192	Mandops
Brack-N	13138	AgriGuard
Brogue	13328	AgriGuard
Buckler	13080	Nufarm UK
Buggy XTG	12951	Sipcam
Bumper Excell	11015	Makhteshim
Cadou Star	13242	Bayer CropScience
Canopy	13181	BASF
Capalo	13170	BASF
Caramba 90	12655	BASF
Carousel	13225	Fine
Caste Off	12832	Nufarm UK
Catalyst	13266	AgriGuard
CDA Vanquish Biactive	12586	Bayer Environ.
Ceando	13271	BASF
Cello	13178	Bayer CropScience
Certis Deal 5	13488	Certis
Certis Dichlobenil Granules	13705	Certis
Certis Metaldehyde 5	13490	Certis
Certis Red 5	13491	Certis
Certis Trigger 5	13493	Certis
Cherokee	13251	Syngenta
Chess WG	13310	Syngenta Bioline
Chimera SX	13171	DuPont
Claymore	13441	BASF
Clayton Belfry	12886	Clayton
Clayton Buzz	12509	Clayton
Clayton Chloron 500 FL	10791	Clayton
Clayton Coldstream	12724	Clayton
Clayton Delilah	12610	Clayton
Clayton Diquat	12739	Clayton
Clayton Edge	12911	Clayton
Clayton Fenican 550	11160	Clayton
Clayton Gantry	09482	Clayton
Clayton Groove	13246	Clayton
Clayton Impress	12912	Clayton
Clayton Krypton	09398	Clayton
Clayton Lambada	13231	Clayton
Clayton Lanark	12942	Clayton
Clayton Lenacil 80 W	09488	Clayton
Clayton Lenaflo	11031	Clayton
Clayton Maximus	12543	Clayton
Clayton Mazarac	11161	Clayton
Clayton Mepiquat	12360	Clayton

Product	Reg. No.	Supplier
Clayton Metazachlor	09688	Clayton
Clayton Metazachlor 50 SC	11719	Clayton
Clayton Mitrex	11921	Clayton
Clayton Oust	11588	Clayton
Clayton Oxen FL	12861	Clayton
Clayton Oxen G	12770	Clayton
Clayton Paradigm	12001	Clayton
Clayton Peak	13547	Clayton
Clayton Pirimicarb 50	12910	Clayton
Clayton Pontoon	13219	Clayton
Clayton Propel 80 WG	11216	Clayton
Clayton Rally	13415	Clayton
Clayton Scabius	13339	Clayton
Clayton Sparta	13457	Clayton
Clayton Spigot	11560	Clayton
Clayton Squat	12271	Clayton
Clayton Standup	11760	Clayton
Clayton Tebucon	08707	Clayton
Clayton Truss	13129	Clayton
Clayton Tunik	10545	Clayton
Clayton Vert	12570	Clayton
Crescent	13330	AgriGuard
Crossfire 480	12516	Dow
Cyd-X	13535	Certis
Cymbal	13312	Belchim
Cypermethrin Lacquer	H3164	Killgerm
Daconil Turf	13602	Syngenta
Dancer Flow	11405	Sipcam
Decimate	13419	Certis
Deft	13263	Sipcam
Depitox	13258	Nufarm UK
Deuce	13240	BASF
Diamant	13253	BASF
Diflufenican GL 500	12531	Globachem
Digital	13561	Interfarm
Diquanet	13413	Hermoo
Dithane 945	12545	Dow
Dithane NT Dry Flowable	12565	Interfarm
Dodifun 400 SC	11657	Hermoo
Draza Forte	13306	Bayer CropScience
Dynamec	13331	Syngenta Bioline
Engage	13690	Interfarm
Equity	12465	Dow
Excelsior	13378	Dow
Exemptor	13122	Scotts
Fenlander 2	13124	Makhteshim
Fernpath Torate XT	13391	AgriGuard
Ferromex Mosskiller Concentrate	13180	Omex
Flagon 400 EC	13470	Makhteshim
Floramite 240 SC	13686	Certis
Flurostar 200	12903	Globachem
Forester	13164	Fargro
Foxtrot EW	13243	Headland
Galaxy	13127	Dow
Galileo	13252	DuPont
Gistar	11397	Hermoo
Glyfos Dakar	13054	Headland
Glyfos Dakar Pro	13147	Headland Amenity
Gly-Plus A	A0377	Greenaway
Govern	13223	Dow
Grail	13385	Nufarm UK
Greenaway Gly-490	12718	Greenaway
Greencrop Doonbeg 2	13025	Greencrop
Greencrop Fenit	A0602	Greencrop
Greencrop Helvick	13052	Greencrop
Grisou	13166	Sipcam
Groslow	13694	Sherriff Amenity
Gro-Stop 100	13742	Certis
Gro-Stop Fog	13741	Certis
Gro-Stop HN	13738	Certis
Guillotine	13562	Interfarm
Gusto	13448	Makhteshim
Hatchet Extra	13213	AgriGuard
Headland Link	11091	Headland
Headland Staff 500	13196	Headland
Helimax S	13109	De Sangosse
Herboxone	13144	Headland
Heritage	13536	Syngenta
Hotspur	13449	AgriGuard
HY-D Super	13198	Agrichem
Ibex	12168	BASF
IMA 100 SL	13559	Q-Chem
Imidasect 5 GR	13581	Fargro
Inka SX	13601	DuPont
Insyst	13414	Certis
Jester	13500	Syngenta
Juventus	13340	BASF
K2	13190	Mandops
Kantor	A0623	Interagro
Karamate Dry Flo Newtec	12691	Landseer
Keel Over	12966	De Sangosse
Klever	13295	Globachem
Kubist Flo	12987	Nufarm UK
Landgold Bentazone SL	13418	Teliton

Product	Reg. No.	Supplier
Landgold Chlorothalonil 50	13473	Teliton
Landgold Clodinafop	13329	Teliton
Landgold Fluroxypyr	13356	Teliton
Landgold Lambda-Z	13533	Teliton
Landgold Metsulfuron	13421	Teliton
Landgold Nicosulfuron	13464	Teliton
Landgold Propyzamide 400 SC	13614	Teliton
Landgold Tepraloxydim	13379	Teliton
Laser	12930	BASF
Lentipur 700	13526	Nufarm UK
Leopard 5 EC	13720	Makhteshim
Lexone 2	12051	DuPont
Lexone 2	13804	DuPont
Liquid Derris	13370	Law Fertilisers
Luxan Vermigon Bait	H7925	Luxan
Luxan Vermigon Blocks	H7926	Luxan
Luxan Vermigon Pellets	H7941	Luxan
Luxan Vermigon Sewer Blocks	H7927	Luxan
Lynx S	13276	De Sangosse
Maccani	13227	BASF
Manipulator	13189	Mandops
Masai	13082	BASF
Mascot Contact	13718	Rigby Taylor
Merit Turf	12415	Bayer Environ.
Midas	13285	DuPont
Minx PX	13337	Nufarm UK
Mitre	12901	Makhteshim
Mixture B NF	A0570	Amega
MM 70 Flo	13318	Nufarm UK
Monkey	12906	Makhteshim
Naprop	13402	Globachem
Nautile DG	13416	Nufarm UK
Neosorexa Gold	H8301	Sorex
Nimrod	13046	Makhteshim
Nirvana	13220	BASF
Nortron Flo	12986	Bayer CropScience
Nugget	13121	AgriGuard
Nu-Shot	13550	Nufarm UK
Oram	13410	Makhteshim
Oryx	13527	BASF
Oskar	13344	Headland
Othello	12695	Bayer CropScience
Pan Proteb	13427	Pan Agriculture
Pan Trilby	13558	Pan Agriculture
Pantheon	13695	Pan Agriculture
PDM 330 EC	13406	BASF
Pelican	12569	Headland
Pennant	13350	Headland
Permasect C	13158	Nufarm UK
Pesta M	13275	De Sangosse
Pharoah	13631	Dow
PicoMax	13456	BASF
Picona	13428	BASF
PicoPro	13454	BASF
PicoStomp	13455	BASF
Pike	12746	Nufarm UK
Pilot Ultra	12929	Nissan
Polymet 5	13480	Unicrop
Prego	13582	Makhteshim
Primer	13226	AgriGuard
Proplant	13359	Fargro
Pyblast	H7485	Agropharm
Pyrethrum Spray	H4636	Killgerm
Quad	13396	Q-Chem
Revus	13484	Syngenta
Rezacur	13717	Sherriff Amenity
Roquat	13474	Q-Chem
Roundup Energy	12945	Monsanto
Rovral WP	13429	BASF
Runner	12940	Bayer CropScience
Sabine	13439	Makhteshim
Sanspor	13313	Syngenta
Scotts Dichlo G Macro	12011	Scotts
Scotts Dichlo G Micro	11857	Scotts
Sipcam Solo-D	12617	Sipcam
Spotlight Plus	12436	Belchim
Spruzit	13438	Certis
SpudGun	13447	AgriGuard
Standon Frontrunner	13308	Standon
Standon Fullstop	13571	Standon
Standon Googly	13281	Standon
Standon Homerun	13177	Standon
Standon Hounddog	13201	Standon
Standon Pirimicarb 50	13290	Standon
Standon Roxx L	13309	Standon
Standon Tonga	13394	Standon
Standon Zero Tolerance	13546	Standon
Steward	13149	DuPont
Stomp 400 SC	13405	BASF
Storite Super	13135	Syngenta
Strength AC	H7469	Certis
Sumimax	13548	Interfarm

Product	Reg. No.	Supplier
Surpass	13590	Headland Amenity
Switch	13185	Syngenta
Synphony	13021	AgriGuard
Tandus	13432	Nufarm UK
TDS Major	13462	De Sangosse
TDS Metarex Amba	13461	De Sangosse
Teliton Ocean	13163	Teliton
Teliton Ocean Power	A0615	Teliton
Templar	10254	Makhteshim
Terbine	11407	Nufarm UK
Thiraflo	13338	Chemtura
Toppel 100 EC	13704	United Phosphorus
Triflurex 48 EC	13445	Makhteshim
Tripod Plus	13168	Makhteshim
Upright	10404	BASF
Vermigon Pasta	H8191	Certis
Vixen	13621	BASF
Volley	13591	Makhteshim
Warbler	13389	Nufarm UK
Wetcol 3 Copper Fungicide	13369	Law Fertilisers
Zimbril	13172	DuPont
Zone 48	13165	AgriGuard

Products Deleted

The appearance of a product name in the following list may not necessarily mean that it is no longer available. In cases of doubt, refer to Section 3 or the supplier.

Product	Reg. No.	Supplier
Acanto	10978	DuPont
Acanto	13043	DuPont
Acanto Prima	11864	DuPont
Aconite 50	10642	DAPT
Actellic D	10509	Syngenta
Adjust	05589	Mandops
Ag-Chem Metribuzin	12007	Ag-Chem
Ag-Chem Prefect	11985	Ag-Chem
Agral	A0421	Syngenta
Agri-50E	-	Fargro
Agriguard Chlorpyrifos	10626	AgriGuard
Agriguard Ethofumesate Flo	09478	AgriGuard
Agritox Dry	10554	Nufarm UK
Aligran	11761	Nufarm UK
Ally SX	12059	DuPont
Alpha Atrazine 50 SC	04877	Makhteshim
Alpha Briotril	04876	Makhteshim
Alpha Briotril Plus 19/19	04740	Makhteshim
Alpha Bromolin 225 EC	08255	Makhteshim
Alpha Bromotril P	07099	Makhteshim
Alpha Chlorpyrifos 48 EC	04821	Makhteshim
Alpha DFF 500 SC	12187	Makhteshim
Alpha Prometryne 50 WP	04871	Makhteshim
Alpha Simazine 50 SC	04801	Makhteshim
Alpha Trifluralin 48 EC	07406	Makhteshim
Amaize	12298	AgriGuard
Amcide	04246	B H & B
Aminotriazole Technical	11137	Nufarm UK
Ardent	09968	Makhteshim
Arelon 500	11639	Nufarm UK
Arizona	11722	Makhteshim
ASAP	11485	Me2
Atlas Brown	07703	Nufarm UK
Atlas Somon	07727	Nufarm UK
Attribut	11221	Bayer CropScience
Auricle	12622	AgriGuard
Autumn Kite	11549	Nufarm UK
Banco	11191	Makhteshim
Barclay Actol	A0126	Barclay
Barclay Barbarian	11362	Barclay
Barclay Busker	11345	Barclay
Barclay Gallup 360	11369	Barclay
Barclay Gallup Amenity	06753	Barclay
Barclay Gallup Biograde 360	11370	Barclay
Barclay Gallup Biograde Amenity	11363	Barclay
Barclay Gallup Hi-Aktiv	11371	Barclay
Barclay Piza 400FL	11495	Barclay
Barleyquat B	07051	Mandops

Product	Reg. No.	Supplier
Batallion	08305	Makhteshim
Baythroid	11663	Makhteshim
Betapal Concentrate	00234	Vitax
Bettaquat B	07050	Mandops
Bison 83 WG	11580	Nufarm UK
Booty	10659	Me2
Buggy SG	08573	Sipcam
Bullet	11204	Makhteshim
Bullion	12050	DuPont
Cabaret	11915	Bayer CropScience
Caddy 240 EC	11256	Bayer CropScience
Carbetamex	11150	Makhteshim
Catalyst	12592	AgriGuard
Certis 2-Aminobutane	11182	Certis
Cherokee	12768	Syngenta
Chess WG	10651	Syngenta
Clarosan	H7692	Scotts
Claymore	11779	BASF
CleanGuard	12055	Amenity Land
Cleanrun 3	11410	Scotts
Clinic	12678	Nufarm UK
Clortosip 500	09320	Sipcam
Contrast	06150	DuPont
Copal 500	10633	DAPT
CoPilot	08042	Interfarm
Corniche	08947	Bayer CropScience
Credo	12368	DuPont
Croptex Bronze	11329	Certis
Croptex Chrome	11180	Certis
Croptex Steel	02418	Certis
Crossfire 480	09929	Bayer Environ.
Delphi	11937	Sipcam
Depitox	11149	Nufarm UK
Dextrone X	00687	Nomix Enviro
Diamant	11557	BASF
Dicotox Extra	09930	Bayer Environ.
Dioweed 50	08050	United Phosphorus
Dithane NT Dry Flowable	09898	Interfarm
Dorac 80 WP	13133	Lanxess
Dormone	09932	Bayer Environ.
Dosaflo	09351	Syngenta
DPX-N7874	12783	DuPont
Dynamec	12539	Syngenta
Elvaron Multi	11422	Bayer CropScience
Embargo G	10367	SumiAgro Amenity
Emrald Berg	12086	Me2
Emrald Eyetort	12034	Me2
Emrald Flumet	12000	Me2
Emrald Wotsit	12060	Me2
Encore	10375	BASF
Enforcer	09288	Scotts
Ethos	11223	Bayer CropScience
Ethosat 500	11501	Makhteshim
Euro	11511	Me2
Evict	11673	Bayer CropScience
Evict	11735	Syngenta
Fenlander 2	12282	Makhteshim
Fernpath Graminite	11646	AgriGuard
Fernpath Pronto	11988	AgriGuard
Fielder	12850	DuPont
Finish SX	12259	DuPont
Flagon 400 EC	08875	Makhteshim
Flex	08885	Syngenta
Flute	08953	Sipcam
FMC Oil	A0435	Belchim
Fortrol	11174	Makhteshim
Fusonil Turf	09695	Rigby Taylor
Galaxy	12228	AgriGuard
Galben M	07220	Sipcam
Garlon 2	10672	Syngenta
Genie 25	10285	DuPont
Gesagard	08410	Syngenta
GEX 353	11474	DuPont
Glyphosate 360	11060	SumiAgro Amenity
Gly-Plus A	A0377	Greenaway
Golmet	12858	Interfarm
Gramoxone 100	10526	Syngenta
Greenaway Gly-490	11064	Greenaway
Greencrop Amaize	11973	Greencrop
Greencrop Gypsy	09432	Greencrop
Greengard	11715	SumiAgro Amenity
Gremlin	09468	Sipcam
Grey Squirrel Liquid Concentrate	06455	Killgerm
Habitat	10498	SumiAgro Amenity
Habitat	12717	Barclay
Harvesan	12238	DuPont
Headland Staff 500	12087	Headland
Herboxone	10032	Headland
Heritage	11383	Syngenta
Holster	10593	SumiAgro Amenity

Product	Reg. No.	Supplier
HY-D Super	11618	Agrichem
Impuls	11780	BASF
Ingot	09997	Bayer CropScience
Intercept 5GR	08126	Scotts
Inter-Metribuzin WG	12235	Lanxess
Inzacur	12014	Nufarm UK
Javelin Gold	12536	Bayer CropScience
Jester	08681	Syngenta
Joules	11047	Nufarm UK
Juggler	09391	Sipcam
K2	10370	Mandops
Karamate Dry Flo Newtec	09759	Landseer
Karathane Liquid	09262	Landseer
Kerb Granules	08917	SumiAgro
Kerb Granules	12946	Barclay
Kerb Pro Granules	09600	SumiAgro Amenity
Kerb Pro Granules	12947	Barclay
KN 540	12009	Monsanto
Kubist Flo	11167	Nufarm UK
Landgold Glyphosate 360	05929	Landgold
Landgold Glyphosate 360	12120	Teliton
Landgold Minoan Two	A0523	Teliton
Landgold Tepraloxydim	11808	Landgold
Landgold Tepraloxydim	12129	Teliton
Lanxess Metribuzin 70 WG	12487	Lanxess
Laser	05251	BASF
Lexone 2	04991	DuPont
Lexone 2	12051	DuPont
Liquid Derris	01213	Law Fertilisers
Lorate	12743	DuPont
Lupus	11291	Bayer CropScience
Luxan Deal	11101	Luxan
Luxan Dichlobenil Granules	09250	Luxan
Luxan Gro Stop 100	12052	Luxan
Luxan Gro-Stop Fog	09388	Luxan
Luxan Gro-Stop HN	07689	Luxan
Luxan Isoproturon 500 Flowable	12426	Luxan
Luxan Metaldehyde 5	12401	Luxan
Luxan Non-Ionic Wetter	A0139	Luxan
Luxan Red 5	12390	Luxan

Product	Reg. No.	Supplier
Luxan Trigger 5	12389	Luxan
Lyric	08252	DuPont
Magic Tandem	11814	Bayer CropScience
Maizon	12833	AgriGuard
Manipulator	05871	Mandops
Maraud	09274	Scotts
Marnoch Clodim	11460	Me2
Marnoch Phorm	11087	Me2
Marshal Soil Insecticide suSCon CR Sachets	11754	Fargro
Masai	10223	BASF
Mazide 25	02067	Vitax
Me2 Aducksbackside	10065	Me2
Me2 Azoxystrobin	09654	Me2
Me2 Booty 2	12010	Me2
Me2 Cindy	10634	Me2
Me2 Cymoxeb	09486	Me2
Me2 Exodus	09786	Me2
Me2 Hooray	12546	Me2
Me2 I-Tora	12209	Me2
Me2 KME	09594	Me2
Me2 Lambda2	12883	Me2
Me2 MCPA	12844	Qdos
Me2 Pendimethalin	12420	Me2
Me2 Penny	10369	Me2
Me2 Puddy	12164	Me2
Me2 Rockall	12229	Me2
Me2 Silthiofam	12513	Me2
Me2 Sylvester	12021	Me2
Me2 Tebuconazole	09751	Me2
Meothrin	10400	BASF
Mixture B	A0161	Amega
MM 70	11582	Nufarm UK
MM 70 Flo	11197	Nufarm UK
Monceren Flowable	11425	Bayer CropScience
Mossicide	09606	SumiAgro Amenity
Mossicide	12948	Barclay
MSS Diuron 500 FL	11493	Nufarm UK
MSS Diuron 80 WP	11503	Nufarm UK
New Estermone	06336	Vitax
Nimrod	10563	Makhteshim
Nomix Mosskiller	12080	Nomix Enviro
Nortron Flo	08154	Bayer CropScience
Nu-Shot	11148	Nufarm UK
Nuturf Carbendazim	11469	Nufarm UK
Opogard	08427	Syngenta

Product	Reg. No.	Supplier
Outrun	10581	SumiAgro Amenity
Outrun	12838	Barclay
Panacide M	05611	Coalite
Panacide M	12413	CMS Chemicals
Panacide TS	05612	Coalite
Panther	10745	Bayer CropScience
PDM 330 EC	11084	BASF
Pennine 50	10638	DAPT
Permasect C	11121	Nufarm UK
PicoMax	10739	BASF
Picona	10715	BASF
PicoPro	10740	BASF
PicoStomp	11987	BASF
Plantvax 75	01601	Fargro
Presite SX	12291	DuPont
Propose 50 W	11170	Interfarm
Pulsar	04002	BASF
Punch C	06801	DuPont
Qdos Brefi	A0539	Me2
Qdos Chikara	A0540	Me2
Qdos Glyfo	12574	Me2
Qdos Rockall 2	12962	Me2
Reflex T	08884	Syngenta
Ringer	10692	SumiAgro Amenity
Ringer	12949	Barclay
Rogor L40	07611	Interfarm
Romany	11000	Greencrop
Ronilan FL	02960	BASF
Roundup Amenity	10319	Monsanto
Roundup Amenity	12672	Monsanto
Roundup Gold	10975	Monsanto
Rovral WP	11694	BASF
Runner	11470	Bayer CropScience
Safeguard	12020	AgriGuard
Sanction 25	10284	DuPont
Scorpio 500 SC	10927	DAPT
Shortcut	09254	Scotts
Sipcam Echo 75	08302	Sipcam
Sipcam Simazine Flowable	07622	Sipcam
Skater	11159	Makhteshim
Sniper 100	12479	AgriGuard
Snitch	12466	Nufarm UK
Solan 40	11897	Nufarm UK
Spasor	09945	Bayer Environ.

Product	Reg. No.	Supplier
Spasor Biactive	09940	Bayer Environ.
Speedway 2	10790	Scotts
Spitfire	12002	AgriGuard
Spotlight 24 EC	11617	Belchim
Spotlight Plus	12100	Belchim
Sprayfast	A0131	Mandops
Sprayfast	A0553	Mandops
Spraygard	A0394	Mandops
Stacato	05892	Sipcam
Standon Mimas WG	12075	Standon
Standon Tepraloxydim	11119	Standon
Standon Vinclozolin	07836	Standon
Stomp 400 SC	11777	BASF
Super Mosstox	09942	Bayer Environ.
Sypex	04650	BASF
Talat	11311	Certis
Target SC	11070	Unicrop
Temik 10G	10021	Bayer CropScience
Terpitz	09634	Me2
Terrazole 35 WP	10468	Scotts
Thiraflo	11526	Chemtura
Tigress Ultra	08946	Bayer CropScience
Toppel 10	08772	United Phosphorus
Torq	08370	Fargro
Totril	10026	Bayer CropScience
Touchdown	10538	Syngenta
Triflurex 48 EC	07947	Makhteshim
Trifolex-Tra	10396	BASF
Trik	11441	Nufarm UK
Trimaran	11400	Nufarm UK
Trump	10376	BASF
Trustee Elite	11556	Barclay
Trustee Elite	11556	Dalgety
Tuberon	11061	Unicrop
Turfclear	07506	Scotts
Twin	11172	Makhteshim
Twist	11230	Bayer CropScience
Uppercut	12591	Nufarm UK
Voodoo	09414	Sipcam
Waterloo	11196	Interfarm
Wetcol 3	02360	Law Fertilisers
Winner	12573	Headland Amenity
XL-All Insecticide	02369	Vitax

Introduction

Purpose

The primary aim of this book is to provide a practical handbook of pesticides, plant growth regulators and adjuvants that the farmer or grower can realistically and legally obtain in the UK, and to indicate the purposes for which they may be used. It is designed to help in the identification of products appropriate for a particular problem. In addition to uses recommended on product labels, details are provided of those uses that do not appear on labels but which have been granted specific off-label approval (SOLA). As well as identifying the products available, the book provides guidance on how to use them safely and effectively, but without giving details of doses, volumes, spray schedules or approved tank mixtures. Sections 5 and 6 provide essential background information on a wide range of pesticide-related issues including legislation, codes of practice, poisons and treatment of poisoning, products for use in special situations, and weed and crop growth stage keys.

While we have tried to cover all other important factors, this book does **not** provide a full statement of product recommendations. **Before using any pesticide product it is essential that the user should read the label carefully and comply strictly with the instructions it contains. This *Guide* is not a substitute for the product label**.

Scope

The *Guide* is confined to pesticides registered in the UK for use in arable agriculture, horticulture (including amenity use), forestry and areas in or near water. Within these fields of use, there are about 2500 products in UK with current approval, and the *Guide* is a reference for virtually all of them, given in two sections. Section 2 gives full details of products notified to the Editor as available on the market. Products are included in this section only if requested by the supplier, and supported by evidence of approval. Section 3 gives brief details of all other registered products with extant approval.

Section 2 lists some 265 individual active ingredients and a further 180 mixtures containing these active ingredients. For each active ingredient, or mixture of active ingredients, a list is shown of available approved products within the fields of use specified above, together with the name of the supplier from whom each may be obtained. All types of pesticides covered by the Control of Pesticides Regulations or the Plant Protection Products Regulations are included. This embraces acaricides, algicides, herbicides, fungicides, insecticides, lumbricides, molluscicides, nematicides and rodenticides, together with plant growth regulators used a straw shorteners, sprout inhibitors and for various horticultural purposes.

The total number of products included in Section 2 (i.e. available on the market for the year of this edition) is about 1300.

The *Guide* also gives information (in Section 4) on some 150 authorised adjuvants which, although not themselves active pesticides, may be added to pesticide products to improve their effectiveness in use.

The *Guide* does **not** include products approved solely for amateur, home garden, domestic, food storage, public health or animal health uses.

Sources of Information

The information for this edition has been drawn from these authoritative sources:

- approved labels and product manuals received from suppliers of pesticides up to October 2007
- websites of the Pesticides Safety Directorate (PSD, www.pesticides.gov.uk) and the Health and Safety Executive (HSE, www.hse.gov.uk).

Criteria for Inclusion

To be included in the *Guide*, a product must meet the following conditions:

- it must have extant approval under UK pesticides legislation
- information on the approved uses must have been provided by the supplier
- it must be expected to be on the UK market during the currency of this edition.

When a company changes its name, whether by merger, take-over or joint venture, it is obliged to re-register its products in the name of the new company, and new Ministry registration numbers are normally assigned. Where stocks of the previously registered products remain legally available, both old and new identities are included in the *Guide* and remain until approval for the former lapses, or stocks are notified as exhausted.

Products that have been withdrawn from the market and whose approval will finally lapse during 2008 are identified in each profile. After the indicated date, sale, storage or use of the product bearing that approval number becomes illegal. Where there is a direct replacement product, this is indicated.

The Voluntary Initiative

The Voluntary Initiative (VI) is a programme of measures, agreed by the crop protection industry and related organisations with Government, to minimise the environmental impacts of crop protection products. The programme has provided a framework of practices and principles to help Government achieve its objective of protection of water quality and enhancement of farmland biodiversity. Many of the environmental protection schemes launched under the Voluntary Initiative represent current best practice. The first 5-year phase of the programme formally concluded in March 2006, but it is continuing, with revised proposals and targets, initially for 2 years, until March 2008 with strong indications that it will continue beyond this date. Anyone involved with the products in this *Guide*, whether as a user or advisor, is likely to be affected directly or indirectly.

This edition of *The UK Pesticide Guide* can help in two specific ways.

- It is illegal to use, or even store, a product for which the approval has expired. While mere storage is unlikely to have any environmental impact, their incorrect or inappropriate disposal could undoubtedly do so. All the products listed in Sections 2 and 3 of this *Guide* have extant approval for all or part of 2008. A date is shown for any whose approval will expire before the end of the year. Therefore any products not listed here are likely to be obsolete, although a check should be made with the manufacturer or supplier before disposal.

- Product labels contain much information that is often too crowded and complicated. Another project in the VI, undertaken by the Crop Protection Association (CPA) together with the Pesticides Safety Directorate and the Health and Safety Executive, has been a review of product labels and the production of Best Practice Labelling Guidance. CPA member companies have adapted their labels to be compliant with these guidelines at the same time as making the mandatory changes required by the Dangerous Preparations Directive. The presentation of information in this *Guide* reflects this improved clarity as far as possible. In the online version we have gone further by using the new symbols for environmental hazard, product activity and LERAP category, as well as providing links to the Voluntary Initiative website (www.voluntaryinitiative.org.uk) to enable access to product Environmental Information Sheets (also produced as a VI project).

Notes on Contents and Layout

The book consists of six main sections:

1 Crop/Pest Guide
2 Pesticide Profiles
3 Products Also Registered
4 Adjuvants
5 Useful Information
6 Appendices

1 Crop/Pest Guide

This section enables the user to identify which active ingredients are approved for a particular crop/pest combination. The crops are grouped as shown in the Crop Index. For convenience, some crops and targets have been grouped into generic units (for example, 'cereals' are handled as a group, not individually). Therefore indications from the Crop/Pest Guide must always be followed up by reference to the specific entry in the pesticide profiles section because a product may not be approved on all crops in the group. Chemicals indicated as having uses in cereals, for example, may be approved for use only on winter wheat and barley, not for other cereals. Because of differences in the wording of product labels it may sometimes be necessary to refer to broader categories of organism in addition to specific organisms (e.g. annual grasses and annual weeds in addition to annual meadow grass).

2 Pesticide Profiles

Each active ingredient has a separate, numbered profile entry, as does each mixture of active ingredients. The entries are arranged in alphabetical order using the common names approved by the British Standards Institution, and used in *The Pesticide Manual* (BCPC, 2006). Where an active ingredient is available only in mixtures, this is stated. The ingredients of the mixtures are themselves ordered alphabetically, and the entries appear at the correct point for the first named ingredient.

Below the profile title, a brief phrase describes the main use of the active ingredient(s). This is followed, where appropriate, by the mode of action code(s) as published by the Fungicide, Herbicide and Insecticide Resistance Action Committees.

Within each profile entry, a table lists the products approved and available on the market, in the following style:

Product name	Main supplier	Active ingredient contents	Formulation type	Registration number
1 Broadsword	United Phosphorus	200:85:65 g/l	EC	09140
2 Greengard	SumiAgro Amenity	200:85:65 g/l	EC	11715
3 Nu-Shot	Nufarm UK	200:85:65 g/l	EC	11148

Many of the **product names** are registered trademarks, but no special indication of this is given. Individual products are indexed under their entry number in the Index of Proprietary Names at the back of the book. The **main supplier** indicates the marketing outlet through which the product may be purchased. Full addresses and telephone/fax numbers of suppliers are listed in Appendix 1. Some website and e-mail addresses are also given. For mixtures, the **active ingredient contents** are given in the same order as they appear in the profile heading. The **formulation types** are listed in full in the Key to Abbreviations and Acronyms (Appendix 5). The **registration number** normally refers to registration with PSD. In cases where products registered with the Health and Safety Executive are included, HSE numbers are quoted, e.g. H0462.

Below the product table, a **Uses** section lists all approved uses notified by suppliers to the Editor by October 2007, giving both the principal target organisms and the recommended crops or situations (identified in ***bold italics***). Where there is an important condition of use, or the approval is off-label, this is shown in parentheses, e.g. (*off-label*). Numbers in square brackets refer to the numbered products in the table above. Thus, in the example shown above, a use approved for Nu-Shot (product 3) but not for Broadsword or Greengard (products 1 and 2) appears as:

Annual dicotyledons in ***rotational grassland*** [3]

Any **Specific Off-Label Approvals (SOLAs)** for products in the profile are detailed below the list of approved uses. Each SOLA has a separate entry and shows the crops to which it applies, the Notice of Approval number, the expiry date (if due to fall during the current edition) and the product reference number in square brackets. SOLAs do not appear on the product label and are undertaken entirely at the risk of the user.

Below the SOLA paragraph, **Notes** are listed under the following headings. Unless otherwise stated, any reference to dose is made in terms of product rather than active ingredient. Where notes refer to particular products, rather than to the entry generally, this is indicated by numbers in square brackets as described above.

Approval information	Information of a general nature about the approval status of the active ingredient or products in the profile is given here. Notes on approval for aerial or ULV application are given. Inclusion of the active ingredient in Annex I under EC Directive 91/414 is shown, as well as any acceptance from the British Beer and Pub Association (BBPA) for its use on barley or hops for brewing or malting. Where a product approval will finally expire in 2008, the expiry date is shown here.
Efficacy	Factors important in making the most effective use of the treatment. Certain factors, such as the need to apply chemicals uniformly, the need to spray at appropriate volume rates, and the need to prevent settling out of active ingredient in spray tanks, are assumed to be important for all treatments and are not emphasised in individual profiles.
Restrictions	Notes are included in this section where products are subject to the Poisons Law, and where the label warns about organophosphorus and/or anticholinesterase compounds. Factors important in optimising performance and minimising the risk of crop damage, including statutory conditions relating to the maximum permitted number of applications, are listed. Any restrictions on crop varieties that may not be sprayed are mentioned here.
Environmental safety	Where the label specifies an environmental hazard classification, it is noted, together with the associated risk phrases. Any other special operator precautions, and any conditions concerning withholding livestock from treated areas, are specified. Where any of the products in the profile are subject to Category A, Category B or broadcast air-assisted buffer zone restrictions under the LERAP scheme, the relevant classification is shown. Other environmental hazards are also noted here, including potential dangers to livestock, game, wildlife, bees and fish. The need to avoid drift onto neighbouring crops and to wash out equipment thoroughly after use are important with all pesticide treatments, but may receive special mention here if of particular significance.
Crop-specific information	Instructions about timing of application or cultivations that are specific to a particular crop, rather than generally applicable, are listed here. The latest permitted use and harvest intervals (if any) are shown for individual crops.
Following crops guidance	Any specific label instructions about what may be grown after a treated crop, whether harvested normally or after failure, are shown here.

Hazard classification and safety precautions	The label hazard classification(s) and precautions are shown using a series of letter or number codes which are explained in Appendix 4. The codes are listed under a number of sub-headings as follows:

- Hazard
- Risk phrases
- Operator protection
- Environmental protection
- Consumer protection
- Storage and disposal
- Treated seed
- Vertebrate/rodent control products
- Medical advice

This section is given for information only and should not be used for the purpose of making a COSHH assessment without reference to the actual label of the product to be used.

3 Products also Registered

Products with extant approval for all or part of the year of the edition in which they appear are listed in this section if they have not been requested by their supplier or manufacturer for inclusion in Section 2. Details shown are the same (apart from formulation) as those in the product tables in Section 2 and, where relevant, an approval expiry date is shown. Not all the products listed here will be available in the market, but this list, together with the products in Section 2, comprises a comprehensive listing of all approved products in the UK for uses within the scope of the *Guide* for the year in question.

4 Adjuvants

Adjuvants are listed in a table ordered alphabetically by product name. For each product, details are shown of the main supplier, the authorisation number and the mode of action (e.g. wetter, sticker, etc.) as shown on the label. A brief statement of the uses of the adjuvant is given. Protective clothing requirements and label precautions are listed using the codes from Appendix 4.

5 Useful Information

This section summarises legislation covering approval, storage, sale and use of pesticides in the UK. There are brief articles on the broader issues of using crop protection chemicals, including resistance management. Lists are provided of products approved for use in or near water, in forestry, as seed treatments, and for aerial application. Chemicals subject to the Poisons Laws are listed, and there is a summary of first aid measures if pesticide poisoning should be suspected. Finally, this section provides guidance on environmental protection issues and covers the protection of bees and the use of pesticides in or near water.

6 Appendices

Appendix 1	Gives names, addresses, telephone and fax numbers of all companies listed as main suppliers in the pesticide profiles. E-mail and website addresses are also listed where available. Where a supplier is no longer trading under that name (usually following a merger or takeover) but products in that name are still listed in the *Guide* (because they are still available in the supply chain), a cross-reference indicates the new 'parent' company from which technical or commercial information can be obtained.
Appendix 2	Gives names, addresses, telephone and fax numbers of useful contacts, including the National Poisons Information Service. Website addresses are included where available.

Appendix 3	Gives details of the keys to crop and weed growth stages, including the publication reference for each. The numeric codes are used in the descriptive sections of the pesticide profiles (Section 2).
Appendix 4	Shows the full text for code letters and numbers used in the pesticide profiles (Section 2) to indicate personal protective equipment requirements and label precautions.
Appendix 5	Shows the full text for the formulation abbreviations used in the pesticide profiles (Section 2). Other abbreviations and acronyms used in the *Guide* are explained here.
Appendix 6	Provides full definitions of officially agreed descriptive phrases for crops or situations used in the pesticide profiles (Section 2) where misunderstandings can occur.
Appendix 7	Shows a list of useful reference publications which amplify the summarised information provided in this section.

SECTION 1
CROP/PEST GUIDE

Crop/Pest Guide Index

Important Note: The Crop/Pest Guide Index refers to pages on which the subject can be located.

Arable and vegetable crops

Flowers and ornamentals

Forestry

Fruit and hops

Crop/Pest Guide

Important note: For convenience, some crops and pests or targets have been brought together into generic groups in this guide, e.g. 'cereals', 'annual grasses'. It is essential to check the profile entry in Section 2 *and* the label to ensure that a product is approved for a specific crop/ pest combination, e.g. winter wheat/blackgrass.

Arable and vegetable crops

Agricultural herbage - Grass

Diseases	Crown rust	propiconazole, propiconazole (*for animal fodder*)
	Drechslera leaf spot	propiconazole, propiconazole (*for animal fodder*)
	Powdery mildew	propiconazole, propiconazole (*for animal fodder*)
	Rhynchosporium	propiconazole, propiconazole (*for animal fodder*)
Pests	Aphids	pirimicarb
	Birds/mammals	aluminium ammonium sulphate, aluminium phosphide, strychnine hydrochloride (commodity substance) (*areas of restricted public access*)
	Flies	chlorpyrifos, cypermethrin
	Leatherjackets	chlorpyrifos, methiocarb (*seed admixture*)
	Slugs/snails	methiocarb (*seed admixture*)
Plant growth regulation	Quality/yield control	sulphur
Weeds	Aquatic weeds	MCPA
	Broad-leaved weeds	2,4-D, 2,4-D + dicamba, 2,4-D + dicamba + triclopyr, 2,4-D + MCPA, 2,4-DB + linuron + MCPA, 2,4-DB + MCPA, amidosulfuron, aminopyralid + fluroxypyr, aminopyralid + triclopyr, asulam, bentazone + MCPA + MCPB, bromoxynil + ethofumesate + ioxynil, bromoxynil + ioxynil + mecoprop-P, citronella oil, clopyralid, clopyralid (*off-label - spot treatment*), clopyralid + fluroxypyr + triclopyr, clopyralid + triclopyr, dicamba + MCPA + mecoprop-P, dicamba + mecoprop-P, dichlorprop-P, fluroxypyr, fluroxypyr + triclopyr, glyphosate, glyphosate (*wiper application*), MCPA, MCPA + MCPB, MCPB, mecoprop-P, thifensulfuron-methyl
	Grass weeds	bromoxynil + ethofumesate + ioxynil, glyphosate, paraquat (*sward destruction/direct drilling*)
	Weeds, miscellaneous	diquat + paraquat, glufosinate-ammonium, glyphosate
	Woody weeds/scrub	asulam, clopyralid + triclopyr

Agricultural herbage - Herbage legumes

Weeds	Broad-leaved weeds	2,4-DB, carbetamide, chlorpropham, isoxaben (*off-label - grown for game cover*), MCPA + MCPB, MCPB (*seed crops*), propyzamide
	Crops as weeds	carbetamide, fluazifop-P-butyl (*off-label*)
	Grass weeds	carbetamide, chlorpropham, fluazifop-P-butyl (*off-label*), propyzamide, tri-allate

	Weeds, miscellaneous	clopyralid (*off-label - game cover*)

Agricultural herbage - Herbage seed

Crop control	Desiccation	diquat
Diseases	Crown rust	propiconazole
	Damping off	thiram (*seed treatment*)
	Disease control/foliar feed	prochloraz (*off-label*), prochloraz + propiconazole (*off-label*), prochloraz + tebuconazole (*off-label*)
	Drechslera leaf spot	propiconazole
	Foliar diseases	azoxystrobin (*off-label*), azoxystrobin + chlorothalonil (*off-label*), azoxystrobin + cyproconazole (*off-label*), azoxystrobin + fenpropimorph (*off-label*), carbendazim (*off-label*), chlorothalonil (*off-label*), chlorothalonil + flutriafol (*off-label*), chlorothalonil + mancozeb (*off-label*), cyproconazole + cyprodinil (*off-label*), difenoconazole (*off-label*), epoxiconazole (*off-label*), epoxiconazole + fenpropimorph (*off-label*), epoxiconazole + fenpropimorph + kresoxim-methyl (*off-label*), epoxiconazole + kresoxim-methyl (*off-label*), epoxiconazole + kresoxim-methyl + pyraclostrobin (*off-label*), epoxiconazole + pyraclostrobin (*off-label*), famoxadone + flusilazole (*off-label*), fenpropimorph (*off-label*), fenpropimorph + flusilazole (*off-label*), fenpropimorph + kresoxim-methyl (*off-label*), fenpropimorph + pyraclostrobin (*off-label*), mancozeb (*off-label*), metconazole (*off-label*), metrafenone (*off-label*), prochloraz (*off-label*), spiroxamine + tebuconazole (*off-label*), tebuconazole (*off-label*), tebuconazole + triadimenol (*off-label*)
	Powdery mildew	cyflufenamid (*off-label*), fenpropimorph + quinoxyfen (*off-label*), propiconazole, sulphur (*off-label*)
	Rhynchosporium	propiconazole
	Rust	azoxystrobin + fenpropimorph (*off-label*)
	Septoria	azoxystrobin + fenpropimorph (*off-label*)
Pests	Aphids	deltamethrin (*off-label*), dimethoate, lambda-cyhalothrin (*off-label*), lambda-cyhalothrin + pirimicarb (*off-label*), tau-fluvalinate (*off-label*)
Plant growth regulation	Growth control	trinexapac-ethyl
Weeds	Broad-leaved weeds	asulam (*off-label*), bifenox (*off-label*), bromoxynil + diflufenican + ioxynil (*off-label*), bromoxynil + ethofumesate + ioxynil, clodinafop-propargyl + trifluralin (*off-label*), clopyralid (*off-label*), dicamba + MCPA + mecoprop-P, dichlorprop-P + MCPA + mecoprop-P (*off-label*), diflufenican + flufenacet (*off-label*), diflufenican + flurtamone (*off-label*), diflufenican + isoproturon (*off-label*), ethofumesate, florasulam (*off-label*), florasulam + fluroxypyr (*off-label*), flufenacet + pendimethalin (*off-label*), linuron + trifluralin (*off-label*), MCPA, MCPA + MCPB, MCPB, mecoprop-P, pendimethalin (*off-label*), propyzamide
	Grass weeds	bromoxynil + ethofumesate + ioxynil, clodinafop-propargyl (*off-label*), clodinafop-propargyl + trifluralin (*off-label*), diflufenican + flufenacet (*off-label*), diflufenican + flurtamone (*off-label*), diflufenican +

isoproturon (*off-label*), ethofumesate, flufenacet + pendimethalin (*off-label*), linuron + trifluralin (*off-label*), pendimethalin (*off-label*), propyzamide

Brassicas - Brassica seed crops

Diseases	Alternaria	iprodione
	Botrytis	iprodione
Weeds	Broad-leaved weeds	carbetamide, propyzamide
	Crops as weeds	carbetamide
	Grass weeds	carbetamide, propyzamide

Brassicas - Brassicas, general

Diseases	Alternaria	iprodione (*seed treatment*)
	Downy mildew	chlorothalonil
	Ring spot	chlorothalonil
Pests	Aphids	pymetrozine (*off-label*)

Brassicas - Fodder brassicas

Diseases	Alternaria	azoxystrobin, iprodione, iprodione (*seed treatment*)
	Black rot	copper oxychloride (*off-label*)
	Downy mildew	fosetyl-aluminium (*off-label - seedlings*)
	Powdery mildew	azoxystrobin + difenoconazole
	Ring spot	azoxystrobin, difenoconazole (*off-label*), tebuconazole (*off-label*)
	Spear rot	copper oxychloride (*off-label*)
	White blister	azoxystrobin, azoxystrobin + difenoconazole, boscalid + pyraclostrobin (*off-label*), mancozeb + metalaxyl-M (*off-label*)
Pests	Aphids	carbosulfan, imidacloprid (*off-label - seed treatment*), nicotine, pirimicarb, pymetrozine (*off-label*)
	Beetles	alpha-cypermethrin, carbosulfan, deltamethrin
	Caterpillars	alpha-cypermethrin, Bacillus thuringiensis (*off-label*), cypermethrin, deltamethrin, nicotine
	Flies	carbosulfan, chlorpyrifos (*off-label*)
	Leaf miners	nicotine
	Pests, miscellaneous	dimethoate (*off-label*)
	Weevils	carbosulfan
Weeds	Broad-leaved weeds	chlorthal-dimethyl, chlorthal-dimethyl + propachlor, clomazone (*off-label*), clopyralid, metazachlor (*off-label*), napropamide, propachlor, trifluralin
	Crops as weeds	fluazifop-P-butyl (*stockfeed only*)
	Grass weeds	chlorthal-dimethyl + propachlor, fluazifop-P-butyl (*stockfeed only*), napropamide, propachlor, trifluralin

Brassicas - Leaf and flowerhead brassicas

Diseases	Alternaria	azoxystrobin, boscalid + pyraclostrobin, chlorothalonil + metalaxyl-M (*moderate control*), difenoconazole,

		iprodione, iprodione (*seed treatment*), tebuconazole, tebuconazole + trifloxystrobin
	Black rot	copper oxychloride (*off-label*)
	Botrytis	chlorothalonil, chlorothalonil (*off-label*)
	Damping off	etridiazole, etridiazole (*seeds and seedlings*), thiram (*seed treatment*), tolclofos-methyl
	Downy mildew	chlorothalonil, chlorothalonil (*off-label*), chlorothalonil + metalaxyl-M, fosetyl-aluminium (*off-label - seedlings*), propamocarb hydrochloride
	Foliar diseases	chlorothalonil (*off-label - seedling*)
	Light leaf spot	tebuconazole, tebuconazole + trifloxystrobin
	Phoma leaf spot	tebuconazole + trifloxystrobin
	Phytophthora	etridiazole, etridiazole (*transplants*), propamocarb hydrochloride
	Powdery mildew	azoxystrobin + difenoconazole, tebuconazole, tebuconazole + trifloxystrobin
	Pythium	propamocarb hydrochloride
	Ring spot	azoxystrobin, boscalid + pyraclostrobin, chlorothalonil, chlorothalonil (*off-label*), chlorothalonil + metalaxyl-M (*reduction*), difenoconazole, difenoconazole (*off-label*), tebuconazole, tebuconazole (*off-label*), tebuconazole + trifloxystrobin
	Spear rot	copper oxychloride (*off-label*)
	Stem canker	flusilazole (*off-label*)
	Storage rots	metalaxyl-M (*off-label*)
	White blister	azoxystrobin, azoxystrobin + difenoconazole, boscalid + pyraclostrobin, boscalid + pyraclostrobin (*off-label*), boscalid + pyraclostrobin (*qualified minor use*), chlorothalonil + metalaxyl-M, mancozeb + metalaxyl-M (*off-label*), tebuconazole + trifloxystrobin
	White rot	thiacloprid (*off-label*)
Pests	Aphids	acetamiprid (*off-label*), bifenthrin, carbosulfan, chlorpyrifos, cypermethrin, deltamethrin (*off-label*), dimethoate, fatty acids, imidacloprid (*off-label - seed treatment*), lambda-cyhalothrin + pirimicarb, nicotine, pirimicarb, pymetrozine (*off-label*), pyrethrins
	Beetles	alpha-cypermethrin, carbosulfan, deltamethrin, lambda-cyhalothrin + pirimicarb, pyrethrins
	Birds/mammals	aluminium ammonium sulphate
	Caterpillars	alpha-cypermethrin, Bacillus thuringiensis, Bacillus thuringiensis (*off-label*), bifenthrin, chlorpyrifos, cyfluthrin, cypermethrin, deltamethrin, diflubenzuron, indoxacarb, lambda-cyhalothrin, lambda-cyhalothrin + pirimicarb, nicotine, pyrethrins, spinosad, spinosad (*off-label*)
	Cutworms	chlorpyrifos
	Flies	carbosulfan, chlorpyrifos, chlorpyrifos (*off-label*), spinosad (*off-label*)
	Leaf miners	nicotine
	Leatherjackets	chlorpyrifos
	Mealybugs	fatty acids
	Pests, miscellaneous	deltamethrin (*off-label*), dimethoate (*off-label*)
	Scale insects	fatty acids
	Slugs/snails	methiocarb, thiodicarb

	Spider mites	fatty acids
	Weevils	carbosulfan
	Whiteflies	acetamiprid (*off-label*), bifenthrin, chlorpyrifos, cypermethrin, fatty acids, lambda-cyhalothrin, lambda-cyhalothrin + pirimicarb, pyrethrins
Weeds	Broad-leaved weeds	carbetamide, chlorthal-dimethyl, chlorthal-dimethyl (*off-label*), chlorthal-dimethyl + propachlor, chlorthal-dimethyl + propachlor (*off-label*), clomazone (*off-label*), clopyralid, metazachlor, metazachlor (*off-label*), napropamide, pendimethalin, propachlor, trifluralin
	Crops as weeds	carbetamide, cycloxydim, pendimethalin, tepraloxydim
	Grass weeds	carbetamide, chlorthal-dimethyl + propachlor, chlorthal-dimethyl + propachlor (*off-label*), cycloxydim, fluazifop-P-butyl (*off-label*), metazachlor, napropamide, pendimethalin, propachlor, tepraloxydim, trifluralin

Brassicas - Mustard

Crop control	Desiccation	diquat (*off-label*), glyphosate
Diseases	Alternaria	iprodione, iprodione (*seed treatment*)
	Botrytis	chlorothalonil (*off-label*), iprodione
	Downy mildew	chlorothalonil (*off-label*)
	Foliar diseases	carbendazim (*off-label*), difenoconazole (*off-label*), tebuconazole (*off-label*)
Pests	Aphids	lambda-cyhalothrin (*off-label*), lambda-cyhalothrin + pirimicarb (*off-label*), pirimicarb (*off-label*), tau-fluvalinate (*off-label*)
	Beetles	deltamethrin, thiacloprid
	Flies	tefluthrin (*off-label - seed treatment*)
	Midges	deltamethrin
	Weevils	deltamethrin
Weeds	Broad-leaved weeds	carbetamide (*off-label*), chlorthal-dimethyl + propachlor, clomazone (*off-label*), clopyralid + picloram (*off-label*), glyphosate, metazachlor (*off-label*), metazachlor + quinmerac (*off-label*), propachlor, trifluralin
	Crops as weeds	glyphosate
	Grass weeds	carbetamide (*off-label*), chlorthal-dimethyl + propachlor, glyphosate, metazachlor (*off-label*), metazachlor + quinmerac (*off-label*), propachlor, propaquizafop (*off-label*), tepraloxydim (*off-label*), trifluralin
	Weeds, miscellaneous	glyphosate

Brassicas - Root brassicas

Diseases	Alternaria	azoxystrobin (*off-label*), azoxystrobin + difenoconazole (*off-label*), fenpropimorph (*off-label*), iprodione (*off-label*), iprodione (*seed treatment*), iprodione + thiophanate-methyl (*off-label*), tebuconazole
	Crown rot	fenpropimorph (*off-label*), iprodione + thiophanate-methyl (*off-label*)
	Damping off	thiram (*off-label - seed treatment*), thiram (*seed treatment*)

	Downy mildew	metalaxyl-M (*off-label*), propamocarb hydrochloride (*off-label*)
	Foliar diseases	tebuconazole (*off-label*), tebuconazole + trifloxystrobin (*off-label*)
	Fungus diseases	tebuconazole (*off-label*)
	Powdery mildew	azoxystrobin (*off-label*), azoxystrobin + difenoconazole (*off-label*), fenpropimorph (*off-label*), sulphur, sulphur (*off-label*), tebuconazole
	Rhizoctonia	azoxystrobin (*off-label*), tolclofos-methyl (*with fleece or mesh covers*), tolclofos-methyl (*without covers*)
	Ring spot	tebuconazole (*off-label*)
	Sclerotinia	boscalid + pyraclostrobin (*off-label*), tebuconazole
	Stem canker	iprodione (*off-label*)
	White blister	mancozeb + metalaxyl-M (*off-label*), metalaxyl-M (*off-label*), propamocarb hydrochloride (*off-label*)
Pests	Aphids	deltamethrin (*off-label*), lambda-cyhalothrin + pirimicarb (*off-label*), nicotine, pirimicarb, pirimicarb (*off-label*)
	Beetles	deltamethrin
	Caterpillars	deltamethrin, nicotine
	Cutworms	Bacillus thuringiensis (*off-label*), lambda-cyhalothrin + pirimicarb (*off-label*)
	Flies	chlorpyrifos (*off-label*), lambda-cyhalothrin (*off-label*)
	Leaf miners	nicotine
	Pests, miscellaneous	deltamethrin (*off-label*), dimethoate (*off-label*)
	Weevils	lambda-cyhalothrin (*off-label*)
Plant growth regulation	Quality/yield control	sulphur
Weeds	Broad-leaved weeds	chlorpropham (*off-label*), chlorthal-dimethyl, chlorthal-dimethyl + propachlor, clomazone (*off-label*), clopyralid, glyphosate, linuron (*off-label*), metamitron (*off-label*), metazachlor, propachlor, propachlor (*off-label*), trifluralin
	Crops as weeds	cycloxydim, fluazifop-P-butyl (*stockfeed only*), glyphosate
	Grass weeds	chlorpropham (*off-label*), chlorthal-dimethyl + propachlor, cycloxydim, fluazifop-P-butyl (*stockfeed only*), glyphosate, metamitron (*off-label*), metazachlor, propachlor, propachlor (*off-label*), propaquizafop, propaquizafop (*off-label*), trifluralin
	Weeds, miscellaneous	linuron (*off-label*), metamitron (*off-label*)

Brassicas - Salad greens

Diseases	Alternaria	azoxystrobin (*off-label - for baby leaf production*), iprodione (*seed treatment*)
	Black rot	copper oxychloride (*off-label*)
	Botrytis	fenhexamid (*off-label - for baby leaf production*), iprodione (*off-label*)
	Damping off	cymoxanil + fludioxonil + metalaxyl-M (*off-label*), tolclofos-methyl
	Downy mildew	azoxystrobin (*off-label - for baby leaf production*), dimethomorph + mancozeb (*off-label*), dimethomorph + mancozeb (*off-label - for baby leaf production*), fosetyl-aluminium (*off-label - for baby leaf production*), fosetyl-

		aluminium (*off-label - seedlings*), propamocarb hydrochloride
	Foliar diseases	thiram (*off-label - for baby leaf production*)
	Phytophthora	propamocarb hydrochloride
	Pythium	propamocarb hydrochloride
	Ring spot	difenoconazole (*off-label*), tebuconazole (*off-label*), tebuconazole (*off-label - for baby leaf production*)
	Spear rot	copper oxychloride (*off-label*)
	White blister	boscalid + pyraclostrobin (*off-label*), boscalid + pyraclostrobin (*off-label - for baby leaf production*), mancozeb + metalaxyl-M (*off-label*)
Pests	Aphids	carbosulfan (*off-label*), chlorpyrifos, deltamethrin (*off-label*), lambda-cyhalothrin (*off-label - for baby leaf production*), nicotine, pirimicarb, pirimicarb (*off-label*), pirimicarb (*off-label - for baby leaf production*), pymetrozine (*off-label - for baby leaf production*)
	Beetles	deltamethrin (*off-label - for baby leaf production*)
	Birds/mammals	aluminium ammonium sulphate
	Caterpillars	alpha-cypermethrin (*off-label - for baby leaf production*), Bacillus thuringiensis (*off-label*), Bacillus thuringiensis (*off-label - for baby leaf production*), chlorpyrifos, deltamethrin (*off-label*), deltamethrin (*off-label - for baby leaf production*), diflubenzuron (*off-label - for baby leaf production*), nicotine
	Cutworms	chlorpyrifos
	Flies	chlorpyrifos, chlorpyrifos (*off-label*), tefluthrin (*off-label - seed treatment*)
	Leaf miners	nicotine
	Leatherjackets	chlorpyrifos
	Pests, miscellaneous	deltamethrin (*off-label*), dimethoate (*off-label*)
	Whiteflies	chlorpyrifos
Weeds	Broad-leaved weeds	chlorthal-dimethyl (*off-label*), chlorthal-dimethyl + propachlor (*off-label*), metazachlor (*off-label*), pendimethalin (*off-label*), pendimethalin (*off-label - for baby leaf production*), phenmedipham (*off-label - for baby leaf production*), propachlor (*off-label*), propachlor (*off-label - for baby leaf production*), propyzamide (*off-label - for baby leaf production outdoors or under crop covers*)
	Grass weeds	chlorthal-dimethyl + propachlor (*off-label*), fluazifop-P-butyl (*off-label*), pendimethalin (*off-label - for baby leaf production*), propachlor (*off-label*), propachlor (*off-label - for baby leaf production*), propyzamide (*off-label - for baby leaf production outdoors or under crop covers*)

Cereals - Barley

Crop control	Desiccation	diquat (*stockfeed only*), glyphosate
Diseases	Covered smut	carboxin + thiram (*seed treatment*), clothianidin + prothioconazole (*seed treatment*), clothianidin + prothioconazole + tebuconazole + triazoxide (*seed treatment*), fludioxonil (*seed treatment*), fludioxonil + flutriafol (*seed treatment*), fludioxonil + tefluthrin, fluquinconazole + prochloraz (*seed treatment*), fuberidazole + imidacloprid + triadimenol (*seed treatment*), fuberidazole + triadimenol (*seed treatment*),

	guazatine + imazalil (*seed treatment*), prochloraz + triticonazole (*seed treatment*), prothioconazole (*seed treatment*), prothioconazole + tebuconazole + triazoxide (*seed treatment*)
Eyespot	azoxystrobin + cyproconazole (*reduction*), boscalid + epoxiconazole (*moderate*), boscalid + epoxiconazole (*moderate control*), carbendazim, carbendazim + flusilazole, chlorothalonil + cyproconazole, cyproconazole + cyprodinil, cyprodinil, cyprodinil + picoxystrobin, epoxiconazole (*reduction*), epoxiconazole + fenpropimorph (*reduction*), epoxiconazole + fenpropimorph + kresoxim-methyl (*reduction*), epoxiconazole + kresoxim-methyl (*reduction*), fluoxastrobin + prothioconazole (*reduction*), fluoxastrobin + prothioconazole + trifloxystrobin (*reduction*), flusilazole, prochloraz, prochloraz + propiconazole, prochloraz + tebuconazole, prothioconazole, prothioconazole + spiroxamine, prothioconazole + spiroxamine + tebuconazole (*reduction*), prothioconazole + spiroxamine + tebuconazole (*reducton*), prothioconazole + tebuconazole (*reduction*), prothioconazole + trifloxystrobin (*reduction*)
Foot rot	carboxin + thiram (*seed treatment*), clothianidin + prothioconazole (*seed treatment*), clothianidin + prothioconazole + tebuconazole + triazoxide (*seed treatment*), fludioxonil (*seed treatment*), fludioxonil + flutriafol (*seed treatment*), fludioxonil + tefluthrin, fluquinconazole + prochloraz (*seed treatment*), fuberidazole + imidacloprid + triadimenol (*seed treatment*), fuberidazole + imidacloprid + triadimenol (*seed treatment - reduction*), fuberidazole + triadimenol (*seed treatment*), guazatine (*seed treatment*), guazatine (*seed treatment - reduction*), guazatine + imazalil (*seed treatment*), guazatine + imazalil (*seed treatment - moderate control*), prochloraz + triticonazole (*seed treatment*), prothioconazole (*seed treatment*), prothioconazole + tebuconazole + triazoxide (*seed treatment*)
Late ear diseases	fluoxastrobin + prothioconazole, fluoxastrobin + prothioconazole + trifloxystrobin, prothioconazole, prothioconazole + spiroxamine + tebuconazole, prothioconazole + tebuconazole, prothioconazole + trifloxystrobin
Leaf stripe	carboxin + thiram (*seed treatment*), carboxin + thiram (*seed treatment - reduction*), clothianidin + prothioconazole (*seed treatment*), clothianidin + prothioconazole + tebuconazole + triazoxide (*seed treatment*), fludioxonil (*seed treatment - reduction*), fludioxonil + flutriafol (*seed treatment*), fludioxonil + tefluthrin (*partial control*), fuberidazole + triadimenol (*seed treatment*), guazatine (*seed treatment*), guazatine + imazalil (*seed treatment*), prochloraz + triticonazole (*seed treatment*), prothioconazole (*seed treatment*), prothioconazole + tebuconazole + triazoxide (*seed treatment*)
Loose smut	carboxin + thiram (*seed treatment - reduction*), clothianidin + prothioconazole (*seed treatment*), clothianidin + prothioconazole + tebuconazole + triazoxide (*seed treatment*), fludioxonil + flutriafol (*seed*

	treatment), fluquinconazole + prochloraz (*seed treatment*), fuberidazole + imidacloprid + triadimenol (*seed treatment*), fuberidazole + triadimenol (*seed treatment*), prochloraz + triticonazole (*seed treatment*), prothioconazole (*seed treatment*), prothioconazole + tebuconazole + triazoxide (*seed treatment*)
Net blotch	azoxystrobin, azoxystrobin + chlorothalonil, azoxystrobin + cyproconazole, azoxystrobin + fenpropimorph, boscalid + epoxiconazole, carboxin + thiram (*seed treatment*), chlorothalonil + cyproconazole, chlorothalonil + cyproconazole + propiconazole (*reduction*), chlorothalonil + picoxystrobin, clothianidin + prothioconazole + tebuconazole + triazoxide (*seed treatment*), cyproconazole + cyprodinil, cyproconazole + propiconazole, cyproconazole + trifloxystrobin, cyprodinil, cyprodinil + picoxystrobin, epoxiconazole, epoxiconazole + fenpropimorph, epoxiconazole + fenpropimorph + kresoxim-methyl, epoxiconazole + fenpropimorph + metrafenone, epoxiconazole + fenpropimorph + pyraclostrobin, epoxiconazole + kresoxim-methyl, epoxiconazole + kresoxim-methyl + pyraclostrobin, epoxiconazole + metrafenone, epoxiconazole + pyraclostrobin, famoxadone + flusilazole, fenpropimorph + flusilazole, fenpropimorph + pyraclostrobin, fluoxastrobin + prothioconazole, fluoxastrobin + prothioconazole + trifloxystrobin, flusilazole, fuberidazole + imidacloprid + triadimenol (*seed treatment - seed-borne only*), guazatine (*seed treatment*), guazatine + imazalil (*seed treatment - moderate control*), iprodione, mancozeb, metconazole (*reduction*), picoxystrobin, prochloraz, prochloraz + tebuconazole, prothioconazole, prothioconazole + spiroxamine, prothioconazole + spiroxamine + tebuconazole, prothioconazole + tebuconazole, prothioconazole + tebuconazole + triazoxide (*seed treatment*), prothioconazole + trifloxystrobin, pyraclostrobin, spiroxamine + tebuconazole, tebuconazole, tebuconazole + triadimenol, trifloxystrobin
Powdery mildew	azoxystrobin, azoxystrobin + cyproconazole, azoxystrobin + fenpropimorph, carbendazim + flusilazole, chlorothalonil + cyproconazole, chlorothalonil + cyproconazole + propiconazole, chlorothalonil + picoxystrobin, cyflufenamid, cyproconazole + cyprodinil, cyproconazole + propiconazole, cyproconazole + trifloxystrobin, cyprodinil, cyprodinil + picoxystrobin, epoxiconazole, epoxiconazole + fenpropimorph, epoxiconazole + fenpropimorph + kresoxim-methyl, epoxiconazole + fenpropimorph + metrafenone, epoxiconazole + fenpropimorph + pyraclostrobin, epoxiconazole + kresoxim-methyl, epoxiconazole + metrafenone, fenpropidin, fenpropimorph, fenpropimorph + flusilazole, fenpropimorph + kresoxim-methyl, fenpropimorph + pyraclostrobin, fenpropimorph + quinoxyfen, fluoxastrobin + prothioconazole, fluoxastrobin + prothioconazole + trifloxystrobin, fluquinconazole + prochloraz, flusilazole, flutriafol, fuberidazole + imidacloprid + triadimenol (*seed treatment*), metconazole, metconazole (*moderate control*), metrafenone, picoxystrobin, prochloraz, prochloraz + tebuconazole, propiconazole, proquinazid,

	prothioconazole, prothioconazole + spiroxamine, prothioconazole + spiroxamine + tebuconazole, prothioconazole + tebuconazole, prothioconazole + trifloxystrobin, quinoxyfen, spiroxamine, spiroxamine + tebuconazole, sulphur, tebuconazole, tebuconazole + triadimenol, tetraconazole
Rhynchosporium	azoxystrobin, azoxystrobin + chlorothalonil, azoxystrobin + cyproconazole (*moderate control*), azoxystrobin + fenpropimorph, boscalid + epoxiconazole (*moderate*), boscalid + epoxiconazole (*moderate control*), carbendazim, chlorothalonil, chlorothalonil + cyproconazole, chlorothalonil + cyproconazole + propiconazole (*reduction*), chlorothalonil + flusilazole, chlorothalonil + picoxystrobin, cyproconazole + cyprodinil, cyproconazole + propiconazole, cyproconazole + trifloxystrobin, cyprodinil, cyprodinil (*moderate control*), cyprodinil + picoxystrobin, epoxiconazole, epoxiconazole + fenpropimorph, epoxiconazole + fenpropimorph + kresoxim-methyl, epoxiconazole + fenpropimorph + metrafenone, epoxiconazole + fenpropimorph + pyraclostrobin, epoxiconazole + kresoxim-methyl, epoxiconazole + kresoxim-methyl + pyraclostrobin, epoxiconazole + metrafenone, epoxiconazole + pyraclostrobin, famoxadone + flusilazole, fenpropidin (*moderate control*), fenpropimorph, fenpropimorph + flusilazole, fenpropimorph + kresoxim-methyl, fenpropimorph + pyraclostrobin, fluoxastrobin + prothioconazole, fluoxastrobin + prothioconazole + trifloxystrobin, fluquinconazole + prochloraz, flusilazole, flutriafol, mancozeb, metconazole, picoxystrobin, prochloraz, prochloraz + propiconazole, prochloraz + tebuconazole, propiconazole, prothioconazole, prothioconazole + spiroxamine, prothioconazole + spiroxamine + tebuconazole, prothioconazole + tebuconazole, prothioconazole + trifloxystrobin, pyraclostrobin (*moderate*), spiroxamine (*reduction*), spiroxamine + tebuconazole, tebuconazole, tebuconazole + triadimenol, trifloxystrobin
Rust	azoxystrobin, azoxystrobin + chlorothalonil, azoxystrobin + cyproconazole, azoxystrobin + fenpropimorph, boscalid + epoxiconazole, carbendazim + flusilazole, chlorothalonil + cyproconazole, chlorothalonil + cyproconazole + propiconazole, chlorothalonil + picoxystrobin, cyproconazole + cyprodinil, cyproconazole + propiconazole, cyproconazole + trifloxystrobin, cyprodinil + picoxystrobin, epoxiconazole, epoxiconazole + fenpropimorph, epoxiconazole + fenpropimorph + kresoxim-methyl, epoxiconazole + fenpropimorph + metrafenone, epoxiconazole + fenpropimorph + pyraclostrobin, epoxiconazole + kresoxim-methyl, epoxiconazole + kresoxim-methyl + pyraclostrobin, epoxiconazole + metrafenone, epoxiconazole + pyraclostrobin, famoxadone + flusilazole, fenpropidin, fenpropidin (*moderate control*), fenpropimorph, fenpropimorph + flusilazole, fenpropimorph + pyraclostrobin, fluoxastrobin + prothioconazole, fluoxastrobin + prothioconazole + trifloxystrobin, fluquinconazole + prochloraz, fluquinconazole +

		prochloraz (*seed treatment*), flusilazole, flutriafol, fuberidazole + imidacloprid + triadimenol (*seed treatment*), mancozeb, metconazole, picoxystrobin, prochloraz + tebuconazole, propiconazole, prothioconazole, prothioconazole + spiroxamine, prothioconazole + spiroxamine + tebuconazole, prothioconazole + tebuconazole, prothioconazole + trifloxystrobin, pyraclostrobin, spiroxamine, spiroxamine + tebuconazole, tebuconazole, tebuconazole + triadimenol, tetraconazole, trifloxystrobin
	Septoria	flusilazole
	Snow mould	fludioxonil (*seed treatment*), fludioxonil + flutriafol (*seed treatment*), fludioxonil + tefluthrin, fuberidazole + imidacloprid + triadimenol (*seed treatment - reduction*), guazatine + imazalil (*seed treatment*)
	Sooty moulds	mancozeb
	Take-all	azoxystrobin (*reduction*), azoxystrobin + chlorothalonil (*reduction*), azoxystrobin + cyproconazole (*reduction*), azoxystrobin + fenpropimorph (*reduction*), fluoxastrobin + prothioconazole (*reduction*), fluquinconazole + prochloraz (*seed treatment - reduction*), silthiofam (*seed treatment*)
Pests	Aphids	alpha-cypermethrin, bifenthrin, clothianidin (*seed treatment*), clothianidin + prothioconazole (*seed treatment*), clothianidin + prothioconazole + tebuconazole + triazoxide (*seed treatment*), cyfluthrin, cypermethrin, cypermethrin (*autumn sown*), deltamethrin, esfenvalerate, fuberidazole + imidacloprid + triadimenol (*seed treatment*), lambda-cyhalothrin, lambda-cyhalothrin + pirimicarb, pirimicarb, tau-fluvalinate, zeta-cypermethrin
	Birds/mammals	aluminium ammonium sulphate
	Flies	alpha-cypermethrin, chlorpyrifos, cypermethrin, cypermethrin (*autumn sown*), deltamethrin, fludioxonil + tefluthrin
	Leafhoppers	clothianidin (*seed treatment*), clothianidin + prothioconazole (*seed treatment*), clothianidin + prothioconazole + tebuconazole + triazoxide (*seed treatment*)
	Leatherjackets	chlorpyrifos
	Midges	chlorpyrifos
	Slugs/snails	clothianidin (*seed treatment*), clothianidin + prothioconazole (*seed treatment*), clothianidin + prothioconazole + tebuconazole + triazoxide (*seed treatment - reduction*), thiodicarb
	Wireworms	clothianidin (*seed treatment*), clothianidin + prothioconazole (*seed treatment*), clothianidin + prothioconazole + tebuconazole + triazoxide (*seed treatment - reduction*), fludioxonil + tefluthrin, fuberidazole + imidacloprid + triadimenol (*reduction of damage*)
Plant growth regulation	Growth control	2-chloroethylphosphonic acid, 2-chloroethylphosphonic acid + mepiquat chloride, chlormequat, chlormequat + 2-chloroethylphosphonic acid, chlormequat + 2-chloroethylphosphonic acid + mepiquat chloride, mepiquat chloride + prohexadione-calcium, trinexapac-ethyl

	Quality/yield control	2-chloroethylphosphonic acid + mepiquat chloride (*low lodging situations*), chlormequat, mepiquat chloride + prohexadione-calcium, sulphur
Weeds	Broad-leaved weeds	2,4-D, 2,4-D + MCPA, 2,4-DB + MCPA, amidosulfuron, amidosulfuron + iodosulfuron-methyl-sodium, bifenox, bifenox + isoproturon, bromoxynil, bromoxynil + diflufenican + ioxynil, bromoxynil + ioxynil, bromoxynil + ioxynil + mecoprop-P, carfentrazone-ethyl, carfentrazone-ethyl + mecoprop-P, carfentrazone-ethyl + metsulfuron-methyl, chlorotoluron, chlorotoluron + diflufenican, chlorotoluron + isoproturon, chlorotoluron + isoproturon (*off-label*), clopyralid, clopyralid + florasulam + fluroxypyr, dicamba + MCPA + mecoprop-P, dicamba + mecoprop-P, dichlorprop-P, dichlorprop-P + MCPA + mecoprop-P, diflufenican, diflufenican + flufenacet, diflufenican + flurtamone, diflufenican + isoproturon, florasulam, florasulam + fluroxypyr, flufenacet + pendimethalin, flupyrsulfuron-methyl, fluroxypyr, fluroxypyr + thifensulfuron-methyl + tribenuron-methyl, glyphosate, iodosulfuron-methyl-sodium, isoproturon, isoproturon (*off-label*), linuron, linuron + trifluralin, MCPA, MCPA + MCPB, MCPB, mecoprop-P, metsulfuron-methyl, metsulfuron-methyl + thifensulfuron-methyl, metsulfuron-methyl + tribenuron-methyl, pendimethalin, pendimethalin + picolinafen, pendimethalin + trifluralin, picolinafen, prosulfocarb, thifensulfuron-methyl + tribenuron-methyl, tribenuron-methyl, trifluralin
	Crops as weeds	amidosulfuron + iodosulfuron-methyl-sodium, diflufenican, diflufenican + flufenacet, diflufenican + flurtamone, florasulam, fluroxypyr, glyphosate, iodosulfuron-methyl-sodium, metsulfuron-methyl + tribenuron-methyl, pendimethalin
	Grass weeds	bifenox + isoproturon, chlorotoluron, chlorotoluron + diflufenican, chlorotoluron + isoproturon, chlorotoluron + isoproturon (*off-label*), diflufenican + flufenacet, diflufenican + flurtamone, diflufenican + isoproturon, fenoxaprop-P-ethyl, flufenacet + pendimethalin, flupyrsulfuron-methyl, glyphosate, isoproturon, isoproturon (*off-label*), linuron, linuron + trifluralin, pendimethalin, pendimethalin + picolinafen, pendimethalin + trifluralin, pinoxaden, prosulfocarb, tralkoxydim, tri-allate, trifluralin
	Weeds, miscellaneous	glyphosate

Cereals - Cereals, general

Crop control	Desiccation	glyphosate (*off-label - for wild bird seed production*)
Pests	Leatherjackets	methiocarb (*reduction*)
	Slugs/snails	methiocarb
Weeds	Broad-leaved weeds	2,4-DB + MCPA, isoxaben (*off-label - grown for game cover*), MCPA + MCPB
	Weeds, miscellaneous	clopyralid (*off-label - game cover*)

Cereals - Maize/sweetcorn

Diseases	Damping off	thiram (*seed treatment*)
Pests	Aphids	nicotine, pirimicarb, pirimicarb (*off-label*), pymetrozine (*off-label*)
	Caterpillars	Bacillus thuringiensis (*off-label*), nicotine
	Flies	chlorpyrifos, lambda-cyhalothrin (*off-label*)
	Leaf miners	nicotine
	Slugs/snails	methiocarb
Weeds	Broad-leaved weeds	bromoxynil, bromoxynil + prosulfuron, bromoxynil + terbuthylazine, bromoxynil + terbuthylazine (*off-label*), clopyralid, flufenacet + isoxaflutole, fluroxypyr, isoxaben (*off-label - grown for game cover*), mesotrione, mesotrione (*off-label*), mesotrione + terbuthylazine, mesotrione + terbuthylazine (*off-label*), nicosulfuron, pendimethalin, pendimethalin (*off-label - under covers*), pendimethalin (*off-label - under covers and mulches*), pendimethalin (*off-label - under crop covers or mulches*), rimsulfuron
	Crops as weeds	fluroxypyr (*off-label*), mesotrione, mesotrione (*off-label*), nicosulfuron, pendimethalin, rimsulfuron
	Grass weeds	flufenacet + isoxaflutole, mesotrione + terbuthylazine, mesotrione + terbuthylazine (*off-label*), nicosulfuron, pendimethalin, pendimethalin (*off-label - under crop covers or mulches*)

Cereals - Oats

Crop control	Desiccation	diquat (*stockfeed only*), glyphosate
Diseases	Covered smut	carboxin + thiram (*seed treatment*), fluoxastrobin + prothioconazole (*seed treatment*), prochloraz + triticonazole (*seed treatment*)
	Crown rust	azoxystrobin, azoxystrobin + cyproconazole, azoxystrobin + fenpropimorph, boscalid + epoxiconazole, epoxiconazole + fenpropimorph + metrafenone, epoxiconazole + fenpropimorph + pyraclostrobin, epoxiconazole + kresoxim-methyl + pyraclostrobin, epoxiconazole + metrafenone, epoxiconazole + pyraclostrobin, epoxiconazole + pyraclostrobin (*qualified minor use*), fenpropimorph + pyraclostrobin, fluoxastrobin + prothioconazole, picoxystrobin, propiconazole, prothioconazole, prothioconazole + spiroxamine, prothioconazole + spiroxamine + tebuconazole, prothioconazole + tebuconazole, pyraclostrobin, tebuconazole, tebuconazole (*reduction*), tebuconazole + triadimenol, tetraconazole
	Eyespot	epoxiconazole + fenpropimorph + kresoxim-methyl (*reduction*), epoxiconazole + kresoxim-methyl (*reduction*), fluoxastrobin + prothioconazole, prothioconazole, prothioconazole + spiroxamine, prothioconazole + spiroxamine + tebuconazole, prothioconazole + tebuconazole
	Foot rot	carboxin + thiram (*seed treatment*), clothianidin + prothioconazole (*seed treatment*), fludioxonil (*seed treatment*), fludioxonil + flutriafol (*seed treatment*), fludioxonil + tefluthrin, fluoxastrobin + prothioconazole

		(*seed treatment*), fuberidazole + imidacloprid + triadimenol (*seed treatment - reduction*), fuberidazole + triadimenol (*seed treatment*), guazatine (*seed treatment - reduction*), prochloraz + triticonazole (*seed treatment*), prothioconazole (*seed treatment*)
	Leaf spot	fludioxonil (*seed treatment*), fludioxonil + flutriafol (*seed treatment*), fludioxonil + tefluthrin, fuberidazole + imidacloprid + triadimenol (*seed treatment*), fuberidazole + triadimenol (*seed treatment*), guazatine (*seed treatment*), guazatine + imazalil (*seed treatment*)
	Loose smut	carboxin + thiram (*seed treatment - reduction*), clothianidin + prothioconazole (*seed treatment*), fluoxastrobin + prothioconazole (*seed treatment*), fuberidazole + imidacloprid + triadimenol (*seed treatment*), fuberidazole + triadimenol (*seed treatment*), prochloraz + triticonazole (*seed treatment*), prothioconazole (*seed treatment*)
	Powdery mildew	azoxystrobin, azoxystrobin + cyproconazole, azoxystrobin + fenpropimorph, cyflufenamid, epoxiconazole, epoxiconazole + fenpropimorph, epoxiconazole + fenpropimorph + kresoxim-methyl, epoxiconazole + fenpropimorph + metrafenone, epoxiconazole + fenpropimorph + pyraclostrobin, epoxiconazole + kresoxim-methyl, epoxiconazole + metrafenone, fenpropimorph, fenpropimorph + kresoxim-methyl, fenpropimorph + pyraclostrobin, fenpropimorph + quinoxyfen, fluoxastrobin + prothioconazole, fuberidazole + imidacloprid + triadimenol (*seed treatment*), metrafenone (*qualified minor use*), picoxystrobin, propiconazole, proquinazid, prothioconazole, prothioconazole + spiroxamine, prothioconazole + spiroxamine + tebuconazole, prothioconazole + tebuconazole, quinoxyfen, sulphur, tebuconazole, tebuconazole + triadimenol
	Snow mould	fludioxonil (*seed treatment*)
Pests	Aphids	bifenthrin, clothianidin (*seed treatment*), clothianidin + prothioconazole (*seed treatment*), cypermethrin, deltamethrin, fuberidazole + imidacloprid + triadimenol (*seed treatment*), lambda-cyhalothrin, lambda-cyhalothrin + pirimicarb, pirimicarb, zeta-cypermethrin
	Birds/mammals	aluminium ammonium sulphate
	Flies	chlorpyrifos, cypermethrin, deltamethrin
	Leafhoppers	clothianidin (*seed treatment*), clothianidin + prothioconazole (*seed treatment*)
	Leatherjackets	chlorpyrifos
	Midges	chlorpyrifos
	Slugs/snails	clothianidin (*seed treatment*), clothianidin + prothioconazole (*seed treatment*), thiodicarb
	Thrips	chlorpyrifos
	Wireworms	clothianidin (*seed treatment*), clothianidin + prothioconazole (*seed treatment*), fludioxonil + tefluthrin, fuberidazole + imidacloprid + triadimenol (*reduction of damage*)
Plant growth regulation	Growth control	chlormequat, trinexapac-ethyl
	Quality/yield control	chlormequat

Weeds	Broad-leaved weeds	2,4-D, 2,4-D + MCPA, 2,4-DB + MCPA, amidosulfuron, bromoxynil, bromoxynil + ioxynil, bromoxynil + ioxynil + mecoprop-P, carfentrazone-ethyl, carfentrazone-ethyl + flupyrsulfuron-methyl, carfentrazone-ethyl + mecoprop-P, carfentrazone-ethyl + metsulfuron-methyl, clopyralid, clopyralid + florasulam + fluroxypyr, dicamba + MCPA + mecoprop-P, dicamba + mecoprop-P, dichlorprop-P, dichlorprop-P + MCPA + mecoprop-P, florasulam, florasulam + fluroxypyr, flupyrsulfuron-methyl + thifensulfuron-methyl, flupyrsulfuron-methyl + tribenuron-methyl, fluroxypyr, glyphosate, linuron, MCPA, MCPA + MCPB, MCPB, mecoprop-P, metsulfuron-methyl, metsulfuron-methyl + tribenuron-methyl, tribenuron-methyl
	Crops as weeds	florasulam, glyphosate, metsulfuron-methyl + tribenuron-methyl
	Grass weeds	carfentrazone-ethyl + flupyrsulfuron-methyl, flupyrsulfuron-methyl + thifensulfuron-methyl, flupyrsulfuron-methyl + tribenuron-methyl, glyphosate, linuron, linuron + trifluralin (*off-label*)
	Weeds, miscellaneous	glyphosate

Cereals - Rye/triticale

Diseases	Blue mould	fuberidazole + triadimenol (*seed treatment*)
	Bunt	clothianidin + prothioconazole (*seed treatment*), fluoxastrobin + prothioconazole (*seed treatment*), prothioconazole (*seed treatment*)
	Eyespot	cyprodinil (*off-label*), epoxiconazole + fenpropimorph + kresoxim-methyl (*reduction*), epoxiconazole + kresoxim-methyl (*reduction*), fluoxastrobin + prothioconazole (*reduction*), prochloraz, prochloraz + tebuconazole, prothioconazole, prothioconazole + spiroxamine, prothioconazole + spiroxamine + tebuconazole (*reducton*), prothioconazole + tebuconazole (*reduction*)
	Foliar diseases	azoxystrobin + chlorothalonil (*off-label*), azoxystrobin + cyproconazole (*off-label*), carbendazim (*off-label*), chlorothalonil (*off-label*), chlorothalonil + flutriafol (*off-label*), chlorothalonil + mancozeb (*off-label*), cyproconazole + cyprodinil (*off-label*), cyproconazole + propiconazole (*off-label*), cyprodinil (*off-label*), difenoconazole (*off-label*), epoxiconazole + kresoxim-methyl + pyraclostrobin (*off-label*), epoxiconazole + pyraclostrobin (*off-label*), famoxadone + flusilazole (*off-label*), fenpropidin (*off-label*), fenpropimorph + flusilazole (*off-label*), fenpropimorph + pyraclostrobin (*off-label*), mancozeb (*off-label*), metconazole (*off-label*), metrafenone (*off-label*), prochloraz (*off-label*), tebuconazole (*off-label*), tebuconazole + triadimenol (*off-label*)
	Foot rot	carboxin + thiram (*seed treatment*), clothianidin + prothioconazole (*seed treatment*), fluoxastrobin + prothioconazole (*seed treatment*), fuberidazole + triadimenol (*seed treatment*), prochloraz + triticonazole (*seed treatment*), prothioconazole (*seed treatment*)
	Late ear diseases	prothioconazole + spiroxamine

	Powdery mildew	azoxystrobin, azoxystrobin + cyproconazole, azoxystrobin + fenpropimorph, cyflufenamid, epoxiconazole, epoxiconazole + fenpropimorph, epoxiconazole + fenpropimorph + kresoxim-methyl, epoxiconazole + fenpropimorph + metrafenone, epoxiconazole + kresoxim-methyl, epoxiconazole + metrafenone, fenpropimorph, fenpropimorph + kresoxim-methyl, fenpropimorph + quinoxyfen, fluoxastrobin + prothioconazole, prochloraz, prochloraz + tebuconazole, propiconazole, proquinazid, prothioconazole, prothioconazole + spiroxamine, prothioconazole + spiroxamine + tebuconazole, prothioconazole + tebuconazole, quinoxyfen, spiroxamine, spiroxamine + tebuconazole, sulphur, sulphur (*off-label*), tebuconazole, tebuconazole + triadimenol
	Rhynchosporium	azoxystrobin, azoxystrobin + cyproconazole (*moderate control*), azoxystrobin + fenpropimorph, azoxystrobin + fenpropimorph (*reduction*), epoxiconazole, epoxiconazole + fenpropimorph, epoxiconazole + fenpropimorph + kresoxim-methyl, epoxiconazole + fenpropimorph + metrafenone, epoxiconazole + kresoxim-methyl, epoxiconazole + metrafenone, fenpropimorph + kresoxim-methyl, fluoxastrobin + prothioconazole, prochloraz, prochloraz + tebuconazole, propiconazole, prothioconazole, prothioconazole + spiroxamine, prothioconazole + spiroxamine + tebuconazole, prothioconazole + tebuconazole, spiroxamine (*reduction*), tebuconazole
	Rust	azoxystrobin, azoxystrobin + cyproconazole, azoxystrobin + fenpropimorph, epoxiconazole, epoxiconazole + fenpropimorph, epoxiconazole + fenpropimorph + kresoxim-methyl, epoxiconazole + fenpropimorph + metrafenone, epoxiconazole + kresoxim-methyl, epoxiconazole + metrafenone, fenpropimorph, fluoxastrobin + prothioconazole, prochloraz + tebuconazole, propiconazole, prothioconazole, prothioconazole + spiroxamine, prothioconazole + spiroxamine + tebuconazole, prothioconazole + tebuconazole, spiroxamine, spiroxamine + tebuconazole, tebuconazole, tebuconazole + triadimenol
	Septoria	epoxiconazole, epoxiconazole + fenpropimorph, epoxiconazole + fenpropimorph + kresoxim-methyl, epoxiconazole + fenpropimorph + metrafenone, epoxiconazole + kresoxim-methyl, epoxiconazole + metrafenone, fenpropimorph + kresoxim-methyl (*reduction*), prochloraz, propiconazole
	Take-all	azoxystrobin + fenpropimorph (*reduction*)
Pests	Aphids	clothianidin (*seed treatment*), clothianidin + prothioconazole (*seed treatment*), cypermethrin, deltamethrin (*off-label*), dimethoate, lambda-cyhalothrin (*off-label*), lambda-cyhalothrin + pirimicarb, pirimicarb, tau-fluvalinate (*off-label*)
	Flies	chlorpyrifos, chlorpyrifos (*off-label*), cypermethrin
	Leafhoppers	clothianidin (*seed treatment*), clothianidin + prothioconazole (*seed treatment*)
	Leatherjackets	chlorpyrifos, chlorpyrifos (*off-label*)
	Midges	chlorpyrifos, chlorpyrifos (*off-label*)

	Slugs/snails	clothianidin (*seed treatment*), clothianidin + prothioconazole (*seed treatment*), thiodicarb
	Wireworms	clothianidin (*seed treatment*), clothianidin + prothioconazole (*seed treatment*)
Plant growth regulation	Growth control	2-chloroethylphosphonic acid, 2-chloroethylphosphonic acid + mepiquat chloride, chlormequat, chlormequat (*off-label*), trinexapac-ethyl
	Quality/yield control	chlormequat, chlormequat (*off-label*)
Weeds	Broad-leaved weeds	2,4-D, amidosulfuron, amidosulfuron + iodosulfuron-methyl-sodium, bifenox, bromoxynil + diflufenican + ioxynil, bromoxynil + ioxynil, bromoxynil + ioxynil + mecoprop-P, bromoxynil + ioxynil + mecoprop-P (*off-label*), carfentrazone-ethyl, carfentrazone-ethyl + flupyrsulfuron-methyl, carfentrazone-ethyl + metsulfuron-methyl, chlorotoluron, chlorotoluron (*off-label*), chlorotoluron + isoproturon (*off-label*), clodinafop-propargyl + trifluralin (*off-label*), dicamba + MCPA + mecoprop-P, dicamba + mecoprop-P, diflufenican, diflufenican + flufenacet (*off-label*), diflufenican + flurtamone (*off-label*), diflufenican + isoproturon, florasulam (*off-label*), florasulam + fluroxypyr (*off-label*), flufenacet + pendimethalin (*off-label*), fluroxypyr, iodosulfuron-methyl-sodium, isoproturon, isoproturon (*off-label*), linuron + trifluralin, linuron + trifluralin (*off-label*), MCPA, metsulfuron-methyl, metsulfuron-methyl + tribenuron-methyl, pendimethalin, tribenuron-methyl
	Crops as weeds	amidosulfuron + iodosulfuron-methyl-sodium, diflufenican, metsulfuron-methyl + tribenuron-methyl, pendimethalin
	Grass weeds	carfentrazone-ethyl + flupyrsulfuron-methyl, chlorotoluron, chlorotoluron (*off-label*), chlorotoluron + isoproturon (*off-label*), clodinafop-propargyl, clodinafop-propargyl + trifluralin (*off-label*), diflufenican + flufenacet (*off-label*), diflufenican + flurtamone (*off-label*), diflufenican + isoproturon, flufenacet + pendimethalin (*off-label*), iodosulfuron-methyl-sodium (*from seed*), isoproturon, isoproturon (*off-label*), linuron + trifluralin, linuron + trifluralin (*off-label*), pendimethalin, tralkoxydim, tri-allate

Cereals - Undersown cereals

Weeds	Broad-leaved weeds	2,4-D, 2,4-DB (*undersown with lucerne*), 2,4-DB + linuron + MCPA (*undersown with clover*), 2,4-DB + MCPA, 2,4-DB + MCPA (*red or white clover*), bentazone + MCPA + MCPB, bromoxynil + ioxynil, dicamba + MCPA + mecoprop-P, dicamba + MCPA + mecoprop-P (*grass only*), dicamba + mecoprop-P, dichlorprop-P, MCPA, MCPA (*red clover or grass*), MCPA + MCPB, MCPB, mecoprop-P

Cereals - Wheat

Crop control	Desiccation	glyphosate
Diseases	Blue mould	fuberidazole + imidacloprid + triadimenol (*seed treatment*), fuberidazole + triadimenol (*seed treatment*)

Bunt	carboxin + thiram (*seed treatment*), clothianidin + prothioconazole (*seed treatment*), fludioxonil (*seed treatment*), fludioxonil + flutriafol (*seed treatment*), fludioxonil + tefluthrin, fluoxastrobin + prothioconazole (*seed treatment*), fluquinconazole (*seed treatment*), fluquinconazole + prochloraz (*seed treatment*), fuberidazole + imidacloprid + triadimenol (*seed treatment*), fuberidazole + triadimenol (*seed treatment*), prochloraz + triticonazole (*off-label - seed treatment*), prochloraz + triticonazole (*seed treatment*), prothioconazole (*seed treatment*)
Disease control/foliar feed	prochloraz (*off-label*)
Drechslera leaf spot	fenpropimorph + pyraclostrobin
Ear blight	dimoxystrobin + epoxiconazole, epoxiconazole, epoxiconazole (*reduction*), epoxiconazole + fenpropimorph, epoxiconazole + fenpropimorph (*reduction*), epoxiconazole + fenpropimorph + kresoxim-methyl (*reduction*), epoxiconazole + fenpropimorph + metrafenone (*reduction*), epoxiconazole + fenpropimorph + pyraclostrobin (*reduction*), epoxiconazole + kresoxim-methyl (*reduction*), epoxiconazole + metrafenone (*reduction*), epoxiconazole + pyraclostrobin (*reduction*), metconazole (*reduction*), spiroxamine + tebuconazole, tebuconazole, tebuconazole + triadimenol
Eyespot	azoxystrobin + cyproconazole (*reduction*), boscalid + epoxiconazole (*moderate*), boscalid + epoxiconazole (*moderate control*), carbendazim, carbendazim + flusilazole, chlorothalonil + cyproconazole, cyproconazole + cyprodinil, cyproconazole + quinoxyfen (*reduction*), cyprodinil, cyprodinil (*off-label*), cyprodinil + picoxystrobin (*moderate control*), epoxiconazole (*reduction*), epoxiconazole + fenpropimorph (*reduction*), epoxiconazole + fenpropimorph + kresoxim-methyl (*reduction*), epoxiconazole + kresoxim-methyl (*reduction*), fluoxastrobin + prothioconazole (*reduction*), fluoxastrobin + prothioconazole + trifloxystrobin (*reduction*), flusilazole, metrafenone (*reduction*), picoxystrobin (*reduction*), prochloraz, prochloraz + propiconazole, prochloraz + tebuconazole, prothioconazole, prothioconazole + spiroxamine, prothioconazole + spiroxamine + tebuconazole (*reduction*), prothioconazole + tebuconazole (*reduction*), prothioconazole + trifloxystrobin
Foliar diseases	azoxystrobin (*off-label*), azoxystrobin + chlorothalonil (*off-label*), azoxystrobin + cyproconazole (*off-label*), azoxystrobin + fenpropimorph (*off-label*), boscalid + epoxiconazole (*off-label*), carbendazim (*off-label*), chlorothalonil (*off-label*), chlorothalonil + flutriafol (*off-label*), chlorothalonil + mancozeb (*off-label*), cyproconazole + cyprodinil (*off-label*), cyproconazole + propiconazole (*off-label*), cyprodinil (*off-label*), difenoconazole (*off-label*), epoxiconazole (*off-label*), epoxiconazole + fenpropimorph (*off-label*), epoxiconazole + fenpropimorph + kresoxim-methyl (*off-label*), epoxiconazole + kresoxim-methyl (*off-label*), epoxiconazole + kresoxim-methyl + pyraclostrobin (*off-label*), epoxiconazole + pyraclostrobin (*off-label*), famoxadone + flusilazole (*off-label*), fenpropidin (*off-*

	label), fenpropimorph (*off-label*), fenpropimorph + flusilazole (*off-label*), fenpropimorph + kresoxim-methyl (*off-label*), mancozeb (*off-label*), metconazole (*off-label*), metrafenone (*off-label*)
Foot rot	carboxin + thiram (*seed treatment*), clothianidin + prothioconazole (*seed treatment*), fludioxonil (*seed treatment*), fludioxonil + flutriafol (*seed treatment*), fludioxonil + tefluthrin, fluoxastrobin + prothioconazole (*seed treatment*), fluquinconazole + prochloraz (*seed treatment*), fuberidazole + imidacloprid + triadimenol (*seed treatment - reduction*), fuberidazole + triadimenol (*seed treatment*), guazatine (*seed treatment - reduction*), prochloraz + triticonazole (*off-label - seed treatment*), prochloraz + triticonazole (*seed treatment*), prothioconazole (*seed treatment*)
Late ear diseases	azoxystrobin, azoxystrobin + fenpropimorph (*moderate control*), chlorothalonil + flutriafol, cyproconazole + propiconazole, fluoxastrobin + prothioconazole, fluoxastrobin + prothioconazole + trifloxystrobin, picoxystrobin, prochloraz + tebuconazole, prothioconazole, prothioconazole + spiroxamine, prothioconazole + spiroxamine + tebuconazole, prothioconazole + tebuconazole, prothioconazole + trifloxystrobin, tebuconazole + trifloxystrobin
Loose smut	clothianidin + prothioconazole (*seed treatment*), fludioxonil + flutriafol (*seed treatment*), fluoxastrobin + prothioconazole (*seed treatment*), fuberidazole + imidacloprid + triadimenol (*seed treatment*), fuberidazole + triadimenol (*seed treatment*), prochloraz + triticonazole (*off-label - seed treatment*), prochloraz + triticonazole (*seed treatment*), prothioconazole (*seed treatment*)
Powdery mildew	azoxystrobin, azoxystrobin + cyproconazole, carbendazim + flusilazole, chlorothalonil + cyproconazole, chlorothalonil + cyproconazole + propiconazole (*moderate*), chlorothalonil + flutriafol, cyflufenamid, cyproconazole + cyprodinil, cyproconazole + propiconazole, cyproconazole + quinoxyfen, cyproconazole + trifloxystrobin, cyprodinil (*moderate control*), epoxiconazole, epoxiconazole + fenpropimorph, epoxiconazole + fenpropimorph + metrafenone, epoxiconazole + kresoxim-methyl, epoxiconazole + metrafenone, fenpropidin, fenpropimorph, fenpropimorph + flusilazole, fenpropimorph + quinoxyfen, fluoxastrobin + prothioconazole, fluoxastrobin + prothioconazole + trifloxystrobin, fluquinconazole, fluquinconazole + prochloraz (*moderate control*), flusilazole, flutriafol, fuberidazole + imidacloprid + triadimenol (*seed treatment*), metconazole (*moderate control*), metrafenone, prochloraz, prochloraz + tebuconazole, propiconazole, proquinazid, proquinazid (*off-label*), prothioconazole, prothioconazole + spiroxamine, prothioconazole + spiroxamine + tebuconazole, prothioconazole + tebuconazole, prothioconazole + trifloxystrobin, quinoxyfen, spiroxamine, spiroxamine + tebuconazole, sulphur, tebuconazole, tebuconazole + triadimenol, tebuconazole + trifloxystrobin, tetraconazole

Rust	azoxystrobin, azoxystrobin + chlorothalonil, azoxystrobin + cyproconazole, azoxystrobin + fenpropimorph, azoxystrobin + fenpropimorph (*off-label*), boscalid + epoxiconazole, carbendazim + flusilazole, chlorothalonil + cyproconazole, chlorothalonil + cyproconazole + propiconazole, chlorothalonil + flutriafol, chlorothalonil + picoxystrobin, cyproconazole + cyprodinil, cyproconazole + propiconazole, cyproconazole + quinoxyfen, cyproconazole + trifloxystrobin, cyprodinil + picoxystrobin, difenoconazole, dimoxystrobin + epoxiconazole, epoxiconazole, epoxiconazole + fenpropimorph, epoxiconazole + fenpropimorph + kresoxim-methyl, epoxiconazole + fenpropimorph + metrafenone, epoxiconazole + fenpropimorph + pyraclostrobin, epoxiconazole + kresoxim-methyl, epoxiconazole + kresoxim-methyl + pyraclostrobin, epoxiconazole + metrafenone, epoxiconazole + pyraclostrobin, famoxadone + flusilazole, fenpropidin (*moderate control*), fenpropimorph, fenpropimorph + flusilazole, fenpropimorph + pyraclostrobin, fluoxastrobin + prothioconazole, fluoxastrobin + prothioconazole + trifloxystrobin, fluquinconazole, fluquinconazole (*seed treatment*), fluquinconazole + prochloraz, fluquinconazole + prochloraz (*seed treatment*), fluquinconazole + prochloraz (*seed treatment - moderate control*), flusilazole, flutriafol, fuberidazole + imidacloprid + triadimenol (*seed treatment*), mancozeb, metconazole, picoxystrobin, prochloraz + tebuconazole, propiconazole, prothioconazole, prothioconazole + spiroxamine, prothioconazole + spiroxamine + tebuconazole, prothioconazole + tebuconazole, prothioconazole + trifloxystrobin, pyraclostrobin, spiroxamine, spiroxamine + tebuconazole, tebuconazole, tebuconazole + triadimenol, tebuconazole + trifloxystrobin, tetraconazole, trifloxystrobin
Septoria	azoxystrobin, azoxystrobin + chlorothalonil, azoxystrobin + cyproconazole, azoxystrobin + fenpropimorph, azoxystrobin + fenpropimorph (*off-label*), boscalid + epoxiconazole, carbendazim + flusilazole, carboxin + thiram (*seed treatment*), chlorothalonil, chlorothalonil + cyproconazole, chlorothalonil + cyproconazole + propiconazole, chlorothalonil + flusilazole, chlorothalonil + flutriafol, chlorothalonil + mancozeb, chlorothalonil + picoxystrobin, cyproconazole + cyprodinil, cyproconazole + propiconazole, cyproconazole + quinoxyfen, cyproconazole + trifloxystrobin, cyprodinil + picoxystrobin, difenoconazole, dimoxystrobin + epoxiconazole, epoxiconazole, epoxiconazole + fenpropimorph, epoxiconazole + fenpropimorph + kresoxim-methyl, epoxiconazole + fenpropimorph + metrafenone, epoxiconazole + fenpropimorph + pyraclostrobin, epoxiconazole + kresoxim-methyl, epoxiconazole + kresoxim-methyl + pyraclostrobin, epoxiconazole + metrafenone, epoxiconazole + pyraclostrobin, famoxadone + flusilazole, fenpropimorph + flusilazole, fenpropimorph + kresoxim-methyl (*reduction*), fenpropimorph + pyraclostrobin,

		fludioxonil (*seed treatment*), fludioxonil + flutriafol (*seed treatment*), fludioxonil + tefluthrin, fluoxastrobin + prothioconazole, fluoxastrobin + prothioconazole + trifloxystrobin, fluquinconazole, fluquinconazole (*seed treatment*), fluquinconazole + prochloraz, fluquinconazole + prochloraz (*seed treatment*), fluquinconazole + prochloraz (*seed treatment - moderate control*), flusilazole, flutriafol, fuberidazole + imidacloprid + triadimenol (*seed treatment*), fuberidazole + imidacloprid + triadimenol (*seed treatment - reduction*), guazatine (*seed treatment*), guazatine (*seed treatment - partial control*), iprodione, mancozeb, metconazole, picoxystrobin, prochloraz, prochloraz + propiconazole, prochloraz + tebuconazole, prochloraz + triticonazole (*off-label - seed treatment*), prochloraz + triticonazole (*seed treatment*), propiconazole, prothioconazole, prothioconazole + spiroxamine, prothioconazole + spiroxamine + tebuconazole, prothioconazole + tebuconazole, prothioconazole + trifloxystrobin, pyraclostrobin, spiroxamine + tebuconazole, tebuconazole, tebuconazole + triadimenol, tebuconazole + trifloxystrobin, tetraconazole, trifloxystrobin
	Snow mould	fludioxonil (*seed treatment*), fludioxonil + flutriafol (*seed treatment*), fludioxonil + tefluthrin
	Sooty moulds	epoxiconazole (*reduction*), epoxiconazole + fenpropimorph (*reduction*), epoxiconazole + fenpropimorph + kresoxim-methyl (*reduction*), epoxiconazole + fenpropimorph + metrafenone (*reduction*), epoxiconazole + kresoxim-methyl (*reduction*), epoxiconazole + metrafenone (*reduction*), mancozeb, propiconazole, spiroxamine + tebuconazole, tebuconazole, tebuconazole + triadimenol, tetraconazole
	Take-all	azoxystrobin (*reduction*), azoxystrobin + chlorothalonil (*reduction*), azoxystrobin + cyproconazole (*reduction*), azoxystrobin + fenpropimorph (*reduction*), fluoxastrobin + prothioconazole (*reduction*), fluquinconazole (*seed treatment - reduction*), fluquinconazole + prochloraz (*seed treatment - reduction*), silthiofam (*seed treatment*)
	Tan spot	boscalid + epoxiconazole (*reduction - qualified minor use*), chlorothalonil + picoxystrobin, dimoxystrobin + epoxiconazole, epoxiconazole + pyraclostrobin, fluoxastrobin + prothioconazole, fluoxastrobin + prothioconazole + trifloxystrobin, prothioconazole, prothioconazole + spiroxamine, prothioconazole + tebuconazole
Pests	Aphids	alpha-cypermethrin, bifenthrin, clothianidin (*seed treatment*), clothianidin + prothioconazole (*seed treatment*), cyfluthrin, cypermethrin, cypermethrin (*autumn sown*), deltamethrin, deltamethrin (*off-label*), dimethoate, dimethoate (*off-label*), esfenvalerate, flonicamid, fuberidazole + imidacloprid + triadimenol (*seed treatment*), lambda-cyhalothrin, lambda-cyhalothrin + pirimicarb, pirimicarb, tau-fluvalinate, tau-fluvalinate (*off-label*), zeta-cypermethrin
	Birds/mammals	aluminium ammonium sulphate
	Flies	alpha-cypermethrin, chlorpyrifos, chlorpyrifos (*off-label*), cypermethrin, cypermethrin (*autumn sown*),

		deltamethrin, dimethoate, fludioxonil + tefluthrin, lambda-cyhalothrin
	Leafhoppers	clothianidin (*seed treatment*), clothianidin + prothioconazole (*seed treatment*)
	Leatherjackets	chlorpyrifos, chlorpyrifos (*off-label*)
	Midges	chlorpyrifos, chlorpyrifos (*off-label*), lambda-cyhalothrin, thiacloprid
	Slugs/snails	clothianidin (*seed treatment*), clothianidin + prothioconazole (*seed treatment*), thiodicarb
	Suckers	lambda-cyhalothrin
	Thrips	chlorpyrifos
	Wireworms	clothianidin (*seed treatment*), clothianidin + prothioconazole (*seed treatment*), fludioxonil + tefluthrin, fuberidazole + imidacloprid + triadimenol (*reduction of damage*)
Plant growth regulation	Growth control	2-chloroethylphosphonic acid, 2-chloroethylphosphonic acid + mepiquat chloride, chlormequat, chlormequat (*off-label*), chlormequat + 2-chloroethylphosphonic acid, chlormequat + 2-chloroethylphosphonic acid + mepiquat chloride, chlormequat + imazaquin, chlormequat + mepiquat chloride, mepiquat chloride + prohexadione-calcium, mepiquat chloride + prohexadione-calcium (*off-label*), trinexapac-ethyl, trinexapac-ethyl (*off-label - cv A C Barrie only*)
	Quality/yield control	chlormequat, chlormequat + imazaquin, mepiquat chloride + prohexadione-calcium, mepiquat chloride + prohexadione-calcium (*off-label*), sulphur
Weeds	Broad-leaved weeds	2,4-D, 2,4-D + MCPA, 2,4-DB + MCPA, amidosulfuron, amidosulfuron + iodosulfuron-methyl-sodium, bifenox, bifenox (*off-label*), bifenox + isoproturon, bromoxynil, bromoxynil + diflufenican + ioxynil, bromoxynil + ioxynil, bromoxynil + ioxynil + mecoprop-P, carfentrazone-ethyl, carfentrazone-ethyl + flupyrsulfuron-methyl, carfentrazone-ethyl + mecoprop-P, carfentrazone-ethyl + metsulfuron-methyl, chlorotoluron, chlorotoluron + diflufenican, chlorotoluron + isoproturon, chlorotoluron + isoproturon (*off-label*), clodinafop-propargyl + trifluralin, clodinafop-propargyl + trifluralin (*off-label*), clopyralid, clopyralid + florasulam + fluroxypyr, dicamba + MCPA + mecoprop-P, dicamba + mecoprop-P, dichlorprop-P, dichlorprop-P + MCPA + mecoprop-P, diflufenican, diflufenican + flufenacet, diflufenican + flufenacet (*off-label*), diflufenican + flupyrsulfuron-methyl, diflufenican + flurtamone, diflufenican + iodosulfuron-methyl-sodium + mesosulfuron-methyl, diflufenican + isoproturon, florasulam, florasulam + fluroxypyr, florasulam + fluroxypyr (*off-label*), flufenacet + pendimethalin, flufenacet + pendimethalin (*off-label*), flumioxazin, flupyrsulfuron-methyl, flupyrsulfuron-methyl + thifensulfuron-methyl, flupyrsulfuron-methyl + tribenuron-methyl, fluroxypyr, fluroxypyr + thifensulfuron-methyl + tribenuron-methyl, glyphosate, iodosulfuron-methyl-sodium, iodosulfuron-methyl-sodium + mesosulfuron-methyl, isoproturon, isoproturon (*autumn sown*), isoproturon (*off-label*), linuron, linuron + trifluralin, linuron + trifluralin (*off-label*), MCPA, MCPA + MCPB, MCPB, mecoprop-P, mecoprop-P (*off-label*), metsulfuron-methyl, metsulfuron-methyl + thifensulfuron-methyl,

		metsulfuron-methyl + tribenuron-methyl, metsulfuron-methyl + tribenuron-methyl (*off-label*), pendimethalin, pendimethalin + picolinafen, pendimethalin + trifluralin, picolinafen, prosulfocarb, sulfosulfuron, thifensulfuron-methyl + tribenuron-methyl, tribenuron-methyl, trifluralin
	Crops as weeds	amidosulfuron + iodosulfuron-methyl-sodium, diflufenican, diflufenican + flufenacet, diflufenican + flurtamone, diflufenican + iodosulfuron-methyl-sodium + mesosulfuron-methyl, florasulam, flumioxazin, fluroxypyr, glyphosate, metsulfuron-methyl + tribenuron-methyl, metsulfuron-methyl + tribenuron-methyl (*off-label*), pendimethalin
	Grass weeds	bifenox + isoproturon, carfentrazone-ethyl + flupyrsulfuron-methyl, chlorotoluron, chlorotoluron + diflufenican, chlorotoluron + isoproturon, chlorotoluron + isoproturon (*off-label*), clodinafop-propargyl, clodinafop-propargyl + pinoxaden, clodinafop-propargyl + pinoxaden (*from seed*), clodinafop-propargyl + trifluralin, clodinafop-propargyl + trifluralin (*off-label*), diflufenican + flufenacet, diflufenican + flufenacet (*off-label*), diflufenican + flupyrsulfuron-methyl, diflufenican + flurtamone, diflufenican + iodosulfuron-methyl-sodium + mesosulfuron-methyl, diflufenican + isoproturon, fenoxaprop-P-ethyl, flufenacet + pendimethalin, flufenacet + pendimethalin (*off-label*), flumioxazin, flupyrsulfuron-methyl, flupyrsulfuron-methyl + thifensulfuron-methyl, flupyrsulfuron-methyl + tribenuron-methyl, glyphosate, iodosulfuron-methyl-sodium (*from seed*), iodosulfuron-methyl-sodium + mesosulfuron-methyl, isoproturon, isoproturon (*autumn sown*), isoproturon (*off-label*), linuron, linuron + trifluralin, linuron + trifluralin (*off-label*), pendimethalin, pendimethalin + picolinafen, pendimethalin + trifluralin, pinoxaden, propoxycarbazone-sodium, prosulfocarb, sulfosulfuron, tralkoxydim, tri-allate, trifluralin
	Weeds, miscellaneous	glyphosate

Edible fungi - Mushrooms

Diseases	Bacterial blotch	sodium hypochlorite (commodity substance)
	Fungus diseases	formaldehyde (commodity substance) (*spray or fumigant*)
	Trichoderma	carbendazim (*off-label - spawn treatment*)
Pests	Flies	deltamethrin (*off-label*), diflubenzuron

Fruiting vegetables - Aubergines

Diseases	Botrytis	azoxystrobin (*off-label*), pyrimethanil (*off-label*)
	Didymella stem rot	azoxystrobin (*off-label*)
	Phytophthora	azoxystrobin (*off-label*), propamocarb hydrochloride
	Powdery mildew	azoxystrobin (*off-label*), sulphur (*off-label*)
	Pythium	propamocarb hydrochloride
Pests	Aphids	nicotine, Verticillium lecanii
	Leaf miners	deltamethrin (*off-label*), oxamyl (*off-label*)

	Leafhoppers	nicotine
	Thrips	deltamethrin (*off-label*), nicotine
	Whiteflies	buprofezin, nicotine, Verticillium lecanii

Fruiting vegetables - Cucurbits

Diseases	Anthracnose	mancozeb (*off-label*)
	Botrytis	mepanipyrim (*off-label*)
	Damping off	etridiazole (*seeds and seedlings*)
	Downy mildew	mancozeb (*off-label*)
	Gummosis	mancozeb (*off-label*)
	Phytophthora	etridiazole (*transplants*), propamocarb hydrochloride
	Powdery mildew	azoxystrobin (*off-label*), bupirimate (*off-label*), bupirimate (*outdoor only*), copper ammonium carbonate, fenarimol (*off-label*), myclobutanil (*off-label*)
	Pythium	propamocarb hydrochloride
Pests	Aphids	fatty acids, nicotine, pirimicarb, pirimicarb (*off-label*)
	Caterpillars	nicotine
	Leaf miners	nicotine
	Mealybugs	fatty acids
	Scale insects	fatty acids
	Spider mites	fatty acids
	Whiteflies	fatty acids
Weeds	Broad-leaved weeds	isoxaben (*off-label*), propyzamide (*off-label - outdoor or under crop covers*)
	Grass weeds	propyzamide (*off-label - outdoor or under crop covers*)
	Weeds, miscellaneous	chlorthal-dimethyl (*off-label*)

Fruiting vegetables - Peppers

Diseases	Damping off	propamocarb hydrochloride (*off-label*)
	Phytophthora	propamocarb hydrochloride
	Pythium	propamocarb hydrochloride
	Root rot	propamocarb hydrochloride (*off-label*)
Pests	Aphids	fatty acids, nicotine, pirimicarb
	Caterpillars	nicotine
	Leaf miners	nicotine
	Mealybugs	fatty acids
	Scale insects	fatty acids
	Spider mites	fatty acids
	Whiteflies	fatty acids

Fruiting vegetables - Tomatoes

Diseases	Botrytis	fenhexamid (*off-label*), iprodione, pyrimethanil (*off-label*), thiram
	Damping off	copper oxychloride, etridiazole (*seeds and seedlings*)
	Foot rot	copper oxychloride

	Phytophthora	Bordeaux mixture, copper ammonium carbonate, copper oxychloride, etridiazole (*transplants*), propamocarb hydrochloride
	Pythium	propamocarb hydrochloride
Pests	Aphids	fatty acids, nicotine, pirimicarb, pyrethrins
	Beetles	pyrethrins
	Caterpillars	nicotine, pyrethrins
	Leaf miners	nicotine
	Mealybugs	fatty acids
	Scale insects	fatty acids
	Spider mites	fatty acids
	Whiteflies	fatty acids, pyrethrins
Plant growth regulation	Fruiting control	2-chloroethylphosphonic acid

Herb crops - Herbs

Diseases	Botrytis	chlorothalonil (*off-label*), fenhexamid (*off-label*), iprodione (*off-label*), propamocarb hydrochloride (*off-label*)
	Damping off	cymoxanil + fludioxonil + metalaxyl-M (*off-label*)
	Downy mildew	chlorothalonil (*off-label*), dimethomorph + mancozeb (*off-label*), fosetyl-aluminium (*off-label*), propamocarb hydrochloride (*off-label*)
	Foliar diseases	difenoconazole (*off-label*), mancozeb (*off-label*), tebuconazole (*off-label*), thiram (*off-label*)
	Fungus diseases	mancozeb + metalaxyl-M (*off-label*)
	Powdery mildew	azoxystrobin (*off-label*), tebuconazole (*off-label*)
	Ring spot	azoxystrobin (*off-label*), prochloraz (*off-label*)
	Rust	azoxystrobin (*off-label*), tebuconazole (*off-label*)
	Sclerotinia	boscalid + pyraclostrobin (*off-label*)
	Seed-borne diseases	thiram (*seed soak*)
Pests	Aphids	acetamiprid (*off-label*), deltamethrin (*off-label*), lambda-cyhalothrin (*off-label*), lambda-cyhalothrin + pirimicarb (*off-label*), nicotine, pirimicarb (*off-label*)
	Beetles	deltamethrin (*off-label*)
	Caterpillars	Bacillus thuringiensis (*off-label*), deltamethrin (*off-label*), diflubenzuron (*off-label*), nicotine
	Cutworms	lambda-cyhalothrin (*off-label*)
	Flies	lambda-cyhalothrin (*off-label*), tefluthrin (*off-label - seed treatment*)
	Leaf miners	nicotine
	Leafhoppers	deltamethrin (*off-label*)
	Pests, miscellaneous	deltamethrin (*off-label*)
Weeds	Broad-leaved weeds	asulam (*off-label*), chloridazon (*off-label*), chloridazon + propachlor (*off-label - post-em. Pre-em on label*), chloridazon + quinmerac (*off-label*), chlorpropham, chlorthal-dimethyl, chlorthal-dimethyl (*off-label*), clomazone (*off-label*), clopyralid (*off-label*), ioxynil (*off-label*), linuron (*off-label*), metamitron (*off-label*), metazachlor (*off-label*), metazachlor + quinmerac (*off-label*), pendimethalin, pendimethalin (*off-label*), phenmedipham (*off-label*), propachlor, propachlor (*off-*

		label), propyzamide (*off-label*), propyzamide (*off-label - outdoor or under crop covers*), trifluralin
	Crops as weeds	pendimethalin, propaquizafop (*off-label*)
	Grass weeds	chloridazon (*off-label*), chloridazon + propachlor (*off-label - post-em. Pre-em on label*), chloridazon + quinmerac (*off-label*), chlorpropham, metazachlor (*off-label*), metazachlor + quinmerac (*off-label*), pendimethalin, pendimethalin (*off-label*), propachlor, propachlor (*off-label*), propaquizafop (*off-label*), propyzamide (*off-label*), propyzamide (*off-label - outdoor or under crop covers*), tepraloxydim (*off-label*), trifluralin
	Weeds, miscellaneous	diquat (*off-label*)

Leafy vegetables - Endives

Diseases	Botrytis	fenhexamid (*off-label*), iprodione (*off-label*)
	Damping off	thiram (*off-label - seed treatment*)
	Downy mildew	dimethomorph + mancozeb (*off-label*), fosetyl-aluminium (*off-label*)
	Foliar diseases	mancozeb (*off-label*), thiram (*off-label*)
	Sclerotinia	iprodione (*off-label*)
Pests	Aphids	acetamiprid (*off-label*), deltamethrin (*off-label*), lambda-cyhalothrin (*off-label*), nicotine, pirimicarb (*off-label*), pymetrozine (*off-label*)
	Caterpillars	Bacillus thuringiensis (*off-label*), diflubenzuron (*off-label*), nicotine
	Leaf miners	nicotine
	Pests, miscellaneous	deltamethrin (*off-label*)
	Thrips	pirimicarb (*off-label*)
Weeds	Broad-leaved weeds	pendimethalin (*off-label*), propachlor (*off-label - under crop covers*), propyzamide (*off-label*), propyzamide (*off-label - outdoor or under crop covers*)
	Grass weeds	pendimethalin (*off-label*), propachlor (*off-label - under crop covers*), propyzamide (*off-label*), propyzamide (*off-label - outdoor or under crop covers*)

Leafy vegetables - Lettuce

Diseases	Big vein	carbendazim (*off-label*)
	Botrytis	fenhexamid (*off-label*), iprodione, iprodione (*off-label*), iprodione (*outdoor crops*), propamocarb hydrochloride, thiram (*outdoor crops*)
	Bottom rot	boscalid + pyraclostrobin
	Damping off	thiram (*off-label - seed treatment*)
	Downy mildew	dimethomorph + mancozeb (*off-label*), fosetyl-aluminium (*off-label*), mancozeb, propamocarb hydrochloride
	Foliar diseases	mancozeb (*off-label*), thiram (*off-label*)
	Fungus diseases	mancozeb + metalaxyl-M (*off-label*)
	Ring spot	prochloraz (*off-label*)
	Sclerotinia	azoxystrobin (*off-label*), boscalid + pyraclostrobin, iprodione (*off-label*)

Pests	Aphids	acetamiprid (*off-label*), cypermethrin, deltamethrin (*off-label*), dimethoate, fatty acids, lambda-cyhalothrin (*off-label*), lambda-cyhalothrin + pirimicarb, nicotine, pirimicarb, pirimicarb (*off-label*), pirimicarb (*outdoor crops*), pymetrozine (*off-label*), pyrethrins
	Caterpillars	Bacillus thuringiensis (*off-label*), cypermethrin, deltamethrin (*off-label*), diflubenzuron (*off-label*), nicotine, pyrethrins
	Cutworms	cypermethrin, deltamethrin, lambda-cyhalothrin, lambda-cyhalothrin + pirimicarb
	Leaf miners	nicotine
	Mealybugs	fatty acids
	Pests, miscellaneous	deltamethrin (*off-label*)
	Scale insects	fatty acids
	Slugs/snails	methiocarb
	Spider mites	fatty acids
	Whiteflies	fatty acids
Weeds	Broad-leaved weeds	chlorpropham, pendimethalin (*off-label*), phenmedipham (*off-label*), propachlor (*off-label*), propachlor (*off-label - under crop covers*), propyzamide, propyzamide (*off-label*), propyzamide (*off-label - outdoor or under crop covers*), propyzamide (*outdoor crops*), trifluralin
	Grass weeds	chlorpropham, pendimethalin (*off-label*), propachlor (*off-label*), propachlor (*off-label - under crop covers*), propyzamide, propyzamide (*off-label*), propyzamide (*off-label - outdoor or under crop covers*), propyzamide (*outdoor crops*), trifluralin

Leafy vegetables - Spinach

Diseases	Botrytis	fenhexamid (*off-label*)
	Damping off	cymoxanil + fludioxonil + metalaxyl-M (*off-label*)
	Downy mildew	boscalid + pyraclostrobin (*off-label - Must be applied between 1 Apr and 31 Oct*), fosetyl-aluminium (*off-label*), metalaxyl-M (*off-label*)
	Foliar diseases	thiram (*off-label*)
Pests	Aphids	deltamethrin (*off-label*), nicotine, pirimicarb (*off-label*), pymetrozine
	Caterpillars	Bacillus thuringiensis (*off-label*), diflubenzuron (*off-label*), nicotine
	Flies	tefluthrin (*off-label - seed treatment*)
	Leaf miners	nicotine
	Pests, miscellaneous	deltamethrin (*off-label*)
	Slugs/snails	methiocarb
Weeds	Broad-leaved weeds	chloridazon (*off-label*), chloridazon + quinmerac (*off-label*), clopyralid (*off-label*), phenmedipham (*off-label*)
	Crops as weeds	propaquizafop (*off-label*)
	Grass weeds	chloridazon (*off-label*), chloridazon + quinmerac (*off-label*), propaquizafop (*off-label*)

Leafy vegetables - Watercress

Diseases	Downy mildew	fosetyl-aluminium (*off-label - during propagation*), propamocarb hydrochloride (*off-label - during propagation*), propamocarb hydrochloride (*off-label - under protection*)
	Phytophthora	fosetyl-aluminium (*off-label - during propagation*), propamocarb hydrochloride (*off-label - during propagation*), propamocarb hydrochloride (*off-label - under protection*)
	Pythium	copper oxychloride (*off-label - during propagation*), fosetyl-aluminium (*off-label - during propagation*), propamocarb hydrochloride (*off-label - during propagation*), propamocarb hydrochloride (*off-label - under protection*)
	Rhizoctonia	copper oxychloride (*off-label - during propagation*)
Pests	Beetles	deltamethrin (*off-label*)
	Caterpillars	Bacillus thuringiensis (*off-label*)
	Leafhoppers	deltamethrin (*off-label*)

Legumes - Beans (Phaseolus)

Crop control	Desiccation	diquat (*off-label*)
Diseases	Botrytis	azoxystrobin (*off-label*), cyprodinil + fludioxonil (*moderate control*), iprodione (*off-label*)
	Damping off	thiram (*seed treatment*)
	Rust	tebuconazole (*off-label*)
	Sclerotinia	cyprodinil + fludioxonil, iprodione (*off-label*)
Pests	Aphids	fatty acids, nicotine, pirimicarb
	Caterpillars	Bacillus thuringiensis (*off-label*), lambda-cyhalothrin (*off-label*), nicotine
	Flies	chlorpyrifos (*off-label - harvested as a dry pulse*), chlorpyrifos (*off-label - harvested dry as a pulse*)
	Leaf miners	nicotine
	Mealybugs	fatty acids
	Scale insects	fatty acids
	Spider mites	fatty acids
	Whiteflies	fatty acids
Weeds	Broad-leaved weeds	bentazone, chlorthal-dimethyl, chlorthal-dimethyl (*off-label*), linuron (*off-label*), pendimethalin (*off-label*), trifluralin
	Crops as weeds	chlorthal-dimethyl (*off-label*), cycloxydim
	Grass weeds	cycloxydim, fluazifop-P-butyl (*off-label*), pendimethalin (*off-label*), trifluralin

Legumes - Beans (Vicia)

Crop control	Desiccation	diquat, diquat (*stock or pigeon feed only*), glufosinate-ammonium, glyphosate
Diseases	Ascochyta	azoxystrobin (*off-label*), thiabendazole + thiram (*seed treatment*)
	Botrytis	azoxystrobin (*off-label*), cyprodinil + fludioxonil (*moderate control*)

	Chocolate spot	boscalid + pyraclostrobin (*moderate*), chlorothalonil, chlorothalonil + cyproconazole, chlorothalonil + pyrimethanil, iprodione, iprodione + thiophanate-methyl, tebuconazole
	Damping off	thiabendazole + thiram (*seed treatment*), thiram (*seed treatment*)
	Downy mildew	chlorothalonil + metalaxyl-M, cymoxanil + fludioxonil + metalaxyl-M (*off-label - seed treatment*), fosetyl-aluminium
	Rust	azoxystrobin, boscalid + pyraclostrobin, chlorothalonil + cyproconazole, metconazole, metconazole (*off-label*), tebuconazole, tebuconazole (*off-label*)
	Sclerotinia	cyprodinil + fludioxonil
Pests	Aphids	fatty acids, nicotine, pirimicarb
	Beetles	lambda-cyhalothrin (*off-label*)
	Birds/mammals	aluminium ammonium sulphate
	Caterpillars	Bacillus thuringiensis (*off-label*), nicotine
	Leaf miners	nicotine
	Mealybugs	fatty acids
	Scale insects	fatty acids
	Spider mites	fatty acids
	Thrips	lambda-cyhalothrin (*off-label*)
	Weevils	alpha-cypermethrin, cypermethrin, deltamethrin, lambda-cyhalothrin, lambda-cyhalothrin + pirimicarb, zeta-cypermethrin
	Whiteflies	fatty acids
Weeds	Broad-leaved weeds	bentazone, carbetamide, clomazone, glyphosate, imazamox + pendimethalin, imazamox + pendimethalin (*off-label*), isoxaben + terbuthylazine, linuron + trifluralin, pendimethalin (*off-label*), pendimethalin + trifluralin (*off-label*), propyzamide, trifluralin
	Crops as weeds	carbetamide, cycloxydim, fluazifop-P-butyl, glyphosate, pendimethalin (*off-label*), propyzamide, quizalofop-P-ethyl, quizalofop-P-tefuryl, tepraloxydim
	Grass weeds	carbetamide, cycloxydim, fluazifop-P-butyl, fluazifop-P-butyl (*off-label*), glyphosate, linuron + trifluralin, pendimethalin (*off-label*), pendimethalin + trifluralin (*off-label*), propaquizafop, propyzamide, quizalofop-P-ethyl, quizalofop-P-tefuryl, tepraloxydim, tri-allate, tri-allate (*off-label*), trifluralin
	Weeds, miscellaneous	glyphosate

Legumes - Lupins

Crop control	Desiccation	glyphosate (*off-label*)
Diseases	Ascochyta	azoxystrobin (*off-label*), metconazole (*qualified minor use*)
	Botrytis	chlorothalonil (*off-label*), cyprodinil + fludioxonil (*off-label*), metconazole (*qualified minor use*)
	Damping off	cymoxanil + fludioxonil + metalaxyl-M (*off-label*), thiram (*off-label - seed treatment*)
	Downy mildew	cymoxanil + fludioxonil + metalaxyl-M (*off-label*), fosetyl-aluminium (*off-label - seed treatment*)

	Foliar diseases	chlorothalonil (*off-label*)
	Pythium	cymoxanil + fludioxonil + metalaxyl-M (*off-label*)
	Rust	azoxystrobin (*off-label*), metconazole (*qualified minor use*)
Pests	Aphids	deltamethrin (*off-label*), lambda-cyhalothrin (*off-label*), lambda-cyhalothrin + pirimicarb (*off-label*)
	Pests, miscellaneous	deltamethrin (*off-label*)
Weeds	Broad-leaved weeds	carbetamide (*off-label*), clomazone (*off-label*), isoxaben + terbuthylazine (*off-label*), linuron + trifluralin (*off-label*), pendimethalin (*off-label*), propyzamide (*off-label*)
	Grass weeds	carbetamide (*off-label*), linuron + trifluralin (*off-label*), pendimethalin (*off-label*), propaquizafop (*off-label*), propyzamide (*off-label*), tepraloxydim (*off-label*), tri-allate (*off-label*)

Legumes - Peas

Crop control	Desiccation	diquat, glufosinate-ammonium, glufosinate-ammonium (*not for seed*), glyphosate, sulphuric acid (commodity substance)
Diseases	Ascochyta	azoxystrobin, chlorothalonil, chlorothalonil + pyrimethanil, cymoxanil + fludioxonil + metalaxyl-M (*seed treatment*), cyprodinil + fludioxonil, metconazole (*reduction*), thiabendazole + thiram (*seed treatment*)
	Botrytis	chlorothalonil, chlorothalonil + cyproconazole, chlorothalonil + pyrimethanil, cyprodinil + fludioxonil (*moderate control*), iprodione (*off-label*), metconazole (*reduction*)
	Damping off	cymoxanil + fludioxonil + metalaxyl-M (*seed treatment*), thiabendazole + thiram (*seed treatment*), thiram (*seed treatment*)
	Downy mildew	cymoxanil + fludioxonil + metalaxyl-M (*seed treatment*), fosetyl-aluminium (*off-label - seed treatment*)
	Mycosphaerella	chlorothalonil, chlorothalonil + pyrimethanil, cyprodinil + fludioxonil, metconazole (*reduction*)
	Pythium	cymoxanil + fludioxonil + metalaxyl-M (*seed treatment*)
	Rust	metconazole
	Sclerotinia	cyprodinil + fludioxonil, iprodione (*off-label*)
Pests	Aphids	alpha-cypermethrin (*reduction*), bifenthrin, cypermethrin, deltamethrin, fatty acids, lambda-cyhalothrin, lambda-cyhalothrin + pirimicarb, nicotine, pirimicarb, zeta-cypermethrin
	Birds/mammals	aluminium ammonium sulphate
	Caterpillars	alpha-cypermethrin, cypermethrin, deltamethrin, lambda-cyhalothrin, lambda-cyhalothrin + pirimicarb, nicotine, zeta-cypermethrin
	Leaf miners	nicotine
	Mealybugs	fatty acids
	Midges	deltamethrin, lambda-cyhalothrin, lambda-cyhalothrin + pirimicarb
	Scale insects	fatty acids
	Spider mites	fatty acids

	Weevils	alpha-cypermethrin, cypermethrin, deltamethrin, lambda-cyhalothrin, lambda-cyhalothrin + pirimicarb, zeta-cypermethrin
	Whiteflies	fatty acids
Weeds	Broad-leaved weeds	bentazone, chlorthal-dimethyl (*off-label*), clomazone, glyphosate, imazamox + pendimethalin, isoxaben + terbuthylazine, linuron + trifluralin, MCPA + MCPB, MCPB, pendimethalin, pendimethalin (*off-label*)
	Crops as weeds	cycloxydim, fluazifop-P-butyl, glyphosate, pendimethalin, quizalofop-P-ethyl, quizalofop-P-tefuryl, tepraloxydim
	Grass weeds	cycloxydim, fluazifop-P-butyl, glyphosate, linuron + trifluralin, pendimethalin, propaquizafop, quizalofop-P-ethyl, quizalofop-P-tefuryl, tepraloxydim, tri-allate
	Weeds, miscellaneous	glyphosate

Miscellaneous arable - Industrial crops

Crop control	Desiccation	glyphosate (*off-label - for wild bird seed production*)

Miscellaneous arable - Miscellaneous arable crops

Crop control	Desiccation	glyphosate (*off-label - for wild bird seed production*)
Diseases	Botrytis	iprodione + thiophanate-methyl (*off-label*)
	Sclerotinia	Coniothyrium minitans
Pests	Aphids	lambda-cyhalothrin + pirimicarb (*off-label - for wild bird seed production*), physical pest control, pyrethrins
	Caterpillars	pyrethrins
	Mealybugs	physical pest control
	Slugs/snails	ferric phosphate, metaldehyde, metaldehyde (*around*)
	Spider mites	physical pest control, pyrethrins
	Thrips	pyrethrins
	Whiteflies	physical pest control
Plant growth regulation	Growth control	trinexapac-ethyl (*off-label - for wild bird seed production*)
Weeds	Broad-leaved weeds	bromoxynil + ioxynil + mecoprop-P (*off-label - for wild bird seed production*), carfentrazone-ethyl (*before planting*), diquat (*around*), diquat + paraquat, glyphosate (*pre-planting or sowing*), isoxaben (*off-label - grown for game cover*), paraquat
	Crops as weeds	carfentrazone-ethyl (*before planting*), diquat + paraquat, paraquat, quizalofop-P-tefuryl
	Grass weeds	diquat + paraquat, glyphosate (*pre-planting or sowing*), paraquat, quizalofop-P-tefuryl, tri-allate (*off-label - for wild bird seed production*)
	Weeds, miscellaneous	clopyralid (*off-label - game cover*), glyphosate (*before planting*), glyphosate (*pre-sowing/planting*)

Miscellaneous arable - Miscellaneous arable situations

Diseases	Soil-borne diseases	dazomet (*soil fumigation*)
Pests	Birds/mammals	aluminium ammonium sulphate (*seed treatment*), ziram

	Free-living nematodes	dazomet (*soil fumigation*)
	Slugs/snails	metaldehyde
	Soil pests	dazomet (*soil fumigation*)
Weeds	Broad-leaved weeds	amitrole, citronella oil, diquat + paraquat, glufosinate-ammonium, glufosinate-ammonium (*uncropped*), glyphosate, paraquat, paraquat (*minimum cultivation*)
	Crops as weeds	amitrole (*barley stubble*), diquat + paraquat, fluazifop-P-butyl, glyphosate, paraquat, paraquat (*minimum cultivation*), tepraloxydim
	Grass weeds	amitrole, diquat + paraquat, fluazifop-P-butyl, glufosinate-ammonium, glufosinate-ammonium (*uncropped*), glyphosate, paraquat, paraquat (*minimum cultivation*), paraquat (*stubble treatment*), tepraloxydim
	Weeds, miscellaneous	2,4-D + dicamba + triclopyr, cycloxydim, dazomet (*soil fumigation*), diquat + paraquat, fluazifop-P-butyl, glufosinate-ammonium, glyphosate, glyphosate (*wiper application*), metsulfuron-methyl, paraquat, tepraloxydim, thifensulfuron-methyl

Miscellaneous field vegetables - All vegetables

Diseases	Soil-borne diseases	dazomet (*soil fumigation*)
Pests	Aphids	physical pest control, rotenone
	Free-living nematodes	dazomet (*soil fumigation*)
	Leafhoppers	physical pest control
	Mites	physical pest control
	Soil pests	dazomet (*soil fumigation*)
	Thrips	physical pest control
	Whiteflies	physical pest control
Weeds	Broad-leaved weeds	glufosinate-ammonium
	Grass weeds	glufosinate-ammonium
	Weeds, miscellaneous	dazomet (*soil fumigation*), glufosinate-ammonium

Oilseed crops - Linseed/flax

Crop control	Desiccation	diquat, glufosinate-ammonium, glyphosate
Diseases	Alternaria	iprodione (*seed treatment*)
	Botrytis	chlorothalonil (*off-label*), metconazole (*off-label*), tebuconazole (*reduction*)
	Disease control/foliar feed	prochloraz + tebuconazole (*off-label*)
	Downy mildew	chlorothalonil (*off-label*)
	Foliar diseases	carbendazim (*off-label*), difenoconazole (*off-label*), prochloraz (*off-label*)
	Powdery mildew	tebuconazole
	Seed-borne diseases	prochloraz (*seed treatment*)
Pests	Aphids	lambda-cyhalothrin (*off-label*), lambda-cyhalothrin + pirimicarb (*off-label*), pirimicarb (*off-label*), tau-fluvalinate (*off-label*)
	Beetles	bifenthrin, zeta-cypermethrin

Weeds	Broad-leaved weeds	amidosulfuron, bentazone, bromoxynil, carbetamide (*off-label*), clomazone (*off-label*), clopyralid, clopyralid + picloram (*off-label*), flupyrsulfuron-methyl (*off-label*), glyphosate, MCPA, metazachlor (*off-label*), metazachlor + quinmerac (*off-label*), metsulfuron-methyl, trifluralin
	Crops as weeds	cycloxydim, fluazifop-P-butyl, glyphosate, quizalofop-P-ethyl, quizalofop-P-tefuryl, tepraloxydim
	Grass weeds	carbetamide (*off-label*), cycloxydim, fluazifop-P-butyl, flupyrsulfuron-methyl (*off-label*), glyphosate, metazachlor (*off-label*), metazachlor + quinmerac (*off-label*), propaquizafop, propaquizafop (*off-label*), quizalofop-P-ethyl, quizalofop-P-tefuryl, tepraloxydim, trifluralin
	Weeds, miscellaneous	glyphosate

Oilseed crops - Miscellaneous oilseeds

Crop control	Desiccation	diquat (*off-label*), glyphosate (*off-label*)
Diseases	Alternaria	boscalid (*off-label*)
	Botrytis	boscalid (*off-label*), chlorothalonil (*off-label*)
	Damping off	cymoxanil + fludioxonil + metalaxyl-M (*off-label - seed treatment*), thiram (*off-label - seed treatment*)
	Downy mildew	chlorothalonil (*off-label*), chlorothalonil + metalaxyl-M (*off-label*), cymoxanil + fludioxonil + metalaxyl-M (*off-label - seed treatment*), dimethomorph + mancozeb (*off-label*), mancozeb (*off-label*)
	Foliar diseases	carbendazim (*off-label*), difenoconazole (*off-label*), tebuconazole (*off-label*)
	Powdery mildew	azoxystrobin (*off-label*), sulphur (*off-label*)
	Pythium	cymoxanil + fludioxonil + metalaxyl-M (*off-label - seed treatment*), thiram (*off-label - seed treatment*)
	Sclerotinia	boscalid (*off-label*)
	Seed-borne diseases	prochloraz (*off-label - seed treatment*)
	White blister	propiconazole (*off-label*)
Pests	Aphids	deltamethrin (*off-label*), lambda-cyhalothrin (*off-label*), lambda-cyhalothrin + pirimicarb (*off-label*), pirimicarb (*off-label*), tau-fluvalinate (*off-label*)
	Beetles	deltamethrin (*off-label*), lambda-cyhalothrin (*off-label*)
Weeds	Broad-leaved weeds	asulam (*off-label*), carbetamide (*off-label*), chlorthal-dimethyl (*off-label*), clomazone (*off-label*), clopyralid (*off-label*), clopyralid + picloram (*off-label*), ethofumesate (*off-label*), fluroxypyr (*off-label*), metazachlor (*off-label*), metazachlor + quinmerac (*off-label*), pendimethalin (*off-label*), propyzamide (*off-label*)
	Crops as weeds	fluroxypyr (*off-label*), propaquizafop (*off-label*)
	Grass weeds	asulam (*off-label*), carbetamide (*off-label*), ethofumesate (*off-label*), metazachlor (*off-label*), metazachlor + quinmerac (*off-label*), pendimethalin (*off-label*), propaquizafop (*off-label*), propyzamide (*off-label*), tepraloxydim (*off-label*)
	Weeds, miscellaneous	diquat (*off-label*)

Oilseed crops - Oilseed rape

Crop control	Desiccation	diquat, glufosinate-ammonium, glyphosate
Diseases	Alternaria	azoxystrobin, azoxystrobin + cyproconazole, boscalid, difenoconazole, iprodione, iprodione (*seed treatment*), iprodione + thiophanate-methyl, metconazole, prochloraz, tebuconazole
	Black scurf and stem canker	difenoconazole, iprodione + thiophanate-methyl, tebuconazole
	Botrytis	chlorothalonil, iprodione, iprodione + thiophanate-methyl, prochloraz
	Damping off	thiram (*seed treatment*)
	Downy mildew	chlorothalonil, mancozeb
	Light leaf spot	carbendazim, carbendazim + flusilazole, difenoconazole, famoxadone + flusilazole, flusilazole, iprodione + thiophanate-methyl, metconazole, prochloraz, prochloraz + propiconazole, propiconazole (*reduction*), prothioconazole + tebuconazole, tebuconazole
	Phoma leaf spot	carbendazim + flusilazole
	Powdery mildew	sulphur
	Ring spot	tebuconazole, tebuconazole (*reduction*)
	Sclerotinia	azoxystrobin, azoxystrobin + cyproconazole, boscalid, iprodione, iprodione + thiophanate-methyl, prochloraz, prochloraz + tebuconazole, prothioconazole, prothioconazole + tebuconazole, tebuconazole
	Stem canker	carbendazim + flusilazole (*reduction*), famoxadone + flusilazole (*suppression*), metconazole (*reduction*), prochloraz, prochloraz + propiconazole, prochloraz + thiram (*seed treatment*), prothioconazole + tebuconazole, tebuconazole
	White leaf spot	prochloraz
Pests	Aphids	deltamethrin, lambda-cyhalothrin, lambda-cyhalothrin + pirimicarb, pirimicarb, tau-fluvalinate
	Beetles	alpha-cypermethrin, beta-cyfluthrin + imidacloprid (*seed treatment*), bifenthrin, cyfluthrin, cypermethrin, deltamethrin, lambda-cyhalothrin, lambda-cyhalothrin + pirimicarb, tau-fluvalinate, thiacloprid, zeta-cypermethrin
	Birds/mammals	aluminium ammonium sulphate
	Midges	alpha-cypermethrin, bifenthrin, cypermethrin, deltamethrin, lambda-cyhalothrin, lambda-cyhalothrin + pirimicarb, zeta-cypermethrin
	Slugs/snails	methiocarb, thiodicarb
	Weevils	alpha-cypermethrin, bifenthrin, cypermethrin, deltamethrin, lambda-cyhalothrin, lambda-cyhalothrin + pirimicarb, zeta-cypermethrin
Plant growth regulation	Growth control	tebuconazole
	Quality/yield control	sulphur
Weeds	Broad-leaved weeds	bifenox (*off-label*), carbetamide, clomazone, clopyralid, clopyralid + picloram, clopyralid + picloram (*off-label*), dimethenamid-p + metazachlor, glyphosate, metazachlor, metazachlor + quinmerac, napropamide, propachlor, propyzamide, trifluralin

	Crops as weeds	carbetamide, cycloxydim, fluazifop-P-butyl, glyphosate, propyzamide, quizalofop-P-ethyl, quizalofop-P-tefuryl, tepraloxydim
	Grass weeds	carbetamide, cycloxydim, fluazifop-P-butyl, glyphosate, metazachlor, metazachlor + quinmerac, napropamide, propachlor, propaquizafop, propyzamide, quizalofop-P-ethyl, quizalofop-P-tefuryl, tepraloxydim, trifluralin
	Weeds, miscellaneous	glyphosate

Oilseed crops - Soya

Diseases	Damping off	thiram (*seed treatment - qualified minor use*), thiram (*seed treatment - qualified minor use*)
Weeds	Broad-leaved weeds	bentazone (*off-label*)

Oilseed crops - Sunflowers

Crop control	Desiccation	diquat (*off-label*)
Pests	Slugs/snails	methiocarb
Weeds	Broad-leaved weeds	isoxaben (*off-label - grown for game cover*), pendimethalin
	Crops as weeds	pendimethalin
	Grass weeds	pendimethalin

Root and tuber crops - Beet crops

Diseases	Aphanomyces	hymexazol (*off-label - seed treatment*)
	Black leg	hymexazol (*seed treatment*)
	Botrytis	iprodione (*off-label*)
	Cercospora leaf spot	epoxiconazole + pyraclostrobin
	Damping off	cymoxanil + fludioxonil + metalaxyl-M (*off-label - seed treatment*)
	Downy mildew	mancozeb (*off-label*)
	Foliar diseases	epoxiconazole + pyraclostrobin (*off-label*)
	Powdery mildew	carbendazim + flusilazole, carbendazim + flusilazole (*off-label*), difenoconazole + fenpropidin, epoxiconazole + pyraclostrobin, flusilazole, quinoxyfen, sulphur, sulphur (*off-label*)
	Ramularia leaf spots	difenoconazole + fenpropidin, epoxiconazole + pyraclostrobin, propiconazole (*reduction*)
	Root malformation disorder	azoxystrobin (*off-label*), mancozeb (*off-label*), metalaxyl-M (*off-label*), propamocarb hydrochloride (*off-label*)
	Rust	carbendazim + flusilazole, difenoconazole + fenpropidin, epoxiconazole + pyraclostrobin, fenpropimorph (*off-label*), flusilazole, propiconazole
	Seed-borne diseases	thiram (*seed soak*)
	Stem canker	iprodione (*off-label*)
Pests	Aphids	beta-cyfluthrin + clothianidin (*seed treatment*), carbosulfan, deltamethrin (*off-label*), dimethoate, dimethoate (*excluding Myzus persicae*), imidacloprid (*seed treatment*), lambda-cyhalothrin + pirimicarb, lambda-cyhalothrin + pirimicarb (*off-label*), nicotine,

		oxamyl, oxamyl (*off-label*), pirimicarb, pirimicarb (*off-label*)
	Beetles	benfuracarb, carbosulfan, chlorpyrifos, chlorpyrifos (*off-label*), deltamethrin, deltamethrin (*off-label*), imidacloprid (*seed treatment*), lambda-cyhalothrin, lambda-cyhalothrin (*off-label*), lambda-cyhalothrin + pirimicarb, lambda-cyhalothrin + pirimicarb (*off-label*), oxamyl, oxamyl (*off-label*), tefluthrin (*seed treatment*), thiamethoxam (*seed treatment*)
	Birds/mammals	aluminium ammonium sulphate
	Caterpillars	cypermethrin, lambda-cyhalothrin (*off-label*), nicotine
	Cutworms	Bacillus thuringiensis (*off-label*), cypermethrin, lambda-cyhalothrin, lambda-cyhalothrin (*off-label*), lambda-cyhalothrin + pirimicarb, lambda-cyhalothrin + pirimicarb (*off-label*), methiocarb (*reduction*), zeta-cypermethrin
	Flies	carbosulfan, imidacloprid (*seed treatment*), oxamyl, oxamyl (*off-label*)
	Free-living nematodes	benfuracarb, carbosulfan, oxamyl, oxamyl (*off-label*)
	Leaf miners	benfuracarb, dimethoate, lambda-cyhalothrin, lambda-cyhalothrin (*off-label*), lambda-cyhalothrin + pirimicarb, lambda-cyhalothrin + pirimicarb (*off-label*), nicotine, thiamethoxam (*seed treatment*)
	Leatherjackets	chlorpyrifos, chlorpyrifos (*off-label*), methiocarb (*reduction*)
	Millipedes	benfuracarb, carbosulfan, imidacloprid (*seed treatment*), methiocarb (*reduction*), oxamyl, oxamyl (*off-label*), tefluthrin (*seed treatment*), thiamethoxam (*seed treatment*)
	Pests, miscellaneous	deltamethrin (*off-label*)
	Slugs/snails	methiocarb
	Springtails	benfuracarb, carbosulfan, imidacloprid (*seed treatment*), tefluthrin (*seed treatment*), thiamethoxam (*seed treatment*)
	Symphylids	benfuracarb, carbosulfan, imidacloprid (*seed treatment*), tefluthrin (*seed treatment*), thiamethoxam (*seed treatment*)
	Wireworms	carbosulfan, thiamethoxam (*seed treatment - reduction*)
Plant growth regulation	Quality/yield control	sulphur
Weeds	Broad-leaved weeds	carbetamide, chloridazon, chloridazon + chlorpropham + metamitron, chloridazon + ethofumesate, chloridazon + lenacil, chloridazon + metamitron, chloridazon + metamitron (*off-label*), chloridazon + quinmerac, chloridazon + quinmerac (*off-label*), clopyralid, desmedipham + ethofumesate + phenmedipham, desmedipham + phenmedipham, diquat, ethofumesate, ethofumesate + metamitron, ethofumesate + phenmedipham, glufosinate-ammonium, glyphosate, lenacil, lenacil + triflusulfuron-methyl, metamitron, paraquat, phenmedipham, propyzamide, sodium chloride (commodity substance), trifluralin, triflusulfuron-methyl, triflusulfuron-methyl (*off-label*)
	Crops as weeds	carbetamide, cycloxydim, fluazifop-P-butyl, glyphosate, glyphosate (*wiper application*), lenacil + triflusulfuron-methyl, paraquat, propaquizafop (*off-label*),

		propyzamide, quizalofop-P-ethyl, quizalofop-P-tefuryl, sodium chloride (commodity substance), tepraloxydim
	Grass weeds	carbetamide, chloridazon, chloridazon + chlorpropham + metamitron, chloridazon + ethofumesate, chloridazon + lenacil, chloridazon + metamitron, chloridazon + metamitron (*off-label*), chloridazon + quinmerac, chloridazon + quinmerac (*off-label*), cycloxydim, desmedipham + ethofumesate + phenmedipham, ethofumesate, ethofumesate + metamitron, ethofumesate + phenmedipham, fluazifop-P-butyl, fluazifop-P-butyl (*off-label*), glufosinate-ammonium, glyphosate, lenacil, metamitron, paraquat, propaquizafop, propaquizafop (*off-label*), propyzamide, quizalofop-P-ethyl, quizalofop-P-tefuryl, tepraloxydim, tri-allate, trifluralin
	Weeds, miscellaneous	glufosinate-ammonium

Root and tuber crops - Carrots/parsnips

Diseases	Alternaria	azoxystrobin, azoxystrobin (*off-label*), azoxystrobin + difenoconazole, azoxystrobin + difenoconazole (*off-label*), boscalid + pyraclostrobin (*moderate*), fenpropimorph (*off-label*), iprodione + thiophanate-methyl (*off-label*), tebuconazole, tebuconazole + trifloxystrobin
	Cavity spot	metalaxyl-M, metalaxyl-M (*off-label*)
	Crown rot	fenpropimorph (*off-label*), iprodione + thiophanate-methyl (*off-label*)
	Foliar diseases	tebuconazole + trifloxystrobin (*off-label*)
	Powdery mildew	azoxystrobin, azoxystrobin (*off-label*), azoxystrobin + difenoconazole, azoxystrobin + difenoconazole (*off-label*), boscalid + pyraclostrobin, fenpropimorph (*off-label*), sulphur (*off-label*), tebuconazole, tebuconazole + trifloxystrobin
	Pythium	cymoxanil + fludioxonil + metalaxyl-M (*off-label - seed treatment*)
	Sclerotinia	boscalid + pyraclostrobin (*moderate*), boscalid + pyraclostrobin (*off-label*), tebuconazole, tebuconazole + trifloxystrobin
	Seed-borne diseases	thiram (*seed soak*)
Pests	Aphids	carbosulfan, deltamethrin (*off-label*), lambda-cyhalothrin + pirimicarb, lambda-cyhalothrin + pirimicarb (*off-label*), nicotine, pirimicarb
	Birds/mammals	aluminium ammonium sulphate
	Caterpillars	deltamethrin (*off-label*), nicotine
	Cutworms	Bacillus thuringiensis (*off-label*), lambda-cyhalothrin, lambda-cyhalothrin + pirimicarb, lambda-cyhalothrin + pirimicarb (*off-label*)
	Flies	lambda-cyhalothrin, lambda-cyhalothrin (*off-label*), tefluthrin (*off-label - seed treatment*)
	Free-living nematodes	carbosulfan
	Leaf miners	nicotine
	Pests, miscellaneous	deltamethrin (*off-label*)
	Stem nematodes	oxamyl

Plant growth regulation	Growth control	maleic hydrazide (*off-label*)
Weeds	Broad-leaved weeds	chlorpropham, clomazone, isoxaben (*off-label*), linuron, metamitron (*off-label*), metribuzin (*off-label*), pendimethalin, trifluralin
	Crops as weeds	cycloxydim, fluazifop-P-butyl, pendimethalin, tepraloxydim
	Grass weeds	chlorpropham, cycloxydim, fluazifop-P-butyl, fluazifop-P-butyl (*off-label*), linuron, metamitron (*off-label*), metribuzin (*off-label*), pendimethalin, propaquizafop, tepraloxydim, tepraloxydim (*off-label*), trifluralin
	Weeds, miscellaneous	metamitron (*off-label*)

Root and tuber crops - Miscellaneous root crops

Diseases	Alternaria	azoxystrobin (*off-label*), azoxystrobin + difenoconazole (*off-label*), fenpropimorph (*off-label*)
	Crown rot	fenpropimorph (*off-label*)
	Foliar diseases	tebuconazole (*off-label*), tebuconazole + trifloxystrobin (*off-label*)
	Powdery mildew	azoxystrobin (*off-label*), azoxystrobin + difenoconazole (*off-label*), fenpropimorph (*off-label*)
	Rust	azoxystrobin (*off-label*)
	Sclerotinia	azoxystrobin (*off-label*)
	Septoria	chlorothalonil (*off-label*), difenoconazole (*off-label*), mancozeb (*off-label*)
Pests	Aphids	deltamethrin (*off-label*), nicotine, pirimicarb (*off-label*), pymetrozine (*off-label*)
	Caterpillars	nicotine
	Cutworms	Bacillus thuringiensis (*off-label*), lambda-cyhalothrin (*off-label*)
	Flies	lambda-cyhalothrin (*off-label*)
	Leaf miners	nicotine
	Pests, miscellaneous	deltamethrin (*off-label*), dimethoate (*off-label*)
Weeds	Broad-leaved weeds	chlorpropham (*off-label*), linuron (*off-label*), metribuzin (*off-label*), pendimethalin (*off-label*), prosulfocarb (*off-label*)
	Grass weeds	chlorpropham (*off-label*), metribuzin (*off-label*), pendimethalin (*off-label*)
	Weeds, miscellaneous	linuron (*off-label*)

Root and tuber crops - Potatoes

Crop control	Desiccation	carfentrazone-ethyl, diquat, glufosinate-ammonium, sulphuric acid (commodity substance)
Diseases	Black dot	azoxystrobin
	Black scurf and stem canker	azoxystrobin, flutolanil (*tuber treatment*), imazalil + pencycuron (*tuber treatment*), imazalil + pencycuron (*tuber treatment - reduction*), iprodione (*seed treatment*), pencycuron (*tuber treatment*), tolclofos-methyl (*tuber treatment*)

	Dry rot	imazalil, imazalil + thiabendazole (*tuber treatment - post-harvest*), thiabendazole (*tuber treatment - post-harvest*)
	Fungus diseases	peroxyacetic acid (commodity substance) (*tuber treatment*)
	Gangrene	imazalil, imazalil + thiabendazole (*tuber treatment - post-harvest*), thiabendazole (*tuber treatment - post-harvest*)
	Phytophthora	benalaxyl + mancozeb, benthiavalicarb-isopropyl + mancozeb, Bordeaux mixture, chlorothalonil, chlorothalonil + mancozeb, chlorothalonil + propamocarb hydrochloride, copper oxychloride, cyazofamid, cymoxanil, cymoxanil + famoxadone, cymoxanil + mancozeb, dimethomorph + mancozeb, fenamidone + propamocarb hydrochloride, fluazinam, fluazinam + metalaxyl-M, fluopicolide + propamocarb hydrochloride, mancozeb, mancozeb + metalaxyl-M, mancozeb + propamocarb hydrochloride, mancozeb + zoxamide, mandipropamid
	Powdery scab	fluazinam (*off-label*)
	Rhizoctonia	tolclofos-methyl (*off-label - chitted seed treatment*)
	Silver scurf	imazalil, imazalil + pencycuron (*tuber treatment - reduction*), imazalil + thiabendazole (*tuber treatment - post-harvest*), thiabendazole (*tuber treatment - post-harvest*)
	Skin spot	imazalil, thiabendazole (*tuber treatment - post-harvest*)
Pests	Aphids	acetamiprid, flonicamid, lambda-cyhalothrin, lambda-cyhalothrin + pirimicarb, nicotine, oxamyl, pirimicarb, pymetrozine, thiacloprid
	Caterpillars	cypermethrin, nicotine
	Cutworms	chlorpyrifos, cypermethrin, lambda-cyhalothrin + pirimicarb, zeta-cypermethrin
	Cyst nematodes	1,3-dichloropropene, ethoprophos, fosthiazate, oxamyl
	Free-living nematodes	1,3-dichloropropene, fosthiazate (*reduction*), oxamyl, peroxyacetic acid (commodity substance) (*tuber treatment*)
	Leaf miners	nicotine
	Leatherjackets	methiocarb (*reduction*)
	Slugs/snails	methiocarb, thiodicarb
	Wireworms	ethoprophos, fosthiazate (*reduction*)
Plant growth regulation	Growth control	chlorpropham, chlorpropham (*thermal fog*), ethylene (commodity substance) (*in store*), maleic hydrazide
	Plant growth regulation, miscellaneous	maleic hydrazide
Weeds	Broad-leaved weeds	bentazone, carfentrazone-ethyl, clomazone, diquat + paraquat, flufenacet + metribuzin, glufosinate-ammonium, linuron, metribuzin, paraquat, pendimethalin, prosulfocarb, rimsulfuron
	Crops as weeds	carfentrazone-ethyl, cycloxydim, diquat + paraquat, metribuzin, paraquat, pendimethalin, quizalofop-P-tefuryl, rimsulfuron
	Grass weeds	cycloxydim, diquat + paraquat, flufenacet + metribuzin, glufosinate-ammonium, linuron, metribuzin, paraquat, pendimethalin, propaquizafop, prosulfocarb, quizalofop-P-tefuryl
	Weeds, miscellaneous	diquat, glufosinate-ammonium (*not for seed*)

Stem and bulb vegetables - Asparagus

Diseases	Botrytis	cyprodinil + fludioxonil (*off-label*)
	Phytophthora	metalaxyl-M (*off-label*)
	Rust	azoxystrobin, azoxystrobin + difenoconazole (*off-label*), difenoconazole (*off-label*), mancozeb (*off-label*)
	Stemphylium	azoxystrobin, azoxystrobin + difenoconazole (*off-label*), cyprodinil + fludioxonil (*off-label*), iprodione (*off-label*), mancozeb (*off-label*)
Pests	Aphids	nicotine
	Caterpillars	nicotine
	Leaf miners	nicotine
Weeds	Broad-leaved weeds	clomazone (*off-label*), clopyralid (*off-label*), diuron (*off-label*), isoxaben (*off-label*), linuron (*off-label*), metamitron (*off-label*), metribuzin (*off-label*), pendimethalin (*off-label*), triclopyr (*off-label - directed treatment*)
	Grass weeds	diuron (*off-label*), linuron (*off-label*), metamitron (*off-label*), metribuzin (*off-label*), pendimethalin (*off-label*), triclopyr (*off-label - directed treatment*)
	Weeds, miscellaneous	glyphosate (*off-label*), metamitron (*off-label*), metribuzin (*off-label*)

Stem and bulb vegetables - Celery/chicory

Diseases	Botrytis	azoxystrobin (*off-label*), iprodione (*off-label*)
	Damping off	etridiazole (*seeds and seedlings*), thiram (*off-label - seed treatment*)
	Downy mildew	dimethomorph + mancozeb (*off-label*), mancozeb (*off-label*)
	Foliar diseases	mancozeb (*off-label*), thiram (*off-label*)
	Phytophthora	etridiazole (*transplants*), fosetyl-aluminium (*off-label - for forcing*)
	Rhizoctonia	azoxystrobin (*off-label*)
	Rust	mancozeb (*off-label*)
	Sclerotinia	azoxystrobin (*off-label*), iprodione (*off-label*)
	Seed-borne diseases	thiram (*seed soak*)
	Septoria	Bordeaux mixture, chlorothalonil (*qualified minor use*), copper ammonium carbonate, copper oxychloride, difenoconazole (*off-label*), mancozeb (*off-label*)
Pests	Aphids	acetamiprid (*off-label*), deltamethrin (*off-label*), nicotine, pirimicarb, pirimicarb (*off-label*), pirimicarb (*off-label - for forcing*), pymetrozine (*off-label*)
	Beetles	deltamethrin (*off-label*)
	Caterpillars	Bacillus thuringiensis (*off-label*), deltamethrin (*off-label*), diflubenzuron (*off-label*), lambda-cyhalothrin (*off-label*), nicotine
	Cutworms	lambda-cyhalothrin (*off-label - for forcing*)
	Flies	lambda-cyhalothrin (*off-label*)
	Leaf miners	nicotine
	Leafhoppers	deltamethrin (*off-label*)
	Pests, miscellaneous	deltamethrin (*off-label*)
Plant growth regulation	Quality/yield control	gibberellins

Weeds	Broad-leaved weeds	chlorpropham, linuron, metamitron (*off-label*), pendimethalin (*off-label*), propachlor (*off-label*), propachlor (*off-label - under crop covers*), propyzamide (*off-label*), triflusulfuron-methyl (*off-label*)
	Crops as weeds	propaquizafop (*off-label*)
	Grass weeds	chlorpropham, linuron, pendimethalin (*off-label*), propachlor (*off-label*), propachlor (*off-label - under crop covers*), propaquizafop (*off-label*), propyzamide (*off-label*)

Stem and bulb vegetables - Globe artichokes/cardoons

Diseases	Downy mildew	azoxystrobin (*off-label*)
	Powdery mildew	azoxystrobin (*off-label*), myclobutanil (*off-label*)
Pests	Aphids	nicotine
	Caterpillars	nicotine
	Leaf miners	nicotine
Weeds	Broad-leaved weeds	metamitron (*off-label*)

Stem and bulb vegetables - Onions/leeks/garlic

Diseases	Bacterial blight	copper oxychloride (*off-label*)
	Botrytis	chlorothalonil, chlorothalonil (*off-label*), cyprodinil + fludioxonil (*off-label*), iprodione, iprodione (*off-label*), thiabendazole + thiram (*off-label - seed treatment*)
	Collar rot	iprodione
	Damping off	thiram (*off-label - seed treatment*), thiram (*seed treatment*)
	Downy mildew	azoxystrobin, azoxystrobin (*off-label*), chlorothalonil + metalaxyl-M (*qualified minor use*), dimethomorph + mancozeb (*off-label*), fosetyl-aluminium (*off-label*), mancozeb (*off-label*), mancozeb + metalaxyl-M (*off-label*)
	Fusarium	thiabendazole + thiram (*off-label - seed treatment*)
	Phytophthora	propamocarb hydrochloride
	Purple blotch	azoxystrobin, azoxystrobin + difenoconazole, tebuconazole + trifloxystrobin
	Pythium	propamocarb hydrochloride
	Rhynchosporium	propiconazole (*off-label*)
	Rust	azoxystrobin, azoxystrobin + difenoconazole, fenpropimorph, propiconazole (*off-label*), tebuconazole, tebuconazole + trifloxystrobin
	Storage rots	copper oxychloride (*off-label*)
	White rot	tebuconazole (*off-label*)
	White tip	azoxystrobin + difenoconazole (*qualified minor use*), boscalid + pyraclostrobin (*off-label*), chlorothalonil + metalaxyl-M (*qualified minor use*), dimethomorph + mancozeb (*off-label*), mancozeb (*off-label*), tebuconazole + trifloxystrobin
Pests	Aphids	deltamethrin (*off-label*), dimethoate (*off-label*), nicotine
	Caterpillars	deltamethrin (*off-label*), nicotine
	Cutworms	Bacillus thuringiensis (*off-label*), chlorpyrifos
	Flies	tefluthrin (*off-label - seed treatment*)

	Free-living nematodes	oxamyl (*off-label*)
	Leaf miners	nicotine
	Pests, miscellaneous	deltamethrin (*off-label*)
	Stem nematodes	oxamyl (*off-label*)
	Thrips	deltamethrin (*off-label*), lambda-cyhalothrin (*off-label*), spinosad, spinosad (*off-label*)
Plant growth regulation	Growth control	maleic hydrazide, maleic hydrazide (*off-label*)
Weeds	Broad-leaved weeds	bentazone (*off-label*), chloridazon (*off-label*), chloridazon + propachlor (*off-label*), chloridazon + propachlor (*off-label - post-em. Pre-em on label*), chlorpropham, chlorpropham (*off-label*), chlorthal-dimethyl, chlorthal-dimethyl (*off-label*), chlorthal-dimethyl + propachlor, chlorthal-dimethyl + propachlor (*off-label*), clopyralid, fluroxypyr (*off-label*), glyphosate, ioxynil, linuron (*off-label*), metazachlor (*off-label*), pendimethalin, pendimethalin (*off-label*), pendimethalin (*off-label - pre + post emergence treatment*), propachlor, prosulfocarb (*off-label*)
	Crops as weeds	cycloxydim, fluazifop-P-butyl, fluroxypyr (*off-label*), glyphosate, pendimethalin, propaquizafop (*off-label*), prosulfocarb (*off-label*), tepraloxydim
	Grass weeds	chloridazon (*off-label*), chloridazon + propachlor (*off-label*), chloridazon + propachlor (*off-label - post-em. Pre-em on label*), chlorpropham, chlorpropham (*off-label*), chlorthal-dimethyl + propachlor, chlorthal-dimethyl + propachlor (*off-label*), cycloxydim, fluazifop-P-butyl, fluazifop-P-butyl (*off-label*), glyphosate, metazachlor (*off-label*), pendimethalin, pendimethalin (*off-label*), propachlor, propaquizafop, propaquizafop (*off-label*), tepraloxydim, tepraloxydim (*off-label*)
	Weeds, miscellaneous	linuron (*off-label*)

Flowers and ornamentals

Flowers - Bedding plants, general

Diseases	Alternaria	iprodione (*seed treatment*)
	Botrytis	carbendazim (*off-label*), thiram
	Petal blight	mancozeb
	Phytophthora	propamocarb hydrochloride
	Powdery mildew	bupirimate, carbendazim (*off-label*), copper ammonium carbonate, dinocap
	Pythium	propamocarb hydrochloride
	Rust	azoxystrobin (*off-label - in pots*), mancozeb, propiconazole (*off-label*), thiram
Pests	Aphids	imidacloprid, malathion, thiacloprid
	Beetles	thiacloprid
	Flies	imidacloprid
	Weevils	imidacloprid, thiacloprid
	Whiteflies	imidacloprid, thiacloprid

Plant growth regulation	Flowering control	paclobutrazol
	Growth control	chlormequat, daminozide, paclobutrazol
Weeds	Broad-leaved weeds	chlorpropham, chlorthal-dimethyl
	Grass weeds	chlorpropham

Flowers - Bulbs/corms

Crop control	Desiccation	sulphuric acid (commodity substance)
Diseases	Botrytis	carbendazim (*off-label*), chlorothalonil (*off-label - for galanthamine production*), tebuconazole (*off-label - for galanthamine production*), thiram
	Crown rot	metalaxyl-M (*off-label*)
	Fire	mancozeb, thiram
	Fungus diseases	formaldehyde (commodity substance) (*dip*), peroxyacetic acid (commodity substance) (*dip*)
	Fusarium	carbendazim (*off-label*), thiabendazole (*off-label - post-lifting or pre-planting dip*), thiabendazole (*post-lifting spray*)
	Ink disease	chlorothalonil (*qualified minor use*)
	Penicillium rot	carbendazim (*off-label*)
	Phytophthora	etridiazole, propamocarb hydrochloride
	Powdery mildew	bupirimate
	Pythium	propamocarb hydrochloride
	Sclerotinia	carbendazim (*off-label*)
	Stagonospora	carbendazim (*off-label*)
	White mould	chlorothalonil (*off-label*), mancozeb (*off-label - for galanthamine production*)
Pests	Aphids	malathion, nicotine
	Birds/mammals	aluminium ammonium sulphate (*bulb treatment*), aluminium ammonium sulphate (*corm treatment*)
	Capsid bugs	nicotine
	Flies	chlorpyrifos (*off-label*)
	Free-living nematodes	peroxyacetic acid (commodity substance) (*dip*)
	Leaf miners	nicotine
	Leafhoppers	nicotine
	Sawflies	nicotine
	Stem nematodes	1,3-dichloropropene
	Thrips	nicotine
Plant growth regulation	Flowering control	paclobutrazol
	Growth control	2-chloroethylphosphonic acid, chlormequat, paclobutrazol
Weeds	Broad-leaved weeds	bentazone, chlorpropham, diuron (*off-label - for galanthamine production*), paraquat, pendimethalin (*off-label - for galanthamine production*)
	Crops as weeds	cycloxydim, pendimethalin (*off-label - for galanthamine production*)
	Grass weeds	chlorpropham, cycloxydim, paraquat

Flowers - Miscellaneous flowers

Diseases	Black spot	captan, kresoxim-methyl, mancozeb, myclobutanil
	Foliar diseases	tebuconazole (*off-label*)
	Powdery mildew	bupirimate, dinocap, fenarimol, kresoxim-methyl, myclobutanil
	Rust	mancozeb, myclobutanil, penconazole
Pests	Aphids	deltamethrin (*off-label*), imidacloprid, lambda-cyhalothrin + pirimicarb (*off-label*), malathion, pirimicarb (*off-label*), rotenone
	Cutworms	lambda-cyhalothrin + pirimicarb (*off-label*)
	Flies	imidacloprid, lambda-cyhalothrin (*off-label*)
	Pests, miscellaneous	deltamethrin (*off-label*)
	Sawflies	rotenone
	Weevils	imidacloprid
	Whiteflies	imidacloprid
Plant growth regulation	Flowering control	paclobutrazol
	Growth control	paclobutrazol
Weeds	Broad-leaved weeds	chlorthal-dimethyl, dichlobenil, lenacil, metribuzin (*off-label*), pendimethalin (*off-label*), propyzamide
	Grass weeds	dichlobenil, lenacil, pendimethalin (*off-label*), propaquizafop (*off-label*), propyzamide, tepraloxydim (*off-label*)
	Weeds, miscellaneous	linuron (*off-label*)

Flowers - Pot plants

Diseases	Botrytis	carbendazim (*off-label*), iprodione
	Phytophthora	fosetyl-aluminium, propamocarb hydrochloride
	Powdery mildew	carbendazim (*off-label*)
	Pythium	propamocarb hydrochloride
	Root rot	fosetyl-aluminium
	Rust	mancozeb
Pests	Aphids	deltamethrin, imidacloprid, thiacloprid
	Beetles	thiacloprid
	Caterpillars	deltamethrin
	Flies	imidacloprid
	Mealybugs	deltamethrin, petroleum oil
	Scale insects	deltamethrin, petroleum oil
	Spider mites	petroleum oil
	Weevils	imidacloprid, thiacloprid
	Whiteflies	deltamethrin, imidacloprid, thiacloprid
Plant growth regulation	Flowering control	2-chloroethylphosphonic acid, paclobutrazol
	Fruiting control	paclobutrazol
	Growth control	2-chloroethylphosphonic acid, chlormequat, daminozide, paclobutrazol

Flowers - Protected bulbs/corms

Pests	Aphids	malathion, pirimicarb

Flowers - Protected flowers

Diseases	Black spot	kresoxim-methyl
	Powdery mildew	kresoxim-methyl
	Ring spot	difenoconazole (*off-label - ground grown (or pot grown - SOLA 20050159)*), difenoconazole (*off-label - ground grown (or pot grown - SOLA 20051488)*)
	Rust	azoxystrobin (*off-label*), difenoconazole (*off-label - ground grown (or pot grown - SOLA 20050159)*), difenoconazole (*off-label - ground grown (or pot grown - SOLA 20051488)*), propiconazole (*off-label*)
Pests	Aphids	malathion, nicotine, pirimicarb, Verticillium lecanii
	Spider mites	tebufenpyrad
Plant growth regulation	Growth control	2-chloroethylphosphonic acid

Miscellaneous flowers and ornamentals - Miscellaneous uses

Diseases	Fungus diseases	potassium bicarbonate (commodity substance)
	Sclerotinia	Coniothyrium minitans
Pests	Aphids	pyrethrins
	Caterpillars	diflubenzuron, pyrethrins
	Leaf miners	deltamethrin (*off-label*)
	Slugs/snails	ferric phosphate, metaldehyde, metaldehyde (*around*), methiocarb (*outdoor only*)
	Spider mites	pyrethrins
	Thrips	deltamethrin (*off-label*), pyrethrins
Weeds	Broad-leaved weeds	carfentrazone-ethyl (*before planting*), diquat (*around*), glyphosate (*pre-planting or sowing*)
	Crops as weeds	carfentrazone-ethyl (*before planting*)
	Grass weeds	glyphosate (*pre-planting or sowing*)
	Weeds, miscellaneous	glyphosate (*before planting*), glyphosate (*pre-sowing/ planting*)

Miscellaneous flowers and ornamentals - Soils

Diseases	Soil-borne diseases	metam-sodium
Pests	Free-living nematodes	metam-sodium
	Leaf miners	deltamethrin (*off-label*), deltamethrin (*off-label - after recovery from wilting*)
	Soil pests	metam-sodium
	Thrips	deltamethrin (*off-label*), deltamethrin (*off-label - after recovery from wilting*)
Weeds	Weeds, miscellaneous	metam-sodium

Ornamentals - Nursery stock

Diseases	Alternaria	iprodione (*seed treatment*)
	Black root rot	carbendazim (*off-label*)
	Black spot	myclobutanil
	Botrytis	chlorothalonil, thiram (*except Hydrangea*)
	Damping off	copper ammonium carbonate, etridiazole (*seeds and seedlings*), tolclofos-methyl
	Downy mildew	benthiavalicarb-isopropyl + mancozeb (*off-label*)
	Foot rot	tolclofos-methyl
	Fungus diseases	prochloraz
	Phytophthora	etridiazole, etridiazole (*rooted cuttings and transplants*), fosetyl-aluminium, metalaxyl-M, propamocarb hydrochloride
	Powdery mildew	cyflufenamid (*off-label*), myclobutanil
	Pruning wounds	tebuconazole
	Pythium	metalaxyl-M, propamocarb hydrochloride
	Root rot	tolclofos-methyl
	Rust	myclobutanil
	Stem canker	tebuconazole
Pests	Ants	pirimiphos-methyl
	Aphids	acetamiprid, cypermethrin, deltamethrin, dimethoate, imidacloprid, imidacloprid (*container grown*), nicotine, physical pest control, pirimicarb, pirimiphos-methyl, pymetrozine, pyrethrins, rotenone, thiacloprid, Verticillium lecanii
	Beetles	thiacloprid
	Birds/mammals	ziram
	Capsid bugs	cypermethrin, deltamethrin, pirimiphos-methyl
	Caterpillars	Bacillus thuringiensis, cypermethrin, deltamethrin, diflubenzuron, indoxacarb, nicotine, pyrethrins, teflubenzuron
	Cutworms	cypermethrin
	Earwigs	pirimiphos-methyl
	Flies	imidacloprid, imidacloprid (*container grown*)
	Leaf miners	deltamethrin (*off-label*), dimethoate, nicotine, oxamyl (*off-label*), pirimiphos-methyl, thiacloprid (*off-label*)
	Leafhoppers	nicotine, physical pest control
	Mealybugs	deltamethrin, petroleum oil, physical pest control
	Mites	physical pest control
	Sawflies	pirimiphos-methyl
	Scale insects	deltamethrin, petroleum oil
	Slugs/snails	metaldehyde
	Spider mites	abamectin, bifenthrin, dimethoate, petroleum oil, physical pest control, pirimiphos-methyl
	Thrips	abamectin, cypermethrin, deltamethrin, deltamethrin (*off-label*), nicotine, physical pest control, pirimiphos-methyl, spinosad, thiacloprid (*off-label*)
	Weevils	chlorpyrifos, fipronil, imidacloprid, imidacloprid (*container grown*), thiacloprid
	Whiteflies	acetamiprid, buprofezin, cypermethrin, deltamethrin, imidacloprid, imidacloprid (*container grown*), nicotine,

		physical pest control, pirimiphos-methyl, teflubenzuron, thiacloprid, thiacloprid (*off-label*), Verticillium lecanii
Plant growth regulation	Growth control	1-naphthylacetic acid, 4-indol-3-ylbutyric acid, 4-indol-3-ylbutyric acid + 1-naphthylacetic acid, chlormequat, indol-3-ylacetic acid
Weeds	Broad-leaved weeds	ammonium sulphamate, clopyralid, dichlobenil, diquat, diquat + paraquat, diuron (*not container grown*), glufosinate-ammonium, isoxaben, isoxaben (*off-label*), isoxaben + trifluralin (*outdoor stock only*), napropamide, oxadiazon, propachlor, propyzamide
	Crops as weeds	cycloxydim
	Grass weeds	ammonium sulphamate, cycloxydim, dichlobenil, diquat + paraquat, diuron, fluazifop-P-butyl (*off-label*), glufosinate-ammonium, isoxaben + trifluralin (*outdoor stock only*), napropamide, oxadiazon, propachlor, propyzamide, quizalofop-P-tefuryl (*off-label*)
	Weeds, miscellaneous	glyphosate

Ornamentals - Trees and shrubs

Diseases	Downy mildew	benthiavalicarb-isopropyl + mancozeb (*off-label*)
	Fungus diseases	prochloraz
	Powdery mildew	cyflufenamid (*off-label*), penconazole
	Rhizoctonia	chloropicrin (*soil fumigation*)
	Scab	penconazole
Pests	Aphids	deltamethrin
	Birds/mammals	aluminium ammonium sulphate, warfarin (*nuts*)
	Capsid bugs	deltamethrin
	Caterpillars	Bacillus thuringiensis, deltamethrin, diflubenzuron
	Mealybugs	deltamethrin
	Scale insects	deltamethrin
	Thrips	deltamethrin
	Whiteflies	deltamethrin
Plant growth regulation	Growth control	maleic hydrazide
Weeds	Broad-leaved weeds	ammonium sulphamate, asulam, chlorthal-dimethyl, dichlobenil, diuron, diuron + glyphosate, glufosinate-ammonium, isoxaben, isoxaben + trifluralin, metazachlor, propyzamide
	Grass weeds	ammonium sulphamate, dichlobenil, diuron, diuron + glyphosate, glufosinate-ammonium, isoxaben + trifluralin, metazachlor, propyzamide, quizalofop-P-tefuryl (*off-label*)
	Weeds, miscellaneous	glyphosate, glyphosate (*wiper application*)
	Woody weeds/scrub	asulam, glyphosate

Ornamentals - Woody ornamentals

Diseases	Fungus diseases	prochloraz
Pests	Aphids	fatty acids
	Mealybugs	fatty acids

	Scale insects	fatty acids
	Spider mites	fatty acids
	Whiteflies	fatty acids
Weeds	Aquatic weeds	propyzamide
	Bindweeds	oxadiazon
	Broad-leaved weeds	dichlobenil, diuron, glufosinate-ammonium, lenacil, napropamide, oxadiazon, propyzamide
	Grass weeds	dichlobenil, diuron, glufosinate-ammonium, lenacil, napropamide, oxadiazon, propyzamide
	Weeds, miscellaneous	glyphosate, propyzamide

Forestry

Forest nurseries, general - Forest nurseries

Diseases	Pruning wounds	tebuconazole
	Stem canker	tebuconazole
Pests	Aphids	pirimicarb
	Birds/mammals	aluminium ammonium sulphate
	Weevils	alpha-cypermethrin (*off-label*), carbosulfan (*off-label*)
Weeds	Broad-leaved weeds	diquat + paraquat, napropamide, paraquat (*stale seedbed*), propyzamide
	Grass weeds	cycloxydim, diquat + paraquat, napropamide, paraquat (*stale seedbed*), propaquizafop, propyzamide

Forestry plantations, general - Forestry plantations

Crop control	Chemical stripping/ thinning	glyphosate, glyphosate (*stump treatment*)
Diseases	Pruning wounds	tebuconazole
	Stem canker	tebuconazole
Pests	Birds/mammals	aluminium ammonium sulphate, warfarin, ziram
	Caterpillars	diflubenzuron
	Weevils	alpha-cypermethrin (*off-label*), carbosulfan, cypermethrin
Weeds	Aquatic weeds	glyphosate, propyzamide
	Broad-leaved weeds	2,4-D + dicamba + triclopyr, clopyralid (*off-label*), diquat + paraquat, glufosinate-ammonium, glyphosate, glyphosate (*wiper application*), isoxaben, metamitron (*off-label*), metazachlor, paraquat, pendimethalin (*off-label*), propyzamide, triclopyr
	Grass weeds	ammonium sulphamate, cycloxydim, diquat + paraquat, glufosinate-ammonium, glyphosate, metamitron (*off-label*), metazachlor, paraquat, propaquizafop, propyzamide
	Weeds, miscellaneous	glufosinate-ammonium, glyphosate, metamitron (*off-label*), paraquat, propyzamide
	Woody weeds/scrub	2,4-D + dicamba + triclopyr, ammonium sulphamate, asulam, asulam (*off-label*), glyphosate, triclopyr

Miscellaneous forestry situations - Cut logs/timber

Diseases	Fungus diseases	urea (commodity substance)
Pests	Beetles	chlorpyrifos
Weeds	Grass weeds	propaquizafop

Woodland on farms - Woodland

Diseases	Pruning wounds	tebuconazole
	Stem canker	tebuconazole
Pests	Aphids	lambda-cyhalothrin (*off-label*)
	Beetles	lambda-cyhalothrin (*off-label*)
	Birds/mammals	aluminium phosphide
	Sawflies	lambda-cyhalothrin (*off-label*)
Plant growth regulation	Plant growth regulation, miscellaneous	gibberellins (*qualified minor use*)
Weeds	Broad-leaved weeds	2,4-D + dicamba + triclopyr, lenacil (*off-label*), metazachlor, pendimethalin (*off-label*)
	Crops as weeds	fluazifop-P-butyl
	Grass weeds	cycloxydim, fluazifop-P-butyl, lenacil (*off-label*), metazachlor, pendimethalin (*off-label*), propaquizafop
	Weeds, miscellaneous	amitrole (*off-label*)
	Woody weeds/scrub	2,4-D + dicamba + triclopyr

Fruit and hops

Bush fruit - Bilberries/blueberries/cranberries

Diseases	Botrytis	fenhexamid (*off-label*), pyrimethanil (*off-label*)
	Downy mildew	copper oxychloride (*off-label*)
	Leaf spot	dodine (*off-label*)
	Powdery mildew	bupirimate (*off-label*), fenpropimorph (*off-label*), penconazole (*off-label*)
Pests	Aphids	pirimicarb (*off-label*), pymetrozine (*off-label*)
	Caterpillars	Bacillus thuringiensis (*off-label*)
	Scale insects	thiacloprid (*off-label*)
Weeds	Bindweeds	oxadiazon (*off-label*)
	Broad-leaved weeds	asulam (*off-label*), chlorthal-dimethyl (*off-label*), dichlobenil (*off-label*), diuron (*off-label*), oxadiazon (*off-label*), pendimethalin (*off-label*), propachlor (*off-label*), propyzamide (*off-label*)
	Grass weeds	dichlobenil (*off-label*), oxadiazon (*off-label*), pendimethalin (*off-label*), propachlor (*off-label*), propyzamide (*off-label*)
	Weeds, miscellaneous	glufosinate-ammonium, glyphosate (*off-label*)

Bush fruit - Currants

Diseases	Botrytis	chlorothalonil, chlorothalonil (*qualified minor use*), fenhexamid, pyrimethanil (*off-label*)
	Downy mildew	copper oxychloride (*off-label*)
	Foliar diseases	mancozeb (*off-label*)
	Leaf spot	Bordeaux mixture, chlorothalonil, chlorothalonil (*qualified minor use*), copper ammonium carbonate, dodine, dodine (*off-label*), mancozeb
	Powdery mildew	bupirimate, bupirimate (*off-label*), chlorothalonil, chlorothalonil (*qualified minor use*), fenarimol, fenpropimorph (*off-label*), kresoxim-methyl, myclobutanil, penconazole, penconazole (*off-label*)
	Rust	copper oxychloride, thiram
Pests	Aphids	chlorpyrifos, pirimicarb, pymetrozine (*off-label*)
	Capsid bugs	chlorpyrifos
	Caterpillars	Bacillus thuringiensis (*off-label*), chlorpyrifos, diflubenzuron
	Gall, rust and leaf & bud mites	sulphur
	Midges	lambda-cyhalothrin (*off-label*)
	Sawflies	lambda-cyhalothrin (*off-label*)
	Scale insects	thiacloprid (*off-label*)
	Spider mites	bifenthrin (*off-label*), chlorpyrifos, clofentezine (*off-label*)
Weeds	Bindweeds	oxadiazon, oxadiazon (*off-label*)
	Broad-leaved weeds	asulam, asulam (*off-label*), chlorpropham, chlorthal-dimethyl, chlorthal-dimethyl (*off-label*), dichlobenil, diuron (*off-label*), isoxaben, lenacil, MCPB, napropamide, oxadiazon, oxadiazon (*off-label*), paraquat, pendimethalin, pendimethalin (*off-label*), propachlor (*off-label*), propyzamide
	Crops as weeds	fluazifop-P-butyl, pendimethalin
	Grass weeds	chlorpropham, dichlobenil, fluazifop-P-butyl, lenacil, napropamide, oxadiazon, oxadiazon (*off-label*), paraquat, pendimethalin, pendimethalin (*off-label*), propachlor (*off-label*), propyzamide
	Weeds, miscellaneous	glyphosate (*off-label*)

Bush fruit - Gooseberries

Diseases	Botrytis	chlorothalonil, chlorothalonil (*qualified minor use*), fenhexamid, pyrimethanil (*off-label*)
	Downy mildew	copper oxychloride (*off-label*)
	Leaf spot	chlorothalonil (*qualified minor use*), dodine (*off-label*), mancozeb
	Powdery mildew	bupirimate, chlorothalonil, chlorothalonil (*qualified minor use*), fenarimol, fenpropimorph (*off-label*), myclobutanil, penconazole (*off-label*), sulphur
Pests	Aphids	chlorpyrifos, pirimicarb, pymetrozine (*off-label*)
	Capsid bugs	chlorpyrifos
	Caterpillars	Bacillus thuringiensis (*off-label*), chlorpyrifos
	Midges	lambda-cyhalothrin (*off-label*)
	Sawflies	lambda-cyhalothrin (*off-label*), nicotine, rotenone

	Scale insects	thiacloprid (*off-label*)
	Spider mites	chlorpyrifos
Weeds	Bindweeds	oxadiazon
	Broad-leaved weeds	asulam (*off-label*), chlorpropham, chlorthal-dimethyl, dichlobenil, diuron (*off-label*), isoxaben, lenacil, MCPB, napropamide, oxadiazon, paraquat, pendimethalin, propachlor (*off-label*), propyzamide
	Crops as weeds	fluazifop-P-butyl, pendimethalin
	Grass weeds	chlorpropham, dichlobenil, fluazifop-P-butyl, lenacil, napropamide, oxadiazon, paraquat, pendimethalin, propachlor (*off-label*), propyzamide
	Weeds, miscellaneous	glyphosate (*off-label*)

Bush fruit - Protected bush fruit

Pests	Aphids	pymetrozine (*off-label*), pyrethrins
	Caterpillars	pyrethrins
	Spider mites	bifenthrin (*off-label*)

Bush fruit - Vines

Diseases	Botrytis	cyprodinil + fludioxonil (*off-label*), fenhexamid, iprodione (*off-label*), pyrimethanil (*off-label*)
	Downy mildew	copper oxychloride, fosetyl-aluminium (*off-label*), metalaxyl-M (*off-label*)
	Foliar diseases	mancozeb (*off-label*)
	Fungus diseases	mancozeb (*off-label*)
	Powdery mildew	dinocap (*off-label*), fenbuconazole (*off-label*), myclobutanil (*off-label*), sulphur
Pests	Aphids	nicotine
	Capsid bugs	lambda-cyhalothrin (*off-label*)
	Caterpillars	lambda-cyhalothrin (*off-label*)
	Mealybugs	petroleum oil
	Scale insects	petroleum oil
	Spider mites	petroleum oil
	Wasps	lambda-cyhalothrin (*off-label*)
Weeds	Bindweeds	oxadiazon
	Broad-leaved weeds	dichlobenil (*off-label*), glufosinate-ammonium, isoxaben, oxadiazon, propyzamide (*off-label*)
	Grass weeds	glufosinate-ammonium, oxadiazon, propyzamide (*off-label*)
	Weeds, miscellaneous	glufosinate-ammonium, glyphosate (*off-label*)

Cane fruit - Outdoor cane fruit

Diseases	Botrytis	chlorothalonil, fenhexamid, iprodione, pyrimethanil (*off-label*), thiram
	Cane blight	carbendazim (*off-label*), tebuconazole (*off-label*)

	Cane spot	Bordeaux mixture, chlorothalonil, copper ammonium carbonate, copper oxychloride, copper oxychloride (*off-label*), thiram
	Powdery mildew	azoxystrobin (*off-label*), bupirimate (*off-label*), bupirimate (*outdoor only*), chlorothalonil, fenarimol, fenpropimorph (*off-label*)
	Purple blotch	copper oxychloride
	Root rot	fluazinam (*off-label*)
	Spur blight	Bordeaux mixture, thiram
	Verticillium wilt	chloropicrin (*off-label - soil fumigation*)
Pests	Aphids	1,3-dichloropropene, chlorpyrifos, chlorpyrifos (*off-label*), deltamethrin (*off-label*), pirimicarb, pirimicarb (*off-label*), pymetrozine (*off-label*), pyrethrins, rotenone
	Beetles	chlorpyrifos, deltamethrin, rotenone
	Birds/mammals	aluminium ammonium sulphate
	Capsid bugs	lambda-cyhalothrin (*off-label*), thiacloprid (*off-label*)
	Caterpillars	Bacillus thuringiensis, Bacillus thuringiensis (*off-label*), pyrethrins
	Free-living nematodes	1,3-dichloropropene
	Mealybugs	petroleum oil
	Midges	chlorpyrifos
	Pests, miscellaneous	deltamethrin (*off-label*)
	Scale insects	petroleum oil
	Spider mites	bifenthrin (*off-label*), chlorpyrifos, chlorpyrifos (*off-label*), clofentezine (*off-label*), petroleum oil
	Weevils	bifenthrin (*off-label*), lambda-cyhalothrin (*off-label*)
Weeds	Bindweeds	oxadiazon, oxadiazon (*off-label*)
	Broad-leaved weeds	asulam (*off-label*), chlorthal-dimethyl, chlorthal-dimethyl (*off-label*), dichlobenil, glufosinate-ammonium, isoxaben, lenacil, lenacil (*off-label*), MCPB, napropamide, oxadiazon, oxadiazon (*off-label*), paraquat, pendimethalin, propachlor (*off-label*), propyzamide, propyzamide (*England only*), trifluralin
	Crops as weeds	fluazifop-P-butyl, pendimethalin
	Grass weeds	dichlobenil, fluazifop-P-butyl, glufosinate-ammonium, lenacil, lenacil (*off-label*), napropamide, oxadiazon, oxadiazon (*off-label*), paraquat, pendimethalin, propachlor (*off-label*), propyzamide, propyzamide (*England only*), trifluralin
	Weeds, miscellaneous	glufosinate-ammonium

Cane fruit - Protected cane fruit

Diseases	Botrytis	pyrimethanil (*off-label*)
	Cane blight	carbendazim (*off-label*)
	Powdery mildew	azoxystrobin (*off-label*), myclobutanil (*off-label*)
	Root rot	fluazinam (*off-label*)
	Verticillium wilt	chloropicrin (*off-label - soil fumigation*)
Pests	Aphids	pymetrozine (*off-label*), pyrethrins
	Beetles	thiacloprid (*off-label*)
	Capsid bugs	thiacloprid (*off-label*)

	Caterpillars	pyrethrins
	Gall, rust and leaf & bud mites	abamectin (*off-label*)
	Spider mites	abamectin (*off-label*), bifenthrin (*off-label*), clofentezine (*off-label*)
	Weevils	bifenthrin (*off-label*)

Hops, general - Hops

Crop control	Chemical stripping/ thinning	diquat, diquat + paraquat, paraquat
Diseases	Downy mildew	Bordeaux mixture, chlorothalonil, copper oxychloride, fosetyl-aluminium, metalaxyl-M (*off-label*)
	Powdery mildew	bupirimate, fenpropimorph (*off-label*), myclobutanil (*off-label*), penconazole, sulphur
Pests	Aphids	1,3-dichloropropene, bifenthrin, cypermethrin, deltamethrin, imidacloprid, pymetrozine (*off-label*), tebufenpyrad
	Free-living nematodes	1,3-dichloropropene
	Mealybugs	petroleum oil
	Scale insects	petroleum oil
	Spider mites	bifenthrin, petroleum oil, tebufenpyrad
Weeds	Bindweeds	oxadiazon
	Broad-leaved weeds	asulam, isoxaben, oxadiazon, paraquat, pendimethalin, propyzamide (*off-label*)
	Crops as weeds	fluazifop-P-butyl, pendimethalin
	Grass weeds	fluazifop-P-butyl, oxadiazon, paraquat, pendimethalin, propyzamide (*off-label*)

Miscellaneous fruit situations - Fruit crops, general

Pests	Aphids	fatty acids, physical pest control
	Leafhoppers	physical pest control
	Mealybugs	fatty acids, petroleum oil
	Mites	physical pest control
	Scale insects	fatty acids, petroleum oil
	Spider mites	fatty acids, petroleum oil
	Thrips	physical pest control
	Whiteflies	fatty acids, physical pest control
Plant growth regulation	Fruiting control	ethylene (commodity substance) (*in store*)
Weeds	Broad-leaved weeds	glufosinate-ammonium
	Grass weeds	glufosinate-ammonium

Miscellaneous fruit situations - Fruit nursery stock

Weeds	Broad-leaved weeds	metazachlor
	Grass weeds	metazachlor

Miscellaneous fruit situations - Orchards

Weeds	Broad-leaved weeds	paraquat
	Grass weeds	paraquat

Other fruit - Protected miscellaneous fruit

Diseases	Black spot	azoxystrobin (*off-label*), boscalid + pyraclostrobin, cyprodinil + fludioxonil (*qualified minor use*)
	Botrytis	boscalid + pyraclostrobin, cyprodinil + fludioxonil, iprodione, iprodione (*off-label*), mepanipyrim
	Crown rot	fosetyl-aluminium (*off-label*)
	Powdery mildew	boscalid + pyraclostrobin, bupirimate, kresoxim-methyl, quinoxyfen (*off-label*)
	Red core	fosetyl-aluminium (*off-label*)
Pests	Aphids	nicotine, physical pest control, pymetrozine (*off-label*)
	Capsid bugs	thiacloprid (*off-label*)
	Leafhoppers	physical pest control
	Mites	physical pest control
	Pests, miscellaneous	deltamethrin (*off-label*)
	Spider mites	abamectin (*off-label*), bifenazate, clofentezine (*off-label*), spiromesifen (*off-label - on inert media or by NFT*)
	Tarsonemid mites	fenpyroximate (*off-label*)
	Thrips	physical pest control
	Whiteflies	physical pest control, spiromesifen (*off-label - on inert media or by NFT*)
Plant growth regulation	Quality/yield control	gibberellins

Other fruit - Rhubarb

Diseases	Bacterial blight	copper oxychloride (*off-label*)
	Damping off	thiram (*off-label - seed treatment*)
Pests	Aphids	pirimicarb (*off-label*)
	Pests, miscellaneous	deltamethrin (*off-label*)
Weeds	Broad-leaved weeds	dichlobenil (*off-label - established*), metamitron (*off-label*), pendimethalin (*off-label*), propachlor (*off-label*), propyzamide, propyzamide (*outdoor*)
	Grass weeds	dichlobenil (*off-label - established*), propachlor (*off-label*), propyzamide, propyzamide (*outdoor*)
	Weeds, miscellaneous	glyphosate (*off-label*)

Other fruit - Strawberries, general

Crop control	Sucker/shoot control	paraquat
Diseases	Black spot	azoxystrobin (*off-label*), boscalid + pyraclostrobin, cyprodinil + fludioxonil (*qualified minor use*)
	Botrytis	boscalid + pyraclostrobin, captan, chlorothalonil, chlorothalonil (*off-label*), chlorothalonil (*outdoor crops only*), cyprodinil + fludioxonil, fenhexamid, iprodione, mepanipyrim, pyrimethanil, thiram

	Crown rot	chloropicrin (*soil fumigation*), fosetyl-aluminium (*off-label*)
	Powdery mildew	boscalid + pyraclostrobin, bupirimate, dinocap, fenarimol, fenpropimorph (*off-label*), kresoxim-methyl, myclobutanil, quinoxyfen (*off-label*), sulphur
	Red core	chloropicrin (*soil fumigation*), fosetyl-aluminium, fosetyl-aluminium (*off-label*)
	Verticillium wilt	chloropicrin (*soil fumigation*)
Pests	Aphids	1,3-dichloropropene, chlorpyrifos, nicotine, pirimicarb, pymetrozine (*off-label*)
	Beetles	methiocarb
	Birds/mammals	aluminium ammonium sulphate
	Capsid bugs	thiacloprid (*off-label*)
	Caterpillars	Bacillus thuringiensis, chlorpyrifos
	Free-living nematodes	1,3-dichloropropene, chloropicrin (*soil fumigation*)
	Slugs/snails	methiocarb
	Spider mites	bifenthrin, chlorpyrifos, clofentezine (*off-label*), tebufenpyrad
	Stem nematodes	1,3-dichloropropene
	Tarsonemid mites	abamectin (*off-label - in propagation*), fenpyroximate (*off-label*)
	Weevils	chlorpyrifos
Weeds	Broad-leaved weeds	asulam (*off-label*), chlorpropham, chlorthal-dimethyl, clopyralid, glufosinate-ammonium, isoxaben, lenacil, metamitron (*off-label*), napropamide, pendimethalin, phenmedipham, propachlor, propachlor (*off-label*), propyzamide, trifluralin
	Crops as weeds	cycloxydim, fluazifop-P-butyl, pendimethalin
	Grass weeds	chlorpropham, cycloxydim, fluazifop-P-butyl, glufosinate-ammonium, lenacil, metamitron (*off-label*), napropamide, pendimethalin, propachlor, propachlor (*off-label*), propyzamide, trifluralin
	Weeds, miscellaneous	glufosinate-ammonium

Soft fruit - Bush fruit

Pests	Aphids	chlorpyrifos, nicotine, pyrethrins, rotenone
	Birds/mammals	aluminium ammonium sulphate
	Capsid bugs	chlorpyrifos, nicotine
	Caterpillars	chlorpyrifos, pyrethrins
	Leafhoppers	nicotine
	Mealybugs	petroleum oil
	Sawflies	nicotine
	Scale insects	petroleum oil
	Spider mites	chlorpyrifos, petroleum oil
Weeds	Broad-leaved weeds	glufosinate-ammonium
	Grass weeds	glufosinate-ammonium
	Weeds, miscellaneous	glufosinate-ammonium

Tree fruit - Miscellaneous top fruit

Diseases	Rhizoctonia	chloropicrin (*soil fumigation*)
Pests	Aphids	nicotine, rotenone
	Birds/mammals	aluminium ammonium sulphate, ziram
	Capsid bugs	nicotine
	Sawflies	nicotine

Tree fruit - Nuts

Diseases	Bacterial canker	copper oxychloride (*off-label*)
	Phytophthora	copper oxychloride (*off-label*)
	Pruning wounds	tebuconazole
	Stem canker	tebuconazole, tebuconazole (*off-label*)
Pests	Aphids	lambda-cyhalothrin (*off-label*)
	Caterpillars	lambda-cyhalothrin (*off-label*)
	Mites	thiacloprid (*off-label*)
Weeds	Bindweeds	oxadiazon (*off-label*)
	Broad-leaved weeds	asulam (*off-label*), dichlobenil (*off-label*), diuron (*off-label*), glufosinate-ammonium, oxadiazon (*off-label*), pendimethalin (*off-label*), propyzamide (*off-label*)
	Grass weeds	dichlobenil (*off-label*), diuron (*off-label*), glufosinate-ammonium, oxadiazon (*off-label*), pendimethalin (*off-label*), propyzamide (*off-label*)
	Weeds, miscellaneous	glufosinate-ammonium, glyphosate (*off-label*), propyzamide (*off-label*)

Tree fruit - Pome fruit

Crop control	Sucker/shoot control	glyphosate
Diseases	Botrytis	iprodione (*off-label*), pyrimethanil (*off-label*)
	Botrytis fruit rot	thiram
	Collar rot	copper oxychloride (*off-label*), fosetyl-aluminium
	Crown rot	fosetyl-aluminium
	Leaf curl	copper oxychloride (*off-label*)
	Phytophthora	mancozeb + metalaxyl-M (*off-label - applied to orchard floor*)
	Powdery mildew	boscalid + pyraclostrobin, bupirimate, dinocap, dithianon + pyraclostrobin, fenarimol, fenbuconazole (*reduction*), kresoxim-methyl (*reduction*), myclobutanil, penconazole, sulphur
	Pruning wounds	tebuconazole
	Scab	Bordeaux mixture, boscalid + pyraclostrobin, captan, dithianon, dithianon + pyraclostrobin, dodine, dodine (*off-label*), fenarimol, fenbuconazole, kresoxim-methyl, mancozeb, myclobutanil, pyrimethanil, pyrimethanil (*off-label*), sulphur, sulphur (*off-label*), thiram
	Stem canker	Bordeaux mixture, copper oxychloride, tebuconazole, tebuconazole (*off-label*)
	Storage rots	captan, thiram

Pests	Aphids	acetamiprid, chlorpyrifos, chlorpyrifos (*off-label*), cyfluthrin, cypermethrin, deltamethrin, nicotine, pirimicarb, thiacloprid
	Capsid bugs	chlorpyrifos, chlorpyrifos (*off-label*), cypermethrin, deltamethrin, nicotine
	Caterpillars	Bacillus thuringiensis (*off-label*), chlorpyrifos, chlorpyrifos (*off-label*), Cydia pomonella GV, cyfluthrin, cypermethrin, deltamethrin, diflubenzuron, fenoxycarb, indoxacarb, methoxyfenozide, spinosad, spinosad (*off-label*)
	Gall, rust and leaf & bud mites	diflubenzuron
	Pests, miscellaneous	deltamethrin (*off-label*)
	Sawflies	chlorpyrifos, cypermethrin, deltamethrin, rotenone
	Spider mites	bifenthrin, chlorpyrifos, clofentezine, fenpyroximate, tebufenpyrad
	Suckers	chlorpyrifos, cyfluthrin, cypermethrin, deltamethrin, diflubenzuron, lambda-cyhalothrin
	Weevils	chlorpyrifos
Plant growth regulation	Fruiting control	1-methylcyclopropene (*post-harvest use*), 2-chloroethylphosphonic acid (*off-label - for cider making*), gibberellins, paclobutrazol
	Growth control	gibberellins (*off-label*), paclobutrazol, prohexadione-calcium
	Quality/yield control	gibberellins, gibberellins (*off-label*)
Weeds	Bindweeds	oxadiazon
	Broad-leaved weeds	2,4-D, 2,4-D + dichlorprop-P + MCPA + mecoprop-P, amitrole, asulam, asulam (*off-label*), clopyralid (*off-label*), dicamba + MCPA + mecoprop-P, dichlobenil, dichlobenil (*off-label*), diuron, fluroxypyr (*off-label*), glufosinate-ammonium, isoxaben, oxadiazon, pendimethalin, pendimethalin (*off-label*), propyzamide, propyzamide (*off-label*)
	Crops as weeds	pendimethalin
	Grass weeds	amitrole, dichlobenil, dichlobenil (*off-label*), diuron, glufosinate-ammonium, oxadiazon, pendimethalin, pendimethalin (*off-label*), propyzamide, propyzamide (*off-label*)
	Weeds, miscellaneous	amitrole, amitrole (*off-label*), glufosinate-ammonium, glyphosate

Tree fruit - Stone fruit

Crop control	Sucker/shoot control	glyphosate
Diseases	Bacterial canker	Bordeaux mixture, copper oxychloride
	Blossom wilt	boscalid + pyraclostrobin (*off-label*), cyprodinil + fludioxonil (*off-label*), fenbuconazole (*off-label*), myclobutanil (*off-label*)
	Botrytis	fenhexamid (*off-label*)
	Leaf curl	Bordeaux mixture, copper ammonium carbonate, copper oxychloride, copper oxychloride (*off-label*)
	Powdery mildew	myclobutanil (*off-label*)
	Pruning wounds	tebuconazole
	Rust	myclobutanil (*off-label*)

	Sclerotinia	boscalid + pyraclostrobin (*off-label*), fenbuconazole (*off-label*), myclobutanil (*off-label*)
	Stem canker	tebuconazole
Pests	Aphids	acetamiprid, chlorpyrifos, cypermethrin, deltamethrin, nicotine, pirimicarb, pirimicarb (*off-label*), thiacloprid (*off-label*), thiacloprid (*off-label - under temporary rain covers*)
	Caterpillars	Bacillus thuringiensis (*off-label*), chlorpyrifos, cypermethrin, deltamethrin, diflubenzuron
	Gall, rust and leaf & bud mites	diflubenzuron
	Sawflies	deltamethrin
	Spider mites	chlorpyrifos, clofentezine, fenpyroximate (*off-label*)
Plant growth regulation	Growth control	paclobutrazol (*off-label*)
Weeds	Broad-leaved weeds	asulam, asulam (*off-label*), glufosinate-ammonium, isoxaben, pendimethalin, propyzamide, propyzamide (*off-label*)
	Crops as weeds	pendimethalin
	Grass weeds	glufosinate-ammonium, pendimethalin, propyzamide, propyzamide (*off-label*)
	Weeds, miscellaneous	amitrole (*off-label*), glufosinate-ammonium, glyphosate

Grain/crop store uses

Stored produce - Food/produce storage

Diseases	Alternaria	iprodione
	Botrytis	iprodione

Stored seed - Stored grain/rape/linseed

Pests	Birds/mammals	aluminium ammonium sulphate
	Food/grain storage pests	bifenthrin + malathion, chlorpyrifos-methyl, d-phenothrin + tetramethrin, permethrin, pirimiphos-methyl
	Pests, miscellaneous	aluminium phosphide, magnesium phosphide

Grass

Turf/amenity grass - Amenity grassland

Diseases	Anthracnose	azoxystrobin, propiconazole (*qualified minor use*)
	Brown patch	azoxystrobin, iprodione, propiconazole (*qualified minor use*)
	Crown rust	azoxystrobin
	Dollar spot	iprodione, propiconazole
	Fairy rings	azoxystrobin
	Fusarium	azoxystrobin, iprodione, propiconazole, trifloxystrobin

	Melting out	azoxystrobin, iprodione
	Red thread	iprodione, trifloxystrobin
	Snow mould	iprodione
	Take-all patch	azoxystrobin
Pests	Flies	chlorpyrifos
	Leatherjackets	chlorpyrifos
	Slugs/snails	metaldehyde
Plant growth regulation	Growth control	maleic hydrazide, trinexapac-ethyl
Weeds	Broad-leaved weeds	2,4-D, 2,4-D + dicamba, 2,4-D + dicamba + triclopyr, asulam, asulam (*not fine turf*), bromoxynil + ioxynil + mecoprop-P, clopyralid + triclopyr, dicamba + MCPA + mecoprop-P, ethofumesate, MCPA
	Grass weeds	ethofumesate
	Mosses	ferrous sulphate
	Weeds, miscellaneous	glyphosate, glyphosate (*wiper application*)
	Woody weeds/scrub	clopyralid + triclopyr

Turf/amenity grass - Managed amenity turf

Crop control	Miscellaneous non-selective situations	glufosinate-ammonium
Diseases	Anthracnose	azoxystrobin, chlorothalonil, propiconazole (*qualified minor use*)
	Brown patch	azoxystrobin, iprodione, propiconazole (*qualified minor use*)
	Crown rust	azoxystrobin
	Dollar spot	carbendazim, chlorothalonil, fenarimol, iprodione, propiconazole, pyraclostrobin (*reduction*), thiophanate-methyl
	Fairy rings	azoxystrobin
	Fusarium	azoxystrobin, carbendazim, carbendazim + epoxiconazole, chlorothalonil, fenarimol, iprodione, myclobutanil, prochloraz + tebuconazole, propiconazole, pyraclostrobin, thiophanate-methyl, trifloxystrobin
	Melting out	azoxystrobin, iprodione
	Pythium	fosetyl-aluminium (*off-label*)
	Red thread	chlorothalonil, fenarimol, iprodione, pyraclostrobin, thiophanate-methyl, trifloxystrobin
	Snow mould	iprodione
	Take-all patch	azoxystrobin
Pests	Birds/mammals	aluminium phosphide
	Chafers	imidacloprid
	Earthworms	carbendazim, thiophanate-methyl
	Flies	chlorpyrifos
	Leatherjackets	chlorpyrifos, imidacloprid
	Slugs/snails	metaldehyde
Plant growth regulation	Growth control	maleic hydrazide, trinexapac-ethyl
Weeds	Broad-leaved weeds	2,4-D, 2,4-D + dicamba, 2,4-D + florasulam, 2,4-D + mecoprop-P, bromoxynil + ioxynil + mecoprop-P,

		chlorthal-dimethyl, clopyralid + 2,4-D + MCPA, clopyralid + diflufenican + MCPA, clopyralid + fluroxypyr + MCPA, dicamba + dichlorprop-P + ferrous sulphate + MCPA, dicamba + dichlorprop-P + MCPA, dicamba + MCPA + mecoprop-P, dichlorprop-P + ferrous sulphate + MCPA, dichlorprop-P + MCPA, ferrous sulphate + MCPA + mecoprop-P, fluroxypyr + mecoprop-P, MCPA, MCPA + mecoprop-P, mecoprop-P
	Crops as weeds	2,4-D + florasulam, dichlorprop-P + MCPA, mecoprop-P
	Grass weeds	amitrole (*off-label - on amitrole resistant turf*), tepraloxydim (*off-label*)
	Mosses	dicamba + dichlorprop-P + ferrous sulphate + MCPA, dichlorprop-P + ferrous sulphate + MCPA, ferrous sulphate, ferrous sulphate + MCPA + mecoprop-P
	Weeds, miscellaneous	glyphosate (*pre-establishment*)

Non-crop pest control

Farm buildings/yards - Farm buildings

Diseases	Fungus diseases	formaldehyde (commodity substance)
Pests	Birds/mammals	aluminium ammonium sulphate, brodifacoum (*indoor use only*), bromadiolone, chlorophacinone, difenacoum, warfarin
	Flies	diflubenzuron, d-phenothrin + tetramethrin, permethrin, tetramethrin
	Food/grain storage pests	alpha-cypermethrin
	Pests, miscellaneous	alpha-cypermethrin, cypermethrin, pyrethrins
	Wasps	d-phenothrin + tetramethrin
Weeds	Weeds, miscellaneous	glyphosate

Farm buildings/yards - Farmyards

Pests	Birds/mammals	bromadiolone, difenacoum

Farmland pest control - Farmland situations

Pests	Birds/mammals	aluminium phosphide
Weeds	Crops as weeds	dichlobenil (*blight prevention*)

Miscellaneous non-crop pest control - Manure/rubbish

Pests	Flies	diflubenzuron

Miscellaneous non-crop pest control - Miscellaneous pest control situations

Pests	Birds/mammals	carbon dioxide (commodity substance), paraffin oil (commodity substance) (*egg treatment*), strychnine hydrochloride (commodity substance) (*areas of restricted public access*)

Protected salad and vegetable crops

Protected brassicas - Protected brassica vegetables

Diseases	Black rot	copper oxychloride (*off-label*)
	Downy mildew	fosetyl-aluminium (*off-label*)
	Spear rot	copper oxychloride (*off-label*)
Pests	Aphids	pyrethrins
	Beetles	pyrethrins
	Caterpillars	Bacillus thuringiensis (*off-label*), pyrethrins
	Pests, miscellaneous	dimethoate (*off-label*)
	Whiteflies	pyrethrins

Protected brassicas - Protected salad brassicas

Diseases	Botrytis	boscalid + pyraclostrobin (*off-label*), iprodione (*off-label - for baby leaf production*)
	Downy mildew	fosetyl-aluminium (*off-label - for baby leaf production*)
	Foliar diseases	thiram (*off-label - for baby leaf production*)
	Ring spot	boscalid + pyraclostrobin (*off-label*)
	Sclerotinia	boscalid + pyraclostrobin (*off-label - for baby leaf production*)
Pests	Aphids	deltamethrin (*off-label*), lambda-cyhalothrin + pirimicarb (*off-label - for baby leaf production*), pirimicarb (*off-label - for baby leaf production*), pymetrozine (*off-label*), pymetrozine (*off-label - for baby leaf production*), thiacloprid (*off-label*), thiacloprid (*off-label - for baby leaf production*)
	Beetles	deltamethrin (*off-label - for baby leaf production*)
	Caterpillars	Bacillus thuringiensis (*off-label*), Bacillus thuringiensis (*off-label - for baby leaf production*), deltamethrin (*off-label*), deltamethrin (*off-label - for baby leaf production*)
	Pests, miscellaneous	deltamethrin (*off-label*), dimethoate (*off-label*)
	Whiteflies	Verticillium lecanii (*off-label*)
Weeds	Broad-leaved weeds	pendimethalin (*off-label*), propyzamide (*off-label - for baby leaf production*)
	Grass weeds	propyzamide (*off-label - for baby leaf production*)

Protected crops, general - All protected crops

Diseases	Fungus diseases	potassium bicarbonate (commodity substance)
	Sclerotinia	Coniothyrium minitans
	Soil-borne diseases	dazomet (*soil fumigation*)

Pests	Aphids	nicotine, pyrethrins, rotenone
	Beetles	pyrethrins
	Capsid bugs	nicotine
	Caterpillars	pyrethrins
	Free-living nematodes	dazomet (*soil fumigation*)
	Leaf miners	nicotine
	Leafhoppers	nicotine
	Mealybugs	pyrethrins
	Sawflies	nicotine
	Scale insects	pyrethrins
	Slugs/snails	ferric phosphate, metaldehyde
	Soil pests	dazomet (*soil fumigation*)
	Spider mites	pyrethrins
	Thrips	nicotine, pyrethrins
	Whiteflies	pyrethrins
Weeds	Weeds, miscellaneous	dazomet (*soil fumigation*)

Protected crops, general - Glasshouses

Diseases	Fungus diseases	formaldehyde (commodity substance) (*spray, dip or fumigant*)
	Phytophthora	fosetyl-aluminium
	Root rot	fosetyl-aluminium
	Soil-borne diseases	metam-sodium
Pests	Free-living nematodes	metam-sodium
	Pests, miscellaneous	pyrethrins
	Soil pests	metam-sodium
Weeds	Weeds, miscellaneous	metam-sodium

Protected crops, general - Protected vegetables, general

Diseases	White blister	propamocarb hydrochloride (*off-label*)
Pests	Aphids	physical pest control, pirimicarb (*off-label*)
	Leafhoppers	physical pest control
	Mites	physical pest control
	Thrips	physical pest control
	Whiteflies	physical pest control

Protected crops, general - Soils and compost

Diseases	Soil-borne diseases	formaldehyde (commodity substance) (*drench*), metam-sodium
Pests	Free-living nematodes	metam-sodium
	Pests, miscellaneous	dimethoate (*off-label - for watercress progagation*), dimethoate (*off-label - for watercress propagation*)
	Soil pests	metam-sodium
Weeds	Weeds, miscellaneous	metam-sodium

Protected fruiting vegetables - Protected aubergines

Diseases	Botrytis	cyprodinil + fludioxonil (*off-label*), fenhexamid (*off-label*), iprodione (*off-label*)
	Powdery mildew	myclobutanil (*off-label*)
Pests	Aphids	acetamiprid, deltamethrin (*off-label*), pirimicarb (*off-label*), pymetrozine (*off-label*)
	Caterpillars	Bacillus thuringiensis (*off-label*), indoxacarb, spinosad (*off-label*)
	Leaf miners	deltamethrin (*off-label*), thiacloprid (*off-label*)
	Pests, miscellaneous	deltamethrin (*off-label*)
	Spider mites	abamectin (*off-label*), spiromesifen (*off-label - on inert media or by NFT*)
	Thrips	abamectin (*off-label*), deltamethrin (*off-label*), spinosad (*off-label*), thiacloprid (*off-label*)
	Whiteflies	acetamiprid, deltamethrin (*off-label*), spiromesifen (*off-label - on inert media or by NFT*), thiacloprid (*off-label*)

Protected fruiting vegetables - Protected cucurbits

Diseases	Botrytis	chlorothalonil, cyprodinil + fludioxonil (*off-label*), fenhexamid (*off-label*), iprodione (*off-label*)
	Damping off	propamocarb hydrochloride (*off-label*)
	Downy mildew	azoxystrobin (*off-label*), metalaxyl-M (*off-label*)
	Fusarium	cyprodinil + fludioxonil (*off-label*)
	Mycosphaerella	cyprodinil + fludioxonil (*off-label*)
	Phytophthora	propamocarb hydrochloride
	Powdery mildew	azoxystrobin (*off-label*), bupirimate, fenarimol (*off-label*), imazalil, myclobutanil (*off-label*), sulphur (*off-label*)
	Pythium	propamocarb hydrochloride
	Root rot	propamocarb hydrochloride (*off-label*)
Pests	Ants	pirimiphos-methyl
	Aphids	acetamiprid, deltamethrin, deltamethrin (*off-label*), nicotine, pirimicarb (*off-label*), pirimiphos-methyl, pymetrozine, pymetrozine (*off-label*), Verticillium lecanii
	Capsid bugs	pirimiphos-methyl
	Caterpillars	Bacillus thuringiensis, Bacillus thuringiensis (*off-label*), deltamethrin, indoxacarb, spinosad (*off-label*)
	Earwigs	pirimiphos-methyl
	Leaf miners	deltamethrin (*off-label*), oxamyl (*off-label*), pirimiphos-methyl, thiacloprid (*off-label*)
	Leafhoppers	nicotine
	Mealybugs	deltamethrin, petroleum oil
	Pests, miscellaneous	deltamethrin (*off-label*)
	Sawflies	pirimiphos-methyl
	Scale insects	deltamethrin, petroleum oil
	Spider mites	abamectin, abamectin (*off-label*), petroleum oil, pirimiphos-methyl, spiromesifen (*off-label - on inert media or by NFT*)
	Thrips	abamectin, deltamethrin (*off-label*), nicotine, pirimiphos-methyl, spinosad, spinosad (*off-label*), thiacloprid (*off-label*)

	Whiteflies	acetamiprid, buprofezin, buprofezin (*off-label*), cypermethrin, deltamethrin, deltamethrin (*off-label*), nicotine, pirimiphos-methyl, spiromesifen (*off-label - on inert media or by NFT*), thiacloprid (*off-label*), Verticillium lecanii, Verticillium lecanii (*off-label*)
Weeds	Broad-leaved weeds	isoxaben (*off-label*)

Protected fruiting vegetables - Protected tomatoes

Diseases	Botrytis	azoxystrobin (*off-label*), chlorothalonil, fenhexamid (*off-label*), iprodione, pyrimethanil (*off-label*), thiram
	Damping off	copper oxychloride, etridiazole (*seeds and seedlings*), propamocarb hydrochloride (*off-label*)
	Didymella stem rot	azoxystrobin (*off-label*), carbendazim (*off-label*)
	Foot rot	copper oxychloride
	Leaf mould	chlorothalonil, copper ammonium carbonate
	Phytophthora	azoxystrobin (*off-label*), chlorothalonil, copper oxychloride, etridiazole (*transplants*), propamocarb hydrochloride
	Powdery mildew	azoxystrobin (*off-label*), bupirimate (*off-label*), fenarimol (*off-label*), imazalil (*off-label - in synthetic/inert media*), myclobutanil (*off-label*), sulphur (*off-label*)
	Pythium	propamocarb hydrochloride
	Root rot	propamocarb hydrochloride (*off-label*)
Pests	Ants	pirimiphos-methyl
	Aphids	acetamiprid, deltamethrin, fatty acids, nicotine, pirimicarb, pirimiphos-methyl, pyrethrins, Verticillium lecanii
	Beetles	pyrethrins
	Capsid bugs	pirimiphos-methyl
	Caterpillars	Bacillus thuringiensis, deltamethrin, indoxacarb, nicotine, pyrethrins, spinosad (*off-label*)
	Earwigs	pirimiphos-methyl
	Leaf miners	abamectin, abamectin (*off-label*), deltamethrin (*off-label*), nicotine, oxamyl (*off-label*), pirimiphos-methyl, thiacloprid (*off-label*)
	Leafhoppers	nicotine
	Mealybugs	deltamethrin, fatty acids, petroleum oil, pyrethrins (*off-label*)
	Mirid bugs	pyrethrins (*off-label*)
	Sawflies	pirimiphos-methyl
	Scale insects	deltamethrin, fatty acids, petroleum oil
	Spider mites	abamectin, fatty acids, petroleum oil, pirimiphos-methyl
	Thrips	deltamethrin (*off-label*), nicotine, pirimiphos-methyl, spinosad (*off-label*), thiacloprid (*off-label*)
	Whiteflies	acetamiprid, buprofezin, deltamethrin, fatty acids, nicotine, pirimiphos-methyl, pymetrozine (*off-label*), pyrethrins, spiromesifen, thiacloprid (*off-label*), Verticillium lecanii
Plant growth regulation	Fruiting control	2-chloroethylphosphonic acid

Protected herb crops - Protected herbs

Diseases	Botrytis	iprodione (*off-label*), propamocarb hydrochloride (*off-label*)
	Downy mildew	fosetyl-aluminium (*off-label*), propamocarb hydrochloride (*off-label*)
	Foliar diseases	mancozeb (*off-label*), thiram (*off-label*)
	Fungus diseases	mancozeb + metalaxyl-M (*off-label*)
	Powdery mildew	sulphur (*off-label*)
	Rhizoctonia	azoxystrobin (*off-label*)
	Ring spot	prochloraz (*off-label*)
	Sclerotinia	boscalid + pyraclostrobin (*off-label*)
Pests	Aphids	acetamiprid (*off-label*), lambda-cyhalothrin + pirimicarb (*off-label*), nicotine, pirimicarb (*off-label*), pymetrozine (*off-label*), thiacloprid (*off-label*)
	Beetles	deltamethrin (*off-label*)
	Caterpillars	Bacillus thuringiensis (*off-label*)
	Leafhoppers	deltamethrin (*off-label*), nicotine
	Pests, miscellaneous	abamectin (*off-label*), deltamethrin (*off-label*)
	Thrips	nicotine, pirimicarb (*off-label*)
	Whiteflies	nicotine, Verticillium lecanii (*off-label*)
Weeds	Broad-leaved weeds	chlorpropham with cetrimide (*off-label*), pendimethalin (*off-label*), propyzamide (*off-label*)
	Grass weeds	propyzamide (*off-label*)

Protected leafy vegetables - Mustard and cress

Diseases	Botrytis	fenhexamid (*off-label*), iprodione (*off-label*), propamocarb hydrochloride (*off-label*)
	Damping off	etridiazole, thiram (*off-label - seed treatment*)
	Downy mildew	dimethomorph + mancozeb (*off-label*), propamocarb hydrochloride (*off-label*)
	Sclerotinia	iprodione (*off-label*)
Pests	Aphids	acetamiprid (*off-label*), nicotine
	Beetles	deltamethrin (*off-label*)
	Caterpillars	nicotine
	Leaf miners	nicotine
	Leafhoppers	deltamethrin (*off-label*)
	Pests, miscellaneous	deltamethrin (*off-label*)
Weeds	Broad-leaved weeds	pendimethalin (*off-label*), propachlor (*off-label*)
	Grass weeds	pendimethalin (*off-label*), propachlor (*off-label*)

Protected leafy vegetables - Protected leafy vegetables

Diseases	Big vein	carbendazim (*off-label*)
	Botrytis	iprodione, iprodione (*off-label*), propamocarb hydrochloride, pyrimethanil (*off-label*), thiram
	Bottom rot	boscalid + pyraclostrobin, tolclofos-methyl
	Downy mildew	fosetyl-aluminium, fosetyl-aluminium (*off-label*), mancozeb, propamocarb hydrochloride, thiram
	Fungus diseases	mancozeb + metalaxyl-M (*off-label*)

	Rhizoctonia	azoxystrobin (*off-label*)
	Ring spot	prochloraz (*off-label*)
	Sclerotinia	boscalid + pyraclostrobin
Pests	Aphids	acetamiprid (*off-label*), lambda-cyhalothrin + pirimicarb (*off-label*), nicotine, pirimicarb, pirimicarb (*off-label*), pymetrozine (*off-label*), pyrethrins, thiacloprid (*off-label*), Verticillium lecanii
	Beetles	deltamethrin (*off-label*)
	Caterpillars	Bacillus thuringiensis (*off-label*), pyrethrins
	Leafhoppers	deltamethrin (*off-label*), nicotine
	Pests, miscellaneous	abamectin (*off-label*), deltamethrin (*off-label*)
	Thrips	nicotine, pirimicarb (*off-label*)
	Whiteflies	cypermethrin, nicotine, Verticillium lecanii
Weeds	Broad-leaved weeds	chlorpropham (*off-label*), chlorpropham with cetrimide, propyzamide (*off-label*)
	Grass weeds	chlorpropham (*off-label*), chlorpropham with cetrimide, propyzamide (*off-label*)

Protected leafy vegetables - Protected spinach

Diseases	Downy mildew	metalaxyl-M (*off-label*)
	Foliar diseases	thiram (*off-label*)
Pests	Aphids	pirimicarb (*off-label*), thiacloprid (*off-label*)
	Pests, miscellaneous	dimethoate (*off-label*)

Protected legumes - Protected peas and beans

Pests	Aphids	Verticillium lecanii
	Leaf miners	oxamyl (*off-label*)
	Whiteflies	Verticillium lecanii, Verticillium lecanii (*off-label*)

Protected root and tuber vegetables - Protected carrots/ parsnips/celeriac

Pests	Pests, miscellaneous	dimethoate (*off-label*)
Weeds	Broad-leaved weeds	isoxaben (*off-label - temporary protection*)

Protected root and tuber vegetables - Protected potatoes

Diseases	Phytophthora	benalaxyl + mancozeb
Pests	Aphids	nicotine
	Leafhoppers	nicotine
	Thrips	nicotine
	Whiteflies	nicotine

Protected root and tuber vegetables - Protected root brassicas

Diseases	Botrytis	propamocarb hydrochloride (*off-label*)
	Downy mildew	propamocarb hydrochloride (*off-label*)

	Rhizoctonia	tolclofos-methyl (*off-label*)
	White blister	mancozeb + metalaxyl-M (*off-label*)
Pests	Aphids	pirimicarb (*off-label*)

Protected stem and bulb vegetables - Protected celery/chicory

Diseases	Botrytis	azoxystrobin (*off-label*)
	Phytophthora	azoxystrobin (*off-label - for forcing*)
	Rhizoctonia	azoxystrobin (*off-label*), tolclofos-methyl (*off-label*)
	Sclerotinia	azoxystrobin (*off-label*)
Pests	Aphids	acetamiprid (*off-label*), deltamethrin (*off-label*), lambda-cyhalothrin + pirimicarb (*off-label*), pirimicarb (*off-label*), pymetrozine (*off-label*)
	Beetles	deltamethrin (*off-label*)
	Caterpillars	Bacillus thuringiensis (*off-label*), deltamethrin (*off-label*)
	Leafhoppers	deltamethrin (*off-label*)
	Pests, miscellaneous	abamectin (*off-label*), deltamethrin (*off-label*)
Weeds	Broad-leaved weeds	propyzamide (*off-label*)
	Grass weeds	propyzamide (*off-label*)

Protected stem and bulb vegetables - Protected onions/leeks/garlic

Pests	Aphids	deltamethrin (*off-label*), nicotine
	Caterpillars	deltamethrin (*off-label*)
	Leafhoppers	nicotine
	Pests, miscellaneous	deltamethrin (*off-label*)
	Stem nematodes	oxamyl (*off-label*)
	Thrips	nicotine
	Whiteflies	nicotine

Total vegetation control

Aquatic situations, general - Aquatic situations

Plant growth regulation	Growth control	maleic hydrazide
Weeds	Aquatic weeds	2,4-D, dichlobenil, glyphosate
	Grass weeds	glyphosate
	Weeds, miscellaneous	glyphosate

Non-crop areas, general - Farm buildings/yards

Plant growth regulation	Growth control	maleic hydrazide

Non-crop areas, general - Miscellaneous non-crop situations

Plant growth regulation	Growth control	maleic hydrazide
Weeds	Broad-leaved weeds	asulam, MCPA, triclopyr
	Weeds, miscellaneous	diuron, glyphosate
	Woody weeds/scrub	glyphosate, triclopyr

Non-crop areas, general - Non-crop farm areas

Weeds	Broad-leaved weeds	2,4-D + dicamba + triclopyr, 2,4-D + picloram, diuron, diuron + glyphosate, glufosinate-ammonium, MCPA, paraquat, picloram, triclopyr
	Grass weeds	diuron, diuron + glyphosate, glufosinate-ammonium, paraquat
	Weeds, miscellaneous	dichlobenil, diuron, glufosinate-ammonium, glyphosate, sodium chlorate
	Woody weeds/scrub	2,4-D + picloram, asulam, glyphosate, picloram, triclopyr

Non-crop areas, general - Paths/roads etc

Pests	Slugs/snails	metaldehyde
Weeds	Broad-leaved weeds	2,4-D + dicamba + triclopyr, diquat + paraquat, diuron, MCPA
	Grass weeds	diquat + paraquat, diuron, quizalofop-P-tefuryl (*off-label*)
	Weeds, miscellaneous	acetic acid, dichlobenil, diflufenican + glyphosate, glyphosate, sodium chlorate
	Woody weeds/scrub	2,4-D + dicamba + triclopyr, glyphosate

SECTION 2
PESTICIDE PROFILES

1 abamectin

A selective acaricide and insecticide for use in ornamentals and other protected crops
IRAC mode of action code: 6

Products

Dynamec	Syngenta Bioline	18 g/l	EC	13331

Uses

- Insect control in ***protected chives*** *(off-label)*, ***protected cress*** *(off-label)*, ***protected frise*** *(off-label)*, ***protected herbs (see appendix 6)*** *(off-label)*, ***protected lamb's lettuce*** *(off-label)*, ***protected lettuce*** *(off-label)*, ***protected parsley*** *(off-label)*, ***protected radicchio*** *(off-label)*, ***protected scarole*** *(off-label)*
- Leaf and bud mite in ***protected blackberries*** *(off-label)*, ***protected raspberries*** *(off-label)*
- Leaf miner in ***protected cherry tomatoes*** *(off-label)*, ***protected tomatoes***
- Red spider mites in ***protected blackberries*** *(off-label)*, ***protected cayenne peppers*** *(off-label)*, ***protected peppers*** *(off-label)*, ***protected raspberries*** *(off-label)*, ***protected strawberries*** *(off-label)*
- Tarsonemid mites in ***strawberries*** *(off-label - in propagation)*
- Two-spotted spider mite in ***ornamental plant production***, ***protected aubergines*** *(off-label)*, ***protected cucumbers***, ***protected ornamentals***, ***protected tomatoes***
- Western flower thrips in ***ornamental plant production***, ***protected aubergines*** *(off-label)*, ***protected cucumbers***, ***protected ornamentals***

Specific Off-Label Approvals (SOLAs)

- ***protected aubergines*** *20070421*
- ***protected blackberries*** *20072290*
- ***protected cayenne peppers*** *20070422*
- ***protected cherry tomatoes*** *20070422*
- ***protected chives*** *20070430*
- ***protected cress*** *20070430*
- ***protected frise*** *20070430*
- ***protected herbs (see appendix 6)*** *20070430*
- ***protected lamb's lettuce*** *20070430*
- ***protected lettuce*** *20070430*
- ***protected parsley*** *20070430*
- ***protected peppers*** *20070422*
- ***protected radicchio*** *20070430*
- ***protected raspberries*** *20072290*
- ***protected scarole*** *20070430*
- ***protected strawberries*** *20070423*
- ***strawberries*** *(in propagation) 20070423*

Efficacy guidance

- Abamectin controls adults and immature stages of the two-spotted spdier mite, the larval stages of leaf miners and the nymphs of Western Flower Thrips, plus a useful reduction in adults
- Treat at first sign of infestation. Repeat sprays may be required
- For effective control total cover of all plant surfaces is essential, but avoid run-off
- Target pests quickly become immobilised but 3-5 d may be required for maximum mortality
- Indoor applications should be made through a hydraulic nozzle applicator or a knapsack applicator. Outdoors suitable high volume hydraulic nozzle applicators should be used

Restrictions

- Number of treatments 6 on protected tomatoes and cucumbers (only 4 of which can be made when flowers or fruit present); not restricted on flowers but rotation with other products advised
- Maximum concentration must not exceed 50 ml per 100 l water
- Do not mix with wetters, stickers or other adjuvants
- Do not use on ferns (*Adiantum* spp) or Shasta daisies
- Do not treat protected crops which are in flower or have set fruit between 1 Nov and 28 Feb
- Do not treat cherry tomatoes (but protected cherry tomatoes may be treated off-label)
- Consult manufacturer for list of plant varieties tested for safety

- There is insufficient evidence to support product compatibility with integrated and biological pest control programmes
- Unprotected persons must be kept out of treated areas until the spray has dried

Crop-specific information

- HI 3 d for protected edible crops
- On tomato or cucumber crops that are in flower, or have started to set fruit, treat only between 1 Mar and 31 Oct. Seedling tomatoes or cucumbers that have not started to flower or set fruit may be treated at any time
- Some spotting or staining may occur on carnation, kalanchoe and begonia foliage

Environmental safety

- Dangerous for the environment
- Very toxic to aquatic organisms
- High risk to bees. Do not apply to crops in flower or to those in which bees are actively foraging. Do not apply when flowering weeds are present
- Keep in original container, tightly closed, in a safe place, under lock and key
- Where bumble bees are used in tomatoes as pollinators, keep them out for 24 h after treatment

Hazard classification and safety precautions

Hazard H03, H11
Risk phrases R22a, R50, R53a
Operator protection A, C, D, H, K, M; U02a, U05a, U20a
Environmental protection E02a (until spray has dried); E12a, E12e, E15b, E34, E38
Storage and disposal D01, D02, D05, D09b, D10c, D12a
Medical advice M04a

2 acetamiprid

A neonicotinoid insecticide for use in top fruit and horticulture
IRAC mode of action code: 4A

Products

1	Gazelle	Certis	20% w/w	SP	12909
2	Insyst	Certis	20% w/w	SP	13414

Uses

- Aphids in ***apples***, ***cherries***, ***chives*** *(off-label)*, ***cress*** *(off-label)*, ***frise*** *(off-label)*, ***herbs (see appendix 6)*** *(off-label)*, ***lamb's lettuce*** *(off-label)*, ***lettuce*** *(off-label)*, ***ornamental plant production***, ***parsley*** *(off-label)*, ***pears***, ***plums***, ***protected aubergines***, ***protected chives*** *(off-label)*, ***protected frise*** *(off-label)*, ***protected herbs (see appendix 6)*** *(off-label)*, ***protected lamb's lettuce*** *(off-label)*, ***protected lettuce*** *(off-label)*, ***protected ornamentals***, ***protected parsley*** *(off-label)*, ***protected peppers***, ***protected radicchio*** *(off-label)*, ***protected scarole*** *(off-label)*, ***protected tomatoes***, ***radicchio*** *(off-label)*, ***scarole*** *(off-label)* [1]; ***brussels sprouts*** *(off-label)*, ***seed potatoes***, ***ware potatoes*** [2]
- Whitefly in ***brussels sprouts*** *(off-label)* [2]; ***protected aubergines***, ***protected ornamentals***, ***protected peppers***, ***protected tomatoes*** [1]

Specific Off-Label Approvals (SOLAs)

- ***brussels sprouts*** *20072866* [2]
- ***chives*** *20070139* [1]
- ***cress*** *20070139* [1]
- ***frise*** *20070139* [1]
- ***herbs (see appendix 6)*** *20070139* [1]
- ***lamb's lettuce*** *20070139* [1]
- ***lettuce*** *20070139* [1]
- ***parsley*** *20070139* [1]
- ***protected chives*** *20070139* [1]
- ***protected frise*** *20070139* [1]
- ***protected herbs (see appendix 6)*** *20070139* [1]
- ***protected lamb's lettuce*** *20070139* [1]
- ***protected lettuce*** *20070139* [1]
- ***protected parsley*** *20070139* [1]

- ***protected radicchio*** *20070139* [1]
- ***protected scarole*** *20070139* [1]
- ***radicchio*** *20070139* [1]
- ***scarole*** *20070139* [1]

Approval information
- Acetamiprid included in Annex I under EC Directive 91/414

Efficacy guidance
- Best results obtained from application at the first sign of pest attack or when appropriate thresholds are reached
- Thorough coverage of foliage is essential to ensure best control. Acetamiprid has contact, systemic and translaminar activity

Restrictions
- Maximum number of treatments 1 per yr for cherries, ware potatoes; 2 per yr or crop for apples, pears, plums, ornamental plant production, protected crops, seed potatoes
- Do not use more than two applications of any neonicotinoid insecticide (e.g. acetamiprid, clothianidin, imidacloprid, thiacloprid) on any crop. Previous soil or seed treatment with a neonicotinoid counts as one such treatment
- 2 treatments are permitted on seed potatoes but must not be used consecutively [2]

Crop-specific information
- HI 3 d for protected crops; 14 d for apples, cherries, pears, plums, potatoes

Environmental safety
- Harmful to aquatic organisms
- LERAP Category B
- Broadcast air-assisted LERAP (20 m) [1]
- Acetamiprid is slightly toxic to predatory mites and generally slightly toxic to other beneficials

Hazard classification and safety precautions

Risk phrases R52, R53a
Operator protection A; U02a, U19a, U20a
Environmental protection E15b, E16a, E38 [1, 2]; E16b [2]; E17b [1] (20 m)
Storage and disposal D05 [1]; D09a, D10b, D12a [1, 2]

3 acetic acid

A non-selective herbicide for non-crop situations

Products

Natural Weed Spray No 1	Headland Amenity	240 g/l	SL	12328

Uses
- General weed control in ***hard surfaces***, ***natural surfaces not intended to bear vegetation***, ***permeable surfaces overlying soil***

Efficacy guidance
- Best results obtained from treatment of young tender weeds less than 10 cm high
- Treat in spring and repeat as necessary throughout the growing season
- Treat survivors as soon as fresh growth is seen
- Ensure complete coverage of foliage to the point of run-off
- Rainfall after treatment may reduce efficacy

Restrictions
- No restriction on number of treatments

Following crops guidance
- There is no residual activity in the soil and sowing or planting may take place as soon as treated weeds have died

Environmental safety
- Harmful to aquatic organisms
- High risk to bees
- Do not apply to crops in flower or to those in which bees are actively foraging. Do not apply when flowering weeds are present

SEE SECTION 3 FOR PRODUCTS ALSO REGISTERED

- Keep people and animals off treated dense weed patches until spray has dried. This is not necessary for areas with occasional low growing prostrate weeds such as on pathways

Hazard classification and safety precautions

Hazard H04
Risk phrases R36, R37, R38, R52
Operator protection A, C, D, H; U05a, U09a, U19a, U20b
Environmental protection E12a, E12e, E15a
Storage and disposal D01, D02, D09a, D12a

4 alpha-cypermethrin

A contact and ingested pyrethroid insecticide for use in arable crops and agricultural buildings
IRAC mode of action code: 3

Products

1	Antec Durakil 1.5 SC	Antec Biosentry	15 g/l	SC	H7560
2	Antec Durakil 6SC	Antec Biosentry	60 g/l	SC	H7559
3	Contest	BASF	15% w/w	WG	10216
4	Strength AC	Certis	15 g/l	SC	H7469

Uses

- Brassica pod midge in ***winter oilseed rape*** [3]
- Cabbage seed weevil in ***spring oilseed rape***, ***winter oilseed rape*** [3]
- Cabbage stem flea beetle in ***winter oilseed rape*** [3]
- Cabbage white butterfly in ***salad brassicas*** *(off-label - for baby leaf production)* [3]
- Caterpillars in ***broccoli***, ***brussels sprouts***, ***cabbages***, ***calabrese***, ***cauliflowers***, ***kale*** [3]
- Cereal aphid in ***spring barley***, ***spring wheat***, ***winter barley***, ***winter wheat*** [3]
- Diamond-back moth in ***salad brassicas*** *(off-label - for baby leaf production)* [3]
- Flea beetle in ***broccoli***, ***brussels sprouts***, ***cabbages***, ***calabrese***, ***cauliflowers***, ***kale*** [3]
- Insect pests in ***farm buildings***, ***poultry houses*** [4]
- Lesser mealworm in ***poultry houses*** [1, 2]
- Pea and bean weevil in ***broad beans***, ***combining peas***, ***spring field beans***, ***vining peas***, ***winter field beans*** [3]
- Pea aphid in ***combining peas*** *(reduction)*, ***vining peas*** *(reduction)* [3]
- Pea moth in ***combining peas***, ***vining peas*** [3]
- Pine weevil in ***forest nurseries*** *(off-label)*, ***forestry transplants*** *(off-label)* [3]
- Pollen beetle in ***spring oilseed rape***, ***winter oilseed rape*** [3]
- Poultry red mite in ***poultry houses*** [1, 2]
- Rape winter stem weevil in ***winter oilseed rape*** [3]
- Small white butterfly in ***salad brassicas*** *(off-label - for baby leaf production)* [3]
- Yellow cereal fly in ***winter barley***, ***winter wheat*** [3]

Specific Off-Label Approvals (SOLAs)

- ***forest nurseries*** *20062391* [3]
- ***forestry transplants*** *20062391* [3]
- ***salad brassicas*** *(for baby leaf production) 20011221* [3]

Approval information

- Alpha-cypermethrin included in Annex I under EC Directive 91/414
- Accepted by BBPA for use on malting barley

Efficacy guidance

- For cabbage stem flea beetle control spray oilseed rape when adult or larval damage first seen and about 1 mth later [3]
- For flowering pests on oilseed rape apply at any time during flowering, on pollen beetle best results achieved at green to yellow bud stage (GS 3,3-3,7), on seed weevil between 20 pods set stage and 80% petal fall (GS 4,7-5,8) [3]
- Spray cereals in autumn for control of cereal aphids, in spring/summer for grain aphids. (See label for details) [3]
- For flea beetle, caterpillar and cabbage aphid control on brassicas apply when the pest or damage first seen or as a preventive spray. Repeat if necessary [3]
- For pea and bean weevil control in peas and beans apply when pest attack first seen and repeat as necessary [3]

- For lesser mealworm control in poultry houses apply a coarse, low-pressure spray as routine treatment after clean-out and before each new crop. Spray vertical surfaces and ensure an overlap onto ceilings. It is not necessary to treat the floor [1, 2, 4]
- Use the highest recommended concentration in poultry houses where extreme residual action is required or surfaces are dirty or highly absorbent [1, 2, 4]

Restrictions

- Maximum number of treatments 4 per crop on edible brassicas, 3 per crop on winter oilseed rape, peas, 2 on beans, spring oilseed rape (only 1 after yellow bud stage - GS 3,7) [3]
- Maximum number of applications in animal husbandry use 2 when used in premises that are occupied by poultry [1, 2]
- Apply up to 2 sprays on cereals in autumn and spring, 1 in summer between 1 Apr and 31 Aug. See label for details of rates and maximum total dose [3]
- Only 1 aphicide treatment may be applied in cereals between 1 Apr and 31 Aug and spray volume must not be reduced in this period [3]
- Do not apply to a cereal crop if any product containing a pyrethroid or dimethoate has been applied after the start of ear emergence (GS 51) [3]

Crop-specific information

- Latest use: before the end of flowering for oilseed rape; before 31 Aug for cereals [3]
- HI vining peas 1 d; brassicas 7 d; combining peas, broad beans, field beans 11 d [3]
- For summer cereal application do not spray within 6 m from edge of crop and do not reduce volume when used after 31 Mar [3]

Environmental safety

- Dangerous for the environment [3]
- Very toxic to aquatic organisms [3]
- Extremely dangerous (harmful [4]) to fish or other aquatic life. Do not contaminate surface waters or ditches with chemical or used container [1, 2]
- Dangerous to bees [1, 2]
- LERAP Category A [3]
- Where possible spray oilseed rape crops in the late evening or early morning or in dull weather [3]
- Do not spray within 6 m of the edge of a cereal crop after 31 Mar in yr of harvest [3]
- Do not apply directly to poultry; collect eggs before application [1, 2]

Hazard classification and safety precautions

Hazard H03, H11 [3]

Risk phrases R22a, R50, R53a [3]

Operator protection A, H; U02a [1, 2]; U02b, U09a, U19a [1, 2, 4]; U05a, U10 [3]; U14, U15 [4]; U20b [1-4]

Environmental protection E02a [1, 2] (until dry); E02d, E05c, E13c [4]; E05a [1, 2, 4]; E12c, E13a [1, 2]; E15a [1-3]; E16c, E16d, E38 [3]

Consumer protection C04, C10 [4]; C05, C06, C07, C09, C11 [1, 2, 4]

Storage and disposal D01, D05 [3, 4]; D02, D10b [3]; D06a, D06c [4]; D09a, D12a [1-4]; D11a [1, 2]

Vertebrate/rodent control products V03b [4]

Medical advice M04a [4]; M05a [3]

5 aluminium ammonium sulphate

An inorganic bird and animal repellent

Products

1	Curb Crop Spray Powder	Sphere	88% w/w	WP	02480
2	Guardsman STP Seed Dressing Powder	Sphere	88% w/w	DS	03606
3	Liquid Curb Crop Spray	Sphere	83 g/l	SC	03164

Uses

- Animal repellent in ***agricultural premises***, ***all bush fruit***, ***all cane fruit***, ***all top fruit***, ***broad beans***, ***carrots***, ***flowerhead brassicas***, ***forest nursery beds***, ***forestry plantations***, ***grain stores***, ***leaf brassicas***, ***peas***, ***permanent grassland***, ***spring barley***, ***spring field beans***, ***spring***

oats, spring oilseed rape, spring wheat, strawberries, sugar beet, winter barley, winter field beans, winter oats, winter oilseed rape, winter wheat [1, 3]; ***amenity vegetation*** [1]
- Bird repellent in ***agricultural premises, all bush fruit, all cane fruit, all top fruit, broad beans, carrots, flowerhead brassicas, forest nursery beds, forestry plantations, grain stores, leaf brassicas, peas, permanent grassland, spring barley, spring field beans, spring oats, spring oilseed rape, spring wheat, strawberries, sugar beet, winter barley, winter field beans, winter oats, winter oilseed rape, winter wheat*** [1, 3]; ***amenity vegetation*** [1]
- Birds in ***corms*** *(corm treatment)*, ***flower bulbs*** *(bulb treatment)*, ***seeds*** *(seed treatment)* [2]
- Damaging mammals in ***corms*** *(corm treatment)*, ***flower bulbs*** *(bulb treatment)*, ***seeds*** *(seed treatment)* [2]

Efficacy guidance
- Apply as overall spray to growing crops before damage starts or mix powder with seed depending on type of protection required
- Spray deposit protects growth present at spraying but gives little protection to new growth
- Product must be sprayed onto dry foliage to be effective and must dry completely before dew or frost forms. In winter this may require some wind

Restrictions
- Maximum number of treatments 1 per batch for corms, flower bulbs, seeds [2]

Crop-specific information
- Latest use: no restriction

Hazard classification and safety precautions

Operator protection U05a, U20a [1, 3]; U20b [2]
Environmental protection E15a [1-3]; E19b [1, 3]
Storage and disposal D01, D02, D05, D12a [1, 3]; D09a [2]; D11a [1-3]
Treated seed S02, S05 [2]
Medical advice M03 [1, 3]

6 aluminium phosphide

A phosphine generating compound used against vertebrates and grain store pests
IRAC mode of action code: 24A

Products

1	Degesch Fumigation Tablets	Rentokil	56% w/w	GE	09313
2	Luxan Talunex	Luxan	57% w/w	GE	06563
3	Phostoxin	Rentokil	56% w/w	GE	09315

Uses
- Insect pests in ***stored grain*** [1]
- Moles in ***farm woodland, lawns, managed amenity turf, permanent grassland*** [2]; ***farmland*** [2, 3]
- Rabbits in ***farm woodland, lawns, managed amenity turf, permanent grassland*** [2]; ***farmland*** [2, 3]
- Rats in ***farm woodland, lawns, managed amenity turf, permanent grassland*** [2]; ***farmland*** [2, 3]

Approval information
- Accepted by BBPA for use in stores for malting barley

Efficacy guidance
- Product releases poisonous hydrogen phosphide gas in contact with moisture
- Place fumigation tablets in grain stores as directed [1]
- Place pellets in burrows or runs and seal hole by heeling in or covering with turf. Do not cover pellets with soil. Inspect daily and treat any new or re-opened holes [2, 3]
- Apply pellets by means of Luxan Topex Applicator [2]

Restrictions
- Aluminium phosphide is subject to the Poisons Rules 1982 and the Poisons Act 1972. See notes in Section 5
- Only to be used by operators instructed or trained in the use of aluminium phosphide and familiar with the precautionary measures to be taken. See label and HSE Guidance Notes for full precautions

- Only open container outdoors [2, 3] and for immediate use. Keep away from liquid or water as this causes immediate release of gas. Do not use in wet weather
- Do not use within 3 m of human or animal habitation. Before application ensure that no humans or domestic animals are in adjacent buildings or structures. Allow a minimum airing-off period of 4 h before re-admission

Environmental safety

- Product liberates very toxic, highly flammable gas
- Dangerous for the environment [2]
- Very toxic to aquatic organisms [2]
- Dangerous to fish or other aquatic life. Do not contaminate surface waters or ditches with chemical or used container [1, 3]
- Prevent access by livestock, pets and other non-target mammals and birds to buildings under fumigation and ventilation [1]
- Pellets must never be placed or allowed to remain on ground surface
- Do not use adjacent to watercourses
- Take particular care to avoid gassing non-target animals, especially those protected under the Wildlife and Countryside Act (e.g. badgers, polecat, reptiles, natterjack toads, most birds). Do not use in burrows where there is evidence of badger or fox activity, or when burrows might be occupied by birds
- Dust remaining after decomposition is harmless and of no environmental hazard
- Keep in original container, tightly closed, in a safe place, under lock and key
- Dispose of empty containers as directed on label

Hazard classification and safety precautions

Hazard H01, H07 [1-3]; H03 [3]; H11 [2]
Risk phrases R21, R26, R28 [1-3]; R36, R37, R50 [2]
Operator protection A, H [1-3]; D [1, 3]; G [2]; U01, U13, U19a, U20a [1-3]; U05a [2, 3]; U05b [1]; U07 [1, 3]; U14, U15 [2]; U18 [1, 2]
Environmental protection E02a [1] (4 h min); E02a [3] (4 h); E02b [1]; E13b [1, 3]; E15a [2]; E34 [1-3]
Storage and disposal D01, D02, D09b, D11b [1-3]; D07 [1, 3]
Vertebrate/rodent control products V04a [2, 3]
Medical advice M04a

7 amidosulfuron

A post-emergence sulfonylurea herbicide for cleavers and other broad-leaved weed control in cereals
HRAC mode of action code: B

Products

1	Eagle	Bayer CropScience	75% w/w	WG	07318
2	Landgold Amidosulfuron	Landgold	75% w/w	WG	09021
3	Landgold Amidosulfuron	Teliton	75% w/w	WG	12110
4	Squire	Bayer CropScience	50% w/w	WG	08715

Uses

- Annual dicotyledons in ***durum wheat***, ***spring barley***, ***spring oats***, ***spring rye***, ***spring wheat***, ***triticale***, ***winter barley***, ***winter oats***, ***winter rye***, ***winter wheat*** [1-3]; ***linseed*** [1]
- Charlock in ***permanent grassland***, ***rotational grass*** [4]
- Cleavers in ***durum wheat***, ***spring barley***, ***spring oats***, ***spring rye***, ***spring wheat***, ***triticale***, ***winter barley***, ***winter oats***, ***winter rye***, ***winter wheat*** [1-3]; ***linseed*** [1]; ***permanent grassland***, ***rotational grass*** [4]
- Docks in ***permanent grassland***, ***rotational grass*** [4]
- Forget-me-not in ***permanent grassland***, ***rotational grass*** [4]
- Shepherd's purse in ***permanent grassland***, ***rotational grass*** [4]

Approval information

- Accepted by BBPA for use on malting barley

SEE SECTION 3 FOR PRODUCTS ALSO REGISTERED

Efficacy guidance

- For best results apply in spring (from 1 Feb) in warm weather when soil moist and weeds growing actively. When used in grassland following cutting or grazing, docks should be allowed to regrow before treatment
- Weed kill is slow, especially under cool, dry conditions. Weeds may sometimes only be stunted but will have little or no competitive effect on crop
- May be used on all soil types unless certain sequences are used on linseed. See label
- Spray is rainfast after 1 h
- Cleavers controlled from emergence to flower bud stage. If present at application charlock (up to flower bud), shepherds purse (up to flower bud) and field forget-me-not (up to 6 leaves) will also be controlled
- Amidosulfuron is a member of the ALS-inhibitor group of herbicides and products should be used in a planned Resistance Management strategy. See Section 5 for more information

Restrictions

- Maximum number of treatments 1 per crop [1-3]
- Maximum total dose equivalent to one full dose treatment on permanent or rotational grass [4]
- Use after 1 Feb and do not apply to rotational grass after 30 Jun, or to permanent grassland after 15 Oct [4]
- Do not apply to crops undersown or due to be undersown with clover or alfalfa [1-3]
- Do not use on swards containing red clover but may be used on swards with white clover. Where clover has been newly established apply from 3-leaf stage of grass and 1-2 trifoliate leaves of clover [4]
- Do not spray crops under stress, suffering drought, waterlogged, grazed, lacking nutrients or if soil compacted
- Do not spray if frost expected
- Do not roll or harrow within 1 wk of spraying
- Hay or silage from treated grass crops must not be cut for at least 21 d after treatment [4]
- Specific restrictions apply to use in sequence or tank mixture with other sulfonylurea or ALS-inhibiting herbicides. See label for details. There are no recommendations for mixtures with metsulfuron-methyl products on linseed
- Certain mixtures with fungicides are expressly forbidden. See label for details

Crop-specific information

- Latest use: before first spikelets just visible (GS 51) for cereals; before flower buds visible for linseed; 15 Oct for grassland
- Broadcast cereal crops should be sprayed post-emergence after plants have a well established root system

Following crops guidance

- If a treated crop fails cereals may be sown after 15 d and thorough cultivation
- After normal harvest of a treated crop only cereals, winter oilseed rape, mustard, turnips, winter field beans or vetches may be sown in the same year as treatment and these must be preceded by ploughing or thorough cultivation
- Only cereals may be sown within 12 mth of application to grassland [4]
- Cereals or potatoes must be sown as the following crop after use of permitted mixtures or sequences with other sulfonylurea herbicides in cereals. Only cereals may be sown after the use of such sequences in linseed

Environmental safety

- Dangerous for the environment [4]
- Very toxic to aquatic organisms [4]
- Keep livestock out of treated areas for at least 7d following treatment and until poisonous weeds, such as ragwort, have died down and become unpalatable [4]
- Take care to wash out sprayers thoroughly. See label for details
- Avoid drift onto neighbouring broad-leaved plants or onto surface waters or ditches

Hazard classification and safety precautions

Hazard H04, H11 [4]
Risk phrases R36, R50, R53a [4]
Operator protection C [4]; U05a, U20b [4]; U20a [1-3]

Environmental protection E06a [4] (7 d); E15a [1-4]; E38 [4]
Storage and disposal D01, D02, D12a [4]; D10a [1-4]

8 amidosulfuron + iodosulfuron-methyl-sodium

A post-emergence sulfonylurea herbicide mixture for cereals
HRAC mode of action code: B + B

Products

1	Chekker	Bayer CropScience	13.4:1.3% w/w	WG	10955
2	Sekator	Bayer CropScience	5:1.25% w/w	WG	12634

Uses

- Annual dicotyledons in ***spring barley, spring rye, spring wheat, triticale, winter barley, winter rye, winter wheat***
- Chickweed in ***spring barley, spring rye, spring wheat, triticale, winter barley, winter rye, winter wheat***
- Cleavers in ***spring barley, spring rye, spring wheat, triticale, winter barley, winter rye, winter wheat***
- Mayweeds in ***spring barley, spring rye, spring wheat, triticale, winter barley, winter rye, winter wheat***
- Volunteer oilseed rape in ***spring barley, spring rye, spring wheat, triticale, winter barley, winter rye, winter wheat***

Approval information

- Iodosulfuron-methyl-sodium included in Annex I under EC Directive 91/414
- Accepted by BBPA for use on malting barley

Efficacy guidance

- Best results obtained from treatment in warm weather when soil is moist and the weeds are growing actively
- Weeds must be present at application to be controlled
- Dry conditions resulting in moisture stress may reduce effectiveness
- Weed control is slow especially under cool dry conditions
- Occasionally weeds may only be stunted but they will normally have little or no competitive effect on the crop
- Amidosulfuron and iodosulfuron are members of the ALS-inhibitor group of herbicides and products should be used in a planned Resistance Management strategy. See Section 5 for more information

Restrictions

- Maximum number of treatments 1 per crop
- Must only be applied between 1 Feb in yr of harvest and specified latest time of application
- Do not apply to crops undersown or to be undersown with grass, clover or alfalfa
- Do not roll or harrow within 1 wk of spraying
- Do not spray crops under stress from any cause or if the soil is compacted
- Do not spray if rain or frost expected
- Do not apply in mixture or in sequence with any other ALS inhibitor

Crop-specific information

- Latest use: before first spikelet of inflorescence just visible (GS 51)
- Treat drilled crops after the 2-leaf stage; treat broadcast crops after the plants have a well-established root system
- Applications to spring barley may cause transient crop yellowing

Following crops guidance

- Cereals, winter oilseed rape and winter field beans may be sown in the same yr as treatment provided they are preceded by ploughing or thorough cultivation. Any crop may be sown in the spring of the yr following treatment
- A minimum of 3 mth must elapse between treatment and sowing winter oilseed rape

Environmental safety

- Dangerous for the environment
- Toxic to aquatic organisms

SEE SECTION 3 FOR PRODUCTS ALSO REGISTERED

- LERAP Category B
- Take extreme care to avoid damage by drift onto broad-leaved plants outside the target area or onto ponds, waterways and ditches
- Observe carefully label instructions for sprayer cleaning

Hazard classification and safety precautions

Hazard H04, H11
Risk phrases R36, R51, R53a
Operator protection A, C, H; U05a, U08, U11, U14, U15, U20b
Environmental protection E15a, E16a, E16b, E38
Storage and disposal D01, D02, D10a, D12a

9 aminopyralid

A pyridine carboxylic acid herbicide available only in mixtures
HRAC mode of action code: O

10 aminopyralid + fluroxypyr

A foliar acting herbicide mixture for use in grassland
HRAC mode of action code: O + O

Products

Forefront	Dow	30:100 g/l	EO	12765

Uses

- Buttercups in ***permanent grassland, rotational grass***
- Chickweed in ***permanent grassland, rotational grass***
- Dandelions in ***permanent grassland, rotational grass***
- Docks in ***permanent grassland, rotational grass***
- Stinging nettle in ***permanent grassland, rotational grass***
- Thistles in ***permanent grassland, rotational grass***

Approval information

- Fluroxypyr included in Annex I under EC Directive 91/414

Efficacy guidance

- For best results and to avoid crop check, grass and weeds must be growing actively
- Allow 2-3 wk after cutting for hay or silage for sufficient regrowth to occur before spraying and leave 7 d afterwards to allow maximum translocation
- Where there is a high reservoir of weed seed or a historically high weed population a programmed approach may be needed involving a second treatment in the following yr
- Control may be reduced if rain falls within 1 h of spraying

Restrictions

- Maximum number of treatments 1 per yr
- Use of an antifoam is compulsory
- Do not use any treated plant material, or manure from animals fed on treated crops, for composting or mulching
- Do not apply to newly sown grass within 8 wk of sowing and preferably not until tillering
- Do not use on crops grown for seed

Crop-specific information

- Treatment will kill clover
- Treatment may occasionally cause transient yellowing of the sward which is quickly outgrown
- Late treatments may lead to a slight transient leaning of grass that does not affect yield

Following crops guidance

- Do not drill clover or other legumes within 4 mth of treatment, or potatoes in the spring following treatment in the previous autumn
- In the event of failure of newly seeded treated grassland, grass may be re-seeded immediately, or wheat may be sown provided 4 mth have elapsed since application

- Residues in plant tissues, including manure, may affect succeeding susceptible crops of peas, beans, other legumes, carrots, other Umbelliferae, potatoes, tomatoes, lettuce and other Compositae. These crops should not be sown within 3 mth of ploughing up treated grassland

Environmental safety

- Dangerous for the environment
- Toxic to aquatic organisms
- Keep livestock out of treated areas for at least 7 d following treatment and until poisonous weeds, such as ragwort, have died down and become unpalatable
- To protect groundwater do not apply between 1Sep and 28 Feb
- Avoid damage by drift onto susceptible crops, non-target plants or waterways

Hazard classification and safety precautions

Hazard H04, H11
Risk phrases R38, R41, R51, R53a, R67
Operator protection A, C; U05a, U08, U11, U14, U23a
Environmental protection E07a, E15b, E38
Consumer protection C01
Storage and disposal D01, D02, D05, D09a, D10b, D12a

11 aminopyralid + triclopyr

A foliar acting herbicide mixture for broad-leaved weed control in grassland
HRAC mode of action code: O + O

Products

Pharoah	Dow	30:240 g/l	EW	13631

Uses

- Annual dicotyledons in ***permanent grassland, rotational grass***
- Perennial dicotyledons in ***permanent grassland, rotational grass***

Approval information

- Triclopyr included in Annex I under EC Directive 91/414

Efficacy guidance

- For best results and to avoid crop check, grass and weeds must be growing actively
- Use adeqate water volume to ensure good weed coverage. Increase water volume if necessary where the weed population is high and where the grass is dense
- Allow 2-3 wk after cutting for hay or silage for sufficient regrowth to occur before spraying and leave 7 d afterwards to allow maximum translocation
- Where there is a high reservoir of weed seed or a historically high weed population a programmed approach may be needed involving a second treatment in the following yr
- Control may be reduced if rain falls within 1 h of spraying

Restrictions

- Maximum number of treatments 1 per yr
- Do not use any treated plant material for composting or mulching
- Do not use manure or slurry from animals fed on treated grass on susceptible crops (see following crops guidance)
- Do not roll or harrow 10 d before or within 7 d after application
- Do not cut grass for 7 d after application
- Do not apply to newly sown grass within 8 wk of sowing and preferably not until tillering
- Do not use on crops grown for seed
- Do not drill clover within 4 mth of application. Clover may be re-introduced 4 mth after application provided treated plant remains have completely decayed

Crop-specific information

- Latest use: 7 d before grazing or harvest for grassland
- Late applications may lead to transient leaning of grass which does not affect final yield

Following crops guidance

- Ensure that all plant remains of a treated crop have completely decayed before planting susceptible crops such as peas, beans and other legumes, sugar beet, carrots and other Umbelliferae, potatoes and tomatoes, lettuce and other Compositae

SEE SECTION 3 FOR PRODUCTS ALSO REGISTERED

- Do not plant potatoes, sugar beet, vegetables, beans or other leguminous crops in the calendar yr following application

Environmental safety

- Dangerous for the environment
- Toxic to aquatic organisms
- Keep livestock out of treated areas for at least 7 d after treatment or until foliage of any poisonous weeds such as ragwort has died and become unpalatable
- To protect groundwater do not apply to grass leys less than 1 yr old
- Take extreme care to avoid drift onto susceptible crops, non-target plants or waterways. All conifers, especially pine and larch, are very sensitive and may be damaged by vapour drift in hot conditions
- LERAP Category B

Hazard classification and safety precautions

Hazard H04, H11
Risk phrases R43, R51, R53a
Operator protection A, H; U05a, U08, U14, U23a
Environmental protection E06a (7 d); E15b, E16a, E38
Consumer protection C01
Storage and disposal D01, D02, D05, D09a, D10b, D12a

12 amitrole

A translocated, foliar-acting, non-selective triazole herbicide
HRAC mode of action code: F3

Products

Weedazol-TL	Nufarm UK	225 g/l	SL	11968

Uses

- Annual and perennial weeds in ***apricots*** *(off-label)*, ***cherries*** *(off-label)*, ***peaches*** *(off-label)*, ***plums*** *(off-label)*, ***quinces*** *(off-label)*
- Annual dicotyledons in ***fallows, headlands, stubbles***
- Annual grasses in ***fallows, headlands, stubbles***
- Barren brome in ***apple orchards, pear orchards***
- Couch in ***apple orchards, fallows, headlands, pear orchards, stubbles***
- Creeping thistle in ***fallows, stubbles***
- Docks in ***fallows, headlands, stubbles***
- General weed control in ***apple orchards, farm woodland*** *(off-label)*, ***pear orchards***
- Grass weeds in ***managed amenity turf*** *(off-label - on amitrole resistant turf)*
- Perennial dicotyledons in ***apple orchards, fallows, headlands, pear orchards***
- Perennial grasses in ***apple orchards, fallows, headlands, pear orchards***
- Volunteer potatoes in ***stubbles*** *(barley stubble)*

Specific Off-Label Approvals (SOLAs)

- ***apricots*** *20051981*
- ***cherries*** *20051981*
- ***farm woodland*** *20051980*
- ***managed amenity turf*** *(on amitrole resistant turf) 20051979*
- ***peaches*** *20051981*
- ***plums*** *20051981*
- ***quinces*** *20051981*

Approval information

- Amitrole included in Annex I under EC Directive 91/414

Efficacy guidance

- In non-crop land may be applied at any time from Apr to Oct. Best results achieved in spring or early summer when weeds growing actively. For coltsfoot, hogweed and horsetail summer and autumn applications are preferred
- Uptake is via foliage and heavy rain immediately after application will reduce efficacy. Amitrole is less affected by drought than some residual herbicides and remains effective for up to 2 mth

- Applications made in summer may not give complete control of couch if past the shooting stage or not actively growing
- Effective crop competition and efficient ploughing are essential for good couch control

Restrictions

- Maximum number of treatments 1 per yr
- Keep off suckers or foliage of desirable trees or shrubs
- Do not spray areas into which the roots of adjacent trees or shrubs extend
- Application to land intended for spring barley should be in the preceding autumn, not the spring
- Do not spray on sloping ground when rain imminent and run-off may occur
- Do not spray if foliage is wet or rain imminent
- Do not use in low temperatures or in drought
- Do not mix product with acids

Crop-specific information

- Latest use: before end Jun or after harvest for apple and pear orchards; end Oct for headlands; end Oct and at least 2 wk before cultivation and drilling for stubbles and fallows

Following crops guidance

- Amitrole breaks down fairly quickly in medium and heavy soils and 3 wk should be allowed between application and sowing or planting. On sandy soils the interval should be 6 wk

Environmental safety

- Harmful to aquatic organisms
- Keep livestock out of treated areas for at least two weeks following treatment and until poisonous weeds, such as ragwort, have died down and become unpalatable
- Harmful to fish or other aquatic life. Do not contaminate surface waters or ditches with chemical or used container

Hazard classification and safety precautions

Hazard H03
Risk phrases R22a, R48, R52, R53a, R63
Operator protection A, C; U05a, U08, U19a, U20b
Environmental protection E07a, E15a
Storage and disposal D01, D02, D09a, D10b

13 ammonium sulphamate

A non-selective, inorganic, general purpose herbicide and tree-killer

Products

Root-Out	Dax	98.5% w/w	CR	03510

Uses

- Annual dicotyledons in ***ornamental plant production*, *trees and shrubs***
- Annual grasses in ***ornamental plant production*, *trees and shrubs***
- Perennial dicotyledons in ***ornamental plant production*, *trees and shrubs***
- Perennial grasses in ***forest*, *ornamental plant production***
- Rhododendrons in ***forest***
- Woody weeds in ***forest***

Approval information

- Ammonium sulphamate will not be included in Annex I of EC Directive 91/414. To comply with this EC decision products listed in this profile were withdrawn from the market in 22 May 2007, and all approvals for their storage and use will expire on 22 May 2008
- Approval expiry 22 May 2008 [1]

Efficacy guidance

- Apply as spray to low scrub and herbaceous weeds from Apr to Sep in dry weather when rain unlikely and cultivate after 3-8 wk
- Apply as crystals in frills or notches in trunks of standing trees at any time of year
- Apply as concentrated solution or crystals to stump surfaces within 48 h of cutting. Rhododendrons must be cut level with ground and sprayed to cover cut surface, bark and immediate root area

SEE SECTION 3 FOR PRODUCTS ALSO REGISTERED

- Stainless steel or plastic sprayers are recommended. Solutions are corrosive to mild steel, galvanised iron, brass and copper

Restrictions

- Keep spray at least 30 cm from growing plants. Low doses may be used under mature trees with undamaged bark

Following crops guidance

- Allow 8-12 wk after treatment before replanting

Environmental safety

- Harmful to fish or other aquatic life. Do not contaminate surface waters or ditches with chemical or used container

Hazard classification and safety precautions

Operator protection U11, U14, U15, U19a, U20a
Environmental protection E13c
Storage and disposal D01, D09a

14 asulam

A translocated carbamate herbicide for control of docks and bracken
HRAC mode of action code: I

Products

1	Asulox	United Phosphorus	400 g/l	SL	13175
2	Brack-N	AgriGuard	400 g/l	SL	13138
3	Greencrop Frond	Greencrop	400 g/l	SL	11912

Uses

- Bracken in ***amenity vegetation*** [3]; ***forest***, ***permanent grassland*** [1-3]; ***forest*** *(off-label)* [1]; ***non-crop areas*** [2]
- Brome grasses in ***poppies for morphine production*** *(off-label)* [1]
- Docks in ***almonds*** *(off-label)*, ***chestnuts*** *(off-label)*, ***hazel nuts*** *(off-label)*, ***walnuts*** *(off-label)* [1, 3]; ***amenity grassland*** *(not fine turf)*, ***apple orchards***, ***blackcurrants***, ***cherries***, ***hops***, ***pear orchards***, ***permanent grassland***, ***plums*** [1-3]; ***amenity vegetation***, ***road verges***, ***waste ground*** [1, 2]; ***blackberries*** *(off-label)*, ***blueberries*** *(off-label)*, ***clover seed crops*** *(off-label)*, ***cranberries*** *(off-label)*, ***damsons*** *(off-label)*, ***gooseberries*** *(off-label)*, ***loganberries*** *(off-label)*, ***mint*** *(off-label)*, ***nectarines*** *(off-label)*, ***parsley*** *(off-label)*, ***poppies for morphine production*** *(off-label)*, ***quinces*** *(off-label)*, ***raspberries*** *(off-label)*, ***redcurrants*** *(off-label)*, ***rotational grass***, ***strawberries*** *(off-label)*, ***tarragon*** *(off-label)*, ***whitecurrants*** *(off-label)* [1]
- Meadow grasses in ***poppies for morphine production*** *(off-label)* [1]

Specific Off-Label Approvals (SOLAs)

- ***almonds*** *20070816* [1], *20060866* [3]
- ***blackberries*** *20070799* [1]
- ***blueberries*** *20070799* [1]
- ***chestnuts*** *20070816* [1], *20060866* [3]
- ***clover seed crops*** *20070825* [1]
- ***cranberries*** *20070799* [1]
- ***damsons*** *20070799* [1]
- ***forest*** *20070823* [1]
- ***gooseberries*** *20070799* [1]
- ***hazel nuts*** *20070816* [1], *20060866* [3]
- ***loganberries*** *20070799* [1]
- ***mint*** *20070822* [1]
- ***nectarines*** *20070799* [1]
- ***parsley*** *20070822* [1]
- ***poppies for morphine production*** *20070826* [1]
- ***quinces*** *20070799* [1]
- ***raspberries*** *20070799* [1]
- ***redcurrants*** *20070799* [1]
- ***strawberries*** *20070824* [1]
- ***tarragon*** *20070822* [1]

- ***walnuts*** *20070816* [1], *20060866* [3]
- ***whitecurrants*** *20070799* [1]

Approval information

- May be applied through CDA equipment
- Approved for aerial application on bracken in agricultural grassland, amenity grassland, forestry and rough upland intended for grazing [09969, 120023]. See notes in Section 5
- Approved for use near surface waters. See notes in Section 5
- Accepted by BBPA for use on hops

Efficacy guidance

- Spray bracken when fronds fully expanded but not senescent, usually Jul-Aug; docks in full leaf before flower stem emergence
- Bracken fronds must not be damaged by stock, frost or cutting before treatment
- Uptake and reliability of bracken control may be improved by use of specified additives - see label. Additives not recommended on forestry land
- To allow adequate translocation do not cut or admit stock for 14 d after spraying bracken or 7 d after spraying docks. Preferably leave undisturbed until late autumn
- Complete bracken control rarely achieved by one treatment. Survivors should be sprayed when they recover to full green frond, which may be in the ensuing year but more likely in the second year following initial application

Restrictions

- Maximum number of treatments 1 per crop or 1 per yr
- Do not apply in drought or hot, dry conditions
- Do not use in pasture before mowing for hay

Crop-specific information

- Latest use: Aug-Sep for all situations. See label for details
- In forestry areas some young trees may be checked if sprayed directly (see label)
- In fruit crops apply as a directed spray
- Do not treat blackcurrant cuttings, hop sets or weak hills
- Some grasses and herbs will be damaged by full dose. Most sensitive are cocksfoot, Yorkshire fog, timothy, bents, annual meadow-grass, daisies, docks, plantains, saxifrage
- Apply as spot treatment in parsley, mint and tarragon, not directly to crop

Following crops guidance

- Allow at least 6 wk between spraying and planting any crop

Environmental safety

- Dangerous for the environment
- Very toxic to aquatic organisms
- Keep livestock out of treated areas for at least two weeks following treatment and until poisonous weeds, such as ragwort, have died down and become unpalatable
- The use of asulam near surface waters has been considered by PSD. Whilst every care should be taken to avoid contamination, any that does occur during the normal course of spraying should offer no harm to operators, to users and consumers of the water, to domestic and farm animals and to wildlife. Before spraying such areas the appropriate regulatory authority should be notified

Hazard classification and safety precautions

Hazard H04, H11

Risk phrases R43, R50 [1-3]; R53a [3]

Operator protection A, C, D, H [1-3]; M [1]; M [2] (for ULV application); M [3]; U08, U19a, U20b [3]; U14 [1-3]; U19a, U20b [1, 2] (ULV use); U20c [1, 2]

Environmental protection E07a, E15a [1-3]; E38 [1, 2]

Storage and disposal D05 [3]; D09a [1-3]; D10a [2, 3]; D10c [1]; D12a [1, 2]

SEE SECTION 3 FOR PRODUCTS ALSO REGISTERED

15 azoxystrobin

A systemic translaminar and protectant strobilurin fungicide for a wide range of crops
FRAC mode of action code: 11

Products

1	Amicron	AgriGuard	250 g/l	SC	13336
2	Amistar	Syngenta	250 g/l	SC	10443
3	Clayton Belfry	Clayton	250 g/l	SC	12886
4	Heritage	Syngenta	50% w/w	WG	13536
5	Landgold Strobilurin 250	Landgold	250 g/l	SC	09595
6	Landgold Strobilurin 250	Teliton	250 g/l	SC	12128
7	Standon Azoxystrobin	Standon	250 g/l	SC	09515

Uses

- Alternaria in ***broccoli***, ***brussels sprouts***, ***cabbages***, ***calabrese***, ***carrots***, ***cauliflowers***, ***collards***, ***kale***, ***spring oilseed rape***, ***winter oilseed rape*** [1-3]; ***chicory root*** *(off-label)*, ***horseradish*** *(off-label)*, ***parsnips*** *(off-label)*, ***salad brassicas*** *(off-label - for baby leaf production)* [2]
- Anthracnose in ***amenity grassland***, ***managed amenity turf*** [4]
- Ascochyta in ***broad beans*** *(off-label)*, ***lupins*** *(off-label)* [2]; ***combining peas***, ***vining peas*** [1-3]
- Black dot in ***potatoes*** [1-3]
- Black scurf and stem canker in ***potatoes*** [1-3]
- Black spot in ***protected strawberries*** *(off-label)*, ***strawberries*** *(off-label)* [2]
- Botrytis in ***broad beans*** *(off-label)*, ***celery (outdoor)*** *(off-label)*, ***dwarf beans*** *(off-label)*, ***navy beans*** *(off-label)*, ***protected celery*** *(off-label)*, ***runner beans*** *(off-label)* [2]
- Brown patch in ***amenity grassland***, ***managed amenity turf*** [4]
- Brown rust in ***spring barley***, ***spring wheat***, ***winter barley***, ***winter wheat*** [1-3, 5-7]; ***spring rye***, ***triticale***, ***winter rye*** [1-3]
- Crown rust in ***amenity grassland***, ***managed amenity turf*** [4]; ***spring oats***, ***winter oats*** [1-3]
- Didymella in ***inert substrate aubergines*** *(off-label)*, ***inert substrate tomatoes*** *(off-label)* [2]
- Didymella stem rot in ***aubergines*** *(off-label)*, ***protected tomatoes*** *(off-label)* [2]
- Downy mildew in ***artichokes*** *(off-label)*, ***garlic*** *(off-label)*, ***inert substrate courgettes*** *(off-label)*, ***inert substrate cucumbers*** *(off-label)*, ***inert substrate gherkins*** *(off-label)*, ***protected courgettes*** *(off-label)*, ***protected cucumbers*** *(off-label)*, ***protected gherkins*** *(off-label)*, ***protected marrows*** *(off-label)*, ***protected melons*** *(off-label)*, ***protected pumpkins*** *(off-label)*, ***protected squashes*** *(off-label)*, ***salad brassicas*** *(off-label - for baby leaf production)*, ***salad onions*** *(off-label)*, ***shallots*** *(off-label)* [2]; ***bulb onions*** [1-3]
- Fairy rings in ***amenity grassland***, ***managed amenity turf*** [4]
- Foliar disease control in ***durum wheat*** *(off-label)* [2]; ***grass seed crops*** *(off-label)* [2, 7]
- Fusarium patch in ***amenity grassland***, ***managed amenity turf*** [4]
- Glume blotch in ***spring wheat***, ***winter wheat*** [1-3, 5-7]
- Grey mould in ***aubergines*** *(off-label)*, ***inert substrate aubergines*** *(off-label)*, ***inert substrate tomatoes*** *(off-label)*, ***protected tomatoes*** *(off-label)* [2]
- Late blight in ***aubergines*** *(off-label)*, ***inert substrate aubergines*** *(off-label)*, ***inert substrate tomatoes*** *(off-label)*, ***protected tomatoes*** *(off-label)* [2]
- Late ear diseases in ***spring wheat***, ***winter wheat*** [1-3, 5-7]
- Melting out in ***amenity grassland***, ***managed amenity turf*** [4]
- Net blotch in ***spring barley***, ***winter barley*** [1-3, 5-7]
- Phytophthora in ***protected chicory*** *(off-label - for forcing)* [2]
- Powdery mildew in ***artichokes*** *(off-label)*, ***aubergines*** *(off-label)*, ***blackberries*** *(off-label)*, ***chicory root*** *(off-label)*, ***chives*** *(off-label)*, ***courgettes*** *(off-label)*, ***herbs (see appendix 6)*** *(off-label)*, ***horseradish*** *(off-label)*, ***inert substrate aubergines*** *(off-label)*, ***inert substrate courgettes*** *(off-label)*, ***inert substrate cucumbers*** *(off-label)*, ***inert substrate gherkins*** *(off-label)*, ***inert substrate tomatoes*** *(off-label)*, ***parsley*** *(off-label)*, ***parsnips*** *(off-label)*, ***poppies for morphine production*** *(off-label)*, ***protected blackberries*** *(off-label)*, ***protected cayenne peppers*** *(off-label)*, ***protected courgettes*** *(off-label)*, ***protected cucumbers*** *(off-label)*, ***protected gherkins*** *(off-label)*, ***protected marrows*** *(off-label)*, ***protected melons*** *(off-label)*, ***protected peppers*** *(off-label)*, ***protected pumpkins*** *(off-label)*, ***protected raspberries*** *(off-label)*, ***protected squashes*** *(off-label)*, ***protected tomatoes*** *(off-label)*, ***raspberries*** *(off-label)* [2]; ***carrots***, ***spring oats***, ***triticale***, ***winter oats***, ***winter rye*** [1-3]; ***spring barley***, ***winter barley*** [1-3, 5-7]; ***spring wheat***, ***winter wheat*** [7]

- Purple blotch in ***leeks*** [1-3]
- Rhizoctonia in ***celery (outdoor)*** *(off-label)*, ***protected celery*** *(off-label)*, ***protected chives*** *(off-label)*, ***protected herbs (see appendix 6)*** *(off-label)*, ***protected lettuce*** *(off-label)*, ***protected parsley*** *(off-label)*, ***swedes*** *(off-label)*, ***turnips*** *(off-label)* [2]
- Rhynchosporium in ***spring barley***, ***winter barley*** [1-3, 5-7]; ***spring rye***, ***triticale***, ***winter rye*** [1-3]
- Ring spot in ***broccoli***, ***brussels sprouts***, ***cabbages***, ***calabrese***, ***cauliflowers***, ***collards***, ***kale*** [1-3]; ***chives*** *(off-label)*, ***herbs (see appendix 6)*** *(off-label)*, ***parsley*** *(off-label)* [2]
- Root malformation disorder in ***red beet*** *(off-label)* [2]
- Rust in ***asparagus***, ***leeks***, ***spring field beans***, ***winter field beans*** [1-3]; ***chicory root*** *(off-label)*, ***chives*** *(off-label)*, ***herbs (see appendix 6)*** *(off-label)*, ***lupins*** *(off-label)*, ***parsley*** *(off-label)* [2]
- Sclerotinia in ***celeriac*** *(off-label)*, ***celery (outdoor)*** *(off-label)*, ***lettuce*** *(off-label)*, ***protected celery*** *(off-label)* [2]
- Sclerotinia stem rot in ***spring oilseed rape***, ***winter oilseed rape*** [1-3]
- Septoria leaf blotch in ***spring wheat***, ***winter wheat*** [1-3, 5-7]
- Stemphylium in ***asparagus*** [1-3]
- Take-all in ***spring barley*** *(reduction)*, ***spring wheat*** *(reduction)*, ***winter barley*** *(reduction)*, ***winter wheat*** *(reduction)* [1-3]
- Take-all patch in ***amenity grassland***, ***managed amenity turf*** [4]
- White blister in ***broccoli***, ***brussels sprouts***, ***cabbages***, ***calabrese***, ***cauliflowers***, ***collards***, ***kale*** [1-3]
- White rust in ***chrysanthemums*** *(off-label - in pots)*, ***protected chrysanthemums*** *(off-label)* [2]
- Yellow rust in ***spring wheat***, ***winter wheat*** [1-3, 5-7]

Specific Off-Label Approvals (SOLAs)

- ***artichokes*** *20051814 expires Jul 2008* [2]
- ***aubergines*** *20021533 expires Jul 2008* [2]
- ***blackberries*** *20030365 expires Jul 2008* [2]
- ***broad beans*** *20032311 expires Jul 2008* [2]
- ***celeriac*** *20031862 expires Jul 2008* [2]
- ***celery (outdoor)*** *20011041 expires Jul 2008* [2]
- ***chicory root*** *20051813 expires Jul 2008* [2]
- ***chives*** *20021293 expires Jul 2008* [2]
- ***chrysanthemums*** *(in pots) 20011684 expires Jul 2008* [2]
- ***courgettes*** *20051985 expires Jul 2008* [2]
- ***durum wheat*** *20061722 expires Jul 2008* [2]
- ***dwarf beans*** *20032311 expires Jul 2008* [2]
- ***garlic*** *20061724 expires Jul 2008* [2]
- ***grass seed crops*** *20061722 expires Jul 2008* [2], *20061095 expires Jul 2008* [7]
- ***herbs (see appendix 6)*** *20021293 expires Jul 2008* [2]
- ***horseradish*** *20061721 expires Jul 2008* [2]
- ***inert substrate aubergines*** *20011685 expires Jul 2008* [2]
- ***inert substrate courgettes*** *20011685 expires Jul 2008* [2]
- ***inert substrate cucumbers*** *20011685 expires Jul 2008* [2]
- ***inert substrate gherkins*** *20011685 expires Jul 2008* [2]
- ***inert substrate tomatoes*** *20011685 expires Jul 2008* [2]
- ***lettuce*** *20011465 expires Jul 2008* [2]
- ***lupins*** *20061723 expires Jul 2008* [2]
- ***navy beans*** *20032311 expires Jul 2008* [2]
- ***parsley*** *20021293 expires Jul 2008* [2]
- ***parsnips*** *20061721 expires Jul 2008* [2]
- ***poppies for morphine production*** *20031137 expires Jul 2008* [2]
- ***protected blackberries*** *20051194 expires Jul 2008* [2]
- ***protected cayenne peppers*** *20021295 expires Jul 2008* [2]
- ***protected celery*** *20011041 expires Jul 2008* [2]
- ***protected chicory*** *(for forcing) 20051813 expires Jul 2008* [2]
- ***protected chives*** *20030659 expires Jul 2008* [2]
- ***protected chrysanthemums*** *20011684 expires Jul 2008* [2]
- ***protected courgettes*** *20021533 expires Jul 2008* [2]
- ***protected cucumbers*** *20021533 expires Jul 2008* [2]
- ***protected gherkins*** *20021533 expires Jul 2008* [2]

- ***protected herbs (see appendix 6)*** *20030659 expires Jul 2008* [2]
- ***protected lettuce*** *20030659 expires Jul 2008* [2]
- ***protected marrows*** *20070371 expires Jul 2008* [2]
- ***protected melons*** *20070371 expires Jul 2008* [2]
- ***protected parsley*** *20030659 expires Jul 2008* [2]
- ***protected peppers*** *20021295 expires Jul 2008* [2]
- ***protected pumpkins*** *20070371 expires Jul 2008* [2]
- ***protected raspberries*** *20051194 expires Jul 2008* [2]
- ***protected squashes*** *20070371 expires Jul 2008* [2]
- ***protected strawberries*** *20021294 expires Jul 2008* [2]
- ***protected tomatoes*** *20021533 expires Jul 2008* [2]
- ***raspberries*** *20030365 expires Jul 2008* [2]
- ***red beet*** *20040614 expires Jul 2008* [2]
- ***runner beans*** *20032311 expires Jul 2008* [2]
- ***salad brassicas*** *(for baby leaf production) 20011465 expires Jul 2008* [2]
- ***salad onions*** *20021687 expires Jul 2008* [2]
- ***shallots*** *20061724 expires Jul 2008* [2]
- ***strawberries*** *20021294 expires Jul 2008* [2]
- ***swedes*** *20040614 expires Jul 2008* [2]
- ***turnips*** *20040614 expires Jul 2008* [2]

Approval information

- Azoxystrobin included in Annex I under EC Directive 91/414
- Accepted by BBPA for use on malting barley
- Approval expiry 01 Jul 2008 [5]

Efficacy guidance

- Best results obtained from use as a protectant or during early stages of disease establishment or when a predictive assessment indicates a risk of disease development
- Azoxystrobin inhibits fungal respiration and should always be used in mixture with fungicides with other modes of action
- Treatment under poor growing conditions may give less reliable results
- For good control of *Fusarium* patch in amenity turf and grass repeat treatment at minimum intervals of 2 wk
- Azoxystrobin is a member of the QoI cross resistance group. Product should be used preventatively and not relied on for its curative potential
- Use product in cereals as part of an Integrated Crop Management strategy incorporating other methods of control, including where appropriate other fungicides with a different mode of action. Do not apply more than two foliar applications of QoI containing products to any cereal crop
- There is a significant risk of widespread resistance occurring in *Septoria tritici* populations in UK. Failure to follow resistance management action may result in reduced levels of disease control
- On cereal crops product must always be used in mixture with another product, recommended for control of the same target disease, that contains a fungicide from a different cross resistance group and is applied at a dose that will give robust control
- Strains of barley powdery mildew resistant to QoIs are common in the UK

Restrictions

- Maximum number of treatments 1 per crop for potatoes; 2 per crop for brassicas, peas, cereals, oilseed rape; 4 per crop or yr for onions, carrots, leeks, amenity turf [1-3]
- Maximum total dose ranges from 2-4 times the single full dose depending on crop and product. See labels for details
- On turf the maximum number of treatments is 4 per yr but they must not exceed one third of the total number of fungicide treatments applied
- Do not use where there is risk of spray drift onto neighbouring apple crops
- The same spray equipment should not be used to treat apples

Crop-specific information

- Latest use: at planting for potatoes; grain watery ripe (GS 71) for cereals; before senescence for asparagus
- HI: 10 d for carrots;14 d for broccoli, Brussels sprouts, bulb onions, cabbages, calabrese, cauliflowers, collards, kale, vining peas; 21 d for leeks, spring oilseed rape, winter oilseed rape; 36 d for combining peas, 35 d for field beans [1-3]

- In cereals control of established infections can be improved by appropriate tank mixtures or application as part of a programme. Always use in mixture with another product from a different cross-resistance group
- In turf use product at full dose rate in a disease control programme, alternating with fungicides of different modes of action
- In potatoes when used as incorporated treatment apply overall to the entire area to be planted, incorporate to 15 cm and plant on the same day. In-furrow spray should be directed at the furrow and not the seed tubers
- Applications to brassica crops must only be made to a developed leaf canopy and not before growth stages specified on the label
- Heavy disease pressure in brassicae and oilseed rape may require a second treatment
- All crops should be treated when not under stress. Check leaf wax on peas if necessary
- Consult processor before treating any crops for processing
- Treat asparagus after the harvest season. Where a new bed is established do not treat within 3 wk of transplanting out the crowns
- Do not apply to turf when ground is frozen or during drought

Environmental safety

- Dangerous for the environment
- Very toxic to aquatic organisms
- Avoid spray drift onto surrounding areas or crops, especially apples, plums or privet
- LERAP Category B (potatoes only) [1-3]

Hazard classification and safety precautions

Hazard H11

Risk phrases R50, R53a

Operator protection A [1, 2]; A [3] (during treatment of potatoes); U05a [1-6]; U09a, U19a [1-3, 5-7]; U20b [1-7]

Environmental protection E15a [1-3, 5-7]; E15b [4]; E16a [1-3] (potatoes only); E16b [2, 3] (potatoes only); E38 [1-4]

Storage and disposal D01, D02 [2-6]; D03 [4]; D05, D09a [1-7]; D07 [1]; D10b [3-7]; D10c [1, 2]; D12a [1-4]

Medical advice M05a [4]

16 azoxystrobin + chlorothalonil

A preventative and systemic fungicide mixture for cereals

FRAC mode of action code: 11 + M5

Products

1	Amistar Opti	Syngenta	100:500 g/l	SC	12515
2	Olympus	Syngenta	80:400 g/l	SC	12517

Uses

- Brown rust in ***spring barley***, ***spring wheat***, ***winter barley***, ***winter wheat***
- Foliar disease control in ***durum wheat*** *(off-label)*, ***grass seed crops*** *(off-label)*, ***spring rye*** *(off-label)*, ***triticale*** *(off-label)*, ***winter rye*** *(off-label)*
- Glume blotch in ***spring wheat***, ***winter wheat***
- Net blotch in ***spring barley***, ***winter barley***
- Rhynchosporium in ***spring barley***, ***winter barley***
- Septoria leaf blotch in ***spring wheat***, ***winter wheat***
- Take-all in ***spring barley*** *(reduction)*, ***spring wheat*** *(reduction)*, ***winter barley*** *(reduction)*, ***winter wheat*** *(reduction)*
- Yellow rust in ***spring wheat***, ***winter wheat***

Specific Off-Label Approvals (SOLAs)

- ***durum wheat*** *20052897* [1], *20061114* [2]
- ***grass seed crops*** *20052897* [1], *20061114* [2]
- ***spring rye*** *20052897* [1], *20061114* [2]
- ***triticale*** *20052897* [1], *20061114* [2]
- ***winter rye*** *20052897* [1], *20061114* [2]

SEE SECTION 3 FOR PRODUCTS ALSO REGISTERED

Approval information

- Azoxystrobin and chlorothalonil included in Annex I under EC Directive 91/414
- Accepted by BBPA for use on malting barley

Efficacy guidance

- Best results obtained from applications made as a protectant treatment or in earliest stages of disease development. Further applications may be needed if disease attack is prolonged
- Reduction of barley spotting occurs when used as part of a programme with other fungicides
- Control of *Septoria* and rust diseases may be improved by mixture with a triazole fungicide
- Azoxystrobin is a member of the QoI cross resistance group. Product should be used preventatively and not relied on for its curative potential
- Use product in cereals as part of an Integrated Crop Management strategy incorporating other methods of control, including where appropriate other fungicides with a different mode of action. Do not apply more than two foliar applications of QoI containing products to any cereal crop
- There is a significant risk of widespread resistance occurring in *Septoria tritici* populations in UK. Failure to follow resistance management action may result in reduced levels of disease control

Restrictions

- Maximum number of treatments 2 per crop
- Maximum total dose on barley equivalent to one full dose treatment
- Do not use where there is risk of spray drift onto neighbouring apple crops
- The same spray equipment should not be used to treat apples

Crop-specific information

- Latest use: before beginning of heading (GS 51) for barley; before caryopsis watery ripe (GS 71) for wheat

Environmental safety

- Dangerous for the environment
- Very toxic to aquatic organisms
- LERAP Category B

Hazard classification and safety precautions

Hazard H02 [1]; H03 [2]; H11 [1, 2]
Risk phrases R20 [2]; R22a, R23 [1]; R37, R40, R41, R43, R50, R53a [1, 2]
Operator protection A, C, H; U02a, U05a, U09a, U11, U14, U15, U19a, U20b
Environmental protection E15a, E16a, E34, E38
Storage and disposal D01, D02, D05, D09a, D10c, D12a
Medical advice M04a

17 azoxystrobin + cyproconazole

A contact and systemic broad spectrum fungicide mixture for cereals and oilseed rape
FRAC mode of action code: 11 + 3

Products

Priori Xtra	Syngenta	200:80 g/l	SC	11518

Uses

- Alternaria in ***spring oilseed rape***, ***winter oilseed rape***
- Brown rust in ***spring barley***, ***spring rye***, ***spring wheat***, ***winter barley***, ***winter rye***, ***winter wheat***
- Crown rust in ***spring oats***, ***winter oats***
- Eyespot in ***spring barley*** *(reduction)*, ***spring wheat*** *(reduction)*, ***winter barley*** *(reduction)*, ***winter wheat*** *(reduction)*
- Foliar disease control in ***durum wheat*** *(off-label)*, ***grass seed crops*** *(off-label)*, ***triticale*** *(off-label)*
- Glume blotch in ***spring wheat***, ***winter wheat***
- Net blotch in ***spring barley***, ***winter barley***
- Powdery mildew in ***spring barley***, ***spring oats***, ***spring rye***, ***spring wheat***, ***winter barley***, ***winter oats***, ***winter rye***, ***winter wheat***
- Rhynchosporium in ***spring barley*** *(moderate control)*, ***spring rye*** *(moderate control)*, ***winter barley*** *(moderate control)*, ***winter rye*** *(moderate control)*
- Sclerotinia stem rot in ***spring oilseed rape***, ***winter oilseed rape***
- Septoria leaf blotch in ***spring wheat***, ***winter wheat***

- Take-all in ***spring barley*** *(reduction)*, ***spring wheat*** *(reduction)*, ***winter barley*** *(reduction)*, ***winter wheat*** *(reduction)*
- Yellow rust in ***spring wheat***, ***winter wheat***

Specific Off-Label Approvals (SOLAs)

- ***durum wheat*** *20052891*
- ***grass seed crops*** *20052891*
- ***triticale*** *20052891*

Approval information

- Azoxystrobin included in Annex I under EC Directive 91/414
- Accepted by BBPA for use on malting barley

Efficacy guidance

- Best results obtained from treatment during the early stages of disease development
- A second application may be needed if disease attack is prolonged
- Azoxystrobin is a member of the QoI cross resistance group. Product should be used preventatively and not relied on for its curative potential
- Use product as part of an Integrated Crop Management strategy incorporating other methods of control, including where appropriate other fungicides with a different mode of action. Do not apply more than two foliar applications of QoI containing products to any cereal crop
- There is a significant risk of widespread resistance occurring in *Septoria tritici* populations in UK. Failure to follow resistance management action may result in reduced levels of disease control
- Strains of wheat and barley powdery mildew resistant to QoIs are common in the UK. Control of wheat powdery mildew can only be relied upon from the triazole component
- Where specific control of wheat mildew is required this should be achieved through a programme of measures including products recommended for the control of mildew that contain a fungicide from a different cross-resistance group and applied at a dose that will give robust control
- Cyproconazole is a DMI fungicide. Resistance to some DMI fungicides has been identified in Septoria leaf blotch which may seriously affect performance of some products. For further advice contact a specialist advisor and visit the Fungicide Resistance Action Group (FRAG)-UK website

Restrictions

- Maximum total dose equivalent to two full dose treatments
- Do not use where there is risk of spray drift onto neighbouring apple crops
- The same spray equipment should not be used to treat apples

Crop-specific information

- Latest use: up to and including anthesis complete (GS 69) for rye and wheat; up to and including emergence of ear complete (GS 59) for barley and oats; before 10% of pods have reached final size (GS 71) for oilseed rape

Environmental safety

- Dangerous for the environment
- Very toxic to aquatic organisms

Hazard classification and safety precautions

Hazard H03, H11
Risk phrases R22a, R50, R53a, R63
Operator protection A; U05a, U09a, U19a, U20b
Environmental protection E15b, E34, E38
Storage and disposal D01, D02, D05, D09a, D10c, D12a
Medical advice M03

18 azoxystrobin + difenoconazole

A broad spectrum fungicide mixture for field crops
FRAC mode of action code: 11 + 3

Products

Amistar Top	Syngenta	200:125 g/l	EC	12761

Uses

- Alternaria blight in ***carrots***, ***horseradish*** *(off-label)*, ***parsley root*** *(off-label)*, ***parsnips*** *(off-label)*, ***salsify*** *(off-label)*

SEE SECTION 3 FOR PRODUCTS ALSO REGISTERED

- Powdery mildew in ***broccoli***, ***brussels sprouts***, ***cabbages***, ***calabrese***, ***carrots***, ***cauliflowers***, ***collards***, ***horseradish*** *(off-label)*, ***kale***, ***parsley root*** *(off-label)*, ***parsnips*** *(off-label)*, ***salsify*** *(off-label)*
- Purple blotch in ***leeks***
- Rust in ***asparagus*** *(off-label)*, ***leeks***
- Stemphylium in ***asparagus*** *(off-label)*
- White blister in ***broccoli***, ***brussels sprouts***, ***cabbages***, ***calabrese***, ***cauliflowers***, ***collards***, ***kale***
- White tip in ***leeks*** *(qualified minor use)*

Specific Off-Label Approvals (SOLAs)

- ***asparagus*** *20070831*
- ***horseradish*** *20061476*
- ***parsley root*** *20061476*
- ***parsnips*** *20061476*
- ***salsify*** *20061476*

Approval information

- Azoxystrobin included in Annex I under EC Directive 91/414

Efficacy guidance

- Best results obtained from applications made in the earliest stages of disease development or as a protectant treatment following a disease risk assessment
- Ensure the crop is free from any stress caused by environmental or agronomic effects
- Azoxystrobin is a member of the QoI cross resistance group. Product should be used preventatively and not relied on for its curative potential
- Use as part of an Integrated Crop Management strategy incorporating other methods of control, including where appropriate other fungicides with a different mode of action. Do not apply more than two foliar applications of QoI containing products

Restrictions

- Maximum number of treatments 2 per crop
- Do not apply where there is a risk of spray drift onto neighbouring apple crops
- Consult processors before treating a crop destined for processing

Crop-specific information

- HI 14 d for carrots; 21 d for brassicas, leeks
- Minimum spray interval of 14 d must be observed on brassicas

Environmental safety

- Dangerous for the environment
- Very toxic to aquatic organisms

Hazard classification and safety precautions

Hazard H04, H11
Risk phrases R20, R43, R50, R53a
Operator protection A, H; U05a, U09a, U19a, U20b
Environmental protection E15b, E38
Storage and disposal D01, D02, D05, D09a, D10c, D12a

19 azoxystrobin + fenpropimorph

A protectant and eradicant fungicide mixture for cereals
FRAC mode of action code: 11 + 5

Products

1	Amistar Pro	Syngenta	100:280 g/l	SE	10513
2	Aspect	Syngenta	100:280 g/l	SE	10516

Uses

- Brown rust in ***durum wheat*** *(off-label)*, ***grass seed crops*** *(off-label)* [2]; ***spring barley***, ***spring rye***, ***spring wheat***, ***triticale***, ***winter barley***, ***winter rye***, ***winter wheat*** [1, 2]
- Crown rust in ***spring oats***, ***winter oats*** [1, 2]
- Foliar disease control in ***durum wheat*** *(off-label)*, ***grass seed crops*** *(off-label)* [1]
- Glume blotch in ***spring wheat***, ***winter wheat*** [1, 2]
- Late ear diseases in ***spring wheat*** *(moderate control)*, ***winter wheat*** *(moderate control)* [1, 2]

- Net blotch in ***spring barley, winter barley*** [1, 2]
- Powdery mildew in ***spring barley, spring oats, spring rye, triticale, winter barley, winter oats, winter rye*** [1, 2]
- Rhynchosporium in ***spring barley, spring rye, triticale*** *(reduction)*, ***winter barley, winter rye*** [1, 2]
- Septoria leaf blotch in ***durum wheat*** *(off-label)*, ***grass seed crops*** *(off-label)* [2]; ***spring wheat, winter wheat*** [1, 2]
- Take-all in ***spring barley*** *(reduction)*, ***spring rye*** *(reduction)*, ***spring wheat*** *(reduction)*, ***triticale*** *(reduction)*, ***winter barley*** *(reduction)*, ***winter rye*** *(reduction)*, ***winter wheat*** *(reduction)* [1, 2]
- Yellow rust in ***durum wheat*** *(off-label)*, ***grass seed crops*** *(off-label)* [2]; ***spring wheat, winter wheat*** [1, 2]

Specific Off-Label Approvals (SOLAs)

- ***durum wheat*** *20060962* [1, 2]
- ***grass seed crops*** *20060962* [1, 2]

Approval information

- Azoxystrobin included in Annex I under EC Directive 91/414
- Accepted by BBPA for use on malting barley

Efficacy guidance

- Best results obtained from application before infection following a disease risk assessment, or when disease first seen in crop
- Results may be less reliable when used on crops under stress
- Treatments for protection against ear disease should be made at ear emergence
- Azoxystrobin is a member of the QoI cross resistance group. Product should be used preventatively and not relied on for its curative potential
- Use product as part of an Integrated Crop Management strategy incorporating other methods of control, including where appropriate other fungicides with a different mode of action. Do not apply more than two foliar applications of QoI containing products to any cereal crop
- There is a significant risk of widespread resistance occurring in *Septoria tritici* populations in UK. Failure to follow resistance management action may result in reduced levels of disease control
- Strains of barley powdery mildew resistant to QoIs are common in the UK
- In wheat product must always be used in mixture with another product, recommended for control of the same target disease, that contains a fungicide from a different cross resistance group and is applied at a dose that will give robust control

Restrictions

- Maximum total dose equivalent to 2 full dose treatments
- Do not use where there is risk of spray drift onto neighbouring apple crops
- The same spray equipment should not be used to treat apples

Crop-specific information

- Latest use: before early milk stage (GS 73) for wheat, barley; up to and including grain watery ripe stage (GS 71) for oats, rye, triticale
- HI 5 wk

Environmental safety

- Dangerous for the environment
- Very toxic to aquatic organisms

Hazard classification and safety precautions

Hazard H03, H11
Risk phrases R20, R38, R43, R50, R53a, R63
Operator protection A, H; U02a, U05a, U09a, U14, U15, U19a, U20b
Environmental protection E15a, E38
Storage and disposal D01, D02, D05, D09a, D10c, D12a
Medical advice M03

SEE SECTION 3 FOR PRODUCTS ALSO REGISTERED

20 Bacillus thuringiensis

A bacterial insecticide for control of caterpillars
IRAC mode of action code: 11

Products

Dipel DF	Fargro	32000 IU/mg	WG	11184

Uses

- Caterpillars in ***amenity vegetation***, ***apples*** *(off-label)*, ***blackberries*** *(off-label)*, ***borage*** *(off-label)*, ***broccoli***, ***brussels sprouts***, ***cabbages***, ***calabrese*** *(off-label)*, ***cauliflowers***, ***celery (outdoor)*** *(off-label)*, ***cherries*** *(off-label)*, ***chinese cabbage*** *(off-label)*, ***chives*** *(off-label)*, ***collards*** *(off-label)*, ***crab apples*** *(off-label)*, ***frise*** *(off-label)*, ***herbs (see appendix 6)*** *(off-label)*, ***kale*** *(off-label)*, ***leaf spinach*** *(off-label)*, ***lettuce*** *(off-label)*, ***ornamental plant production***, ***parsley*** *(off-label)*, ***pears*** *(off-label)*, ***protected aubergines*** *(off-label)*, ***protected broccoli*** *(off-label)*, ***protected brussels sprouts*** *(off-label)*, ***protected cabbages*** *(off-label)*, ***protected calabrese*** *(off-label)*, ***protected cauliflowers*** *(off-label)*, ***protected cayenne peppers*** *(off-label)*, ***protected celery*** *(off-label)*, ***protected chinese cabbage*** *(off-label)*, ***protected chives*** *(off-label)*, ***protected cucumbers***, ***protected frise*** *(off-label)*, ***protected herbs (see appendix 6)*** *(off-label)*, ***protected kale*** *(off-label)*, ***protected lettuce*** *(off-label)*, ***protected ornamentals***, ***protected parsley*** *(off-label)*, ***protected peppers***, ***protected salad brassicas*** *(off-label - for baby leaf production)*, ***protected scarole*** *(off-label)*, ***protected tomatoes***, ***protected watercress*** *(off-label)*, ***radicchio*** *(off-label)*, ***raspberries***, ***rubus hybrids*** *(off-label)*, ***salad brassicas*** *(off-label - for baby leaf production)*, ***scarole*** *(off-label)*, ***strawberries***, ***watercress*** *(off-label)*
- Cutworms in ***bulb onions*** *(off-label)*, ***carrots*** *(off-label)*, ***garlic*** *(off-label)*, ***horseradish*** *(off-label)*, ***leeks*** *(off-label)*, ***parsley root*** *(off-label)*, ***parsnips*** *(off-label)*, ***red beet*** *(off-label)*, ***salsify*** *(off-label)*, ***shallots*** *(off-label)*
- Silver Y moth in ***broad beans*** *(off-label)*, ***dwarf beans*** *(off-label)*, ***french beans*** *(off-label)*, ***runner beans*** *(off-label)*, ***sweetcorn*** *(off-label)*
- Winter moth in ***bilberries*** *(off-label)*, ***blackcurrants*** *(off-label)*, ***blueberries*** *(off-label)*, ***cranberries*** *(off-label)*, ***gooseberries*** *(off-label)*, ***redcurrants*** *(off-label)*, ***vaccinium spp.*** *(off-label)*, ***whitecurrants*** *(off-label)*

Specific Off-Label Approvals (SOLAs)

- ***apples*** *20042063*
- ***bilberries*** *20040358*
- ***blackberries*** *20061554*
- ***blackcurrants*** *20040358*
- ***blueberries*** *20040358*
- ***borage*** *20042063*
- ***broad beans*** *20070431*
- ***bulb onions*** *20070434*
- ***calabrese*** *20040739*
- ***carrots*** *20070433*
- ***celery (outdoor)*** *20040739*
- ***cherries*** *20042063*
- ***chinese cabbage*** *20040739*
- ***chives*** *20042063*
- ***collards*** *20040739*
- ***crab apples*** *20042063*
- ***cranberries*** *20040358*
- ***dwarf beans*** *20070431*
- ***french beans*** *20070431*
- ***frise*** *20042063*
- ***garlic*** *20070434*
- ***gooseberries*** *20040358*
- ***herbs (see appendix 6)*** *20042063*
- ***horseradish*** *20070433*
- ***kale*** *20040739*
- ***leaf spinach*** *20042063*
- ***leeks*** *20070436*

- ***lettuce*** *20042063*
- ***parsley*** *20042063*
- ***parsley root*** *20070433*
- ***parsnips*** *20070433*
- ***pears*** *20042063*
- ***protected aubergines*** *20061555*
- ***protected broccoli*** *20040739*
- ***protected brussels sprouts*** *20040739*
- ***protected cabbages*** *20040739*
- ***protected calabrese*** *20040739*
- ***protected cauliflowers*** *20040739*
- ***protected cayenne peppers*** *20061556*
- ***protected celery*** *20040739*
- ***protected chinese cabbage*** *20040739*
- ***protected chives*** *20042063*
- ***protected frise*** *20042063*
- ***protected herbs (see appendix 6)*** *20042063*
- ***protected kale*** *20040739*
- ***protected lettuce*** *20042063*
- ***protected parsley*** *20042063*
- ***protected salad brassicas*** *(for baby leaf production) 20042063*
- ***protected scarole*** *20042063*
- ***protected watercress*** *20031071*
- ***radicchio*** *20042063*
- ***red beet*** *20070435*
- ***redcurrants*** *20040358*
- ***rubus hybrids*** *20061554*
- ***runner beans*** *20070431*
- ***salad brassicas*** *(for baby leaf production) 20042063*
- ***salsify*** *20070433*
- ***scarole*** *20042063*
- ***shallots*** *20070434*
- ***sweetcorn*** *20070432*
- ***vaccinium spp.*** *20040358*
- ***watercress*** *20031071*
- ***whitecurrants*** *20040358*

Efficacy guidance

- Pest control achieved by ingestion by caterpillars of the treated plant vegetation. Caterpillars cease feeding and die in 1-3 d
- Apply as soon as larvae appear on crop and repeat every 7-10 d until the end of the hatching period
- Good coverage is essential, especially of undersides of leaves. Spray onto dry foliage and do not apply if rain expected within 6 h

Restrictions

- No restriction on number of treatments on edible crops
- Apply spray mixture as soon as possible after preparation

Crop-specific information

- HI zero

Environmental safety

- Store out of direct sunlight

Hazard classification and safety precautions

Operator protection A, C; U05a, U15, U20c
Environmental protection E15a
Storage and disposal D01, D02, D05, D09a, D11a

SEE SECTION 3 FOR PRODUCTS ALSO REGISTERED

21 benalaxyl

A phenylamide (acylalanine) fungicide available only in mixtures
FRAC mode of action code: 4

22 benalaxyl + mancozeb

A systemic and protectant fungicide mixture
FRAC mode of action code: 4 + M3

Products

1	Intro Plus	Interfarm	8:65% w/w	WP	11630
2	Tairel	Sipcam	8:65% w/w	WB	07767

Uses

- Blight in ***potatoes***, ***protected potatoes***

Approval information

- Benalaxyl and mancozeb included in Annex I under EC Directive 91/414
- Approved for aerial application on potatoes [2]. See notes in Section 5

Efficacy guidance

- Apply to potatoes at blight warning prior to crop becoming infected and repeat at 10-21 d intervals depending on risk of infection
- Spray irrigated potatoes after irrigation and at 14 d intervals, crops in polythene tunnels at 10 d intervals
- When active potato growth ceases use a non-systemic fungicide to end of season, starting not more than 10 d after last application
- For reduction of downy mildew in oilseed rape apply at seedling to 7-leaf stage, before end Nov as soon as infection seen

Restrictions

- Maximum number of treatments 5 per crop (including other phenylamide-based fungicides) for potatoes
- Do not treat potatoes showing active blight infection

Crop-specific information

- HI 7 d

Environmental safety

- Dangerous for the environment
- Very toxic to aquatic organisms

Hazard classification and safety precautions

Hazard H04, H11
Risk phrases R37, R43, R50, R53a
Operator protection A; U05a, U14, U15, U19a
Environmental protection E15a
Storage and disposal D01, D02, D12b
Medical advice M04a

23 benfuracarb

A soil-applied carbamate insecticide and nematicide for beet crops
IRAC mode of action code: 1A

Products

Oncol 10G	Nufarm UK	10% w/w	GR	12139

Uses

- Docking disorder vectors in ***fodder beet***, ***mangels***, ***sugar beet***
- Leaf miner in ***fodder beet***, ***mangels***, ***sugar beet***
- Millipedes in ***fodder beet***, ***mangels***, ***sugar beet***
- Pygmy beetle in ***fodder beet***, ***mangels***, ***sugar beet***

- Springtails in ***fodder beet***, ***mangels***, ***sugar beet***
- Symphylids in ***fodder beet***, ***mangels***, ***sugar beet***

Approval information

- In 2006 PSD required that all products containing this active ingredient should carry the following warning in the main area of the container label: "Benfuracarb is an anticholinesterase carbamate. Handle with care"
- Benfuracarb will not be included in Annex I of EC Directive 91/414. All approvals for the use of products listed in this profile will expire in March 2009

Efficacy guidance

- Apply at sowing with suitable applicator so that the granules are mixed with the moving soil closing the seed furrow. See label for recommended applicators and calibration

Restrictions

- Contains an anticholinesterase carbamate compound. Do not use if under medical advice not to work with such compounds
- Maximum number of treatments 1 per crop

Crop-specific information

- Latest use: at planting

Environmental safety

- Dangerous for the environment
- Toxic to aquatic organisms
- Do not empty into drains

Hazard classification and safety precautions

Hazard H03, H11
Risk phrases R20, R51, R53a
Operator protection A, B, C, D, E, H, K, M; U02a, U05a, U19a, U20a
Environmental protection E15a, E19b
Storage and disposal D01, D02, D09a, D11a, D12a
Medical advice M02, M05a

24 bentazone

A post-emergence contact benzothiadiazinone herbicide
HRAC mode of action code: C3

Products

1	Basagran SG	BASF	87% w/w	SG	08360
2	Landgold Bentazone SL	Teliton	480 g/l	SL	13418
3	Pulse	AgriGuard	480 g/l	SL	12764
4	Zone 48	AgriGuard	480 g/l	SL	13165

Uses

- Annual dicotyledons in ***broad beans***, ***french beans***, ***linseed***, ***narcissi***, ***navy beans***, ***peas***, ***potatoes***, ***runner beans***, ***spring field beans***, ***winter field beans*** [1-4]; ***bulb onions*** *(off-label)*, ***garlic*** *(off-label)*, ***leeks*** *(off-label)*, ***shallots*** *(off-label)*, ***soya beans*** *(off-label)* [1]
- Common storksbill in ***leeks*** *(off-label)* [1]
- Fool's parsley in ***leeks*** *(off-label)* [1]
- Mayweeds in ***bulb onions*** *(off-label)*, ***garlic*** *(off-label)*, ***shallots*** *(off-label)* [1]

Specific Off-Label Approvals (SOLAs)

- ***bulb onions*** *20061631* [1]
- ***garlic*** *20061631* [1]
- ***leeks*** *20061630* [1]
- ***shallots*** *20061631* [1]
- ***soya beans*** *20061629* [1]

Approval information

- Bentazone included in Annex I under EC Directive 91/414
- Approval expiry 30 Apr 2008 [3]

Efficacy guidance

- Most effective control obtained when weeds are growing actively and less than 5 cm high or across. Good spray cover is essential
- Various recommendations are made for use in spray programmes with other herbicides or in tank mixes. See label for details
- The addition of specified adjuvant oils is recommended for use on some crops to improve fat hen control. Do not use under hot or humid conditions. See label for details
- Split dose application may be made in all recommended crops except peas and generally gives better weed control. See label for details

Restrictions

- Maximum number of treatments normally 2 per crop but check label
- Crops must be treated at correct stage of growth to avoid danger of scorch. See label for details
- Not all varieties of recommended crops are fully tolerant. Use only on tolerant varieties named in label. Do not use on forage or mange-tout varieties of peas
- Do not use on crops which have been affected by drought, waterlogging, frost or other stress conditions
- Do not apply insecticides within 7 d of treatment
- Leave 14 d after using a post-emergence grass herbicide and carry out a leaf wax test where relevant or 7 d where treatment precedes the grass herbicide
- Do not spray at temperatures above 21 °C. Delay spraying until evening if necessary
- Do not apply if rain or frost expected, if unseasonably cold, if foliage wet or in drought
- A minimum of 6 h (preferably 12 h) free from rain is required after application
- May be used on selected varieties of maincrop and second early potatoes (see label for details), not on seed crops or first earlies

Crop-specific information

- Latest use: before shoots exceed 15 cm high for potatoes and spring field beans (or 6-7 leaf pairs); 4 leaf pairs (6 pairs or 15 cm high with split dose) for broad beans; before flower buds visible for French, navy, runner and winter field beans, and linseed; before flower buds can be found enclosed in terminal shoot for peas
- Best results in narcissi obtained by using a suitable pre-emergence herbicide first
- Consult processor before using on crops for processing
- A satisfactory wax test must be carried out before use on peas
- Do not treat narcissi during flower bud initiation

Environmental safety

- Dangerous for the environment
- Harmful to aquatic organisms

Hazard classification and safety precautions

Hazard H03 [1-4]; H11 [3, 4]
Risk phrases R22a, R43 [1-4]; R36 [3, 4]; R41 [1]; R52, R53a [1, 2]
Operator protection A, C [1-4]; H [2]; U05a, U08, U19a, U20b [1-4]; U11 [1]; U14 [1, 2]
Environmental protection E15a [3]; E15b [1, 2, 4]; E34 [1-4]
Storage and disposal D01, D02, D09a [1-4]; D05 [2-4]; D07 [3, 4]; D10b [1, 3]; D10c [2, 4]
Medical advice M03 [1, 2, 4]; M05a [1]

25 bentazone + MCPA + MCPB

A post-emergence herbicide for undersown spring cereals and grass
HRAC mode of action code: C3 + O + O

Products

Acumen	BASF	200:80:200 g/l	SL	00028

Uses

- Annual dicotyledons in ***rotational grass***, ***undersown spring cereals***

Approval information

- Bentazone, MCPA and MCPB included in Annex I under EC Directive 91/414

Efficacy guidance

- Best results when weeds small and actively growing provided crop is at correct growth stage
- A minimum of 6 h free from rain is required after treatment

Restrictions

- Maximum number of treatments 1 per crop on undersown cereals; 1 per yr on grassland
- Do not treat red clover after 3-trifoliate leaf stage
- Do not use on crops suffering from herbicide damage or physical stress or in temperatures above 21 °C
- Do not use on seed crops or on cereals undersown with lucerne or seed mixtures containing lucerne
- Do not roll or harrow for 7 d before or after spraying
- Clovers may be scorched and undersown crop checked but effects likely to be outgrown. Clover stand in grassland may be reduced
- Do not apply if frost expected, if crop wet or when temperatures at or above 21 °C

Crop-specific information

- Latest use: before 1st node detectable for undersown cereals (GS 31), 2 wk before grazing for grassland
- Apply to cereals from 2-fully expanded leaf stage but before first node detectable (GS 12-30) provided clover has reached 1-trifoliate leaf stage
- On first year grass leys use on seedling stage weeds before end of Sep
- Apply to newly sown leys after grass has reached 2-leaf stage provided clovers have at least 1 trifoliate leaf and red clover has not passed 3-trifoliate leaf stage

Environmental safety

- Dangerous for the environment
- Toxic to aquatic organisms
- Keep livestock out of treated areas for at least 2 wk and until foliage of poisonous weeds such as ragwort has died and become unpalatable

Hazard classification and safety precautions

Hazard H03, H11
Risk phrases R22a, R36, R38, R43, R51, R53a
Operator protection A, C; U05a, U08, U14, U19a, U20b
Environmental protection E07a, E15a, E34, E38
Storage and disposal D01, D02, D08, D09a, D10c, D12a
Medical advice M03, M05a

26 benthiavalicarb-isopropyl

An amino acid amide carbamate fungicide available only in mixtures
FRAC mode of action code: 40

27 benthiavalicarb-isopropyl + mancozeb

A fungicide mixture for potatoes
FRAC mode of action code: 40 + M3

Products

1	En-Garde	Certis	1.75:70% w/w	WG	12985
2	Valbon	Certis	1.75:70% w/w	WG	12603

Uses

- Blight in ***potatoes*** [1, 2]
- Downy mildew in ***amenity vegetation*** *(off-label)*, ***ornamental plant production*** *(off-label)* [2]

Specific Off-Label Approvals (SOLAs)

- ***amenity vegetation*** *20070240* [2]
- ***ornamental plant production*** *20070240* [2]

Approval information

- Mancozeb included in Annex I under EC Directive 91/414

SEE SECTION 3 FOR PRODUCTS ALSO REGISTERED

Efficacy guidance

- Apply as a protectant spray commencing before blight enters the crop irrespective of growth stage following a blight warning or where there is a local source of infection
- In the absence of a warning commence the spray programme before the foliage meets in the rows
- Repeat treatment every 7-10 d depending on disease pressure
- Increase water volume as necessary to ensure thorough coverage of the plant
- Use of air assisted sprayers or drop legs may help to improve coverage

Restrictions

- Maximum number of treatments 6 per crop
- Do not use more than three consecutive treatments. Include products with a different mode of action in the programme

Crop-specific information

- HI 7 d for potatoes

Environmental safety

- Very toxic to aquatic organisms
- Risk to non-target insects or other arthropods. Avoid spraying within 6 m of field boundary
- LERAP Category B

Hazard classification and safety precautions

Hazard H04
Risk phrases R40, R43, R50, R53a
Operator protection A, H; U14
Environmental protection E15a, E16a, E22c
Storage and disposal D01, D02, D05, D09a, D11a, D12b

28 beta-cyfluthrin

A non-systemic pyrethroid insecticide available only in mixtures
IRAC mode of action code: 3

29 beta-cyfluthrin + clothianidin

An insecticidal seed treatment mixture for beet
IRAC mode of action code: 3 + 4A

Products

Poncho Beta	Bayer CropScience	53:400 g/l	FS	12076

Uses

- Beet virus yellows vectors in ***fodder beet*** *(seed treatment)*, ***sugar beet*** *(seed treatment)*

Approval information

- Beta-cyfluthrin and clothianidin included in Annex I under EC Directive 91/414

Efficacy guidance

- In addition to control of aphid virus vectors, product improves crop establishment by reducing damage caused by symphylids, springtails, millipedes, wireworms and leatherjackets
- Additional control measures should be taken where very high populations of soil pests are present
- Product reduces direct feeding damage by foliar pests such as pygmy mangold beetle, beet flea beetle, mangold fly
- Product is not active against nematodes
- Treatment does not alter physical characteristics of pelleted seed and no change to standard drill settings should be necessary

Restrictions

- Maximum number of treatments: 1 per seed batch
- Product must be co-applied with a colouring dye
- Product must only be applied as a coating to pelleted seed using special treatment machinery

Crop-specific information

- Latest use: pre-drilling

Environmental safety

- Extremely dangerous to fish or other aquatic life. Do not contaminate surface waters or ditches with chemical or used container

Hazard classification and safety precautions

Hazard H03
Risk phrases R22a
Operator protection A, H; U04a, U05a, U07, U13, U14, U20b, U24
Environmental protection E13a, E34, E36a
Storage and disposal D01, D02, D05, D09a, D14
Treated seed S02, S03, S04a, S04b, S05, S06a, S06b, S07
Medical advice M03

30 beta-cyfluthrin + imidacloprid

An insecticide mixture for seed treatment of oilseed rape
IRAC mode of action code: 3 + 4A

Products

1	Chinook Blue	Bayer CropScience	100:100 g/l	LS	11262
2	Chinook Colourless	Bayer CropScience	100:100 g/l	LS	11206

Uses

- Cabbage stem flea beetle in ***spring oilseed rape*** *(seed treatment)* [1]; ***winter oilseed rape*** *(seed treatment)* [1, 2]
- Flea beetle in ***spring oilseed rape*** *(seed treatment)*, ***winter oilseed rape*** *(seed treatment)* [1]

Approval information

- Beta-cyfluthrin included in Annex I under EC Directive 91/414

Efficacy guidance

- Drill treated seed as soon as possible after treatment
- Product will reduce damage by early attacks of flea beetle and may also reduce subsequent larval damage. However a follow-up spray treatment may be required if pest activity is heavy and prolonged

Restrictions

- Maximum number of treatments 1 per batch of seed
- Do not use on seed with more than 9% moisture content, on sprouted seed or on cracked, split or otherwise damaged seed
- Product must be co-applied with a colouring agent [2]
- Must be applied using seed treatment machine recommended by manufacturer
- Allow treated seed to dry before packaging. Drying can be assisted by subsequent application of talc
- Product can cause a transient tingling or numbing sensation to exposed skin. Avoid skin contact with the product, treated seed or dust throughout all operations in the treatment plant and during drilling

Crop-specific information

- Latest use: before drilling

Environmental safety

- Dangerous for the environment
- Toxic to aquatic organisms
- Extremely dangerous to fish or other aquatic life. Do not contaminate surface waters or ditches with chemical or used container
- Do not use treated seed as food or feed
- Dangerous to birds, game and other wildlife. Treated seed should not be left on the soil surface. Bury spillages
- Do not broadcast treated seed

Hazard classification and safety precautions

Hazard H03, H11
Risk phrases R22a, R51, R53a
Operator protection A, H; U04a, U05a, U07, U20b

Environmental protection E13a, E34, E36a, E38
Storage and disposal D01, D02, D05, D09a, D12a, D14
Treated seed S01, S02, S03, S04c, S05, S06a, S06b, S07, S08
Medical advice M03

31 bifenazate

A bifenazate acaricide for use on protected strawberries
IRAC mode of action code: 25

Products

Floramite 240 SC	Certis	240 g/l	SC	13686

Uses

- Two-spotted spider mite in ***protected strawberries***

Approval information

- Bifenazate included in Annex I under EC Directive 91/414

Efficacy guidance

- Bifenazate acts by contact knockdown after approximately 4 d and a subsequent period of residual control
- All mobile stages of mites are controlled with occasionally some ovicidal activity
- Best results obtained from a programme of two treatments started as soon as first spider mites are seen
- To minimise possible development of resistance use in a planned Resistance Management strategy. Alternate at least two products with different modes of action between treatment programmes of bifenazate. See Section 5 for more information

Restrictions

- Maximum number of treatments 2 per yr for protected strawberries
- After use in protected environments avoid re-entry for at least 2 h while full ventilation is carried out allowing spray to dry on leaves
- Personnel working among treated crops should wear suitable long-sleeved garments and gloves for 14 d after treatment
- Do not apply as a low volume spray

Crop-specific information

- HI: 7 d for protected strawberries
- Carry out small scale tolerance test on the strawberry cultivar before large scale use

Environmental safety

- Dangerous for the environment
- Toxic to aquatic organisms
- Harmful to non-target predatory mites. Avoid spray drift onto field margins, hedges, ditches, surface water and neighbouring crops

Hazard classification and safety precautions

Hazard H04, H11
Risk phrases R43, R51, R53a
Operator protection A, H; U02a, U04a, U05a, U10, U14, U15, U19a, U20b
Environmental protection E15b, E38
Storage and disposal D01, D02, D05, D09a, D10b, D12b
Medical advice M03

32 bifenox

A diphenyl ether herbicide for cereals
HRAC mode of action code: E

Products

1	Fox	Makhteshim	480 g/l	SC	11981
2	Oram	Makhteshim	480 g/l	SC	13410
3	Sabine	Makhteshim	480 g/l	SC	13439

Uses

- Annual dicotyledons in ***durum wheat*** *(off-label)*, ***grass seed crops*** *(off-label)* [1]
- Cleavers in ***triticale*** [1, 2]; ***winter barley***, ***winter rye***, ***winter wheat*** [1-3]
- Field pansy in ***triticale*** [1, 2]; ***winter barley***, ***winter rye***, ***winter wheat*** [1-3]
- Field speedwell in ***triticale*** [1, 2]; ***winter barley***, ***winter rye***, ***winter wheat*** [1-3]
- Forget-me-not in ***triticale*** [1, 2]; ***winter barley***, ***winter rye***, ***winter wheat*** [1-3]
- Geranium species in ***spring oilseed rape*** *(off-label)*, ***winter oilseed rape*** *(off-label)* [1]
- Ivy-leaved speedwell in ***triticale*** [1, 2]; ***winter barley***, ***winter rye***, ***winter wheat*** [1-3]
- Poppies in ***triticale*** [1, 2]; ***winter barley***, ***winter rye***, ***winter wheat*** [1-3]
- Red dead-nettle in ***triticale*** [1, 2]; ***winter barley***, ***winter rye***, ***winter wheat*** [1-3]

Specific Off-Label Approvals (SOLAs)

- ***durum wheat*** *20061238* [1]
- ***grass seed crops*** *20061238* [1]
- ***spring oilseed rape*** *20060321* [1]
- ***winter oilseed rape*** *20060321* [1]

Approval information

- Accepted by BBPA for use on malting barley

Efficacy guidance

- Bifenox is absorbed by foliage and emerging roots of susceptible species
- Best results obtained when weeds are growing actively with adequate soil moisture

Restrictions

- Maximum number of treatments: one per crop
- Do not apply to crops suffering from stress from whatever cause
- Do not apply if the crop is wet or if rain or frost is expected
- Avoid drift onto broad-leaved plants outside the target area

Crop-specific information

- Latest use: before 2nd node detectable (GS 32) for all crops

Environmental safety

- Dangerous for the environment
- Very toxic to aquatic organisms
- Do not empty into drains

Hazard classification and safety precautions

Hazard H11
Risk phrases R50, R53a
Operator protection A; U05a, U08, U20b
Environmental protection E15a, E19b, E34, E38
Storage and disposal D01, D02, D09a, D12a
Medical advice M03

33 bifenox + isoproturon

A contact and residual herbicide mixture for cereals
HRAC mode of action code: E + C2

Products

Banco	Makhteshim	160:400 g/l	SC	13019

Uses

- Annual dicotyledons in ***winter barley***, ***winter wheat***
- Annual grasses in ***winter barley***, ***winter wheat***
- Blackgrass in ***winter barley***, ***winter wheat***
- Field pansy in ***winter barley***, ***winter wheat***
- Wild oats in ***winter barley***, ***winter wheat***

Approval information

- Isoproturon included in Annex I under EC Directive 91/414.; However, see isoproturon entry for information on phasing out of approvals of all products containing isoproturon by 30 June 2009
- Accepted by BBPA for use on malting barley

Efficacy guidance

- Best results obtained from treatment when crop and weeds are growing actively in moist soil. Weeds are controlled by leaf and root uptake
- Weed control may be reduced on loose or cloddy seedbeds and where trash or straw is not dispersed and buried before treatment
- Rolling or harrowing after treatment will reduce control
- Grass weeds should not be treated before 1 Jan
- Heavy infestations of ryegrass (e.g. after a seed crop) may not be satisfactorily controlled
- Always follow WRAG guidelines for preventing and managing herbicide resistant weeds. See Section 5 for more information

Restrictions

- Maximum number of treatments 1 per crop for wheat and barley
- Do not treat durum wheat, oats, triticale or rye
- Do not treat undersown crops, or those to be undersown
- Do not spray crops under stress from water logging or drought, or suffering from pest or disease attack
- Do not roll or harrow during the 7 d prior to post-emergence treatment
- Do not spray if frost imminent or after prolonged frosty weather

Crop-specific information

- Latest use: before 2nd node detectable (GS 32) for barley and wheat
- Early sown winter cereals may be damaged if application takes place during rapid growth in the autumn

Environmental safety

- Dangerous for the environment
- Very toxic to aquatic organisms
- Do not empty into drains
- Do not apply to dry, cracked or waterlogged soils where heavy rain may lead to contamination of drains by isoproturon

Hazard classification and safety precautions

Hazard H03, H11
Risk phrases R40, R50, R53a
Operator protection A, C, H; U05a, U08, U13, U19a, U20b
Environmental protection E15a, E19b
Storage and disposal D01, D02, D05, D09a, D10b, D12a
Medical advice M05a

34 bifenthrin

A contact and residual pyrethroid acaricide/insecticide for use in agricultural and horticultural crops
IRAC mode of action code: 3

Products

1	Brigade 80 SC	Belchim	80 g/l	SC	12853
2	Gyro	Certis	100 g/l	EC	13026
3	Starion Flo	Belchim	80 g/l	SC	12455
4	Talstar 80 Flo	Belchim	80 g/l	SC	12352
5	UPL Bifenthrin	United Phosphorus	100 g/l	EC	12653

Uses

- Aphids in ***broccoli***, ***brussels sprouts***, ***cabbages***, ***calabrese***, ***cauliflowers*** [1-5]; ***combining peas***, ***vining peas***, ***winter barley***, ***winter oats***, ***winter wheat*** [1-3, 5]; ***spring barley***, ***spring oats***, ***spring wheat*** [1, 3]
- Cabbage seed weevil in ***spring oilseed rape*** [1, 3]; ***winter oilseed rape*** [1-3, 5]
- Cabbage stem flea beetle in ***winter oilseed rape*** [1-3, 5]
- Caterpillars in ***broccoli***, ***brussels sprouts***, ***cabbages***, ***calabrese***, ***cauliflowers*** [1-5]
- Clay-coloured weevil in ***blackberries*** *(off-label)*, ***protected blackberries*** *(off-label)*, ***protected raspberries*** *(off-label)*, ***raspberries*** *(off-label)* [1, 3]
- Damson-hop aphid in ***hops*** [1-5]
- Flea beetle in ***linseed***, ***winter oilseed rape*** [1-3, 5]; ***spring oilseed rape*** [1, 3]

- Fruit tree red spider mite in ***apples, pears*** [1-5]
- Pod midge in ***spring oilseed rape*** [1, 3]; ***winter oilseed rape*** [1-3, 5]
- Pollen beetle in ***spring oilseed rape*** [1, 3]; ***winter oilseed rape*** [1-3, 5]
- Rape winter stem weevil in ***winter oilseed rape*** [1-3, 5]
- Two-spotted spider mite in ***blackberries*** *(off-label)*, ***protected blackberries*** *(off-label)*, ***protected raspberries*** *(off-label)*, ***raspberries*** *(off-label)* [1, 3]; ***blackcurrants*** *(off-label)*, ***protected blackcurrants*** *(off-label)* [3, 4]; ***hops, ornamental plant production, strawberries*** [1-5]
- Virus vectors in ***winter barley, winter oats, winter wheat*** [1-3, 5]
- Whitefly in ***broccoli, brussels sprouts, cabbages, calabrese, cauliflowers*** [1-5]

Specific Off-Label Approvals (SOLAs)

- ***blackberries*** *20061628* [1], *20060109* [3]
- ***blackcurrants*** *20063125* [3], *20063033* [4]
- ***protected blackberries*** *20061628* [1], *20060109* [3]
- ***protected blackcurrants*** *20063125* [3], *20063033* [4]
- ***protected raspberries*** *20061628* [1], *20060109* [3]
- ***raspberries*** *20061628* [1], *20060109* [3]

Approval information

- Accepted by BBPA for use on malting barley and hops

Efficacy guidance

- Timing of application varies with crop and pest. See label for details
- Good spray cover of upper and lower plant surfaces essential to achieve effective pest control. Spray volumes should be increased where crop is dense

Restrictions

- Maximum number of treatments 5 per yr for hops; 4 per crop on brassicas; 2 per yr for apples, pears, strawberries. Other crops either 2 per yr or a maximum total dose is stipulated. See label
- Do not apply to a cereal crop if any product containing a pyrethroid insecticide or dimethoate has been applied after the start of ear emergence (GS 51)
- Consult processor before use on crops for processing
- Do not treat ornamental specimens in blossom or under stress

Crop-specific information

- Latest use: before 31 Mar in yr of harvest for cereals, before 30 Nov in yr of sowing for winter oilseed rape [2, 5]; before late milk stage (GS 77) for wheat, before early milk stage (GS 73) for barley, oats [1, 3];
- HI 2 d for brassicas, 3 d for peas; 8 wk for linseed [1-3, 5], oilseed rape [1, 3]

Environmental safety

- Dangerous for the environment
- Very toxic to aquatic organisms
- High risk to bees. Do not apply to crops in flower, or to those in which bees are actively foraging, except as directed on named crops. Do not apply when flowering weeds are present [1, 3, 4]
- Extremely dangerous to bees. Do not apply to crops in flower or to those in which bees are actively foraging. Do not apply when flowering weeds are present [2, 5]
- Extremely dangerous to fish or other aquatic life. Do not contaminate surface waters or ditches with chemical or used container
- High risk to non-target insects or other arthropods. Not compatible with IPM systems
- Do not spray cereal crops within 6 m of the field boundary or after 31 Mar in the yr of harvest
- Keep in original container, tightly closed, in a safe place, under lock and key [1, 3, 4]
- LERAP Category A and broadcast air-assisted LERAP (30 m [1, 3, 4]) (50 m [2])

Hazard classification and safety precautions

Hazard H03, H11

Risk phrases R20, R22a, R50, R53a [1-5]; R36, R37 [2, 5]; R43 [1, 3, 4]

Operator protection A, H [1-5]; C [2, 5]; U02a, U04a, U08, U19a [1, 3, 4]; U05a, U20b [1-5]; U11, U14, U15 [2, 5]

Environmental protection E12a, E15b, E34, E38 [1, 3, 4]; E12b, E12e [2, 5]; E12f [1, 3, 4] (named crops - see label); E13a, E16c, E16d [1-5]; E17b [1] (30 m); E17b [2] (50 m); E17b [3, 4] (30 m); E17b [5] (50 m); E22a [1] (or after 31 Mar); E22a [2]; E22a [3, 4] (or after 31 Mar); E22a [5]

Storage and disposal D01, D02, D05 [1-5]; D09a, D10c [2, 5]; D09b, D10b, D12a [1, 3, 4]

Medical advice M03 [1-5]; M05a [2, 5]; M05b [1, 3, 4]

SEE SECTION 3 FOR PRODUCTS ALSO REGISTERED

35 bifenthrin + malathion

An insecticide mixture for pest control in stored grain and grain stores
IRAC mode of action code: 3 + 1B

Products

1	Prostore 157 UL	Nickerson	7.5:150 g/l	UL	12017
2	Prostore 420EC	Nickerson	20:400 g/l	EC	12210

Uses

- Grain storage pests in ***grain stores*** [2]; ***stored grain*** [1, 2]

Approval information

- Accepted by BBPA for use in stores for malting barley
- Malathion will not be included in Annex I of EC Directive 91/414. To comply with this EC decision all products listed in this profile will need to be withdrawn from the market by about September 2007, and all approvals for their storage and use will finally expire in about September 2008
- Approval expiry 06 Dec 2008 [1, 2]

Efficacy guidance

- Best results in stored grain obtained from treatment at the start of storage after high-temperature drying. Cool warm grain as soon as possible to prevent moisture migration and ensure optimum protection from reinfestation
- For guidance on cleaning grain stores refer to HGCA publication "The Grain Storage Guide [2]
- Controls a wide range of stored grain insect pests and cosmopolitan food mites
- Under normal storage conditions a single treatment will give 3-6 mth protection
- Knock-down is complete within 24 h at normal harvest grain temperatures
- Control may not be satisfactory if pest strains resistant to bifenthrin or malathion are present

Restrictions

- Maximum number of treatments 1 per store [2] or per batch of grain
- Do not treat wheat (including durum), oats, rye or triticale intended for use as seed
- Must only be applied with appropriate equipment designed to treat grain
- Do not apply with hand-held equipment

Environmental safety

- Dangerous for the environment
- Very toxic to aquatic organisms

Hazard classification and safety precautions

Hazard H03, H11
Risk phrases R22a, R38, R41, R43 [2]; R22b, R50, R53a, R66, R67 [1, 2]
Operator protection A, C, H [1, 2]; D, J, M [2]; U05a, U23a [1, 2]; U14 [2]
Environmental protection E15a, E34, E38
Storage and disposal D01, D02, D11a [1, 2]; D12a [2]
Medical advice M03, M05b

36 Bordeaux mixture

A protectant copper sulphate/lime complex fungicide
FRAC mode of action code: M1

Products

Wetcol 3 Copper Fungicide	Law Fertilisers	30 g/l (copper)	SC	13369

Uses

- Bacterial canker in ***cherries***
- Blight in ***potatoes***, ***tomatoes (outdoor)***
- Cane spot in ***loganberries***, ***raspberries***
- Canker in ***apples***, ***pears***
- Celery leaf spot in ***celery (outdoor)***
- Currant leaf spot in ***blackcurrants***
- Downy mildew in ***hops***
- Leaf curl in ***apricots***, ***nectarines***, ***peaches***

- Scab in ***apples, pears***
- Spur blight in ***raspberries***

Efficacy guidance

- Spray interval normally 7-14 d but varies with disease and crop. See label for details
- Commence spraying potatoes before crop meets in row or immediately first blight period occurs
- For canker control spray monthly from Aug to Oct
- For peach leaf curl control spray at leaf fall in autumn and again in Feb
- Spray when crop foliage dry. Do not spray if rain imminent

Restrictions

- Do not use on copper sensitive cultivars, including Doyenne du Comice pears

Environmental safety

- Harmful to fish or other aquatic life. Do not contaminate surface waters or ditches with chemical or used container
- Harmful to livestock. Keep all livestock out of treated areas for at least 3 wk

Hazard classification and safety precautions

Hazard H03, H04
Risk phrases R22a, R36, R37, R38
Operator protection U13, U15, U20a
Environmental protection E06c (3 wk); E13c
Storage and disposal D09a, D11a

37 boscalid

A translocated and translaminar anilide fungicide
FRAC mode of action code: 7

Products

Filan	BASF	50% w/w	WG	11449

Uses

- Alternaria in ***poppies for morphine production*** *(off-label)*, ***spring oilseed rape***, ***winter oilseed rape***
- Botrytis in ***poppies for morphine production*** *(off-label)*
- Sclerotinia stem rot in ***poppies for morphine production*** *(off-label)*, ***spring oilseed rape***, ***winter oilseed rape***

Specific Off-Label Approvals (SOLAs)

- ***poppies for morphine production*** *20040163 expires Oct 2008*

Efficacy guidance

- For best results apply as a protectant spray before symptoms are visible
- Applications against *Sclerotinia* should be made in high disease risk situations at early to full flower
- *Sclerotinia* control may be reduced when high risk conditions occur after flowering, leading to secondary disease spread
- Applications against *Alternaria* should be made at full flowering. Later applications may result in reduced levels of control
- Ensure adequate spray penetration and good coverage

Restrictions

- Maximum total dose equivalent to two full dose treatments
- Do no treat crops to be used for seed production later than full flowering stage
- Avoid spray drift onto neighbouring crops

Crop-specific information

- Latest use: up to and including when 50% of pods have reached final size (GS 75) for oilseed rape

Environmental safety

- Dangerous for the environment
- Toxic to aquatic organisms
- Product represents minimal hazard to bees when used as directed. However local bee-keepers should be notified if crops are to be sprayed when in flower

Hazard classification and safety precautions

Hazard H11

SEE SECTION 3 FOR PRODUCTS ALSO REGISTERED

Risk phrases R51, R53a
Operator protection U05a, U20c
Environmental protection E15a, E38
Storage and disposal D01, D02, D05, D09a, D10c, D12a

38 boscalid + epoxiconazole

A broad spectrum fungicide mixture for cereals
FRAC mode of action code: 7 + 3

Products

1	Deuce	BASF	233:67 g/l	SC	13240
2	Splice	BASF	233:67 g/l	SC	12315
3	Totem	BASF	233:67 g/l	SC	12692
4	Tracker	BASF	233:67 g/l	SC	12295
5	Venture	BASF	233:67 g/l	SC	12316

Uses

- Brown rust in ***spring barley***, ***spring wheat***, ***winter barley***, ***winter wheat*** [1-5]
- Crown rust in ***spring oats***, ***winter oats*** [1-5]
- Eyespot in ***spring barley*** *(moderate)*, ***spring wheat*** *(moderate)*, ***winter barley*** *(moderate)*, ***winter wheat*** *(moderate)* [1, 2, 4, 5]; ***spring barley*** *(moderate control)*, ***spring wheat*** *(moderate control)*, ***winter barley*** *(moderate control)*, ***winter wheat*** *(moderate control)* [3]
- Foliar disease control in ***durum wheat*** *(off-label)* [2-5]
- Net blotch in ***spring barley***, ***winter barley*** [1-5]
- Rhynchosporium in ***spring barley*** *(moderate)*, ***winter barley*** *(moderate)* [1, 2, 4, 5]; ***spring barley*** *(moderate control)*, ***winter barley*** *(moderate control)* [3]
- Septoria leaf blotch in ***spring wheat***, ***winter wheat*** [1-5]
- Tan spot in ***spring wheat*** *(reduction - qualified minor use)*, ***winter wheat*** *(reduction - qualified minor use)* [3]

Specific Off-Label Approvals (SOLAs)

- ***durum wheat*** *20070460* [2], *20070458* [3], *20070462* [4], *20070461* [5]

Approval information

- Accepted by BBPA for use on malting barley (before ear emergence only)

Efficacy guidance

- Apply at the start of foliar or stem based disease attack
- Optimum effect against eyespot achieved by spraying between leaf-sheath erect and second node detectable stages (GS 30-32)
- Epoxiconazole is a DMI fungicide. Resistance to some DMI fungicides has been identified in Septoria leaf blotch which may seriously affect performance of some products. For further advice contact a specialist advisor and visit the Fungicide Resistance Action Group (FRAG)-UK website

Restrictions

- Maximum number of treatments 1 per crop [3]
- Maximum total dose equivalent to two full dose treatments [1, 2, 4, 5]

Crop-specific information

- Latest use: before cereal ear emergence

Environmental safety

- Dangerous for the environment
- Toxic to aquatic organisms
- LERAP Category B
- Avoid drift onto neighbouring crops. May cause damage to broad-leaved plant species

Hazard classification and safety precautions

Hazard H03, H11
Risk phrases R36, R40, R51, R53a, R62, R63
Operator protection A, C, H; U05a, U20b
Environmental protection E15a, E16a, E38
Storage and disposal D01, D02, D08, D10c, D12a
Medical advice M03

39 boscalid + pyraclostrobin

A protectant and systemic fungicide mixture
FRAC mode of action code: 7 + 11

Products

1	Bellis	BASF	25.2:12.8% w/w	WG	12522
2	Signum	BASF	26.7:6.7% w/w	WG	11450

Uses

- Alternaria in ***cabbages***, ***carrots*** *(moderate)*, ***cauliflowers*** [2]
- Black spot in ***protected strawberries***, ***strawberries*** [2]
- Blossom wilt in ***cherries*** *(off-label)*, ***mirabelles*** *(off-label)*, ***plums*** *(off-label)*, ***protected cherries*** *(off-label)*, ***protected mirabelles*** *(off-label)*, ***protected plums*** *(off-label)* [2]
- Botrytis in ***protected chinese cabbage*** *(off-label)*, ***protected choi sum*** *(off-label)*, ***protected pak choi*** *(off-label)* [2]
- Brown rot in ***cherries*** *(off-label)*, ***mirabelles*** *(off-label)*, ***plums*** *(off-label)*, ***protected cherries*** *(off-label)*, ***protected mirabelles*** *(off-label)*, ***protected plums*** *(off-label)* [2]
- Chocolate spot in ***spring field beans*** *(moderate)*, ***winter field beans*** *(moderate)* [2]
- Downy mildew in ***leaf spinach*** *(off-label - Must be applied between 1 Apr and 31 Oct)* [2]
- Grey mould in ***protected strawberries***, ***strawberries*** [2]
- Powdery mildew in ***apples*** [1]; ***carrots***, ***protected strawberries***, ***strawberries*** [2]
- Rhizoctonia rot in ***lettuce***, ***protected lettuce*** [2]
- Ring spot in ***brussels sprouts***, ***cabbages***, ***cauliflowers***, ***protected chinese cabbage*** *(off-label)*, ***protected choi sum*** *(off-label)*, ***protected pak choi*** *(off-label)* [2]
- Rust in ***spring field beans***, ***winter field beans*** [2]
- Scab in ***apples***, ***pears*** [1]
- Sclerotinia in ***carrots*** *(moderate)*, ***herbs (see appendix 6)*** *(off-label)*, ***horseradish*** *(off-label)*, ***lettuce***, ***parsnips*** *(off-label)*, ***protected herbs (see appendix 6)*** *(off-label)*, ***protected lettuce***, ***protected salad brassicas*** *(off-label - for baby leaf production)* [2]
- White blister in ***brussels sprouts***, ***cabbages*** *(qualified minor use)*, ***chinese cabbage*** *(off-label)*, ***collards*** *(off-label)*, ***kale*** *(off-label)*, ***komatsuna*** *(off-label)*, ***pak choi*** *(off-label)*, ***salad brassicas*** *(off-label - for baby leaf production)* [2]
- White tip in ***leeks*** *(off-label)* [2]

Specific Off-Label Approvals (SOLAs)

- ***cherries*** *20052089 expires Oct 2008* [2]
- ***chinese cabbage*** *20031595 expires Oct 2008* [2]
- ***collards*** *20031595 expires Oct 2008* [2]
- ***herbs (see appendix 6)*** *20041984 expires Oct 2008* [2]
- ***horseradish*** *20051317 expires Oct 2008* [2]
- ***kale*** *20031595 expires Oct 2008* [2]
- ***komatsuna*** *20031595 expires Oct 2008* [2]
- ***leaf spinach*** *(Must be applied between 1 Apr and 31 Oct) 20052378 expires Oct 2008* [2]
- ***leeks*** *20062848 expires Oct 2008* [2]
- ***mirabelles*** *20052089 expires Oct 2008* [2]
- ***pak choi*** *20031595 expires Oct 2008* [2]
- ***parsnips*** *20051317 expires Oct 2008* [2]
- ***plums*** *20052089 expires Oct 2008* [2]
- ***protected cherries*** *20052089 expires Oct 2008* [2]
- ***protected chinese cabbage*** *20051349 expires Oct 2008* [2]
- ***protected choi sum*** *20051349 expires Oct 2008* [2]
- ***protected herbs (see appendix 6)*** *20041984 expires Oct 2008* [2]
- ***protected mirabelles*** *20052089 expires Oct 2008* [2]
- ***protected pak choi*** *20051349 expires Oct 2008* [2]
- ***protected plums*** *20052089 expires Oct 2008* [2]
- ***protected salad brassicas*** *(for baby leaf production) 20041984 expires Oct 2008* [2]
- ***salad brassicas*** *(for baby leaf production) 20031595 expires Oct 2008* [2]

Approval information

- Pyraclostrobin included in Annex I under EC Directive 91/414
- Accepted by BBPA for use on malting barley (before ear emergence only)

Efficacy guidance

- On brassicas apply as a protectant spray or at the first sign of disease and repeat at 3-4 wk intervals depending on disease pressure [2]
- Ensure adequate spray penetration and coverage by increasing water volume in dense crops [2]
- For best results on strawberries apply as a protectant spray at the white bud stage. Applications should be made in sequence with other products as part of a fungicide spray programme during flowering at 7-10 day intervals [2]
- On carrots and field beans apply as a protectant spray or at the first sign of disease with a repeat treatment if needed, as directed on the label [2]
- On lettuce apply as a protectant spray 1-2 wk after planting [2]
- Optimum results on apples and pears obtained from a protectant treatment from bud burst [1]
- Application as the final 2 sprays on apples and pears gives reduction in storage rots [1]
- Pyraclostrobin is a member of the QoI cross-resistance group of fungicides and should be used in programmes with fungicides with a different mode of action

Restrictions

- Maximum total dose equivalent to three full dose treatments on brassicas; two full dose treatments on all other field crops [2]
- Maximum number of treatments (including other QoI treatments) on apples and pears 4 per yr if total number of applications is 12 or more, or 3 per yr if the total number is fewer than 12 [1]
- Do not use more than 2 consecutive treatments on apples and pears, and these must be separated by a minimum of 2 applications of a fungicide with a different mode of action [1]
- Use a maximum of three applications per yr on brassicas and no more than two per yr on other field crops [2]
- Do not use consecutive treatments; apply in alternation with fungicides from a different cross resistance group and effective against the target diseases [2]
- Do not apply more than 6 kg product to the same area of land per yr [2]
- Consult processor before use on crops for processing
- Applications to lettuce and protected salad crops may only be made between 1 Apr and 31 Oct [2]

Crop-specific information

- HI 21 d for field beans; 14 d for brassica crops, carrots, lettuce; 7d for apples, pears; 3 d for strawberries

Environmental safety

- Dangerous for the environment
- Very toxic to aquatic organisms
- LERAP Category B [2]
- Broadcast air-assisted LERAP (40 m) [1]

Hazard classification and safety precautions

Hazard H03, H11
Risk phrases R22a, R50, R53a
Operator protection A; U05a, U20c [1, 2]; U13 [1]
Environmental protection E15a, E38 [1, 2]; E16a [2]; E17b [1] (40 m); E34 [1]
Storage and disposal D01, D02, D10c, D12a [1, 2]; D08 [1]; D09a [2]
Medical advice M03 [1]; M05a [1, 2]

40 brodifacoum

An anticoagulant coumarin rodenticide

Products

Sorexa Checkatube	Sorex	0.2% w/w	XX	H6702

Uses

- Mice in ***farm buildings*** *(indoor use only)*

Efficacy guidance

- Product is a self-contained control device which requires no bait handling
- Best results obtained by placing tubes where mice are active and will readily enter the device

- To acquire a lethal dose mice must pass through the tube a number of times
- Where a continuous source of infestation is present, or where sustained protection is needed, tubes can be installed permanently and replaced at monthly intervals
- For complete eradication other control methods should also be used bearing in mind the resistance status of the target population

Restrictions

- For use only by professional pest contractors
- Do not use outdoors. Products must be used in situations where baits are placed within a building or other enclosed structure, and the target is living or feeding predominantly within that building or structure
- Do not attempt to open or recharge the tubes

Environmental safety

- Prevent access to baits by children, birds and other animals
- Do not place tubes where food, feed or water could become contaminated
- Remove all remains of tubes after use and burn or bury
- Search for and burn or bury all rodent bodies. Do not place in refuse bins or on open rubbish tips
- Keep in original container, tightly closed, in a safe place, under lock and key

Hazard classification and safety precautions

Operator protection U13, U20b
Storage and disposal D02, D09b
Vertebrate/rodent control products V01b, V02, V03a, V04a
Medical advice M03

41 bromadiolone

An anti-coagulant coumarin-derivative rodenticide

Products

1	Endorats Premium Rat Killer	Irish Drugs	0.005% w/w	GB	H6725
2	Luxan Vermigon Bait	Luxan	0.005% w/w	RB	H7925
3	Luxan Vermigon Blocks	Luxan	0.005% w/w	BB	H7926
4	Luxan Vermigon Pellets	Luxan	0.005% w/w	PT	H7941
5	Luxan Vermigon Sewer Blocks	Luxan	0.005% w/w	BB	H7927
6	Sakarat Bromabait	Killgerm	0.005% w/w	RB	H7902
7	Tomcat 2	Antec Biosentry	0.005% w/w	PT	H6736
8	Tomcat 2 Blox	Antec Biosentry	0.005% w/w	BB	H6731
9	Vermigon Pasta	Certis	0.005% w/w	RB	H8191

Uses

- Mice in ***farm buildings*** [2-9]; ***farmyards*** [2-6, 9]
- Rats in ***farm buildings*** [1-9]; ***farmyards*** [2-6, 9]

Efficacy guidance

- Ready-to-use baits are formulated on a mould-resistant, whole-wheat base
- Use in baiting programme. Place baits in protected situations, sufficient for continuous feeding between treatments
- Chemical is effective against warfarin- and coumatetralyl-resistant rats and mice and does not induce bait shyness
- Use bait bags where loose baiting inconvenient (eg behind ricks, silage clamps etc)
- The resistance status of the rodent population should be assessed when considering the choice of product to use

Restrictions

- For use only by professional operators

Environmental safety

- Access to baits by children, birds and animals, particularly cats, dogs, pigs and poultry, must be prevented
- Baits must not be placed where food, feed or water could become contaminated
- Remains of bait and bait containers must be removed after treatment and burned or buried

SEE SECTION 3 FOR PRODUCTS ALSO REGISTERED

- Rodent bodies must be searched for and burned or buried. They must not be placed in refuse bins or on rubbish tips
- Take extreme care to prevent domestic animals having access to the bait

Hazard classification and safety precautions

Operator protection U13 [1-4, 6-9]; U20b [2-4, 6-9]

Environmental protection E10b [2-4, 9]; E15a [7, 8]

Storage and disposal D05 [2, 3, 6, 9]; D06a, D06c, D09c [2, 9]; D07 [2-4, 6, 9]; D09a [1-4, 6-9]; D11a [2-4, 6-9]

Treated seed S06a [1]

Vertebrate/rodent control products V01a, V03a, V04a [1, 7, 8]; V01b, V03b, V04b [2-4, 6, 9]; V02 [1-4, 6-9]

Medical advice M03 [2-4, 6-9]

42 bromoxynil

A contact acting HBN herbicide
HRAC mode of action code: C3

Products

1	Alpha Bromolin 225 EC	Makhteshim	225 g/l	EC	13431
2	Alpha Bromotril P	Makhteshim	240 g/l	SC	13326
3	Barclay Mutiny	Barclay	250 g/l	SC	11379
4	Emblem	Nufarm UK	20% w/w	WB	11559
5	Flagon 400 EC	Makhteshim	400 g/l	EC	13470
6	Greencrop Tassle	Greencrop	240 g/l	SC	09659

Uses

- Annual dicotyledons in ***forage maize*** [2-4, 6]; ***linseed***, ***spring barley***, ***spring oats***, ***spring wheat***, ***winter barley***, ***winter oats***, ***winter wheat*** [1, 5]

Approval information

- Bromoxynil included in Annex I under EC Directive 91/414
- Accepted by BBPA for use on malting barley
- Approval expiry 31 Aug 2008 [3, 6]

Efficacy guidance

- Spray when main weed flush has germinated and the largest are at the 4 leaf stage
- Weed control can be enhanced by using a split treatment spraying each application when the weeds are seedling to 2 true leaves. Apply the second treatment before the crop canopy covers the ground [2, 3]

Restrictions

- Maximum number of treatments 1 per crop or yr or maximum total dose equivalent to one full dose treatment
- Do not apply with oils or other adjuvants
- Do not apply using hand-held equipment or at concentrations higher than those recommended
- Do not apply during frosty weather, drought, when soil is waterlogged, when rain expected within 4 h or to crops under any stress
- Take particular care to avoid drift onto neighbouring susceptible crops or open water surfaces

Crop-specific information

- Latest use: before 10 fully expanded leaf stage of crop for maize, sweetcorn; before crop 20 cm tall and before flower buds visible for linseed; before 2nd node detectable (GS 32) for cereals
- Apply to spring sown maize from 2 fully expanded leaves up to 9 fully expanded leaves [2, 3, 6]
- Apply to spring sown linseed from 1 fully expanded leaf to 20 cm tall [1, 5]
- Foliar scorch, which rapidly disappears without affecting growth, will occur if treatment made in hot weather or during rapid growth

Environmental safety

- Dangerous for the environment [1, 3-6]
- Very toxic to aquatic organisms
- Flammable [5]
- Keep livestock out of treated areas for at least 6 wk after treatment
- High risk to bees. Do not apply to crops in flower or to those in which bees are actively foraging

- Do not contaminate surface waters or ditches with chemical or used container
- LERAP Category B [4]

Hazard classification and safety precautions

Hazard H03 [1-6]; H08 [5]; H11 [1, 3-6]

Risk phrases R20 [1, 2, 5]; R22a, R53a [1-6]; R22b, R38 [1, 5]; R36 [3, 5, 6]; R37 [5]; R43 [4, 5]; R50 [1, 4, 5]; R51 [6]; R52 [2, 3]; R63 [1, 2, 4-6]

Operator protection A, C [1-6]; H [1, 5]; U02a, U20a [2, 3, 5, 6]; U03, U11 [5]; U05a, U19a [1-6]; U08, U13 [1-3, 5, 6]; U10 [1, 5]; U14, U15 [1, 2, 4, 5]; U20b [1]; U22a [4]; U23a [1, 3-6]

Environmental protection E06a [1, 5] (6 wk); E12a, E16a [4]; E12d [1-3, 5, 6]; E12e [1-4, 6]; E13b [2]; E15a [1, 3-6]; E34 [1, 3, 5, 6]; E38 [1, 2, 4, 5]

Consumer protection C02a [1, 5] (6 wk)

Storage and disposal D01, D02, D09a [1-6]; D05, D10b [1-3, 5, 6]; D12a [1, 2, 4, 5]

Medical advice M03 [2, 3, 6]; M04a [2, 5]; M04b [1]; M05b [5]

SECTION 2

43 bromoxynil + diflufenican + ioxynil

A selective contact and translocated herbicide for cereals

HRAC mode of action code: C3 + F1 + C3

Products

Capture	Nufarm UK	300:50:200 g/l	SC	12514

Uses

- Annual dicotyledons in ***grass seed crops*** *(off-label)*, ***spring barley***, ***spring wheat***, ***triticale***, ***winter barley***, ***winter rye***, ***winter wheat***
- Chickweed in ***spring barley***, ***spring wheat***, ***triticale***, ***winter barley***, ***winter rye***, ***winter wheat***
- Field speedwell in ***spring barley***, ***spring wheat***, ***triticale***, ***winter barley***, ***winter rye***, ***winter wheat***
- Ivy-leaved speedwell in ***spring barley***, ***spring wheat***, ***triticale***, ***winter barley***, ***winter rye***, ***winter wheat***
- Knotgrass in ***spring barley***, ***spring wheat***, ***triticale***, ***winter barley***, ***winter rye***, ***winter wheat***
- Mayweeds in ***spring barley***, ***spring wheat***, ***triticale***, ***winter barley***, ***winter rye***, ***winter wheat***

Specific Off-Label Approvals (SOLAs)

- ***grass seed crops*** *20052240*

Approval information

- Bromoxynil and ioxynil included in Annex I under EC Directive 91/414
- Accepted by BBPA for use on malting barley

Efficacy guidance

- Best results obtained on small weeds and when competition removed early
- Good spray coverage essential for good activity

Restrictions

- Maximum number of treatments 1 per crop
- Do not treat crops undersown or to be undersown
- Do not treat frosted crops or those that are under stress from any cause
- Do not apply by hand-held equipment or at concentrations higher than those recommended

Crop-specific information

- Latest use: before 2nd node detectable (GS 32)

Following crops guidance

- In the event of crop failure winter wheat may be drilled immediately after normal cultivations. Winter barley may be re-drilled after ploughing
- Land must be ploughed and an interval of 12 wk elapse after treatment before planting spring crops of wheat, barley, oilseed rape, peas, field beans, sugar beet, potatoes, carrots, edible brassicas or onions
- Successive treatments with any products containing diflufenican can lead to soil build-up and inversion ploughing to 150 mm must precede sowing any following non-cereal crop. Even where ploughing occurs some crops may be damaged

Environmental safety

- Dangerous for the environment

SEE SECTION 3 FOR PRODUCTS ALSO REGISTERED

- Toxic to aquatic organisms
- Keep livestock out of treated areas for at least 2 wk after treatment
- Harmful to bees. Do not apply to crops in flower or to those in which bees are actively foraging
- LERAP Category B

Hazard classification and safety precautions

Hazard H03, H11
Risk phrases R22a, R51, R53a, R63
Operator protection A, C, H; U05a, U08, U13, U19a, U20b
Environmental protection E06a (2 wk); E12d, E12e, E15a, E16a, E34, E38
Consumer protection C02a (2 wk)
Storage and disposal D01, D02, D05, D09a, D10b, D12a
Medical advice M03

44 bromoxynil + ethofumesate + ioxynil

A post-emergence herbicide for new grass leys
HRAC mode of action code: C3 + N + C3

Products

Leyclene	Interfarm	50:200:25 g/l	EC	07263

Uses

- Annual dicotyledons in ***grass seed crops*, *rotational grass***
- Annual meadow grass in ***grass seed crops*, *rotational grass***

Approval information

- Bromoxynil, ethofumesate and ioxynil included in Annex I under EC Directive 91/414
- Approval expiry 31 Aug 2008 [1]

Efficacy guidance

- Best results when weeds small and growing actively in a vigorous crop, soil moist and further rain within 10 d. Mid Oct to end Dec normally suitable
- Annual meadow grass controlled during early crop establishment. Spray weed grasses before fully tillered
- Ash or trash should be burned, buried or removed before spraying

Restrictions

- Maximum number of treatments 2 per yr
- Do not apply by hand-held equipment or at concentrations higher than those recommended
- Do not use on crops under stress, during periods of very dry weather, prolonged frost or waterlogging
- Do not use where clovers or other legumes are valued components of ley
- Do not roll for 7 d before or after spraying
- Do not spray in cold conditions or when heavy rain or frost imminent
- Do not use on soils with more than 10% organic matter
- Do not cut for 14 d after spraying or graze in Jan-Feb after spraying in Oct-Dec

Crop-specific information

- Latest use: 6 wk before cutting or grazing
- Apply to healthy ryegrasses or tall fescue after 2-3 leaf stage, to cocksfoot, timothy and meadow fescue at least 60 d after emergence and after 2-3 leaf stage

Following crops guidance

- Any crop may be sown 5 mth after application following mould board ploughing to at least 15 cm

Environmental safety

- Dangerous for the environment
- Toxic to aquatic organisms
- Flammable [1]
- Keep livestock out of treated areas for at least 6 wk after treatment
- Harmful to bees. Do not apply to crops in flower or to those in which bees are actively foraging

Hazard classification and safety precautions

Hazard H04, H08, H11
Risk phrases R20, R21, R22a, R38, R51, R53a, R63

Operator protection A, C; U05a, U08, U13, U19a, U20b
Environmental protection E06a (6 wk); E12d, E12e, E15a, E34, E38
Consumer protection C02a (6 wk)
Storage and disposal D01, D02, D09a, D10b, D12a
Medical advice M03

45 bromoxynil + ioxynil

A contact acting post-emergence HBN herbicide for cereals
HRAC mode of action code: C3 + C3

Products

1	Alpha Briotril	Makhteshim	240:160 g/l	EC	13580
2	Alpha Briotril Plus 19/19	Makhteshim	190:190 g/l	EC	13579
3	Mextrol Biox	Nufarm UK	200:200 g/l	EC	09470
4	Oxytril CM	Bayer CropScience	200:200 g/l	EC	10005
5	Stellox	Nufarm UK	200:200 g/l	EC	11551

Uses

- Annual dicotyledons in ***spring barley***, ***spring oats***, ***spring rye***, ***spring wheat***, ***triticale***, ***winter barley***, ***winter oats***, ***winter rye***, ***winter wheat*** [1-5]; ***undersown barley***, ***undersown oats***, ***undersown rye***, ***undersown triticale***, ***undersown wheat*** [1, 2]; ***undersown spring cereals***, ***undersown winter cereals*** [3-5]

Approval information

- Bromoxynil and ioxynil included in Annex I under EC Directive 91/414
- Accepted by BBPA for use on malting barley

Efficacy guidance

- Best results achieved on young weeds growing actively in a highly competitive crop
- Do not apply during periods of drought or when rain imminent (some labels say 'if likely within 4 or 6 h')
- Recommended for tank mixture with hormone herbicides to extend weed spectrum. See labels for details

Restrictions

- Maximum number of treatments 1 per crop
- Do not spray crops stressed by drought, waterlogging or other factors
- Do not roll or harrow for several days before or after spraying. Number of days specified varies with product. See label for details
- Do not apply by hand-held equipment or at concentrations higher than those recommended

Crop-specific information

- Latest use: before 2nd node detectable stage (GS 31)
- HI (animal consumption) 6 wk
- Apply to winter or spring cereals from 1-2 fully expanded leaf stage, but before second node detectable (GS 32)
- Spray oats in spring when danger of frost past. Do not spray winter oats in autumn
- Apply to undersown cereals pre-sowing or pre-emergence of legume provided cover crop is at correct stage. Only spray trefoil pre-sowing
- On crops undersown with grasses alone or on direct re-seeds apply from 2-leaf stage of grass

Environmental safety

- Dangerous for the environment
- Very toxic to aquatic organisms
- Flammable [1, 2]
- Keep livestock out of treated areas for at least 6 wk after treatment and until poisonous weeds, such as ragwort, have died down and become unpalatable
- Harmful to bees. Do not apply to crops in flower or to those in which bees are actively foraging. Do not apply when flowering weeds are present [3-5]

Hazard classification and safety precautions

Hazard H03, H11 [1-5]; H08 [1, 2]
Risk phrases R22a, R22b, R43, R50, R53a, R63, R67 [1-5]; R36, R37, R38 [1, 2]; R66 [3-5]

Operator protection A, C; U02a, U14, U15, U20a [1, 2]; U05a, U20b [3-5]; U08, U13, U19a, U23a [1-5]
Environmental protection E06a [1, 2] (6 wk); E07a, E12d, E12e [3-5]; E13b [1, 2, 4, 5]; E15a [3]; E34, E38 [1-5]
Consumer protection C02a [1, 2] (6 wk); C02a [3-5] (14 d)
Storage and disposal D01, D02, D12a [1-5]; D05 [1, 2]; D09a, D11a [3-5]
Medical advice M03, M05b

46 bromoxynil + ioxynil + mecoprop-P

A post-emergence contact and translocated herbicide
HRAC mode of action code: C3 + C3 + O

Products

1	Astalavista	Nufarm UK	56:56:224 g/l	EW	12335
2	Swipe P	Nufarm UK	56:56:224 g/l	EW	11955

Uses

- Annual dicotyledons in ***amenity grassland***, ***managed amenity turf*** [1]; ***canary seed*** *(off-label - for wild bird seed production)*, ***durum wheat***, ***rotational grass***, ***spring barley***, ***spring oats***, ***spring rye*** *(off-label)*, ***spring wheat***, ***triticale***, ***winter barley***, ***winter oats***, ***winter rye*** *(off-label)*, ***winter wheat*** [2]

Specific Off-Label Approvals (SOLAs)

- ***canary seed*** *(for wild bird seed production) 20070868* [2]
- ***spring rye*** *20052220* [2]
- ***winter rye*** *20052220* [2]

Approval information

- Bromoxynil, ioxynil and mecoprop-P included in Annex I under EC Directive 91/414
- Accepted by BBPA for use on malting barley

Efficacy guidance

- Best results when weeds small and growing actively in a strongly competitive crop
- Do not spray in rain or when rain imminent. Control may be reduced by rain within 6 h
- Application to wet crops or weeds may reduce control. Rainfall in the 6 hour period following spraying may reduce weed control
- To achieve optimum control of large over-wintered weeds or in advanced crops increase water volume to aid spray penetration and cover
- Recommended for tank-mixing with approved MCPA-amine for hemp nettle control

Restrictions

- Maximum number of treatments 1 per crop or per yr. The total amount of mecoprop-P applied in a single yr must not exceed the maximum total dose approved for any single product for the crop/ situation
- Do not treat durum wheat or winter oats in autumn
- Do not spray crops undersown with legumes or use on winter oats or durum wheat in autumn
- Winter and spring cereal crops should not be sprayed in the spring until the risk of frost is over
- Do not treat fine grasses such as those found on bowling or putting greens [1]
- Do not use mowings from the first cut of recently treated turf for mulching and allow compost made from subsequent cuts to rot completely before use. Susceptible crops such as brassicas should not be mulched with such compost at any time [1]
- Do not spray crops under stress from frost, waterlogging, drought or other causes
- Do not roll within 5 d after spraying
- Do not mix with manganese sulphate
- Do not apply by hand-held equipment or at concentrations higher than those recommended

Crop-specific information

- Latest use: before first node detectable (GS31) for spring oats; before second node detectable (GS32) for durum wheat, spring barley, spring wheat, triticale, winter barley, winter oats, winter wheat
- HI (animal consumption) 6 wk

- Spray cereals from 3 leaves unfolded (GS 13) to before second node detectable (GS 32) for winter sown cereals, spring wheat and spring barley and before first node detectable (GS 31) for spring oats
- Apply to direct sown ryegrass or cereals undersown with ryegrass from 2-3 leaf stage of grass
- Yield of barley may be reduced if frost occurs within 3-4 wk of treatment of low vigour crops on light soils or subject to stress
- Some cereal crop yellowing may follow treatment but yield not normally affected
- Amenity grass and turf must be well established (tillered) before treatment

Environmental safety

- Dangerous for the environment
- Very toxic to aquatic organisms
- Keep livestock out of treated areas for at least two weeks following treatment and until poisonous weeds, such as ragwort, have died down and become unpalatable
- Harmful to bees. Do not apply to crops in flower or to those in which bees are actively foraging. Do not apply when flowering weeds are present
- LERAP Category B
- Do not harvest crops for animal consumption for at least 6 wk after last application
- Avoid drift onto neighbouring crops, especially beans, beet, brassicas (including oilseed rape), carrots, lettuce, legumes, tomatoes and other horticultural crops, which are all very susceptible

Hazard classification and safety precautions

Hazard H03, H11
Risk phrases R20, R22a, R38, R41, R43, R50, R53a, R63
Operator protection A, C, H, M; U09a, U11, U13, U14, U15, U19a, U20b, U23a
Environmental protection E07a [2]; E12d, E12e, E15a, E16a, E34, E38 [1, 2]
Consumer protection C02a [2]
Storage and disposal D05, D09a, D10c, D12a
Medical advice M03

47 bromoxynil + prosulfuron

A contact and residual herbicide mixture for maize
HRAC mode of action code: C3 + B

Products

Jester	Syngenta	60:3% w/w	WG	13500

Uses

- Annual dicotyledons in ***forage maize, grain maize***
- Black bindweed in ***forage maize, grain maize***
- Chickweed in ***forage maize, grain maize***
- Hemp-nettle in ***forage maize, grain maize***
- Knotgrass in ***forage maize, grain maize***
- Mayweeds in ***forage maize, grain maize***

Approval information

- Bromoxynil and prosulfuron included in Annex I under EC Directive 91/414

Efficacy guidance

- Prosulfuron is a member of the ALS-inhibitor group of herbicides and products should be used in a planned Resistance Management strategy. See Section 5 for more information

Restrictions

- Maximum total dose equivalent to one full dose treatment
- Do not treat crops grown for seed production
- Consult before use on crops intended for processing
- Do not use in frosty weather or on crops under stress
- Do not tank mix with organophosphate insecticides or apply in tank mix or sequence with any other sulfonylurea
- Do not apply by knapsack sprayer or in volumes less than those recommended
- Take care to wash out sprayers thoroughly. See label for details

Crop-specific information

- Latest use: before 7 crop leaves unfolded for maize
- Apply post-emergence up to when the crop has six unfolded leaves
- Product must be used with Agral
- Apply in cool conditions, or during the evening, to avoid scorch

Following crops guidance

- Winter or spring cereals, winter or spring beans, spring sown peas or oilseed rape may be sown following normal harvest of a treated crop. Mould-board ploughing to 20 cm is recommended in some cases

Environmental safety

- Dangerous for the environment
- Very toxic to aquatic organisms
- LERAP Category B
- Take special care to avoid drift outside the target area

Hazard classification and safety precautions

Hazard H02, H11
Risk phrases R22a, R23, R36, R43, R50, R53a, R63
Operator protection A, C, D; U05a, U11, U15, U19a, U20c, U23a
Environmental protection E15b, E16a, E34, E38
Storage and disposal D01, D02, D05, D09a, D10c, D12a
Medical advice M04a

48 bromoxynil + terbuthylazine

A post-emergence contact and residual herbicide for maize
HRAC mode of action code: C3 + C1

Products

Templar	Makhteshim	200:300 g/l	SC	10254

Uses

- Annual dicotyledons in ***forage maize***, ***sweetcorn*** *(off-label)*

Specific Off-Label Approvals (SOLAs)

- ***sweetcorn*** *20072179*

Approval information

- Bromoxynil included in Annex I under EC Directive 91/414

Efficacy guidance

- Best results obtained from treatment when main flush of weeds has germinated and largest at 4 leaf stage
- Control can be enhanced by use of split dose treatment but second spray must be applied before crop canopy covers ground
- Soils should be moist at time of treatment and should not be disturbed afterwards
- Residual activity may be reduced on soils with high OM content, or where high amounts of slurry or manure have been applied

Restrictions

- Maximum total dose equivalent to one full dose treatment
- Foliar scorch may occur after treatment in hot conditions or when growth particularly rapid. In such conditions spray in the evening
- Do not apply during frosty weather, drought, when soil is waterlogged or when rain expected within 12 hr
- Do not treat crops under any stress whatever
- Product should be used only where damaging populations of competitive weeds have emerged. Yield may be reduced when used in absence of significant weed populations
- Succeeding crops may not be planted until the spring following application. Mould board plough to 150 mm before doing so

Crop-specific information

- Before 9 fully expanded leaf stage

Following crops guidance

- Keep livestock out of treated areas for at least 6 wk

Environmental safety

- Dangerous for the environment
- Very toxic to aquatic organisms
- Keep livestock out of treated areas for at least 6 wk after treatment
- Harmful to bees. Do not apply to crops in flower or to those in which bees are actively foraging. Do not apply when flowering weeds are present
- Do not apply by hand-held equipment or at concentrations higher than those recommended

Hazard classification and safety precautions

Hazard H03, H11
Risk phrases R20, R22a, R50, R53a
Operator protection A, C, H; U02a, U05a, U08, U13, U14, U15, U19a, U20a, U23a
Environmental protection E06a (6 wk); E12d, E12e, E13b, E38
Storage and disposal D01, D02, D12a
Medical advice M04a

49 bupirimate

A systemic aminopyrimidinol fungicide active against powdery mildew
FRAC mode of action code: 8

Products

Nimrod	Makhteshim	250 g/l	EC	13046

Uses

- Powdery mildew in ***apples***, ***begonias***, ***bilberries*** *(off-label)*, ***blackberries*** *(off-label)*, ***blackcurrants***, ***blueberries*** *(off-label)*, ***chrysanthemums***, ***courgettes*** *(outdoor only)*, ***cranberries*** *(off-label)*, ***gooseberries***, ***hops***, ***marrows*** *(outdoor only)*, ***pears***, ***protected cucumbers***, ***protected strawberries***, ***protected tomatoes*** *(off-label)*, ***pumpkins*** *(off-label)*, ***raspberries*** *(outdoor only)*, ***redcurrants*** *(off-label)*, ***roses***, ***rubus hybrids*** *(off-label)*, ***squashes*** *(off-label)*, ***strawberries***, ***vaccinium spp.*** *(off-label)*, ***whitecurrants*** *(off-label)*

Specific Off-Label Approvals (SOLAs)

- ***bilberries*** *20070995*
- ***blackberries*** *20070996*
- ***blueberries*** *20070995*
- ***cranberries*** *20070995*
- ***protected tomatoes*** *20070997*
- ***pumpkins*** *20070994*
- ***redcurrants*** *20070995*
- ***rubus hybrids*** *20070996*
- ***squashes*** *20070994*
- ***vaccinium spp.*** *20070995*
- ***whitecurrants*** *20070995*

Approval information

- Accepted by BBPA for use on hops

Efficacy guidance

- On apples during periods that favour disease development lower doses applied weekly give better results than higher rates fortnightly
- Not effective in protected crops against strains of mildew resistant to bupirimate

Restrictions

- Maximum number of treatments or maximum total dose depends on crop and dose (see label for details)

Crop-specific information

- HI: 1 d for apples, pears, strawberries; 2 d for cucurbits; 7 d for blackcurrants; 8 d for raspberries; 14 d for gooseberries, hops
- Apply before or at first signs of disease and repeat at 7-14 d intervals. Timing and maximum dose vary with crop. See label for details

- With apples, hops and ornamentals cultivars may vary in sensitivity to spray. See label for details
- If necessary to spray cucurbits in winter or early spring spray a few plants 10-14 d before spraying whole crop to test for likelihood of leaf spotting problem
- On roses some leaf puckering may occur on young soft growth in early spring or under low light intensity. Avoid use of high rates or wetter on such growth
- Never spray flowering begonias (or buds showing colour) as this can scorch petals
- Do not mix with other chemicals for application to begonias, cucumbers or gerberas

Environmental safety

- Dangerous for the environment
- Toxic to aquatic organisms
- Flammable
- Product has negligible effect on *Phytoseiulus* and *Encarsia* and may be used in conjunction with biological control of red spider mite

Hazard classification and safety precautions

Hazard H03, H08, H11
Risk phrases R22b, R38, R51, R53a
Operator protection A, C; U05a, U20b
Environmental protection E15a
Storage and disposal D01, D02, D09a, D10c
Medical advice M05b

50 buprofezin

A moulting inhibitor, thiadiazine insecticide for whitefly control
IRAC mode of action code: 16

Products

Applaud	Certis	250 g/l	SC	11532

Uses

- Glasshouse whitefly in ***aubergines***, ***protected cayenne peppers*** *(off-label)*, ***protected cucumbers***, ***protected ornamentals***, ***protected peppers***, ***protected tomatoes***
- Tobacco whitefly in ***aubergines***, ***protected cayenne peppers*** *(off-label)*, ***protected cucumbers***, ***protected ornamentals***, ***protected peppers***, ***protected tomatoes***

Specific Off-Label Approvals (SOLAs)

- ***protected cayenne peppers*** *20061563*

Efficacy guidance

- Product has contact, residual and some vapour activity
- Whitefly most susceptible at larval stages but residual effect can also kill nymphs emerging from treated eggs and application to pupae reduces emergence
- Adult whitefly not directly affected. Resistant strains of tobacco whitefly are known and where present control likely to be reduced or ineffective

Restrictions

- Maximum number of treatments 8 per crop for tomatoes and cucumbers; 4 per crop on protected ornamentals; 2 per crop for aubergines and peppers
- Do not apply more than 2 sprays within a 65 d period on tomatoes, or within a 45 d period on cucumbers
- Do not treat *Dieffenbachia* or *Closmoplictrum*
- Do not apply to crops under stress
- Do not leave spray liquid in sprayer for long periods
- Do not apply as fog or mist

Crop-specific information

- HI for all crops 3 d
- In IPM programme apply as single application and allow at least 60 d before re-applying
- In All Chemical programme apply twice at 7-14 d interval and allow at least 60 d before re-applying
- See label for list of ornamentals successfully treated but small scale test advised to check varietal tolerance. This is especially important if spraying flowering ornamentals with buds showing colour

Environmental safety

- Product may be used either in IPM programme in association with *Encarsia formosa* or in All Chemical programme

Hazard classification and safety precautions

Hazard H04
Risk phrases R36, R53a
Operator protection U05a, U14, U15, U20c
Environmental protection E15a, E34
Storage and disposal D01, D02, D09a, D11a

51 captan

A protectant phthalimide fungicide with horticultural uses
FRAC mode of action code: M4

Products

1	Alpha Captan 80 WDG	Makhteshim	80% w/w	WG	07096
2	Alpha Captan 83 WP	Makhteshim	83% w/w	WP	04806

Uses

- Black spot in ***roses***
- Botrytis in ***strawberries***
- Gloeosporium rot in ***apples***
- Scab in ***apples, pears***

Approval information

- Captan included in Annex I under EC Directive 91/414

Restrictions

- Maximum number of treatments 12 per yr on apples and pears as pre-harvest sprays
- Product must not be used as dip or drench on apples or pears [2]
- Do not use on apple cultivars Bramley, Monarch, Winston, King Edward, Spartan, Kidd's Orange or Red Delicious or on pear cultivar D'Anjou
- Do not mix with alkaline materials or oils
- Do not use on fruit for processing
- Powered visor respirator with hood and neck cape must be used when handling concentrate

Crop-specific information

- HI apples, pears 14 d; strawberries 7 d
- For control of scab apply at bud burst and repeat at 10-14 d intervals until danger of scab infection ceased
- For suppression of fruit storage rots apply from late Jul and repeat at 2-3 wk intervals
- For black spot control in roses apply after pruning with 3 further applications at 14 d intervals or spray when spots appear and repeat at 7-10 d intervals
- For grey mould in strawberries spray at first open flower and repeat every 7-10 d [1]
- Do not leave diluted material for more than 2 h. Agitate well before and during spraying

Environmental safety

- Dangerous for the environment
- Very toxic to aquatic organisms

Hazard classification and safety precautions

Hazard H02 [2]; H03 [1]; H11 [1, 2]
Risk phrases R23, R37, R38 [2]; R36, R40, R43, R50, R53a [1, 2]; R41 [1]
Operator protection A [1, 2]; D, E, H [1]; U05a, U11, U19a [1, 2]; U09a, U20c [1]; U14, U15, U20b [2]
Environmental protection E15a
Storage and disposal D01, D02, D09a, D11a, D12a

SEE SECTION 3 FOR PRODUCTS ALSO REGISTERED

52 carbendazim

A systemic benzimidazole fungicide with curative and protectant activity
FRAC mode of action code: 1

Products

1	AgriGuard Pro-Turf	AgriGuard	500 g/l	SC	11837
2	Caste Off	Nufarm UK	500 g/l	SC	12832
3	Delsene 50 Flo	Nufarm UK	500 g/l	SC	11452
4	Mascot Systemic	Rigby Taylor	500 g/l	SC	11335

Uses

- Big vein in ***lettuce*** *(off-label)*, ***protected lettuce*** *(off-label)* [3]
- Black root rot in ***ornamental plant production*** *(off-label)*, ***protected ornamentals*** *(off-label)* [3]
- Botrytis in ***bedding plants*** *(off-label)*, ***bulbs/corms*** *(off-label)*, ***chrysanthemums*** *(off-label)*, ***pot plants*** *(off-label)* [3]
- Cane blight in ***blackberries*** *(off-label)*, ***loganberries*** *(off-label)*, ***protected blackberries*** *(off-label)*, ***protected loganberries*** *(off-label)*, ***protected raspberries*** *(off-label)*, ***protected rubus hybrids*** *(off-label)*, ***raspberries*** *(off-label)*, ***rubus hybrids*** *(off-label)* [3]
- Didymella stem rot in ***protected tomatoes*** *(off-label)* [3]
- Dollar spot in ***managed amenity turf*** [1, 2, 4]
- Eyespot in ***spring barley***, ***winter barley***, ***winter wheat*** [3]
- Foliar disease control in ***durum wheat*** *(off-label)*, ***grass seed crops*** *(off-label)*, ***honesty*** *(off-label)*, ***mustard*** *(off-label)*, ***spring linseed*** *(off-label)*, ***spring rye*** *(off-label)*, ***triticale*** *(off-label)*, ***winter linseed*** *(off-label)*, ***winter rye*** *(off-label)* [3]
- Fusarium in ***bulbs/corms*** *(off-label)*, ***freesias*** *(off-label)* [3]
- Fusarium patch in ***managed amenity turf*** [1, 2, 4]
- Light leaf spot in ***spring oilseed rape***, ***winter oilseed rape*** [3]
- Penicillium rot in ***bulbs/corms*** *(off-label)* [3]
- Powdery mildew in ***bedding plants*** *(off-label)*, ***chrysanthemums*** *(off-label)*, ***pot plants*** *(off-label)* [3]
- Rhynchosporium in ***spring barley***, ***winter barley*** [3]
- Sclerotinia in ***bulbs/corms*** *(off-label)* [3]
- Stagonospora in ***bulbs/corms*** *(off-label)* [3]
- Trichoderma in ***mushroom compost*** *(off-label - spawn treatment)* [3]
- Wormcast formation in ***managed amenity turf*** [1, 2, 4]

Specific Off-Label Approvals (SOLAs)

- ***bedding plants*** *20041004* [3]
- ***blackberries*** *20041009 expires Jul 2008* [3]
- ***bulbs/corms*** *20041004* [3]
- ***chrysanthemums*** *20041004* [3]
- ***durum wheat*** *20063557* [3]
- ***freesias*** *20041004* [3]
- ***grass seed crops*** *20063557* [3]
- ***honesty*** *20063556* [3]
- ***lettuce*** *20041011* [3]
- ***loganberries*** *20041009 expires Jul 2008* [3]
- ***mushroom compost*** *(spawn treatment) 20041007* [3]
- ***mustard*** *20063556* [3]
- ***ornamental plant production*** *20041004* [3]
- ***pot plants*** *20041004* [3]
- ***protected blackberries*** *20041009 expires Jul 2008* [3]
- ***protected lettuce*** *20041011* [3]
- ***protected loganberries*** *20041009 expires Jul 2008* [3]
- ***protected ornamentals*** *20041004* [3]
- ***protected raspberries*** *20041009 expires Jul 2008* [3]
- ***protected rubus hybrids*** *20041009 expires Jul 2008* [3]
- ***protected tomatoes*** *20041010* [3]
- ***raspberries*** *20041009 expires Jul 2008* [3]
- ***rubus hybrids*** *20041009 expires Jul 2008* [3]

- ***spring linseed*** *20063556* [3]
- ***spring rye*** *20063557* [3]
- ***triticale*** *20063557* [3]
- ***winter linseed*** *20063556* [3]
- ***winter rye*** *20063557* [3]

Approval information

- Following implementation of Directive 98/82/EC, approval for use of carbendazim on numerous crops was revoked in 1999. UK approvals for use of carbendazim on strawberries revoked in 2001 as a result of implementation of the MRL Directives. Some further MRL reductions were imposed in 2006 resulting in the immediate revocation of uses on apples, pears, celery, cucmbers and rhubarb
- Carbendazim has been included in Annex I under Directive 91/414 for 3 years from January 2007 and all approvals for products containing this active ingredient will expire in December 2009
- Accepted by BBPA for use on malting barley
- Approval expiry 31 Aug 2008 [4]

Efficacy guidance

- Products vary in the diseases listed as controlled for several crops. Labels must be consulted for full details and for rates and timings
- Mostly applied as spray or drench. Spray treatments normally applied at first sign of disease and repeated after 1 mth if required
- Apply as drench rather than spray where red spider mite predators are being used
- Do not apply during drought conditions. Rain or irrigation after treatment may improve control
- Leave 4 mth intervals where used only for worm cast control [10692, 12949, 1, 2, 4]
- To delay appearance of resistant strains alternate treatment with non-MBC fungicide. Eyespot in cereals and *Botrytis cinerea* in many crops are now widely resistant

Restrictions

- Maximum number of treatments (including applications of any product containing benomyl, carbendazim or thiophanate-methyl) varies with crop treated and product used - see labels for details
- Do not treat crops or turf suffering from drought or other physical or chemical stress
- Consult processors before using on crops for processing
- Not compatible with alkaline products such as lime sulphur

Crop-specific information

- Latest use: before 2nd node detectable (GS32) for cereals
- HI 2 d for inert substrate cucumbers, protected tomatoes; 7-14 d (depends on dose) for apples, pears; 14 d for mushrooms, protected celery; 21 d for oilseed rape, dwarf beans, lettuce, 6 mth for cane fruit
- On turf apply as preventative treatment in spring or autumn during periods of high disease risk [10692, 12949, 1, 2, 4]
- Apply as a drench to control soil-borne diseases in cucumbers and tomatoes and as a pre-planting dip treatment for bulbs

Environmental safety

- Dangerous for the environment
- Very toxic to aquatic organisms
- After use dipping suspension must not be discharged directly into ditches or drains

Hazard classification and safety precautions

Hazard H03, H11
Risk phrases R22a [1]; R46, R60, R61 [3]; R50, R53a [2-4]; R68 [2, 4]
Operator protection A, C, H, M; U19a, U20c [2, 4]; U20a [1]; U20b [3]
Environmental protection E13c [3]; E15a [1, 2, 4]; E38 [2, 4]
Storage and disposal D01, D02, D11a [3]; D03, D12a [2, 4]; D05, D09a [1-4]; D07 [1]; D10b [1, 2, 4]

SEE SECTION 3 FOR PRODUCTS ALSO REGISTERED

53 carbendazim + epoxiconazole

A systemic fungicide mixture for managed amenity turf
FRAC mode of action code: 1 + 3

Products

Capricorn	Bayer Environ.	125:125 g/l	SC	10035

Uses
- Fusarium patch in ***managed amenity turf***

Approval information
- Carbendazim has been included in Annex I under Directive 91/414 for 3 years from January 2007 and all approvals for products containing this active ingredient will expire in December 2009
- Approval expiry 30 Jun 2008 [1]

Efficacy guidance
- Treat at the first sign of disease and repeat as necessary
- Apply after cutting and delay further mowing for at least 48 h to allow systemic distribution within the plant
- Only moderate control of Fusarium patch claimed

Restrictions
- Maximum number of treatments 2 per yr
- Do not apply during drought conditions or to frozen turf
- Use only on established turf

Environmental safety
- Harmful to fish or other aquatic life. Do not contaminate surface waters or ditches with chemical or used container
- LERAP Category B

Hazard classification and safety precautions

Hazard H04
Risk phrases R43
Operator protection A, C, H, M; U04a, U05a, U09b, U20c
Environmental protection E13c, E16a, E16b, E34
Storage and disposal D01, D02, D05, D09a, D10b

54 carbendazim + flusilazole

A broad-spectrum systemic and protectant fungicide for cereals, oilseed rape and sugar beet
FRAC mode of action code: 1 + 3

Products

1 Contrast	DuPont	125:250 g/l	SC	13107
2 Harvesan	DuPont	125:250 g/l	SC	13106
3 Punch C	DuPont	125:250 g/l	SC	13023

Uses
- Brown rust in ***spring barley***, ***spring wheat***, ***winter barley***, ***winter wheat*** [1-3]
- Canker in ***spring oilseed rape*** *(reduction)*, ***winter oilseed rape*** *(reduction)* [1-3]
- Eyespot in ***spring barley***, ***spring wheat***, ***winter barley***, ***winter wheat*** [1-3]
- Glume blotch in ***spring wheat***, ***winter wheat*** [1-3]
- Light leaf spot in ***spring oilseed rape***, ***winter oilseed rape*** [1-3]
- Phoma leaf spot in ***spring oilseed rape***, ***winter oilseed rape*** [1-3]
- Powdery mildew in ***fodder beet*** *(off-label)* [3]; ***spring barley***, ***spring wheat***, ***sugar beet***, ***winter barley***, ***winter wheat*** [1-3]
- Rust in ***sugar beet*** [1-3]
- Septoria leaf blotch in ***spring wheat***, ***winter wheat*** [1-3]
- Yellow rust in ***spring barley***, ***spring wheat***, ***winter barley***, ***winter wheat*** [1-3]

Specific Off-Label Approvals (SOLAs)
- ***fodder beet*** *20063163* [3]

Approval information

- Carbendazim has been included in Annex I under Directive 91/414 for 3 years from January 2007 and all approvals for products containing this active ingredient will expire in December 2009
- Flusilazole has been included in Annex I under Directive 91/414 for 18 months from January 2007 and all approvals for products containing this active ingredient will expire in June 2008
- Accepted by BBPA for use on malting barley

Efficacy guidance

- Apply at early stage of disease development or in routine preventive programme
- Most effective timing of treatment on cereals varies with disease. See label for details
- Higher rate active against both MBC-sensitive and MBC-resistant eyespot
- Rain occurring within 2 h of spraying may reduce effectiveness
- To prevent build-up of resistant strains of cereal mildew tank mix with approved morpholine fungicide
- Flusilazole is a DMI fungicide. Resistance to some DMI fungicides has been identified in Septoria leaf blotch which may seriously affect performance of some products. For further advice contact a specialist advisor and visit the Fungicide Resistance Action Group (FRAG)-UK website

Restrictions

- Maximum number of treatments 1 per crop on sugar beet; 2 per crop on cereals and oilseed rape
- Do not apply to crops under stress or during frosty weather

Crop-specific information

- Latest use on cereals varies with crop and dose used - see label for details; before first flower opened stage for oilseed rape
- HI 7 wk for sugar beet
- Treat oilseed rape in autumn when leaf lesions first appear and spring from the start of stem extension when disease appears

Environmental safety

- Dangerous for the environment
- Toxic to aquatic organisms
- Dangerous to fish or other aquatic life. Do not contaminate surface waters or ditches with chemical or used container

Hazard classification and safety precautions

Hazard H02, H11
Risk phrases R22a, R40, R41, R43, R46, R51, R53a, R60, R61
Operator protection A, C, H, M; U05a, U11, U19a, U20b
Environmental protection E13b, E34, E38
Storage and disposal D01, D02, D05, D09a, D10b, D12a
Medical advice M04a

55 carbetamide

A residual pre- and post-emergence carbamate herbicide for a range of field crops
HRAC mode of action code: K2

Products

1	Carbetamex	Makhteshim	70% w/w	WP	13045
2	Crawler	Makhteshim	60% w/w	WG	11840

Uses

- Annual dicotyledons in ***cabbage seed crops***, ***collards***, ***fodder rape seed crops***, ***kale seed crops***, ***lucerne***, ***red clover***, ***sainfoin***, ***spring cabbage***, ***sugar beet seed crops***, ***swede seed crops***, ***turnip seed crops***, ***white clover***, ***winter field beans***, ***winter oilseed rape*** [1, 2]; ***evening primrose*** *(off-label)*, ***honesty*** *(off-label)*, ***lupins*** *(off-label)*, ***mustard*** *(off-label)*, ***spring linseed*** *(off-label)* [2]
- Annual grasses in ***cabbage seed crops***, ***collards***, ***fodder rape seed crops***, ***kale seed crops***, ***lucerne***, ***red clover***, ***sainfoin***, ***spring cabbage***, ***sugar beet seed crops***, ***swede seed crops***, ***turnip seed crops***, ***white clover***, ***winter field beans***, ***winter oilseed rape*** [1, 2]; ***evening primrose*** *(off-label)*, ***honesty*** *(off-label)*, ***lupins*** *(off-label)*, ***mustard*** *(off-label)*, ***spring linseed*** *(off-label)* [2]

- Volunteer cereals in ***cabbage seed crops***, ***collards***, ***fodder rape seed crops***, ***kale seed crops***, ***lucerne***, ***red clover***, ***sainfoin***, ***spring cabbage***, ***sugar beet seed crops***, ***swede seed crops***, ***turnip seed crops***, ***white clover***, ***winter field beans***, ***winter oilseed rape*** [1, 2]

Specific Off-Label Approvals (SOLAs)

- ***evening primrose*** *20061236* [2]
- ***honesty*** *20061236* [2]
- ***lupins*** *20061237* [2]
- ***mustard*** *20061236* [2]
- ***spring linseed*** *20061236* [2]

Efficacy guidance

- Best results pre- or early post-emergence of weeds under cool, moist conditions. Adequate soil moisture is essential
- Dicotyledons controlled include chickweed, cleavers and speedwell
- Weed growth stops rapidly after treatment but full effects may take 6-8 wk to develop
- Various tank mixes are recommended to broaden the weed spectrum. See label for details
- Always follow WRAG guidelines for preventing and managing herbicide resistant weeds. Section 5 for more information

Restrictions

- Maximum number of treatments 1 per crop for all crops
- Do not treat any crop on waterlogged soil
- Do not use on soils with more than 10% organic matter as residual activity is impaired
- Do not apply during prolonged periods of cold weather when weeds are dormant

Crop-specific information

- HI 6 wk for all crops
- Apply to brassicas from mid-Oct to end-Feb provided crop has at least 4 true leaves (spring cabbage, spring greens), 3-4 true leaves (seed crops, oilseed rape)
- Apply to established lucerne or sainfoin from Nov to end-Feb
- Apply to established red or white clover from Feb to mid-Mar

Following crops guidance

- Succeeding crops may be sown 2 wk after treatment for brassicas, field beans, 8 wk after treatment for peas, runner beans, 16 wk after treatment for cereals, maize
- Ploughing is not necessary before sowing subsequent crops

Environmental safety

- Do not graze crops for at least 6 wk after treatment

Hazard classification and safety precautions

Operator protection U19a [1]; U20c [1, 2]
Environmental protection E15a
Storage and disposal D09a, D11a

56 carbon dioxide (commodity substance)

A gas for the control of trapped rodents and other vertebrates

Products

carbon dioxide	various	99.9%	GA

Uses

- Birds in ***traps***
- Mice in ***traps***
- Rats in ***traps***

Approval information

- Approval for the use of carbon dioxide as a commodity substance was granted on 8 October 1993 by Ministers under regulation 5 of the Control of Pesticides Regulations 1986
- Only to be used where a licence has been issued in accordance with Section 16(1) of the Wildlife and Countryside Act 1981
- Carbon dioxide is being supported for review in the fourth stage of the EC Review Programme under Directive 91/414. Whether or not it is included in Annex I, the existing commodity chemical approval will have to be revoked because substances listed in Annex I must be approved for

marketing and use under the EC regime. If carbon dioxide is included in Annex I, products containing it will need to gain approval in the normal way if they carry label claims for pesticidal activity

Efficacy guidance

- Use to destroy trapped rodent pests
- Use to control birds covered by general licences issued by the Agriculture and Environment Departments under Section 16(1) of the Wildlife and Countryside Act (1981) for the control of opportunistic bird species, where birds have been trapped or stupefied with alphachloralose/ seconal

Restrictions

- Operators must wear self-contained breathing apparatus when carbon dioxide levels are greater than 0.5% v/v
- Operators must be suitably trained and competent

Environmental safety

- Unprotected persons and non-target animals must be excluded from the treatment enclosures and surrounding areas unless the carbon dioxide levels are below 0.5% v/v

Hazard classification and safety precautions

Operator protection G

57 carbosulfan

A systemic carbamate insecticide for control of soil and stem pests

IRAC mode of action code: 1A

Products

1	Marshal Soil Insecticide suSCon CR granules	Fargro	10% w/w	GR	06978
2	Posse 10G	Belchim	10% w/w	GR	11640

Uses

- Aphids in ***broccoli***, ***brussels sprouts***, ***cabbages***, ***calabrese***, ***carrots***, ***cauliflowers***, ***chinese cabbage*** *(off-label)*, ***choi sum*** *(off-label)*, ***collards***, ***fodder beet***, ***kale***, ***mangels***, ***pak choi*** *(off-label)*, ***parsnips***, ***sugar beet*** [2]
- Cabbage root fly in ***broccoli***, ***brussels sprouts***, ***cabbages***, ***calabrese***, ***cauliflowers***, ***collards***, ***kale*** [2]
- Cabbage stem weevil in ***broccoli***, ***brussels sprouts***, ***cabbages***, ***calabrese***, ***cauliflowers***, ***collards***, ***kale*** [2]
- Flea beetle in ***broccoli***, ***brussels sprouts***, ***cabbages***, ***calabrese***, ***cauliflowers***, ***collards***, ***fodder beet***, ***kale***, ***mangels***, ***sugar beet*** [2]
- Free-living nematodes in ***carrots***, ***fodder beet***, ***mangels***, ***parsnips***, ***sugar beet*** [2]
- Large pine weevil in ***forest***, ***forest nurseries*** *(off-label)* [1]
- Mangold fly in ***fodder beet***, ***mangels***, ***sugar beet*** [2]
- Millipedes in ***fodder beet***, ***mangels***, ***sugar beet*** [2]
- Pygmy mangold beetle in ***fodder beet***, ***mangels***, ***sugar beet*** [2]
- Springtails in ***fodder beet***, ***mangels***, ***sugar beet*** [2]
- Symphylids in ***fodder beet***, ***mangels***, ***sugar beet*** [2]
- Wireworm in ***fodder beet***, ***mangels***, ***sugar beet*** [2]

Specific Off-Label Approvals (SOLAs)

- ***chinese cabbage*** *20061299* [2]
- ***choi sum*** *20061299* [2]
- ***forest nurseries*** *992383* [1]
- ***pak choi*** *20061299* [2]

Approval information

- In 2006 PSD required that all products containing this active ingredient should carry the following warning in the main area of the container label: "Carbosulfan is an anticholinesterase carbamate. Handle with care"

- Carbosulfan will not be included in Annex I of EC Directive 91/414. To comply with this EC decision all approvals for the storage and use of all products listed in this profile will expire on 13 December 2008
- Approval expiry 13 Dec 2008 [1, 2]

Efficacy guidance

- At recommended rates seed of drilled arable crops is not damaged by contact with product [1]
- Products formulated as controlled release granules to give pine weevil control throughout the 2-year establishment phase [1]

Restrictions

- Contains an anticholinesterase carbamate compound. Do not use if under medical advice not to work with such compounds
- Maximum number of treatments 1 per crop or tree

Crop-specific information

- Latest use: before or at planting for forestry uses [1]
- HI 100 d for beet crops, carrots, parsnips; 12 wk for broccoli, cabbages, calabrese, cauliflowers; 14 wk for Brussels sprouts; 23 wk for collards [2]
- Apply to drilled crops with suitable granule applicator feeding directly into seed furrow or immediately behind drill coulter (behind seed drill boot for brassicas) or use bow-wave technique [2]
- See label for details of suitable applicators and settings. Correct calibration is essential [2]
- Where used in forestry, granules should be placed in the planting hole by hand or metered applicator before or after placing the tree [1]
- Safe to Douglas Fir and Sitka Spruce. Other conifers may vary in their sensitivity [1]
- Forest trees may take 10-15 d to achieve full protection. An additional pre-planting insecticide dip or spray is advised, but post-planting sprays during the 2-yr establishment should not be necessary [1]

Environmental safety

- Dangerous for the environment
- Very toxic to aquatic organisms
- Dangerous to game, wild birds and animals
- Do not empty into drains [2]
- Keep in original container, tightly closed, in a safe place, under lock and key

Hazard classification and safety precautions

Hazard H03, H11
Risk phrases R20, R51 [1]; R22a, R50 [2]; R43, R53a [1, 2]
Operator protection A, C, D, E [1]; B [2]; C [2] (or D+E); H, K, M; U02a, U04a, U05a, U13, U14, U20a [1, 2]; U09a, U19a [2]; U10, U12 [1]
Environmental protection E10a, E13b [1, 2]; E19b, E34 [2]; E38 [1]
Storage and disposal D01, D02 [1, 2]; D09a, D12a [1]; D09b, D11a [2]
Treated seed S04a [1]
Medical advice M02 [1, 2]; M03, M05a [2]

58 carboxin

A carboxamide fungicide available only in mixtures
FRAC mode of action code: 7

59 carboxin + thiram

A fungicide seed dressing for cereals
FRAC mode of action code: 7 + M3

Products

Anchor	Chemtura	200:200 g/l	FS	08684

Uses

- Bunt in ***winter wheat*** *(seed treatment)*

- Covered smut in ***spring barley*** *(seed treatment)*, ***spring oats*** *(seed treatment)*, ***winter barley*** *(seed treatment)*, ***winter oats*** *(seed treatment)*
- Fusarium foot rot and seedling blight in ***spring barley*** *(seed treatment)*, ***spring oats*** *(seed treatment)*, ***spring rye*** *(seed treatment)*, ***spring wheat*** *(seed treatment)*, ***triticale*** *(seed treatment)*, ***winter barley*** *(seed treatment)*, ***winter oats*** *(seed treatment)*, ***winter rye*** *(seed treatment)*, ***winter wheat*** *(seed treatment)*
- Leaf stripe in ***spring barley*** *(seed treatment)*, ***winter barley*** *(seed treatment - reduction)*
- Loose smut in ***spring barley*** *(seed treatment - reduction)*, ***spring oats*** *(seed treatment - reduction)*, ***winter barley*** *(seed treatment - reduction)*, ***winter oats*** *(seed treatment - reduction)*
- Net blotch in ***spring barley*** *(seed treatment)*
- Septoria seedling blight in ***spring wheat*** *(seed treatment)*, ***winter wheat*** *(seed treatment)*
- Stinking smut in ***spring wheat*** *(seed treatment)*

Approval information
- Thiram included in Annex I under EC Directive 91/414
- Accepted by BBPA for use on malting barley

Efficacy guidance
- Apply through suitable liquid flowable seed treating equipment of the batch treatment or continuous flow type where a secondary mixing auger is fitted
- Drill flow may be affected by treatment. Always re-calibrate seed drill before use

Restrictions
- Maximum number of treatments 1 per batch of seed
- Do not treat seed with moisture content above 16%
- Do not apply to cracked, split or sprouted seed
- Do not store treated seed from one season to the next

Crop-specific information
- Latest use: pre-drilling

Environmental safety
- Dangerous for the environment
- Very toxic to aquatic organisms
- Do not use treated seed as food or feed
- Treated seed harmful to game and wildlife

Hazard classification and safety precautions

Hazard H03, H11
Risk phrases R22a, R48, R50, R53a
Operator protection A, D, H; U05a, U09a, U20c
Environmental protection E15a, E34, E38
Storage and disposal D01, D02, D05, D06a, D09a, D10a
Treated seed S01, S02, S04b, S05, S06a, S07
Medical advice M03

60 carfentrazone-ethyl

A triazolinone contact herbicide
HRAC mode of action code: E

Products

1	Aurora 50 WG	Belchim	50% w/w	WG	11613
2	Carfen	AgriGuard	50% w/w	WG	13005
3	Shark	Belchim	60 g/l	ME	12762
4	Spotlight Plus	Belchim	60 g/l	ME	12436

Uses
- Annual dicotyledons in ***all edible crops (outdoor)*** *(before planting)*, ***all non-edible crops (outdoor)*** *(before planting)*, ***potatoes*** [3]
- Black bindweed in ***potatoes*** [3]
- Cleavers in ***durum wheat***, ***spring barley***, ***spring oats***, ***spring wheat***, ***triticale***, ***winter barley***, ***winter oats***, ***winter wheat*** [1, 2]; ***potatoes*** [3]
- Fat hen in ***potatoes*** [3]
- Haulm destruction in ***seed potatoes***, ***ware potatoes*** [4]

SEE SECTION 3 FOR PRODUCTS ALSO REGISTERED

- Ivy-leaved speedwell in ***durum wheat***, ***spring barley***, ***spring oats***, ***spring wheat***, ***triticale***, ***winter barley***, ***winter oats***, ***winter wheat*** [1, 2]; ***potatoes*** [3]
- Knotgrass in ***potatoes*** [3]
- Redshank in ***potatoes*** [3]
- Volunteer oilseed rape in ***all edible crops (outdoor)*** *(before planting)*, ***all non-edible crops (outdoor)*** *(before planting)*, ***potatoes*** [3]

Approval information

- Carfentrazone-ethyl included in Annex I under EC Directive 91/414
- Accepted by BBPA for use on malting barley and hops

Efficacy guidance

- Best weed control results achieved from good spray cover applied to small actively growing weeds [1-3]
- Carfentrazone-ethyl acts by contact only; see label for optimum timing on specified weeds. Weeds emerging after application will not be controlled [1-3]
- For weed control in cereals use as two spray programme with one application in autumn and one in spring [1, 2]
- Efficacy of haulm destruction will be reduced where flailed haulm covers the stems at application [4]
- For potato crops with very dense vigorous haulm or where regrowth occurs following a single application a second application may be necessary to achieve satisfactory desiccation. A minimum interval between applications of 7 d should be observed to achieve optimum performance [4]

Restrictions

- Maximum number of treatments 2 per crop for cereals (1 in Autumn and 1 in Spring) [1, 2]; 1 per crop for weed control in potatoes [3]; 1 per yr for pre-planting treatments [3]
- Maximum total dose for potato haulm destruction equivalent to 1.6 full dose treatments [4]
- Do not treat cereal crops under stress from drought, waterlogging, cold, pests, diseases, nutrient or lime deficiency or any factors reducing plant growth [1, 2]
- Do not treat cereals undersown with clover or other legumes [1, 2]
- Allow at least 2-3 wk between application and lifting potatoes to allow skins to set if potatoes are to be stored [4]
- Follow label instructions for sprayer cleaning
- Do not apply through knapsack sprayers
- Contact processor before using a split dose on potatoes for processing

Crop-specific information

- Latest use: 1 mth before planting edible and non-edible crops; before 3rd node detectable (GS 33) on cereals [1, 2]
- HI 7 d for potatoes [4]
- For weed control in cereals treat from 2 leaf stage [1, 2]
- When used as a treatment prior to planting a subsequent crop apply before weeds exceed maximum sizes indicated in the label [3]
- If treated potato tubers are to be stored then allow at least 2-3 wk between the final application and lifting to allow skins to set [4]

Following crops guidance

- No restrictions apply on the planting of succeeding crops 1 mth after application to potatoes for haulm destruction or as a pre-planting treatment, or 3 mth after application to cereals for weed control
- In the event of failure of a treated cereal crop, all cereals, ryegrass, maize, oilseed rape, peas, sunflowers, *Phacelia*, vetches, carrots or onions may be planted within 1 mth of treatment

Environmental safety

- Dangerous for the environment
- Very toxic to aquatic organisms
- Some non-target crops are sensitive. Avoid drift onto broad-leaved plants outside the treated area, or onto ponds waterways or ditches

Hazard classification and safety precautions

Hazard H04, H11

Risk phrases R38 [4]; R43, R50, R53a [1-4]

Operator protection A, H; U05a, U14 [1-4]; U08, U13 [1, 2]; U20a [3, 4]

Environmental protection E15a [1, 3]; E15b [2, 4]; E34 [1, 2]; E38 [1-4]
Storage and disposal D01, D02, D09a, D12a [1-4]; D10b [1, 2]; D10c [3, 4]
Medical advice M05a [1]

61 carfentrazone-ethyl + flupyrsulfuron-methyl

A foliar and residual acting herbicide for cereals
HRAC mode of action code: E + B

Products

Lexus Class	DuPont	33.3:16.7% w/w	WG	10809

Uses

- Annual dicotyledons in ***triticale***, ***winter oats***, ***winter rye***, ***winter wheat***
- Blackgrass in ***triticale***, ***winter oats***, ***winter rye***, ***winter wheat***

Approval information

- Carfentrazone-ethyl and flupyrsulfuron-methyl included in Annex I under EC Directive 91/414

Efficacy guidance

- Best results obtained when applied to small actively growing weeds
- Good spray cover of weeds must be obtained
- Increased degradation of active ingredient in high soil temperatures reduces residual activity
- Weed control may be reduced in dry soil conditions but susceptible weeds germinating soon after treatment will be controlled if adequate soil moisture present
- Symptoms may not be apparent on some weeds for up to 4 wk, depending on weather conditions
- Blackgrass should be treated from 1 leaf but before first node
- Flupyrsulfuron-methyl is a member of the ALS-inhibitor group of herbicides. To avoid the build up of resistance do not use any product containing an ALS-inhibitor herbicide with claims for control of grass weeds more than once on any crop
- Use these products as part of a resistance management strategy that includes cultural methods of control and does not use ALS inhibitors as the sole chemical method of grass weed control

Restrictions

- Maximum number of treatments 1 per crop
- Do not use on barley, on crops undersown with grasses or legumes, or any other broad-leaved crop
- Do not apply within 7 d of rolling
- Do not treat any crop suffering from drought, waterlogging, pest or disease attack, nutrient deficiency, or any other stress factors
- Specific restrictions apply to use in sequence or tank mixture with other sulfonylurea or ALS-inhibiting herbicides. See label for details

Crop-specific information

- Latest use: before 31 Dec in yr of sowing for winter oats, winter rye, triticale; before 1st node detectable (GS 31) for winter wheat
- Slight chlorosis and stunting may occur in certain conditions. Recovery is rapid and yield not affected

Following crops guidance

- Only cereals, oilseed rape, field beans, clover or grass may be sown in the yr of harvest of a treated crop
- In the event of crop failure only winter or spring wheat may be sown 1-3 mth after treatment. Land should be ploughed and cultivated to 15 cm minimum before resowing

Environmental safety

- Dangerous for the environment
- Very toxic to aquatic organisms
- Take extreme care to avoid damage by drift outside the target area or onto surface waters or ditches
- Spraying equipment should not be drained or flushed onto land planted, or to be planted, with trees or crops other than cereals and should be thoroughly cleansed after use - see label for instructions

SEE SECTION 3 FOR PRODUCTS ALSO REGISTERED

Hazard classification and safety precautions

Hazard H04, H11
Risk phrases R43, R50, R53a
Operator protection A, H; U05a, U08, U14, U19a, U20b
Environmental protection E15b, E38
Storage and disposal D01, D02, D09a, D10b, D11a, D12a

62 carfentrazone-ethyl + mecoprop-P

A foliar applied herbicide for cereals
HRAC mode of action code: E + O

Products

Platform S	Belchim	1.5:60% w/w	WG	11611

Uses

- Charlock in ***spring barley***, ***spring oats***, ***spring wheat***, ***winter barley***, ***winter oats***, ***winter wheat***
- Chickweed in ***spring barley***, ***spring oats***, ***spring wheat***, ***winter barley***, ***winter oats***, ***winter wheat***
- Cleavers in ***spring barley***, ***spring oats***, ***spring wheat***, ***winter barley***, ***winter oats***, ***winter wheat***
- Field speedwell in ***spring barley***, ***spring oats***, ***spring wheat***, ***winter barley***, ***winter oats***, ***winter wheat***
- Ivy-leaved speedwell in ***spring barley***, ***spring oats***, ***spring wheat***, ***winter barley***, ***winter oats***, ***winter wheat***
- Red dead-nettle in ***spring barley***, ***spring oats***, ***spring wheat***, ***winter barley***, ***winter oats***, ***winter wheat***

Approval information

- Carfentrazone-ethyl and mecoprop-P included in Annex I under EC Directive 91/414
- Accepted by BBPA for use on malting barley

Efficacy guidance

- Best results obtained when weeds have germinated and growing vigorously in warm moist conditions
- Treatment of large weeds and poor spray coverage may result in reduced weed control

Restrictions

- Maximum number of treatments 2 per crop. The total amount of mecoprop-P applied in a single yr must not exceed the maximum total dose approved for any single product for the crop/situation
- Do not treat crops suffering from stress from any cause
- Do not treat crops undersown or to be undersown

Crop-specific information

- Latest use: before 3rd node detectable (GS 33)
- Can be used on all varieties of wheat and barley in autumn or spring from the beginning of tillering
- Early sown crops may be prone to damage if treated after period of rapid growth in autumn

Following crops guidance

- In the event of crop failure, any cereal, maize, oilseed rape, peas, vetches or sunflowers may be sown 1 mth after a spring treatment. Any crop may be planted 3 mth after treatment

Environmental safety

- Dangerous for the environment
- Very toxic to aquatic organisms
- Keep livestock out of treated areas for at least two weeks following treatment and until poisonous weeds, such as ragwort, have died down and become unpalatable

Hazard classification and safety precautions

Hazard H03, H11
Risk phrases R22a, R41, R43, R50, R53a
Operator protection A, C, H, M; U05a, U08, U11, U13, U14, U20b
Environmental protection E07a, E15a, E34, E38

Storage and disposal D01, D02, D09a, D10b, D12a
Medical advice M05a

63 carfentrazone-ethyl + metsulfuron-methyl

A foliar applied herbicide mixture for cereals
HRAC mode of action code: E + B

Products

Ally Express	DuPont	40:10% w/w	WG	08640

Uses

- Annual dicotyledons in ***durum wheat***, ***spring barley***, ***spring oats***, ***spring wheat***, ***triticale***, ***winter barley***, ***winter oats***, ***winter wheat***

Approval information

- Carfentrazone-ethyl and metsulfuron-methyl included in Annex I under EC Directive 91/414
- Accepted by BBPA for use on malting barley

Efficacy guidance

- Best results achieved from applications made in good growing conditions
- Good spray cover of weeds must be obtained
- Growth of weeds is inhibited within hours of treatment but the time taken for visible colour changes to appear will vary according to species and weather
- Product has short residual life in soil. Under normal moisture conditions susceptible weeds germinating soon after treatment will be controlled
- Metsulfuron-methyl is a member of the ALS-inhibitor group of herbicides and products should be used in a planned Resistance Management strategy. See Section 5 for more information

Restrictions

- Maximum number of treatments 1 per crop
- Product must only be used after 1 Feb
- Do not use on any crop suffering stress from drought, waterlogging, cold, pest or disease attack, or nutrient deficiency
- Do not use on crops undersown with grass or legumes or on any broad-leaved crop
- Do not apply within 7 d of rolling
- Specific restrictions apply to use in sequence or tank mixture with other sulfonylurea or ALS-inhibiting herbicides. See label for details
- Do not tank mix with products containing chlorpyrifos, diclofop-methyl, difenzoquat, flamprop-methyl, flutriafol, propiconazole or tralkoxydim

Crop-specific information

- Latest use: before 3rd node detectable (GS 33)
- Can be used on all soil types
- Apply in the spring from the 3-leaf stage on all crops
- Slight necrotic spotting of crops can occur under certain crop and soil conditions. Recovery is rapid and there is no effect on grain yield or quality

Following crops guidance

- Only cereals, oilseed rape, field beans or grass may be sown as a following crop in the same calendar yr. Any crop may follow in the next spring
- In the event of failure of a treated crop, only winter wheat may be sown 1-3 mth later after ploughing and cultivating to at least 15 cm

Environmental safety

- Dangerous for the environment
- Very toxic to aquatic organisms
- Take extreme care to avoid drift onto broad-leaved plants outside the target area or onto ponds, waterways or ditches, or onto land intended for cropping
- Spraying equipment should not be drained or flushed onto land planted, or to be planted, with trees or crops other than cereals and should be thoroughly cleansed after use - see label for instructions

Hazard classification and safety precautions

Hazard H04, H11

Risk phrases R43, R50, R53a
Operator protection A, H; U05a, U08, U14, U19a, U20b
Environmental protection E15b, E38
Storage and disposal D01, D02, D09a, D11a, D12a

64 chloridazon

A residual pyridazinone herbicide for beet crops
HRAC mode of action code: C1

Products

1	Better DF	Sipcam	65% w/w	SG	06250
2	Better Flowable	Sipcam	430 g/l	SC	04924
3	Burex 430 SC	United Phosphorus	430 g/l	SC	12619
4	Globazone 430 SC	Globachem	430 g/l	SC	12878
5	Lidazone 65 WG	Globachem	65% w/w	WG	10847
6	Parador	United Phosphorus	430 g/l	SC	12372
7	Pyramin DF	BASF	65% w/w	WG	03438
8	Takron	BASF	430 g/l	SC	11627
9	Tempest	AgriGuard	430 g/l	SC	12688

Uses

- Annual dicotyledons in ***bulb onions*** *(off-label)*, ***leeks*** *(off-label)*, ***salad onions*** *(off-label)* [1, 5, 7]; ***chard*** *(off-label)*, ***chives*** *(off-label)*, ***herbs (see appendix 6)*** *(off-label)*, ***leaf spinach*** *(off-label)*, ***parsley*** *(off-label)*, ***spinach beet*** *(off-label)* [7]; ***fodder beet***, ***mangels***, ***sugar beet*** [1-9]; ***garlic*** *(off-label)*, ***shallots*** *(off-label)* [1, 5]
- Annual meadow grass in ***bulb onions*** *(off-label)*, ***leeks*** *(off-label)*, ***salad onions*** *(off-label)* [1, 5, 7]; ***chard*** *(off-label)*, ***chives*** *(off-label)*, ***herbs (see appendix 6)*** *(off-label)*, ***leaf spinach*** *(off-label)*, ***parsley*** *(off-label)*, ***spinach beet*** *(off-label)* [7]; ***fodder beet***, ***mangels***, ***sugar beet*** [1, 2, 4-9]; ***garlic*** *(off-label)*, ***shallots*** *(off-label)* [1, 5]

Specific Off-Label Approvals (SOLAs)

- ***bulb onions*** *20012570* [1], *20021819* [5], *970732* [7]
- ***chard*** *20062979* [7]
- ***chives*** *20062979* [7]
- ***garlic*** *20012570* [1], *20021819* [5]
- ***herbs (see appendix 6)*** *20062979* [7]
- ***leaf spinach*** *20062979* [7]
- ***leeks*** *20012570* [1], *20021819* [5], *970732* [7]
- ***parsley*** *20062979* [7]
- ***salad onions*** *20012570* [1], *20021819* [5], *970732* [7]
- ***shallots*** *20012570* [1], *20021819* [5]
- ***spinach beet*** *20062979* [7]

Approval information

- Chloridazon included in Annex I under EC Directive 91/414

Efficacy guidance

- Absorbed by roots of germinating weeds and best results achieved pre-emergence of weeds or crop when soil moist and adequate rain falls after application
- Application rate for some products depends on soil type. Check label for details

Restrictions

- Maximum number of treatments generally 1 per crop (pre-emergence) for fodder beet and mangels; 1 (pre-emergence) + 3 (post-emergence) per crop for sugar beet, but labels vary slightly. Check labels for maximum total dose for the crop to be treated
- Maximum total dose for onions and leeks equivalent to one full dose recommended for these crops [7]
- Do not use on Coarse Sands, Sands or Fine Sands or where organic matter exceeds 5%

Crop-specific information

- Latest use: pre-emergence for fodder beet and mangels; normally before leaves of crop meet in row for sugar beet, but labels vary slightly; up to and including second true leaf stage for onions and leeks [7]

- Where used pre-emergence spray as soon as possible after drilling in mid-Mar to mid-Apr on fine, firm, clod-free seedbed
- Where crop drilled after mid-Apr or soil dry apply pre-drilling and incorporate to 2.5 cm immediately afterwards
- Various tank mixes recommended on sugar beet for pre- and post-emergence use and as repeated low dose treatments. See label for details
- Crop vigour may be reduced by treatment of crops growing under unfavourable conditions including poor tilth, drilling at incorrect depth, soil capping, physical damage, pest or disease damage, excess seed dressing, trace-element deficiency or a sudden rise in temperature after a cold spell

Following crops guidance

- Winter cereals or any spring sown crop may follow a treated beet crop harvested at the normal time and after ploughing to at least 150 mm
- In the event of failure of a treated crop only a beet crop or maize may be drilled, after cultivation

Environmental safety

- Dangerous for the environment
- Very toxic to aquatic organisms

Hazard classification and safety precautions

Hazard H03 [1, 7]; H04 [2-6, 8, 9]; H11 [1-9]
Risk phrases R22a [1, 7]; R36, R37, R38 [9]; R43 [2-6, 8, 9]; R50 [1, 2, 4-9]; R51 [3]; R53a [1-9]
Operator protection A [3, 4, 6, 8, 9]; U05a, U08 [1-9]; U14 [2, 4-6, 8, 9]; U15 [9]; U19a [2, 6, 8, 9]; U20a [3, 5]; U20b [1, 2, 4, 6-9]
Environmental protection E13b [3, 6]; E13c [5]; E15a [1, 2, 4-9]; E38 [1, 2, 4, 6-8]
Storage and disposal D01, D02, D09a [1-9]; D05 [3, 4]; D10b [2, 5, 6, 8, 9]; D10c [3, 4, 6]; D11a [1, 7]; D12a [1, 2, 4, 6-8]; D12b [3]
Medical advice M05a [1, 2, 6-9]

65 chloridazon + chlorpropham + metamitron

A residual herbicide mixture for beet crops
HRAC mode of action code: C1 + K2 + C1

Products

Newtron	Whyte Agrochemicals	83:42:333 g/l	SE	11208

Uses

- Annual dicotyledons in ***fodder beet***, ***mangels***, ***sugar beet***
- Annual meadow grass in ***fodder beet***, ***mangels***, ***sugar beet***

Approval information

- Chloridazob and chlorpropham included in Annex I under EC Directive 91/414
- Approval expiry 31 Aug 2008 [1]

Efficacy guidance

- Apply as part of a spray programme. For best results a full programme of pre- and post-emergence treatments is required
- Pre-emergence weed control will be reduced under very dry conditions but subsequent rain may activate the chemical
- Best post-emergence performance requires tank mxture with phenmedipham. See label for details
- Correct post-emergence timing important for good control. First treatment should be made when weeds at early cotyledon stage and subsequent applications made when new weed flushes appear

Restrictions

- Subject to the maximum total dose, maximum number of treatments 1 pre-emergence followed by 3 post-emergence, or 5 post-emergence treatments unless another product has been used pre-emergence, in which case 4 post-emergence sprays may be applied
- Must be tank mixed when used post-emergence. See label
- Do not roll or harrow treated soil. Spray after rolling

SEE SECTION 3 FOR PRODUCTS ALSO REGISTERED

Crop-specific information

- Latest use: before crop leaves meet between rows

Following crops guidance

- Only sugar beet, fodder beet or mangels should be sown within 4 mth of last application
- After harvesting treated crops land should be mould-board ploughed to 15 cm and any spring crop may then be drilled

Environmental safety

- Dangerous for the environment
- Very toxic to aquatic organisms

Hazard classification and safety precautions

Hazard H03, H11
Risk phrases R22a, R36, R38, R43, R50, R53a
Operator protection A, C; U05a, U08, U11, U14, U15, U19a, U20a
Environmental protection E15a, E34
Storage and disposal D01, D02, D09a, D10c, D12b

66 chloridazon + ethofumesate

A contact and residual herbicide for beet crops
HRAC mode of action code: C1 + N

Products

Magnum	BASF	275:170 g/l	SC	11727

Uses

- Annual dicotyledons in ***fodder beet***, ***sugar beet***
- Annual meadow grass in ***fodder beet***, ***sugar beet***

Approval information

- Chloridazon and ethofumesate included in Annex I under EC Directive 91/414

Efficacy guidance

- Best results achieved on a fine firm clod-free seedbed when soil moist and adequate rain falls after spraying. Efficacy and crop safety may be reduced if heavy rain falls just after incorporation
- Effectiveness may be reduced under conditions of low pH
- May be applied by conventional or repeat low dose method. See label for details

Restrictions

- Maximum number of treatments 1 pre-emergence for fodder beet; 1 pre-emergence plus 3 post-emergence for sugar beet
- May be used on soil classes Loamy Sand - Silty Clay Loam. Additional restrictions apply for some tank mixtures
- Do not treat beet post-emergence with recommended tank mixtures when temperature is, or is likely to be, above 21 °C on day of spraying or under conditions of high light intensity
- If a mixture with phenmedipham is applied to a crop previously treated with a pre-emergence herbicide and the crop is suffering from stress from whatever cause, no further such post-emergence applications may be made

Crop-specific information

- Latest use: pre-crop emergence for fodder beet; before crop leaves meet between rows for sugar beet
- Apply up to cotyledon stage of weeds
- Crop vigour may be reduced by treatment of crops growing under unfavourable conditions including poor tilth, drilling at incorrect depth, soil capping, physical damage, excess nitrogen, excess seed dressing, trace element deficiency or a sudden rise in temperature after a cold spell. Frost after pre-emergence treatment may check crop growth

Following crops guidance

- In the event of crop failure only sugar beet, fodder beet or mangels may be re-drilled
- Any crop may be sown 3 mth after spraying following ploughing to 15 cm

Environmental safety

- Dangerous for the environment
- Very toxic to aquatic organisms

Hazard classification and safety precautions

Hazard H03, H11
Risk phrases R22a, R50, R53a
Operator protection A, H; U05a, U08, U19a, U20a
Environmental protection E15a, E38
Storage and disposal D01, D02, D09a, D10b, D12a
Medical advice M05a

67 chloridazon + lenacil

A residual pre- and post-emergence herbicide for beet crops
HRAC mode of action code: C1 + C1

Products

Advisor S	DuPont	200:133 g/l	SC	10948

Uses

- Annual dicotyledons in ***fodder beet*, *mangels*, *sugar beet***
- Annual meadow grass in ***fodder beet*, *mangels*, *sugar beet***

Approval information

- Chloridazon included in Annex I under EC Directive 91/414

Efficacy guidance

- Best pre-emergence results achieved from application to firm, moist, weed-free seedbed when adequate rain falls afterwards. Weed spectrum can be increased by tank mixture with other herbicides. See label
- If dry conditions prevail after application results can be improved by irrigation
- Season long weed control achieved by a programme of pre- and post-emergence treatments
- After pre-drilling incorporated treatment sugar beet should be drilled within 3 wk

Restrictions

- Maximum number of treatments 1 per crop pre-drilling (sugar beet) or 1 pre-emergence (sugar beet, fodder beet, mangels) followed in either case by 1 post-emergence (or 3 post-emergence at low dose) per crop
- Crops should be drilled to at least 15 mm and the seed well covered
- Do not use on Sands, stony or gravelly soils, Heavy soils or those with more than 10% organic matter
- Only use in post-emergence mixtures on crops growing vigorously and not stressed by drought, pest attack, deficiency or other factors

Crop-specific information

- Latest use: pre-emergence or before crop meets across rows post-emergence
- Heavy rainfall after spraying may reduce crop stand especially when such rain is followed by very hot weather

Following crops guidance

- In the event of crop failure only beet crops may be re-drilled on treated land and no further application should be made for at least 4 mth.
- Any succeeding crop may be sown 4 mth after treatment following ploughing to 15 cm

Environmental safety

- Dangerous for the environment
- Very toxic to aquatic organisms
- Harmful to fish or other aquatic life. Do not contaminate surface waters or ditches with chemical or used container

Hazard classification and safety precautions

Hazard H04, H11
Risk phrases R43, R50, R53a
Operator protection U05a, U08, U14, U19a, U20b

SEE SECTION 3 FOR PRODUCTS ALSO REGISTERED

Environmental protection E13c, E38
Storage and disposal D01, D02, D05, D09a, D10a, D12a

68 chloridazon + metamitron

A contact and residual herbicide mixture for sugar beet
HRAC mode of action code: C1 + C1

Products

Volcan Combi	Sipcam	300:280 g/l	SC	10256

Uses

- Annual dicotyledons in ***fodder beet*** *(off-label)*, ***sugar beet***
- Annual meadow grass in ***fodder beet*** *(off-label)*, ***sugar beet***

Specific Off-Label Approvals (SOLAs)

- ***fodder beet*** *20061540*

Approval information

- Chloridazon included in Annex I under EC Directive 91/414

Efficacy guidance

- Best results achieved from a sequential programmme of sprays
- Pre-emergence use improves efficacy of post-emergence programme. Best results obtained from application to a moist seed bed
- First post-emergence treatment should be made when weeds at early cotyledon stage and subsequent applications made when new weed flushes reach this stage
- Weeds surviving an earlier treatment should be treated again after 7-10 d even if no new weeds have appeared

Restrictions

- Maximum number of treaments 1 pre-emergence followed by 3 post-emergence
- Take advice if light soils have a high proportion of stones

Crop-specific information

- Latest use: when leaves of crop meet in rows
- Tolerance of crops growing under stress from any cause may be reduced
- Crops treated pre-emergence and subsequently subjected to frost may be checked and recovery may not be complete

Following crops guidance

- After the last application only sugar beet or mangels may be sown within 4 mth; cereals may be sown after 16 wk. Land should be mouldboard ploughed to 15 cm and thoroughly cultivated before any succeeding crop

Environmental safety

- Harmful to fish or other aquatic life. Do not contaminate surface waters or ditches with chemical or used container

Hazard classification and safety precautions

Hazard H04
Risk phrases R36, R38
Operator protection U08, U14, U15, U20a
Environmental protection E13c, E34
Storage and disposal D09a, D10b

69 chloridazon + propachlor

A residual pre-emergence herbicide for use in onions and leeks
HRAC mode of action code: C1 + K3

Products

Ashlade CP	Nufarm UK	86:400 g/l	SC	06481

Uses

- Annual dicotyledons in ***bulb onions*** *(off-label - post-em. Pre-em on label)*, ***chives*** *(off-label - post-em. Pre-em on label)*, ***garlic*** *(off-label)*, ***leeks*** *(off-label - post-em. Pre-em on label)*, ***salad onions*** *(off-label - post-em. Pre-em on label)*, ***shallots*** *(off-label)*
- Annual grasses in ***bulb onions*** *(off-label - post-em. Pre-em on label)*, ***chives*** *(off-label - post-em. Pre-em on label)*, ***garlic*** *(off-label)*, ***leeks*** *(off-label - post-em. Pre-em on label)*, ***salad onions*** *(off-label - post-em. Pre-em on label)*, ***shallots*** *(off-label)*

Specific Off-Label Approvals (SOLAs)

- ***bulb onions*** *(post-em. Pre-em on label) 981362*
- ***chives*** *(post-em. Pre-em on label) 951233*
- ***garlic*** *20061296*
- ***leeks*** *(post-em. Pre-em on label) 981362*
- ***salad onions*** *(post-em. Pre-em on label) 951233*
- ***shallots*** *20061296*

Approval information

- Chloridazon included in Annex I under EC Directive 91/414

Efficacy guidance

- Best results achieved from application to firm, moist, weed-free seedbed when adequate rain falls afterwards

Restrictions

- Maximum number of treatments 1 per crop
- Do not use on soils with more than 10% organic matter
- Ensure crops are drilled to 20 mm depth

Crop-specific information

- Latest use: pre-emergence of crop; before 2 true leaf stage for onions and leeks
- HI 12 wk for chives, salad onions
- Apply pre-emergence of sown crops, preferably soon after drilling, before weeds emerge. Loose or fluffy seedbeds must be consolidated before application
- Apply to transplanted crops when soil has settled after planting
- Crops stressed by nutrient deficiency, pests or diseases, poor growing conditions or pesticide damage may be checked by treatment, especially on sandy or gravelly soils

Following crops guidance

- In the event of crop failure only onions, leeks or maize should be planted
- Any crop can follow a treated onion or leek crop harvested normally as long as the ground is cultivated thoroughly before drilling

Environmental safety

- Harmful to fish or other aquatic life. Do not contaminate surface waters or ditches with chemical or used container

Hazard classification and safety precautions

Hazard H03
Risk phrases R22a, R36, R43
Operator protection A, C; U02a, U04a, U05a, U08, U11, U13, U14, U19a, U20a
Environmental protection E13c
Storage and disposal D01, D02, D09a, D10b
Medical advice M03

70 chloridazon + quinmerac

A herbicide mixture for use in beet crops
HRAC mode of action code: C1 + O

Products

Fiesta T	BASF	360:60 g/l	SC	11734

Uses

- Annual dicotyledons in ***beet leaves*** *(off-label)*, ***chard*** *(off-label)*, ***chives*** *(off-label)*, ***fodder beet***, ***herbs (see appendix 6)*** *(off-label)*, ***leaf spinach*** *(off-label)*, ***mangels***, ***parsley*** *(off-label)*, ***spinach beet*** *(off-label)*, ***sugar beet***

- Annual grasses in ***beet leaves*** *(off-label)*, ***chard*** *(off-label)*, ***chives*** *(off-label)*, ***herbs (see appendix 6)*** *(off-label)*, ***leaf spinach*** *(off-label)*, ***parsley*** *(off-label)*, ***spinach beet*** *(off-label)*
- Annual meadow grass in ***fodder beet***, ***mangels***, ***sugar beet***
- Chickweed in ***fodder beet***, ***mangels***, ***sugar beet***
- Cleavers in ***fodder beet***, ***mangels***, ***sugar beet***
- Field speedwell in ***fodder beet***, ***mangels***, ***sugar beet***
- Ivy-leaved speedwell in ***fodder beet***, ***mangels***, ***sugar beet***
- Mayweeds in ***fodder beet***, ***mangels***, ***sugar beet***
- Poppies in ***fodder beet***, ***mangels***, ***sugar beet***

Specific Off-Label Approvals (SOLAs)

- ***beet leaves*** *20063164*
- ***chard*** *20063164*
- ***chives*** *20063164*
- ***herbs (see appendix 6)*** *20063164*
- ***leaf spinach*** *20063164*
- ***parsley*** *20063164*
- ***spinach beet*** *20063164*

Approval information

- Chloridazon included in Annex I under EC Directive 91/414

Efficacy guidance

- Best results obtained when adequate soil moisture is present at application and afterwards to form an active herbicidal layer in the soil
- A programme of pre-emergence treatment followed by post-emergence application(s) in mixture with a contact herbicide optimises weed control and is essential for some difficult weed species such as cleavers
- Treatment pre-emergence only may not provide sufficient residual activity to give season-long weed control
- Effectiveness may be reduced under conditions of low pH
- Always follow WRAG guidelines for preventing and managing herbicide resistant weeds. Section 5 for more information

Restrictions

- Maximum total dose on sugar beet equivalent to one treatment pre-emergence plus two treatments post-emergence, all at maximum individual dose; maximum total dose on fodder beet and mangels equivalent to one full dose pre-emergence treatment
- Heavy rain shortly after treatment may check crop growth particularly when it leaves water standing in surface depressions
- Treatment of stressed crops or those growing in unfavourable conditions may depress crop vigour and possibly reduce stand
- Do not use on Sands or soils of high organic matter content
- Where rates of nitrogen higher than those generally recommended are considered necessary, apply at least 3 wk before drilling

Crop-specific information

- Latest use: before plants meet between the rows for sugar beet; pre-emergence for fodder beet, mangels

Following crops guidance

- In the spring following normal harvest of a treated crop sow only cereals, beet or mangel crops, potatoes or field beans, after ploughing
- In the event of failure of a treated crop only a beet crop may be drilled, after cultivation

Environmental safety

- Dangerous for the environment
- Toxic to aquatic organisms
- To reduce movement to groundwater do not apply to dry soil or when heavy rain is forecast

Hazard classification and safety precautions

Hazard H04, H11
Risk phrases R43, R51, R53a
Operator protection A; U05a, U08, U14, U19a, U20b
Environmental protection E15a, E38

Storage and disposal D01, D02, D08, D09a, D10c, D12a
Medical advice M03, M05a

71 chlormequat

A plant-growth regulator for reducing stem growth and lodging

Products

1	Adjust	Mandops	620 g/l	SL	13148
2	Agriguard Chlormequat 700	AgriGuard	700 g/l	SL	09782
3	Agriguard Chlormequat 720	AgriGuard	720 g/l	SL	09919
4	Barclay Take 5	Barclay	645 g/l	SL	11368
5	Barleyquat B	Mandops	620 g/l	SL	13187
6	BASF 3C Chlormequat 720	BASF	720 g/l	SL	06514
7	Bettaquat B	Mandops	637 g/l	SL	13192
8	Clayton Coldstream	Clayton	400 g/l	SL	12724
9	Clayton Squat	Clayton	400 g/l	SL	12271
10	Clayton Standup	Clayton	700 g/l	SL	11760
11	Fargro Chlormequat	Fargro	460 g/l	SL	02600
12	Fernpath Tangent	AgriGuard	460 g/l	SL	10341
13	Greencrop Carna	Greencrop	600 g/l	SL	09403
14	Hive	Nufarm UK	730 g/l	SL	11392
15	K2	Mandops	620 g/l	SL	13190
16	Mandops Chlormequat 700	Mandops	700 g/l	SL	06002
17	Manipulator	Mandops	620 g/l	SL	13189
18	Mirquat	Nufarm UK	730 g/l	SL	11406
19	New 5C Cycocel	BASF	645 g/l	SL	01482
20	New 5C Quintacel	Nufarm UK	645 g/l	SL	12074
21	Sigma PCT	Nufarm UK	460 g/l	SL	11209
22	Stabilan 700	Nufarm UK	700 g/l	SL	11393
23	Terbine	Nufarm UK	730 g/l	SL	11407

Uses

- Increasing yield in ***spring barley*** [1, 5, 15-17]; ***spring oats, winter oats*** [1, 5, 9]; ***spring rye*** *(off-label)*, ***triticale*** *(off-label)*, ***winter rye*** *(off-label)* [5]; ***spring wheat, winter rye, winter wheat*** [9]; ***winter barley*** [1, 2, 5, 6, 9, 12, 13, 15-17, 19, 20]
- Lodging control in ***durum wheat*** *(off-label)* [1, 7, 15, 17]; ***spring barley*** [1, 5, 15, 17]; ***spring oats, winter oats*** [1-6, 8-10, 12-20, 22, 23]; ***spring rye*** [6, 19, 20]; ***spring rye*** *(off-label)*, ***triticale*** *(off-label)*, ***winter rye*** *(off-label)* [1, 5, 7, 15, 17]; ***spring wheat, winter wheat*** [1-4, 6-10, 12-23]; ***triticale*** [4, 6, 12, 19, 20]; ***winter barley*** [1-6, 8-10, 14, 15, 17, 18, 21-23]; ***winter rye*** [2-4, 6, 8-10, 14, 18-20, 22]
- Stem shortening in ***bedding plants, camellias, hibiscus trionum, lilies*** [10, 11, 14, 18, 22, 23]; ***geraniums, poinsettias*** [6, 10, 11, 14, 18-20, 22, 23]

Specific Off-Label Approvals (SOLAs)

- ***durum wheat*** *20070390* [1], *20070391* [7], *20070392* [15], *20070393* [17]
- ***spring rye*** *20070390* [1], *20070394* [5], *20070391* [7], *20070392* [15], *20070393* [17]
- ***triticale*** *20070390* [1], *20070394* [5], *20070391* [7], *20070392* [15], *20070393* [17]
- ***winter rye*** *20070390* [1], *20070394* [5], *20070391* [7], *20070392* [15], *20070393* [17]

Approval information

- Approved for aerial application on wheat and oats [2, 6, 13, 16, 18, 21]; on winter barley [6, 13]; on triticale, rye [6]. See notes in Section 5
- Accepted by BBPA for use on malting barley

Efficacy guidance

- Most effective results on cereals normally achieved from application from Apr onwards, on wheat and rye from leaf sheath erect to first node detectable (GS 30-31), on oats at second node detectable (GS 32), on winter barley from mid-tillering to leaf sheath erect (GS 25-30). However, recommendations vary with product. See label for details
- Influence on growth varies with crop and growth stage. Risk of lodging reduced by application at early stem extension. Root development and yield can be improved by earlier treatment
- Results on barley can be variable
- In tank mixes with other pesticides on cereals optimum timing for herbicide action may differ from that for growth reduction. See label for details of tank mix recommendations

- Most products recommended for use on oats require addition of approved non-ionic wetter. Check label
- Some products are formulated with trace elements to help compensate for increased demand during rapid growth

Restrictions

- Maximum number of treatments or maximum total dose varies with crop and product and whether split dose treatments are recommended. Check labels
- Do not use on very late sown spring wheat or oats or on crops under stress
- Mixtures with liquid nitrogen fertilizers may cause scorch and are specifically excluded on some labels
- Do not use on soils of low fertility unless such crops regularly receive adequate dressings of nitrogen
- At least 6 h, preferably 24 h, required before rain for maximum effectiveness. Do not apply to wet crops
- Check labels for tank mixtures known to be incompatible

Crop-specific information

- Latest use varies with crop and product. See label for details
- May be used on cereals undersown with grass or clovers
- Ornamentals to be treated must be well established and growing vigorously. Do not treat in strong sunlight or when temperatures are likely to fall below 10 °C
- Temporary yellow spotting may occur on poinsettias. It can be minimised by use of a non-ionic wetting agent - see label

Environmental safety

- Harmful to aquatic organisms [1, 5, 10, 11, 14, 17, 18, 22]
- Wash equipment thoroughly with water and wetting agent immediately after use and spray out. Spray out again before storing or using for another product. Traces can cause harm to susceptible crops sprayed later
- Do not use straw from treated cereals as horticultural growth medium or mulch [6, 19]

Hazard classification and safety precautions

Hazard H03

Risk phrases R21 [2, 3, 8, 9]; R22a [1-23]; R36 [17]; R52 [1, 5, 10, 11, 14, 17, 18, 22]; R53a [1, 5, 10, 11, 14, 22]

Operator protection A [1-23]; C [17]; U05a, U19a [1-23]; U08 [1-13, 15-17, 19, 21, 23]; U09a [14, 18, 20, 22]; U13 [11, 16]; U20a [16, 20]; U20b [1-15, 17-19, 21-23]

Environmental protection E15a [1, 3-23]; E34 [1-23]

Consumer protection C01 [16]

Storage and disposal D01, D02, D09a [1-23]; D05 [2-4, 8-11, 13, 14, 16, 18, 20, 22]; D10a [10, 16, 22]; D10b [1-5, 7-9, 11-15, 17, 18, 21, 23]; D10c [6, 19, 20]; D12a [8, 9]

Medical advice M03 [1-23]; M05a [4, 6, 10, 11, 14, 18-20, 22]

72 chlormequat + 2-chloroethylphosphonic acid

A plant growth regulator for use in cereals

Products

1	Greencrop Tycoon	Greencrop	305:155 g/l	SL	09571
2	Strate	Bayer CropScience	360:180 g/l	SL	10020
3	Upgrade	Bayer CropScience	360:180 g/l	SL	10029

Uses

- Lodging control in ***spring barley***, ***winter barley***, ***winter wheat***

Approval information

- 2-chloroethylphosphonic acid (ethephon) included in Annex I under EC Directive 91/414
- Accepted by BBPA for use on malting barley
- In 2006 PSD required that all products containing 2-chloroethylphosphonic acid should carry the following warning in the main area of the container label: "2-chloroethylphosphonic acid is an anticholinesterase organophosphate. Handle with care"

Efficacy guidance

- Best results obtained when crops growing vigorously

- Recommended dose varies with growth stage. See labels for details and recommendations for use of sequential treatments

Restrictions

- 2-chloroethylphosphonic acid is an anticholinesterase organophosphorus compound. Do not use if under medical advice not to work with such compounds
- Maximum number of treatments 1 per crop; maximum total dose equivalent to one full dose treatment
- Product must always be used with specified non-ionic wetter - see labels
- Do not use on any crop in sequence with any other product containing 2-chloroethylphosphonic acid
- Do not spray when crop wet or rain imminent
- Do not spray during cold weather or periods of night frost, when soil is very dry, when crop diseased or suffering pest damage, nutrient deficiency or herbicide stress
- If used on seed crops grown for certification inform seed merchant beforehand
- Do not use on wheat variety Moulin or on any winter varieties sown in spring [1]
- Do not use on spring barley Triumph [1]
- Do not treat barley on soils with more than 10% organic matter [1]
- Do not use in programme with any other product containing 2-chloroethylphosphonic acid [1]
- Only crops growing under conditions of high fertility should be treated

Crop-specific information

- Latest use: before flag leaf ligule/collar just visible (GS 39) or 1st spikelet visible (GS 51) for wheat or barley at top dose; or before flag leaf sheath opening (GS 47) for winter wheat at reduced dose
- Apply before lodging has started

Environmental safety

- Dangerous for the environment [2, 3]
- Harmful to aquatic organisms [1]
- Harmful to fish or other aquatic life. Do not contaminate surface waters or ditches with chemical or used container [2, 3]
- Do not use straw from treated cereals as a horticultural growth medium or as a mulch [1]

Hazard classification and safety precautions

Hazard H03 [1-3]; H11 [2, 3]
Risk phrases R22a, R37 [1-3]; R41 [2, 3]; R52 [1]
Operator protection A [1-3]; C [2, 3]; U05a, U08, U19a, U20b [1-3]; U11, U13 [2, 3]
Environmental protection E13c [2, 3]; E15a [1]; E34 [1-3]
Storage and disposal D01, D02, D09a, D10b [1-3]; D05 [2, 3]
Medical advice M01, M03 [1-3]; M05a [1]

73 chlormequat + 2-chloroethylphosphonic acid + mepiquat chloride

A plant growth regulator for reducing lodging in cereals

Products

Cyclade	BASF	230:155:75 g/l	SL	08958

Uses

- Lodging control in ***spring barley***, ***winter barley***, ***winter wheat***

Approval information

- 2-chloroethylphosphonic acid (ethephon) included in Annex I under EC Directive 91/414
- Accepted by BBPA for use on malting barley
- In 2006 PSD required that all products containing 2-chloroethylphosphonic acid should carry the following warning in the main area of the container label: "2-chloroethylphosphonic acid is an anticholinesterase organophosphate. Handle with care"

Efficacy guidance

- Best results achieved in a vigorous, actively growing crop with adequate fertility and moisture
- Optimum timing on all crops is from second node detectable stage (GS 32)
- Recommended for use as part of an intensive growing system which includes provision for optimum fertilizer treatment and disease control

Restrictions

- 2-chloroethylphosphonic acid is an anticholinesterase organophosphorus compound. Do not use if under medical advice not to work with such compounds
- Maximum number of treatments 1 per crop
- Maximum total dose depends on spraying regime adopted. See label
- Must be used with a non-ionic wetting agent
- Do not apply to stressed crops or those on soils of low fertility unless receiving adequate dressings of fertilizer
- Do not apply in temperatures above 21 °C or if crop is wet or if rain expected
- Do not treat variety Moulin nor any winter varieties sown in spring
- Do not use in a programme with any other product containing 2-chloroethylphosphonic acid
- Do not apply to barley on soils with more than 10% organic matter (winter wheat may be treated)
- Notify seed merchant in advance if use on a seed crop is proposed

Crop-specific information

- Latest use: before first spikelet of ear visible (GS 51) using reduced dose on winter barley; before flag leaf sheath opening (GS 47) using reduced dose on winter wheat; before flag leaf just visible on spring barley
- May be applied to crops undersown with grasses or clovers
- Treatment may cause some delay in ear emergence

Environmental safety

- Harmful to aquatic organisms
- Do not use straw from treated crops as a horticultural growth medium

Hazard classification and safety precautions

Hazard H03
Risk phrases R22a, R37, R52
Operator protection A; U05a, U08, U19a, U20b
Environmental protection E15a, E34
Storage and disposal D01, D02, D09a, D10c
Medical advice M01, M03, M05a

74 chlormequat + imazaquin

A plant growth regulator mixture for winter wheat

Products

1	Meteor	BASF	368:0.8 g/l	SL	10403
2	Standon Imazaquin 5C	Standon	368:0.8 g/l	SL	08813
3	Upright	BASF	368:0.8 g/l	SL	10404

Uses

- Increasing yield in ***winter wheat***
- Lodging control in ***winter wheat***

Efficacy guidance

- Apply to crops during good growing conditions or to those at risk from lodging
- On soils of low fertility, best results obtained where adequate nitrogen fertilizer used

Restrictions

- Maximum number of applications 1 per crop (2 per crop at split dose)
- Do not treat durum wheat
- Do not apply to undersown crops
- Do not apply when crop wet or rain imminent

Crop-specific information

- Latest use: before second node detectable (GS 31)
- Apply as single dose from leaf sheath lengthening up to and including 1st node detectable or as split dose, the first from tillers formed to leaf sheath lengthening, the second from leaf sheath erect up to and including 1st node detectable

Environmental safety

- Dangerous for the environment

- Toxic to aquatic organisms
- Do not use straw from treated cereals as horticultural growth medium or mulch

Hazard classification and safety precautions

Hazard H03 [1-3]; H11 [2]
Risk phrases R20, R52 [1, 3]; R22a, R36, R53a [1-3]; R51 [2]
Operator protection A, C; U05a, U08, U13, U19a, U20b [1-3]; U11, U15 [1, 3]
Environmental protection E15a, E34
Storage and disposal D01, D02, D05, D09a, D10b [1-3]; D06c, D12a [1, 3]
Medical advice M03 [1-3]; M05a [1, 3]

75 chlormequat + mepiquat chloride

A plant growth regulator for reducing lodging in wheat

Products

Stronghold	BASF	345:115 g/l	SL	09134

Uses
- Lodging control in ***winter wheat***

Efficacy guidance
- Optimum timing is when leaf sheaths erect (GS 30)
- Benefit will vary according to crop and stage of growth at application

Restrictions
- Maximum total dose equivalent to one full dose treatment
- Do not apply to stressed crops or those on soils of low fertility unless receiving adequate dressings of fertilizer
- Do not treat crops where significant foot diseases, especially take-all, are expected
- Do not treat crops on soils of low fertility
- Do not apply in temperatures above 21 °C or if crop is wet or if rain expected
- Do not treat any winter varieties sown in spring
- Notify seed merchant in advance if use on a seed crop is proposed

Crop-specific information
- Latest use: before 3rd node detectable (GS 33)
- Apply during good growing conditions at the correct timings - see label
- May be applied to crops undersown with grasses or clovers
- Treatment may cause some delay in ear emergence
- Mixtures with liquid fertilizers may cause scorching in some circumstances

Environmental safety
- Do not use straw from treated crops as a horticultural growth medium or mulch

Hazard classification and safety precautions

Hazard H03
Risk phrases R22a
Operator protection A; U05a, U08, U19a, U20b
Environmental protection E15a, E34
Storage and disposal D01, D02, D09a, D10c
Medical advice M03, M05a

76 2-chloroethylphosphonic acid

A plant growth regulator for cereals and various horticultural crops

See also *chlormequat + 2-chloroethylphosphonic acid*
chlormequat + 2-chloroethylphosphonic acid + mepiquat chloride

Products

1	Agriguard Cerusite	AgriGuard	480 g/l	SL	11494
2	Cerone	Bayer CropScience	480 g/l	SL	09985
3	Ethrel C	Certis	480 g/l	SL	11387
4	Pan Ethephon	Pan Agriculture	480 g/l	SL	11020
5	Pan Stiffen	Pan Agriculture	480 g/l	SL	11644

SEE SECTION 3 FOR PRODUCTS ALSO REGISTERED

Uses

- Basal bud stimulation in ***protected roses*** [3]
- Fruit ripening in ***apples*** *(off-label - for cider making)*, ***protected tomatoes, tomatoes (outdoor)*** [3]
- Increasing branching in ***geraniums*** [3]
- Inducing flowering in ***bromeliads*** [3]
- Lodging control in ***spring barley*** [1, 2]; ***triticale***, ***winter barley***, ***winter rye***, ***winter wheat*** [1, 2, 4, 5]
- Stem shortening in ***narcissi*** [3]

Specific Off-Label Approvals (SOLAs)

- ***apples*** *(for cider making) 20030875* [3]

Approval information

- 2-chloroethylphosphonic acid (ethephon) included in Annex I under EC Directive 91/414
- Approved for aerial application on winter barley [1, 2]. See notes in Section 5
- In 2006 PSD required that all products containing this active ingredient should carry the following warning in the main area of the container label: "2-chloroethylphosphonic acid is an anticholinesterase organophosphate. Handle with care"

Efficacy guidance

- Best results achieved on crops growing vigorously under conditions of high fertility
- Optimum timing varies between crops and products. See labels for details
- Do not spray crops when wet or if rain imminent
- Best results on horticultural crops when temperature does not fall below 10 °C [3]
- Use on tomatoes 17 d before planned pulling date [3]
- Apply as drench to daffodils when stems average 15 cm [3]
- Apply to glasshouse roses when new growth started after pruning [3]

Restrictions

- 2-chloroethylphosphonic acid is an anticholinesterase organophosphorus compound. Do not use if under medical advice not to work with such compounds
- Maximum number of treatments 1 per crop or yr
- Do not spray crops suffering from stress caused by any factor, during cold weather or period of night frost nor when soil very dry
- Do not apply to cereals within 10 d of herbicide or liquid fertilizer application
- Do not spray wheat or triticale where the leaf sheaths have split and the ear is visible

Crop-specific information

- Latest use: before 1st spikelet visible (GS 51) for spring barley, winter barley, winter rye; before flag leaf sheath opening (GS 47) for triticale, winter wheat
- HI cider apples, tomatoes 5 d

Environmental safety

- Dangerous for the environment [1, 4, 5]
- Harmful to aquatic organisms
- Avoid accidental deposits on painted objects such as cars, trucks, aircraft

Hazard classification and safety precautions

Hazard H04 [1-5]; H11 [1, 4, 5]

Risk phrases R36 [1]; R37, R41, R52, R53a [2-5]; R38 [1-5]

Operator protection A, C; U02a, U09a, U20a [3]; U05a, U13 [1-5]; U08, U20b [1, 2, 4, 5]; U11 [2, 3]

Environmental protection E13c, E38 [2, 3]; E15a [1, 4, 5]; E34 [1, 3]

Storage and disposal D01, D02, D09a, D10b [1-5]; D12a [2, 3]

Medical advice M01 [2-5]

77 2-chloroethylphosphonic acid + mepiquat chloride

A plant growth regulator for reducing lodging in cereals

Products

1	Barclay Banshee	Barclay	155:305 g/l	SL	11343
2	Clayton Mepiquat	Clayton	155:305 g/l	SL	12360
3	Guilder	Nufarm UK	155:305 g/l	SL	11894
4	Standon Mepiquat Plus	Standon	155:305 g/l	SL	09373
5	Terpal	BASF	155:305 g/l	SL	02103

Uses

- Increasing yield in ***winter barley*** *(low lodging situations)* [1-3, 5]
- Lodging control in ***spring barley*** [1-3, 5]; ***triticale***, ***winter barley***, ***winter rye***, ***winter wheat*** [1-5]

Approval information

- 2-chloroethylphosphonic acid (ethephon) included in Annex I under EC Directive 91/414
- In 2006 PSD required that all products containing 2-chloroethylphosphonic acid should carry the following warning in the main area of the container label: "2-chloroethylphosphonic acid is an anticholinesterase organophosphate. Handle with care"

Efficacy guidance

- Best results achieved on crops growing vigorously under conditions of high fertility
- Recommended dose and timing vary with crop, cultivar, growing conditions, previous treatment and desired degree of lodging control. See label for details
- May be applied to crops undersown with grass or clovers
- Do not apply to crops if wet or rain expected as efficacy will be impaired

Restrictions

- 2-chloroethylphosphonic acid is an anticholinesterase organophosphorus compound. Do not use if under medical advice not to work with such compounds
- Maximum number of treatments 2 per crop
- Add an authorised non-ionic wetter to spray solution. See label for recommended product and rate
- Do not treat crops damaged by herbicides or stressed by drought, waterlogging etc
- Do not treat crops on soils of low fertility unless adequately fertilized
- Do not use in a programme with any other product containing 2-chloroethylphosphonic acid
- Do not apply to winter cultivars sown in spring or treat winter barley, triticale or winter rye on soils with more than 10% organic matter (winter wheat may be treated)
- Do not apply at temperatures above 21 °C

Crop-specific information

- Latest use: before ear visible (GS 49) for winter barley, spring barley, winter wheat and triticale; flag leaf just visible (GS 37) for winter rye
- Late tillering may be increased with crops subject to moisture stress and may reduce quality of malting barley

Environmental safety

- Do not use straw from treated cereals as a mulch or growing medium

Hazard classification and safety precautions

Hazard H03
Risk phrases R22a, R37, R53a [1-5]; R36 [1, 3-5]; R41, R53b [2]; R52 [3]
Operator protection A, C [3, 5]; U05a, U11 [2]; U20b [1-5]
Environmental protection E15a
Storage and disposal D01, D02, D09a [1-5]; D05, D10b, D12a [2]; D08, D10c [1, 3-5]
Medical advice M01 [1-5]; M05a [1, 3-5]

78 chlorophacinone

An indandione anticoagulant rodenticide

Products

1	Drat Rat Bait	B H & B	0.005% w/w	RB	H6743
2	Endorats	Irish Drugs	0.005% w/w	RB	H6744

SEE SECTION 3 FOR PRODUCTS ALSO REGISTERED

Uses

- Rats in ***farm buildings*** [1, 2]
- Voles in ***farm buildings*** [1]

Efficacy guidance

- Chemical formulated with oil, thus improving weather resistance of bait
- Use in baiting programme
- Bait stations should be sited where rats active, by rat holes, along runs or in harbourages. Place bait in suitable containers
- Replenish baits every few days and remove unused bait when take ceases or after 7-10 d

Restrictions

- For use only by professional operators
- Resistance status of target population should be taken into account when considering choice of rodenticide
- Bait stations may be sited conveniently but bait should be inaccessible to non-target animals and protected from prevailing weather

Environmental safety

- Prevent access to baits by children, domestic animals and birds; see label for other precautions required
- Harmful to game, wild birds and animals

Hazard classification and safety precautions

Operator protection U13 [1, 2]; U20a [2]; U20b [1]
Storage and disposal D09a [1, 2]; D10a [1]; D11a [2]
Vertebrate/rodent control products V01a, V02, V03a, V04a

79 chloropicrin

A highly toxic horticultural soil fumigant
IRAC mode of action code: 8B

Products

1	Chloropicrin Fumigant	Dewco-Lloyd	99.5% w/w	VP	04216
2	K & S Chlorofume	K & S Fumigation	99.3% w/w	VP	08722

Uses

- Crown rot in ***strawberries*** *(soil fumigation)*
- Nematodes in ***strawberries*** *(soil fumigation)*
- Red core in ***strawberries*** *(soil fumigation)*
- Replant disease in ***all top fruit*** *(soil fumigation)*, ***hardy ornamentals*** *(soil fumigation)*
- Verticillium wilt in ***blackberries*** *(off-label - soil fumigation)*, ***protected blackberries*** *(off-label - soil fumigation)*, ***protected raspberries*** *(off-label - soil fumigation)*, ***protected rubus hybrids*** *(off-label - soil fumigation)*, ***raspberries*** *(off-label - soil fumigation)*, ***rubus hybrids*** *(off-label - soil fumigation)*, ***strawberries*** *(soil fumigation)*

Specific Off-Label Approvals (SOLAs)

- ***blackberries*** *(soil fumigation) 20051948* [1], *(soil fumigation) 20051579* [2]
- ***protected blackberries*** *(soil fumigation) 20051948* [1], *(soil fumigation) 20051579* [2]
- ***protected raspberries*** *(soil fumigation) 20051948* [1], *(soil fumigation) 20051579* [2]
- ***protected rubus hybrids*** *(soil fumigation) 20051948* [1], *(soil fumigation) 20051579* [2]
- ***raspberries*** *(soil fumigation) 20051948* [1], *(soil fumigation) 20051579* [2]
- ***rubus hybrids*** *(soil fumigation) 20051948* [1], *(soil fumigation) 20051579* [2]

Efficacy guidance

- Treat pre-planting
- Apply with specialised injection equipment
- For treating small areas or re-planting a single tree, a hand-operated injector may be used. Mark the area to be treated and inject to 22 cm at intervals of 22 cm
- Polythene sheeting (150 gauge) should be progressively laid over soil as treatment proceeds. The margin of the sheeting around the treated area must be embedded or covered with treated soil. Remove progressively after at least 4 d provided good air movement conditions prevail. Aerate soil for 15 d before planting
- Carry out a cress test before planting

Restrictions

- Chloropicrin is subject to the Poisons Rules 1982 and the Poisons Act 1972. See notes in Section 5
- Before use, consult the code of practice for the fumigation of soil with chloropicrin. 2 fumigators must be present
- Maximum number of treatments 1 per crop
- Remove contaminated gloves, boots or other clothing immediately and ventilate them in the open air until all odour is eliminated
- Keep unprotected persons out of treated areas and adjacent premises until advised otherwise by professional operator
- Keep treated areas covered with sheets of low gas permeability for at least 4 d after treatment
- Avoid releasing vapour into the air when persistent still air conditions prevail

Crop-specific information

- Latest use: at least 20 d before planting

Following crops guidance

- Carry out a cress test before replanting treated soil

Environmental safety

- Avoid treatment or vapour release when persistent still air conditions prevail
- Dangerous to game, wild birds and animals
- Dangerous to bees
- Dangerous to fish or other aquatic life. Do not contaminate surface waters or ditches with chemical or used container
- Dangerous to livestock. Keep all livestock out of treated areas until advised otherwise by fumigator
- Keep in original container, tightly closed, in a safe place, under lock and key

Hazard classification and safety precautions

Hazard H01 [1, 2]; H04 [1]
Risk phrases R26, R27, R28, R37 [1, 2]; R34 [2]; R36, R38 [1]
Operator protection A, G, H, K, M; U01, U19b [2]; U04a, U05a, U10, U19a, U20a [1, 2]; U13, U14, U15 [1]
Environmental protection E02a, E06b [1, 2] (until advised); E10a, E12c, E13b [1]; E15a, E34 [1, 2]
Storage and disposal D01, D02, D09b, D11a [1, 2]; D14 [2]
Medical advice M04a

80 chlorothalonil

A protectant chlorophenyl fungicide for use in many crops and turf
FRAC mode of action code: M5

See also azoxystrobin + chlorothalonil

Products

1	Agriguard Chlorothalonil	AgriGuard	500 g/l	SC	12201
2	Bravo 500	Syngenta	500 g/l	SC	10518
3	Daconil Turf	Scotts	500 g/l	SC	09265
4	Daconil Turf	Syngenta	500 g/l	SC	13602
5	Daconil Weatherstik	Syngenta	500 g/l	SC	12636
6	Expert-Turf	AgriGuard	500 g/l	SC	12571
7	Greencrop Orchid 2	Greencrop	500 g/l	SC	12769
8	Greencrop Orchid B	Greencrop	500 g/l	SC	12251
9	Landgold Chlorothalonil 50	Teliton	500 g/l	SC	13473
10	Mascot Contact	Rigby Taylor	500 g/l	SC	13718
11	Pan Trilby	Pan Agriculture	500 g/l	SC	13558
12	Repulse	Certis	500 g/l	SC	11328
13	Rezacur	Sherriff Amenity	500 g/l	SC	13717
14	Sanspor	Syngenta	500 g/l	SC	13313
15	Sonar	Syngenta	720 g/l	SC	13083
16	Supreme	AgriGuard	500 g/l	SC	12233

Uses

- American gooseberry mildew in ***gooseberries*** [2, 14]
- Anthracnose in ***managed amenity turf*** [3-6, 10, 11, 13]
- Ascochyta in ***combining peas*** [1, 2, 7-9, 12, 14-16]
- Blight in ***potatoes***, ***protected tomatoes*** [1, 2, 7-9, 12, 14-16]
- Botrytis in ***blackberries***, ***raspberries*** [1, 7-9, 15, 16]; ***blackcurrants***, ***gooseberries*** *(qualified minor use)*, ***redcurrants*** *(qualified minor use)*, ***strawberries*** *(outdoor crops only)* [1, 9, 16]; ***borage*** *(off-label)*, ***daffodils*** *(off-label - for galanthamine production)*, ***evening primrose*** *(off-label)*, ***honesty*** *(off-label)*, ***linseed*** *(off-label)*, ***mustard*** *(off-label)*, ***strawberries*** *(off-label)* [2]; ***broccoli***, ***brussels sprouts***, ***cabbages***, ***cauliflowers***, ***protected cucumbers***, ***protected ornamentals*** [1, 2, 7-9, 12, 14-16]; ***bulb onions***, ***combining peas*** [1, 2, 9, 12, 14, 16]; ***calabrese*** *(off-label)*, ***lupins*** *(off-label)* [2, 12]; ***garlic*** *(off-label)*, ***shallots*** *(off-label)* [1, 12]; ***gooseberries*** [2, 14]; ***protected tomatoes***, ***strawberries*** [2, 7, 8, 12, 14, 15]; ***spring oilseed rape***, ***winter oilseed rape*** [1, 2, 7-9, 14-16]
- Cane spot in ***blackberries***, ***raspberries*** [7, 8, 12, 15]
- Celery leaf spot in ***celeriac*** *(off-label)* [2]; ***celery (outdoor)*** *(qualified minor use)* [1, 2, 7-9, 12, 14-16]
- Chocolate spot in ***spring field beans***, ***winter field beans*** [1, 2, 7-9, 14-16]
- Currant leaf spot in ***blackcurrants***, ***gooseberries*** *(qualified minor use)*, ***redcurrants*** *(qualified minor use)* [1, 2, 7-9, 12, 14-16]
- Dollar spot in ***managed amenity turf*** [3-6, 10, 11, 13]
- Downy mildew in ***borage*** *(off-label)*, ***evening primrose*** *(off-label)*, ***honesty*** *(off-label)*, ***linseed*** *(off-label)*, ***mustard*** *(off-label)* [2]; ***brassica seed beds*** [2, 7-9, 14, 15]; ***broccoli***, ***brussels sprouts***, ***cabbages***, ***cauliflowers***, ***hops*** [1, 2, 7-9, 12, 14-16]; ***calabrese*** *(off-label)* [2, 12]; ***edible brassicas*** [12]; ***spring oilseed rape***, ***winter oilseed rape*** [1, 2, 7-9, 14-16]
- Foliar disease control in ***calabrese*** *(off-label - seedling)*, ***lupins*** *(off-label)* [1]; ***durum wheat*** *(off-label)*, ***spring rye*** *(off-label)*, ***triticale*** *(off-label)*, ***winter rye*** *(off-label)* [2]; ***grass seed crops*** *(off-label)* [1, 2]
- Fusarium patch in ***managed amenity turf*** [3-6, 10, 11, 13]
- Glume blotch in ***spring wheat*** [1, 2, 9, 14, 16]; ***winter wheat*** [1, 2, 7-9, 14-16]
- Ink disease in ***irises*** *(qualified minor use)* [1, 2, 7-9, 14-16]
- Leaf mould in ***protected tomatoes*** [1, 9, 12, 16]
- Leaf rot in ***bulb onions*** [1, 2, 7-9, 12, 14-16]; ***garlic*** *(off-label)*, ***shallots*** *(off-label)* [2]
- Mycosphaerella in ***combining peas*** [1, 2, 7-9, 12, 14-16]
- Neck rot in ***bulb onions*** [1, 2, 7-9, 12, 14-16]; ***garlic*** *(off-label)*, ***shallots*** *(off-label)* [2]
- Powdery mildew in ***blackberries***, ***raspberries*** [7, 8, 15]; ***blackcurrants***, ***gooseberries*** *(qualified minor use)*, ***redcurrants*** *(qualified minor use)* [1, 9, 16]
- Red thread in ***managed amenity turf*** [3-6, 10, 11, 13]
- Rhynchosporium in ***spring barley***, ***winter barley*** [2, 7, 8, 14]
- Ring spot in ***broccoli***, ***brussels sprouts***, ***cabbages***, ***cauliflowers*** [7, 8, 12, 15]; ***calabrese*** *(off-label)* [2, 12]; ***edible brassicas*** [12]
- Septoria leaf blotch in ***spring wheat*** [1, 2, 9, 14, 16]; ***winter wheat*** [1, 2, 7-9, 14-16]
- White mould in ***daffodils*** *(off-label)* [2]

Specific Off-Label Approvals (SOLAs)

- ***borage*** *20052941* [2]
- ***calabrese*** *(seedling) 20052922* [1], *20052940* [2], *20052888* [12]
- ***celeriac*** *20032083* [2]
- ***daffodils*** *(for galanthamine production) 20041518* [2], *20041518* [2]
- ***durum wheat*** *20052937* [2]
- ***evening primrose*** *20052941* [2]
- ***garlic*** *20052923* [1], *20052939* [2], *20052887* [12]
- ***grass seed crops*** *20052921* [1], *20052937* [2]
- ***honesty*** *20052941* [2]
- ***linseed*** *20052941* [2]
- ***lupins*** *20052924* [1], *20052938* [2], *20052886* [12]
- ***mustard*** *20052941* [2]
- ***shallots*** *20052923* [1], *20052939* [2], *20052887* [12]
- ***spring rye*** *20052937* [2]
- ***strawberries*** *20011635* [2]

- ***triticale*** *20052937* [2]
- ***winter rye*** *20052937* [2]

Approval information

- Chlorothalonil included in Annex I under EC Directive 91/414. However a Standing Committee decision in July 2007 to reduce to the limit of determination the MRL set for chlorothalonil in blackberries and raspberries led to revocation of approvals for use on these crops
- Approval for aerial spraying on potatoes [1, 2, 7-9, 12, 14-16]. See notes in Section 5
- Accepted by BBPA for use on malting barley and hops

Efficacy guidance

- For some crops products differ in diseases listed as controlled. See label for details and for application rates, timing and number of sprays
- Apply as protective spray or as soon as disease appears and repeat as directed
- On cereals activity against Septoria may be reduced where serious mildew or rust present. In such conditions mix with suitable mildew or rust fungicide
- May be used for preventive and curative treatment of turf but treatment at a late stage of disease development may not be so successful and can leave bare patches of soil requiring renovation. Products are not recommended for curative control of Fusarium patch or Anthracnose [3, 5, 6, 10, 11, 13]
- Optimum results on managed amenity turf obtained from treatment in spring or autumn when conditions most favourable for disease development [3, 5, 6, 10, 11, 13]

Restrictions

- Maximum number of treatments and maximum total doses vary with crop and product - see labels for details
- Operators of vehicle mounted equipment must use a vehicle fitted with a cab and a forced air filtration unit with a pesticide filter complying with HSE Guidance Note PM74 or to an equivalent or higher standard when making broadcast or air-assisted applications
- Must only be used on established turf [3-6, 10, 11, 13]
- On managed amenity turf must only be applied by a pedestrian controlled sprayer or vehicle mounted/drawn equipment [3-6, 10, 11, 13]
- Do not mow or water turf for 24 h after treatment. Do not add surfactant or mix with liquid fertilizer [3-6, 10, 11, 13]

Crop-specific information

- Latest time of application to cereals varies with product. Consult label
- Latest use for winter oilseed rape before flowering; end Aug in yr of harvest for blackcurrants, gooseberries, redcurrants, blueberries, loganberries
- HI 8 wk for field beans; 6 wk for combining peas; 28 d or before 31 Aug for post-harvest treatment for blackcurrants, gooseberries, redcurrants; 14 d for onions, strawberries; 10 d for hops; 7-14 d for broccoli, Brussels sprouts, cabbages, cauliflowers, celery, onions, potatoes; 3 d for cane fruit; 12-48 h for protected cucumbers, protected tomatoes
- For Botrytis control in strawberries important to start spraying early in flowering period and repeat at least 3 times at 10 d intervals
- On strawberries some scorching of calyx may occur with protected crops

Environmental safety

- Dangerous for the environment
- Very toxic to aquatic organisms
- Do not spray from the air within 250 m horizontal distance of surface waters or ditches
- LERAP Category B
- Broadcast air-assisted LERAP (18 m) [1, 2, 7-9, 12, 14-16]

Hazard classification and safety precautions

Hazard H03, H11

Risk phrases R20, R40, R43, R50, R53a [1-16]; R36 [1-11, 13-16]; R37 [1-7, 9-16]; R38 [8, 12]; R41 [12]

Operator protection A, C, H, M [1-16]; J [1, 2, 7-9, 14, 16]; U02a [2, 4, 5, 7-9, 14, 15]; U05a, U20a [1-16]; U09a, U19a [1-9, 12, 14-16]; U14 [12]

Environmental protection E15a [1-8, 10-16]; E15b [9]; E16a [1-16]; E16b [1-3, 7-16]; E17b [1, 2, 7-9, 12, 14-16] (18 m); E18 [1, 2, 7-9, 14-16]; E34 [1, 3-7, 9-11, 13, 16]; E38 [1-6, 8, 12, 14-16]

SEE SECTION 3 FOR PRODUCTS ALSO REGISTERED

Storage and disposal D01, D02, D09a [1-16]; D05 [1-11, 13-16]; D07 [6]; D10a [3, 6, 10-13]; D10b [7]; D10c [1, 2, 4, 5, 8, 9, 14-16]; D12a [1-5, 7-9, 12, 14-16]
Medical advice M03 [6]; M05a [7]

81 chlorothalonil + cyproconazole

A systemic protectant and curative fungicide mixture for cereals
FRAC mode of action code: M5 + 3

Products

Alto Elite	Syngenta	375:40 g/l	SC	08467

Uses
- Brown rust in ***spring barley, winter barley, winter wheat***
- Chocolate spot in ***spring field beans, winter field beans***
- Eyespot in ***winter barley, winter wheat***
- Glume blotch in ***winter wheat***
- Grey mould in ***combining peas***
- Net blotch in ***spring barley, winter barley***
- Powdery mildew in ***spring barley, winter barley, winter wheat***
- Rhynchosporium in ***spring barley, winter barley***
- Rust in ***spring field beans, winter field beans***
- Septoria leaf blotch in ***winter wheat***
- Yellow rust in ***spring barley, winter barley, winter wheat***

Approval information
- Chlorothalonil included in Annex I under EC Directive 91/414
- Accepted by BBPA for use on malting barley

Efficacy guidance
- Apply at first signs of infection or as soon as disease becomes active
- A repeat application may be made if re-infection occurs
- For established mildew tank-mix with an approved mildewicide
- When applied prior to third node detectable (GS 33) a useful reduction of eyespot will be obtained
- Cyproconazole is a DMI fungicide. Resistance to some DMI fungicides has been identified in Septoria leaf blotch which may seriously affect performance of some products. For further advice contact a specialist advisor and visit the Fungicide Resistance Action Group (FRAG)-UK website

Restrictions
- Maximum total dose equivalent to 2 full dose treatments
- Do not apply at concentrations higher than recommended

Crop-specific information
- Latest use: before beginning of anthesis (GS 60) for barley; before caryopsis watery ripe (GS 71) for wheat
- HI peas, field beans 6 wk
- If applied to winter wheat in spring at GS 30-33 straw shortening may occur but yield is not reduced

Environmental safety
- Dangerous for the environment
- Very toxic to aquatic organisms
- LERAP Category B

Hazard classification and safety precautions

Hazard H03, H11
Risk phrases R20, R37, R40, R41, R43, R50, R53a
Operator protection A, C, H, M; U05a, U11, U19a, U20a
Environmental protection E15a, E16a, E16b, E38
Storage and disposal D01, D02, D05, D09a, D10c, D12a
Medical advice M04a

82 chlorothalonil + cyproconazole + propiconazole

A broad-spectrum fungicide mixture for cereals
FRAC mode of action code: M5 + 3 + 3

Products

Cherokee	Syngenta	375:50:62.5 g/l	SE	13251

Uses

- Brown rust in ***spring barley***, ***winter barley***, ***winter wheat***
- Net blotch in ***spring barley*** *(reduction)*, ***winter barley*** *(reduction)*
- Powdery mildew in ***spring barley***, ***winter barley***, ***winter wheat*** *(moderate)*
- Rhynchosporium in ***spring barley*** *(reduction)*, ***winter barley*** *(reduction)*
- Septoria leaf blotch in ***winter wheat***
- Yellow rust in ***winter wheat***

Approval information

- Chlorothalonil and propiconazole included in Annex I under EC Directive 91/414

Efficacy guidance

- Best results obtained from treatment at the early stages of disease development
- Cyproconazole and propiconazole are DMI fungicides. Resistance to some DMI fungicides has been identified in Septoria leaf blotch which may seriously affect performance of some products. For further advice contact a specialist advisor and visit the Fungicide Resistance Action Group (FRAG)-UK website
- Product should be used as part of an Integrated Crop Management strategy incorporating other methods of control and including, where appropriate, other fungicides with a different mode of action
- Product should be used preventatively and not relied on for its curative potential
- Always follow FRAG guidelines for resistance management. See Section 5 for more information

Restrictions

- Maximum total dose equivalent to two full dose treatments on wheat and barley
- Do not treat crops under stress
- Spray solution must be used as soon as possible after mixing

Crop-specific information

- Latest use: before caryopsis watery ripe (GS 71) for winter wheat; before first spikelet just visible (GS 51) for barley
- When applied to winter wheat in the spring straw shortening may occur which causes no loss of yield

Environmental safety

- Dangerous for the environment
- Very toxic to aquatic organisms
- LERAP Category B

Hazard classification and safety precautions

Hazard H03, H11
Risk phrases R20, R36, R37, R40, R43, R50, R53a
Operator protection A, C, H; U02a, U05a, U09a, U10, U11, U14, U15, U19a, U20b
Environmental protection E15b, E16a, E16b, E34, E38
Storage and disposal D01, D02, D05, D07, D09a, D10c, D12a
Medical advice M04a

83 chlorothalonil + flusilazole

A fungicide mixture for wheat and barley
FRAC mode of action code: M5 + 3

Products

Midas	DuPont	200:80 g/l	SC	13285

Uses

- Glume blotch in ***spring wheat***, ***winter wheat***

SEE SECTION 3 FOR PRODUCTS ALSO REGISTERED

- Rhynchosporium in ***spring barley***, ***winter barley***
- Septoria leaf blotch in ***spring wheat***, ***winter wheat***

Approval information

- Chlorothalonil included in Annex I under EC Directive 91/414. Flusilazole has been included in Annex I under Directive 91/414 for 18 months from January 2007 and all approvals for products containing this active ingredient will expire in June 2008

Efficacy guidance

- Best results achieved from treatment when diseases are at low levels and before they spread to new growth
- Under heavy disease pressure a second application may be made 3-4 wk later
- Flusilazole is a DMI fungicide. Resistance to some DMI fungicides has been identified in Septoria leaf blotch which may seriously affect performance of some products. For further advice contact a specialist advisor and visit the Fungicide Resistance Action Group (FRAG)-UK website

Restrictions

- Maximum number of treatments 2 per crop
- Do not apply to crops under stress
- Do not apply during frosty weather

Crop-specific information

- Latest use: end of flowering (GS 69) for wheat; end of heading (GS 59) for barley

Environmental safety

- Dangerous for the environment
- Very toxic to aquatic organisms
- LERAP Category B

Hazard classification and safety precautions

Hazard H02, H11
Risk phrases R37, R40, R41, R43, R50, R53a, R61
Operator protection A, C, H, M; U05a, U11, U20b, U23a
Environmental protection E15a, E16a, E34, E38
Storage and disposal D01, D02, D05, D09a, D10b, D12a
Medical advice M04a

84 chlorothalonil + flutriafol

A systemic eradicant and protectant fungicide for winter wheat
FRAC mode of action code: M5 + 3

Products

1	Halo	Headland	375:47 g/l	SC	11546
2	Impact Excel	Headland	300:47 g/l	SC	11547
3	Prospa	Headland	300:47 g/l	SC	11548

Uses

- Brown rust in ***winter wheat*** [1-3]
- Foliar disease control in ***durum wheat*** *(off-label)*, ***grass seed crops*** *(off-label)*, ***spring rye*** *(off-label)*, ***triticale*** *(off-label)*, ***winter rye*** *(off-label)* [2, 3]
- Late ear diseases in ***winter wheat*** [1-3]
- Powdery mildew in ***winter wheat*** [1-3]
- Septoria leaf blotch in ***winter wheat*** [1-3]
- Yellow rust in ***winter wheat*** [1-3]

Specific Off-Label Approvals (SOLAs)

- ***durum wheat*** *20052890* [2], *20052889* [3]
- ***grass seed crops*** *20052890* [2], *20052889* [3]
- ***spring rye*** *20052890* [2], *20052889* [3]
- ***triticale*** *20052890* [2], *20052889* [3]
- ***winter rye*** *20052890* [2], *20052889* [3]

Approval information

- Chlorothalonil included in Annex I under EC Directive 91/414

Efficacy guidance
- Generally disease control and yield benefit will be optimised when application is made at an early stage of disease development
- Apply as soon as disease is seen establishing in the crop

Restrictions
- Maximum number of treatments 2 per crop on winter wheat

Crop-specific information
- Latest use: before early milk stage (GS 73)
- On certain cultivars with erect leaves high transpiration can result in flag leaf tip scorch. This may be increased by treatment but does not affect yield

Environmental safety
- Dangerous for the environment [2, 3]
- Very toxic to aquatic organisms [2, 3]
- Dangerous to fish or other aquatic life. Do not contaminate surface waters or ditches with chemical or used container [1, 3]
- LERAP Category B

Hazard classification and safety precautions

Hazard H02, H11 [2, 3]; H04 [1]
Risk phrases R23, R40, R50, R53a [2, 3]; R38 [1]; R41, R43 [1-3]
Operator protection A, C, H; U02a, U05a, U09a, U19a, U20b [1-3]; U11 [2, 3]
Environmental protection E13b [1, 3]; E15a [2]; E16a [1-3]; E38 [2, 3]
Storage and disposal D01, D02, D09a, D10b [1-3]; D05 [3]

85 chlorothalonil + mancozeb

A multi-site protectant fungicide mixture
FRAC mode of action code: M5 + M3

Products

1	Adagio	Interfarm	201:274 g/l	SC	10796
2	Guru	Interfarm	286: 194 g/l	SC	10801

Uses
- Blight in ***potatoes*** [1]
- Foliar disease control in ***durum wheat*** *(off-label)*, ***grass seed crops*** *(off-label)*, ***spring rye*** *(off-label)*, ***winter rye*** *(off-label)* [2]
- Glume blotch in ***winter wheat*** [2]
- Septoria leaf blotch in ***winter wheat*** [2]

Specific Off-Label Approvals (SOLAs)
- ***durum wheat*** *20052892* [2]
- ***grass seed crops*** *20052892* [2]
- ***spring rye*** *20052892* [2]
- ***winter rye*** *20052892* [2]

Approval information
- Chlorothalonil and mancozeb included in Annex I under EC Directive 91/414

Efficacy guidance
- Start spray treatments on potatoes immediately after a blight warning or just before the haulm meets in the row [1]
- It is essential to start the blight spray programme before the disease appears in the crop [1]
- Repeat treatments for blight at 7, 10 or 14 d intervals depending on blight risk and continue until haulm is to be destroyed [1]
- Best results on winter wheat achieved from protective applications to the flag leaf. If disease already present on lower leaves treat as soon as flag leaf is just visible (GS 37) [2]
- Activity on Septoria may be reduced in presence of severe mildew infection [2]

Restrictions
- Maximum number of treatments 5 per crop for potatoes [1]
- Maximum total dose on winter wheat equivalent to 1.5 full dose treatments [2]

SEE SECTION 3 FOR PRODUCTS ALSO REGISTERED

- Do not treat crops under stress from frost, drought, waterlogging, trace element deficiency or pest attack [2]
- Broadcast air assisted applications must only be made by equipment fitted with a cab with a forced air filtration unit plus a pesticide filter complying with HSE Guidance Note PM 74 or an equivalent or higher standard

Crop-specific information

- Latest use: before grain watery ripe (GS 71) for winter wheat
- HI potatoes 7 d
- Irrigated potato crops should be sprayed immediately after irrigation [1]

Environmental safety

- Dangerous for the environment
- Very toxic to aquatic organisms
- LERAP Category B

Hazard classification and safety precautions

Hazard H04, H11
Risk phrases R37, R43, R50, R53a [1, 2]; R40 [2]
Operator protection A, C, H, M; U02a, U05a, U08, U10, U11, U13, U14, U19a [1, 2]; U20a [1]; U20b [2]
Environmental protection E15a [1]; E16a, E16b, E38 [1, 2]
Storage and disposal D01, D02, D05, D09a, D10b, D12a
Medical advice M04a [1]

86 chlorothalonil + metalaxyl-M

A systemic and protectant fungicide for various crops
FRAC mode of action code: M5 + 4

Products

Folio Gold	Syngenta	500:37.5 g/l	SC	10704

Uses

- Alternaria in ***brussels sprouts*** *(moderate control)*, ***calabrese*** *(moderate control)*, ***cauliflowers*** *(moderate control)*
- Downy mildew in ***broad beans***, ***brussels sprouts***, ***bulb onions*** *(qualified minor use)*, ***calabrese***, ***cauliflowers***, ***poppies for morphine production*** *(off-label)*, ***shallots*** *(qualified minor use)*, ***spring field beans***, ***winter field beans***
- Ring spot in ***brussels sprouts*** *(reduction)*, ***calabrese*** *(reduction)*, ***cauliflowers*** *(reduction)*
- White blister in ***brussels sprouts***, ***calabrese***, ***cauliflowers***
- White tip in ***leeks*** *(qualified minor use)*

Specific Off-Label Approvals (SOLAs)

- ***poppies for morphine production*** *20040469*

Approval information

- Chlorothalonil and metalaxyl-M included in Annex I under EC Directive 91/414

Efficacy guidance

- Apply at first signs of disease or when weather conditions favourable for disease pressure
- Best results obtained when used in a full and well-timed programme. Repeat treatment at 14-21 d intervals if necessary
- Treatment of established disease will be less effective
- Evidence of effectiveness in bulb onions, shallots and leeks is limited

Restrictions

- Maximum total dose equivalent to 2 full doses on broad beans, field beans, cauliflowers, calabrese; 3 full doses on Brussels sprouts, leeks, onions and shallots

Crop-specific information

- HI 14 d for all crops

Environmental safety

- Dangerous for the environment
- Very toxic to aquatic organisms
- LERAP Category B

Hazard classification and safety precautions

Hazard H03, H11

Risk phrases R20, R36, R37, R38, R40, R43, R50, R53a

Operator protection A, C, H; U05a, U09a, U15, U20a

Environmental protection E15a, E16a, E38

Consumer protection C02a (14 d)

Storage and disposal D01, D02, D05, D09a, D10c, D12a

87 chlorothalonil + picoxystrobin

A broad spectrum fungicide mixture for cereals
FRAC mode of action code: M5 + 11

Products

1	Credo	DuPont	500:100 g/l	SC	13042
2	Zimbrail	DuPont	500:100 g/l	SC	13172

Uses

- Brown rust in ***spring barley, spring wheat, winter barley, winter wheat***
- Glume blotch in ***spring wheat, winter wheat***
- Net blotch in ***spring barley, winter barley***
- Powdery mildew in ***spring barley, winter barley***
- Rhynchosporium in ***spring barley, winter barley***
- Septoria leaf blotch in ***spring wheat, winter wheat***
- Tan spot in ***spring wheat, winter wheat***
- Yellow rust in ***spring wheat, winter wheat***

Approval information

- Chlorothalonil and picoxystrobin included in Annex I under EC Directive 91/414
- Accepted by BBPA for use on malting barley

Efficacy guidance

- Best results obtained as a protectant treatment, or when disease first seen in crop, applied in good growing conditions with adequate soil moisture
- Disease control for 4-6 wk normally achieved during stem elongation
- Results may be less reliable when used on crops under stress
- Treatments for protection against ear disease should be made at ear emergence
- Picoxystrobin is a member of the QoI cross resistance group. Product should be used preventatively and not relied on for its curative potential
- Use product as part of an Integrated Crop Management strategy incorporating other methods of control, including where appropriate other fungicides with a different mode of action. Do not apply more than two foliar applications of QoI containing products to any cereal crop
- Control of many diseases may be improved by use in mixture with an appropriate triazole
- There is a significant risk of widespread resistance occurring in *Septoria tritici* populations in UK. Failure to follow resistance management action may result in reduced levels of disease control
- Strains of barley powdery mildew resistant to QoIs are common in the UK

Restrictions

- Maximum number of treatments 2 per crop of wheat or barley

Crop-specific information

- Latest use: before first spikelet just visible (GS 51) for barley; before grain watery ripe (GS 71) for wheat

Environmental safety

- Dangerous for the environment
- Very toxic to aquatic organisms
- LERAP Category B

Hazard classification and safety precautions

Hazard H03, H11

Risk phrases R20, R37, R40, R43, R50, R53a

Operator protection A, H; U05a, U09a, U14, U15, U19a, U20a

Environmental protection E15a, E16a, E16b, E38
Storage and disposal D01, D02, D05, D09a, D10c, D12a

88 chlorothalonil + propamocarb hydrochloride

A contact and systemic fungicide mixture for blight control in potatoes
FRAC mode of action code: M5 + 28

Products

1 Merlin	Bayer CropScience	375:375 g/l	SC	07943
2 Pan Magician	Pan Agriculture	375:375 g/l	SC	11992

Uses
- Blight in ***potatoes***

Approval information
- Chlorothalonil and propamocarb hydrochloride included in Annex I under EC Directive 91/414

Efficacy guidance
- Commence treatment early in the season as soon as there is risk of infection
- In the absence of a blight warning treatment should start just before potatoes meet along the row
- Use only as a protectant. Stop use when blight readily visible (1% leaf area destroyed)
- Repeat sprays at 10-14 d intervals depending on blight infection risk. See label for details
- Complete blight spray programme after end Aug up to haulm destruction with protectant fungicides preferably fentin based

Restrictions
- Maximum total dose equivalent to 6 full dose treatments on potatoes
- Apply to dry foliage. Do not apply if rainfall or irrigation imminent

Crop-specific information
- HI 7 d for potatoes

Environmental safety
- Dangerous for the environment
- Very toxic to aquatic organisms
- LERAP Category B

Hazard classification and safety precautions
Hazard H03, H11
Risk phrases R20, R40, R41, R43, R50, R53a
Operator protection A, C, H; U05a, U08, U11, U14, U19a, U20a
Environmental protection E15a, E16a, E16b, E34, E38
Storage and disposal D01, D02, D05, D09a, D10b, D12a

89 chlorothalonil + pyrimethanil

A fungicide mixture for use in combining peas
FRAC mode of action code: M5 + 9

Products

Walabi	BASF	375:150 g/l	SC	12265

Uses
- Ascochyta in ***combining peas***
- Botrytis in ***combining peas***
- Chocolate spot in ***spring field beans***, ***winter field beans***
- Mycosphaerella in ***combining peas***

Approval information
- Chlorothalonil and pyrimethanil included in Annex I under EC Directive 91/414.

Efficacy guidance
- Use only as a protectant spray in a two spray programme
- For best results ensure good foliar cover

Restrictions

- Maximum total dose equivalent to two full dose treatments
- Consult processors before use on crops for processing

Crop-specific information

- HI 6 wk for combining peas; 8 wk for field beans

Environmental safety

- Dangerous for the environment
- Very toxic to aquatic organisms
- LERAP Category B

Hazard classification and safety precautions

Hazard H03, H11
Risk phrases R36, R40, R43, R50, R53a
Operator protection A, C, H; U02a, U11
Environmental protection E15a, E16a
Storage and disposal D01, D02, D05, D09a, D10c, D12a
Medical advice M03

90 chlorotoluron

A contact and residual urea herbicide for cereals
HRAC mode of action code: C2

Products

1	Alpha Chlorotoluron 500	Makhteshim	500 g/l	SC	04848
2	Atol	Nufarm UK	700 g/l	SC	11138
3	Clayton Chloron 500 FL	Clayton	500 g/l	SC	10791
4	CTU Minrinse	Makhteshim	90% w/w	WG	12458
5	Headland Tolerate	Headland	500 g/l	SC	10774
6	Lentipur 700	Nufarm UK	700 g/l	SC	13526
7	Lentipur CL 500	Nufarm UK	500 g/l	SC	08743
8	Tolugan 700	Makhteshim	700 g/l	SC	08064
9	Tolurex 90 WDG	Makhteshim	90% w/w	WG	11403

Uses

- Annual dicotyledons in ***durum wheat*** [4, 7, 9]; ***spring rye*** *(off-label)* [1, 7]; ***triticale***, ***winter barley***, ***winter wheat*** [1-9]; ***winter rye*** *(off-label)* [1, 7, 8]
- Annual grasses in ***durum wheat*** [4, 7, 9]; ***spring rye*** *(off-label)* [1, 7]; ***triticale***, ***winter barley***, ***winter wheat*** [1-9]; ***winter rye*** *(off-label)* [1, 7, 8]
- Blackgrass in ***durum wheat*** [4, 7, 9]; ***triticale***, ***winter barley***, ***winter wheat*** [1-9]
- Rough meadow grass in ***durum wheat*** [4, 7, 9]; ***triticale***, ***winter barley***, ***winter wheat*** [1-9]
- Wild oats in ***durum wheat*** [4, 7, 9]; ***triticale***, ***winter barley***, ***winter wheat*** [1-9]

Specific Off-Label Approvals (SOLAs)

- ***spring rye*** *20052337* [1], *20052336* [7]
- ***winter rye*** *20052337* [1], *20052336* [7], *20052338* [8]

Approval information

- Chlorotoluron included in Annex I under EC Directive 91/414
- May be applied through CDA equipment. See labels for details [10633, 2, 4, 7]
- Approved for aerial application on winter wheat, winter barley, durum wheat, triticale [7]. See notes in Section 5
- Accepted by BBPA for use on malting barley

Efficacy guidance

- Best results achieved by application soon after drilling. Application in autumn controls most weeds germinating in early spring
- For wild oat control apply within 1 wk of drilling, not after 2-leaf stage. Blackgrass and meadow grasses controlled to 5 leaf, ryegrasses to 3 leaf stage
- Any trash or burnt straw should be buried and dispersed during seedbed preparation
- Control may be reduced if prolonged dry conditions follow application
- Harrowing after treatment may reduce weed control
- Always follow WRAG guidelines for preventing and managing herbicide resistant weeds. Section 5 for more information

Restrictions

- Maximum number of treatments 1 per crop
- Use only on listed crop varieties. See label. Ensure seed well covered at drilling
- Apply only as pre-emergence spray in durum wheat, pre- or post-emergence in wheat, barley or triticale
- Do not apply pre-emergence to crops sown after 30 Nov
- Do not apply to crops severely checked by waterlogging, pests, frost or other factors
- Do not use on undersown crops or those due to be undersown
- Do not apply post-emergence in mixture with liquid fertilizers
- Do not roll for 7 d before or after application to an emerged crop
- Do not use on soils with more than 10% organic matter
- Crops on stony or gravelly soils may be damaged, especially after heavy rain

Crop-specific information

- Latest use: pre-emergence for durum wheat. Post emergence timings on other crops vary - see labels
- Early sown crops may be damaged if applied before or during a period of rapid autumn growth
- Autumn treated crops must be drilled by 1 Dec and spring treated crops must be drilled by 1 Feb

Environmental safety

- Dangerous for the environment
- Very toxic to aquatic organisms. May cause long-term adverse effects in the aquatic environment

Hazard classification and safety precautions

Hazard H03 [1-3, 5-7]; H04 [1]; H11 [1-9]
Risk phrases R22a, R36, R38 [1, 3, 5]; R40 [3, 7]; R50, R53a [1-9]; R63 [2, 3, 6, 7]; R68 [2, 6, 7]
Operator protection A, C; U05a [1-6, 8, 9]; U08, U19a [1-9]; U10 [5]; U11 [3, 5]; U20a [2, 6, 8]; U20b [1, 3-5, 7, 9]
Environmental protection E15a [1-9]; E34 [1-6, 8, 9]
Storage and disposal D01, D02, D09a, D10b [1-9]; D05 [3, 5]; D12a [2, 5-7]; D12b [1, 3, 4, 8, 9]
Medical advice M03 [1-6, 8, 9]; M05a [3]

91 chlorotoluron + diflufenican

A soil and foliar acting herbicide mixture for winter cereals
HRAC mode of action code: C2 + F1

Products

Buckler	Nufarm UK	600:25 g/l	SC	13080

Uses

- Annual dicotyledons in ***winter barley***, ***winter wheat***
- Annual meadow grass in ***winter barley***, ***winter wheat***

Approval information

- Chlorotoluron included in Annex I under EC Directive 91/414

Efficacy guidance

- Best results obtained in crops sown in a fine firm seedbed with clods no greater than fist size and adequate moisture present at spraying. Good even spray coverage of the soil is essential
- Good weed control depends on burial and dispersal of any trash or straw before drilling
- Weed control may be reduced in prolonged periods of below average rainfall
- In direct drilled crops soil surface should be cultivated and the drill slots closed before spraying
- Crops should be treated pre-emergence and before end Feb
- Spring gerrminating weeds will not be controlled
- Fluffy seedbeds should be rolled before treatment
- Harrowing a treated crop may reduce the level of weed control

Restrictions

- Maximum number of treatments 1 per crop for wheat, barley
- Do not apply if heavy rain is expected within 4 h or if crop is stressed
- Do not roll for 7 d before or after application. Do not roll autumn crops until spring
- Do not treat undersown crops, or those to be undersown

- Do not apply on soils with more than 10% organic matter
- Treat only listed varieties of wheat and barley

Crop-specific information

- Latest use: before end Feb for winter barley, winter wheat
- Transient leaf discolouration occasionally occurs after treatment of wheat or barley
- Treatment of wheat or barley on stony or gravelly soils may cause damage especially if heavy rain falls after application
- Early sown crops of wheat or barley may be damaged if treated during a period of rapid autumn growth

Following crops guidance

- Any crop may be drilled or planted following normal harvest of a treated crop provided the soil is ploughed to 150 mm
- In the event of failure of a treated crop only winter wheat or winter barley may be drilled immediately after ploughing. A period of 12 wk must elapse after ploughing before spring sowing of wheat, barley, oilseed rape, peas, beans, sugar beet, potatoes, carrots, edible brassicas or onions
- Successive treatments with any products containing diflufenican can lead to soil build-up and inversion ploughing to 150 mm must precede sowing any following non-cereal crop. Even where ploughing occurs some crops may be damaged

Environmental safety

- Dangerous for the environment
- Very toxic to aquatic organisms
- LERAP Category B
- Do not apply to dry, cracked or waterlogged soils

Hazard classification and safety precautions

Hazard H03, H11
Risk phrases R40, R43, R50, R53a, R63
Operator protection A, H; U05a, U08, U13, U19a, U20b
Environmental protection E15a, E16a, E38
Storage and disposal D01, D02, D05, D09a, D10b, D12a

92 chlorotoluron + isoproturon

A urea herbicide mixture for cereals
HRAC mode of action code: C2 + C2

Products

Tolugan Extra	Makhteshim	300:300 g/l	SC	09393

Uses

- Annual dicotyledons in ***durum wheat*** *(off-label)*, ***spring barley*** *(off-label)*, ***triticale*** *(off-label)*, ***winter barley***, ***winter rye*** *(off-label)*, ***winter wheat***
- Annual grasses in ***durum wheat*** *(off-label)*, ***spring barley*** *(off-label)*, ***triticale*** *(off-label)*, ***winter barley***, ***winter rye*** *(off-label)*, ***winter wheat***
- Blackgrass in ***winter barley***, ***winter wheat***
- Rough meadow grass in ***winter barley***, ***winter wheat***
- Wild oats in ***winter barley***, ***winter wheat***

Specific Off-Label Approvals (SOLAs)

- ***durum wheat*** *20052993*
- ***spring barley*** *20070111*
- ***triticale*** *20052993*
- ***winter rye*** *20052993*

Approval information

- Chlorotoluron and isoproturon included in Annex I under EC Directive 91/414. However, see isoproturon entry for information on phasing out of approvals of all products containing isoproturon by 30 June 2009
- Accepted by BBPA for use on malting barley

SEE SECTION 3 FOR PRODUCTS ALSO REGISTERED

Efficacy guidance

- Best results achieved by application soon after drilling. Application in autumn controls most weeds germinating in early spring
- For wild oat control apply within 1 wk of drilling, not after 2-leaf stage. Blackgrass and meadow grasses controlled to 5 leaf, ryegrasses to 3 leaf stage
- Any trash or burnt straw should be buried and dispersed during seedbed preparation
- Control may be reduced if prolonged dry conditions follow application
- Harrowing after treatment may reduce weed control
- Where strains of herbicide-resistant blackgrass occur control may be unsatisfactory
- Always follow WRAG guidelines for preventing and managing herbicide resistant weeds. Section 5 for more information

Restrictions

- Maximum number of treatments 1 per crop
- Use only on listed crop varieties. See label. Ensure seed well covered at drilling
- Apply only as post-emergence treatment
- Do not apply to crops severely checked by waterlogging, pests, frost or other factors
- Do not use on undersown crops or those due to be undersown
- Do not apply post-emergence in mixture with liquid fertilizers
- Do not roll for 7 d before or after application to an emerged crop
- Do not use on soils with more than 10% organic matter
- Crops on stony or gravelly soils may be damaged, especially after heavy rain

Crop-specific information

- Latest use: before end of tillering (GS 29) for winter barley, winter wheat

Environmental safety

- Dangerous for the environment
- Very toxic to aquatic organisms
- Do not apply to dry, cracked or waterlogged soils where heavy rain may lead to contamination of drains by isoproturon

Hazard classification and safety precautions

Hazard H03, H11
Risk phrases R40, R43, R50, R53a
Operator protection A, C, H; U05a, U08, U19a, U20b
Environmental protection E15a, E34
Storage and disposal D01, D02, D05, D09a, D10c, D12b

93 chlorpropham

A residual carbamate herbicide and potato sprout suppressant
HRAC mode of action code: K2

See also chloridazon + chlorpropham + metamitron

Products

1	BL 500	Whyte Agrochemicals	500 g/l	HN	00279
2	Comrade	United Phosphorus	400 g/l	EC	10181
3	Gro-Stop 100	Certis	300 g/l	HN	13742
4	Gro-Stop Fog	Certis	300 g/l	HN	13741
5	Gro-Stop HN	Certis	300 g/l	HN	13738
6	Jupiter 40 EC	United Phosphorus	400 g/l	EC	12852
7	MSS CIPC 5 G	Whyte Agrochemicals	5% w/w	GR	01402
8	MSS CIPC 50 M	Whyte Agrochemicals	500 g/l	HN	11165
9	MSS Sprout Nip	Whyte Agrochemicals	100% w/w	HN	11804
10	Pro-Long	Whyte Agrochemicals	500 g/l	HN	12600
11	Warefog 25	Whyte Agrochemicals	600 g/l	HN	06776

Uses

- Annual dicotyledons in ***acroclinium***, ***african marigolds***, ***bulb onions***, ***calendula***, ***carrots***, ***china asters***, ***chrysanthemums***, ***coreopsis***, ***flower bulbs***, ***french marigolds***, ***lettuce***, ***parsley***, ***straw flower***, ***sweet sultan*** [2, 6]; ***blackcurrants***, ***celery (outdoor)***, ***gooseberries***, ***lucerne***, ***strawberries*** [2]; ***garlic*** *(off-label)*, ***horseradish*** *(off-label)*, ***leeks***, ***parsley root*** *(off-label)*, ***protected frise*** *(off-label)*, ***shallots*** *(off-label)* [6]

- Annual grasses in ***acroclinium, african marigolds, bulb onions, calendula, carrots, china asters, chrysanthemums, coreopsis, flower bulbs, french marigolds, lettuce, parsley, straw flower, sweet sultan*** [2, 6]; ***blackcurrants, celery (outdoor), gooseberries, lucerne, strawberries*** [2]; ***garlic*** *(off-label)*, ***horseradish*** *(off-label)*, ***leeks, parsley root*** *(off-label)*, ***protected frise*** *(off-label)*, ***shallots*** *(off-label)* [6]
- Chickweed in ***acroclinium, african marigolds, calendula, carrots, china asters, chrysanthemums, coreopsis, french marigolds, lettuce, parsley, straw flower, sweet sultan*** [2, 6]; ***blackcurrants, celery (outdoor), gooseberries, lucerne, strawberries*** [2]; ***bulb onions, flower bulbs, leeks*** [6]
- Polygonums in ***acroclinium, african marigolds, calendula, carrots, china asters, chrysanthemums, coreopsis, french marigolds, lettuce, parsley, straw flower, sweet sultan*** [2, 6]; ***blackcurrants, celery (outdoor), gooseberries, lucerne, strawberries*** [2]; ***bulb onions, flower bulbs, leeks*** [6]
- Sprout suppression in ***ware potatoes*** [7, 8]; ***ware potatoes*** *(thermal fog)* [1, 3-5, 9-11]

Specific Off-Label Approvals (SOLAs)

- ***garlic*** *20070827* [6]
- ***horseradish*** *20070828* [6]
- ***parsley root*** *20070828* [6]
- ***protected frise*** *20070829* [6]
- ***shallots*** *20070827* [6]

Approval information

- Chlorpropham included in Annex I under EC Directive 91/414
- Some products are formulated for application by thermal fogging. See labels for details [07689, 09388, 120521, 9, 11]
- Approval expiry 31 Aug 2008 [2]

Efficacy guidance

- Apply weed control sprays to freshly cultivated soil. Adequate rainfall must occur after spraying. Activity is greater in cold, wet than warm, dry conditions
- For sprout suppression apply with suitable fogging or rotary atomiser equipment or sprinkle granules over dry tubers before sprouting commences. Repeat applications may be needed. See labels for details
- Best results on potatoes obtained in purpose-built box stores with suitable forced draft ventilation. Potatoes in bulk stores should not be stacked more than 3 m high
- Blockage of air spaces between tubers prevents circulation of vapour and consequent loss of efficacy
- It is important to treat potatoes before the eyes open to obtain best results
- Effectiveness of fogging reduced in non-dedicated stores without proper insulation and temperature controls. Best results obtained at 5-10 °C and 75-80% humidity

Restrictions

- Maximum number of treatments normally 1 per batch for sprout suppression of potatoes; 1 per crop or yr for other named crops or situations
- Not to be used on grass seed crops if grass to be grazed or cut for fodder before 31 May following treatment [2]
- Excess rainfall after application may result in crop damage
- Do not use on Sands, Very Light soils or soils low in organic matter
- Poor conditions at drilling or planting, soil compaction, surface capping, waterlogging or attack by pests may result in crop damage
- On crops under glass high temperatures and poor ventilation may cause crop damage
- Only clean, mature, disease-free potatoes should be treated for sprout suppression. Use of chlorpropham can inhibit tuber wound healing and the severity of skin spot infection in store may be increased if damaged tubers are treated
- Do not fog potatoes with a high level of skin spot
- Do not use on potatoes for seed. Do not handle, dry or store seed potatoes or any other seed or bulbs in boxes or buildings in which potatoes are being or have been treated
- Do not remove treated potatoes from store for sale or processing for at least 21 d after application

Crop-specific information

- Latest use: 3 d before drilling carrots; 7d before planting out celery; 2 wk before planting or at least 10 d after planting for chrysanthemums [2]; 21 d before removal for sale or processing for potatoes [1, 7, 8, 11]; before tulip leaves unfurl or flower bulbs 5 cm high
- Cure potatoes according to label instructions before treatment and allow 3 wk between completion of loading into store and first treatment
- Apply weed control sprays to seeded crops pre-emergence of crop or weeds, to onions as soon as first crop seedlings visible, to planted crops a few days before planting, to bulbs immediately after planting, to fruit crops in late autumn-early winter. See label for further details

Following crops guidance

- There is a risk of damage to seed potatoes which are handled or stored in boxes or buildings previously treated with chlorpropham

Environmental safety

- Dangerous for the environment
- Toxic to aquatic organisms
- Flammable [1, 2, 6, 8-10]
- Keep unprotected persons out of treated stores for at least 24 h after application
- Keep in original container, tightly closed, in a safe place, under lock and key

Hazard classification and safety precautions

Hazard H02, H07 [1, 8-10]; H03 [2-6, 11]; H08 [2, 6]; H11 [1, 3-11]

Risk phrases R20, R22a [2, 6]; R21 [2]; R22b, R37, R66 [6]; R23, R24, R39 [1, 8-10]; R25 [1, 10]; R36 [1, 6, 8-11]; R38 [2, 6, 11]; R40, R42, R43 [3-5]; R48 [11]; R51 [1, 3-6, 8, 10, 11]; R52 [7]; R53a [1, 3-8, 10, 11]

Operator protection A [1-6, 8-10]; C [1, 2, 8-10]; D, E, H, M [1, 3-5, 8-11]; J [1, 8-11]; U02a, U04a, U20a [3-5]; U05a [1-11]; U08 [1-5, 8-11]; U09a [2, 6]; U14, U15 [3-5, 11]; U19a [1, 3-5, 8-11]; U20b [1, 2, 6-10]; U20c [11]

Environmental protection E02a [1, 8-10] (24 h); E02a [11] (12 h); E13b [6]; E15a [1-5, 7-11]; E34 [1-11]

Consumer protection C02b [1, 3-5, 7, 9-11]; C12 [3-5]

Storage and disposal D01, D02 [1-11]; D05 [1-6, 8-10]; D06b [1, 2, 8, 9]; D07, D12a [3-5]; D08 [2]; D09a [2, 6, 7, 11]; D09b [1, 3-5, 8-10]; D10a [1, 8-11]; D10c [2, 6]; D11a [3-5, 7]; D12b [1, 2, 7-11]

Medical advice M03 [2, 6]; M04a [1, 3-5, 8-10]

94 chlorpropham with cetrimide

A soil-acting herbicide for lettuce under cold glass

HRAC mode of action code: K2

Products

Croptex Pewter	Certis	80:80 g/l	SC	11181

Uses

- Annual dicotyledons in ***protected chives*** *(off-label)*, ***protected herbs (see appendix 6)*** *(off-label)*, ***protected lettuce***, ***protected parsley*** *(off-label)*
- Annual grasses in ***protected lettuce***
- Chickweed in ***protected lettuce***
- Polygonums in ***protected lettuce***

Specific Off-Label Approvals (SOLAs)

- ***protected chives*** *20053031*
- ***protected herbs (see appendix 6)*** *20053031*
- ***protected parsley*** *20053031*

Approval information

- Chlorpropham included in Annex I under EC Directive 91/414

Efficacy guidance

- Adequate irrigation must be applied before or after treatment
- Best results achieved on firm soil of fine tilth, free from clods and weeds
- Control of susceptible weeds is achieved at, or shortly after, germination and residual control lasts for 6-8 wk depending on weather conditions

Restrictions

- Maximum number of treatments 1 per crop
- Apply within 24 h of drilling
- Do not apply to crop foliage or use where seed has germinated
- Excess irrigation may cause temporary check to crop under certain circumstances
- Do not apply where tomatoes, brassicas or other sensitive crops are growing in the same house

Crop-specific information

- Latest use: before crop emergence or pre-planting
- Apply to drilled lettuce under cold glass within 24 h post-drilling, to transplanted crops pre-planting and treat up to 7 d later

Following crops guidance

- In the event of crop failure only lettuce should be grown within 2 mth
- After cutting a normally harvested crop the treated area should be cultivated to a minimum depth of 15 cm before sowing any succeeding crop

Environmental safety

- Dangerous for the environment
- Toxic to aquatic organisms
- Highly flammable

Hazard classification and safety precautions

Hazard H03, H07, H11
Risk phrases R21, R22a, R36, R37, R38, R51, R53a
Operator protection A, C; U05a, U08, U14, U15, U19a, U20a
Environmental protection E15a, E34
Storage and disposal D01, D02, D12a
Medical advice M03

95 chlorpyrifos

A contact and ingested organophosphorus insecticide and acaricide
IRAC mode of action code: 1B

Products

1	Alpha Chlorpyrifos 48 EC	Makhteshim	480 g/l	EC	13532
2	Ballad	Headland	480 g/l	EC	11659
3	Chevron 48	DAPT	480 g/l	EC	10645
4	Clayton Pontoon	Clayton	480 g/l	EC	13219
5	Crossfire 480	Dow	480 g/l	EC	12516
6	Cyren	Headland	480 g/l	EC	11028
7	Dursban WG	Dow	75% w/w	WG	09153
8	Equity	Dow	480 g/l	EC	12465
9	Govern	Dow	75% w/w	WG	13223
10	Greencrop Pontoon	Greencrop	480 g/l	EC	09667
11	Parapet	Dow	75 % w/w	WG	12773
12	Spannit Granules	SumiAgro	6% w/w	GR	10935
13	Spannit Granules	Barclay	6% w/w	GR	12950
14	Standon Chlorpyrifos 48	Standon	480 g/l	EC	08286
15	suSCon Green Soil Insecticide	Fargro	10% w/w	CG	06312

Uses

- Ambrosia beetle in ***cut logs*** [1-4, 6-11]
- Aphids in ***all currants*** [2, 4, 6, 8, 10, 14]; ***apples***, ***gooseberries***, ***pears***, ***plums***, ***raspberries***, ***strawberries*** [1-4, 6, 8-11, 14]; ***blackberries*** *(off-label)* [8, 11]; ***blackcurrants*** [1, 3, 9, 11]; ***broccoli***, ***cabbages***, ***calabrese***, ***cauliflowers***, ***chinese cabbage*** [1-4, 6-11, 14]; ***brussels sprouts*** [1-3, 6]; ***quinces*** *(off-label)* [8]; ***redcurrants***, ***whitecurrants*** [4, 9-11]
- Apple blossom weevil in ***apples*** [1-4, 6, 8-11, 14]
- Apple sucker in ***apples*** [9, 11, 14]
- Bean seed fly in ***green beans*** *(off-label - harvested as a dry pulse)* [6, 7]
- Cabbage root fly in ***broccoli***, ***cabbages***, ***cauliflowers*** [1-4, 6-14]; ***brussels sprouts*** [1-4, 6-9, 11]; ***calabrese***, ***chinese cabbage*** [1-4, 6-11, 14]; ***choi sum*** *(off-label)*, ***collards*** *(off-label)*, ***kale***

SEE SECTION 3 FOR PRODUCTS ALSO REGISTERED

(off-label), ***kohlrabi*** *(off-label)*, ***mooli*** *(off-label)*, ***pak choi*** *(off-label)*, ***radishes*** *(off-label)* [6, 7, 11]; ***green beans*** *(off-label - harvested dry as a pulse)* [11]; ***tatsoi*** *(off-label)* [6]

- Capsids in ***all currants*** [2, 4, 6, 8, 10, 14]; ***apples***, ***gooseberries***, ***pears*** [1-4, 6, 8-11, 14]; ***blackcurrants*** [1, 3, 9, 11]; ***quinces*** *(off-label)* [8]; ***redcurrants***, ***whitecurrants*** [4, 9-11]
- Caterpillars in ***all currants*** [2, 4, 6, 8, 10, 14]; ***apples***, ***pears*** [1-4, 6, 8, 10]; ***blackcurrants*** [1, 3, 9, 11]; ***broccoli***, ***cabbages***, ***calabrese***, ***cauliflowers***, ***chinese cabbage*** [1-4, 6-11, 14]; ***brussels sprouts*** [1-3, 6]; ***gooseberries*** [1-4, 6, 8-11, 14]; ***quinces*** *(off-label)* [8]; ***redcurrants***, ***whitecurrants*** [4, 9-11]
- Codling moth in ***apples***, ***pears*** [1-4, 6, 8-11, 14]
- Cutworms in ***broccoli***, ***cabbages***, ***calabrese***, ***cauliflowers***, ***chinese cabbage*** [1, 4, 7-11, 14]; ***brussels sprouts*** [1]; ***bulb onions*** [1, 2, 4, 6-11, 14]; ***seed potatoes*** [1, 2, 4, 7-11, 14]
- Damson-hop aphid in ***plums*** [1, 4, 9-11, 14]
- Elm bark beetle in ***cut logs*** [2, 3, 6]
- Frit fly in ***amenity grassland*** [2-6, 10, 14]; ***durum wheat***, ***spring rye***, ***triticale***, ***winter rye*** [7]; ***durum wheat*** *(off-label)*, ***spring rye*** *(off-label)*, ***triticale*** *(off-label)*, ***winter rye*** *(off-label)* [8, 11]; ***forage maize***, ***permanent grassland***, ***spring oats***, ***spring wheat***, ***winter oats***, ***winter wheat*** [1-4, 6-11, 14]; ***managed amenity turf*** [1-6, 10, 14]; ***rotational grass*** [1-3, 6, 7, 9, 11]; ***spring barley***, ***winter barley*** [8]
- Larch shoot beetle in ***cut logs*** [1-4, 6-11]
- Leatherjackets in ***amenity grassland*** [2-6, 10, 14]; ***broccoli***, ***cabbages***, ***calabrese***, ***cauliflowers***, ***chinese cabbage*** [2, 3, 6-9, 11, 14]; ***brussels sprouts*** [2, 3, 6]; ***durum wheat***, ***spring rye***, ***triticale***, ***winter rye*** [7]; ***durum wheat*** *(off-label)*, ***spring rye*** *(off-label)*, ***triticale*** *(off-label)*, ***winter rye*** *(off-label)* [8, 11]; ***fodder beet*** *(off-label)* [7, 8, 11]; ***managed amenity turf*** [1-6, 10, 14]; ***permanent grassland***, ***spring oats***, ***spring wheat***, ***sugar beet***, ***winter oats***, ***winter wheat*** [1-4, 6-11, 14]; ***rotational grass*** [1-3, 6, 7, 9, 11]; ***spring barley***, ***winter barley*** [1-4, 6-9, 11]
- Mealy aphid in ***plums*** [1, 4, 10]
- Narcissus fly in ***flower bulbs*** *(off-label)* [6]
- Pear sucker in ***pears*** [9, 11, 14]
- Pine shoot beetle in ***cut logs*** [1-4, 6-11]
- Pygmy mangold beetle in ***fodder beet*** *(off-label)* [7, 8, 11]; ***sugar beet*** [2, 3, 6-9, 11]
- Raspberry beetle in ***raspberries*** [1-4, 6, 8-11, 14]
- Raspberry cane midge in ***raspberries*** [1-4, 6, 8-11, 14]
- Red spider mites in ***all currants*** [2, 4, 6, 8, 10, 14]; ***apples***, ***gooseberries***, ***pears***, ***plums***, ***raspberries***, ***strawberries*** [1-4, 6, 8-11, 14]; ***blackberries*** *(off-label)* [8]; ***blackcurrants*** [1, 3, 9, 11]; ***redcurrants***, ***whitecurrants*** [4, 9-11]
- Sawflies in ***apples*** [1-4, 6, 8-11, 14]; ***pears*** [8]
- Strawberry blossom weevil in ***strawberries*** [8]
- Suckers in ***apples***, ***pears*** [1-4, 6, 8, 10]
- Summer-fruit tortrix moth in ***apples*** [9, 11, 14]
- Thrips in ***spring oats***, ***spring wheat***, ***winter oats***, ***winter wheat*** [2, 3, 6]
- Tortrix moths in ***apples***, ***pears***, ***plums***, ***strawberries*** [1-4, 6, 8-11, 14]
- Vine weevil in ***ornamental plant production*** [15]; ***strawberries*** [1-4, 6, 8-11, 14]
- Wheat bulb fly in ***durum wheat***, ***spring rye***, ***triticale***, ***winter rye*** [7]; ***durum wheat*** *(off-label)*, ***spring rye*** *(off-label)*, ***triticale*** *(off-label)*, ***winter rye*** *(off-label)* [8, 11]; ***spring barley***, ***winter barley*** [4, 7-9, 11]; ***spring oats***, ***spring wheat***, ***winter oats***, ***winter wheat*** [1-4, 6-11, 14]
- Wheat-blossom midge in ***durum wheat***, ***spring rye***, ***triticale***, ***winter rye*** [7]; ***durum wheat*** *(off-label)*, ***spring rye*** *(off-label)*, ***triticale*** *(off-label)*, ***winter rye*** *(off-label)* [8, 11]; ***spring barley***, ***winter barley*** [2, 3, 6]; ***spring oats***, ***spring wheat***, ***winter oats***, ***winter wheat*** [2, 3, 6-9, 11]
- Whitefly in ***broccoli***, ***cabbages***, ***calabrese***, ***cauliflowers***, ***chinese cabbage*** [2, 3, 6-9, 11]; ***brussels sprouts*** [2, 3, 6]
- Winter moth in ***apples***, ***pears***, ***plums*** [1-4, 6, 8-11, 14]
- Woolly aphid in ***apples*** [1-4, 6, 8-11, 14]; ***pears*** [2, 3, 6, 8]

Specific Off-Label Approvals (SOLAs)

- ***blackberries*** *20052964* [8], *20061585* [11]
- ***choi sum*** *20050237* [6], *20031390* [7], *20061587* [11]
- ***collards*** *20050237* [6], *20031390* [7], *20061587* [11]
- ***durum wheat*** *20061583* [11]
- ***flower bulbs*** *20050236* [6]
- ***fodder beet*** *20052979* [7], *20052965* [8], *20061584* [11]

- ***green beans*** *(harvested as a dry pulse) 20050235* [6], *(harvested as a dry pulse) 20031391* [7], *(harvested dry as a pulse) 20061586* [11]
- ***kale*** *20050237* [6], *20031390* [7], *20061587* [11]
- ***kohlrabi*** *20050235* [6], *20031391* [7], *20061586* [11]
- ***mooli*** *20050235* [6], *20031391* [7], *20061586* [11]
- ***pak choi*** *20050237* [6], *20031390* [7], *20061587* [11]
- ***quinces*** *20052974* [8]
- ***radishes*** *20050235* [6], *20031391* [7], *20061586* [11]
- ***spring rye*** *20061481* [8], *20061583* [11]
- ***tatsoi*** *20050237* [6]
- ***triticale*** *20061583* [11]
- ***winter rye*** *20061583* [11]

Approval information

- Chlorpyrifos included in Annex I under EC Directive 91/414
- Accepted by BBPA for use on malting barley
- Following implementation of Directive 98/82/EC, approval for use of chlorpyrifos on numerous crops was revoked in 1999
- Following the environmental review of chlorpyrifos uses on ware potatoes, carrots, conifers (drench), forestry trees (dip) and cereals (for aphids and yellow cereal fly) were revoked in September 2005. In addition the maximum permitted number of treatments, or the maximum total dose, on several other crops was reduced.
- In 2006 PSD required that all products containing this active ingredient should carry the following warning in the main area of the container label: "Chlorpyrifos is an anticholinesterase organophosphate. Handle with care"
- Approval expiry 31 Aug 2008 [3, 10, 14]
- Approval expiry 30 Apr 2008 [12]

Efficacy guidance

- Brassicas raised in plant-raising beds may require retreatment at transplanting
- Activity on field crops may be reduced when soil temperature below 5 °C or on organic soils
- In dry conditions the effect of granules applied as a surface band may be reduced [12, 13]
- Where pear suckers resistant to chlorpyrifos occur control is unlikely to be satisfactory
- Efficacy against frit fly and leatherjackets in grass may be reduced if applied during periods of frost
- Thorough incorporation by hand or mixing machine is essential when used in potting media for ornamental plant production
- When used for control of vine weevil in ornamental plant production the recently emerged and young larvae are controlled by incorporation into the growing media. A single application will control successive generations for the growing cycle of plants kept in the same containers for 2 yr, with partial control for a third yr. Where plants are potted-on into larger containers the fresh growing media must also be treated to obtain control [15]

Restrictions

- Contains an anticholinesterase organophosphate compound. Do not use if under medical advice not to work with such compounds
- Maximum number of treatments and timing vary with crop, product and pest. See label for details
- Do not apply to sugar beet under stress or within 4 d of applying a herbicide
- Once incorporated, treated growing media must be used within 30 d
- Do not use product in growing media used for aquatic or semi-aquatic plants, and not for propagation of any edible crop
- Do not graze lactating cows on treated pasture for 14 d after treatment

Crop-specific information

- Latest use 4d after transplanting edible brassica crops; end Jul for sugar beet; varies for other crops. See individual labels
- HI range from 7 d to 6 wk depending on crop. See labels
- For vine weevil control in ornamental nursery stock incorporate in growing medium when plants first potted from rooted cutting stage. Treat the fresh growing medium when plants are potted on into larger containers [15]
- For turf pests apply from Nov where high larval populations detected or damage seen
- On lettuce apply only to strong well developed plants when damage first seen, or on professional advice

- In apples use pre-blossom up to pink/white bud and post-blossom after petal fall
- Test tolerance of glasshouse ornamentals before widescale use for propagating unrooted cuttings, or potting unusual plants and new species, and when using any media with a high content of non-peat ingredients

Environmental safety

- Dangerous for the environment
- Very toxic to aquatic organisms
- Flammable [3-5, 8, 10, 14]
- Keep livestock out of treated areas for at least 14 d after treatment [1-8]
- High risk to bees. Do not apply to crops in flower or to those in which bees are actively foraging. Do not apply when flowering weeds are present [1-11, 14]
- Chlorpyrifos is dangerous to some beneficial arthropods, especially parasitoid wasps, ground beetles, rove beetles and hoverflies. Environmental protection advice is not to spray summer applications within 12 m of the edge of the growing crop
- LERAP Category A [1-11, 14]
- Broadcast air-assisted LERAP (18 m) [1, 3-5, 7-11, 14]

Hazard classification and safety precautions

Hazard H03 [1-15]; H04 [3]; H08 [3-5, 8, 10, 14]; H11 [1, 2, 4-15]

Risk phrases R20 [1, 2, 4-6, 8]; R21 [10, 12-14]; R22a [1-15]; R22b, R38 [1-6, 8, 10, 14]; R36 [1, 4, 5, 8]; R37 [4, 5, 8, 10, 14]; R41, R43 [1, 3]; R50 [1, 2, 4-11, 14]; R51 [12, 13, 15]; R53a [1, 2, 4-15]; R67 [10, 14]

Operator protection A [1-15]; C [1-3, 6]; H [1-4, 6, 10, 14]; P [8]; U02a [1, 3, 12, 13, 15]; U05a [1-6, 8, 10, 14]; U08, U19a [1-11, 14]; U11 [4, 5, 8]; U14 [1, 2, 6]; U20a [12, 13]; U20b [1-11, 14, 15]

Environmental protection E06a [1-8] (14 d); E12a [2, 6, 9, 11]; E12c [1, 3-5, 7, 8, 10, 14]; E12e, E16c [1-11, 14]; E13a [1-3, 6]; E15a [4, 5, 7-14]; E16d [1-3, 5-11, 14]; E17a [2, 6] (18 m); E17b [1, 3-5, 7-11, 14] (18 m); E34 [1-4, 6-11, 14]; E38 [1, 5, 7-9, 11-13, 15]

Consumer protection C01 [15]; C02a [12, 13] (6 wk)

Storage and disposal D01 [1-15]; D02 [1, 3-5, 7, 8, 10, 14]; D05 [1-4, 6, 10, 14, 15]; D09a [1-8, 10, 12-15]; D09c [9, 11]; D10b [1-11, 14]; D11a [12, 13, 15]; D12a [1, 4, 5, 7-9, 11-13, 15]; D12b [2, 6]

Medical advice M01 [3-5, 7-15]; M03 [2-11, 14]; M04a [15]; M05b [1-6, 8, 10, 14]

96 chlorpyrifos-methyl

An organophosphorus insecticide and acaricide for grain store use
IRAC mode of action code: 1B

Products

Reldan 22	Dow	225 g/l	EC	12404

Uses

- Grain storage pests in ***stored grain***
- Pre-harvest hygiene in ***grain stores***

Approval information

- Chlorpyrifos-methyl included in Annex I under EC Directive 91/414
- Accepted by BBPA for use in stores for malting barley
- In 2006 PSD required that all products containing this active ingredient should carry the following warning in the main area of the container label: "Chlorpyrifos-methyl is an anticholinesterase organophosphate. Handle with care"

Efficacy guidance

- Apply to grain after drying to moisture content below 14%, cooling and cleaning
- Insecticide may become depleted at grain surface if grain is being cooled by continuous extraction of air from the base leading to reduced control of grain store pests especially mites
- Resistance to organophosphorus compounds sometimes occurs in insect and mite pests of stored products

Restrictions

- Contains an anticholinesterase organophosphorus compound. Do not use if under medical advice not to work with such compounds
- Maximum number of treatments 1 per batch or 1 per store, prior to storage
- Only treat grain in good condition

- Do not treat grain intended for sowing
- Minimum of 90 d must elapse between treatment of grain and removal from store for consumption or processing

Crop-specific information

- May be applied pre-harvest to surfaces of empty store and grain handling machinery and as admixture with grain
- May be used in wheat, barley, oats, rye or triticale stores

Environmental safety

- Dangerous for the environment
- Very toxic to aquatic organisms

Hazard classification and safety precautions

Hazard H03, H11
Risk phrases R22b, R41, R50, R53a
Operator protection A, C, H, M; U05a, U08, U19a, U20b
Environmental protection E15a, E34
Storage and disposal D01, D02, D05, D09a, D10b
Medical advice M01, M03, M05b

97 chlorthal-dimethyl

A residual benzenedicarboxylic acid herbicide for use in horticulture
HRAC mode of action code: K1

Products

Dacthal W-75	Certis	75% w/w	WP	11323

Uses

- Annual dicotyledons in ***bedding plants***, ***bilberries*** *(off-label)*, ***blackberries*** *(off-label)*, ***blackcurrants***, ***blueberries*** *(off-label)*, ***broccoli***, ***brussels sprouts***, ***bulb onions***, ***cabbages***, ***calabrese***, ***cauliflowers***, ***chinese cabbage*** *(off-label)*, ***chives*** *(off-label)*, ***choi sum*** *(off-label)*, ***collards*** *(off-label)*, ***cranberries*** *(off-label)*, ***edible podded peas*** *(off-label)*, ***fodder rape***, ***garlic*** *(off-label)*, ***gooseberries***, ***herbs (see appendix 6)*** *(off-label)*, ***kale***, ***leeks***, ***pak choi*** *(off-label)*, ***parsley*** *(off-label)*, ***raspberries***, ***redcurrants*** *(off-label)*, ***roses***, ***rubus hybrids*** *(off-label)*, ***runner beans***, ***sage***, ***shallots*** *(off-label)*, ***shrubs***, ***strawberries***, ***swedes***, ***trees***, ***turnips***, ***whitecurrants*** *(off-label)*
- Black nightshade in ***dwarf beans*** *(off-label)*, ***navy beans*** *(off-label)*, ***poppies for morphine production*** *(off-label)*
- Fat hen in ***poppies for morphine production*** *(off-label)*
- General weed control in ***courgettes*** *(off-label)*, ***marrows*** *(off-label)*
- Slender speedwell in ***managed amenity turf***
- Volunteer potatoes in ***dwarf beans*** *(off-label)*, ***navy beans*** *(off-label)*

Specific Off-Label Approvals (SOLAs)

- ***bilberries*** *20061551*
- ***blackberries*** *20061553*
- ***blueberries*** *20061551*
- ***chinese cabbage*** *20061552*
- ***chives*** *20030873*
- ***choi sum*** *20061552*
- ***collards*** *20061552*
- ***courgettes*** *20041561*
- ***cranberries*** *20061551*
- ***dwarf beans*** *20030874*
- ***edible podded peas*** *20061549*
- ***garlic*** *20061550*
- ***herbs (see appendix 6)*** *20030873*
- ***marrows*** *20041561*
- ***navy beans*** *20030874*
- ***pak choi*** *20061552*
- ***parsley*** *20030873*

- ***poppies for morphine production*** *20032407*
- ***redcurrants*** *20061551*
- ***rubus hybrids*** *20061553*
- ***shallots*** *20061550*
- ***whitecurrants*** *20061551*

Efficacy guidance

- Best results on fine firm weed-free soil when adequate rain or irrigation follows
- Apply after drilling or planting prior to weed emergence. Rates and timing vary with crop and soil type. See label for details
- For control of slender speedwell in turf apply when weeds growing actively. Do not mow for at least 3 d after treatment

Restrictions

- Maximum number of treatments 1 per crop or 1 per season for all crops
- Do not apply mixture with propachlor to newly planted strawberries after rolling or application of other herbicides
- Do not use on strawberries between flowering and harvest
- Many types of ornamental have been treated successfully. See label for details. For species of unknown susceptibility treat a small number of plants first
- Do not use on turf where bent grasses form a major constituent of sward
- Do not plant lettuce within 6 mth of application, seeded turf within 2 mth, other crops within 3 mth. In the event of crop failure deep plough before re-drilling or planting
- Do not use on dwarf French beans
- Do not use on organic soils

Crop-specific information

- Latest use: before crop emergence on runner beans, sage; end of flowering for strawberries. Some restrictions apply to other crops. See label for details
- Recommended alone on roses, runner beans, various ornamentals, strawberries and turf, in tank-mix with propachlor on brassicas, onions, leeks, sage, ornamentals, established strawberries and newly planted soft fruit

Following crops guidance

- The interval between treatment and planting a following crop should be 6 mth for lettuce, 2 mth for seeded turf, 3 mth for other crops
- In the event of failure of a treated crop the land must be deep mould-board ploughed before re-drilling or re-planting

Environmental safety

- May cause long-term adverse effects in the aquatic environment

Hazard classification and safety precautions

Hazard H03
Risk phrases R40, R53a
Operator protection U19a, U20a
Environmental protection E15a
Storage and disposal D01, D02, D09a, D11a
Medical advice M04a

98 chlorthal-dimethyl + propachlor

A residual herbicide mixture for a range of vegetable crops
HRAC mode of action code: K1 + K3

Products

1	Decimate	Certis	225:216 g/l	SC	11008
2	Decimate	Certis	225:216 g/l	SC	13419
3	Greencrop Vegex	Greencrop	225:216 g/l	SC	12439

Uses

- Annual dicotyledons in ***broccoli***, ***brussels sprouts***, ***bulb onions***, ***cabbages***, ***calabrese***, ***cauliflowers***, ***fodder rape***, ***kale***, ***leeks***, ***mustard***, ***salad onions***, ***swedes***, ***turnips*** [1-3]; ***chinese cabbage*** *(off-label)*, ***choi sum*** *(off-label)*, ***collards*** *(off-label)*, ***garlic*** *(off-label)*, ***pak choi*** *(off-label)*, ***shallots*** *(off-label)* [1]

- Annual meadow grass in ***broccoli***, ***brussels sprouts***, ***bulb onions***, ***cabbages***, ***calabrese***, ***cauliflowers***, ***fodder rape***, ***kale***, ***leeks***, ***mustard***, ***salad onions***, ***swedes***, ***turnips*** [1-3]; ***chinese cabbage*** *(off-label)*, ***choi sum*** *(off-label)*, ***collards*** *(off-label)*, ***garlic*** *(off-label)*, ***pak choi*** *(off-label)*, ***shallots*** *(off-label)* [1]

Specific Off-Label Approvals (SOLAs)

- ***chinese cabbage*** *20072148 expires Sep 2008* [1]
- ***choi sum*** *20072148 expires Sep 2008* [1]
- ***collards*** *20072148 expires Sep 2008* [1]
- ***garlic*** *20072147 expires Sep 2008* [1]
- ***pak choi*** *20072148 expires Sep 2008* [1]
- ***shallots*** *20072147 expires Sep 2008* [1]

Approval information

- Approval expiry 30 Sep 2008 [1]

Efficacy guidance

- Product may be applied pre- or post-emergence of the crop but pre-emergence of the weeds. Emerged weeds are not controlled
- Soils should be fine, firm and free of weed growth
- For best results apply to moist soil and before rain. In dry weather irrigate lightly after application and before seedling weeds appear
- Weed control may be reduced under very wet, dry or windy conditions
- Do not use on soils with more than 10% organic matter as weed control will be impaired

Restrictions

- Maximum number of treatments 1 per crop for all crops

Crop-specific information

- Use on brassica crops pre-crop emergence may result in a check to growth but this is usually outgrown within 8 wk without effect on yield or quality
- Post-emergence treatment on brassica crops should not be made until they have 3-4 true leaves. Post-emergence treatment on onions should be made post-crook stage
- Young transplanted plants, particularly cabbages and cauliflowers, raised under glass should be hardened off before treatment
- Care should be taken with modular transplants to ensure that that roots are not exposed to spray

Following crops guidance

- The interval between treatment and planting succeeding crops should be 2 mth for seeded turf, 6 mth for lettuce, 3 mth for all other crops
- In the event of failure of a treated crop the land must be deep mouldboard ploughed before redrilling or replanting

Environmental safety

- Dangerous for the environment
- Toxic to aquatic organisms

Hazard classification and safety precautions

Hazard H03, H11
Risk phrases R36 [1, 2]; R38 [3]; R40, R43, R51, R53a [1-3]
Operator protection A, C; U02a, U04a, U05a, U09b, U11, U14, U15, U19a, U20a
Environmental protection E15a
Storage and disposal D01, D02, D09a, D10a
Medical advice M04a

99 citronella oil

A natural plant extract herbicide

Products

Barrier H	Barrier	22.9% w/w	EW	10136

Uses

- Ragwort in ***land temporarily removed from production***, ***permanent grassland***

Efficacy guidance

- Best results obtained from spot treatment of ragwort in the rosette stage, during dry still conditions
- Aerial growth of ragwort is rapidly destroyed. Longer term control depends on overall management strategy
- Check for regrowth after 28 d and re-apply as necessary

Crop-specific information

- Contact with grasses will result in transient scorch which is outgrown in good growing conditions

Environmental safety

- Apply away from bees
- Harmful to fish or other aquatic life. Do not contaminate surface waters or ditches with chemical or used container
- Keep livestock out of treated areas for at least 2 wk and until foliage of any poisonous weeds such as ragwort has died and become unpalatable

Hazard classification and safety precautions

Hazard H04
Risk phrases R36, R38
Operator protection A, C; U05a, U20c
Environmental protection E07b (2 wk); E12g, E13c
Storage and disposal D01, D02, D05, D09a, D11a

100 clodinafop-propargyl

A foliar acting herbicide for annual grass weed control in wheat, triticale and rye
HRAC mode of action code: A

Products

1	Landgold Clodinafop	Teliton	240 g/l	EC	13329
2	Topik	Syngenta	240 g/l	EC	12333

Uses

- Blackgrass in ***durum wheat***, ***spring rye***, ***spring wheat***, ***triticale***, ***winter rye***, ***winter wheat*** [1, 2]; ***grass seed crops*** *(off-label)* [2]
- Rough meadow grass in ***durum wheat***, ***spring rye***, ***spring wheat***, ***triticale***, ***winter rye***, ***winter wheat*** [1, 2]; ***grass seed crops*** *(off-label)* [2]
- Wild oats in ***durum wheat***, ***spring rye***, ***spring wheat***, ***triticale***, ***winter rye***, ***winter wheat*** [1, 2]; ***grass seed crops*** *(off-label)* [2]

Specific Off-Label Approvals (SOLAs)

- ***grass seed crops*** *20060956* [2]

Approval information

- Clodinafop-propargyl included in Annex I under EC Directive 91/414

Efficacy guidance

- Spray when majority of weeds have germinated but before competition reduces yield
- Products contain a herbicide safener (cloquintocet-mexyl) that improves crop tolerance to clodinafop-propargyl
- Optimum control achieved when all grass weeds emerged. Wait for delayed germination on dry or cloddy seedbed
- A mineral oil additive is recommended to give more consistent control of very high blackgrass populations or for late season treatments. See label for details
- Weed control not affected by soil type, organic matter or straw residues
- Control may be reduced if rain falls within 1 h of treatment
- Clodinafop-propargyl is an ACCase inhibitor herbicide. To avoid the build up of resistance do not apply products containing an ACCase inhibitor herbicide more than twice to any crop. In addition do not use any product containing clodinafop-propargyl in mixture or sequence with any other product containing the same ingredient
- Use these products as part of a resistance management strategy that includes cultural methods of control and does not use ACCase inhibitors as the sole chemical method of grass weed control

- Applying a second product containing an ACCase inhibitor to a crop will increase the risk of resistance development; only use a second ACCase inhibitor to control different weeds at a different timing
- Always follow WRAG guidelines for preventing and managing herbicide resistant weeds. See Section 5

Restrictions

- Maximum number of treatments 1 per crop
- Do not use on barley or oats
- Do not treat crops under stress or suffering from waterlogging, pest attack, disease or frost
- Do not treat crops undersown with grass mixtures
- Do not mix with products containing MCPA, mecoprop-P, 2,4-D or 2,4-DB
- MCPA, mecoprop, 2,4-D or 2,4-DB should not be applied within 21 d before, or 7 d after, treatment

Crop-specific information

- Latest use: before second node detectable stage (GS 32) for durum wheat, triticale, rye; before flag leaf sheath extending (GS 41) for wheat
- Spray in autumn, winter or spring from 1 true leaf stage (GS 11) to before second node detectable (GS 32) on durum, rye, triticale; before flag leaf sheath extends (GS 41) on wheat

Following crops guidance

- Any broad leaved crop or cereal (except oats) may be sown after failure of a treated crop provided that min 3 wk have elapsed between application and drilling a cereal
- After normal harvest of a treated crop any broad leaved crop or wheat, durum wheat, rye, triticale or barley should be sown. Oats and grass should not be sown until the following spring

Environmental safety

- Dangerous for the environment
- Very toxic to aquatic organisms

Hazard classification and safety precautions

Hazard H11
Risk phrases R50, R53a
Operator protection A, C, H, K; U02a, U05a, U09a, U20a
Environmental protection E15a, E38 [2]; E15b [1]
Storage and disposal D01, D02, D05, D09a, D10c, D12a

101 clodinafop-propargyl + pinoxaden

A foliar acting herbicide mixture for winter wheat
HRAC mode of action code: A + A

Products

Traxos	Syngenta	100:100 g/l	EC	12742

Uses

- Italian ryegrass in ***winter wheat***
- Perennial ryegrass in ***winter wheat*** *(from seed)*
- Wild oats in ***winter wheat***

Approval information

- Clodinafop-propargyl included in Annex I under EC Directive 91/414

Efficacy guidance

- Products contain a herbicide safener (cloquintocet-mexyl) that improves crop tolerance to clodinafop-propargyl
- Best results obtained from treatment when all grass weeds have emerged. There is no residual activity
- Broad-leaved weeds are not controlled
- Treat before emerged weed competition reduces yield
- For ryegrass control use as part of a programme including other products with activity against ryegrass
- For blackgrass control must be used as part of an integrated control strategy
- Grass weed control may be reduced if rain falls within 1 hr of application

- Clodinafop-propargyl and pinoxaden are ACCase inhibitor herbicides. To avoid the build up of resistance do not apply products containing an ACCase inhibitor herbicide more than twice to any crop. In addition do not use any product containing clodinafop-propargyl or pinoxaden in mixture or sequence with any other product containing the same ingredient
- Use as part of a resistance management strategy that includes cultural methods of control and does not use ACCase inhibitors as the sole chemical method of grass weed control
- Applying a second product containing an ACCase inhibitor to a crop will increase the risk of resistance development; only use a second ACCase inhibitor to control different weeds at a different timing
- Always follow WRAG guidelines for preventing and managing herbicide resistant weeds. See Section 5

Restrictions

- Maximum number of treatments 1 per crop
- Product must always be used with specified adjuvant. See label
- Do not use on other cereals
- Do not treat crops under stress from any cause
- Do not treat crops undersown with grass or grass mixtures
- Avoid use of hormone-containing herbicides in mixture or in sequence. See label for restrictions on tank mixes

Crop-specific information

- Latest use: before flag leaf sheath extended (before GS 41) for winter wheat
- All varieties of winter wheat may be treated

Following crops guidance

- There are no restrictions on succeeding crops in a normal rotation
- In the event of failure of a treated crop ryegrass, maize, oats or any broad-leaved crop may be planted after a minimum interval of 4 wk from application

Environmental safety

- Dangerous for the environment
- Toxic to aquatic organisms

Hazard classification and safety precautions

Hazard H04, H11
Risk phrases R36, R38, R43, R51, R53a
Operator protection A, C, H, K; U05a, U09a, U20b
Environmental protection E15b, E38
Storage and disposal D01, D02, D05, D09a, D10c, D11a, D12a

102 clodinafop-propargyl + trifluralin

A contact and residual herbicide for winter wheat
HRAC mode of action code: A + K1

Products

Hawk	Syngenta	12:383 g/l	EC	12507

Uses

- Annual dicotyledons in ***durum wheat*** *(off-label)*, ***grass seed crops*** *(off-label)*, ***spring rye*** *(off-label)*, ***triticale*** *(off-label)*, ***winter rye*** *(off-label)*, ***winter wheat***
- Blackgrass in ***durum wheat*** *(off-label)*, ***grass seed crops*** *(off-label)*, ***spring rye*** *(off-label)*, ***triticale*** *(off-label)*, ***winter rye*** *(off-label)*, ***winter wheat***
- Rough meadow grass in ***durum wheat*** *(off-label)*, ***grass seed crops*** *(off-label)*, ***spring rye*** *(off-label)*, ***triticale*** *(off-label)*, ***winter rye*** *(off-label)*, ***winter wheat***
- Wild oats in ***durum wheat*** *(off-label)*, ***grass seed crops*** *(off-label)*, ***spring rye*** *(off-label)*, ***triticale*** *(off-label)*, ***winter rye*** *(off-label)*, ***winter wheat***

Specific Off-Label Approvals (SOLAs)

- ***durum wheat*** *20060967*
- ***grass seed crops*** *20060967*
- ***spring rye*** *20060967*

- ***triticale*** *20060967*
- ***winter rye*** *20060967*

Approval information

- Clodinafop-propargyl included in Annex I under EC Directive 91/414
- Trifluralin will not be included in Annex I of EC Directive 91/414. All approvals for the storage and use of all products listed in this profile will expire on 20 March 2009 unless an earlier date has been set

Efficacy guidance

- Product contains a herbicide safener (cloquintocet-mexyl) that improves crop tolerance to clodinafop-propargyl
- Optimum weed control obtained when most grass weeds emerged and broad leaved weeds at susceptible stage before they compete with the crop. Pre-emergence control of annual meadow-grass and broad leaved weeds may not be satisfactory on medium and heavy soils
- Delay treatment if dry or cloddy seedbeds favour late weed germination
- Optimum weed control obtained in crops growing in a fine, firm seedbed cleared of trash and straw
- Wild oats and other weeds germinating from beneath treated zone not controlled
- Speed of action depends on temperature and growing conditions and may appear slow in dry or cold weather
- Rain within 1 h of application may reduce grass weed control
- Clodinafop-propargyl is an ACCase inhibitor herbicide. To avoid the build up of resistance do not apply products containing an ACCase inhibitor herbicide more than twice to any crop. In addition do not use any product containing clodinafop-propargyl in mixture or sequence with any other product containing the same ingredient
- Use these products as part of a resistance management strategy that includes cultural methods of control and does not use ACCase inhibitors as the sole chemical method of grass weed control
- Applying a second product containing an ACCase inhibitor to a crop will increase the risk of resistance development; only use a second ACCase inhibitor to control different weeds at a different timing
- Always follow WRAG guidelines for preventing and managing herbicide resistant weeds. See Section 5

Restrictions

- Maximum number of treatments 1 per crop
- Do not use on barley or oats
- Do not treat crops under stress or suffering from waterlogging, pest attack, disease or frost
- Do not treat crops undersown with grass mixtures
- Do not mix with products containing MCPA, mecoprop-P, 2,4-D or 2,4-DB
- MCPA, mecoprop, 2,4-D or 2,4-DB should not be applied within 21 d before, or 7 d after, treatment
- Do not use on Sands, stony or gravelly soils or soils with over 10% organic matter
- Product must be applied with specified adjuvant - see label

Crop-specific information

- Latest use: before ear at 1 cm stage (GS 30) for winter wheat
- Spray in autumn, winter or spring from first to third leaf unfolded stage (GS 11-13) but before ear at 1 cm stage (GS 30)

Following crops guidance

- After normal harvest any broad leaved crop (except sugar beet), barley, wheat, durum wheat, rye or triticale may be sown. Before drilling or planting subsequent crops soil must be mouldboard ploughed to 15 cm
- See label for details of crops that may be safely drilled or planted within 5 mth of treatment. If a treated crop should fail after 5 mth but before normal harvest, sow only a broad-leaved crop (not sugar beet)
- 12 mth must elapse after treatment before sugar beet is drilled

Environmental safety

- Dangerous for the environment
- Very toxic to aquatic organisms
- Crops must not be treated if there is risk of run-off (eg from frozen foliage)

SEE SECTION 3 FOR PRODUCTS ALSO REGISTERED

Hazard classification and safety precautions

Hazard H04, H11

Risk phrases R36, R38, R50, R53a

Operator protection A, C, H, K; U02a, U05a, U09a, U11, U20a

Environmental protection E15a, E38

Storage and disposal D01, D02, D05, D09a, D10c, D12a

103 clofentezine

A selective ovicidal tetrazine acaricide for use in top fruit

IRAC mode of action code: 10A

Products

Apollo 50 SC	Makhteshim	500 g/l	SC	10590

Uses

- Red spider mites in ***apples***, ***cherries***, ***pears***, ***plums***
- Spider mites in ***blackcurrants*** *(off-label)*, ***protected blackberries*** *(off-label)*, ***protected raspberries*** *(off-label)*, ***protected strawberries*** *(off-label)*, ***raspberries*** *(off-label)*, ***strawberries*** *(off-label)*

Specific Off-Label Approvals (SOLAs)

- ***blackcurrants*** *20012269*
- ***protected blackberries*** *20012268*
- ***protected raspberries*** *20012268*
- ***protected strawberries*** *20012271*
- ***raspberries*** *20012269*
- ***strawberries*** *20012269*

Efficacy guidance

- Acts on eggs and early motile stages of mites. For effective control total cover of plants is essential, particular care being needed to cover undersides of leaves

Restrictions

- Maximum number of treatments 1 per yr for apples, pears, cherries, plums

Crop-specific information

- HI apples, pears 28 d; cherries, plums 8 wk
- For red spider mite control spray apples and pears between bud burst and pink bud, plums and cherries between white bud and first flower. Rust mite is also suppressed
- On established infestations apply in conjunction with an adult acaricide

Environmental safety

- Harmful to aquatic organisms
- Product safe on predatory mites, bees and other predatory insects

Hazard classification and safety precautions

Risk phrases R52

Operator protection U05a, U08, U20b

Environmental protection E15a

Storage and disposal D01, D02, D05, D09a, D11a

104 clomazone

An isoxazolidinone residual herbicide for oilseed rape and field beans

HRAC mode of action code: F3

Products

1	Centium 360 CS	Belchim	360 g/l	CS	11607
2	Cirrus CS	Belchim	360 g/l	CS	11751
3	Gamit 36 CS	Belchim	360 g/l	CS	12598
4	Greencrop Dingle 360	Greencrop	360 g/l	CS	12760
5	Landgold Clomazone 360	Teliton	360 g/l	CS	13074

Uses

- Annual dicotyledons in ***broccoli*** *(off-label)*, ***brussels sprouts*** *(off-label)*, ***cabbages*** *(off-label)*, ***calabrese*** *(off-label)*, ***cauliflowers*** *(off-label)*, ***collards*** *(off-label)*, ***kale*** *(off-label)* [1]
- Chickweed in ***asparagus*** *(off-label)* [1, 3]; ***borage*** *(off-label)*, ***canary flower (echium spp.)*** *(off-label)*, ***honesty*** *(off-label)*, ***lupins*** *(off-label)*, ***mustard*** *(off-label)*, ***spring linseed*** *(off-label)* [2]; ***broccoli*** *(off-label)*, ***brussels sprouts*** *(off-label)*, ***cabbages*** *(off-label)*, ***calabrese*** *(off-label)*, ***carrots***, ***cauliflowers*** *(off-label)*, ***chives*** *(off-label)*, ***collards*** *(off-label)*, ***herbs (see appendix 6)*** *(off-label)*, ***kale*** *(off-label)*, ***parsley*** *(off-label)*, ***potatoes***, ***spring cabbage*** *(off-label)*, ***swedes*** *(off-label)* [3]; ***combining peas***, ***spring field beans***, ***spring oilseed rape***, ***vining peas***, ***winter field beans***, ***winter oilseed rape*** [1, 2, 4, 5]
- Cleavers in ***asparagus*** *(off-label)* [1, 3]; ***borage*** *(off-label)*, ***canary flower (echium spp.)*** *(off-label)*, ***honesty*** *(off-label)*, ***lupins*** *(off-label)*, ***mustard*** *(off-label)*, ***spring linseed*** *(off-label)* [2]; ***broccoli*** *(off-label)*, ***brussels sprouts*** *(off-label)*, ***cabbages*** *(off-label)*, ***calabrese*** *(off-label)*, ***carrots***, ***cauliflowers*** *(off-label)*, ***chives*** *(off-label)*, ***collards*** *(off-label)*, ***herbs (see appendix 6)*** *(off-label)*, ***kale*** *(off-label)*, ***parsley*** *(off-label)*, ***potatoes***, ***spring cabbage*** *(off-label)*, ***swedes*** *(off-label)* [3]; ***combining peas***, ***spring field beans***, ***spring oilseed rape***, ***vining peas***, ***winter field beans***, ***winter oilseed rape*** [1, 2, 4, 5]
- Fool's parsley in ***asparagus*** *(off-label)* [1, 3]; ***borage*** *(off-label)*, ***canary flower (echium spp.)*** *(off-label)*, ***honesty*** *(off-label)*, ***lupins*** *(off-label)*, ***mustard*** *(off-label)*, ***spring linseed*** *(off-label)* [2]; ***broccoli*** *(off-label)*, ***brussels sprouts*** *(off-label)*, ***cabbages*** *(off-label)*, ***calabrese*** *(off-label)*, ***carrots***, ***cauliflowers*** *(off-label)*, ***chives*** *(off-label)*, ***collards*** *(off-label)*, ***herbs (see appendix 6)*** *(off-label)*, ***kale*** *(off-label)*, ***parsley*** *(off-label)*, ***potatoes***, ***spring cabbage*** *(off-label)*, ***swedes*** *(off-label)* [3]; ***combining peas***, ***spring field beans***, ***spring oilseed rape***, ***vining peas*** [1, 2, 4, 5]
- Red dead-nettle in ***asparagus*** *(off-label)* [1, 3]; ***borage*** *(off-label)*, ***canary flower (echium spp.)*** *(off-label)*, ***honesty*** *(off-label)*, ***lupins*** *(off-label)*, ***mustard*** *(off-label)*, ***spring linseed*** *(off-label)* [2]; ***broccoli*** *(off-label)*, ***brussels sprouts*** *(off-label)*, ***cabbages*** *(off-label)*, ***calabrese*** *(off-label)*, ***carrots***, ***cauliflowers*** *(off-label)*, ***chives*** *(off-label)*, ***collards*** *(off-label)*, ***herbs (see appendix 6)*** *(off-label)*, ***kale*** *(off-label)*, ***parsley*** *(off-label)*, ***potatoes***, ***spring cabbage*** *(off-label)*, ***swedes*** *(off-label)* [3]; ***combining peas***, ***spring field beans***, ***spring oilseed rape***, ***vining peas***, ***winter field beans***, ***winter oilseed rape*** [1, 2, 4, 5]
- Shepherd's purse in ***asparagus*** *(off-label)* [1, 3]; ***borage*** *(off-label)*, ***canary flower (echium spp.)*** *(off-label)*, ***honesty*** *(off-label)*, ***lupins*** *(off-label)*, ***mustard*** *(off-label)*, ***spring linseed*** *(off-label)* [2]; ***broccoli*** *(off-label)*, ***brussels sprouts*** *(off-label)*, ***cabbages*** *(off-label)*, ***calabrese*** *(off-label)*, ***carrots***, ***cauliflowers*** *(off-label)*, ***chives*** *(off-label)*, ***collards*** *(off-label)*, ***herbs (see appendix 6)*** *(off-label)*, ***kale*** *(off-label)*, ***parsley*** *(off-label)*, ***potatoes***, ***spring cabbage*** *(off-label)*, ***swedes*** *(off-label)* [3]; ***combining peas***, ***spring field beans***, ***spring oilseed rape***, ***vining peas***, ***winter field beans***, ***winter oilseed rape*** [1, 2, 4, 5]

Specific Off-Label Approvals (SOLAs)

- ***asparagus*** *20060530* [1], *20060658* [3]
- ***borage*** *20061214* [2]
- ***broccoli*** *20041031* [1], *20060659* [3]
- ***brussels sprouts*** *20041031* [1], *20060659* [3]
- ***cabbages*** *20041031* [1], *20060659* [3]
- ***calabrese*** *20041031* [1], *20060659* [3]
- ***canary flower (echium spp.)*** *20061214* [2]
- ***cauliflowers*** *20041031* [1], *20060659* [3]
- ***chives*** *20061609* [3]
- ***collards*** *20041031* [1], *20060659* [3]
- ***herbs (see appendix 6)*** *20061609* [3]
- ***honesty*** *20061214* [2]
- ***kale*** *20041031* [1], *20060659* [3]
- ***lupins*** *20061213* [2]
- ***mustard*** *20061214* [2]
- ***parsley*** *20061609* [3]
- ***spring cabbage*** *20060659* [3]
- ***spring linseed*** *20061214* [2]
- ***swedes*** *20070933* [3]

Approval information

- Clomazone included in Annex I under EC Directive 91/414
- Approval expiry 30 Jun 2008 [4]

Efficacy guidance

- Best results obtained from application as soon as possible after sowing crop and before emergence of crop or weeds
- Uptake is via roots and shoots. Seedbeds should be firm, level and free from clods. Loose puffy seedbeds should be consolidated before spraying
- Efficacy is reduced on organic soils, on dry cloddy seedbeds and if prolonged dry weather follows application
- Clomazone acts by inhibiting synthesis of chlorophyll pigments. Susceptible weeds emerge but are chlorotic and die shortly afterwards
- Season-long control of weeds may not be achieved
- Always follow WRAG guidelines for preventing and managing herbicide resistant weeds. Section 5 for more information

Restrictions

- Maximum number of treatments one per crop
- Crops must be covered by a minimum of 20 mm settled soil. Do not apply to broadcast crops. Direct-drilled crops should be harrowed across the slits to cover seed before spraying
- Do not use on compacted soils or soils of poor structure that may be liable to waterlogging
- Do not use on Sands or Very Light soils or those with more than 10% organic matter
- Do not treat two consecutive crops of carrots with clomazone in one calendar yr [3]
- Consult manufacturer or your advisor before use on potato seed crops [3]
- Do not overlap spray swaths. Crop plants emerged at time of treatment may be severely damaged

Crop-specific information

- Latest use: pre-emergence of crop
- Severe, but normally transient, crop damage may occur in overlaps on field beans
- Some transient crop bleaching may occur under certain climatic conditions and can be severe where heavy rain follows application. This is normally rapidly outgrown and has no effect on final crop yield. Overlapping spray swaths may cause severe damage to field beans

Following crops guidance

- Following normal harvest of a spring or autumn treated crop, cereals, oilseed rape, field beans, combining peas, potatoes, maize, turnips, linseed or sugar beet may be sown
- In the event of failure of an autumn treated crop, winter cereals or winter beans may be sown in the autumn if 6 wk have elapsed since treatment. In the spring following crop failure combining peas, field beans or potatoes may be sown if 6 wk have elapsed since treatment, and spring cereals, maize, turnips, onions, carrots or linseed may be sown if 7 mth have elapsed since treatment
- In the event of a failure of a spring treated crop a wide range of crops may be sown provided intervals of 6-9 wk have elapsed since treatment. See label for details
- Prior to resowing any listed replacement crop the soil should be ploughed and cultivated to 15 cm

Environmental safety

- Take extreme care to avoid drift outside the target area, or on to ponds, waterways or ditches as considerable damage may occur. Apply using a coarse quality spray

Hazard classification and safety precautions

Hazard H04 [4, 5]

Risk phrases R43 [4, 5]

Operator protection A [1-5]; H [4, 5]; U05a, U14, U20b

Environmental protection E15a [4, 5]; E15b [1-3]; E34 [5]

Storage and disposal D01, D02, D09a, D10b [1-5]; D05 [4]; D12a [5]

105 clopyralid

A foliar translocated picolinic herbicide for a wide range of crops
HRAC mode of action code: O

Products

1	Dow Shield	Dow	200 g/l	SL	10988
2	Fernpath Torate	AgriGuard	200 g/l	SL	11033
3	Fernpath Torate XT	AgriGuard	200g/l	SL	13391
4	Glopyr 200 SL	Globachem	200 g/l	SL	10979
5	Greencrop Champion	Greencrop	200 g/l	SL	11755
6	Landgold Clopyralid 200	Teliton	200 g/l	SL	12359
7	Lontrel 200	Dow	18.02% w/w	SL	11558
8	Pirlid	AgriGuard	200 g/l	SL	11946
9	Torate	AgriGuard	200 g/l	SL	12611

Uses

- Annual dicotyledons in ***broccoli***, ***brussels sprouts***, ***bulb onions***, ***cabbages***, ***calabrese***, ***fodder beet***, ***fodder rape***, ***forage maize***, ***kale***, ***linseed***, ***mangels***, ***permanent grassland***, ***red beet***, ***salad onions***, ***spring oilseed rape***, ***strawberries***, ***sugar beet***, ***swedes***, ***sweetcorn***, ***turnips***, ***winter barley***, ***winter oats***, ***winter oilseed rape*** [1-9]; ***cauliflowers***, ***conifers and broadleaved trees*** *(off-label)* [1, 7]; ***ornamental plant production*** [1-5, 7-9]; ***rotational grass*** [5-7]; ***spring barley***, ***spring oats*** [1, 3-7]; ***spring wheat***, ***winter wheat*** [1, 3-7, 9]
- Corn marigold in ***broccoli***, ***brussels sprouts***, ***bulb onions***, ***cabbages***, ***calabrese***, ***cauliflowers***, ***fodder beet***, ***fodder rape***, ***forage maize***, ***kale***, ***linseed***, ***mangels***, ***permanent grassland***, ***red beet***, ***salad onions***, ***spring oilseed rape***, ***strawberries***, ***sugar beet***, ***swedes***, ***sweetcorn***, ***turnips***, ***winter barley***, ***winter oats***, ***winter oilseed rape*** [1-9]; ***ornamental plant production*** [1-5, 7-9]; ***rotational grass*** [5-7]; ***spring barley***, ***spring oats*** [1, 3-7]; ***spring wheat***, ***winter wheat*** [1, 3-7, 9]
- Creeping thistle in ***broccoli***, ***brussels sprouts***, ***bulb onions***, ***cabbages***, ***calabrese***, ***cauliflowers***, ***fodder beet***, ***fodder rape***, ***forage maize***, ***kale***, ***linseed***, ***mangels***, ***permanent grassland***, ***red beet***, ***salad onions***, ***spring oilseed rape***, ***strawberries***, ***sugar beet***, ***swedes***, ***sweetcorn***, ***turnips***, ***winter barley***, ***winter oats***, ***winter oilseed rape*** [1-9]; ***ornamental plant production*** [1-5, 7-9]; ***rotational grass*** [5-7]; ***spring barley***, ***spring oats*** [1, 3-7]; ***spring wheat***, ***winter wheat*** [1, 3-7, 9]
- General weed control in ***canary grass*** *(off-label - game cover)*, ***quinoa*** *(off-label - game cover)*, ***sweet clover*** *(off-label - game cover)*, ***tanka millet*** *(off-label - game cover)*, ***white millet*** *(off-label - game cover)* [1, 7]
- Groundsel in ***chives*** *(off-label)*, ***herbs (see appendix 6)*** *(off-label)*, ***leaf spinach*** *(off-label)*, ***parsley*** *(off-label)*, ***spinach beet*** *(off-label)* [1, 7]
- Mayweeds in ***broccoli***, ***brussels sprouts***, ***bulb onions***, ***cabbages***, ***calabrese***, ***cauliflowers***, ***fodder beet***, ***fodder rape***, ***forage maize***, ***kale***, ***linseed***, ***mangels***, ***permanent grassland***, ***red beet***, ***salad onions***, ***spring oilseed rape***, ***strawberries***, ***sugar beet***, ***swedes***, ***sweetcorn***, ***turnips***, ***winter barley***, ***winter oats***, ***winter oilseed rape*** [1-9]; ***chives*** *(off-label)*, ***herbs (see appendix 6)*** *(off-label)*, ***honesty*** *(off-label)*, ***leaf spinach*** *(off-label)*, ***parsley*** *(off-label)*, ***poppies for morphine production*** *(off-label)*, ***spinach beet*** *(off-label)* [1, 7]; ***ornamental plant production*** [1-5, 7-9]; ***rotational grass*** [5-7]; ***spring barley***, ***spring oats*** [1, 3-7]; ***spring wheat***, ***winter wheat*** [1, 3-7, 9]
- Thistles in ***apple orchards*** *(off-label)*, ***asparagus*** *(off-label)*, ***grass seed crops*** *(off-label)*, ***honesty*** *(off-label)*, ***pear orchards*** *(off-label)*, ***permanent grassland*** *(off-label - spot treatment)*, ***poppies for morphine production*** *(off-label)* [1, 7]

Specific Off-Label Approvals (SOLAs)

- ***apple orchards*** *20050471* [1], *20062444* [7]
- ***asparagus*** *20050470* [1], *20062445* [7]
- ***canary grass*** *(game cover) 20050474* [1], *(game cover) 20062450* [7]
- ***chives*** *20050473* [1], *20062442* [7]
- ***conifers and broadleaved trees*** *20050468* [1], *20062449* [7]
- ***grass seed crops*** *20050467* [1], *20062448* [7]
- ***herbs (see appendix 6)*** *20050473* [1], *20062442* [7]
- ***honesty*** *20050469* [1], *20062446* [7]
- ***leaf spinach*** *20050472* [1], *20062443* [7]

- ***parsley*** *20050473* [1], *20062442* [7]
- ***pear orchards*** *20050471* [1], *20062444* [7]
- ***permanent grassland*** *(spot treatment) 20050467* [1], *(spot treatment) 20062448* [7]
- ***poppies for morphine production*** *20040161* [1], *20062447* [7]
- ***quinoa*** *(game cover) 20050474* [1], *(game cover) 20062450* [7]
- ***spinach beet*** *20050472* [1], *20062443* [7]
- ***sweet clover*** *(game cover) 20050474* [1], *(game cover) 20062450* [7]
- ***tanka millet*** *(game cover) 20050474* [1], *(game cover) 20062450* [7]
- ***white millet*** *(game cover) 20050474* [1], *(game cover) 20062450* [7]

Approval information

- Clopyralid included in Annex I under EC Directive 91/414
- Accepted by BBPA for use on malting barley

Efficacy guidance

- Best results achieved by application to young actively growing weed seedlings. Treat creeping thistle at rosette stage and repeat 3-4 wk later as directed
- High activity on weeds of Compositae family. For most crops recommended for use in tank mixes. See label for details

Restrictions

- Maximum total dose varies between the equivalent of one and two full dose treatments, depending on the crop treated. See labels for details
- Do not apply to cereals later than the second node detectable stage (GS 32)
- Do not apply when crop damp or when rain expected within 6 h
- Do not use straw from treated cereals in compost or any other form for glasshouse crops. Straw may be used for strawing down strawberries
- Straw from treated grass seed crops or linseed should be baled and carted away. If incorporated do not plant winter beans in same year
- Do not use on onions at temperatures above 20 °C or when under stress
- Do not treat maiden strawberries or runner beds or apply to early leaf growth during blossom period or within 4 wk of picking. Aug or early Sep sprays may reduce yield

Crop-specific information

- Latest use: 7 d before cutting grass for hay or silage; before 3rd node detectable (GS 33) for cereals; before flower buds visible from above for oilseed rape, linseed
- HI grassland 7 d; apples, pears, strawberries 4 wk; maize, sweetcorn, onions, Brussels sprouts, broccoli, cabbage, cauliflowers, calabrese, kale, fodder rape, oilseed rape, swedes, turnips, sugar beet, red beet, fodder beet, mangels, sage, honesty 6 wk
- Timing of application varies with weed problem, crop and other ingredients of tank mixes. See labels for details
- Apply as directed spray in woody ornamentals, avoiding leaves, buds and green stems. Do not apply in root zone of families Compositae or Leguminosae

Following crops guidance

- Do not plant susceptible autumn-sown crops in same year as treatment. Do not apply later than Jul where susceptible crops are to be planted in spring. See label for details

Environmental safety

- Dangerous for the environment [2-4, 8, 9]
- Harmful to aquatic organisms [1, 3, 4, 7, 9]
- Wash spray equipment thoroughly with water and detergent immediately after use. Traces of product can damage susceptible plants sprayed later
- Keep livestock out of treated areas for at least 7 d and until foliage of any poisonous weeds such as ragwort has died and become unpalatable

Hazard classification and safety precautions

Hazard H11 [2-4, 8, 9]
Risk phrases R50 [4]; R52 [1, 3, 7, 9]; R53a [1, 3, 4, 7, 9]
Operator protection A, C; U08, U19a [1-9]; U20a [5, 6]; U20b [1-4, 7-9]
Environmental protection E06a [3] (7 d); E07a [1, 2, 4-9]; E15a [1-9]
Storage and disposal D01, D12a [1, 7]; D05, D09a, D10b [1-9]

106 clopyralid + 2,4-D + MCPA

A translocated herbicide mixture for managed amenity turf
HRAC mode of action code: O + O + O

Products

Esteem	Vitax	35:150:175 g/l	SL	12555

Uses

- Annual dicotyledons in ***managed amenity turf***
- Perennial dicotyledons in ***managed amenity turf***

Approval information

- Clopyralid, 2,4-D and MCPA included in Annex I under EC Directive 91/414

Efficacy guidance

- Apply when soil is moist and weeds growing actively, normally between Apr and Oct
- Ensure sufficient leaf area for uptake and select appropriate water volume to achieve good spray coverage
- To allow maximum translocation do not cut grass for 3 d after treatment

Restrictions

- Maximum number of treatments on managed amenity turf 1 per yr
- Do not spray in periods of drought unless irrigation is applied
- Do not apply if night temrperatures are low, if ground frost is imminent or in periods of prolonged cold weather
- Do not treat turf less than 1 yr old
- Do not use any treated plant materials for composting or mulching
- Do not mow for 3 d before or after treatment

Crop-specific information

- Consult manufacturer or carry out small scale safety test on turf grass cultivars not previously treated

Following crops guidance

- If reseeding of treated turf is required allow at least 6 wk after treatment before drilling grasses into the sward
- Where treated land is subsequently to be sown with broad-leaved plants allow an interval of at least 6 mth

Environmental safety

- Dangerous for the environment
- Harmful to aquatic organisms
- Wash spray equipment thoroughly with water and detergent immediately after use. Traces of product can damage susceptible plants sprayed later
- Do not spray outside the target area and avoid drift onto non-target plants

Hazard classification and safety precautions

Hazard H03, H11
Risk phrases R22a, R37, R41, R52, R53a
Operator protection A, C, H; U05a, U11, U15
Environmental protection E15a, E34
Storage and disposal D01, D02, D12a

107 clopyralid + diflufenican + MCPA

A selective herbicide for use in managed amenity turf
HRAC mode of action code: O + F1 + O

Products

Spearhead	Bayer Environ.	20:15:300 g/l	SL	09941

Uses

- Annual dicotyledons in ***managed amenity turf***
- Perennial dicotyledons in ***managed amenity turf***

SEE SECTION 3 FOR PRODUCTS ALSO REGISTERED

Approval information

- Clopyralid and MCPA included in Annex I under EC Directive 91/414

Efficacy guidance

- Best results achieved by application when grass and weeds are actively growing
- Treatment during early part of the season is recommended, but not during drought

Restrictions

- Maximum number of treatments 1 per yr
- Only use when sward is satisfactorily established and regular mowing has begun
- Turf sown in spring or early summer may be ready for treatment after 2 mth. Later sown turf should not be sprayed until growth is resumed in the following spring
- Avoid mowing within 3-4 d before or after treatment
- Do not use cuttings from treated area as a mulch for any crop

Environmental safety

- Extremely dangerous to fish or other aquatic life. Do not contaminate surface waters or ditches with chemical or used container
- Keep livestock out of treated areas/away from treated water for at least 1 wk and until foliage of any poisonous weeds, such as ragwort, has died and become unpalatable
- LERAP Category B
- Avoid drift. Small amounts of spray can cause serious injury to herbaceous plants, vegetables, fruit and glasshouse crops
- Keep livestock out of treated areas

Hazard classification and safety precautions

Hazard H03, H04
Risk phrases R20, R21, R22a, R36, R38
Operator protection A, C; U02a, U05a, U08, U19a, U20b
Environmental protection E06a (unstipulated period); E13a, E16a, E16b, E34
Storage and disposal D01, D02, D05, D09a, D10b
Medical advice M03

108 clopyralid + florasulam + fluroxypyr

A translocated herbicide mixture for cereals
HRAC mode of action code: O + B + O

Products

Galaxy	Dow	80:144:2.5 g/l	EC	13127

Uses

- Annual dicotyledons in ***spring barley***, ***spring oats***, ***spring wheat***, ***winter barley***, ***winter oats***, ***winter wheat***
- Chickweed in ***spring barley***, ***spring oats***, ***spring wheat***, ***winter barley***, ***winter oats***, ***winter wheat***
- Cleavers in ***spring barley***, ***spring oats***, ***spring wheat***, ***winter barley***, ***winter oats***, ***winter wheat***

Approval information

- Clopyralid, florasulam and fluroxypyr are all included in Annex I under Directive 91/414

Efficacy guidance

- Best results obtained when weeds are small and actively growing
- Effectiveness may be reduced when soil is very dry
- Use adequate water volume to achieve complete spray coverage of the weeds
- Florasulam is a member of the ALS-inhibitor group of herbicides

Restrictions

- Maximum number of treatments 1 per yr for all crops
- Do not spray when crops are under stress from any cause
- Do not apply through CDA applicators
- Do not roll or harrow for 7d before or after application
- Do not use any treated plant material for composting or mulching
- Do not use manure from animals fed on treated crops for composting

- Specific restrictions apply to use in sequence or tank mixture with other sulfonylurea or ALS-inhibiting herbicides. See label for details

Crop-specific information

- Latest use: before second node detectable for oats; before third node detectable for barley and wheat

Following crops guidance

- Where residues of a treated crop have not completely decayed by the time of planting a succeeding crop, avoid planting peas, beans and other legumes, carrots and other Umbelliferae, potatoes, lettuce and other Compositae, glasshouse and protected crops
- Where the product has been used in mixture with certain named products (see label) only cereals or grass may be sown in the autumn following harvest. Otherwise cereals, oilseed rape, grass or vegetable brassicas as transplants may be sown as a following crop in the same calendar yr as treatment. Oilseed rape may show some temporary reduction of vigour after a dry summer, but yields are not affected
- In addition to the above, field beans, linseed, peas, sugar beet, potatoes, maize, clover (for use in grass/clover mixtures) or carrots may be sown in the calendar yr following treatment
- In the event of failure of a treated crop in spring, only spring wheat, spring barley, spring oats, maize or ryegrass may be sown

Environmental safety

- Dangerous for the environment
- Very toxic to aquatic organisms
- Take extreme care to avoid drift outside the target area

Hazard classification and safety precautions

Hazard H03, H11
Risk phrases R20, R36, R38, R50, R53a
Operator protection A, H; U05a, U11, U19a, U20a
Environmental protection E15b, E34, E38
Storage and disposal D01, D02, D05, D09a, D10b, D12a
Medical advice M03

SECTION 2

109 clopyralid + fluroxypyr + MCPA

A translocated herbicide mixture for use in sports and amenity turf
HRAC mode of action code: O + O + O

Products

Greenor	Rigby Taylor	20:40:200 g/l	ME	10909

Uses

- Annual dicotyledons in ***managed amenity turf***

Approval information

- Clopyralid, fluroxypyr and MCPA included in Annex I under EC Directive 91/414

Efficacy guidance

- Best results achieved when weeds actively growing and turf grass competitive
- Treatment should normally be between Apr-Sep when the soil is moist
- Do not apply during drought unless irrigation is applied
- Allow 3 d before or after mowing established turf to ensure sufficient weed leaf surface present to allow uptake and movement

Restrictions

- Maximum number of treatments 2 per yr
- Do not treat grass under stress from frost, drought, waterlogging, trace element deficiency, disease or pest attack
- Do not treat if night temperatures are low, when frost is imminent or during prolonged cold weather

Crop-specific information

- Treat young turf only in spring when at least 2 mth have elapsed since sowing
- Allow 5 d after mowing young turf before treatment

SEE SECTION 3 FOR PRODUCTS ALSO REGISTERED

- Product selective on a number of turf grass species (see label) but consultation or testing recommended before treatment of any cultivar

Environmental safety

- Dangerous for the environment
- Very toxic to aquatic organisms
- Wash spray equipment thoroughly with water and detergent immediately after use. Traces of product can damage susceptible plants sprayed later

Hazard classification and safety precautions

Hazard H04, H11
Risk phrases R36, R43, R50, R53a
Operator protection A, C; U05a, U08, U14, U19a, U20b
Environmental protection E15a, E38
Storage and disposal D01, D02, D05, D09a, D10b, D12a

110 clopyralid + fluroxypyr + triclopyr

A foliar acting herbicide mixture for grassland
HRAC mode of action code: O + O + O

Products

1	Pastor	Dow	50:75:100 g/l	EC	11168
2	Trinity	AgriGuard	50:75:100 g/l	EC	12738

Uses

- Annual dicotyledons in ***rotational grass***
- Docks in ***permanent grassland***, ***rotational grass***
- Stinging nettle in ***permanent grassland***
- Thistles in ***permanent grassland***

Approval information

- Clopyralid, fluroxypyr and triclopyr included in Annex I under EC Directive 91/414

Efficacy guidance

- Apply in spring or autumn depending on weeds present or as a split treatment in spring followed by autumn
- Treatment must be made when weeds and grass are actively growing
- It is important to ensure sufficient leaf area is present for uptake especially on established docks and thistles
- On large well established docks and where there is a large soil seed reservoir further treatment in the following year may be needed
- To allow maximum translocation in the weeds do not cut grass for 4 wk after treatment

Restrictions

- Maximum number of treatments 1 per yr at full dose or 2 per yr at half dose
- Do not spray in drought, very hot or very cold weather
- Do not treat sports or amenity turf
- Product kills or severely checks clover and should not be used where clover is an important constituent of the sward
- Do not roll or harrow 10 d before or 7 d after treatment
- Do not allow spray or drift to reach other crops, amenity plantings, gardens, ponds, lakes or water courses

Crop-specific information

- HI 7 d
- Application during active growth ensures minimal check to grass. Newly sown grass may be treated from the third leaf visible stage
- Product may be used in established grassland which is under non-rotational setaside arrangements
- Very occasionally some yellowing of the sward may occur after treatment but is quickly outgrown

Following crops guidance

- Residues in incompletely decayed plant tissue may affect succeeding susceptible crops such as peas, beans and other legumes, carrots and related crops, potatoes, tomatoes, lettuce

- Do not plant susceptible autumn-sown crops in the same yr as treatment with product. Spring sown crops may follow if treatment was before end Jul in the previous yr

Environmental safety

- Dangerous for the environment
- Toxic to aquatic organisms
- Keep livestock out of treated areas for at least 7 d following treatment and until foliage of poisonous weeds such as ragwort has died and become unpalatable
- Wash spray equipment thoroughly with water and detergent immediately after use. Traces of product can damage susceptible plants sprayed later

Hazard classification and safety precautions

Hazard H03, H08, H11
Risk phrases R22b, R37, R38, R41, R43, R51, R53a, R67
Operator protection A, C; U02a, U05a, U08, U11, U19a, U20b [1, 2]; U14 [1]
Environmental protection E07a, E15a, E34 [1, 2]; E38 [1]
Consumer protection C01 [1]
Storage and disposal D01, D02, D05, D09a, D10b, D12a [1, 2]; D07 [2]
Medical advice M05b

111 clopyralid + picloram

A post-emergence herbicide mixture for oilseed rape
HRAC mode of action code: O + O

Products

1	Galera	Dow	267:67 g/l	SL	11961
2	Nugget	AgriGuard	267:67 g/l	SL	13121

Uses

- Cleavers in ***evening primrose*** *(off-label)*, ***honesty*** *(off-label)*, ***linseed*** *(off-label)*, ***mustard*** *(off-label)*, ***spring oilseed rape*** *(off-label)* [1]; ***winter oilseed rape*** [1, 2]
- Mayweeds in ***evening primrose*** *(off-label)*, ***honesty*** *(off-label)*, ***linseed*** *(off-label)*, ***mustard*** *(off-label)*, ***spring oilseed rape*** *(off-label)* [1]; ***winter oilseed rape*** [1, 2]

Specific Off-Label Approvals (SOLAs)

- ***evening primrose*** *20060960* [1]
- ***honesty*** *20060960* [1]
- ***linseed*** *20060960* [1]
- ***mustard*** *20060960* [1]
- ***spring oilseed rape*** *20061437* [1]

Approval information

- Clopyralid included in Annex I under EC Directive 91/414

Efficacy guidance

- Best results obtained from treatment when weeds are small and actively growing
- Cleavers that germinate after treatment will not be controlled

Restrictions

- Maximum total dose equivalent to one full dose treatment
- Do not treat crops under stress from cold, drought, pest damage, nutrient deficiency or any other cause
- Do not roll or harrow for 7 d before or after spraying
- Do not apply through CDA applicators
- Do not use any treated plant material for composting or mulching
- Do not use manure from animals fed on treated crops for composting
- Chop and incorporate all treated plant remains in early autumn, or as soon as possible after harvest, to release any residues into the soil. Ensure that all treated plant remains have completely decayed before planting susceptible crops

Crop-specific information

- Latest Use: before flower buds visible above crop canopy for winter oilseed rape

Following crops guidance

- Any crop may be sown in the calendar yr following treatment

- Ploughing or thorough cultivation should be carried out before planting leguminous crops
- Do not attempt to plant peas, beans, other legumes, carrots, other umbelliferous crops, potatoes, lettuce, other Compositae, or any glasshouse or protected crops if treated crop remains have not fully decayed by the time of planting
- In the event of failure of an autumn treated crop only oilseed rape, wheat, barley, oats, maize or ryegrass may be sown in the spring and only after ploughing or thorough cultivation

Environmental safety

- Dangerous for the environment
- Toxic to aquatic organisms
- Take extreme care to avoid drift onto crops and non-target plants outside the target area

Hazard classification and safety precautions

Hazard H11
Risk phrases R51, R53a
Operator protection A, C; U05a
Environmental protection E15a, E34
Storage and disposal D01, D02, D05, D07, D09a

112 clopyralid + triclopyr

A perennial and woody weed herbicide for use in grassland
HRAC mode of action code: O + O

Products

1	Blaster	Headland Amenity	60:240 g/l	EC	10571
2	Grazon 90	Dow	60:240 g/l	EC	05456
3	Greencrop Hi-Grass 90	Greencrop	60:240 g/l	EC	12429
4	Thistlex	Dow	200:200 g/l	EC	11533

Uses

- Brambles in ***amenity grassland*** [1]; ***permanent grassland*** [2, 3]
- Broom in ***amenity grassland*** [1]; ***permanent grassland*** [2, 3]
- Creeping thistle in ***permanent grassland***, ***rotational grass*** [4]
- Docks in ***amenity grassland*** [1]; ***permanent grassland*** [2, 3]
- Gorse in ***amenity grassland*** [1]; ***permanent grassland*** [2, 3]
- Perennial dicotyledons in ***amenity grassland*** [1]; ***permanent grassland*** [2, 3]
- Stinging nettle in ***amenity grassland*** [1]; ***permanent grassland*** [2, 3]
- Thistles in ***amenity grassland*** [1]; ***permanent grassland*** [2, 3]

Approval information

- Clopyralid and triclopyr included in Annex I under EC Directive 91/414
- Approval expiry 31 Aug 2008 [1-3]

Efficacy guidance

- Must be applied to actively growing weeds
- Correct timing crucial for good control. Spray stinging nettle before flowering, docks in rosette stage in spring, creeping thistle before flower stems 15-20 cm high, brambles, broom and gorse in Jun-Aug
- Allow 2-3 wk regrowth after grazing or mowing before spraying perennial weeds
- Where there is a large reservoir of weed seed in the soil further treatment in the following yr may be needed

Restrictions

- Maximum number of treatments 1 per yr
- Only use on permanent pasture or rotational grassland established for at least 1 yr
- Do not apply where clover is an important constituent of sward
- Do not roll or harrow within 10 d before or 7 d after spraying
- Do not cut grass for 21 d before or 28 d after spraying
- Do not use any treated plant material for composting or mulching, and do not use manure for composting from animals fed on treated crops
- Do not apply by hand-held rotary atomiser equipment
- Do not allow drift onto other crops, amenity plantings or gardens, ponds, lakes or water courses. All conifers, especially pine and larch, are very sensitive

Crop-specific information

- Latest use: 7 d before grazing or cutting grass
- Some transient yellowing of treated swards may occur but is quickly outgrown

Following crops guidance

- Residues in plant tissues which have not completely decayed may affect succeeding susceptible crops such as peas, beans, other legumes, carrots, parsnips, potatoes, tomatoes, lettuce, glasshouse and protected crops
- Do not plant susceptible autumn-sown crops (eg winter beans) in same year as treatment and allow at least 9 mth from treatment before planting a susceptible crop in the following yr
- Do not direct drill kale, swedes, turnips, grass or grass mixtures within 6 wk of spraying
- Do not spray after end Jul where susceptible crops are to be planted in the next spring

Environmental safety

- Dangerous for the environment
- Very toxic to aquatic organisms
- Keep livestock out of treated areas for at least 7 d after spraying and until foliage of any poisonous weeds such as ragwort or buttercup has died down and become unpalatable

Hazard classification and safety precautions

Hazard H03 [1-3]; H04 [4]; H11 [1-4]
Risk phrases R22a, R38, R43, R50 [1-3]; R41, R53a [1-4]; R52 [4]
Operator protection A, C [1-4]; H, M [1-3]; U02a, U05a, U11, U20b [1-4]; U08, U19a [1-3]; U14, U23b [1, 2]; U15, U23a [4]
Environmental protection E07a, E15a [1-4]; E23, E34 [1-3]; E38 [1, 2]
Consumer protection C01 [1, 2]
Storage and disposal D01, D02, D05, D09a, D10b, D12a
Medical advice M03 [4]; M05a [3]

113 clothianidin

A nitromethylene neonicotinoid insecticide
IRAC mode of action code: 4A

See also *beta-cyfluthrin + clothianidin*

Products

Deter	Bayer CropScience	250 g/l	FS	12411

Uses

- Leafhoppers in ***durum wheat*** *(seed treatment)*, ***triticale*** *(seed treatment)*, ***winter barley*** *(seed treatment)*, ***winter oats*** *(seed treatment)*, ***winter rye*** *(seed treatment)*, ***winter wheat*** *(seed treatment)*
- Slugs in ***durum wheat*** *(seed treatment)*, ***triticale*** *(seed treatment)*, ***winter barley*** *(seed treatment)*, ***winter oats*** *(seed treatment)*, ***winter rye*** *(seed treatment)*, ***winter wheat*** *(seed treatment)*
- Virus vectors in ***durum wheat*** *(seed treatment)*, ***triticale*** *(seed treatment)*, ***winter barley*** *(seed treatment)*, ***winter oats*** *(seed treatment)*, ***winter rye*** *(seed treatment)*, ***winter wheat*** *(seed treatment)*
- Wireworm in ***durum wheat*** *(seed treatment)*, ***triticale*** *(seed treatment)*, ***winter barley*** *(seed treatment)*, ***winter oats*** *(seed treatment)*, ***winter rye*** *(seed treatment)*, ***winter wheat*** *(seed treatment)*

Approval information

- Clothianidin included in Annex I under EC Directive 91/414
- Accepted by BBPA for use on malting barley

Efficacy guidance

- May only be used in conjunction with manufacturer's approved seed treatment application equipment or by following procedures given in the operating instructions
- Treated seed should preferably be drilled in the same season
- Evenness of seed cover improved by simultaneous application with a small volume (1.5-3 litres/ tonne) of water
- Calibrate drill for treated seed and drill at 2.5-4 cm into firm, well prepared seedbed

- Use minimum 125 kg treated seed per ha
- Seed should be drilled to a depth of 40 mm into a well prepared and firm seedbed. If seed is present on the surface, or if spills have occurred, the field should be harrowed and rolled if conditions are appropriate
- When aphid activity is unusually late, or is heavy and prolonged in areas of high risk, and mild weather predominates, follow-up foliar aphicide may be required
- Incidental suppression of leafhoppers in early spring (Mar/Apr) may be achieved but if a specific attack develops an additional foliar insecticide may be required

Restrictions

- Maximum number of seed treatments one per batch
- Product must be fully re-dispersed and homogeneous before use
- Do not use on seed with more than 16% moisture content, or on sprouted, cracked or skinned seed
- All seed batches should be tested to ensure they are suitable for treatment

Crop-specific information

- Latest use: pre-drilling

Environmental safety

- Harmful to game and wildlife. Treated seed should not be left on the surface. Bury spillages.

Hazard classification and safety precautions

Hazard H04
Risk phrases R43
Operator protection A, H; U04a, U05a, U07, U13, U14, U20b
Environmental protection E03, E15b, E34, E36a
Storage and disposal D01, D02, D05, D09a, D14
Treated seed S01, S02, S03, S04b, S05, S06a, S06b, S07, S08
Medical advice M03

114 clothianidin + prothioconazole

A combined fungicide and insecticide seed dressing for cereals
IRAC mode of action code: 4A

Products

Redigo Deter	Bayer CropScience	250:50 g/l	FS	12423

Uses

- Bunt in ***durum wheat*** *(seed treatment)*, ***triticale*** *(seed treatment)*, ***winter rye*** *(seed treatment)*, ***winter wheat*** *(seed treatment)*
- Covered smut in ***winter barley*** *(seed treatment)*
- Fusarium foot rot and seedling blight in ***durum wheat*** *(seed treatment)*, ***triticale*** *(seed treatment)*, ***winter barley*** *(seed treatment)*, ***winter oats*** *(seed treatment)*, ***winter rye*** *(seed treatment)*, ***winter wheat*** *(seed treatment)*
- Leaf stripe in ***winter barley*** *(seed treatment)*
- Leafhoppers in ***durum wheat*** *(seed treatment)*, ***triticale*** *(seed treatment)*, ***winter barley*** *(seed treatment)*, ***winter oats*** *(seed treatment)*, ***winter rye*** *(seed treatment)*, ***winter wheat*** *(seed treatment)*
- Loose smut in ***durum wheat*** *(seed treatment)*, ***winter barley*** *(seed treatment)*, ***winter oats*** *(seed treatment)*, ***winter wheat*** *(seed treatment)*
- Slugs in ***durum wheat*** *(seed treatment)*, ***triticale*** *(seed treatment)*, ***winter barley*** *(seed treatment)*, ***winter oats*** *(seed treatment)*, ***winter rye*** *(seed treatment)*, ***winter wheat*** *(seed treatment)*
- Virus vectors in ***durum wheat*** *(seed treatment)*, ***triticale*** *(seed treatment)*, ***winter barley*** *(seed treatment)*, ***winter oats*** *(seed treatment)*, ***winter rye*** *(seed treatment)*, ***winter wheat*** *(seed treatment)*
- Wireworm in ***durum wheat*** *(seed treatment)*, ***triticale*** *(seed treatment)*, ***winter barley*** *(seed treatment)*, ***winter oats*** *(seed treatment)*, ***winter rye*** *(seed treatment)*, ***winter wheat*** *(seed treatment)*

Approval information
- Clothianidin included in Annex I under EC Directive 91/414
- Accepted by BBPA for use on malting barley

Efficacy guidance
- May only be used in conjunction with manufacturer's approved seed treatment application equipment or by following procedures given in the operating instructions
- Treated wheat seed should preferably be drilled in the same season. Treated seed of other cereals must be used in the same season
- Evenness of seed cover improved by simultaneous application with a small volume (1.5-3 litres/ tonne) of water
- Calibrate drill for treated seed and drill at 2.5-4 cm into firm, well prepared seedbed
- Use minimum 125 kg treated seed per ha
- When aphid activity unusually late, or is heavy and prolonged in areas of high risk, and mild weather predominates, follow-up foliar aphicide may be required
- Incidental suppression of leafhoppers in early spring (Mar/Apr) may be achieved but if a specific attack develops an additional foliar insecticide may be required
- Protection against slugs only applies to germinating seeds and not to aerial parts after emergence
- Where very high populations of aphids, wireworms or slugs occur additional specific control measures may be needed

Restrictions
- Maximum number of seed treatments one per batch for all cereals
- Product must be fully re-dispersed and homogeneous before use
- Do not use on seed with more than 16% moisture content, or on sprouted, cracked or skinned seed
- All seed batches should be tested to ensure they are suitable for treatment

Crop-specific information
- Latest use: pre-drilling of all cereals

Environmental safety
- Harmful to aquatic organisms

Hazard classification and safety precautions

Hazard H04
Risk phrases R43, R52, R53a
Operator protection A, H; U14, U19a, U20b
Environmental protection E15a, E34, E38
Storage and disposal D01, D02, D09a, D11a, D12a
Treated seed S01, S02, S03, S04b, S05, S06a, S07, S08

115 clothianidin + prothioconazole + tebuconazole + triazoxide

A fungicide and insecticide seed treatment mixture for winter barley
IRAC mode of action code: 4A

Products

Raxil Deter	Bayer CropScience	333.3:33.3:20:13.3 g/l	FS	13030

Uses
- Barley yellow dwarf virus vectors in ***winter barley*** *(seed treatment)*
- Covered smut in ***winter barley*** *(seed treatment)*
- Fusarium foot rot and seedling blight in ***winter barley*** *(seed treatment)*
- Leaf stripe in ***winter barley*** *(seed treatment)*
- Leafhoppers in ***winter barley*** *(seed treatment)*
- Loose smut in ***winter barley*** *(seed treatment)*
- Net blotch in ***winter barley*** *(seed treatment)*
- Slugs in ***winter barley*** *(seed treatment - reduction)*
- Wireworm in ***winter barley*** *(seed treatment - reduction)*

Approval information
- Clothianidin included in Annex I under EC Directive 91/414
- Accepted by BBPA for use on malting barley

Efficacy guidance

- May only be used in conjunction with manufacturer's approved seed treatment application equipment or by following procedures given in the operating instructions
- Treated barley seed must be used in the same season
- Evenness of seed cover improved by simultaneous application with a small volume (1.5-3 litres/ tonne) of water
- Calibrate drill for treated seed and drill at 4 cm into firm, well prepared seedbed
- Use minimum 125 kg treated seed per ha
- When aphid activity unusually late, or is heavy and prolonged in areas of high risk, and mild weather predominates, follow-up foliar aphicide may be required
- Leafhoppers can be suppressed in early spring (Mar/Apr) but if a specific attack develops an additional foliar insecticide may be required
- Protection against slugs only applies to germinating seeds and not to aerial parts after emergence
- Where very high populations of aphids, wireworms or slugs occur additional specific control measures may be needed

Restrictions

- Maximum number of seed treatments one per batch
- Product must be fully re-dispersed and homogeneous before use
- Do not use on seed with more than 16% moisture content, or on sprouted, cracked or skinned seed
- All seed batches should be tested to ensure they are suitable for treatment

Crop-specific information

- Latest use: pre-drilling of winter barley

Environmental safety

- Harmful to aquatic organisms

Hazard classification and safety precautions

Hazard H04
Risk phrases R43, R52, R53a
Operator protection A, D, H; U07, U14, U20a
Environmental protection E15a, E34, E36a, E38
Storage and disposal D01, D02, D09a, D12a, D14
Treated seed S01, S02, S03, S04b, S05, S07, S08

116 Coniothyrium minitans

A fungal parasite of sclerotia in soil

Products

Contans	Belchim	1000 IU/mg	WG	12616

Uses

- Sclerotinia in ***all edible crops (outdoor)***, ***all non-edible crops (outdoor)***, ***protected crops***

Efficacy guidance

- *C.minitans* is a soil acting biological fungicide with specific action against the resting bodies (sclerotia) of *Sclerotinia sclerotiorum* and *S.minor*
- Treat 3 mth before disease protection is required to allow time for the infective sclerotia in the soil to be reduced
- A post-harvest treatment of soil and debris prevents further contamination of the soil with sclerotia produced by the previous crop
- For best results the soil should be moist and the temperature between 12-20 °C
- Application should be followed by soil incorporation into the surface layer to 10 cm with a rotovator or rotary harrow. Thorough spray coverage of the soil is essential to ensure uniform distribution
- If soil temperature drops below 0 °C or rises above 27 °C fungicidal activity is suspended but restarts when soil temperature returns within this range
- In glasshouses untreated areas at the margins should be covered by a film to avoid spread of spores from untreated sclerotia

Restrictions
- Maximum number of treatments 1 per crop or situation
- Do not plough or cultivate after treatment
- Do not mix with other pesticides, acids, alkalines or any product that attacks organic material
- Store product under cool dry conditions away from direct sunlight and away from heat sources

Crop-specific information
- Latest use: pre-planting
- HI: zero

Following crops guidance
- There are no restrictions on following crops and no waiting interval is specified

Hazard classification and safety precautions

Hazard H03
Risk phrases R42
Operator protection A, C, D; U05a, U19a
Environmental protection E15b, E34
Storage and disposal D01, D02, D12a
Medical advice M03

117 copper ammonium carbonate

A protectant copper fungicide
FRAC mode of action code: M1

Products

Croptex Fungex	Certis	8% w/w (copper)	SL	11049

Uses
- Blight in ***tomatoes (outdoor)***
- Cane spot in ***loganberries***, ***raspberries***
- Celery leaf spot in ***celery (outdoor)***
- Currant leaf spot in ***blackcurrants***
- Damping off in ***seedlings of ornamentals***
- Leaf curl in ***peaches***
- Leaf mould in ***protected tomatoes***
- Powdery mildew in ***chrysanthemums***, ***cucumbers***

Approval information
- Accepted by BBPA for use on hops

Efficacy guidance
- Apply spray to both sides of foliage
- With protected crops keep foliage dry before and after spraying
- Ventilate glasshouse immediately after spraying

Restrictions
- Maximum number of treatments 5 per crop for celery; 3 per yr for blackcurrants and loganberries; 2 per yr for peaches and raspberries; 1 per tray when used as a drench for ornamental seedlings
- Do not spray plants which are dry at the roots

Environmental safety
- Harmful to fish or other aquatic life. Do not contaminate surface waters or ditches with chemical or used container
- Keep all livestock out of treated areas for at least 3 wk. Bury or remove spillages

Hazard classification and safety precautions

Hazard H03
Risk phrases R22a, R41
Operator protection U05a, U11, U15, U20c
Environmental protection E06c (3 wk); E13c, E34
Storage and disposal D01, D02, D09a, D10a, D12a
Medical advice M03

118 copper oxychloride

A protectant copper fungicide and bactericide
FRAC mode of action code: M1

Products

1	Cuprokylt	Unicrop	50% w/w (copper)	WP	00604
2	Cuprokylt FL	Unicrop	270 g/l (copper)	SC	08299
3	Headland Inorganic Liquid Copper	Headland	256 g/l (copper)	SC	13009

Uses

- Bacterial canker in ***cherries*** [1, 3]; ***cob nuts*** *(off-label)*, ***hazel nuts*** *(off-label)*, ***walnuts*** *(off-label)* [2]; ***plums*** [1, 2]
- Bacterial rot in ***bulb onions*** *(off-label)*, ***garlic*** *(off-label)*, ***leeks*** *(off-label)*, ***salad onions*** *(off-label)*, ***shallots*** *(off-label)* [1]; ***rhubarb*** *(off-label)* [1, 2]
- Black rot in ***broccoli*** *(off-label)*, ***brussels sprouts*** *(off-label)*, ***cabbages*** *(off-label)*, ***calabrese*** *(off-label)*, ***cauliflowers*** *(off-label)*, ***chinese cabbage*** *(off-label)*, ***collards*** *(off-label)*, ***kale*** *(off-label)* [1]; ***protected brassica seedlings*** *(off-label)* [2]
- Blight in ***cob nuts*** *(off-label)*, ***hazel nuts*** *(off-label)*, ***walnuts*** *(off-label)* [2]; ***potatoes***, ***tomatoes (outdoor)*** [1-3]; ***protected tomatoes*** [1, 2]
- Buck-eye rot in ***protected tomatoes***, ***tomatoes (outdoor)*** [1-3]
- Cane spot in ***blackberries*** *(off-label)*, ***loganberries***, ***raspberries***, ***rubus hybrids*** *(off-label)* [1, 2]
- Canker in ***apples***, ***pears*** [1-3]
- Celery leaf spot in ***celery (outdoor)*** [1-3]
- Collar rot in ***apples*** *(off-label)* [2]
- Damping off in ***protected tomatoes***, ***tomatoes (outdoor)*** [1, 2]
- Downy mildew in ***bilberries*** *(off-label)*, ***blueberries*** *(off-label)*, ***cranberries*** *(off-label)*, ***gooseberries*** *(off-label)*, ***redcurrants*** *(off-label)*, ***whitecurrants*** *(off-label)* [1, 2]; ***hops*** [1-3]; ***wine grapes*** [3]
- Foot rot in ***protected tomatoes***, ***tomatoes (outdoor)*** [1, 2]
- Leaf curl in ***apricots*** *(off-label)*, ***peaches*** [1, 2]; ***quinces*** *(off-label)* [1]
- Pseudomonas storage rots in ***bulb onions*** *(off-label)*, ***garlic*** *(off-label)*, ***leeks*** *(off-label)*, ***salad onions*** *(off-label)*, ***shallots*** *(off-label)* [1]
- Purple blotch in ***blackberries*** [3]
- Pythium in ***watercress*** *(off-label - during propagation)* [2]
- Rhizoctonia in ***watercress*** *(off-label - during propagation)* [2]
- Rust in ***blackcurrants*** [1-3]
- Spear rot in ***broccoli*** *(off-label)*, ***brussels sprouts*** *(off-label)*, ***cabbages*** *(off-label)*, ***calabrese*** *(off-label)*, ***cauliflowers*** *(off-label)*, ***chinese cabbage*** *(off-label)*, ***collards*** *(off-label)*, ***kale*** *(off-label)* [1]; ***protected brassica seedlings*** *(off-label)* [2]

Specific Off-Label Approvals (SOLAs)

- ***apples*** *982336* [2]
- ***apricots*** *20063134* [1], *20063140* [2]
- ***bilberries*** *20063131* [1], *20063137* [2]
- ***blackberries*** *20063132* [1], *20063139* [2]
- ***blueberries*** *20063131* [1], *20063137* [2]
- ***broccoli*** *20010115* [1]
- ***brussels sprouts*** *20010115* [1]
- ***bulb onions*** *991127* [1]
- ***cabbages*** *20010115* [1]
- ***calabrese*** *20010115* [1]
- ***cauliflowers*** *20010115* [1]
- ***chinese cabbage*** *20010115* [1]
- ***cob nuts*** *990385* [2]
- ***collards*** *20010115* [1]
- ***cranberries*** *20063131* [1], *20063137* [2]
- ***garlic*** *991127* [1]
- ***gooseberries*** *20063131* [1], *20063137* [2]
- ***hazel nuts*** *990385* [2]

- ***kale*** *20010115* [1]
- ***leeks*** *991127* [1]
- ***protected brassica seedlings*** *20010117* [2]
- ***quinces*** *20063133* [1]
- ***redcurrants*** *20063131* [1], *20063137* [2]
- ***rhubarb*** *20063135* [1], *20063138* [2]
- ***rubus hybrids*** *20063132* [1], *20063139* [2]
- ***salad onions*** *991127* [1]
- ***shallots*** *991127* [1]
- ***walnuts*** *990385* [2]
- ***watercress*** *(during propagation) 20001538* [2]
- ***whitecurrants*** *20063131* [1], *20063137* [2]

Approval information

- Approved for aerial application on potatoes [1]. See notes in Section 5
- Accepted by BBPA for use on hops

Efficacy guidance

- Spray crops at high volume when foliage dry but avoid run off. Do not spray if rain expected soon
- Spray interval commonly 10-14 d but varies with crop, see label for details
- If buck-eye rot occurs, spray soil surface and lower parts of tomato plants to protect unaffected fruit [1, 2]
- A follow-up spray in the following spring should be made to top fruit severely infected with bacterial canker

Restrictions

- Maximum number of treatments 3 per crop for apples, blackberries, grapevines, pears; 2 per crop for watercress. Not specified for other crops
- Some peach cultivars are sensitive to copper. Treat non-sensitive varieties only [1, 2]

Crop-specific information

- HI hops 7 d
- Slight damage may occur to leaves of cherries and plums

Environmental safety

- Dangerous for the environment
- Very toxic to aquatic organisms
- Keep all livestock out of treated areas for at least 3 wk

Hazard classification and safety precautions

Hazard H03 [2]; H11 [1-3]
Risk phrases R22a, R50 [2]; R51 [1, 3]; R53a [1-3]
Operator protection A [3]; U20a [1, 3]; U20c [2]
Environmental protection E06a [1-3] (3 wk); E13c, E34, E38 [1, 3]; E15a [2]
Storage and disposal D01, D09a [1-3]; D05, D10c, D12a [2]; D10b [1, 3]

119 copper sulphate

See Bordeaux Mixture
FRAC mode of action code: M1

120 cyazofamid

A cyanoimidazolesulfonamide protectant fungicide for potatoes
FRAC mode of action code: 21

Products

Ranman Twinpack	Belchim	400 g/l	KL	11851

Uses

- Blight in ***potatoes***

Approval information

- Cyazofamid included in Annex I under EC Directive 91/414

SEE SECTION 3 FOR PRODUCTS ALSO REGISTERED

Efficacy guidance

- Apply as a protectant treatment before blight enters the crop and repeat every 7-10 d depending on severity of disease pressure
- Commence spray programme immediately the risk of blight in the locality occurs, usually when the crop meets along the rows
- Product must always be used with organosilicone adjuvant provided in the twin pack
- To minimise the chance of development of resistance no more than three applications should be made consecutively (out of a permissible total of six) in the blight control programme. For more information on Resistance Management see Section 5

Restrictions

- Maximum number of treatments 6 per crop (no more than three of which should be consecutive)
- Mixed product must not be allowed to stand overnight
- Consult processor before using on crops intended for processing

Crop-specific information

- HI 7 d

Environmental safety

- Dangerous for the environment
- Very toxic to aquatic organisms
- Do not empty into drains

Hazard classification and safety precautions

Hazard H03 (adjuvant); H11
Risk phrases R20, R36, R48 (adjuvant); R41, R50, R53a
Operator protection A, C; U02a, U05a, U11, U15, U20a; U19a (adjuvant)
Environmental protection E15a, E19b, E34, E38
Storage and disposal D01, D02, D09a, D10c, D12a
Medical advice M03

121 cycloxydim

A translocated post-emergence cyclohexanedione oxime herbicide for grass weed control
HRAC mode of action code: A

Products

1	Greencrop Pomeroy	Greencrop	200 g/l	EC	11769
2	Greencrop Valentia	Greencrop	200 g/l	EC	10197
3	Landgold Cycloxydim	Teliton	200 g/l	EC	12113
4	Laser	BASF	200 g/l	EC	12930
5	Standon Cycloxydim	Standon	200 g/l	EC	08830

Uses

- Annual grasses in ***brussels sprouts, cabbages, cauliflowers, fodder beet, mangels, spring field beans, sugar beet, swedes, winter field beans, winter oilseed rape*** [1-5]; ***bulb onions, carrots, dwarf beans, farm forestry, forest, leeks, linseed, parsnips, salad onions, spring oilseed rape, strawberries*** [1, 2, 4, 5]; ***calabrese, peas*** [2]; ***combining peas, vining peas*** [3, 4]; ***early potatoes*** [1, 3, 4]; ***flower bulbs*** [2, 4, 5]; ***forest nurseries, ornamental plant production*** [1]; ***maincrop potatoes*** [2-4]
- Annual meadow grass in ***combining peas, maincrop potatoes, vining peas*** [1, 5]
- Black bent in ***brussels sprouts, cabbages, cauliflowers, fodder beet, maincrop potatoes, mangels, spring field beans, sugar beet, swedes, winter field beans, winter oilseed rape*** [1-5]; ***bulb onions, carrots, dwarf beans, flower bulbs, leeks, linseed, parsnips, salad onions, spring oilseed rape, strawberries*** [1, 2, 4, 5]; ***calabrese, peas*** [2]; ***combining peas, vining peas*** [1, 3-5]; ***early potatoes*** [1, 3, 4]; ***ornamental plant production*** [1]
- Blackgrass in ***brussels sprouts, cabbages, cauliflowers, fodder beet, maincrop potatoes, mangels, spring field beans, sugar beet, swedes, winter field beans, winter oilseed rape*** [1-5]; ***bulb onions, carrots, dwarf beans, flower bulbs, leeks, linseed, parsnips, salad onions, spring oilseed rape, strawberries*** [1, 2, 4, 5]; ***calabrese, peas*** [2]; ***combining peas, vining peas*** [1, 3-5]; ***early potatoes*** [1, 3, 4]; ***ornamental plant production*** [1]
- Couch in ***brussels sprouts, cabbages, cauliflowers, fodder beet, maincrop potatoes, mangels, spring field beans, sugar beet, swedes, winter field beans, winter oilseed rape***

[1-5]; ***bulb onions, carrots, dwarf beans, flower bulbs, leeks, linseed, parsnips, salad onions, spring oilseed rape, strawberries*** [1, 2, 4, 5]; ***calabrese, peas*** [2]; ***combining peas, vining peas*** [1, 3-5]; ***early potatoes*** [1, 3, 4]; ***ornamental plant production*** [1]
- Creeping bent in ***brussels sprouts, cabbages, cauliflowers, fodder beet, maincrop potatoes, mangels, spring field beans, sugar beet, swedes, winter field beans, winter oilseed rape*** [1-5]; ***bulb onions, carrots, dwarf beans, flower bulbs, leeks, linseed, parsnips, salad onions, spring oilseed rape, strawberries*** [1, 2, 4, 5]; ***calabrese, peas*** [2]; ***combining peas, vining peas*** [1, 3-5]; ***early potatoes*** [1, 3, 4]; ***ornamental plant production*** [1]
- Green cover in ***land temporarily removed from production*** [1, 2, 4]
- Onion couch in ***brussels sprouts, bulb onions, cabbages, carrots, cauliflowers, dwarf beans, flower bulbs, fodder beet, leeks, linseed, maincrop potatoes, mangels, parsnips, salad onions, spring field beans, spring oilseed rape, strawberries, sugar beet, swedes, winter field beans, winter oilseed rape*** [1, 2, 4, 5]; ***calabrese, peas*** [2]; ***combining peas, vining peas*** [1, 4, 5]; ***early potatoes*** [1, 4]; ***ornamental plant production*** [1]
- Perennial grasses in ***farm forestry, forest*** [1, 2, 4, 5]; ***forest nurseries*** [1]
- Volunteer cereals in ***brussels sprouts, cabbages, cauliflowers, fodder beet, maincrop potatoes, mangels, spring field beans, sugar beet, swedes, winter field beans, winter oilseed rape*** [1-5]; ***bulb onions, carrots, dwarf beans, flower bulbs, leeks, linseed, parsnips, salad onions, spring oilseed rape, strawberries*** [1, 2, 4, 5]; ***calabrese, peas*** [2]; ***combining peas, vining peas*** [1, 3-5]; ***early potatoes*** [1, 3, 4]; ***ornamental plant production*** [1]
- Wild oats in ***brussels sprouts, cabbages, cauliflowers, fodder beet, maincrop potatoes, mangels, spring field beans, sugar beet, swedes, winter field beans, winter oilseed rape*** [1-5]; ***bulb onions, carrots, dwarf beans, flower bulbs, leeks, linseed, parsnips, salad onions, spring oilseed rape, strawberries*** [1, 2, 4, 5]; ***calabrese, peas*** [2]; ***combining peas, vining peas*** [1, 3-5]; ***early potatoes*** [1, 3, 4]; ***ornamental plant production*** [1]

Approval information
- Approval expiry 31 Aug 2008 [1-3, 5]

Efficacy guidance
- Best results achieved when weeds small and have not begun to compete with crop. Effectiveness reduced by drought, cool conditions or stress. Weeds emerging after application are not controlled
- Foliage death usually complete after 3-4 wk but longer under cool conditions, especially late treatments to winter oilseed rape
- Perennial grasses should have sufficient foliage to absorb spray and should not be cultivated for at least 14 d after treatment
- On established couch pre-planting cultivation recommended to fragment rhizomes and encourage uniform emergence
- Split applications to volunteer wheat and barley at GS 12-14 will often give adequate control in winter oilseed rape. See label for details
- Apply to dry foliage when rain not expected for at least 2 h
- Cycloxydim is an ACCase inhibitor herbicide. To avoid the build up of resistance do not apply products containing an ACCase inhibitor herbicide more than twice to any crop. In addition do not use any product containing cycloxydim in mixture or sequence with any other product containing the same ingredient
- Use these products as part of a resistance management strategy that includes cultural methods of control and does not use ACCase inhibitors as the sole chemical method of grass weed control
- Applying a second product containing an ACCase inhibitor to a crop will increase the risk of resistance development; only use a second ACCase inhibitor to control different weeds at a different timing
- Always follow WRAG guidelines for preventing and managing herbicide resistant weeds. See Section 5

Restrictions
- Maximum number of treatments 1 per crop or yr in most situations. See label
- Must be used with authorised adjuvant oil. See label
- Do not apply to crops damaged or stressed by adverse weather, pest or disease attack or other pesticide treatment
- Prevent drift onto other crops, especially cereals and grass

Crop-specific information

- HI cabbage, cauliflower, calabrese, salad onions 4 wk; peas, dwarf beans 5 wk; bulb onions, carrots, parsnips, strawberries 6 wk; sugar and fodder beet, leeks, mangels, potatoes, field beans, swedes, Brussels sprouts, winter field beans 8 wk; oilseed rape, soya beans, linseed 12 wk
- Recommended time of application varies with crop. See label for details
- On peas a crystal violet wax test should be done if leaf wax likely to have been affected by weather conditions or other chemical treatment. The wax test is essential if other products are to be sprayed before or after treatment
- May be used on ornamental bulbs when crop 5-10 cm tall. Product has been used on tulips, narcissi, hyacinths and irises but some subjects may be more sensitive and growers advised to check tolerance on small number of plants before treating the rest of the crop
- May be applied to land temporarily removed from production where the green cover is made up predominantly of tolerant crops listed on label. Use on industrial crops of linseed and oilseed rape on land temporarily removed from production also permitted.

Following crops guidance

- Guideline intervals for sowing succeeding crops after failed treated crop: field beans, peas, sugar beet, rape, kale, swedes, radish, white clover, lucerne 1 wk; dwarf French beans 4 wk; wheat, barley, maize 8 wk
- Oats should not be sown after failure of a treated crop

Environmental safety

- Dangerous for the environment
- Toxic to aquatic organisms
- Harmful to fish or other aquatic life. Do not contaminate surface waters or ditches with chemical or used container

Hazard classification and safety precautions

Hazard H03, H11
Risk phrases R22b, R38, R51, R53a [1-5]; R36 [1-3, 5]
Operator protection A, C; U05a, U08, U20b [1-5]; U19a [1, 2]
Environmental protection E15a [1-5]; E38 [4]
Storage and disposal D01, D02, D09a [1-5]; D05, D10b [1-3, 5]; D10c, D12a [4]
Medical advice M05b

122 Cydia pomonella GV

An baculovirus insecticide for codling moth control

Products

Cyd-X	Certis	10 g/l granulovirus	SC	13535

Uses

- Codling moth in ***apples, pears***

Efficacy guidance

- Optimum results require a precise spraying schedule commencing soon after egg laying and before first hatch
- First generation of codling moth normally occurs in first 2 wk of Jun in UK
- Apply in sufficient water to achieve good coverage of whole tree
- Normally 3 applications at intervals of 8 sunny days are needed for each codling moth generation
- Fruit damage is reduced significantly in the first yr and will reduce the population of codling moth in the following yr

Restrictions

- Maximum number of treatments 3 per season
- Do not tank mix with copper or any products with a pH lower than 5 or higher than 8
- Do not use as the exclusive measure for codling moth control. Adopt an anti-resistance strategy. See Section 5
- Consult processor before use on crops for processing

Crop-specific information

- HI: 14 d for apples, pears
- Evidence of efficacy and crop safety on pears is limited

Environmental safety

- Product has no adverse effect on predatory insects and is fully compatible with IPM programmes

Hazard classification and safety precautions

Hazard H03
Risk phrases R42, R43
Operator protection A, D, H; U14, U19a
Environmental protection E15a
Storage and disposal D12a

123 cyflufenamid

An amidoxime fungicide for cereals
FRAC mode of action code: U6

Products

Cyflamid	Certis	50 g/l	EW	12403

Uses

- Powdery mildew in ***amenity vegetation*** *(off-label)*, ***durum wheat***, ***grass seed crops*** *(off-label)*, ***ornamental plant production*** *(off-label)*, ***spring barley***, ***spring oats***, ***spring wheat***, ***triticale***, ***winter barley***, ***winter oats***, ***winter rye***, ***winter wheat***

Specific Off-Label Approvals (SOLAs)

- ***amenity vegetation*** *20070512*
- ***grass seed crops*** *20060602*
- ***ornamental plant production*** *20070512*

Efficacy guidance

- Best results obtained from treatment at the first visible signs of infection by powdery mildew
- Sustained disease presssure may require a second treatment
- Disease spectrum may be broadened by appropriate tank mixtures. See label
- Product is rainfast within 1 h
- Must be used as part of an Integrated Crop Management programme that includes alternating use, or mixture, with fungicides with a different mode of action effective against powdery mildew

Restrictions

- Maximum number of treatments 2 per crop on all recommended cereals
- Apply only in the spring
- Do not apply to crops under stress from drought, waterlogging, cold, pests or diseases, lime or nutrient deficiency or other factors affecting crop growth

Crop-specific information

- Latest use: before start of flowering of cereal crops

Following crops guidance

- No restrictions

Environmental safety

- Dangerous for the environment
- Very toxic to aquatic organisms
- Avoid drift onto ponds, waterways or ditches

Hazard classification and safety precautions

Hazard H11
Risk phrases R50, R53a
Operator protection A; U05a
Environmental protection E15b, E34
Storage and disposal D01, D02, D05, D10c, D12b

124 cyfluthrin

A non-systemic pyrethroid insecticide
IRAC mode of action code: 3

Products

Baythroid	Makhteshim	50 g/l	EC	13221

Uses

- Aphids in ***apples***
- Apple sucker in ***apples***
- Barley yellow dwarf virus vectors in ***winter barley***, ***winter wheat***
- Cabbage stem flea beetle in ***winter oilseed rape***
- Caterpillars in ***broccoli***, ***brussels sprouts***, ***cabbages***, ***cauliflowers***
- Winter moth in ***apples***

Approval information

- Cyfluthrin included in Annex I under EC Directive 91/414
- Approval expiry 31 Dec 2008 [1]

Efficacy guidance

- Best control of virus vectors in winter cereals and cabbage stem flea beetle in oilseed rape obtained from treatment normally in Oct or early Nov. Follow specialist or ADAS advice
- Complete control of cabbage stem flea beetle in oilseed rape may require a second treatment to control later hatches
- Brassica pests should be treated during first stages of attack. A second treatment may be needed if reinfestation occurs
- Apple pests should be treated at green cluster

Restrictions

- Maximum number of treatments 1 per crop or yr for barley, wheat, apples; 2 per crop for brassicas, oilseed rape

Crop-specific information

- Latest use: before 31 Mar in yr of harvest for barley, winter oilseed rape, wheat; before flowerhead formation for broccoli, cauliflowers; before pink bud for apples
- HI 7 d for hops

Environmental safety

- Dangerous for the environment
- Toxic to aquatic organisms
- Extremely dangerous to bees. Do not apply to crops in flower or to those in which bees are actively foraging. Do not apply when flowering weeds are present
- Extremely dangerous to fish or other aquatic life. Do not contaminate surface waters or ditches with chemical or used container
- LERAP Category A
- Broadcast air-assisted LERAP (18 m)

Hazard classification and safety precautions

Hazard H03, H08, H11
Risk phrases R20, R22a, R22b, R36, R51, R53a, R66
Operator protection A, C, H; U04a, U05a, U10, U11, U14, U15, U19a, U20b
Environmental protection E12b, E13a, E16c, E16d, E34, E38; E17b (18 m)
Storage and disposal D01, D02, D05, D09a, D10b, D12a
Medical advice M03, M05b

125 cymoxanil

A cyanoacetamide oxime fungicide for potatoes
FRAC mode of action code: 27

Products

1	Cymbal	Belchim	45% w/w	WG	13312
2	Option	DuPont	60% w/w	WG	11834
3	Sipcam C 50	Sipcam	50% w/w	WP	10610

Uses

- Blight in ***potatoes***

Efficacy guidance

- Product to be used in mixture with specified mixture partners in order to combine systemic and protective activity. See labels for details
- Commence spray programme as soon as weather conditions favourable for disease development occur and before infection appears. At latest the first treatment should be made as the foliage meets along the rows
- Repeat treatments at 7-14 day intervals according to disease incidence and weather conditions
- Treat irrigated crops as soon as possible after irrigation and repeat at 10 d intervals

Restrictions

- Maximum total dose equivalent to four full dose treatments [3]. Not specified for other products
- Minimum spray interval 7 d
- Do not allow packs to become wet during storage and do not keep opened packs from one season to another
- Cymoxanil must be used in tank mixture with other fungicides

Crop-specific information

- HI for potatoes 7 d [3]; 14 d [1, 2]
- Consult processor before treating potatoes grown for processing

Environmental safety

- Dangerous for the environment
- Very toxic to aquatic organisms

Hazard classification and safety precautions

Hazard H03 [2, 3]; H04 [1]; H11 [1, 2]
Risk phrases R22a [2, 3]; R43 [1]; R50, R53a [1, 2]
Operator protection A [1-3]; H [1]; U05a [1-3]; U08, U20b [2, 3]; U09a, U14, U20a [1]; U19a [2]
Environmental protection E13a, E34 [2]; E13c [3]; E15a [1]; E38 [1, 2]
Storage and disposal D01, D02, D09a, D11a [1-3]; D12a [1, 2]
Medical advice M03 [2, 3]

126 cymoxanil + famoxadone

A preventative and curative fungicide mixture for potatoes
FRAC mode of action code: 27 + 11

Products

Tanos	DuPont	25:25% w/w	WG	10677

Uses

- Blight in ***potatoes***

Approval information

- Famoxadone included in Annex I under EC Directive 91/414

Efficacy guidance

- Commence spray programme before infection appears as soon as weather conditions favourable for disease development occur. At latest the first treatment should be made as the foliage meets along the rows
- Repeat treatments at 7-10 day intervals according to disease incidence and weather conditions
- Reduce spray interval if conditions are conducive to the spread of blight
- Spray as soon as possible after irrigation
- Increase water volume in dense crops
- Product combines active substances with different modes of action and is effective against strains of potato blight that are insensitive to phenylamide fungicides
- To minimise the likelihood of development of resistance to QoI fungicides these products should be used in a planned Resistance Management strategy. See Section 5 for more information

Restrictions

- Maximum number of treatments 6 per crop
- Consult processors before using on crops for processing
- Do not apply more than 3 consecutive treatments containing QoI active fungicides

Crop-specific information

- HI 14 d for potatoes

Environmental safety

- Dangerous for the environment
- Very toxic to aquatic organisms
- Dangerous to fish or other aquatic life. Do not contaminate surface waters or ditches with chemical or used container
- LERAP Category B

Hazard classification and safety precautions

Hazard H03, H11
Risk phrases R22a, R50, R53a
Operator protection A, D; U05a, U20b
Environmental protection E13b, E16a, E16b, E34, E38
Storage and disposal D01, D02, D09a, D10b, D12a
Medical advice M03

127 cymoxanil + fludioxonil + metalaxyl-M

A fungicide seed dressing for peas
FRAC mode of action code: 27 + 12 + 4

Products

Wakil XL	Syngenta	10:5:17.5% w/w	WS	10562

Uses

- Ascochyta in ***combining peas*** *(seed treatment)*, ***vining peas*** *(seed treatment)*
- Damping off in ***chard*** *(off-label)*, ***chives*** *(off-label)*, ***combining peas*** *(seed treatment)*, ***herbs (see appendix 6)*** *(off-label)*, ***leaf spinach*** *(off-label)*, ***lupins*** *(off-label)*, ***parsley*** *(off-label)*, ***poppies for morphine production*** *(off-label - seed treatment)*, ***red beet*** *(off-label - seed treatment)*, ***salad brassicas*** *(off-label)*, ***spinach beet*** *(off-label)*, ***vining peas*** *(seed treatment)*
- Downy mildew in ***broad beans*** *(off-label - seed treatment)*, ***combining peas*** *(seed treatment)*, ***lupins*** *(off-label)*, ***poppies for morphine production*** *(off-label - seed treatment)*, ***spring field beans*** *(off-label - seed treatment)*, ***vining peas*** *(seed treatment)*, ***winter field beans*** *(off-label - seed treatment)*
- Pythium in ***carrots*** *(off-label - seed treatment)*, ***combining peas*** *(seed treatment)*, ***lupins*** *(off-label)*, ***parsnips*** *(off-label - seed treatment)*, ***poppies for morphine production*** *(off-label - seed treatment)*, ***vining peas*** *(seed treatment)*

Specific Off-Label Approvals (SOLAs)

- ***broad beans*** *(seed treatment) 20023932*
- ***carrots*** *(seed treatment) 20021191*
- ***chard*** *20070791*
- ***chives*** *20070791*
- ***herbs (see appendix 6)*** *20070791*
- ***leaf spinach*** *20070791*
- ***lupins*** *20060614*
- ***parsley*** *20070791*
- ***parsnips*** *(seed treatment) 20021191*
- ***poppies for morphine production*** *(seed treatment) 20040683*
- ***red beet*** *(seed treatment) 20032313*
- ***salad brassicas*** *20070791*
- ***spinach beet*** *20070791*
- ***spring field beans*** *(seed treatment) 20023932*
- ***winter field beans*** *(seed treatment) 20023932*

Approval information

- Fludioxonil and metalaxyl-M included in Annex I under EC Directive 91/414

Efficacy guidance

- Apply through continuous flow seed treaters which should be calibrated before use

Restrictions

- Max number of treatments 1 per seed batch
- Ensure moisture content of treated seed satisfactory and store in a dry place
- Check calibration of seed drill with treated seed before drilling and sow as soon as possible after treatment
- Consult before using on crops for processing

Crop-specific information

- Latest use: pre-drilling

Environmental safety

- Harmful to aquatic organisms
- Do not use treated seed as food or feed

Hazard classification and safety precautions

Risk phrases R52, R53a
Operator protection A, D, H; U02a, U05a, U20b
Environmental protection E03, E15a
Storage and disposal D01, D02, D05, D07, D09a, D11a, D12a
Treated seed S02, S04b, S05, S07

128 cymoxanil + mancozeb

A protectant and systemic fungicide for potato blight control
FRAC mode of action code: 27 + M3

Products

1	Agriguard Cymoxanil Plus	AgriGuard	4.5:68% w/w	WP	10893
2	Besiege	DuPont	4.5:68% w/w	WP	08086
3	Clayton Krypton	Clayton	4.5:68% w/w	WP	09398
4	Curzate M WG	DuPont	4.5:68% w/w	WG	11901
5	Globe	Sipcam	6:70% w/w	WP	10339
6	Matilda	Nufarm UK	4.5:68% w/w	WP	12006
7	Nautile DG	Nufarm UK	5:68% w/w	WG	13416
8	Rhapsody	DuPont	4.5:68% w/w	WG	11958
9	Rhythm	Interfarm	4.5:68% w/w	WP	09636
10	Rhythm CM	Interfarm	4.5:68% w/w	WP	11825
11	Solace	Nufarm UK	4.5:68% w/w	WP	11936

Uses

- Blight in ***potatoes***

Approval information

- Mancozeb included in Annex I under EC Directive 91/414

Efficacy guidance

- Apply immediately after blight warning or as soon as local conditions dictate and repeat at 7-14 d intervals until haulm dies down or is burnt off
- Spray interval should not be more than 7-10 d in irrigated crops (see product label). Apply treatment after irrigation
- To minimise the likelihood of development of resistance these products should be used in a planned Resistance Management strategy. See Section 5 for more information

Restrictions

- Maximum number of treatments 6 per crop [5]. Not specified for other products
- Do not apply at less than 7 d intervals
- Do not allow packs to become wet during storage

Crop-specific information

- HI 7 d [5]
- Destroy and remove any haulm that remains after harvest of early varieties to reduce blight pressure on neighbouring maincrop potatoes

Environmental safety

- Dangerous for the environment
- Very toxic to aquatic organisms
- Keep product away from fire or sparks

SEE SECTION 3 FOR PRODUCTS ALSO REGISTERED

Hazard classification and safety precautions

Hazard H03 [3]; H04 [1, 2, 4-11]; H11 [1-4, 7-10]

Risk phrases R36, R42 [1-4, 6, 8-11]; R37 [1-4, 6-11]; R38 [1, 6, 11]; R43 [1-11]; R50, R53a [2-4, 7-10]

Operator protection A, C [1-11]; H [5]; U02a [5]; U05a, U08 [1-11]; U11 [2, 3, 6, 7, 9, 10]; U14 [2, 4, 7-10]; U19a [1-4, 6-11]; U20a [1, 11]; U20b [2-10]

Environmental protection E13b [5]; E13c [2, 4, 6, 8]; E15a [1-5, 7-11]; E38 [2, 4, 7-10]

Storage and disposal D01, D02, D09a [1-11]; D06b [7]; D11a [1-6, 8-11]; D12a [2, 7, 9, 10]

Medical advice M04a, M04b [3]

129 cypermethrin

A contact and stomach acting pyrethroid insecticide
IRAC mode of action code: 3

Products

1	Agriguard Cypermethrin EC	AgriGuard	100 g/l	EC	12134
2	Cypermethrin Lacquer	Killgerm	7 g/l	LA	H3164
3	Forester	Fargro	100 g/l	EW	13164
4	Permasect C	Nufarm UK	100 g/l	EC	13158
5	Toppel 100 EC	United Phosphorus	100 g/l	EC	13704

Uses

- Aphids in ***apples***, ***vining peas*** [1, 4, 5]; ***broccoli***, ***brussels sprouts***, ***cabbages***, ***calabrese***, ***cauliflowers***, ***cherries***, ***durum wheat***, ***hops***, ***lettuce***, ***ornamental plant production***, ***pears***, ***plums***, ***triticale***, ***winter barley***, ***winter oats***, ***winter rye***, ***winter wheat*** [1, 5]; ***spring barley*** *(autumn sown)*, ***spring wheat*** *(autumn sown)* [5]
- Barley yellow dwarf virus vectors in ***durum wheat*** [1, 5]; ***spring barley*** *(autumn sown)*, ***spring wheat*** *(autumn sown)* [4, 5]; ***triticale***, ***winter barley***, ***winter oats***, ***winter rye***, ***winter wheat*** [1, 4, 5]
- Bladder pod midge in ***spring oilseed rape***, ***winter oilseed rape*** [1, 5]
- Cabbage seed weevil in ***spring oilseed rape***, ***winter oilseed rape*** [4]
- Cabbage stem flea beetle in ***winter oilseed rape*** [1, 4, 5]
- Capsids in ***apples***, ***ornamental plant production*** [1, 5]; ***pears*** [1, 4, 5]
- Caterpillars in ***apples***, ***cherries***, ***lettuce***, ***ornamental plant production***, ***pears***, ***plums*** [1, 5]; ***broccoli***, ***brussels sprouts***, ***cabbages***, ***calabrese***, ***cauliflowers*** [1, 4, 5]; ***fodder beet***, ***kale***, ***mangels***, ***potatoes***, ***red beet***, ***sugar beet*** [4]
- Codling moth in ***apples*** [1, 4, 5]
- Cutworms in ***fodder beet***, ***mangels***, ***red beet*** [4]; ***lettuce***, ***ornamental plant production*** [1, 5]; ***potatoes***, ***sugar beet*** [1, 4, 5]
- Frit fly in ***permanent grassland*** [5]; ***rotational grass*** [1, 4, 5]
- Insect pests in ***farm buildings*** [2]
- Large pine weevil in ***forest*** [3]
- Pea and bean weevil in ***spring field beans***, ***vining peas***, ***winter field beans*** [1, 4, 5]
- Pea moth in ***vining peas*** [1, 4, 5]
- Pod midge in ***spring oilseed rape***, ***winter oilseed rape*** [4]
- Pollen beetle in ***spring oilseed rape***, ***winter oilseed rape*** [1, 4, 5]
- Rape winter stem weevil in ***winter oilseed rape*** [1, 5]
- Sawflies in ***apples*** [1, 5]
- Seed weevil in ***spring oilseed rape***, ***winter oilseed rape*** [1, 5]
- Suckers in ***apples*** [4]; ***pears*** [1, 5]
- Thrips in ***ornamental plant production*** [1, 5]
- Tortrix moths in ***apples*** [1, 4, 5]; ***pears*** [4]
- Whitefly in ***broccoli***, ***brussels sprouts***, ***cabbages***, ***calabrese***, ***cauliflowers***, ***ornamental plant production***, ***protected cucumbers***, ***protected lettuce***, ***protected ornamentals*** [1, 5]
- Winter moth in ***apples***, ***pears*** [4]
- Yellow cereal fly in ***durum wheat*** [1, 5]; ***spring barley*** *(autumn sown)*, ***spring wheat*** *(autumn sown)* [4, 5]; ***triticale***, ***winter barley***, ***winter oats***, ***winter rye***, ***winter wheat*** [1, 4, 5]

Approval information

- Cypermethrin included in Annex I under EC Directive 91/414

- Following implementation of Directive 98/82/EC, approval for use of cypermethrin on numerous crops was revoked in 1999
- Accepted by BBPA for use on malting barley and hops
- Approval expiry 31 Aug 2008 [1]

Efficacy guidance

- Products combine rapid action, good persistence, and high activity on Lepidoptera.
- As effect is mainly via contact, good coverage is essential for effective action. Spray volume should be increased on dense crops [1, 3-5]
- A repeat spray after 10-14 d is needed for some pests of outdoor crops, several sprays at shorter intervals for whitefly and other glasshouse pests [1, 3-5]
- Rates and timing of sprays vary with crop and pest. See label for details [1, 3-5]
- Where aphids in hops, pear suckers or glasshouse whitefly resistant to cypermethrin occur control is unlikely to be satisfactory [1, 3-5]
- Apply lacquer undiluted in a 5 cm wide band at strategic places. Ensure surfaces are clean and dry before application [2]
- Lacquer normally remains active for 12 mth [2]
- Temporary loss of activity for about 2 d will occur if lacquer surface is washed, until cypermethrin has migrated to the surface layer [2]

Restrictions

- Maximum number of treatments varies with crop and product. See label or approval notice for details [1, 3-5]
- Apply by forest drench gun or metered dose knapsack sprayer [3]

Crop-specific information

- Latest use: end of July for forestry; normally before end Mar in yr of harvest of field crops, but varies with product and crop; white bud stage for cherries, plums. See labels for details
- HI vining peas 7 d; other crops 0 d
- Test spray sample of new or unusual ornamentals or trees before committing whole batches [1, 3, 5]
- Post-planting forestry applications should be made before damage is seen or at the onset of damage during the first 2 yrs after transplanting [3]

Environmental safety

- Dangerous for the environment [1, 3-5]
- Very toxic to aquatic organisms
- High risk to bees. Do not apply to crops in flower, or to those in which bees are actively foraging, except as directed. Do not apply when flowering weeds are present [1]
- Flammable [1, 2, 4, 5]
- LERAP Category A [1, 3-5]
- Broadcast air-assisted LERAP (18 m) [4, 5]
- Do not spray cereals after 31 Mar within 6 m of the edge of the growing crop [1, 3-5]

Hazard classification and safety precautions

Hazard H03 [1-4]; H04 [5]; H08 [1, 2, 4, 5]; H11 [1, 3-5]

Risk phrases R20 [2, 5]; R21, R52 [2]; R22a, R50 [3-5]; R22b, R37 [1, 4, 5]; R36, R51 [1]; R38 [1, 3]; R41 [5]; R43 [3]; R53a [1-5]; R66 [4, 5]; R67 [4]

Operator protection A [1-5]; C [1, 2, 4, 5]; H, M [3]; P [2]; U02a [3, 4]; U04a [2-4]; U05a [1-5]; U08 [5]; U09a, U15 [2]; U10 [1, 3, 4]; U11 [4]; U14 [3]; U16b [2, 3]; U19a [1, 3-5]; U20b [1, 4, 5]

Environmental protection E02a [2]; E12a [1]; E12f [1] (vining peas, oilseed rape); E13a [5]; E15a [1-4]; E16c [1, 4, 5]; E16d [1, 3-5]; E17b [4, 5] (18 m); E34 [3-5]; E38 [3]

Consumer protection C08, C09, C11, C12 [2]

Storage and disposal D01, D02, D09a [1-5]; D05, D10b [1, 3-5]; D12a [3-5]; D12b [2]

Medical advice M03 [3]; M04a [2, 4]; M05b [1, 4, 5]

130 cyproconazole

A contact and systemic triazole fungicide available only in mixtures
FRAC mode of action code: 3

See also *azoxystrobin + cyproconazole*
chlorothalonil + cyproconazole
chlorothalonil + cyproconazole + propiconazole

131 cyproconazole + cyprodinil

A broad spectrum fungicide mixture for cereals
FRAC mode of action code: 3 + 9

Products

Radius	Syngenta	5.33:40 % w/w	WG	09387

Uses

- Brown rust in ***spring barley***, ***winter barley***, ***winter wheat***
- Eyespot in ***spring barley***, ***winter barley***, ***winter wheat***
- Foliar disease control in ***durum wheat*** *(off-label)*, ***grass seed crops*** *(off-label)*, ***spring rye*** *(off-label)*, ***triticale*** *(off-label)*, ***winter rye*** *(off-label)*
- Glume blotch in ***winter wheat***
- Net blotch in ***spring barley***, ***winter barley***
- Powdery mildew in ***spring barley***, ***winter barley***, ***winter wheat***
- Rhynchosporium in ***spring barley***, ***winter barley***
- Septoria leaf blotch in ***winter wheat***
- Yellow rust in ***spring barley***, ***winter barley***, ***winter wheat***

Specific Off-Label Approvals (SOLAs)

- ***durum wheat*** *20060613*
- ***grass seed crops*** *20060613*
- ***spring rye*** *20060613*
- ***triticale*** *20060613*
- ***winter rye*** *20060613*

Approval information

- Cyprodinil included in Annex I under EC Directive 91/414
- Accepted by BBPA for use on malting barley

Efficacy guidance

- Best results obtained from treatment at early stages of development. Later treatments may be required if disease pressure remains high
- Eyespot is fully controlled by treatment in spring during stem extension. Control may be reduced when very dry conditions follow application
- Cyproconazole is a DMI fungicide. Resistance to some DMI fungicides has been identified in Septoria leaf blotch which may seriously affect performance of some products. For further advice contact a specialist advisor and visit the Fungicide Resistance Action Group (FRAG)-UK website

Restrictions

- Maximum total dose equivalent to two full dose treatments

Crop-specific information

- Latest use: before first spikelet of inflorescence visible (GS 51) for barley; before caryopsis watery ripe (GS 71) for winter wheat

Environmental safety

- Dangerous for the environment
- Very toxic to aquatic organisms
- Risk to certain non-target insects or other arthropods. For advice on risk management and use in Integrated Pest Management (IPM) see directions for use

Hazard classification and safety precautions

Hazard H03, H11
Risk phrases R38, R43, R50, R53a, R63

Operator protection A, H; U05a, U20a
Environmental protection E15a, E22b, E38
Storage and disposal D01, D02, D05, D07, D09a, D12a

132 cyproconazole + propiconazole

A broad-spectrum mixture of conazole fungicides for wheat and barley
FRAC mode of action code: 3 + 3

Products

1	Alto Xtra	Syngenta	160:250 g/l	EC	13058
2	Menara	Syngenta	160:250 g/l	EC	13035

Uses
- Brown rust in ***spring barley***, ***winter barley***, ***winter wheat*** [1, 2]
- Foliar disease control in ***durum wheat*** *(off-label)*, ***spring rye*** *(off-label)*, ***triticale*** *(off-label)*, ***winter rye*** *(off-label)* [2]
- Glume blotch in ***winter wheat*** [1, 2]
- Late ear diseases in ***winter wheat*** [1, 2]
- Net blotch in ***spring barley***, ***winter barley*** [1, 2]
- Powdery mildew in ***spring barley***, ***winter barley*** [1, 2]; ***winter wheat*** [1]
- Rhynchosporium in ***spring barley***, ***winter barley*** [1, 2]
- Septoria leaf blotch in ***winter wheat*** [1, 2]
- Yellow rust in ***spring barley***, ***winter barley*** [1]; ***winter wheat*** [1, 2]

Specific Off-Label Approvals (SOLAs)
- ***durum wheat*** *20064168* [2]
- ***spring rye*** *20064168* [2]
- ***triticale*** *20064168* [2]
- ***winter rye*** *20064168* [2]

Approval information
- Propiconazole included in Annex I under EC Directive 91/414
- Accepted by BBPA for use on malting barley

Efficacy guidance
- Treatment should be made at first sign of disease
- Useful reduction of eyespot when applied in spring during stem extension (GS 30-32)
- Cyproconazole and propiconazole are DMI fungicides. Resistance to some DMI fungicides has been identified in Septoria leaf blotch which may seriously affect performance of some products. For further advice contact a specialist advisor and visit the Fungicide Resistance Action Group (FRAG)-UK website

Restrictions
- Maximum total dose equivalent to one full dose treatment for barley; two full dose treatments for winter wheat

Crop-specific information
- Latest use: before beginning of anthesis (GS 60) for barley; before grain watery ripe (GS 71) for wheat
- Application to winter wheat in spring may cause straw shortening but does not cause loss of yield

Environmental safety
- Dangerous for the environment
- Very toxic to aquatic organisms

Hazard classification and safety precautions

Hazard H03, H11
Risk phrases R50, R53a, R63
Operator protection A, C, H; U05a, U08, U20b
Environmental protection E15a, E38
Storage and disposal D01, D02, D05, D07, D09a, D10c, D12a

133 cyproconazole + quinoxyfen

A contact and systemic fungicide mixture for cereals
FRAC mode of action code: 3 + 13

Products

Excelsior	Dow	80 g/l	SC	13378

Uses

- Brown rust in ***winter wheat***
- Eyespot in ***winter wheat*** *(reduction)*
- Glume blotch in ***winter wheat***
- Powdery mildew in ***winter wheat***
- Septoria leaf blotch in ***winter wheat***
- Yellow rust in ***winter wheat***

Approval information

- Quinoxyfen included in Annex I under Directive 91/414
- Accepted by BBPA for use on malting barley

Efficacy guidance

- Best results are achieved when disease is first detected and seen to be active. A second treatment may be needed under prolonged disease pressure
- Treatment for Septoria control should be made as soon as possible after weather that favours infection to prevent spread to the second and flag leaves
- Useful reduction of eyespot when applied between GS 30-34 to control other diseases
- Systemic activity may be reduced under conditions of severe drought stress

Restrictions

- Maximum number of treatments equivalent to a total of two full doses
- Application to winter wheat in spring may cause straw shortening but does not cause loss of yield

Crop-specific information

- Latest use: before ear emergence for wheat

Environmental safety

- Dangerous for the environment
- Very toxic to aquatic organisms
- LERAP Category B

Hazard classification and safety precautions

Hazard H03, H11
Risk phrases R43, R50, R53a, R63
Operator protection A, C, H; U05a, U20b
Environmental protection E15a, E16a, E16b, E38
Storage and disposal D01, D02, D09a, D10b, D12a

134 cyproconazole + trifloxystrobin

A conazole and strobilurin fungicide mixture for wheat and barley
FRAC mode of action code: 3 + 11

Products

Sphere	Bayer CropScience	80:187.5 g/l	EC	11429

Uses

- Brown rust in ***spring barley***, ***winter barley***, ***winter wheat***
- Glume blotch in ***winter wheat***
- Net blotch in ***spring barley***, ***winter barley***
- Powdery mildew in ***spring barley***, ***winter barley***, ***winter wheat***
- Rhynchosporium in ***spring barley***, ***winter barley***
- Septoria leaf blotch in ***winter wheat***
- Yellow rust in ***spring barley***, ***winter barley***, ***winter wheat***

Approval information

- Trifloxystrobin included in Annex I under EC Directive 91/414
- Accepted by BBPA for use on malting barley

Efficacy guidance

- Best results obtained from treatment at early stages of disease development. Further treatment may be needed if disease attack is prolonged
- Cyproconazole is a DMI fungicide. Resistance to some DMI fungicides has been identified in Septoria leaf blotch which may seriously affect performance of some products. For further advice contact a specialist advisor and visit the Fungicide Resistance Action Group (FRAG)-UK website
- Trifloxystrobin is a member of the QoI cross resistance group. Product should be used preventatively and not relied on for its curative potential
- Use product as part of an Integrated Crop Management strategy incorporating other methods of control, including where appropriate other fungicides with a different mode of action. Do not apply more than two foliar applications of QoI containing products to any cereal crop
- There is a significant risk of widespread resistance occurring in *Septoria tritici* populations in UK. Failure to follow resistance management action may result in reduced levels of disease control
- Strains of barley powdery mildew resistant to QoIs are common in the UK

Restrictions

- Maximum total dose per crop equivalent to two full dose treatments

Crop-specific information

- HI 35 d

Environmental safety

- Dangerous for the environment
- Very toxic to aquatic organisms

Hazard classification and safety precautions

Hazard H03, H11
Risk phrases R36, R50, R53a, R63
Operator protection A, C, H; U05a, U09a, U19a, U20b
Environmental protection E15a, E38
Storage and disposal D01, D02, D09a, D10b, D12a

135 cyprodinil

An anilinopyrimidine systemic broad spectrum fungicide for cereals
FRAC mode of action code: 9

See also cyproconazole + cyprodinil

Products

1	Kayak	Syngenta	300 g/l	EC	13119
2	Unix	Syngenta	75% w/w	WG	11512

Uses

- Eyespot in ***durum wheat*** *(off-label)*, ***spring rye*** *(off-label)*, ***spring wheat***, ***triticale*** *(off-label)*, ***winter rye*** *(off-label)*, ***winter wheat*** [2]; ***spring barley***, ***winter barley*** [1, 2]
- Foliar disease control in ***durum wheat*** *(off-label)*, ***spring rye*** *(off-label)*, ***triticale*** *(off-label)*, ***winter rye*** *(off-label)* [2]
- Net blotch in ***spring barley***, ***winter barley*** [1, 2]
- Powdery mildew in ***spring barley***, ***winter barley*** [1, 2]; ***spring wheat*** *(moderate control)*, ***winter wheat*** *(moderate control)* [2]
- Rhynchosporium in ***spring barley***, ***winter barley*** [1]; ***spring barley*** *(moderate control)*, ***winter barley*** *(moderate control)* [2]

Specific Off-Label Approvals (SOLAs)

- ***durum wheat*** *20060600* [2]
- ***spring rye*** *20060600* [2]
- ***triticale*** *20060600* [2]
- ***winter rye*** *20060600* [2]

Approval information

- Cyprodinil included in Annex I under EC Directive 91/414
- Accepted by BBPA for use on malting barley

Efficacy guidance

- Best results obtained from treatment at early stages of disease development
- For best control of eyespot spray before or during the period of stem extension in spring. Control may be reduced if very dry conditions follow treatment

Restrictions

- Maximum total dose equivalent to one and two thirds full dose on wheat and twice full dose on barley [2]
- Maximum number of treatments 2 per crop on wheat and barley [1]

Crop-specific information

- Latest use: up to and including first awns visible (GS 49) for spring barley, winter barley [1, 2]; up to and including grain watery ripe (GS 71) for spring wheat, winter wheat [2]

Environmental safety

- Dangerous for the environment
- Very toxic to aquatic organisms

Hazard classification and safety precautions

Hazard H04 [1]; H11 [1, 2]
Risk phrases R38, R43 [1]; R50, R53a [1, 2]
Operator protection A, H; U05a, U20a [1, 2]; U09a [1]
Environmental protection E15a [2]; E15b, E34 [1]; E38 [1, 2]
Storage and disposal D01, D02, D05, D07, D09a, D12a [1, 2]; D10c [1]

136 cyprodinil + fludioxonil

A broad-spectrum fungicide mixture for legumes and strawberries
FRAC mode of action code: 9 + 12

Products

Switch	Syngenta	37.5:25 % w/w	WG	13185

Uses

- Ascochyta in ***mange-tout peas***, ***sugar snap peas***, ***vining peas***
- Black spot in ***protected strawberries*** *(qualified minor use)*, ***strawberries*** *(qualified minor use)*
- Blossom wilt in ***apricots*** *(off-label)*, ***cherries*** *(off-label)*, ***peaches*** *(off-label)*, ***plums*** *(off-label)*
- Botrytis in ***asparagus*** *(off-label)*, ***broad beans*** *(moderate control)*, ***green beans*** *(moderate control)*, ***lupins*** *(off-label)*, ***mange-tout peas*** *(moderate control)*, ***protected aubergines*** *(off-label)*, ***protected cucumbers*** *(off-label)*, ***protected peppers*** *(off-label)*, ***protected strawberries***, ***runner beans*** *(moderate control)*, ***salad onions*** *(off-label)*, ***strawberries***, ***sugar snap peas*** *(moderate control)*, ***table grapes*** *(off-label)*, ***vining peas*** *(moderate control)*, ***wine grapes*** *(off-label)*
- Fusarium in ***protected cucumbers*** *(off-label)*
- Mycosphaerella in ***mange-tout peas***, ***protected cucumbers*** *(off-label)*, ***sugar snap peas***, ***vining peas***
- Sclerotinia in ***broad beans***, ***green beans***, ***mange-tout peas***, ***runner beans***, ***sugar snap peas***, ***vining peas***
- Stemphylium in ***asparagus*** *(off-label)*

Specific Off-Label Approvals (SOLAs)

- ***apricots*** *20070735*
- ***asparagus*** *20070235*
- ***cherries*** *20070735*
- ***lupins*** *20070237*
- ***peaches*** *20070735*
- ***plums*** *20070735*
- ***protected aubergines*** *20070238*
- ***protected cucumbers*** *20070234*
- ***protected peppers*** *20070238*

- ***salad onions*** *20070239*
- ***table grapes*** *20070236*
- ***wine grapes*** *20070236*

Approval information

- Cyprodinil and fludioxonil included in Annex I under EC Directive 91/414

Efficacy guidance

- Best results obtained from application at the earliest stage of disease development or as a protective treatment following disease risk assessment
- Subsequent tratments may follow after a minimum of 10 d if disease pressure remains high
- To minimise the likelihood of development of resistance product should be used in a planned Resistance Management strategy. See Section 5 for more information

Restrictions

- Maximum number of treatments 2 per crop
- Consult processor before treating any crop for processing

Crop-specific information

- HI broad beans, green beans, mange-tout peas, runner beans, sugar snap peas, vining peas 14 d; protected strawberries, strawberries 3 d
- First signs of disease infection likely to be seen from early flowering for all crops
- Ensure peas are free from any stress before treatment. If necessary check wax with a crystal violet test

Environmental safety

- Dangerous for the environment
- Very toxic to aquatic organisms
- LERAP Category B

Hazard classification and safety precautions

Hazard H11
Risk phrases R50, R53a
Operator protection A, H; U05a, U20a
Environmental protection E15b, E16a, E16b, E38
Storage and disposal D01, D02, D05, D07, D09a, D10c, D12a

137 cyprodinil + picoxystrobin

A broad spectrum fungicide mixture for cereals
FRAC mode of action code: 9 + 11

Products

Acanto Prima	DuPont	30.8% w/w	WG	13041

Uses

- Brown rust in ***spring barley***, ***spring wheat***, ***winter barley***, ***winter wheat***
- Eyespot in ***spring barley***, ***spring wheat*** *(moderate control)*, ***winter barley***, ***winter wheat*** *(moderate control)*
- Glume blotch in ***spring wheat***, ***winter wheat***
- Net blotch in ***spring barley***, ***winter barley***
- Powdery mildew in ***spring barley***, ***winter barley***
- Rhynchosporium in ***spring barley***, ***winter barley***
- Septoria leaf blotch in ***spring wheat***, ***winter wheat***
- Yellow rust in ***spring wheat***, ***winter wheat***

Approval information

- Cyprodinil and picoxystrobin included in Annex I under EC Directive 91/414
- Accepted by BBPA for use on malting barley
- Approval expiry 31 Jul 2007 [11864]

Efficacy guidance

- Best results obtained from use as a protectant treatment or in the earliest stages of disease development. Further applications may be needed if disease attack is prolonged

- Picoxystrobin is a member of the QoI cross resistance group. Product should be used preventatively and not relied on for its curative potential
- Use product as part of an Integrated Crop Management strategy incorporating other methods of control, including where appropriate other fungicides with a different mode of action. Do not apply more than two foliar applications of QoI containing products to any cereal crop
- There is a significant risk of widespread resistance occurring in *Septoria tritici* populations in UK. Failure to follow resistance management action may result in reduced levels of disease control
- In wheat product must always be used in mixture with another product, recommended for control of the same target disease, that contains a fungicide from a different cross resistance group and is applied at a dose that will give robust control
- Strains of barley powdery mildew resistant to QoIs are common in the UK

Restrictions

- Maximum number of treatments 2 per crop

Crop-specific information

- Latest use: before early milk stage (GS 73) for wheat; before ear just visible (GS 51) for barley

Environmental safety

- Dangerous for the environment
- Very toxic to aquatic organisms

Hazard classification and safety precautions

Hazard H11
Risk phrases R50, R53a
Operator protection A, C, H; U05a, U09a, U11, U14, U15, U20a
Environmental protection E15a, E34, E38
Storage and disposal D01, D02, D09a, D10c, D12a

138 2,4-D

A translocated phenoxycarboxylic acid herbicide for cereals, grass and amenity use
HRAC mode of action code: O

See also clopyralid + 2,4-D + MCPA

Products

1	Depitox	Nufarm UK	490 g/l	SL	13258
2	Headland Staff 500	Headland	500 g/l	SL	13196
3	Herboxone	Headland	500 g/l	SL	13144
4	HY-D Super	Agrichem	500 g/l	SL	13198

Uses

- Annual dicotyledons in ***amenity grassland***, ***managed amenity turf***, ***permanent grassland***, ***spring barley***, ***spring wheat***, ***winter barley***, ***winter oats***, ***winter rye***, ***winter wheat*** [1-4]; ***apple orchards***, ***pear orchards***, ***spring rye*** [1]; ***undersown barley***, ***undersown oats***, ***undersown rye***, ***undersown wheat*** [4]
- Aquatic weeds in ***enclosed waters***, ***land immediately adjacent to aquatic areas***, ***open waters*** [1]
- Perennial dicotyledons in ***amenity grassland***, ***managed amenity turf***, ***permanent grassland***, ***spring barley***, ***spring wheat***, ***winter barley***, ***winter oats***, ***winter rye***, ***winter wheat*** [1-4]; ***apple orchards***, ***pear orchards***, ***spring rye*** [1]; ***undersown barley***, ***undersown oats***, ***undersown rye***, ***undersown wheat*** [4]

Approval information

- 2,4-D included in Annex I under EC Directive 91/414
- Approved for aquatic weed control [11149]. See notes in Section 5 on use of herbicides in or near water
- Accepted by BBPA for use on malting barley

Efficacy guidance

- Best results achieved by spraying weeds in seedling to young plant stage when growing actively in a strongly competing crop
- Most effective stage for spraying perennials varies with species. See label for details
- Spray aquatic weeds when in active growth between May and Sep

Restrictions

- Maximum number of treatments normally 1 per crop and in forestry, 1 per yr in grassland and 2 or 3 per yr in amenity turf. Check individual labels
- Do not use on newly sown leys containing clover
- Do not spray grass seed crops after ear emergence
- Do not spray within 6 mth of laying turf or sowing fine grass
- Do not dump surplus herbicide in water or ditch bottoms
- Do not plant conifers until at least 1 mth after treatment
- Do not spray crops stressed by cold weather or drought or if frost expected
- Do not roll or harrow within 7 d before or after spraying
- Do not spray if rain falling or imminent
- Do not mow or roll turf or amenity grass 4 d before or after spraying. The first 4 mowings after treatment must be composted for at least 6 mth before use
- Do not mow grassland or graze for at least 10 d after spraying

Crop-specific information

- Latest use: before 1st node detectable in cereals; end Aug for conifer plantations; before established grassland 25 cm high
- Spray winter cereals in spring when leaf-sheath erect but before first node detectable (GS 31), spring cereals from 5-leaf stage to before first node detectable (GS 15-31)
- Cereals undersown with grass and/or clover, but not with lucerne, may be treated
- Selective treatment of resistant conifers can be made in Aug when growth ceased and plants hardened off, spray must be directed if applied earlier. See label for details

Following crops guidance

- Do not use shortly before or after sowing any crop
- Do not plant succeeding crops within 3 mth [2]
- Do not direct drill brassicas or grass/clover mixtures within 3 wk of application

Environmental safety

- Dangerous for the environment
- Toxic (very toxic [2, 3]) to aquatic organisms
- Dangerous to aquatic higher plants. Do not contaminate surface waters or ditches with chemical or used container
- 2,4-D is active at low concentrations. Take extreme care to avoid drift onto neighbouring crops, especially beet crops, brassicas, most market garden crops including lettuce and tomatoes under glass, pears and vines
- May be used to control aquatic weeds in presence of fish if used in strict accordance with directions for waterweed control and precautions needed for aquatic use [1]
- Keep livestock out of treated areas for at least 2 wk following treatment and until poisonous weeds, such as ragwort, have died down and become unpalatable
- Water containing the herbicide must not be used for irrigation purposes within 3 wk of treatment or until the concentration in water is below 0.05 ppm

Hazard classification and safety precautions

Hazard H03, H11 [2-4]; H04 [1]
Risk phrases R22a [2-4]; R38 [1]; R41, R53a [1-4]; R43, R51 [4]; R50 [2, 3]
Operator protection A, C, H [1-4]; M [1]; U05a, U08, U11, U20b [1-4]; U14, U15 [4]
Environmental protection E07a, E34 [1-4]; E15a [1]; E15b, E38 [2-4]
Storage and disposal D01, D02, D05, D09a [1-4]; D10a, D12a [2-4]; D10b [1]
Medical advice M03 [1-4]; M05a [4]

139 2,4-D + dicamba

A translocated herbicide for use on turf
HRAC mode of action code: O + O

Products

1	Magneto	Nufarm UK	344:120 g/l	SL	12339
2	Scotts Feed and Weed	Scotts	0.8:0.12% w/w	GR	13076
3	Thrust	Nufarm UK	344:120 g/l	SL	12230

SEE SECTION 3 FOR PRODUCTS ALSO REGISTERED

Uses

- Annual dicotyledons in ***amenity grassland*, *permanent grassland*** [1, 3]; ***managed amenity turf*** [2]
- Perennial dicotyledons in ***amenity grassland*, *permanent grassland*** [1, 3]; ***managed amenity turf*** [2]

Approval information

- 2,4-D included in Annex I under EC Directive 91/414

Efficacy guidance

- Best results achieved by application when weeds growing actively in spring or early summer (later with irrigation and feeding)
- More resistant weeds may need repeat treatment after 3 wk
- Improved control of some weeds can be obtained by use of specifed oil adjuvant [1, 3]
- Do not use during drought conditions

Restrictions

- Maximum number of treatments 2 per yr on amenity grass and established grassland [1]
- Do not treat newly sown or turfed areas or grass less than 1 yr old
- Do not mow less than 3 d before treatment and at least 3-4 d afterwards [2]
- Do not treat grass crops intended for seed production [1, 3]
- Do not treat grass suffering from drought, disease or other adverse factors [1, 3]
- Do not roll or harrow for 7 d before or after treatment [1, 3]
- Do not apply when grassland is flowering [1, 3]
- Do not re-seed for 6-8 wk after application [2]
- Avoid spray drift onto cultivated crops or ornamentals

Crop-specific information

- Latest use: before grass 25 cm high for amenity grass, established grassland [1, 3]
- The first four mowings after treatment must be composted for at least 6 mth before use

Environmental safety

- Dangerous for the environment [1, 3]
- Toxic to aquatic organisms
- Keep livestock out of treated areas for at least two weeks following treatment and until poisonous weeds, such as ragwort, have died down and become unpalatable [1, 3]

Hazard classification and safety precautions

Hazard H03, H11 [1, 3]
Risk phrases R22a, R41, R43, R51 [1, 3]; R52 [2]; R53a [1-3]
Operator protection A, C, H [1-3]; D [2]; M [1, 3]; U05a, U11 [1, 3]; U20b [2]
Environmental protection E07a, E15a, E34, E38 [1, 3]; E15b [2]
Storage and disposal D01, D09a [1-3]; D02, D05, D10a, D12a [1, 3]; D11a [2]
Medical advice M03 [1, 3]

140 2,4-D + dicamba + triclopyr

A translocated herbicide for perennial and woody weed control
HRAC mode of action code: O + O + O

Products

1	Broadsword	United Phosphorus	200:85:65 g/l	EC	09140
2	Nu-Shot	Nufarm UK	200:85:65 g/l	EC	13550

Uses

- Annual dicotyledons in ***amenity grassland*** [2]; ***land not intended to bear vegetation*** [1]; ***permanent grassland*, *rotational grass*** [1, 2]
- Brambles in ***farm forestry*, *forest*, *natural surfaces not intended to bear vegetation*** [1, 2]
- Docks in ***amenity grassland*** [2]; ***land not intended to bear vegetation*** [1]; ***permanent grassland*, *rotational grass*** [1, 2]
- Gorse in ***farm forestry*, *forest*, *natural surfaces not intended to bear vegetation*** [1, 2]
- Green cover in ***land temporarily removed from production*** [1]
- Japanese knotweed in ***farm forestry*, *forest*, *natural surfaces not intended to bear vegetation*** [1, 2]

- Perennial dicotyledons in ***amenity grassland*** [2]; ***farm forestry, forest, natural surfaces not intended to bear vegetation, permanent grassland, rotational grass*** [1, 2]; ***land not intended to bear vegetation*** [1]
- Plantains in ***amenity grassland*** [2]; ***land not intended to bear vegetation*** [1]; ***permanent grassland, rotational grass*** [1, 2]
- Rhododendrons in ***farm forestry, forest, natural surfaces not intended to bear vegetation*** [1, 2]
- Stinging nettle in ***amenity grassland*** [2]; ***permanent grassland, rotational grass*** [1, 2]
- Thistles in ***amenity grassland*** [2]; ***land not intended to bear vegetation*** [1]; ***permanent grassland, rotational grass*** [1, 2]
- Woody weeds in ***farm forestry, forest, natural surfaces not intended to bear vegetation*** [1, 2]

Approval information

- 2,4-D and triclopyr included in Annex I under EC Directive 91/414

Efficacy guidance

- Apply as foliar spray to herbaceous or woody weeds. Timing and growth stage for best results vary with species. See label for details
- Dilute with water for stump treatment and apply after felling up to the start of regrowth. Treat any regrowth with a spray to the growing foliage
- May be applied at 1/3 dilution in weed wipers or 1/8 dilution with ropewick applicators

Restrictions

- Maximum total dose per yr equivalent to two full dose treatments
- Do not use on pasture established less than 1 yr or on grass grown for seed
- Where clover a valued constituent of sward only use as a spot treatment
- Do not graze for 7 d or mow for 14 d after treatment
- Do not direct drill grass, clover or brassicas for at least 6 wk after grassland treatment
- Do not plant trees for 1-3 mth after spraying depending on dose applied. See label
- Avoid spray drift into greenhouses or onto crops or ornamentals. Vapour drift may occur in hot conditions

Crop-specific information

- Latest use: before weed flower buds open in grassland, non-cropped land, land temprorarily removed from production [1]; end of summer (spray); after felling up nto the start of regrowth (cut stump spray) for farm forestry [1]
- Sprays may be applied in pines, spruce and fir providing drift is avoided. Optimum time is mid-autumn when tree growth ceased but weeds not yet senescent [1]

Environmental safety

- Dangerous for the environment
- Toxic (very toxic [1]) to aquatic organisms [2]
- Flammable
- Docks and other weeds may become increasingly palatable after treatment and may be preferentially grazed
- Keep livestock out of treated areas for at least two weeks following treatment and until poisonous weeds, such as ragwort, have died down and become unpalatable
- Take extreme care to avoid drift onto neighbouring crops, especially beet crops, brassicas, most market garden crops including lettuce and tomatoes under glass, pears and vines

Hazard classification and safety precautions

Hazard H03, H08, H11

Risk phrases R22a, R22b, R36, R37, R53a [1, 2]; R38, R50, R67 [1]; R43, R51 [2]

Operator protection A, C, H, M [1, 2]; D [2]; U02a, U04a, U13, U20a [1]; U05a, U08, U11, U19a [1, 2]; U14, U20b [2]

Environmental protection E07a, E34 [1, 2]; E13b [1]; E15a, E38 [2]

Consumer protection C01 [1]

Storage and disposal D01, D02, D05, D09a, D10b, D12a

Medical advice M03, M05b

SEE SECTION 3 FOR PRODUCTS ALSO REGISTERED

141 2,4-D + dichlorprop-P + MCPA + mecoprop-P

A translocated herbicide for use in apple and pear orchards
HRAC mode of action code: O + O + O + O

Products

UPL Camppex	United Phosphorus	34.5:66.6:52.8:82.1 g/l	SL	11661

Uses

- Annual dicotyledons in ***apple orchards, pear orchards***
- Chickweed in ***apple orchards, pear orchards***
- Cleavers in ***apple orchards, pear orchards***
- Perennial dicotyledons in ***apple orchards, pear orchards***

Approval information

- 2,4-D, dichlorprop-P, MCPA and mecoprop-P included in Annex I under EC Directive 91/414

Efficacy guidance

- Use on emerged weeds in established apple and pear orchards (from 1 yr after planting) as directed application
- Spray when weeds in active growth and at growth stage recommended on label
- Effectiveness may be reduced by rain within 12 h

Restrictions

- Maximum number of treatments 2 per yr
- Applications must be made around, and not directly to, trees. Do not allow drift onto trees
- Do not spray during blossom period. Do not spray to run-off
- Do not roll, harrow or cut grass crops on orchard floor within at least 3 d before or after spraying

Environmental safety

- Harmful to aquatic organisms
- Harmful to fish or other aquatic life. Do not contaminate surface waters or ditches with chemical or used container
- Keep livestock out of treated areas for at least two weeks following treatment and until poisonous weeds, such as ragwort, have died down and become unpalatable
- Take extreme care to avoid drift onto neighbouring sensitive crops

Hazard classification and safety precautions

Hazard H03
Risk phrases R20, R21, R22a, R43, R52, R53a
Operator protection A, C, H, M; U05a, U08, U20b
Environmental protection E07a, E13c, E34
Storage and disposal D01, D02, D05, D09a, D10c
Medical advice M03, M05a

142 2,4-D + florasulam

A post-emergence herbicide mixture for control of broad leaved weeds in managed amenity turf
HRAC mode of action code: O + B

Products

Junction	Rigby Taylor	452:6.25 g/l	ME	12493

Uses

- Clover in ***managed amenity turf***
- Daisies in ***managed amenity turf***
- Dandelions in ***managed amenity turf***
- Plantains in ***managed amenity turf***
- Sticky mouse-ear in ***managed amenity turf***

Approval information

- 2,4-D and florasulam included in Annex I under EC Directive 91/414

Efficacy guidance

- Best results obtained from treatment of actively growing weeds between Mar and Oct when soil is moist
- Do not apply when rain is imminent or during periods of drought unless irrigation is applied

Restrictions

- Maximum number of treatments on managed amenity turf: 1 per yr

Crop-specific information

- Avoid mowing turf 3 d before and after spraying
- Ensure newly sown turf has established before treating. Turf sown in late summer or autumn should not be sprayed until growth is resumed in the following spring

Following crops guidance

- An interval of 4 wk must elapse bewteen application and re-seeding turf

Environmental safety

- Dangerous for the environment
- Toxic to aquatic organisms

Hazard classification and safety precautions

Hazard H03, H11
Risk phrases R22a, R43, R51, R53a
Operator protection A, H; U02a, U05a, U08, U14, U20a
Environmental protection E15a, E34, E38
Storage and disposal D01, D02, D09a, D10b, D12a

143 2,4-D + MCPA

A translocated herbicide mixture for cereals and grass
HRAC mode of action code: O + O

Products

1	Headland Polo	Headland	360:315 g/l	SL	10283
2	Lupo	Nufarm UK	345:345 g/l	SL	12725

Uses

- Annual dicotyledons in ***permanent grassland***, ***rotational grass*** [1, 2]; ***spring barley***, ***spring wheat***, ***winter barley***, ***winter oats***, ***winter wheat*** [1]
- Perennial dicotyledons in ***permanent grassland***, ***rotational grass*** [1, 2]; ***spring barley***, ***spring wheat***, ***winter barley***, ***winter oats***, ***winter wheat*** [1]

Approval information

- 2,4-D and MCPA included in Annex I under EC Directive 91/414
- Accepted by BBPA for use on malting barley

Efficacy guidance

- Best results achieved by spraying weeds in seedling to young plant stage when growing actively in a strongly competing crop
- Most effective stage for spraying perennials varies with species. See label for details

Restrictions

- Maximum number of treatments 1 per crop in cereals, permanent grassland; 2 per yr in rotational grassland [1]; 1 per crop in permanent and rotational grassland [2]
- Do not spray if rain falling or imminent
- Do not cut grass or graze for at least 10 d after spraying
- Do not use on newly sown leys containing clover or other legumes
- Do not spray crops stressed by cold weather or drought or if frost expected
- Do not use shortly before or after sowing any crop
- Do not roll or harrow within 7 d before or after spraying

Crop-specific information

- Spray winter cereals in spring when leaf-sheath erect but before first node detectable (GS 31), spring cereals from 5-leaf stage to before first node detectable (GS 15-31)
- Latest use: before first node detectable (GS 31) in cereals [1]; before crop is 25 cm high in grassland [2]

SEE SECTION 3 FOR PRODUCTS ALSO REGISTERED

Environmental safety
- Dangerous for the environment
- Toxic to aquatic organisms
- Keep livestock out of treated areas for at least two weeks following treatment and until poisonous weeds, such as ragwort, have died down and become unpalatable
- 2,4-D and MCPA are active at low concentrations. Take extreme care to avoid drift onto neighbouring crops, especially beet crops, brassicas, most market garden crops including lettuce and tomatoes under glass, pears and vines

Hazard classification and safety precautions

Hazard H03, H11
Risk phrases R20, R21, R22a, R41, R43, R53a [1, 2]; R50 [2]; R51 [1]
Operator protection A, C, H, M; U05a, U08, U11, U14, U20b [1, 2]; U15 [1]
Environmental protection E07a, E15a, E34 [1, 2]; E38 [2]
Storage and disposal D01, D02, D05, D09a, D10a [1, 2]; D12a [2]
Medical advice M03 [1, 2]; M05a [1]

144 2,4-D + mecoprop-P

A translocated herbicide for use in amenity turf
HRAC mode of action code: O + O

Products

Supertox 30	Bayer Environ.	93.5:95 g/l	SL	09946

Uses
- Annual dicotyledons in ***managed amenity turf***
- Perennial dicotyledons in ***managed amenity turf***

Approval information
- 2,4-D and mecoprop-P included in Annex I under EC Directive 91/414

Efficacy guidance
- May be applied from Apr to Sep, best results in May-Jun when weeds in active growth. Less susceptible weeds may require second treatment not less than 6 wk later
- For best results apply fertilizer 1-2 wk before treatment
- Do not spray during drought conditions or when rain imminent as efficacy will be impaired
- Do not close mow infrequently mown turf for 3-4 d before or after treatment. On fine turf do not mow for 24 h before or after treatment

Restrictions
- Maximum total dose equivalent to one full dose treatment in first yr followed by half full dose treatment in second yr. The total amount of mecoprop-P applied in a single yr must not exceed the maximum total dose approved for any single product for the crop
- Do not use first 4 mowings for mulching unless composted for at least 6 mth
- Do not apply during frosty weather
- Do not roll or harrow within 7 d before or after treatment

Crop-specific information
- Newly laid turf or newly sown grass should not be treated for at least 6 mth

Environmental safety
- Harmful to aquatic organisms
- Keep livestock out of treated areas for at least two weeks following treatment and until poisonous weeds, such as ragwort, have died down and become unpalatable
- Take extreme care to avoid drift onto neighbouring crops, especially beet crops, brassicas, most market garden crops including lettuce and tomatoes under glass, pears and vines

Hazard classification and safety precautions

Hazard H03, H04
Risk phrases R36, R41, R43, R52, R53a
Operator protection A, C, H, M; U05a, U08, U11, U14, U15, U20b
Environmental protection E07a, E15a, E23, E34
Storage and disposal D01, D02, D05, D09a, D10c, D12a

145 2,4-D + picloram

A persistent translocated herbicide for non-crop land
HRAC mode of action code: O + O

Products

Tordon 101	Dow	240:65 g/l	SL	05816

Uses

- Annual dicotyledons in ***land not intended for cropping***
- Brambles in ***land not intended for cropping***
- Creeping thistle in ***land not intended for cropping***
- Docks in ***land not intended for cropping***
- Perennial dicotyledons in ***land not intended for cropping***
- Ragwort in ***land not intended for cropping***
- Scrub clearance in ***land not intended for cropping***
- Woody weeds in ***land not intended for cropping***

Approval information

- 2,4-D included in Annex I under EC Directive 91/414
- Approval expiry 31 Dec 2008 [1]

Efficacy guidance

- Apply as overall foliar spray during period of active growth when foliage well developed

Restrictions

- Maximum number of treatments 1 per yr
- Do not apply around desirable trees or shrubs where roots may absorb chemical
- Prevent leaching into areas where desirable plants are present
- Avoid drift of spray onto desirable plants
- Do not use cuttings from treated grass for mulching or composting

Environmental safety

- Harmful to aquatic organisms
- Keep livestock out of treated areas for at least 2 wk following treatment and until poisonous weeds, such as ragwort, have died down and become unpalatable

Hazard classification and safety precautions

Hazard H04
Risk phrases R37, R43, R52, R53a
Operator protection A, C; U05a, U08, U14, U15, U20b
Environmental protection E07a, E15a
Storage and disposal D01, D02, D09a, D10a, D12a

146 daminozide

A hydrazide plant growth regulator for use in certain ornamentals

Products

1	B-Nine SG	Certis	85% w/w	SG	12734
2	Dazide Enhance	Fine	85% w/w	SG	11943

Uses

- Internode reduction in ***azaleas***, ***bedding plants***, ***chrysanthemums***, ***hydrangeas***

Approval information

- Daminozide included in Annex I under EC Directive 91/414

Efficacy guidance

- Best results obtained by application in late afternoon when glasshouse has cooled down
- Apply a fine spray using compressed air or power sprayers to give good coverage of dry foliage without run off
- Response to treatment differs widely depending on variety, stage of growth and physiological condition of the plant. It is recommended that any new variety is first tested on a small scale to observe if adverse effects occur

- For plants grown in modules or for seedlings prior to pricking out apply at reduced dose on a 7 d regime, starting at cotyledon stage

Restrictions

- Maximum number of treatments 2 per crop
- Apply only to turgid, well watered plants. Do not water for 24 h after spraying
- Do not mix with other spray chemicals unless specifically recommended
- Do not store product in metal containers

Crop-specific information

- Do not use on chrysanthemum variety Fandango
- Evidence of effectiveness on Azaleas and Hydrangeas is limited
- See label for guidance on a range of bedding plant species

Hazard classification and safety precautions

Operator protection A [1, 2]; H [2]; U05a [1, 2]; U08, U19a, U20b [2]
Environmental protection E15a
Storage and disposal D01, D02, D09a [1, 2]; D08, D10b, D12a [2]
Medical advice M05a [2]

147 dazomet

A methyl isothiocyanate releasing soil fumigant
HRAC mode of action code: Z

Products

Basamid	Certis	97% w/w	GR	11324

Uses

- Nematodes in ***field crops*** *(soil fumigation)*, ***protected crops*** *(soil fumigation)*, ***vegetables*** *(soil fumigation)*
- Soil insects in ***field crops*** *(soil fumigation)*, ***protected crops*** *(soil fumigation)*, ***vegetables*** *(soil fumigation)*
- Soil-borne diseases in ***field crops*** *(soil fumigation)*, ***protected crops*** *(soil fumigation)*, ***vegetables*** *(soil fumigation)*
- Weed seeds in ***field crops*** *(soil fumigation)*, ***protected crops*** *(soil fumigation)*, ***vegetables*** *(soil fumigation)*

Efficacy guidance

- Dazomet acts by releasing methyl isothiocyanate in contact with moist soil
- Soil sterilization is carried out after harvesting one crop and before planting the next
- The soil must be of fine tilth, free of clods and evenly moist to the depth of sterilization
- Soil moisture must not be less than 50% of water-holding capacity or oversaturated. If too dry, water at least 7-14 d before treatment
- In order to obtain short treatment times it is recommended to treat soils when soil temperature is above 7 °C. Treatment should be used outdoors before winter rains make soil too wet to cultivate - usually early Nov
- For club root control treat only in summer when soil temperature above 10 °C
- Where onion white rot a problem unlikely to give effective control where inoculum level high or crop under stress
- Apply granules with suitable applicators, mix into soil immediately to desired depth and seal surface with polythene sheeting, by flooding or heavy rolling. See label for suitable application and incorporation machinery
- With 'planting through' technique polythene seal is left in place to form mulch into which new crop can be planted

Restrictions

- Maximum number of treatments 1 per batch of soil for protected crops or field crop situation
- Do not treat ground where water table may rise into treated layer

Crop-specific information

- Latest use: pre-planting

Following crops guidance

- With 'planting through' technique no gas release cultivations are made and safety test with cress is particularly important. Conduct cress test on soil samples from centre as well as edges of bed. Observe minimum of 30 d from application to cress test in soils at 10 °C or above, at least 50 d in soils below 10 °C. See label for details
- In all other situations 14-28 d after treatment cultivate lightly to allow gas to disperse and conduct cress test after a further 14-28 d (timing depends on soil type and temperature). Do not treat structures containing live plants or any ground within 1 m of live plants

Environmental safety

- Dangerous for the environment
- Very toxic to aquatic organisms

Hazard classification and safety precautions

Hazard H03, H11
Risk phrases R22a, R50, R53a
Operator protection A, M; U04a, U05a, U09a, U19a, U20a
Environmental protection E15a, E34, E38
Storage and disposal D01, D02, D07, D09a, D11a, D12a
Medical advice M03, M05a

148 2,4-DB

A translocated phenoxycarboxylic acid herbicide for use in lucerne and undersown cereals
HRAC mode of action code: O

Products

DB Straight	United Phosphorus	300 g/l	SL	07523

Uses

- Annual dicotyledons in ***lucerne***, ***undersown spring cereals*** *(undersown with lucerne)*, ***undersown winter cereals*** *(undersown with lucerne)*
- Thistles in ***lucerne***, ***undersown spring cereals*** *(undersown with lucerne)*, ***undersown winter cereals*** *(undersown with lucerne)*

Approval information

- 2,4-DB included in Annex I under EC Directive 91/414
- Accepted by BBPA for use on malting barley
- Approval expiry 31 Dec 2008 [1]

Efficacy guidance

- Best results achieved on young seedling weeds under good growing conditions. Treatment less effective in cold weather and dry soil conditions
- Rain within 12 h may reduce effectiveness

Restrictions

- Do not allow spray drift onto neighbouring crops

Crop-specific information

- Latest use: before first node detectable (GS 31) for undersown cereals; fourth trifoliate leaf for lucerne
- In direct sown lucerne spray when seedlings have reached first trifoliate leaf stage. Optimum time 3-4 trifoliate leaves
- Do not treat any lucerne after fourth trifoliate leaf
- In spring barley and spring oats undersown with lucerne spray from when cereal has 1 leaf unfolded and lucerne has first trifoliate leaf
- In spring wheat undersown with lucerne spray from when cereal has 3 leaves unfolded and lucerne has first trifoliate leaf

Environmental safety

- Dangerous for the environment
- Toxic to aquatic organisms
- Keep livestock out of treated areas for at least two weeks following treatment and until poisonous weeds, such as ragwort, have died down and become unpalatable

Hazard classification and safety precautions

Hazard H03, H11
Risk phrases R22a, R38, R41, R51, R53a
Operator protection U08, U11, U20a
Environmental protection E07a, E13c, E19b
Storage and disposal D01, D02, D05, D09a, D10c, D12a
Medical advice M03, M05a

149 2,4-DB + linuron + MCPA

A translocated herbicide for undersown cereals and grass
HRAC mode of action code: O + C2 + O

Products

Alistell	United Phosphorus	220:30:30 g/l	EC	11053

Uses

- Annual dicotyledons in ***rotational grass***, ***undersown spring cereals*** *(undersown with clover)*, ***undersown winter cereals*** *(undersown with clover)*

Approval information

- 2,4-DB, linuron and MCPA included in Annex I under EC Directive 91/414
- Accepted by BBPA for use on malting barley

Efficacy guidance

- Best results achieved on young seedling weeds growing actively in warm, moist weather
- May be applied at any time of year provided crop at correct stage and weather suitable
- Avoid spraying if rain falling or imminent

Restrictions

- Maximum number of treatments 1 per crop or yr
- Spray winter cereals when fully tillered but before first node detectable (GS 29-30)
- Spray spring wheat from 5-fully expanded leaf stage (GS 15), barley and oats from 2-fully expanded leaves (GS 12)
- Do not spray cereals undersown with lucerne, peas or beans
- Apply to clovers after 1-trifoliate leaf, to grasses after 2-fully expanded leaf stage
- Do not spray in conditions of drought, waterlogging or extremes of temperature
- In frosty weather clover leaf scorch may occur but damage normally outgrown
- Do not use on sand or soils with more than 10% organic matter
- Do not roll or harrow within 7 d before or after spraying
- Avoid drift of spray or vapour onto susceptible crops
- Do not apply by hand-held sprayers

Crop-specific information

- Latest use: before first node detectable (GS 31) for undersown cereals; 2 wk before grazing for grassland

Environmental safety

- Dangerous for the environment
- Toxic to aquatic organisms
- Keep livestock out of treated areas for at least two weeks following treatment and until poisonous weeds, such as ragwort, have died down and become unpalatable
- LERAP Category B

Hazard classification and safety precautions

Hazard H03, H11
Risk phrases R22a, R22b, R40, R41, R51, R53a, R66
Operator protection A, C, H; U05a, U08, U11, U13, U19a, U20b
Environmental protection E07a, E13b, E16a, E34, E38
Storage and disposal D01, D02, D09a, D10c, D12a
Medical advice M03, M05a

150 2,4-DB + MCPA

A translocated herbicide for cereals, clovers and leys
HRAC mode of action code: O + O

Products

1	Agrichem DB Plus	Agrichem	243:40 g/l	SL	00044
2	Headland Cedar	Headland	240:40 g/l	SL	12180
3	Redlegor	United Phosphorus	244:44 g/l	SL	07519

Uses

- Annual dicotyledons in ***all cereals*** [1]; ***rotational grass*** [1-3]; ***spring barley, spring oats, undersown barley*** *(red or white clover)*, ***undersown oats*** *(red or white clover)*, ***undersown wheat*** *(red or white clover)*, ***winter barley, winter oats*** [2]; ***spring wheat, winter wheat*** [2, 3]; ***undersown spring cereals, undersown winter cereals*** [1, 3]
- Perennial dicotyledons in ***all cereals*** [1]; ***rotational grass*** [1-3]; ***spring barley, spring oats, undersown barley*** *(red or white clover)*, ***undersown oats*** *(red or white clover)*, ***undersown wheat*** *(red or white clover)*, ***winter barley, winter oats*** [2]; ***spring wheat, winter wheat*** [2, 3]; ***undersown spring cereals, undersown winter cereals*** [1, 3]
- Polygonums in ***rotational grass, spring wheat, undersown spring cereals, undersown winter cereals, winter wheat*** [3]

Approval information

- 2,4-DB and MCPA included in Annex I under EC Directive 91/414
- Accepted by BBPA for use on malting barley

Efficacy guidance

- Best results achieved on young seedling weeds under good growing conditions
- Spray thistles and other perennials when 10-20 cm high provided clover at correct stage
- Effectiveness may be reduced by rain within 12 h, by very cold conditions or drought

Restrictions

- Maximum number of treatments 1 per crop
- Do not spray established clover crops or lucerne
- Do not roll or harrow within 7 d before or after spraying
- Do not spray immediately before or after sowing any crop
- Avoid drift onto neighbouring sensitive crops

Crop-specific information

- Latest use: before first node detectable (GS 31) for cereals; 4th trifoliate leaf stage of clover for grass re-seeds
- Apply in spring to winter cereals from leaf sheath erect stage, to spring barley or oats from 2-leaf stage (GS 12), to spring wheat from 5-leaf stage (GS 15)
- Spray clovers as soon as possible after first trifoliate leaf, grasses after 2-3 leaf stage
- Red clover may suffer temporary distortion after treatment

Environmental safety

- Harmful to aquatic organisms
- Harmful to fish or other aquatic life. Do not contaminate surface waters or ditches with chemical or used container
- Keep livestock out of treated areas for at least two weeks following treatment and until poisonous weeds, such as ragwort, have died down and become unpalatable

Hazard classification and safety precautions

Hazard H03
Risk phrases R22a, R41, R52 [1-3]; R38 [2, 3]; R53a [1, 3]
Operator protection A, C [1, 2]; U05a, U08, U11, U20b [1-3]; U13, U14 [3]; U15 [2, 3]
Environmental protection E07a, E13c [1-3]; E19b [1]; E34 [1, 2]; E38 [2]
Storage and disposal D01, D05, D09a [1-3]; D02, D10c [1, 3]; D10a [2]
Medical advice M03 [1-3]; M05a [1, 3]

151 deltamethrin

A pyrethroid insecticide with contact and residual activity
IRAC mode of action code: 3

Products

1	Agriguard Deltamethrin	AgriGuard	25 g/l	EC	10770
2	Bandu	Headland	25 g/l	EC	10994
3	Decis	Bayer CropScience	25 g/l	EC	07172
4	Decis Protech	Bayer CropScience	15 g/l	EW	11502
5	Landgold Deltaland	Landgold	25 g/l	EC	09906
6	Landgold Deltaland	Teliton	25 g/l	EC	12114
7	Pearl Micro	Bayer CropScience	6.25% w/w	GR	08620

Uses

- American serpentine leaf miner in ***aubergines*** *(off-label)*, ***non-edible ornamentals*** *(off-label)*, ***rooting beds*** *(off-label)* [3]; ***non-edible glasshouse crops*** *(off-label)*, ***protected aubergines*** *(off-label)* [4]; ***protected courgettes*** *(off-label)*, ***protected cucumbers*** *(off-label)*, ***protected gherkins*** *(off-label)* [2-4]; ***protected peppers*** *(off-label)*, ***protected tomatoes*** *(off-label)* [3, 4]; ***rooting beds*** *(off-label - after recovery from wilting)* [2]
- Aphids in ***amenity vegetation*** [1-4]; ***apples***, ***spring barley***, ***spring oats***, ***spring wheat***, ***winter barley***, ***winter oats***, ***winter wheat*** [1-7]; ***blackberries*** *(off-label)*, ***leaf spinach*** *(off-label)*, ***lupins*** *(off-label)*, ***protected aubergines*** *(off-label)*, ***protected cayenne peppers*** *(off-label)* [1, 2]; ***bulb onions*** *(off-label)*, ***celery (outdoor)*** *(off-label)*, ***garlic*** *(off-label)*, ***leeks*** *(off-label)*, ***parsnips*** *(off-label)*, ***peas***, ***protected celery*** *(off-label)*, ***protected chinese cabbage*** *(off-label)*, ***protected salad onions*** *(off-label)*, ***salad onions*** *(off-label)* [7]; ***carrots*** *(off-label)*, ***nursery stock***, ***radicchio*** *(off-label)* [3, 7]; ***chinese cabbage*** *(off-label)*, ***durum wheat*** *(off-label)*, ***spring rye*** *(off-label)*, ***triticale*** *(off-label)*, ***winter rye*** *(off-label)* [2-4, 7]; ***chives*** *(off-label)*, ***choi sum*** *(off-label)*, ***frise*** *(off-label)*, ***herbs (see appendix 6)*** *(off-label)*, ***lamb's lettuce*** *(off-label)*, ***scarole*** *(off-label)* [2]; ***collards*** *(off-label)*, ***pak choi*** *(off-label)* [1]; ***evening primrose*** *(off-label)*, ***protected ornamentals*** [4]; ***fodder beet*** *(off-label)*, ***horseradish*** *(off-label)*, ***mallow (althaea spp.)*** *(off-label)*, ***parsley root*** *(off-label)* [3]; ***grass seed crops*** *(off-label)*, ***lettuce*** *(off-label)* [3, 4, 7]; ***ornamental plant production***, ***protected cucumbers***, ***protected peppers***, ***protected pot plants***, ***protected tomatoes*** [1-4, 7]; ***parsley*** *(off-label)* [2, 3]
- Apple sawfly in ***apples*** [7]
- Apple sucker in ***apples*** [1-7]
- Barley yellow dwarf virus vectors in ***spring barley***, ***spring oats***, ***spring wheat***, ***winter oats*** [7]; ***winter barley***, ***winter wheat*** [1-7]
- Beet virus yellows vectors in ***winter oilseed rape*** [1-7]
- Brassica pod midge in ***mustard***, ***spring oilseed rape***, ***winter oilseed rape*** [7]
- Cabbage seed weevil in ***mustard***, ***spring oilseed rape***, ***winter oilseed rape*** [1-7]
- Cabbage stem flea beetle in ***winter oilseed rape*** [1-7]
- Cabbage stem weevil in ***mustard***, ***spring oilseed rape***, ***winter oilseed rape*** [1-6]
- Capsids in ***amenity vegetation*** [1-4]; ***apples*** [1-7]; ***nursery stock*** [3, 7]; ***ornamental plant production*** [1-4, 7]
- Caterpillars in ***amenity vegetation*** [1-4]; ***apples***, ***broccoli***, ***brussels sprouts***, ***cabbages***, ***cauliflowers***, ***kale***, ***plums***, ***swedes***, ***turnips*** [1-7]; ***bulb onions*** *(off-label)*, ***carrots*** *(off-label)*, ***celery (outdoor)*** *(off-label)*, ***garlic*** *(off-label)*, ***leeks*** *(off-label)*, ***parsnips*** *(off-label)*, ***protected celery*** *(off-label)*, ***protected chinese cabbage*** *(off-label)*, ***protected salad onions*** *(off-label)*, ***salad onions*** *(off-label)* [7]; ***chinese cabbage*** *(off-label)*, ***lettuce*** *(off-label)*, ***radicchio*** *(off-label)* [3, 4, 7]; ***marjoram*** *(off-label)*, ***protected ornamentals***, ***sorrel*** *(off-label)* [4]; ***nursery stock*** [3, 7]; ***ornamental plant production***, ***protected cucumbers***, ***protected peppers***, ***protected pot plants***, ***protected tomatoes*** [1-4, 7]; ***protected salad brassicas*** *(off-label - for baby leaf production)*, ***salad brassicas*** *(off-label - for baby leaf production)* [2-4, 7]
- Codling moth in ***apples*** [1-7]
- Cutworms in ***lettuce*** [1, 2, 4, 7]
- Damson-hop aphid in ***hops***, ***plums*** [1-7]
- Flea beetle in ***broccoli***, ***brussels sprouts***, ***cabbages***, ***cauliflowers***, ***kale***, ***swedes***, ***turnips*** [3, 7]; ***cress*** *(off-label)* [3]; ***evening primrose*** *(off-label)*, ***sorrel*** *(off-label)* [3, 4, 7]; ***fodder beet*** *(off-label)* [2]; ***marjoram*** *(off-label)*, ***protected lamb's lettuce*** *(off-label)*, ***protected scarole*** *(off-label)*, ***protected watercress*** *(off-label)*, ***radicchio*** *(off-label)* [4]; ***protected chives*** *(off-label)*, ***protected***

parsley *(off-label)* [2, 4]; ***protected herbs (see appendix 6)*** *(off-label)*, ***protected lettuce*** *(off-label)* [2, 4, 7]; ***protected radicchio*** *(off-label)* [4, 7]; ***protected salad brassicas*** *(off-label - for baby leaf production)*, ***salad brassicas*** *(off-label - for baby leaf production)* [2-4, 7]; ***sugar beet*** [1-7]

- Fruit tree tortrix moth in ***plums*** [1-6]
- Insect control in ***carrots*** *(off-label)*, ***celery (outdoor)*** *(off-label)*, ***horseradish*** *(off-label)*, ***mallow (althaea spp.)*** *(off-label)*, ***parsley root*** *(off-label)*, ***parsnips*** *(off-label)*, ***protected celery*** *(off-label)*, ***protected chinese cabbage*** *(off-label)*, ***protected salad onions*** *(off-label)*, ***salad onions*** *(off-label)* [2]
- Insect pests in ***blackberries*** *(off-label)*, ***bulb onions*** *(off-label)*, ***celery (outdoor)*** *(off-label)*, ***chives*** *(off-label)*, ***frise*** *(off-label)*, ***garlic*** *(off-label)*, ***herbs (see appendix 6)*** *(off-label)*, ***lamb's lettuce*** *(off-label)*, ***leeks*** *(off-label)*, ***parsley*** *(off-label)*, ***protected celery*** *(off-label)*, ***protected chinese cabbage*** *(off-label)*, ***protected salad onions*** *(off-label)*, ***salad onions*** *(off-label)*, ***scarole*** *(off-label)* [3, 4]; ***carrots*** *(off-label)* [4]; ***choi sum*** *(off-label)*, ***cress*** *(off-label)*, ***protected chives*** *(off-label)*, ***protected frise*** *(off-label)*, ***protected herbs (see appendix 6)*** *(off-label)*, ***protected lamb's lettuce*** *(off-label)*, ***protected lettuce*** *(off-label)*, ***protected parsley*** *(off-label)*, ***protected radicchio*** *(off-label)*, ***protected rhubarb*** *(off-label)*, ***rhubarb*** *(off-label)*, ***shallots*** *(off-label)* [3]; ***collards*** *(off-label)*, ***fodder beet*** *(off-label)* [7]; ***crab apples*** *(off-label)* [6]; ***leaf spinach*** *(off-label)*, ***lupins*** *(off-label)* [3, 4, 7]; ***protected aubergines*** *(off-label)* [3, 6, 7]; ***protected cayenne peppers*** *(off-label)* [4, 7]
- Leafhoppers in ***cress*** *(off-label)* [3]; ***marjoram*** *(off-label)* [3, 4, 7]; ***protected chives*** *(off-label)*, ***protected parsley*** *(off-label)* [2, 4]; ***protected herbs (see appendix 6)*** *(off-label)*, ***protected lettuce*** *(off-label)* [2, 4, 7]; ***protected lamb's lettuce*** *(off-label)*, ***protected scarole*** *(off-label)*, ***protected watercress*** *(off-label)*, ***radicchio*** *(off-label)*, ***sorrel*** *(off-label)* [4]; ***protected radicchio*** *(off-label)* [4, 7]
- Mealybugs in ***amenity vegetation*** [1-4]; ***nursery stock*** [3, 7]; ***ornamental plant production***, ***protected cucumbers***, ***protected peppers***, ***protected pot plants***, ***protected tomatoes*** [1-4, 7]; ***protected ornamentals*** [4]
- Pea and bean weevil in ***broad beans***, ***peas***, ***spring field beans***, ***winter field beans*** [1-7]
- Pea midge in ***peas*** [3, 4, 7]
- Pea moth in ***peas*** [1-7]
- Pear sucker in ***pears*** [1-7]
- Phorid flies in ***mushrooms*** *(off-label)* [2, 4]; ***protected edible fungi*** *(off-label)*, ***protected mushrooms*** *(off-label)* [3]
- Plum fruit moth in ***plums*** [7]
- Plum sawfly in ***plums*** [7]
- Pollen beetle in ***evening primrose*** *(off-label)* [3, 4, 7]; ***mustard***, ***spring oilseed rape***, ***winter oilseed rape*** [1-7]
- Rape winter stem weevil in ***winter oilseed rape*** [7]
- Raspberry beetle in ***raspberries*** [1-7]
- Sawflies in ***apples***, ***plums*** [1-6]
- Scale insects in ***amenity vegetation*** [1-4]; ***nursery stock*** [3, 7]; ***ornamental plant production***, ***protected cucumbers***, ***protected peppers***, ***protected pot plants***, ***protected tomatoes*** [1-4, 7]; ***protected ornamentals*** [4]
- Sciarid flies in ***mushrooms*** *(off-label)* [2, 4]; ***protected edible fungi*** *(off-label)*, ***protected mushrooms*** *(off-label)* [3]
- Thrips in ***amenity vegetation*** [1-4]; ***bulb onions*** *(off-label)*, ***garlic*** *(off-label)*, ***leeks*** *(off-label)*, ***shallots*** *(off-label)* [2]; ***nursery stock*** [3, 7]; ***ornamental plant production*** [1-4, 7]
- Tortrix moths in ***apples*** [1-7]
- Western flower thrips in ***aubergines*** *(off-label)*, ***non-edible ornamentals*** *(off-label)*, ***rooting beds*** *(off-label)* [3]; ***non-edible glasshouse crops*** *(off-label)*, ***protected aubergines*** *(off-label)* [4]; ***protected courgettes*** *(off-label)*, ***protected cucumbers*** *(off-label)*, ***protected gherkins*** *(off-label)* [2-4]; ***protected peppers*** *(off-label)*, ***protected tomatoes*** *(off-label)* [3, 4]; ***rooting beds*** *(off-label - after recovery from wilting)* [2]
- Whitefly in ***amenity vegetation*** [1-4]; ***nursery stock*** [3, 7]; ***ornamental plant production***, ***protected cucumbers***, ***protected peppers***, ***protected pot plants***, ***protected tomatoes*** [1-4, 7]; ***protected aubergines*** *(off-label)*, ***protected cayenne peppers*** *(off-label)* [1]; ***protected ornamentals*** [4]
- Yellow cereal fly in ***spring barley***, ***spring oats***, ***spring wheat***, ***winter barley***, ***winter oats***, ***winter wheat*** [7]

Specific Off-Label Approvals (SOLAs)

- ***aubergines*** *20040511* [3]
- ***blackberries*** *20052308* [1], *20052296* [2], *20052319* [3], *20052321* [4]
- ***bulb onions*** *20070364* [2], *20040528* [3], *20031140* [4], *20040504* [7]
- ***carrots*** *20070361* [2], *20040527* [3], *20031140* [4], *20040504* [7]
- ***celery (outdoor)*** *20041187* [2], *20040529* [3], *20031140* [4], *20040504* [7]
- ***chinese cabbage*** *20052297* [2], *20040526* [3], *20031140* [4], *20040504* [7]
- ***chives*** *20052295* [2], *20052317* [3], *20052322* [4]
- ***choi sum*** *20052297* [2], *20052318* [3]
- ***collards*** *20052307* [1], *20052333* [7]
- ***crab apples*** *20052313* [6]
- ***cress*** *20063972* [3]
- ***durum wheat*** *20052293* [2], *20052315* [3], *20052323* [4], *20052329* [7]
- ***evening primrose*** *20040512* [3], *20031140* [4], *20040504* [7]
- ***fodder beet*** *20070360* [2], *20052316* [3], *20052332* [7]
- ***frise*** *20052295* [2], *20052317* [3], *20052322* [4]
- ***garlic*** *20070364* [2], *20040528* [3], *20031140* [4], *20040504* [7]
- ***grass seed crops*** *20040512* [3], *20031140* [4], *20040504* [7]
- ***herbs (see appendix 6)*** *20052295* [2], *20052317* [3], *20052322* [4]
- ***horseradish*** *20070361* [2], *20040527* [3]
- ***lamb's lettuce*** *20052295* [2], *20052317* [3], *20052322* [4]
- ***leaf spinach*** *20052305* [1], *20052295* [2], *20052317* [3], *20052322* [4], *20060006* [7]
- ***leeks*** *20070365* [2], *20040530* [3], *20031140* [4], *20040504* [7]
- ***lettuce*** *20040526* [3], *20031140* [4], *20040504* [7]
- ***lupins*** *20052304* [1], *20052290* [2], *20052314* [3], *20052320* [4], *20052330* [7]
- ***mallow (althaea spp.)*** *20070361* [2], *20040527* [3]
- ***marjoram*** *20040512* [3], *20031140* [4], *20040504* [7]
- ***mushrooms*** *20041189* [2], *20031138* [4]
- ***non-edible glasshouse crops*** *20031139* [4]
- ***non-edible ornamentals*** *20040511* [3]
- ***pak choi*** *20052307* [1]
- ***parsley*** *20052295* [2], *20040527* [3], *20052317* [3], *20052322* [4]
- ***parsley root*** *20070361* [2], *20040527* [3]
- ***parsnips*** *20070361* [2], *20040504* [7]
- ***protected aubergines*** *20063690* [1], *20063657* [2], *20063660* [3], *20031139* [4], *20063666* [6], *20063661* [7]
- ***protected cayenne peppers*** *20052306* [1], *20052298* [2], *20052324* [4], *20052334* [7]
- ***protected celery*** *20041187* [2], *20040529* [3], *20031140* [4], *20040535* [7]
- ***protected chinese cabbage*** *20041187* [2], *20040529* [3], *20031140* [4], *20040535* [7]
- ***protected chives*** *20041188* [2], *20063972* [3], *20031140* [4]
- ***protected courgettes*** *20070362* [2], *20040511* [3], *20031139* [4]
- ***protected cucumbers*** *20070362* [2], *20040511* [3], *20031139* [4]
- ***protected edible fungi*** *20030040* [3]
- ***protected frise*** *20063972* [3]
- ***protected gherkins*** *20070362* [2], *20040511* [3], *20031139* [4]
- ***protected herbs (see appendix 6)*** *20041188* [2], *20063972* [3], *20031140* [4], *20040740* [7]
- ***protected lamb's lettuce*** *20063972* [3], *20031140* [4]
- ***protected lettuce*** *20041188* [2], *20063972* [3], *20031140* [4], *20040740* [7]
- ***protected mushrooms*** *20030040* [3]
- ***protected parsley*** *20041188* [2], *20063972* [3], *20031140* [4]
- ***protected peppers*** *20040511* [3], *20031139* [4]
- ***protected radicchio*** *20063972* [3], *20031140* [4], *20040740* [7]
- ***protected rhubarb*** *20040529* [3]
- ***protected salad brassicas*** *(for baby leaf production) 20041190* [2], *(for baby leaf production) 20063971* [3], *(for baby leaf production) 20031140* [4], *(for baby leaf production) 20040535* [7]
- ***protected salad onions*** *20041187* [2], *20040529* [3], *20031140* [4], *20040535* [7]
- ***protected scarole*** *20031140* [4]
- ***protected tomatoes*** *20040511* [3], *20031139* [4]
- ***protected watercress*** *20031140* [4]

- ***radicchio*** *20040512* [3], *20031140* [4], *20040504* [7]
- ***rhubarb*** *20040529* [3]
- ***rooting beds*** *(after recovery from wilting) 20070362* [2], *20040511* [3]
- ***salad brassicas*** *(for baby leaf production) 20041190* [2], *(for baby leaf production) 20063971* [3], *(for baby leaf production) 20031140* [4], *(for baby leaf production) 20040504* [7]
- ***salad onions*** *20041187* [2], *20040529* [3], *20031140* [4], *20040504* [7]
- ***scarole*** *20052295* [2], *20052317* [3], *20052322* [4]
- ***shallots*** *20070364* [2], *20040528* [3]
- ***sorrel*** *20040512* [3], *20031140* [4], *20040504* [7]
- ***spring rye*** *20052293* [2], *20052315* [3], *20052323* [4], *20052329* [7]
- ***triticale*** *20052293* [2], *20052315* [3], *20052323* [4], *20052329* [7]
- ***winter rye*** *20052293* [2], *20052315* [3], *20052323* [4], *20052329* [7]

Approval information

- Deltamethrin included in Annex I under EC Directive 91/414
- Accepted by BBPA for use on malting barley and hops
- Approval expiry 01 Nov 2008 [1-4, 6, 7]

Efficacy guidance

- A contact and stomach poison with 3-4 wk persistence, particularly effective on caterpillars and sucking insects
- Normally applied at first signs of damage with follow-up treatments where necessary at 10-14 d intervals. Rates, timing and recommended combinations with other pesticides vary with crop and pest. See label for details
- Spray is rainfast within 1 h
- May be applied in frosty weather provided foliage not covered in ice
- Temperatures above 35 °C may reduce effectiveness or persistence

Restrictions

- Maximum number of treatments varies with crop and pest, 4 per crop for wheat and barley, only 1 application between 1 Apr and 31 Aug. See label for other crops
- Do not apply more than 1 aphicide treatment to cereals in summer
- Do not spray crops suffering from drought or other physical stress
- Consult processer before treating crops for processing
- Do not apply to a cereal crop if any product containing a pyrethroid insecticide or dimethoate has been applied to that crop after the start of ear emergence (GS 51)
- Do not spray cereals after 31 Mar in the year of harvest within 6 m of the outside edge of the crop
- Reduced volume spraying must not be used on cereals after 31 Mar in yr of harvest

Crop-specific information

- Latest use: early dough (GS 83) for barley, oats, wheat; before flowering for mustard, oilseed rape; before 31 Mar for grass seed crops [1-6]; end of flowering for mustard, oilseed rape [7]

Environmental safety

- Dangerous for the environment
- Toxic to aquatic organisms
- Flammable [1-3, 5, 6]
- Extremely dangerous to fish or other aquatic life. Do not contaminate surface waters or ditches with chemical or used container
- High risk to bees. Do not apply to crops in flower or to those in which bees are actively foraging. Do not apply when flowering weeds are present
- Do not apply in tank-mixture with a triazole-containing fungicide when bees are likely to be actively foraging in the crop
- High risk to non-target insects or other arthropods. Do not spray within 6 m of the field boundary
- LERAP Category A
- Broadcast air-assisted LERAP (18 m) [3-7]

Hazard classification and safety precautions

Hazard H02 [7]; H03, H08 [1-3, 5, 6]; H11 [1-7]

Risk phrases R20 [2, 3]; R21 [1, 5, 6]; R22a [1-3, 5, 6]; R22b, R41 [2, 3, 5, 6]; R25 [7]; R36 [1, 7]; R38, R51 [1-3, 5-7]; R50 [4]; R53a [1-7]

Operator protection A, C, H; U04a, U05a, U08, U19a [1-3, 5-7]; U10 [2]; U11, U14, U15 [4]; U20b [1-7]

Environmental protection E12a [3, 4, 7]; E12c [1, 2, 5, 6]; E12f [1, 2] (cereals, oilseed rape, peas, beans); E12f [3, 4] (cereals, oilseed rape, peas, beans - see label for guidance); E12f [5, 6] (oilseed rape, cereals, peas, beans); E12f [7] (cereals, oilseed rape, peas, beans - see label for guidance); E13a [1, 7]; E15a [2-6]; E16c, E16d [1-7]; E17b [3-7] (18 m); E22a [1, 4]; E34 [1-3, 5-7]; E38 [2-4, 7]

Storage and disposal D01, D02, D09a [1-7]; D05 [1-3, 5-7]; D08, D11a [7]; D10b [5, 6]; D10c [1-4]; D12a [3, 4, 7]

Medical advice M03 [1-3, 5-7]; M04a [7]; M05b [2, 3, 5, 6]

152 desmedipham

A contact phenyl carbamate herbicide available only in mixtures
HRAC mode of action code: C1

153 desmedipham + ethofumesate + phenmedipham

A selective contact and residual herbicide for beet
HRAC mode of action code: C1 + N + C1

Products

1	Beetaweed	AgriGuard	25:151:75 g/l	EC	12260
2	Betanal Expert	Bayer CropScience	25:151:75 g/l	EC	10592
3	Landgold Deputy	Landgold	25:151:75 g/l	EC	11975
4	Landgold Deputy	Teliton	25:151:75 g/l	EC	12115

Uses

- Annual dicotyledons in ***sugar beet***
- Annual meadow grass in ***sugar beet***

Approval information

- Desmedipham, ethofumesate and phenmedipham included in Annex I under EC Directive 91/414

Efficacy guidance

- Product recommended for low-volume overall application in a planned spray programme
- Product acts mainly by contact action. A full programme also gives some residual control but this may be reduced on soils with more than 5% organic matter
- Best results achieved from treatments applied at fully expanded cotyledon stage of largest weeds present. Occasional larger weeds will usually be controlled from a full programme of sprays
- Where a pre-emergence band spray has been applied treatment must be timed according to size of the untreated weeds between the rows
- Susceptible weeds may not all be killed by the first spray. Repeat applications as each flush of weeds reaches cotyledon size normally necessary for season long control
- Sequential treatments should be applied when the previous one is still showing an effect on the weeds
- Various mixtures with other beet herbicides are recommended. See label for details

Restrictions

- Maximum total dose equivalent to three full dose treatments
- Do not spray crops stressed by nutrient deficiency, wind damage, pest or disease attack, or previous herbicide treatments. Stressed crops treated under conditions of high light intensity may be checked and not recover fully
- If temperature likely to exceed 21 °C spray after 5 pm
- Crystallisation may occur if spray volume exceeds that recommended or spray mixture not used within 2 h, especially if the water temperature is below 5 °C
- Before use, wash out sprayer to remove all traces of previous products, especially hormone and sulfonyl urea weedkillers

Crop-specific information

- Latest use: before crop meets between rows
- Apply first treatment when majority of crop plants have reached the fully expanded cotyledon stage
- Frost within 7 d of treatment may cause check from which the crop may not recover

Following crops guidance

- Beet crops may be sown at any time after treatment. Any other crop may be sown 3 mth after treatment following mouldboard ploughing to 15 cm minimum

Environmental safety

- Dangerous for the environment
- Toxic to aquatic organisms

Hazard classification and safety precautions

Hazard H11
Risk phrases R51, R53a
Operator protection A, H; U08, U20a
Environmental protection E15a, E38 [1, 2]
Storage and disposal D05, D09a, D10b [1-4]; D07 [3, 4]; D12a [1, 2]
Medical advice M03

154 desmedipham + phenmedipham

A mixture of contact herbicides for use in sugar beet
HRAC mode of action code: C1 + C1

Products

Betanal Carrera	Bayer CropScience	40:125 g/l	SE	11029

Uses

- Annual dicotyledons in ***sugar beet***

Approval information

- Desmedipham and phenmedipham included in Annex I under EC Directive 91/414

Efficacy guidance

- Product is recommended for low volume overall spraying in a planned programme involving pre and/or post-emergence treatments at doses recommended for low volume programmes
- Best results obtained from treatment when earliest germinating weeds have reached cotyledon stage
- Further treatments must be applied as each flush of weeds reaches cotyledon stage but allowing a minimum of 7 d between each spray
- Where a pre-emergence band spray has been applied, the first treatment should be timed according to the size of the weeds in the untreated area between the rows
- Product is absorbed by leaves of emerged weeds which are killed by scorching action in 2-10 d. Apply overall as a fine spray to optimise weed cover and spray retention
- Various tank mixtures and sequences recommended to widen weed spectrum and add residual activity - see label for details

Restrictions

- Maximum total dose equivalent to three full dose treatments
- Do not spray crops stressed by nutrient deficiency, frost, wind damage, pest or disease attack, or previous herbicide treatments. Stressed crops may be checked and not recover fully
- If temperature likely to exceed 21 °C spray after 5 pm
- Before use, wash out sprayer to remove all traces of previous products, especially hormone and sulfonyl urea weedkillers
- Crystallisation may occur if spray volume exceeds that recommended or spray mixture not used within 2 h, especially if the water temperature is below 5 °C
- Product may cause non-reinforced PVC pipes and hoses to soften and swell. Wherever possible, use reinforced PVC or synthetic rubber hoses

Crop-specific information

- Latest use: before crop leaves meet between rows
- Product safe to use on all soil types

Environmental safety

- Dangerous for the environment
- Toxic to aquatic organisms
- Risk to certain non-target insects or other arthropods. Avoid spraying within 6 m of field boundary

Hazard classification and safety precautions

Hazard H04, H11
Risk phrases R43, R51, R53a
Operator protection A, H; U05a, U08, U19a, U20b
Environmental protection E15a, E22b, E38
Storage and disposal D01, D02, D05, D09a, D10b, D12a

155 dicamba

A translocated benzoic herbicide available only in mixtures
HRAC mode of action code: O

See also *2,4-D + dicamba*
2,4-D + dicamba + triclopyr

156 dicamba + dichlorprop-P + ferrous sulphate + MCPA

A herbicide/fertilizer combination for moss and weed control in turf
HRAC mode of action code: O + O + O

Products

Renovator 2	Scotts	0.03:0.22:16.3:0.22% w/w	GR	11411

Uses

- Annual dicotyledons in ***managed amenity turf***
- Moss in ***managed amenity turf***
- Perennial dicotyledons in ***managed amenity turf***

Approval information

- Dichlorprop-P and MCPA included in Annex I under EC Directive 91/414

Efficacy guidance

- Apply from mid-Apr to mid-Aug when weeds are growing
- For best control of moss scarify vigorously after 2 wk to remove dead moss
- Where regrowth of moss or weeds occurs a repeat treatment may be made after 6 wk
- Avoid treatment of wet grass or during drought. If no rain falls within 48 h water in thoroughly

Restrictions

- Do not treat new turf until established for 6-9 mth
- The first 4 mowings after treatment should not be used to mulch cultivated plants unless composted at least 6 mth
- Avoid walking on treated areas until it has rained or they have been watered
- Do not re-seed or turf within 8 wk of last treatment
- Do not cut grass for at least 3 d before and at least 4 d after treatment
- Do not apply during freezing conditions or when rain imminent

Crop-specific information

- Apply with a suitable calibrated fertilizer distributor

Hazard classification and safety precautions

Operator protection U09a, U20a
Environmental protection E15a
Storage and disposal D09a, D11a

157 dicamba + dichlorprop-P + MCPA

A translocated herbicide mixture for turf
HRAC mode of action code: O + O + O

Products

Intrepid 2	Scotts	20.8:167:167 g/l	SL	11594

Uses

- Annual dicotyledons in ***managed amenity turf***
- Perennial dicotyledons in ***managed amenity turf***

Approval information

- Dichlorprop-P and MCPA included in Annex I under EC Directive 91/414

Efficacy guidance

- Apply as directed on label between Apr and Sep when weeds growing actively
- Do not mow for at least 3 d before and 3-4 d after treatment so that there is sufficient leaf growth for spray uptake and sufficient time for translocation. See label for details
- If re-growth occurs or new weeds germinate re-treatment recommended

Restrictions

- Do not use during drought unless irrigation is carried out before and after treatment
- Do not apply during freezing conditions or when heavy rain is imminent
- Do not treat new turf until established for about 6 mth after seeding or turfing
- The first 4 mowings after treatment should not be used to mulch cultivated plants unless composted for at least 6 mth
- Do not re-seed turf within 8 wk of last treatment

Crop-specific information

- Latest use: end Sep for managed amenity turf

Environmental safety

- Prevent spray running or drifting onto cultivated plants, including shrubs and trees

Hazard classification and safety precautions

Hazard H03
Risk phrases R22a, R38, R41, R43
Operator protection A, C; U05a, U08, U11, U19a, U20a
Environmental protection E15a, E34
Storage and disposal D01, D02, D05, D09a, D10b
Medical advice M03, M05a

158 dicamba + MCPA + mecoprop-P

A translocated herbicide for cereals, grassland, amenity grass and orchards
HRAC mode of action code: O + O + O

Products

1	Banlene Super	Certis	18:252:42 g/l	SL	10053
2	Field Marshal	United Phosphorus	18:360:80 g/l	SL	08956
3	Greencrop Triathlon	Greencrop	25:200:200 g/l	SL	10956
4	Headland Relay P	Headland	25:200:200 g/l	SL	08580
5	Headland Relay Turf	Headland Amenity	25:200:200 g/l	SL	08935
6	Headland Transfer	Headland	18:315:50 g/l	SL	11010
7	Headland Trinity	Headland	18:315:50 g/l	SL	10842
8	Hycamba Plus	Agrichem	20.4:125.9:240.2 g/l	SL	10180
9	Hyprone-P	Agrichem	16:101:92 g/l	SL	09125
10	Hysward-P	Agrichem	16:101:92 g/l	SL	09052
11	Mircam Plus	Nufarm UK	19.5:245:43.3 g/l	SL	11525
12	Nocweed	Bayer Environ.	18:252:42 g/l	SL	10609
13	Pasturol Plus	Headland	25:200:200 g/l	SL	10278
14	Pierce	Nufarm UK	31:256:238 g/l	SL	11924
15	Re-act	Scotts	31:256:237 g/l	SL	12231
16	Super Selective Plus	Rigby Taylor	31.25:256:238 g/l	SL	11928
17	T2 Green	Nufarm UK	31: 256: 237 g/l	SL	11925
18	Tribute	Nomix Enviro	18:252:42 g/l	SL	06921
19	Tribute Plus	Nomix Enviro	18:252:42 g/l	SL	09493
20	Tritox	Scotts	15:178:54 g/l	SL	07764
21	UPL Grassland Herbicide	United Phosphorus	25:200:200 g/l	SL	08934

Uses

- Annual dicotyledons in ***amenity grassland*** [3, 7, 8, 10, 13, 21]; ***apple orchards*** [1, 7, 11, 12]; ***grass seed crops*** [1, 6, 7, 9, 11, 12, 14]; ***managed amenity turf*** [3, 5, 7, 8, 10-13, 15-20]; ***pear orchards, spring rye, winter rye*** [1, 7, 12]; ***permanent grassland*** [1, 3, 4, 6, 7, 10-14, 21];

SEE SECTION 3 FOR PRODUCTS ALSO REGISTERED

rotational grass [1-4, 6, 7, 9-14, 21]; ***spring barley, spring oats, spring wheat, winter barley, winter oats, winter wheat*** [1, 2, 6, 7, 9, 11, 12]; ***undersown barley, undersown oats, undersown wheat*** [9, 12]; ***undersown rye*** [12]; ***undersown spring cereals, undersown winter cereals*** [1, 7]; ***undersown spring cereals*** *(grass only)*, ***undersown winter cereals*** *(grass only)* [2, 6]

- Docks in ***amenity grassland*** [10, 21]; ***apple orchards, pear orchards*** [1, 7, 12]; ***grass seed crops*** [1, 7, 9, 12, 14]; ***managed amenity turf*** [10]; ***permanent grassland, rotational grass*** [1, 7, 10, 12, 14, 21]
- Perennial dicotyledons in ***amenity grassland*** [3, 7, 8, 10, 13, 21]; ***apple orchards*** [1, 7, 11, 12]; ***grass seed crops*** [1, 6, 7, 9, 11, 12, 14]; ***managed amenity turf*** [3, 5, 7, 8, 10, 11, 13, 15-20]; ***pear orchards, spring rye, winter rye*** [1, 7, 12]; ***permanent grassland*** [1, 3, 4, 6, 7, 10-14, 21]; ***rotational grass*** [1-4, 6, 7, 9-14, 21]; ***spring barley, spring oats, spring wheat, winter barley, winter oats, winter wheat*** [1, 2, 6, 7, 9, 11, 12]; ***undersown barley, undersown oats, undersown wheat*** [9, 12]; ***undersown rye*** [12]; ***undersown spring cereals, undersown winter cereals*** [1, 7]; ***undersown spring cereals*** *(grass only)*, ***undersown winter cereals*** *(grass only)* [2, 6]

Approval information

- MCPA and mecoprop-P included in Annex I under EC Directive 91/414
- Accepted by BBPA for use on malting barley

Efficacy guidance

- Treatment should be made when weeds growing actively. Weeds hardened by winter weather may be less susceptible
- For best results apply in fine warm weather, preferably when soil is moist. Do not spray if rain expected within 6 h or in drought
- Application of fertilizer 1-2 wk before spraying aids weed control in turf
- Where a second treatment later in the season is needed in amenity situations and on grass allow 4-6 wk between applications to permit sufficient foliage regrowth for uptake

Restrictions

- Maximum number of treatments (including other mecoprop-P products) or maximum total dose varies with crop and product. See label for details. The total amount of mecoprop-P applied in a single yr must not exceed the maximum total dose approved for any single product for the crop/situation
- Do not apply to cereals after the first node is detectable (GS 31), or to grass under stress from drought or cold weather
- Do not spray cereals undersown with clovers or legumes, to be undersown with grass or legumes or grassland where clovers or other legumes are important
- Do not spray leys established less than 18 mth or orchards established less than 3 yr
- Do not roll or harrow within 7 d before or after treatment, or graze for at least 7 d afterwards (longer if poisonous weeds present)
- Do not use on turf or grass in year of establishment. Allow 6-8 wk after treatment before seeding bare patches
- The first mowings after use should not be used for mulching unless composted for 6 mth
- Turf should not be mown for 24 h before or after treatment (3-4 d for closely mown turf)
- Avoid drift onto all broad-leaved plants outside the target area

Crop-specific information

- Latest use: before first node detectable (GS 31) for cereals; 5-6 wk before head emergence for grass seed crops; mid-Oct for established grass
- HI 7-14 d before cutting or grazing for leys, permanent pasture
- Apply to winter cereals from the leaf sheath erect stage (GS 30), and to spring cereals from the 5 expanded leaf stage (GS 15)
- Spray grass seed crops 4-6 wk before flower heads begin to emerge (timothy 6 wk)
- Turf containing bulbs may be treated once the foliage has died down completely [4, 10, 19, 20]

Environmental safety

- Harmful to aquatic organisms
- Keep livestock out of treated areas for at least 2 wk following treatment and until poisonous weeds, such as ragwort, have died down and become unpalatable
- Harmful to fish or other aquatic life. Do not contaminate surface waters or ditches with chemical or used container

Hazard classification and safety precautions

Hazard H03 [1-8, 11-19]; H04 [9, 10, 20, 21]

Risk phrases R20 [4, 6, 7, 13]; R21 [1, 3-7, 11-13, 18, 19]; R22a [1-8, 11-17]; R36 [3, 4, 9-11, 13, 18-21]; R38 [6-8, 11, 21]; R41 [1, 2, 4-8, 12-17]; R43 [21]; R52 [2, 4-11, 13, 14, 17-21]; R53a [2, 6-11, 14, 17, 20, 21]

Operator protection A, C [1-21]; H, M [2-11, 13-21]; U05a [1, 3-21]; U08 [1-11, 13, 18-21]; U09a [12]; U11 [1, 2, 4, 6-17, 20, 21]; U15 [4, 13]; U19a [3, 8, 10, 20]; U20b [1-13, 18-21]

Environmental protection E07a [1-4, 6-20]; E13c [1-10, 12-17, 21]; E15a [2, 11, 14, 15, 17-20]; E23 [18, 19]; E34 [2-7, 11-20]; E38 [4, 13, 18, 19]

Storage and disposal D01, D02, D09a [1-21]; D05 [1-19, 21]; D10a [6, 7, 16, 20]; D10b [1, 3-5, 11-15, 17-19]; D10c [2, 8-10, 21]; D12a [2, 18, 19, 21]

Medical advice M03 [2-7, 11, 13-17, 21]; M05a [2, 7-10, 21]

SECTION 2

159 dicamba + mecoprop-P

A translocated post-emergence herbicide for cereals and grassland

HRAC mode of action code: O + O

Products

1	Camber	Headland	42:319 g/l	SL	09901
2	Di-Farmon R	Headland	42:319 g/l	SL	08472
3	Dockmaster	Nufarm UK	18.7:150 g/l	SL	11651
4	Foundation	Headland	84:600 g/l	SL	11708
5	Headland Saxon	Headland	84:600 g/l	SL	11947
6	High Load Mircam	Nufarm UK	80:600 g/l	SL	11930
7	Hyban-P	Agrichem	18.7:150 g/l	SL	09129
8	Hygrass-P	Agrichem	18.7:150 g/l	SL	09130
9	Mircam	Nufarm UK	18.7:150 g/l	SL	11707
10	Prompt	Headland	84:600 g/l	SL	11948

Uses

- Annual dicotyledons in ***permanent grassland*** [1, 2, 4-6, 8-10]; ***rotational grass*** [6, 8, 9]; ***spring barley, spring oats, spring wheat, winter barley, winter oats, winter wheat*** [1-7, 9, 10]; ***spring rye, triticale, undersown rye, winter rye*** [7]; ***undersown barley, undersown oats, undersown wheat*** [3, 7]
- Buttercups in ***permanent grassland*** [3, 7]; ***rotational grass*** [7]
- Chickweed in ***permanent grassland*** [1-3, 7]; ***rotational grass*** [6, 7]; ***spring barley, spring oats, spring wheat, winter barley, winter oats, winter wheat*** [1-7, 9, 10]; ***spring rye, triticale, undersown rye, winter rye*** [7]; ***undersown barley, undersown oats, undersown wheat*** [3, 7]
- Cleavers in ***permanent grassland*** [1, 2]; ***rotational grass*** [6]; ***spring barley, spring oats, spring wheat, winter barley, winter oats, winter wheat*** [1-7, 9, 10]; ***spring rye, triticale, undersown rye, winter rye*** [7]; ***undersown barley, undersown oats, undersown wheat*** [3, 7]
- Creeping thistle in ***permanent grassland*** [3, 7]; ***rotational grass*** [7]
- Docks in ***permanent grassland*** [3, 7-9]; ***rotational grass*** [7-9]
- Mayweeds in ***permanent grassland*** [1, 2]; ***rotational grass*** [6]; ***spring barley, spring oats, spring wheat, winter barley, winter oats, winter wheat*** [1-7, 9, 10]; ***spring rye, triticale, undersown rye, winter rye*** [7]; ***undersown barley, undersown oats, undersown wheat*** [3, 7]
- Perennial dicotyledons in ***permanent grassland*** [4-6, 8-10]; ***rotational grass*** [8, 9]; ***spring barley, spring oats, spring wheat, winter barley, winter oats, winter wheat*** [3, 7, 9]; ***spring rye, triticale, undersown rye, winter rye*** [7]; ***undersown barley, undersown oats, undersown wheat*** [3, 7]
- Plantains in ***spring barley, spring oats, spring rye, spring wheat, triticale, undersown barley, undersown oats, undersown rye, undersown wheat, winter barley, winter oats, winter rye, winter wheat*** [7]
- Polygonums in ***permanent grassland*** [1, 2]; ***rotational grass*** [6]; ***spring barley, spring oats, spring wheat, winter barley, winter oats, winter wheat*** [1-7, 9, 10]; ***spring rye, triticale, undersown rye, winter rye*** [7]; ***undersown barley, undersown oats, undersown wheat*** [3, 7]
- Scentless mayweed in ***permanent grassland*** [3, 7]; ***rotational grass*** [7]
- Stinging nettle in ***permanent grassland, rotational grass*** [7]
- Thistles in ***permanent grassland, rotational grass*** [8, 9]

SEE SECTION 3 FOR PRODUCTS ALSO REGISTERED

Approval information

- Mecoprop-P included in Annex I under EC Directive 91/414
- Accepted by BBPA for use on malting barley

Efficacy guidance

- Best results by application in warm, moist weather when weeds are actively growing

Restrictions

- Maximum number of treatments 1 per crop for cereals and 1 or 2 per yr on grass depending on label. The total amount of mecoprop-P applied in a single yr must not exceed the maximum total dose approved for any single product for the crop/situation
- Do not spray in cold or frosty conditions
- Do not spray if rain expected within 6 h
- Do not treat undersown grass until tillering begins
- Do not spray cereals undersown with clover or legume mixtures
- Do not roll or harrow within 7 d before or after spraying
- Do not treat crops suffering from stress from any cause
- Use product immediately following dilution; do not allow diluted product to stand before use [5]
- Avoid treatment when drift may damage neighbouring susceptible crops

Crop-specific information

- Latest use: before 1st node detectable for cereals; 7 d before cutting or 14 d before grazing grass
- Apply to winter sown crops from 5 expanded leaf stage (GS 15)
- Apply to spring sown cereals from 5 expanded leaf stage but before first node is detectable (GS 15-31)
- Treat grassland just before perennial weeds flower
- Transient crop prostration may occur after spraying but recovery is rapid

Environmental safety

- Dangerous for the environment [1, 2, 4-6, 10]
- Toxic to aquatic organisms
- Harmful to fish or other aquatic life. Do not contaminate surface waters or ditches with chemical or used container
- Keep livestock out of treated areas for at least 2 wk and until foliage of poisonous weeds such as ragwort has died and become unpalatable

Hazard classification and safety precautions

Hazard H03 [1-6, 9, 10]; H04 [7, 8]; H11 [1, 2, 4-6, 10]

Risk phrases R21, R36 [3, 9]; R22a, R38 [1-6, 9, 10]; R41 [1, 2, 4-8, 10]; R51 [1, 2, 4-6, 10]; R52 [3, 7-9]; R53a [1-10]

Operator protection A, C [1-10]; H, M [1-5, 7-10]; U05a, U11 [1-10]; U08 [1, 2, 4, 5, 7, 10]; U09a [3, 6, 8, 9]; U19a [8]; U20a [2]; U20b [1, 3-10]

Environmental protection E07a [1-5, 7-10]; E13c [3, 6-9]; E15a [1, 2, 4, 5, 10]; E34 [1-10]; E38 [1, 2, 4]

Storage and disposal D01, D02, D05, D09a [1-10]; D10b [1-5, 9, 10]; D10c [7, 8]

Medical advice M03 [1-10]; M05a [7, 8]

160 dichlobenil

A residual benzonitrile herbicide for woody crops and non-crop uses
HRAC mode of action code: L

Products

	Product	Company	Content	Form	Reg. no.
1	Casoron G	Nomix Enviro	6.75% w/w	GR	09023
2	Casoron G	Scotts	6.75% w/w	GR	10709
3	Casoron G	Certis	6.75% w/w	GR	11632
4	Casoron G4	Scotts	4% w/w	GR	10708
5	Certis Dichlobenil Granules	Certis	6.75%	GR	13705
6	Midstream GSR	Scotts	20% w/w	GR	11674
7	Osorno	Nufarm UK	6.75% w/w	GR	12486
8	Scotts Dichlo G Macro	Scotts	6.75% w/w	GR	12011
9	Scotts Dichlo G Micro	Scotts	6.75% w/w	GR	11857

Uses

- Annual dicotyledons in ***almonds*** *(off-label)*, ***bilberries*** *(off-label)*, ***blueberries*** *(off-label)*, ***chestnuts*** *(off-label)*, ***cranberries*** *(off-label)*, ***hazel nuts*** *(off-label)*, ***quinces*** *(off-label)*, ***rhubarb*** *(off-label - established)*, ***walnuts*** *(off-label)*, ***wine grapes*** *(off-label)* [3]; ***amenity vegetation*** [1, 2, 5, 7-9]; ***apple orchards, blackberries, blackcurrants, gooseberries, loganberries, pear orchards, raspberries, redcurrants*** [3, 5, 7]; ***established woody ornamentals*** [4]; ***ornamental plant production*** [1, 2, 8, 9]; ***ornamental trees*** [4, 5, 7]; ***roses*** [1, 2, 4, 5, 7-9]
- Annual grasses in ***almonds*** *(off-label)*, ***bilberries*** *(off-label)*, ***blueberries*** *(off-label)*, ***chestnuts*** *(off-label)*, ***cranberries*** *(off-label)*, ***hazel nuts*** *(off-label)*, ***quinces*** *(off-label)*, ***rhubarb*** *(off-label - established)*, ***walnuts*** *(off-label)* [3]; ***amenity vegetation*** [1, 2, 5, 7-9]; ***apple orchards, blackberries, blackcurrants, gooseberries, loganberries, pear orchards, raspberries, redcurrants*** [3, 5, 7]; ***established woody ornamentals*** [4]; ***ornamental plant production*** [1, 2, 8, 9]; ***ornamental trees*** [4, 5, 7]; ***roses*** [1, 2, 4, 5, 7-9]
- Aquatic weeds in ***aquatic areas*** [3]; ***enclosed waters, land immediately adjacent to aquatic areas, open waters*** [1, 2, 5-9]
- Perennial dicotyledons in ***almonds*** *(off-label)*, ***bilberries*** *(off-label)*, ***blueberries*** *(off-label)*, ***chestnuts*** *(off-label)*, ***cranberries*** *(off-label)*, ***hazel nuts*** *(off-label)*, ***quinces*** *(off-label)*, ***rhubarb*** *(off-label - established)*, ***walnuts*** *(off-label)*, ***wine grapes*** *(off-label)* [3]; ***amenity vegetation, ornamental plant production*** [1, 2, 8, 9]; ***apple orchards, blackberries, blackcurrants, gooseberries, loganberries, pear orchards, raspberries, redcurrants*** [3, 5, 7]; ***established woody ornamentals*** [4]; ***ornamental trees*** [4, 5, 7]; ***roses*** [1, 2, 4, 5, 7-9]
- Perennial grasses in ***almonds*** *(off-label)*, ***bilberries*** *(off-label)*, ***blueberries*** *(off-label)*, ***chestnuts*** *(off-label)*, ***cranberries*** *(off-label)*, ***hazel nuts*** *(off-label)*, ***quinces*** *(off-label)*, ***rhubarb*** *(off-label - established)*, ***walnuts*** *(off-label)* [3]; ***amenity vegetation, ornamental plant production*** [1, 2, 8, 9]; ***apple orchards, blackberries, blackcurrants, gooseberries, loganberries, pear orchards, raspberries, redcurrants*** [3, 5, 7]; ***established woody ornamentals*** [4]; ***ornamental trees*** [4, 5, 7]; ***roses*** [1, 2, 4, 5, 7-9]
- Total vegetation control in ***hard surfaces, natural surfaces not intended to bear vegetation, permeable surfaces overlying soil*** [1, 2, 5, 7-9]; ***land not intended to bear vegetation*** [3]
- Volunteer potatoes in ***potato dumps*** *(blight prevention)* [3, 5, 7]

Specific Off-Label Approvals (SOLAs)

- ***almonds*** *20060585* [3]
- ***bilberries*** *20060587* [3]
- ***blueberries*** *20060587* [3]
- ***chestnuts*** *20060585* [3]
- ***cranberries*** *20060587* [3]
- ***hazel nuts*** *20060585* [3]
- ***quinces*** *20060586* [3]
- ***rhubarb*** *(established) 20041404* [3]
- ***walnuts*** *20060585* [3]
- ***wine grapes*** *20041405 expires Jan 2008* [3]

Approval information

- Approved for aquatic weed control [1-3, 6-9]. See notes in Section 5 on use of herbicides in or near water

Efficacy guidance

- Best results from soil applications achieved by application in winter to moist soil during cool weather, particularly if rain follows soon after. Do not disturb treated soil by hoeing
- In aquatic situations apply as soon as active growth commences to reduce the likelihood of subsequent water deoxygenation caused by die-back of heavy weed stands. Later partial treatment for rooted species may be effective [1, 3, 5-9]
- Where mulching is practised, apply beforehand
- Lower rates control annuals, higher rates perennials, see label for details
- Residual activity in soil lasts 3-6 mth with selective, up to 12 mth with non-selective rates
- Do not use on fen peat or moss soils as efficacy will be severely impaired
- For control of emergent, floating and submerged aquatics apply to water surface in early spring. Intended for use in still or sluggish flowing water [1, 3, 5-9]
- Control in water is effective where flow does not exceed 90 m/hr (2.5 cm/second). Treatment of faster flowing water will only be effective if the flow can be checked for a minimum of 7 d after application [1, 3, 5-9]

Restrictions

- Maximum number of soil treatments 1 per yr in all situations
- Do not treat crops established for less than 2 yr. See label for lists of resistant and sensitive species. Do not treat stone fruit or Norway spruce (Christmas trees)
- Do not apply within 300 m of glasshouses or hops or near areas underplanted with bulbs, annuals or herbaceous stock
- Do not apply to frozen, snow-covered or waterlogged ground or when crop foliage wet
- Do not apply to sites less than 18 mth before replanting or sowing
- Store well away from corms, bulbs, tubers and seed

Crop-specific information

- Latest use: before end of spring for aquatic situations; 4 wk before budding for grapevines; before onset of spring growth for potato dumps and all other crops. See labels for precise details
- Apply to crops in dormant period. See label for details of timing and rates
- Apply evenly by hand applicator or suitable mechanical spreader
- Apply to potato dump sites before emergence of potatoes for blight prevention
- Aquatic treatments intended for use either in still waterbodies or in sluggishly flowing water courses [1, 3, 5-9]

Environmental safety

- Dangerous for the environment [1-3, 5, 7]
- Toxic to aquatic organisms
- If fish are known to be present in a water body intended for treatment for aquatic weed control, partial treatment should be adopted. See label for details [1, 3, 5-9]
- Do not dump surplus herbicide in water or ditch bottoms
- Use not permitted on reservoirs that form part of a water supply system [1, 3, 5-9]
- Do not use treated water for irrigation purposes within 2 wk of treatment or until concentration of dichlobenil falls below 0.3 ppm [1, 3, 5-9]
- Use as an aquatic herbicide subject to requirements set out in *Guidelines for the use of herbicides in or near watercourses or lakes* available from Defra [1, 3, 5-9]

Hazard classification and safety precautions

Hazard H03, H11 [1-3, 5, 7]
Risk phrases R21, R51 [1-3, 5, 7]; R52 [4, 6, 8, 9]; R53a [1-9]
Operator protection A, H [1-3, 5, 7]; U05a [1-3, 5, 7]; U20b [8, 9]; U20c [1-7]
Environmental protection E13c [1-3, 5-7]; E15a, E38 [4, 8, 9]; E19a [6, 8, 9]; E21 [6] (2 wk)
Storage and disposal D01, D02 [1-3, 5, 7]; D04 [8, 9]; D07 [4, 6, 8, 9]; D09a, D11a [1-9]
Medical advice M05b [1-3, 5, 7]

161 1,3-dichloropropene

A halogenated hydrocarbon soil nematicide

Products

Telone II	Dow	94% w/w	VP	05749

Uses

- Free-living nematodes in ***potatoes***
- Nematodes in ***hops, raspberries, strawberries***
- Potato cyst nematode in ***potatoes***
- Stem and bulb nematodes in ***narcissi***
- Stem nematodes in ***strawberries***
- Virus vectors in ***hops, raspberries, strawberries***

Approval information

- Dichloropropene will not be included in Annex I of EC Directive 91/414. All approvals for the storage and use of all products listed in this profile will expire by March 2009 unless an earlier date has been set
- Accepted by BBPA for use on hops

Efficacy guidance

- Before treatment soil should be in friable seed bed condition above 5 °C, with adequate moisture and all crop remains decomposed. Tilling to 30 cm improves results

- Apply with sub-surface A-blade injector combined with soil-sealing roller or, in hops, with a hollow tined injector. See label for details of suitable machinery
- Leave soil undisturbed for 21 d after treatment. Wet or cold soils need longer exposure
- For control of potato-cyst nematode contact firm's representative, check nematode level and use in integrated control programme
- For use in hops apply in May or Jun following the yr of grubbing and leave as long as possible before replanting
- Do not use water to clean out apparatus nor use aluminium or magnesium alloy containers which may corrode

Restrictions

- Maximum number of treatments 2 per yr for hops, 1 per yr for other crops
- Do not drill or plant until odour of fumigant is eliminated
- Do not use fertilizer containing ammonium salts after treatment. Use only nitrate nitrogen fertilizers until crop well established and soil temperature above 18 °C
- Do not use on extremely heavy clays or soils with many large stones

Crop-specific information

- Latest use: 6 wk before planting for raspberries, strawberries; pre-planting of crop for hops, narcissi, potatoes

Following crops guidance

- Allow at least 6 wk after treatment before planting raspberries or strawberries

Environmental safety

- Dangerous for the environment
- Toxic to aquatic organisms
- Keep in original container, tightly closed, in a safe place, under lock and key

Hazard classification and safety precautions

Hazard H02, H08, H11
Risk phrases R20, R21, R22b, R24, R25, R37, R38, R43, R51, R53a
Operator protection A, C, H, M; U02a, U04a, U05a, U09a, U14, U15, U19a, U20a
Environmental protection E15a, E34, E38
Storage and disposal D01, D02, D03, D06b, D09b, D12a, D14
Medical advice M04a, M05b

162 dichlorprop-P

A translocated phenoxy carboxylic acid herbicide for use in cereals
HRAC mode of action code: O

See also *2,4-D + dichlorprop-P + MCPA + mecoprop-P*
dicamba + dichlorprop-P + ferrous sulphate + MCPA
dicamba + dichlorprop-P + MCPA

Products

Headland Link	Headland	500 g/l	SL	11091

Uses

- Annual dicotyledons in ***permanent grassland, rotational grass, spring barley, spring oats, spring wheat, undersown spring cereals, undersown winter cereals, winter barley, winter oats, winter wheat***
- Charlock in ***permanent grassland, rotational grass, spring barley, spring oats, spring wheat, undersown spring cereals, undersown winter cereals, winter barley, winter oats, winter wheat***
- Chickweed in ***permanent grassland, rotational grass, spring barley, spring oats, spring wheat, undersown spring cereals, undersown winter cereals, winter barley, winter oats, winter wheat***
- Fat hen in ***permanent grassland, rotational grass, spring barley, spring oats, spring wheat, undersown spring cereals, undersown winter cereals, winter barley, winter oats, winter wheat***

- Perennial dicotyledons in ***permanent grassland***, ***rotational grass***, ***spring barley***, ***spring oats***, ***spring wheat***, ***undersown spring cereals***, ***undersown winter cereals***, ***winter barley***, ***winter oats***, ***winter wheat***
- Redshank in ***permanent grassland***, ***rotational grass***, ***spring barley***, ***spring oats***, ***spring wheat***, ***undersown spring cereals***, ***undersown winter cereals***, ***winter barley***, ***winter oats***, ***winter wheat***

Approval information

- Dichlorprop-P included in Annex I under EC Directive 91/414
- Accepted by BBPA for use on malting barley

Efficacy guidance

- Best results achieved by application to young seedling weeds in good growing conditions in a strongly competing crop
- Where cleavers are to be controlled increase water volume if crop or weed is dense
- Control of redshank is best achieved in warm temperatures even if this means allowing the weed to grow beyond the seedling stage

Restrictions

- Maximum number of treatments 1 per crop
- Do not apply during cold weather, drought, rain or when rain is expected
- Do not roll or harrow within 7 d before or after spraying
- Do not apply immediately before sowing any crop

Crop-specific information

- Latest use: before second node detectable (GS 32)
- Spray winter cereals in the spring from the leaf sheath erect stage; spray spring cereals from the one leaf stage

Following crops guidance

- Keep livestock out of treated areas until foliage of any poisonous weeds such as ragwort has died and become unpalatable

Environmental safety

- Harmful to fish or other aquatic life. Do not contaminate surface waters or ditches with chemical or used container
- Keep livestock out of treated areas for at least 2 wk and until foliage of any poisonous weeds, such as ragwort, has died and become unpalatable
- Do not spray in windy conditions when drift may cause damage to neighbouring sensitive crops

Hazard classification and safety precautions

Hazard H03, H04
Risk phrases R22a, R41, R43
Operator protection A, C, H; U05a, U08, U20b
Environmental protection E07a, E13c, E34
Storage and disposal D01, D02, D05, D09a, D10a
Medical advice M03

163 dichlorprop-P + ferrous sulphate + MCPA

A herbicide/fertilizer combination for moss and weed control in turf
HRAC mode of action code: O + O

Products

1	SHL Granular Feed, Weed & Mosskiller	Sinclair	0.2:10.9:0.3% w/w	GR	10972
2	SHL Turf Feed, Weed & Mosskiller	Sinclair	0.2:10.9:0.3% w/w	DP	10973

Uses

- Annual dicotyledons in ***managed amenity turf***
- Buttercups in ***managed amenity turf***
- Moss in ***managed amenity turf***
- Perennial dicotyledons in ***managed amenity turf***

Approval information
- Dichlorprop-P and MCPA included in Annex I under EC Directive 91/414

Efficacy guidance
- Apply between Mar and Sep when grass in active growth and soil moist
- A repeat treatment may be needed after 4-6 wk to control perennial weeds or if moss regrows
- Water in if rainfall does not occur within 48 h [2]

Restrictions
- Do not treat newly sown grass for 6 mth after establishment [1]
- Do not apply during drought or freezing conditions or when rain imminent
- Avoid walking on treated areas until it has rained or turf has been watered
- Do not mow for 3-4 d before or after application
- Do not use first 4 mowings after treatment for mulching. Mowings should be composted for 6 mth before use
- Do not treat areas of fine turf such as golf or bowling greens [1]
- Avoid contact with tarmac surfaces as staining may occur

Environmental safety
- Avoid drift onto nearby plants and borders [1]

Hazard classification and safety precautions

Operator protection U20c

Storage and disposal D01, D09a, D11a

164 dichlorprop-P + MCPA

A translocated herbicide for use in managed amenity turf
HRAC mode of action code: O + O

Products

1	SHL Granular Feed and Weed	Sinclair	0.2:0.3% w/w	GR	10970
2	SHL Turf Feed and Weed	Sinclair	0.2:0.3% w/w	DP	10963

Uses
- Annual dicotyledons in ***managed amenity turf*** [1, 2]
- Buttercups in ***managed amenity turf*** [1, 2]
- Clovers in ***managed amenity turf*** [2]
- Daisies in ***managed amenity turf*** [2]
- Dandelions in ***managed amenity turf*** [2]
- Perennial dicotyledons in ***managed amenity turf*** [1, 2]

Approval information
- Dichlorprop-P and MCPA included in Annex I under EC Directive 91/414

Efficacy guidance
- Best results achieved by application to young seedling weeds in active growth between Mar and Sep
- Water in if rainfall does not occur within 48 h [2]

Restrictions
- Do not use on newly sown turf for 6 mth after establishment
- Do not cut or mow for 2-3 d before and after treatment
- Do not spray during cold weather, if rain or frost expected, if crop wet or in drought
- The first four mowings should not be used for mulching and should be composted for 6 mth before use
- Avoid close mowing for 3-4 d before or after treatment
- Do not treat areas of fine turf such as golf or bowling greens [1]

Environmental safety
- Avoid drift onto nearby plants and borders [2]
- Avoid contact with tarmac surfaces as staining may occur

Hazard classification and safety precautions

Operator protection U20c

Storage and disposal D01, D09a, D11a

SEE SECTION 3 FOR PRODUCTS ALSO REGISTERED

165 dichlorprop-P + MCPA + mecoprop-P

A translocated herbicide mixture for winter and spring cereals
HRAC mode of action code: O + O + O

Products

Hymec Triple	Agrichem	310:160:130 g/l	SL	09949

Uses

- Annual dicotyledons in ***durum wheat, grass seed crops*** *(off-label)*, ***spring barley, spring oats, spring wheat, winter barley, winter oats, winter wheat***
- Chickweed in ***durum wheat, spring barley, spring oats, spring wheat, winter barley, winter oats, winter wheat***
- Cleavers in ***durum wheat, spring barley, spring oats, spring wheat, winter barley, winter oats, winter wheat***
- Field pansy in ***durum wheat, spring barley, spring oats, spring wheat, winter barley, winter oats, winter wheat***
- Mayweeds in ***durum wheat, spring barley, spring oats, spring wheat, winter barley, winter oats, winter wheat***
- Poppies in ***durum wheat, spring barley, spring oats, spring wheat, winter barley, winter oats, winter wheat***

Specific Off-Label Approvals (SOLAs)

- ***grass seed crops*** *20060918*

Approval information

- Dichlorprop-P, MCPA and mecoprop-P included in Annex I under EC Directive 91/414
- Accepted by BBPA for use on malting barley

Efficacy guidance

- Best results obtained if application is made while majority of weeds are at seedling stage but not if temperatures are too low
- Optimum results achieved by spraying when temperature is above 10 °C. If temperatures are lower delay spraying until growth becomes more active

Restrictions

- Maximum number of treatments 1 per crop
- Do not spray in windy conditions where spray drift may cause damage to neighbouring crops, especially sugar beet, oilseed rape, peas, turnips and most horticultural crops including lettuce and tomatoes under glass

Crop-specific information

- Latest use: before second node detectable (GS 32) for all crops

Environmental safety

- Harmful to aquatic organisms
- Harmful to fish or other aquatic life. Do not contaminate surface waters or ditches with chemical or used container

Hazard classification and safety precautions

Hazard H03
Risk phrases R22a, R41, R52, R53a
Operator protection A, C, H, M; U05a, U08, U11, U20b
Environmental protection E13c, E34
Storage and disposal D01, D02, D05, D09a, D10c
Medical advice M03

166 difenacoum

An anticoagulant coumarin rodenticide

Products

1	Neokil	Sorex	0.005% w/w	RB	H6769
2	Neosorexa	Sorex	0.005% w/w	RB	H6773
3	Neosorexa Gold	Sorex	0.005% w/w	RB	H8301
4	Sakarat D (Whole Wheat)	Killgerm	0.005% w/w	RB	H7109

Products – continued

5	Sakarat D Wax Bait	Killgerm	0.005% w/w	RB	H7489
6	Sorexa Gel	Sorex	0.005% w/w	RB	H6790

Uses
- Mice in ***farm buildings***, ***farmyards*** [1-6]
- Rats in ***farm buildings***, ***farmyards*** [1-4]

Efficacy guidance
- Difenacoum is a chronic poison and rodents need to feed several times before accumulating a lethal dose. Effective against rodents resistant to other commonly used anticoagulants
- Best results achieved by placing baits at points between nesting and feeding places, at entry points, in holes and where droppings are seen
- A minimum of five baiting points normally required for a small infestation; more than 40 for a large infestation
- Inspect bait sites frequently and top up as long as there is evidence of feeding
- Product formulated for application through a skeleton or caulking gun [6]
- Maintain a few baiting points to guard against reinfestation after a successful control campaign
- Products contain an insecticide and fungistat to prevent insect and mould attack [1, 2]

Restrictions
- Only for use by farmers, horticulturists and other professional users
- When working in rodent infested areas wear synthetic rubber/PVC gloves to protect against rodent-borne diseases

Environmental safety
- Harmful to wildlife
- Cover baits by placing in bait boxes, drain pipes or under boards to prevent access by children, animals or birds
- Products contain human taste deterrent

Hazard classification and safety precautions

Operator protection A [6]; U13, U20b

Environmental protection E10b [3]; E15a [6]

Storage and disposal D09a, D11a

Vertebrate/rodent control products V01a, V03a, V04a [1, 2, 6]; V01b, V04b [3, 5]; V02 [1-3, 5, 6]; V03b [5]; V04c [3]

Medical advice M03

167 difenoconazole

A diphenyl-ether triazole protectant and curative fungicide
FRAC mode of action code: 3

See also azoxystrobin + difenoconazole

Products

1	Difcor 250 EC	Nufarm UK	250 g/l	EC	12361
2	Plover	Syngenta	250 g/l	EC	11763

Uses
- Alternaria in ***broccoli***, ***brussels sprouts***, ***cabbages***, ***calabrese***, ***cauliflowers***, ***spring oilseed rape***, ***winter oilseed rape*** [1, 2]
- Brown rust in ***winter wheat*** [1, 2]
- Celery leaf spot in ***celeriac*** *(off-label)*, ***celery (outdoor)*** *(off-label)* [2]
- Foliar disease control in ***borage*** *(off-label)*, ***canary flower (echium spp.)*** *(off-label)*, ***durum wheat*** *(off-label)*, ***evening primrose*** *(off-label)*, ***grass seed crops*** *(off-label)*, ***honesty*** *(off-label)*, ***linseed*** *(off-label)*, ***mustard*** *(off-label)*, ***spring rye*** *(off-label)*, ***triticale*** *(off-label)*, ***winter rye*** *(off-label)* [2]
- Light leaf spot in ***spring oilseed rape***, ***winter oilseed rape*** [1, 2]
- Ring spot in ***broccoli***, ***brussels sprouts***, ***cabbages***, ***calabrese***, ***cauliflowers***, ***chinese cabbage*** *(off-label)*, ***choi sum*** *(off-label)*, ***collards*** *(off-label)*, ***kale*** *(off-label)*, ***pak choi*** *(off-label)* [1, 2]; ***protected pinks*** *(off-label - ground grown (or pot grown - SOLA 20050159))*, ***protected sweet williams*** *(off-label - ground grown (or pot grown - SOLA 20050159))* [2]; ***protected pinks*** *(off-label*

- ground grown (or pot grown - SOLA 20051488)), ***protected sweet williams*** *(off-label - ground grown (or pot grown - SOLA 20051488))* [1]
- Rust in ***asparagus*** *(off-label)* [1, 2]; ***protected pinks*** *(off-label - ground grown (or pot grown - SOLA 20050159)),* ***protected sweet williams*** *(off-label - ground grown (or pot grown - SOLA 20050159))* [2]; ***protected pinks*** *(off-label - ground grown (or pot grown - SOLA 20051488)),* ***protected sweet williams*** *(off-label - ground grown (or pot grown - SOLA 20051488))* [1]
- Septoria leaf blotch in ***winter wheat*** [1, 2]
- Stem canker in ***spring oilseed rape***, ***winter oilseed rape*** [1, 2]

Specific Off-Label Approvals (SOLAs)

- ***asparagus*** *20051493* [1], *20050158* [2]
- ***borage*** *20060559* [2]
- ***canary flower (echium spp.)*** *20060559* [2]
- ***celeriac*** *20060261* [2]
- ***celery (outdoor)*** *20060260 expires Dec 2008* [2]
- ***chinese cabbage*** *20051490 expires Mar 2008* [1], *20050558* [2]
- ***choi sum*** *20051490 expires Mar 2008* [1], *20050558* [2]
- ***collards*** *20051490 expires Mar 2008* [1], *20050558* [2]
- ***durum wheat*** *20060558* [2]
- ***evening primrose*** *20060559* [2]
- ***grass seed crops*** *20060558* [2]
- ***honesty*** *20060559* [2]
- ***kale*** *20051492* [1], *20050559* [2]
- ***linseed*** *20060559* [2]
- ***mustard*** *20060559* [2]
- ***pak choi*** *20051490 expires Mar 2008* [1], *20050558* [2]
- ***protected pinks*** *(ground grown (or pot grown - SOLA 20051488)) 20051486* [1], *(ground grown (or pot grown - SOLA 20050159)) 20050156* [2]
- ***protected sweet williams*** *(ground grown (or pot grown - SOLA 20051488)) 20051486* [1], *(ground grown (or pot grown - SOLA 20050159)) 20050156* [2]
- ***spring rye*** *20060558* [2]
- ***triticale*** *20060558* [2]
- ***winter rye*** *20060558* [2]

Efficacy guidance

- Adequate control of yellow rust may require an earlier appropriate treatment
- Improved control of established infections on oilseed rape achieved by mixture with carbendazim. See label
- In brassicas a 3-spray programme should be used starting at the first sign of disease and repeated at 14-21 d intervals
- Product is fully rainfast 2 h after application
- For most effective control of Septoria, apply as part of a programme of sprays which includes a suitable flag leaf treatment
- Difenoconazole is a DMI fungicide. Resistance to some DMI fungicides has been identified in Septoria leaf blotch which may seriously affect performance of some products. For further advice contact a specialist advisor and visit the Fungicide Resistance Action Group (FRAG)-UK website

Restrictions

- Maximum number of treatments 3 per crop for brassicas; 2 per crop for oilseed rape; 1 per crop for wheat [1]
- Maximum total dose equivalent to 3 full dose treatments on brassicas; 2 full dose treatments on oilseed rape; 1 full dose treatment on wheat [2]
- Apply to wheat any time from ear fully emerged stage but before early milk-ripe stage (GS 59-73)

Crop-specific information

- Latest use: before grain early milk-ripe stage (GS 73) for cereals; end of flowering for oilseed rape
- HI brassicas 21 d
- Recommended for use on all varieties of winter and spring sown oilseed rape and cabbage
- Treat oilseed rape in autumn from 4 expanded true leaf stage (GS 1,4). A repeat spray may be made in spring at the beginning of stem extension (GS 2,0) if visible symptoms develop

Environmental safety

- Dangerous for the environment

- Very toxic to aquatic organisms
- Product supplied in returnable refillable containers. Follow label instructions

Hazard classification and safety precautions

Hazard H04 [1]; H11 [1, 2]
Risk phrases R38, R41, R51 [1]; R50 [2]; R53a [1, 2]
Operator protection A, C, H; U05a, U09a, U20b [1, 2]; U11 [1]
Environmental protection E15a, E38
Storage and disposal D01, D02, D09a, D12a [1, 2]; D05, D07, D10c [2]; D10b [1]

168 difenoconazole + fenpropidin

A triazole morpholine fungicide mixture for beet crops
FRAC mode of action code: 3 + 5

Products

Spyrale	Syngenta	375:100 g/l	EC	12566

Uses

- Powdery mildew in ***fodder beet***, ***sugar beet***
- Ramularia leaf spots in ***fodder beet***, ***sugar beet***
- Rust in ***fodder beet***, ***sugar beet***

Efficacy guidance

- For best results apply as a preventative treatment or as soon as first symptoms of disease are seen
- Product gives prolonged protection against re-infection but a second application may be needed where crops remain under heavy disease pressure

Restrictions

- Maximum number of treatments 2 per crop of sugar beet or fodder beet

Crop-specific information

- HI: 28 d for fodder beet, sugar beet

Environmental safety

- Dangerous for the environment
- Very toxic to aquatic organisms

Hazard classification and safety precautions

Hazard H03, H11
Risk phrases R22a, R36, R37, R38, R50, R53a
Operator protection A, C, H; U02a, U04a, U05a, U09a, U10, U11, U14, U15, U20a
Environmental protection E15a, E34, E38
Storage and disposal D01, D02, D05, D07, D09a, D10c, D12a
Medical advice M03

169 diflubenzuron

A selective, persistent, contact and stomach acting insecticide
IRAC mode of action code: 15

Products

1	Dimilin Flo	Chemtura	480 g/l	SC	08769
2	Dimilin Flo	Certis	480 g/l	SC	11056

Uses

- Browntail moth in ***amenity vegetation***, ***hedges***, ***nursery stock***, ***ornamental plant production*** [2]
- Bud moth in ***apples***, ***pears*** [2]
- Carnation tortrix moth in ***amenity vegetation***, ***hedges***, ***nursery stock***, ***ornamental plant production*** [2]
- Caterpillars in ***broccoli***, ***brussels sprouts***, ***cabbages***, ***calabrese***, ***cauliflowers***, ***chives*** *(off-label)*, ***endives*** *(off-label)*, ***frise*** *(off-label)*, ***herbs (see appendix 6)*** *(off-label)*, ***leaf spinach*** *(off-label)*,

lettuce *(off-label)*, ***parsley*** *(off-label)*, ***radicchio*** *(off-label)*, ***salad brassicas*** *(off-label - for baby leaf production)*, ***scarole*** *(off-label)* [2]

- Clouded drab moth in ***apples, pears*** [2]
- Codling moth in ***apples, pears*** [2]
- Fruit tree tortrix moth in ***apples, pears*** [2]
- Houseflies in ***livestock houses, manure heaps, refuse tips*** [2]
- Lackey moth in ***amenity vegetation, hedges, nursery stock, ornamental plant production*** [2]
- Oak leaf roller moth in ***forest*** [2]
- Pear sucker in ***pears*** [2]
- Pine beauty moth in ***forest*** [2]
- Pine looper in ***forest*** [2]
- Plum fruit moth in ***plums*** [2]
- Rust mite in ***apples, pears, plums*** [2]
- Sciarid flies in ***mushrooms*** [1]
- Small ermine moth in ***amenity vegetation, hedges, nursery stock, ornamental plant production*** [2]
- Tortrix moths in ***plums*** [2]
- Winter moth in ***amenity vegetation, apples, blackcurrants, forest, hedges, nursery stock, ornamental plant production, pears, plums*** [2]

Specific Off-Label Approvals (SOLAs)

- ***chives*** *20051321* [2]
- ***endives*** *20051321* [2]
- ***frise*** *20051321* [2]
- ***herbs (see appendix 6)*** *20051321* [2]
- ***leaf spinach*** *20051321* [2]
- ***lettuce*** *20051321* [2]
- ***parsley*** *20051321* [2]
- ***radicchio*** *20051321* [2]
- ***salad brassicas*** *(for baby leaf production) 20051321* [2]
- ***scarole*** *20051321* [2]

Approval information

- Approved for aerial application in forestry when average wind velocity does not exceed 18 knots and gusts do not exceed 20 knots [2]. See notes in Section 5

Efficacy guidance

- Most active on young caterpillars and most effective control achieved by spraying as eggs start to hatch
- Dose and timing of spray treatments vary with pest and crop. See label for details
- Addition of wetter recommended for use on brassicas and for pear sucker control in pears

Restrictions

- Maximum number of treatments 3 per yr for apples, pears; 2 per yr for plums, blackcurrants; 2 per crop for brassicas; 1 per yr for forest [2]
- Maximum number of treatments 1 per spawning for mushrooms [1]
- Before treating ornamentals check varietal tolerance on a small sample
- Do not use as a compost drench or incorporated treatment on ornamental crops
- Do not spray protected plants in flower or with flower buds showing colour
- For use only on the food crops specified on the label
- Do not apply directly to livestock/poultry

Crop-specific information

- HI apples, pears, plums, blackcurrants, brassicas 14 d
- Apply as casing mixing treatment (use immediately) on mushrooms or as post-casing drench [1]

Environmental safety

- Dangerous for the environment
- Very toxic to aquatic organisms
- LERAP Category B
- Broadcast air-assisted LERAP (20 m in orchards, 10 m in blackcurrants, forestry or ornamentals)

Hazard classification and safety precautions

Hazard H11

Risk phrases R50
Operator protection U20c
Environmental protection E05a, E16a, E16b, E18 [2]; E15a, E38 [1, 2]; E17b [2] (20 m when used in orchards; 10 m when used in blackcurrants, forestry or ornamentals)
Storage and disposal D05, D12b [2]; D09a, D11a [1, 2]

170 diflufenican

A shoot absorbed pyridinecarboxamide herbicide for winter cereals
HRAC mode of action code: F1

See also *bromoxynil + diflufenican + ioxynil*
chlorotoluron + diflufenican
clopyralid + diflufenican + MCPA

Products

1	Diflanil 500 SC	Q-Chem	500 g/l	SC	12489
2	Diflufenican GL 500	Globachem	500 g/l	SC	12531
3	Hurricane SC	Makhteshim	500 g/l	SC	12424
4	Pelican	Headland	500 g/l	SC	12569
5	Sipcam Solo-D	Sipcam	500 g/l	SC	12617

Uses

- Annual dicotyledons in ***durum wheat***, ***triticale***, ***winter rye*** [1, 2]; ***winter barley***, ***winter wheat*** [1, 2, 4, 5]
- Chickweed in ***durum wheat***, ***triticale***, ***winter rye*** [1, 2]; ***winter barley***, ***winter wheat*** [1, 2, 4, 5]
- Cleavers in ***durum wheat***, ***triticale***, ***winter rye*** [1-3]; ***winter barley***, ***winter wheat*** [1-5]
- Field pansy in ***durum wheat***, ***triticale***, ***winter barley***, ***winter rye***, ***winter wheat*** [3]
- Field speedwell in ***durum wheat***, ***triticale***, ***winter rye*** [1-3]; ***winter barley***, ***winter wheat*** [1-5]
- Ivy-leaved speedwell in ***durum wheat***, ***triticale***, ***winter rye*** [1-3]; ***winter barley***, ***winter wheat*** [1-5]
- Mayweeds in ***durum wheat***, ***triticale***, ***winter barley***, ***winter rye***, ***winter wheat*** [3]
- Poppies in ***durum wheat***, ***triticale***, ***winter barley***, ***winter rye***, ***winter wheat*** [3]
- Red dead-nettle in ***durum wheat***, ***triticale***, ***winter barley***, ***winter rye***, ***winter wheat*** [3]
- Volunteer oilseed rape in ***durum wheat***, ***triticale***, ***winter rye*** [1, 2]; ***winter barley***, ***winter wheat*** [1, 2, 4, 5]

Approval information

- Accepted by BBPA for use on malting barley

Efficacy guidance

- Best results achieved from treatment of small actively growing weeds in early autumn or spring
- Good weed control depends on efficient burial of trash or straw before or during seedbed preparation
- Loose or fluffy seedbeds should be rolled before application
- The final seedbed should be moist, fine and firm with clods no bigger than fist size
- Ensure good even spray coverage and increase spray volume for post-emergence treatments where the crop or weed foliage is dense
- Activity may be slow under cool conditions and final level of weed control may take some time to appear
- Where cleavers are a particular problem a separate specific herbicide treatment may be required
- Efficacy may be impaired on soils with a Kd factor greater than 6
- Always follow WRAG guidelines for preventing and managing herbicide resistant weeds. Section 5 for more information

Restrictions

- Maximum number of treatments 1 per crop
- Do not treat broadcast crops [1-3]
- Do not use on crops being grown for seed [4]
- Do not roll treated crops or harrow at any time after treatment
- Do apply to soils with more than 10% organic matter or on Sands, or very stony or gravelly soils
- Do not treat after a period of cold frosty weather
- Apply before end Feb in yr of harvest [4]

Crop-specific information

- Latest use: before end of tillering (GS 29) for wheat and barley [1-3], end Feb in yr of harvest for wheat and barley [4]; pre crop emergence for triticale and winter rye [1-3]
- Treat only named varieties of rye or triticale [1-3]

Following crops guidance

- Labels vary slightly but in general ploughing to 150 mm and thoroughly mixing the soil is recommended before drilling or planting any succeeding crops either after crop failure or after normal harvest
- In the event of crop failure only winter wheat or winter barley may be re-drilled immediately after ploughing. Spring crops of wheat, barley, oilseed rape, peas, field beans, sugar beet [only1, 3], potatoes, carrots, edible brassicas or onions may be sown provided an interval of 12 wk has elapsed after ploughing
- After normal harvest of a treated crop winter cereals, oilseed rape, field beans, leaf brassicas, sugar beet seed crops and winter onions may be drilled in the following autumn. Other crops listed above for crop failure may be sown in the spring after normal harvest
- Successive treatments with any products containing diflufenican can lead to soil build-up and inversion ploughing to 150 mm must precede sowing any following non-cereal crop. Even where ploughing occurs some crops may be damaged

Environmental safety

- Dangerous for the environment
- Very toxic to aquatic organisms
- LERAP Category B

Hazard classification and safety precautions

Hazard H11
Risk phrases R50, R53a
Operator protection A [1-4]; C [1-3]; U05a [3-5]; U08, U19a [3, 4]; U13, U20a [3]; U20c [4]
Environmental protection E13c, E34 [3]; E15a [1, 2, 5]; E16a [1-5]; E38 [3, 4]
Storage and disposal D01, D02 [3-5]; D05, D09a, D10b [1-5]; D12a [1, 2, 4, 5]
Medical advice M03 [4]

171 diflufenican + flufenacet

A contact and residual herbicide mixture for cereals
HRAC mode of action code: F1 + K3

Products

1	Firebird	Bayer CropScience	200:400 g/l	SC	12421
2	Liberator	Bayer CropScience	100:400 g/l	SC	12032
3	Regatta	Bayer CropScience	100:400 g/l	SC	12054

Uses

- Annual dicotyledons in ***durum wheat*** *(off-label)*, ***triticale*** *(off-label)*, ***winter rye*** *(off-label)* [2]; ***grass seed crops*** *(off-label)* [2, 3]
- Annual grasses in ***durum wheat*** *(off-label)*, ***triticale*** *(off-label)*, ***winter rye*** *(off-label)* [2]; ***grass seed crops*** *(off-label)* [2, 3]
- Annual meadow grass in ***winter barley***, ***winter wheat*** [1-3]
- Blackgrass in ***winter barley***, ***winter wheat*** [1-3]
- Chickweed in ***winter barley***, ***winter wheat*** [1-3]
- Field pansy in ***winter barley***, ***winter wheat*** [1-3]
- Field speedwell in ***winter barley***, ***winter wheat*** [1-3]
- Ivy-leaved speedwell in ***winter barley***, ***winter wheat*** [1-3]
- Mayweeds in ***winter barley***, ***winter wheat*** [1-3]
- Red dead-nettle in ***winter barley***, ***winter wheat*** [2, 3]
- Shepherd's purse in ***winter barley***, ***winter wheat*** [1]
- Volunteer oilseed rape in ***winter barley***, ***winter wheat*** [1]

Specific Off-Label Approvals (SOLAs)

- ***durum wheat*** *20063650* [2]
- ***grass seed crops*** *20063650* [2], *20061029* [3]

- ***triticale*** *20063650* [2]
- ***winter rye*** *20063650* [2]

Approval information

- Flufenacet included in Annex I under EC Directive 91/414
- Accepted by BBPA for use on malting barley

Efficacy guidance

- Best results obtained when there is moist soil at and after application and rain falls within 7 d
- Residual control may be reduced under prolonged dry conditions
- Activity may be slow under cool conditions and final level of weed control may take some time to appear
- Good weed control depends on burying any trash or straw before or during seedbed preparation
- Established perennial grasses and broad-leaved weeds will not be controlled
- Do not use as a stand-alone treatment for blackgrass control. Always follow WRAG guidelines for preventing and managing herbicide resistant weeds. Section 5 for more information

Restrictions

- Maximum number of treatments 1 per crop
- Do not treat undersown cereals or those to be undersown
- Do not use on waterlogged soils or soils prone to waterlogging
- Do not use on Sands or Very Light soils, or very stony or gravelly soils, or on soils containing more than 10% organic matter
- Do not treat broadcast crops and treat shallow-drilled crops post-emergence only
- Do not incorporate into the soil or disturb the soil after application by rolling or harrowing
- Avoid treating crops under stress from whatever cause and avoid treating during periods of prolonged or severe frosts

Crop-specific information

- Latest use: before 31 Dec in yr of sowing and before 3rd tiller stage (GS 23) for wheat or 4th tiller stage (GS 24) for barley
- For pre-emergence treatments the seed should be covered with a minimum of 32 mm settled soil

Following crops guidance

- In the event of crop failure wheat, barley or potatoes may be sown provided the soil is ploughed to 15 cm, and a minimum of 12 weeks elapse between treatment and sowing spring wheat or spring barley
- After normal harvest wheat, barley or potatoes my be sown without special cultivations. Soil must be ploughed or cultivated to 15 cm before sowing oilseed rape, field beans, peas, sugar beet, carrots, onions or edible brassicae
- Successive treatments with any products containing diflufenican can lead to soil build-up and inversion ploughing to 150 mm must precede sowing any following non-cereal crop. Even where ploughing occurs some crops may be damaged

Environmental safety

- Dangerous for the environment
- Very toxic to aquatic organisms
- Risk to non-target insects or other arthropods. Avoid spraying within 6 m of the field boundary to reduce the effects on non-target insects or other arthropods
- LERAP Category B

Hazard classification and safety precautions

Hazard H03, H11
Risk phrases R22a, R43, R48, R50, R53a
Operator protection A, H; U05a, U14
Environmental protection E15a [1, 3]; E16a, E34, E38 [1-3]; E22c [1, 2]
Storage and disposal D01, D02, D09a, D10b, D12a
Medical advice M03, M05a

SEE SECTION 3 FOR PRODUCTS ALSO REGISTERED

172 diflufenican + flupyrsulfuron-methyl

A pyridinecarboxamide and sulfonylurea herbicide mixture for wheat
HRAC mode of action code: F1 + B

Products

Absolute	DuPont	41.7:8.3% w/w	WG	12558

Uses

- Annual dicotyledons in ***winter wheat***
- Blackgrass in ***winter wheat***

Approval information

- Flupyrsulfuron-methyl included in Annex I under EC Directive 91/414

Efficacy guidance

- Best results achieved from applications made in good growing conditions
- Good spray cover of weeds must be obtained
- Best control of blackgrass obtained from application from the 2-leaf stage but tank mixing or sequencing with a partner product is recommended as a resistance management strategy (see below)
- Growth of weeds is inhibited within hours of treatment but visible symptoms may not be apparent for up to 4 wk
- Product has moderate residual life in soil. Under normal moisture conditions susceptible weeds germinating soon after treatment will be controlled
- Product may be used on all soil types except those with more than 10% organic matter, but residual activity and weed control is reduced by high soil temperatures and when soil is very dry and on highly alkaline soils
- Flupyrsulfuron-methyl is a member of the ALS-inhibitor group of herbicides. To avoid the build up of resistance do not use any product containing an ALS-inhibitor herbicide with claims for control of grass weeds more than once on any crop
- Use these products as part of a resistance management strategy that includes cultural methods of control and does not use ALS inhibitors as the sole chemical method of grass weed control

Restrictions

- Maximum number of treatments 1 per crop
- Do not use on wheat undersown with grasses, clover or other legumes, or any other broad-leaved crop
- Do not apply within 7 d of rolling
- Do not treat broadcast crops. Seed should be covered by at least 25 mm of settled soil
- Do not treat any crop suffering from drought, waterlogging, pest or disease attack, nutrient deficiency, or any other stress factors
- Do not use on soils with more than 10% organic matter
- Specific restrictions apply to use in sequence or tank mixture with other sulfonylurea or ALS-inhibiting herbicides. See label for details

Crop-specific information

- Latest use: before 1st node detectable stage (GS 31) for winter wheat
- Under certain conditions slight chlorosis and stunting may occur from which recovery is rapid

Following crops guidance

- Only cereals, oilseed rape or field beans may be sown in the yr of harvest of a treated crop. Ploughing must be carried out before drilling any non-cereal crop
- An interval of at least 20 wk must elapse after treatment before planting wheat, barley, oilseed rape, peas, field beans, sugar beet, potatoes, carrots, brassicas or onions in the following spring, and only after ploughing
- In the event of failure of a treated crop only winter wheat or spring wheat may be sown within 3 mth of treatment, and only after ploughing and cultivating to at least 15 cm. After 3 mth the normal guidance for following crops may be observed
- Successive treatments with any products containing diflufenican can lead to soil build-up and inversion ploughing to 150 mm must precede sowing any following non-cereal crop. Even where ploughing occurs some crops may be damaged

Environmental safety
- Dangerous for the environment
- Very toxic to aquatic organisms
- LERAP Category B
- Take extreme care to avoid damage by drift onto broad-leaved plants outside the target area
- Do not drain or flush spraying equipment onto land planted, or intended for planting, with trees or crops other than cereals
- Poor sprayer cleanout practices may result in inadequate removal of chemical deposits which may lead to damage of subsequently treated non-cereal crops. See label for guidance on cleaning

Hazard classification and safety precautions

Hazard H11
Risk phrases R50, R53a
Operator protection U08, U19a, U20b
Environmental protection E15a, E16a, E38
Storage and disposal D09a, D12a

173 diflufenican + flurtamone

A contact and residual herbicide mixture for cereals
HRAC mode of action code: F1 + F1

Products

1	Bacara	Bayer CropScience	100:250 g/l	SC	10744
2	Graduate	Bayer CropScience	80:400 g/l	SC	10776

Uses
- Annual dicotyledons in ***grass seed crops*** *(off-label)*, ***spring rye*** *(off-label)*, ***triticale*** *(off-label)*, ***winter barley***, ***winter rye*** *(off-label)*, ***winter wheat*** [1, 2]
- Annual grasses in ***grass seed crops*** *(off-label)*, ***spring rye*** *(off-label)*, ***triticale*** *(off-label)*, ***winter rye*** *(off-label)* [1, 2]
- Annual meadow grass in ***winter barley***, ***winter wheat*** [1, 2]
- Blackgrass in ***winter barley***, ***winter wheat*** [1]
- Loose silky bent in ***winter barley***, ***winter wheat*** [1]
- Volunteer oilseed rape in ***winter barley***, ***winter wheat*** [1, 2]

Specific Off-Label Approvals (SOLAs)
- ***grass seed crops*** *20060813* [1], *20060812* [2]
- ***spring rye*** *20060813* [1], *20060812* [2]
- ***triticale*** *20060813* [1], *20060812* [2]
- ***winter rye*** *20060813* [1], *20060812* [2]

Approval information
- Flurtamone included in Annex I under EC Directive 91/414
- Accepted by BBPA for use on malting barley

Efficacy guidance
- Apply pre-emergence or from when crop has first leaf unfolded before susceptible weeds pass recommended size
- Best results obtained on firm, fine seedbeds with adequate soil moisture present at and after application. Increase water volume where crop or weed foliage is dense
- Good weed control requires ash, trash and burnt straw to be buried during seed bed preparation
- Loose fluffy seedbeds should be rolled before application and the final seed bed should be fine and firm without large clods
- Speed of control depends on weather conditions and activity can be slow under cool conditions
- Always follow WRAG guidelines for preventing and managing herbicide resistant weeds. Section 5 for more information

Restrictions
- Maximum number of treatments 1 per crop
- Crops should be drilled to a normal depth of 25 mm and the seed well covered. Do not treat broadcast crops as uncovered seed may be damaged
- Do not treat spring sown cereals, durum wheat, oats, undersown cereals or those to be undersown

- Do not treat frosted crops or when frost is imminent. Severe frost after application, or any other stress, may lead to transient discoloration or scorch
- Do not use on Sands or Very Light soils or those that are very stony or gravelly
- Do not use on waterlogged soils, or on crops subject to temporary waterlogging by heavy rainfall, as there is risk of persistent crop damage which may result in yield loss
- Do not use on soils with more than 10% organic matter
- Do not harrow at any time after application and do not roll autumn treated crops until spring

Crop-specific information

- Latest use: before 2nd node detectable (GS32)
- Take particular care to match spray swaths otherwise crop discoloration and biomass reduction may occur which may lead to yield reduction

Following crops guidance

- In the event of crop failure winter wheat may be redrilled immediately after normal cultivation, and winter barley may be sown after ploughing. Fields must be ploughed to a depth of 15 cm and 20 wk must elapse before sowing spring crops of wheat, barley, oilseed rape, peas, field beans or potatoes
- After normal harvest autumn cereals can be drilled after ploughing. Thorough mixing of the soil must take place before drilling field beans, leaf brassicae or winter oilseed rape. For sugar beet seed crops and winter onions complete inversion of the furrow slice is essential
- Do not broadcast or direct drill oilseed rape or other brassica crops as a following crop on treated land. See label for detailed advice on preparing land for subsequent autumn cropping in the normal rotation
- Successive treatments with any products containing diflufenican can lead to soil build-up and inversion ploughing to 150 mm must precede sowing any following non-cereal crop. Even where ploughing occurs some crops may be damaged

Environmental safety

- Dangerous for the environment
- Very toxic to aquatic organisms
- Extremely dangerous to fish or other aquatic life. Do not contaminate surface waters or ditches with chemical or used container
- LERAP Category B

Hazard classification and safety precautions

Hazard H11
Risk phrases R50, R53a
Operator protection A, H; U20c
Environmental protection E13a, E16a, E16b, E38
Storage and disposal D05, D09a, D10b, D12a

174 diflufenican + glyphosate

A foliar non-selective herbicide mixture
HRAC mode of action code: F1 + G

Products

Pistol	Bayer Environ.	40:250 g/l	SC	12173

Uses

- Annual and perennial weeds in ***natural surfaces not intended to bear vegetation***, ***permeable surfaces overlying soil***

Approval information

- Glyphosate included in Annex I under EC Directive 91/414

Efficacy guidance

- Treat when weeds actively growing from Mar to end Sep and before they begin to senesce
- Performance may be reduced if application is made to plants growing under stress, such as drought or water-logging
- Pre-emergence activity is reduced on soils containing more than 10% organic matter or where organic debris has collected

- Perennial weeds such as docks, perennial sowthistle and willowherb are best treated before flowering or setting seed
- Perennial weeds emerging from established rootstocks after treatment will not be controlled
- A rainfree period of at least 6 h (preferably 24 h) should follow spraying for optimum control
- For optimum control do not cultivate or rake after treatment

Restrictions
- Maximum number of treatments 1 per yr
- May only be used on porous surfaces overlying soil. Must not be used if an impermeable membrane lies between the porous surface and the soil, and must not be used on any non-porous man-made surfaces
- Do not add any wetting agent or adjuvant oil
- Do not spray in windy weather

Following crops guidance
- A period of at least 6 mth must be allowed after treatment of sites that are to be cleared and grubbed before sowing and planting. Soil should be ploughed or dug first to ensure thorough mixing and dilution of any herbicide residues

Environmental safety
- Dangerous for the environment
- Very toxic to aquatic organisms
- LERAP Category B
- Avoid drift onto non-target plants
- Heavy rain after application may wash product onto sensitive areas such as newly sown grass or areas about to be planted

Hazard classification and safety precautions

Hazard H11
Risk phrases R50, R53a
Operator protection A, C; U20b
Environmental protection E15a, E16a, E16b, E38
Storage and disposal D01, D02, D09a, D11a, D12a

175 diflufenican + iodosulfuron-methyl-sodium + mesosulfuron-methyl

A contact and residual herbicide mixture containing sulfonyl ureas for winter wheat
HRAC mode of action code: F1 + B + B

Products

Othello	Bayer CropScience	50:2.5:7.5 g/l	OD	12695

Uses
- Annual dicotyledons in ***winter wheat***
- Annual meadow grass in ***winter wheat***
- Cleavers in ***winter wheat***
- Mayweeds in ***winter wheat***
- Rough meadow grass in ***winter wheat***
- Volunteer oilseed rape in ***winter wheat***

Approval information
- Iodosulfuron-methyl-sodium and mesosulfuron-methyl included in Annex I under EC Directive 91/414

Efficacy guidance
- Optimum control obtained when all weeds are emerged at spraying. Activity is primarily via foliar uptake and good spray coverage of the target weeds is essential
- Translocation occurs readily within the target weeds and growth is inhibited within hours of treatment but symptoms may not be apparent for up to 4 wk, depending on weed species, timing of treatment and weather conditions
- Iodosulfuron-methyl and mesosulfuron-methyl are both members of the ALS-inhibitor group of herbicides. To avoid the build up of resistance do not use any product containing an ALS-inhibitor herbicide with claims for control of grass weeds more than once on any crop

- Use this product as part of a Resistance Management Strategy that includes cultural methods of control and does not use ALS inhibitors as the sole chemical method of weed control in successive crops. See Section 5 for more information

Restrictions

- Maximum number of treatments 1 per crop
- Do not use on crops undersown with grasses, clover or other legumes or any other broad-leaved crop
- Do not use where annual grass weeds other than annual meadow grass and rough meadow grass are present
- Do not use as a stand-alone treatment for control of common chickweed or common poppy. Only use mixtures with non ALS-inhibitor herbicides for these weeds
- Do not use as the sole means of weed control in successive crops
- Specific restrictions apply to use in sequence or tank mixture with other sulfonylurea or ALS-inhibiting herbicides. See label for details
- Do not apply to crops under stress from any cause
- Do not apply when rain is imminent or during periods of frosty weather
- Specified adjuvant must be used. See label

Crop-specific information

- Latest use: before 2nd node detectable for winter wheat
- Transitory crop effects may occur, particularly on overlaps and after late season/spring applications. Recovery is normally complete and yield not affected

Following crops guidance

- In the event of crop failure winter or spring wheat may be drilled after normal cultivation and an interval of 6 wk
- Winter wheat, winter barley or winter oilseed rape may be drilled in the autumn following normal harvest of a treated crop. Spring wheat, spring barley, spring oilseed rape or sugar beet may be drilled in the following spring
- Where the product has been applied in sequence with a permitted ALS-inhibitor herbicide (see label) only winter or spring wheat or barley, or sugar beet may be sown as following crops
- Successive treatments with any products containing diflufenican can lead to soil build-up and inversion ploughing to 150 mm must precede sowing any following non-cereal crop. Even where ploughing occurs some crops may be damaged

Environmental safety

- Dangerous for the environment
- Very toxic to aquatic organisms
- Take extreme care to avoid drift outside the target area
- LERAP Category B

Hazard classification and safety precautions

Hazard H04, H11
Risk phrases R36, R50, R53a
Operator protection A, C; U05a, U11, U20b
Environmental protection E15b, E16a, E38
Storage and disposal D01, D02, D05, D09a, D10a, D12a

176 diflufenican + isoproturon

A contact and residual herbicide for use in winter cereals
HRAC mode of action code: F1 + C2

Products

1	Clayton Fenican 550	Clayton	50:500 g/l	SC	11160
2	Fernpath Ipex	AgriGuard	50:500 g/l	SC	11391
3	Headland Javelin	Headland	62.5:500 g/l	SC	12451
4	Hermolance	Hermoo	62.5:500 g/l	SC	12459
5	Koala	Nufarm UK	97.5:785 g/l	WG	12556
6	Standon Diflufenican-IPU	Standon	50:500 g/l	SC	11109

Uses

- Annual dicotyledons in ***grass seed crops*** *(off-label)* [1, 2, 6]; ***triticale***, ***winter rye*** [1-4, 6]; ***winter barley***, ***winter wheat*** [1-6]
- Annual grasses in ***grass seed crops*** *(off-label)* [1, 2, 6]; ***triticale***, ***winter barley***, ***winter rye***, ***winter wheat*** [1-4, 6]
- Annual meadow grass in ***winter barley***, ***winter wheat*** [5]
- Blackgrass in ***triticale***, ***winter barley***, ***winter rye***, ***winter wheat*** [1-4, 6]
- Rough meadow grass in ***winter barley***, ***winter wheat*** [5]
- Wild oats in ***triticale***, ***winter barley***, ***winter rye***, ***winter wheat*** [1-4, 6]

Specific Off-Label Approvals (SOLAs)

- ***grass seed crops*** *20060028* [1], *20060021* [2], *20060020* [6]

Approval information

- Isoproturon included in Annex I under EC Directive 91/414. However, see isoproturon entry for information on phasing out of approvals of all products containing isoproturon by 30 June 2009
- Accepted by BBPA for use on malting barley

Efficacy guidance

- May be applied in autumn or spring (but only post-emergence on wheat or barley). Best control normally achieved by early post-emergence treatment
- Best results by application to fine, firm seedbed moist at or after application
- Weeds controlled from before emergence to 6 true leaf stage
- Apply to moist, but not waterlogged, soils
- Any trash or ash should be buried during seedbed preparation
- Always follow WRAG guidelines for preventing and managing herbicide resistant weeds. Section 5 for more information

Restrictions

- Maximum number of treatments 1 per crop. Maximum total dose of isoproturon 2.5 kg a.i./ha per yr
- On triticale and winter rye only treat named varieties and apply as pre-emergence spray
- Do not use on other cereals, broadcast or undersown crops or crops to be undersown
- Do not use on Sands or Very Light soils, or those that are very stony or gravelly or on soils with more than 10% organic matter
- Do not spray when heavy rain is forecast or on crops suffering from stress, frost, deficiency, pest or disease attack
- Do not harrow after application nor roll autumn-treated crops until spring

Crop-specific information

- Latest use: before 2nd node detectable (GS 32) (low dose) or before end Feb (high dose) for wheat and barley; pre-emergence of crop for triticale and rye [3, 4]; post-emergence and before end Feb for winter wheat and barley [2, 5, 6]
- Drill crop to normal depth (25 mm) and ensure seed well covered
- Early sown crops may be damaged if application is made during a period of rapid growth in the autumn

Following crops guidance

- In the event of crop failure winter wheat may be redrilled immediately after normal cultivation, and winter barley may be sown after ploughing. Fields must be ploughed to a depth of 15 cm and 20 wk must elapse before sowing spring crops of wheat, barley, oilseed rape, peas, field beans, sugar beet, potatoes, carrots, edible brassicas or onions
- After normal harvest autumn cereals can be drilled after ploughing. Thorough mixing of the soil must take place before drilling field beans, leaf brassicae or winter oilseed rape. For sugar beet seed crops and winter onions complete inversion of the furrow slice is essential
- Do not broadcast or direct drill oilseed rape or other brassica crops as a following crop on treated land. See label for detailed advice on preparing land for subsequent autumn cropping in the normal rotation
- Successive treatments with any products containing diflufenican can lead to soil build-up and inversion ploughing to 150 mm must precede sowing any following non-cereal crop. Even where ploughing occurs some crops may be damaged

Environmental safety

- Dangerous for the environment
- Very toxic to aquatic organisms
- Extremely dangerous to fish or other aquatic life. Do not contaminate surface waters or ditches with chemical or used container
- Do not apply to dry, cracked or waterlogged soils where heavy rain may lead to contamination of drains by isoproturon
- LERAP Category B

Hazard classification and safety precautions

Hazard H03 [1-4, 6]; H04 [5]; H11 [1-6]
Risk phrases R36 [5]; R40, R50, R53a [1-6]
Operator protection A, C, H [1-6]; D [5]; U08, U19a [1]; U20a [2]; U20c [1, 3-6]
Environmental protection E13a [3, 5]; E15a [1, 2, 4, 6]; E16a [1-6]; E16b [1-3, 5, 6]; E38 [3-5]
Storage and disposal D05 [1-5]; D09a [1-6]; D10b [1-4, 6]; D11a [5]; D12a [3-5]

177 dimethenamid-p

A chloroacetamide herbicide available only in mixtures
HRAC mode of action code: K3

178 dimethenamid-p + metazachlor

A soil acting herbicide mixture for oilseed rape
HRAC mode of action code: K3 + K3

Products

Springbok	BASF	200:200 g/l	EC	12983

Uses

- Annual dicotyledons in ***winter oilseed rape***
- Chickweed in ***winter oilseed rape***
- Cleavers in ***winter oilseed rape***
- Common storksbill in ***winter oilseed rape***
- Field speedwell in ***winter oilseed rape***
- Poppies in ***winter oilseed rape***
- Scented mayweed in ***winter oilseed rape***
- Shepherd's purse in ***winter oilseed rape***

Efficacy guidance

- Best results obtained from treatments to fine, firm and moist seedbeds
- Apply pre- or post-emergence of the crop and ideally before weed emergence
- Residual weed control may be reduced under prolonged dry conditions
- Weeds germinating from depth may not be controlled

Restrictions

- Maximum total dose on winter oilseed rape equivalent to one full dose treatment
- Do not disturb soil after application
- Do not treat broadcast crops until they have attained two fully expanded cotyledons
- Do not use on Sands, Very Light soils, or soils containing >10% organic matter
- Do not apply when heavy rain forecast or on soils waterlogged or prone to waterlogging
- Do not treat crops suffering from stress from any cause
- Do not use pre-emergence when crop seed has started to germinate or if not covered with 15 mm of soil

Crop-specific information

- Latest use: before 7th true leaf for winter oilseed rape
- All varieties of winter oilseed rape may be treated

Following crops guidance

- Any crop may follow a normally harvested treated winter oilseed rape crop. Ploughing is not essential before a following cereal crop but is required for all other crops

- In the event of failure of a treated crop winter wheat (excluding durum) or winter barley may be drilled in the same autumn, and any cereal (excluding durum wheat), spring oilseed rape, peas or field beans may be sown in the following spring. Ploughing to at least 150 mm should precede in all cases

Environmental safety

- Dangerous for the environment
- Very toxic to aquatic organisms
- LERAP Category B
- Take extreme care to avoid spray drift onto non-crop plants outside the target area

Hazard classification and safety precautions

Hazard H03, H11
Risk phrases R22a, R36, R38, R43, R50, R53a
Operator protection A, C, H; U05a, U08, U20c
Environmental protection E15a, E16a, E34, E38
Storage and disposal D01, D02, D09a, D10c, D12a
Medical advice M03, M05a

179 dimethoate

A contact and systemic organophosphorus insecticide and acaricide
IRAC mode of action code: 1B

Products

1	BASF Dimethoate 40	BASF	400 g/l	EC	00199
2	Danadim Progress	Headland	400 g/l	EC	12208

Uses

- Aphids in ***bulb onions*** *(off-label)*, ***garlic*** *(off-label)*, ***leeks*** *(off-label)*, ***salad onions*** *(off-label)*, ***shallots*** *(off-label)* [2]; ***durum wheat*** *(off-label)*, ***grass seed crops***, ***mangel seed crops***, ***sugar beet seed crops*** *(excluding Myzus persicae)* [1]; ***fodder beet*** *(excluding Myzus persicae)*, ***lettuce***, ***mangels*** *(excluding Myzus persicae)*, ***ornamental plant production***, ***red beet*** *(excluding Myzus persicae)*, ***spring rye***, ***spring wheat***, ***sugar beet*** *(excluding Myzus persicae)*, ***triticale***, ***winter rye***, ***winter wheat*** [1, 2]
- Cabbage aphid in ***broccoli***, ***brussels sprouts***, ***calabrese***, ***cauliflowers*** [1, 2]
- General insect control in ***calabrese*** *(off-label)*, ***celeriac*** *(off-label)*, ***chinese cabbage*** *(off-label)*, ***collards*** *(off-label)*, ***compost*** *(off-label - for watercress propagation)*, ***kale*** *(off-label)*, ***kohlrabi*** *(off-label)*, ***protected broccoli*** *(off-label)*, ***protected cabbages*** *(off-label)*, ***protected calabrese*** *(off-label)*, ***protected cauliflowers*** *(off-label)*, ***protected celeriac*** *(off-label)*, ***protected chinese cabbage*** *(off-label)*, ***protected collards*** *(off-label)*, ***protected kale*** *(off-label)*, ***protected kohlrabi*** *(off-label)*, ***protected roscoff cauliflowers*** *(off-label)*, ***protected savoy cabbage*** *(off-label)*, ***protected spinach*** *(off-label)*, ***roscoff cauliflowers*** *(off-label)*, ***savoy cabbage*** *(off-label)* [2]; ***compost*** *(off-label - for watercress progagation)* [1]
- Insect pests in ***calabrese*** *(off-label)*, ***celeriac*** *(off-label)*, ***chinese cabbage*** *(off-label)*, ***collards*** *(off-label)*, ***kale*** *(off-label)*, ***kohlrabi*** *(off-label)*, ***protected broccoli*** *(off-label)*, ***protected brussels sprouts*** *(off-label)*, ***protected cabbages*** *(off-label)*, ***protected calabrese*** *(off-label)*, ***protected cauliflowers*** *(off-label)*, ***protected celeriac*** *(off-label)*, ***protected chinese cabbage*** *(off-label)*, ***protected collards*** *(off-label)*, ***protected kale*** *(off-label)*, ***protected kohlrabi*** *(off-label)*, ***protected roscoff cauliflowers*** *(off-label)*, ***protected savoy cabbage*** *(off-label)*, ***protected spinach*** *(off-label)*, ***roscoff cauliflowers*** *(off-label)*, ***savoy cabbage*** *(off-label)* [1]
- Leaf miner in ***fodder beet***, ***mangels***, ***ornamental plant production***, ***red beet***, ***sugar beet*** [1, 2]
- Red spider mites in ***ornamental plant production*** [1, 2]
- Wheat bulb fly in ***spring wheat***, ***winter wheat*** [1, 2]

Specific Off-Label Approvals (SOLAs)

- ***bulb onions*** *20061348* [2]
- ***calabrese*** *940389* [1], *20050682* [2]
- ***celeriac*** *940389* [1], *20050682* [2]
- ***chinese cabbage*** *940389* [1], *20050682* [2]
- ***collards*** *940389* [1], *20050682* [2]
- ***compost*** *(for watercress progagation) 940394* [1], *(for watercress propagation) 20050683* [2]

- ***durum wheat*** *20060807* [1]
- ***garlic*** *20061348* [2]
- ***kale*** *940389* [1], *20050682* [2]
- ***kohlrabi*** *940389* [1], *20050682* [2]
- ***leeks*** *20061348* [2]
- ***protected broccoli*** *940389* [1], *20050682* [2]
- ***protected brussels sprouts*** *940389* [1]
- ***protected cabbages*** *940389* [1], *20050682* [2]
- ***protected calabrese*** *940389* [1], *20050682* [2]
- ***protected cauliflowers*** *940389* [1], *20050682* [2]
- ***protected celeriac*** *940389* [1], *20050682* [2]
- ***protected chinese cabbage*** *940389* [1], *20050682* [2]
- ***protected collards*** *940389* [1], *20050682* [2]
- ***protected kale*** *940389* [1], *20050682* [2]
- ***protected kohlrabi*** *940389* [1], *20050682* [2]
- ***protected roscoff cauliflowers*** *940389* [1], *20050682* [2]
- ***protected savoy cabbage*** *940389* [1], *20050682* [2]
- ***protected spinach*** *940389* [1], *20050682* [2]
- ***roscoff cauliflowers*** *940389* [1], *20050682* [2]
- ***salad onions*** *20061348* [2]
- ***savoy cabbage*** *940389* [1], *20050682* [2]
- ***shallots*** *20061348* [2]

Approval information

- Dimethoate included in Annex I under EC Directive 91/414
- Approved for aerial application on cereals, sugar beet [1]. See notes in Section 5
- In 2006 PSD required that all products containing this active ingredient should carry the following warning in the main area of the container label: "Dimethoate is an anticholinesterase organophosphate. Handle with care"

Efficacy guidance

- Chemical has quick knock-down effect and systemic activity lasts for up to 14 d
- With some crops, products differ in range of pests listed as controlled. Uses section above provides summary. See labels for details
- For most pests apply when pest first seen and repeat 2-3 wk later or as necessary. Timing and number of sprays varies with crop and pest. See labels for details
- Best results achieved when crop growing vigorously. Systemic activity reduced when crops suffering from drought or other stress
- In hot weather apply in early morning or late evening
- Where aphids or spider mites resistant to organophosphorus compounds occur control is unlikely to be satisfactory and repeat treatments may result in lower levels of control

Restrictions

- Contains an anticholinesterase organophosphorus compound. Do not use if under medical advice not to work with such compounds
- Maximum number of treatments 6 per crop for Brussels sprouts, broccoli, cauliflower, calabrese, lettuce; 4 per crop for grass seed crops, cereals; 2 per crop for beet crops, triticale, rye; 1 per crop for mangel and sugar beet seed crops
- On beet crops only one treatment per crop may be made for control of leaf miners (max 84 g a.i./ha) and black bean aphid (max 400 g a.i./ha)
- In beet crops resistant strains of peach-potato aphid (*Myzus persicae*) are common and dimethoate products must not be used to control this pest
- Test for varietal susceptibility on all unusual plants or new cultivars
- Do not tank mix with alkaline materials. See label for recommended tank-mixes
- Consult processor before spraying crops grown for processing
- Must not be applied to cereals if any product containing a pyrethroid insecticide or dimethoate has been sprayed after the start of ear emergence (GS 51)

Crop-specific information

- Latest use: before 30 Jun in yr of harvest for beet crops; flowering just complete stage (GS 30) for wheat, rye, triticale [1]; before 31 Mar in yr of harvest for aerial application to cereals
- HI brassicas 21 d; grass seed crops, lettuce, cereals 14 d; ornamental plant production 7 d

Environmental safety

- Dangerous for the environment
- Harmful to aquatic organisms
- Flammable
- Harmful to game, wild birds and animals
- Harmful to livestock. Keep all livestock out of treated areas for at least 7 d
- Likely to cause adverse effects on beneficial arthropods
- High risk to non-target insects or other arthropods. Do not treat cereals after 1 Apr within 6 m of edge of crop
- High risk to bees. Do not apply to crops in flower or to those in which bees are actively foraging. Do not apply when flowering weeds are present
- Surface residues may also cause bee mortality following spraying
- Dangerous to fish or other aquatic life. Do not contaminate surface waters or ditches with chemical or used container
- LERAP Category A
- Keep in original container, tightly closed, in a safe place, under lock and key

Hazard classification and safety precautions

Hazard H03, H08
Risk phrases R20, R21, R48 [1]; R22a, R43 [1, 2]; R52 [2]
Operator protection A, H, M [1, 2]; C, J [1]; U02a, U04a, U19a [1, 2]; U05a, U10, U13, U20a [1]; U08, U14, U15, U20b [2]
Environmental protection E06c [1, 2] (7 d); E10b, E12a, E12e, E13b, E16c, E16d, E22a, E34 [1, 2]; E18 [1]
Storage and disposal D01, D02, D05, D09b [1, 2]; D08, D10c [1]; D12b [2]
Medical advice M01, M03 [1, 2]; M05a [1]

180 dimethomorph

A cinnamic acid fungicide with translaminar activity available only in mixtures
FRAC mode of action code: 40

181 dimethomorph + mancozeb

A systemic and protectant fungicide for potato blight control
FRAC mode of action code: 40 + M3

Products

Invader	BASF	7.5:66.7% w/w	WG	11978

Uses

- Blight in ***potatoes***
- Downy mildew in ***bulb onions*** *(off-label)*, ***chinese cabbage*** *(off-label)*, ***chives*** *(off-label)*, ***choi sum*** *(off-label)*, ***cress*** *(off-label)*, ***frise*** *(off-label)*, ***garlic*** *(off-label)*, ***herbs (see appendix 6)*** *(off-label)*, ***lamb's lettuce*** *(off-label)*, ***lettuce*** *(off-label)*, ***pak choi*** *(off-label)*, ***poppies for morphine production*** *(off-label)*, ***radicchio*** *(off-label)*, ***salad brassicas*** *(off-label - for baby leaf production)*, ***salad onions*** *(off-label)*, ***scarole*** *(off-label)*, ***shallots*** *(off-label)*, ***tatsoi*** *(off-label)*
- White tip in ***leeks*** *(off-label)*

Specific Off-Label Approvals (SOLAs)

- ***bulb onions*** *20042334*
- ***chinese cabbage*** *20063044*
- ***chives*** *20063044*
- ***choi sum*** *20063044*
- ***cress*** *20063044*
- ***frise*** *20063044*
- ***garlic*** *20042334*
- ***herbs (see appendix 6)*** *20063044*
- ***lamb's lettuce*** *20063044*
- ***leeks*** *20063056*
- ***lettuce*** *20063044*

- ***pak choi*** *20063044*
- ***poppies for morphine production*** *20050909*
- ***radicchio*** *20063044*
- ***salad brassicas*** *(for baby leaf production) 20063044*
- ***salad onions*** *20062953*
- ***scarole*** *20063044*
- ***shallots*** *20042334*
- ***tatsoi*** *20063044*

Approval information
- Dimethomorph and mancozeb included in Annex I under EC Directive 91/414

Efficacy guidance
- Commence treatment as soon as there is a risk of blight infection
- In the absence of a warning treatment should start before the crop meets along the rows
- Repeat treatments every 7-14 d depending in the degree of infection risk
- Irrigated crops should be regarded as at high risk and treated every 7 d
- For best results good spray coverage of the foliage is essential
- To minimise the likelihood of development of resistance these products should be used in a planned Resistance Management strategy. See Section 5 for more information

Restrictions
- Maximum total dose equivalent to eight full dose treatments on potatoes

Crop-specific information
- HI 7 d for all crops

Environmental safety
- Dangerous for the environment
- Very toxic to aquatic organisms
- LERAP Category B

Hazard classification and safety precautions

Hazard H04, H11
Risk phrases R36, R37, R38, R43, R50, R53a
Operator protection A; U02a, U04a, U05a, U08, U13, U19a, U20b
Environmental protection E16a, E16b, E38
Storage and disposal D01, D02, D05, D09a, D12a
Medical advice M05a

182 dimoxystrobin

A protectant strobilurin fungicide for cereals available only in mixtures
FRAC mode of action code: 11

183 dimoxystrobin + epoxiconazole

A protectant and curative fungicide mixture for cereals
FRAC mode of action code: 11 + 3

Products

Swing Gold	BASF	133:50 g/l	SC	11658

Uses
- Brown rust in ***winter wheat***
- Fusarium ear blight in ***winter wheat***
- Glume blotch in ***winter wheat***
- Septoria leaf blotch in ***winter wheat***
- Tan spot in ***winter wheat***

Approval information
- Dimoxystrobin included in Annex I under EC Directive 91/414

Efficacy guidance

- For best results apply from the start of ear emergence
- Dimoxystrobin is a member of the QoI cross resistance group. Product should be used preventatively and not relied on for its curative potential
- Use product as part of an Integrated Crop Management strategy incorporating other methods of control, including where appropriate other fungicides with a different mode of action. Do not apply more than two foliar applications of QoI containing products to any cereal crop
- There is a significant risk of widespread resistance occurring in *Septoria tritici* populations in UK. Failure to follow resistance management action may result in reduced levels of disease control
- Epoxiconazole is a DMI fungicide. Resistance to some DMI fungicides has been identified in Septoria leaf blotch which may seriously affect performance of some products. For further advice contact a specialist advisor and visit the Fungicide Resistance Action Group (FRAG)-UK website

Restrictions

- Maximum number of treatments 1 per crop
- Do not apply before the start of ear emergence

Crop-specific information

- Latest use: up to and including flowering (GS 69)

Environmental safety

- Dangerous for the environment
- Very toxic to aquatic organisms
- LERAP Category B
- Avoid drift onto neighbouring crops

Hazard classification and safety precautions

Hazard H03, H11
Risk phrases R20, R22a, R40, R50, R53a, R63
Operator protection A; U05a, U20b
Environmental protection E15a, E16a, E34, E38
Storage and disposal D01, D02, D05, D09a, D10c, D12a
Medical advice M05a

184 dinocap

A protectant dinitrophenyl fungicide for powdery mildew control
FRAC mode of action code: 29

Products

Karathane Liquid	Landseer	350 g/l	EC	12576

Uses

- Powdery mildew in ***apples***, ***chrysanthemums***, ***roses***, ***strawberries***, ***wine grapes*** *(off-label)*

Specific Off-Label Approvals (SOLAs)

- ***wine grapes*** *20060488*

Approval information

- Dinocap has been included in Annex I under Directive 91/414 for 3 years from January 2007 and all approvals for products containing this active ingredient will expire in December 2009
- Approval expiry 30 Jun 2008 [1]

Efficacy guidance

- Spray at 7-14 d intervals to maintain protective film
- Product must be applied to roses before disease becomes established
- Regular use on apples suppresses red spider mites and rust mites

Restrictions

- Maximum total dose equivalent to 10 full dose treatments on apples
- Maximum number of treatments on strawberries 5 per yr
- When applied to apples during blossom period may cause spotting but no adverse effect on pollination or fruit set. Do not apply to Golden Delicious during blossom period
- Do not apply when temperature is above 24 °C. Do not apply with white oils
- Certain chrysanthemum cultivars may be susceptible
- May cause petal spotting on white roses

Crop-specific information

- HI grapevines 21 d, apples 14 d, strawberries 7 d

Environmental safety

- Dangerous for the environment
- Very toxic to aquatic organisms
- LERAP Category B
- Broadcast air-assisted LERAP (18 m)

Hazard classification and safety precautions

Hazard H02, H08, H11
Risk phrases R20, R22b, R36, R38, R43, R48, R50, R53a, R61, R67
Operator protection A, C, H, J, M; U04a, U05a, U10, U11, U19a, U20a
Environmental protection E15a, E16a, E34, E38; E17b (18 m)
Storage and disposal D01, D02, D05, D09a, D10a, D12a
Medical advice M04a

185 diquat

A non-residual bipyridyl contact herbicide and crop desiccant
HRAC mode of action code: D

Products

1	Brogue	AgriGuard	200 g/l	SL	13328
2	Clayton Diquat	Clayton	200 g/l	SL	12739
3	Diquanet	Hermoo	200 g/l	SL	13413
4	Quad	Q-Chem	200 g/l	SL	13396
5	Reglone	Syngenta	200 g/l	SL	10534
6	Roquat	Q-Chem	200 g/l	SL	13474
7	Standon Googly	Standon	200 g/l	SL	12995
8	Standon Googly	Standon	200 g/l	SL	13281

Uses

- Annual dicotyledons in ***all edible crops (outdoor)*** *(around)*, ***all non-edible crops (outdoor)*** *(around)*, ***ornamental plant production***, ***sugar beet*** [1-8]
- Chemical stripping in ***hops*** [3, 4, 6, 7]
- General weed control in ***chives*** *(off-label)*, ***herbs (see appendix 6)*** *(off-label)*, ***parsley*** *(off-label)*, ***poppies for morphine production*** *(off-label)* [5]; ***potatoes*** [1-8]
- Haulm destruction in ***potatoes*** [1-8]
- Pre-harvest desiccation in ***borage for oilseed production*** *(off-label)*, ***brown mustard*** *(off-label)*, ***french beans*** *(off-label)*, ***hemp for oilseed production*** *(off-label)*, ***poppies for morphine production*** *(off-label)*, ***poppies for oilseed production*** *(off-label)*, ***sesame*** *(off-label)*, ***sunflowers*** *(off-label)*, ***white mustard*** *(off-label)* [5]; ***broad beans*** [2, 5, 8]; ***clover seed crops***, ***combining peas***, ***laid barley*** *(stockfeed only)*, ***laid oats*** *(stockfeed only)*, ***linseed***, ***spring field beans*** *(stock or pigeon feed only)*, ***spring oilseed rape***, ***winter field beans*** *(stock or pigeon feed only)*, ***winter oilseed rape*** [1-8]

Specific Off-Label Approvals (SOLAs)

- ***borage for oilseed production*** *20062390* [5]
- ***brown mustard*** *20062387* [5]
- ***chives*** *20062388* [5]
- ***french beans*** *20062389* [5]
- ***hemp for oilseed production*** *20062386* [5]
- ***herbs (see appendix 6)*** *20062388* [5]
- ***parsley*** *20062388* [5]
- ***poppies for morphine production*** *20062385* [5]
- ***poppies for oilseed production*** *20062387* [5]
- ***sesame*** *20062387* [5]
- ***sunflowers*** *20062387* [5]
- ***white mustard*** *20062387* [5]

Approval information

- Diquat included in Annex I under EC Directive 91/414
- All approvals for aquatic weed control expired in 2004

Efficacy guidance

- Acts rapidly on green parts of plants and rainfast in 15 min
- Best results for potato desiccation achieved by spraying in bright light and low humidity conditions
- For weed control in row crops apply as overall spray before crop emergence or before transplanting or as an inter-row treatment using a sprayer designed to prevent contamination of crop foliage by spray
- Treatments for weed control normally require co-application with paraquat. Use of authorised non-ionic wetter for improved weed control essential for most uses except potato desiccation

Restrictions

- Maximum number of treatments 1 per crop in most situations - see labels for details
- Do not apply potato haulm destruction treatment when soil dry. Tubers may be damaged if spray applied during or shortly after dry periods. See label for details of maximum allowable soil-moisture deficit and varietal drought resistance scores
- Do not add wetters to desiccant sprays for potatoes
- Consult processor before adding Agral or other wetter on peas. Staining may be increased
- Do not use treated straw or haulm as animal feed or bedding within 4 d of spraying
- Do not use on barley or oats intended for human consumption
- Do not apply through ULV or mistblower equipment

Crop-specific information

- Latest use: before crop emergence for sugar beet and before transplanting ornamentals. Not specified for other crops where use varies according to situation
- HI zero when used as a crop desiccant but see label for advisory intervals
- For pre-harvest desiccation, apply to potatoes when tubers the desired size and to other crops when mature or approaching maturity. See label for details of timing and of period to be left before harvesting potatoes and for timing of pea and bean desiccation sprays
- Spray linseed when seed matured evenly over whole field; direct combining can normally begin 10-20 d after spraying
- Apply as a chemical stripping treatment in hops when shoots have reached top wire [3, 4, 6, 7]
- Treatment of hops under stress or when leaves are wet with a heavy dew may cause damage [3, 4, 6, 7]
- Treated laid barley and oats may be used only for stock feed
- Treated peas must be harvested dry
- If potato tubers are to be stored leave 14 d after treatment before lifting
- Use on thin or stressed crops of oilseed rape may increase seed losses and decrease oil yield to an unacceptable level

Environmental safety

- Dangerous for the environment
- Very toxic to aquatic organisms
- Keep all livestock out of treated areas and away from treated water for at least 24 h
- Do not feed treated straw or haulm to livestock within 4 days of spraying
- Do not use on crops if the straw is to be used as animal feed/bedding
- Do not dump surplus herbicide in water or ditch bottoms

Hazard classification and safety precautions

Hazard H01 [2]; H02 [1, 3-8]; H11 [1-8]

Risk phrases R22a, R37, R38, R48, R50, R53a [1-8]; R23 [1, 3-8]; R26 [2]; R36, R43 [4, 6]

Operator protection A [1-8]; C [1, 3, 4, 6, 7]; D [5, 8]; H [2, 5, 8]; M [1-4, 6, 7]; U02a, U04a, U05a, U08, U19a, U20b [1-8]; U10 [1, 3-8]; U15 [4, 6]

Environmental protection E06c [1-8] (24 h); E08a [1, 3, 4, 6, 7] (4 d); E09 [2] (4 d); E09 [5, 8]; E15b, E34 [1-8]; E38 [1, 3-8]

Storage and disposal D01, D02, D05, D09a, D12a [1-8]; D10b [2]; D10c [1, 3-8]

Medical advice M04a

SEE SECTION 3 FOR PRODUCTS ALSO REGISTERED

186 diquat + paraquat

A non-selective non-residual bipyridyl contact herbicide
HRAC mode of action code: D + D

Products

1	Clayton Paradigm	Clayton	80:120 g/l	SL	12001
2	PDQ	Syngenta	80:120 g/l	SL	10532

Uses

- Annual dicotyledons in ***all edible crops (outdoor), forest, forest nurseries, natural surfaces not intended to bear vegetation, ornamental plant production, potatoes, stubbles*** [1, 2]
- Annual grasses in ***all edible crops (outdoor), forest, forest nurseries, natural surfaces not intended to bear vegetation, ornamental plant production, potatoes, stubbles*** [1, 2]
- Chemical stripping in ***hops*** [1]
- Green cover in ***land temporarily removed from production*** [1, 2]
- Sward destruction in ***permanent grassland, rotational grass*** [1, 2]
- Volunteer cereals in ***all edible crops (outdoor), potatoes, stubbles*** [1, 2]

Approval information

- Diquat included in Annex I under EC Directive 91/414. Paraquat inclusion annulled in 2007. As a result all remaining approvals for use of products containing paraquat will expire in July 2008
- Accepted by BBPA for use on hops
- Approval expiry 31 Dec 2008 [1, 2]

Efficacy guidance

- Rapid kill obtained under bright conditions but most effective results on difficult weeds obtained from slower action in winter
- Apply to young emerged weeds less than 15 cm high, annual grasses and volunteer cereals must have at least 2 leaves when sprayed
- Addition of approved non-ionic wetter recommended for control of certain species and with low dose rates. See label for details
- Minimum interval between spraying and cultivation varies. See label for details
- Chemical rapidly inactivated in moist soil, activity reduced in dirty or muddy water
- Spray is rainfast within 10 min of spray drying on weed leaves

Restrictions

- Paraquat is subject to the Poisons Rules 1982 and the Poisons Act 1972. See notes in Section 5
- Do not use on straw or other artificial growing media
- Do not put in a food or drinks container

Crop-specific information

- Latest use normally before planting or emergence, except potatoes - see below - see label
- In potatoes spray earlies up to 10% emergence, maincrop to 40% emergence, provided plants are less than 15 cm high. Do not use post-emergence on potatoes from diseased or small tubers or under very hot, dry conditions
- Use guarded no-drift sprayers to kill inter-row weeds and strawberry runners
- Apply to fruit crops as a directed spray, preferably in dormant season
- If spraying bulbs at end of season ensure all crop foliage is detached from bulbs. Do not use on very sandy soils

Following crops guidance

- Diquat and paraquat are quickly inactivated on contact with soil. Crops can normally be sown or planted from 4 h after spraying. However on sandy or immature peat soils and on forest nursery seedbeds allow 3 d between spraying and planting
- Where trash or dying weeds are left on surface allow at least 3 d before planting

Environmental safety

- Dangerous for the environment
- Very toxic to aquatic organisms
- Paraquat can be harmful to hares; spray stubbles early in the day
- Keep all livestock out of treated areas for at least 24 h
- Keep in original container, tightly closed, in a safe place, under lock and key

Hazard classification and safety precautions

Hazard H02, H11
Risk phrases R21, R22a, R37, R38, R41, R48, R50, R53a
Operator protection A, C, H, M; U02a, U04a, U08, U20a [1]; U04b, U04c, U20b [2]; U05a, U09a, U19a [1, 2]
Environmental protection E06c [1, 2] (24 h); E11, E15a, E34 [1, 2]; E38 [2]
Storage and disposal D01, D02, D05, D09b [1, 2]; D10a [1]; D10c, D12a [2]
Medical advice M03 [2]; M04a, M05c [1, 2]

187 dithianon

A protectant and eradicant dicarbonitrile fungicide for scab control
FRAC mode of action code: M9

Products

1	Dithianon Flowable	BASF	750 g/l	SC	10219
2	Dithianon WG	BASF	70% w/w	WG	12538

Uses

- Scab in ***apples***, ***pears***

Efficacy guidance

- Apply at bud-burst and repeat every 7-14 d until danger of scab infection ceases
- Application at high rate within 48 h of a Mills period prevents new infection
- Spray programme also reduces summer infection with apple canker

Restrictions

- Maximum number of treatments on apples and pears 8 per crop [1]; 12 per crop [2]
- Do not use on Golden Delicious apples after green cluster
- Do not mix with lime sulphur or highly alkaline products

Crop-specific information

- HI 4 wk for apples, pears

Environmental safety

- Dangerous for the environment
- Very toxic to aquatic organisms

Hazard classification and safety precautions

Hazard H03, H11
Risk phrases R20 [1]; R22a, R50, R53a [1, 2]; R41 [2]
Operator protection A [1, 2]; D [2]; H [1]; U05a [1, 2]; U08, U19a, U20a [1]; U11 [2]
Environmental protection E15a, E34, E38
Storage and disposal D01, D02, D05 [1, 2]; D09a, D10b [1]; D12a [2]
Medical advice M03 [1]; M05a [1, 2]

188 dithianon + pyraclostrobin

A protectant fungicide mixture for apples and pears
FRAC mode of action code: M9 + 11

Products

Maccani	BASF	12:4% w/w	WG	13227

Uses

- Powdery mildew in ***apples***
- Scab in ***apples***, ***pears***

Approval information

- Pyraclostrobin included in Annex I under EC Directive 91/414

Efficacy guidance

- Best resullts obtained from treatments applied from bud burst before disease development and repeated at 10 d intervals
- Water volume should be adjusted according to size of tree and leaf area to ensure good coverage

- Pyraclostrobin is a member of the QoI cross resistance group. Product should be used preventatively at the full recommended dose and not relied on for its curative potential
- Where field performance may be adversely affected apply QoI containing fungicides in mixtures in strict alternation with fungicides from a different cross-resistance group

Restrictions

- Maximum number of treatments 4 per crop. However where the total number of fungicide treatments is less than 12, the maximum number of applications of any QoI containing products should be three
- Consult processors before use on crops destined for fruit preservation

Crop-specific information

- HI 35 d for apples, pears
- Some apple varieties may be considered to be susceptible to russetting caused by the dithianon component during fruit development

Environmental safety

- Dangerous for the environment
- Very toxic to aquatic organisms
- Broadcast air-assisted LERAP (40 m)

Hazard classification and safety precautions

Hazard H03, H11
Risk phrases R20, R22a, R36, R50, R53a
Operator protection A, C, D; U05a, U11, U20b
Environmental protection E15b, E34, E38; E17b (40 m)
Storage and disposal D01, D02, D05, D08, D11a, D12a
Medical advice M03, M05a

189 diuron

A residual urea herbicide for non-crop areas and woody crops
HRAC mode of action code: C2

Products

1	Diurex 50SC	Makhteshim	500 g/l	SC	11472
2	Diuron 80 WP	Bayer Environ.	80% w/w	WP	09931
3	Divide	Headland	80% w/w	WP	12206
4	Divide 50 SC	Headland	500 g/l	SC	12377
5	Freeway	Bayer Environ.	500 g/l	SC	11129
6	Nomix Diuron Flowable	Nomix Enviro	500 g/l	SC	12078
7	Unicrop Flowable Diuron	Unicrop	500 g/l	SC	02270
8	UPL Diuron 80	United Phosphorus	80% w/w	WP	07619

Uses

- Annual dicotyledons in ***almonds*** *(off-label)*, ***chestnuts*** *(off-label)*, ***hazel nuts*** *(off-label)*, ***walnuts*** *(off-label)* [1, 7]; ***amenity vegetation*** [1, 4-6]; ***apple orchards***, ***pear orchards*** [1, 5, 7]; ***asparagus*** *(off-label)* [7]; ***established woody ornamentals*** [2]; ***hard surfaces***, ***permeable surfaces overlying soil*** [1, 4, 6]; ***land not intended to bear vegetation*** [2-6]; ***natural surfaces not intended to bear vegetation*** [1]; ***trees and shrubs***, ***woody nursery stock*** *(not container grown)* [5]
- Annual grasses in ***almonds*** *(off-label)*, ***chestnuts*** *(off-label)*, ***hazel nuts*** *(off-label)*, ***walnuts*** *(off-label)* [1, 7]; ***amenity vegetation*** [1, 5]; ***apple orchards***, ***pear orchards*** [1, 5, 7]; ***asparagus*** *(off-label)* [7]; ***established woody ornamentals*** [2]; ***hard surfaces***, ***natural surfaces not intended to bear vegetation***, ***permeable surfaces overlying soil*** [1]; ***land not intended to bear vegetation*** [2, 5]; ***trees and shrubs***, ***woody nursery stock*** [5]
- Annual meadow grass in ***land not intended to bear vegetation*** [3]
- Black nightshade in ***daffodils*** *(off-label - for galanthamine production)* [7]
- Perennial dicotyledons in ***amenity vegetation***, ***hard surfaces***, ***land not intended to bear vegetation***, ***permeable surfaces overlying soil*** [4, 6]
- Perennial grasses in ***amenity vegetation***, ***hard surfaces***, ***land not intended to bear vegetation***, ***permeable surfaces overlying soil*** [4, 6]
- Stinging nettle in ***daffodils*** *(off-label - for galanthamine production)* [7]
- Total vegetation control in ***industrial sites***, ***land not intended for cropping*** [8]

- Willowherb in ***asparagus*** *(off-label)*, ***gooseberries*** *(off-label)*, ***redcurrants*** *(off-label)*, ***whitecurrants*** *(off-label)* [1]; ***bilberries*** *(off-label)*, ***blackcurrants*** *(off-label)*, ***blueberries*** *(off-label)*, ***cranberries*** *(off-label)* [1, 7]; ***daffodils*** *(off-label - for galanthamine production)* [7]

Specific Off-Label Approvals (SOLAs)

- ***almonds*** *20061396* [1], *20061395* [7]
- ***asparagus*** *20063727* [1], *982352* [7]
- ***bilberries*** *20063727* [1], *20061394* [7]
- ***blackcurrants*** *20063727* [1], *20061394* [7]
- ***blueberries*** *20063727* [1], *20061394* [7]
- ***chestnuts*** *20061396* [1], *20061395* [7]
- ***cranberries*** *20063727* [1], *20061394* [7]
- ***daffodils*** *(for galanthamine production) 20041521* [7]
- ***gooseberries*** *20063727* [1]
- ***hazel nuts*** *20061396* [1], *20061395* [7]
- ***redcurrants*** *20063727* [1]
- ***walnuts*** *20061396* [1], *20061395* [7]
- ***whitecurrants*** *20063727* [1]

Approval information

- Diuron will not be included in Annex I of EC Directive 91/414. To comply with this EC decision all approvals for the storage and use of all products listed in this profile will expire on 13 December 2008
- Approval expiry 13 Dec 2008 [1-4, 6-8]
- Approval expiry 31 Aug 2008 [5]

Efficacy guidance

- Best results when applied to moist soil and rain falls soon afterwards
- Length of residual activity may be reduced on heavy or highly organic soils or those with ash substrates
- Selective rates must be applied to weed-free soil and activity persists for 2-3 mth
- Application for total vegetation control may be at any time of year, best results obtained from start of Feb to end Apr (end May in Scotland). See label for list of resistant species
- Application to frozen ground not recommended

Restrictions

- Maximum number of treatments or maximum total dose varies according to product, dose and situation. See label
- Do not treat trees and shrubs less than 5 cm tall or established less than 12 mth or in areas of container grown trees and shrubs. See label for list of sensitive ornamental species that should not be treated
- Do not use around trees and shrubs on Sands or Very Light soils or those that are gravelly or where less than 1% organic matter
- Do not use on lawns, grass tennis courts or similar areas of turf
- Do not apply non-selective rates on or near desirable plants where chemical may be washed into contact with roots. Some labels give details of species which should not be treated
- Application for amenity use should only take place between the beginning of Feb and end of Apr (May in Scotland)

Crop-specific information

- Latest use: not specified
- Around trees and shrubs spray in late winter or early spring
- Treat hard surfaces between Feb and end May
- Apply to weed-free soil in apple and pear orchards established for at least 1 yr during Feb-Mar

Following crops guidance

- Succeeding crops should not be planted for at least 12 mth after treatment. Do not plant susceptible crops such as vegetables for at least 2 yr

Environmental safety

- Dangerous for the environment
- Very toxic to aquatic organisms
- Harmful to fish or other aquatic life. Do not contaminate surface waters or ditches with chemical or used container
- Do not apply over drains or in drainage channels or gullies

- Avoid run-off when using on paved and similar surfaces
- Do not apply, or allow to drift visibly, on to water
- Do not mix or load within 20 m of well waters unless on an impervious pad to prevent ground water infiltration

Hazard classification and safety precautions

Hazard H03 [1, 2, 4-8]; H04 [2, 3]; H11 [1-8]

Risk phrases R22a [1, 8]; R36 [2-4]; R38 [2, 4]; R40 [1-5, 7, 8]; R48, R50, R53a [1-8]; R68 [6]

Operator protection A [1, 3-6, 8]; C [3, 4]; D [8]; H [1, 5, 6, 8]; M [1, 5, 6]; N [1, 6]; U05a [1-7]; U08 [2-8]; U14, U15 [1]; U19a, U20b [2-5, 7, 8]; U20a [1, 6]

Environmental protection E13b [8]; E13c [3, 4]; E15a [1, 2, 5-7]; E34 [2, 4-7]; E38 [2-7]

Storage and disposal D01, D02, D09a [1-8]; D05 [5, 7, 8]; D06a, D08 [8]; D10b [1, 4, 6]; D10c [5, 7]; D11a [2-4, 8]; D12a [2-7]; D12b [1, 8]

Medical advice M03 [2, 5, 7]; M04a [1]; M05a [3, 6, 8]

190 diuron + glyphosate

A non-selective residual herbicide mixture for non-crop and amenity use

HRAC mode of action code: C2 + G

Products

1	Touché	Nomix Enviro	217.6:109 g/l	RH	07913
2	Xanadu	Bayer Environ.	125:100 g/l	EW	09943

Uses

- Annual dicotyledons in ***amenity vegetation***, ***land not intended to bear vegetation***
- Annual grasses in ***amenity vegetation***, ***land not intended to bear vegetation***
- Perennial dicotyledons in ***amenity vegetation***, ***land not intended to bear vegetation***
- Perennial grasses in ***amenity vegetation***, ***land not intended to bear vegetation***

Approval information

- Glyphosate included in Annex I under EC Directive 91/414
- Diuron will not be included in Annex I of EC Directive 91/414. To comply with this EC decision all products listed in this profile will need to be withdrawn from the market by about November 2007, and all approvals for their storage and use will finally expire in about November 2008
- Approved for use through ULV applicators [1]
- Approval expiry 13 Dec 2008 [1, 2]

Efficacy guidance

- Best results achieved from treatment when weeds are green and actively growing, usually in early spring. Symptoms may be slow to appear in poor growing conditions
- Perennial grasses are susceptible when tillering and making new rhizome growth, normally when plants have 4-5 new leaves
- Perennial dicotyledons most susceptible if treated at or near flowering but will be severely checked if treated at other times when growing actively
- Most species of germinating weeds are controlled. See label for details of more resistant species
- Weed control may be reduced on heavy or highly organic soils and when weeds are suffering stress in any situation
- At least 6 h and preferably 24 h without rain must follow spraying. Rain soon after treatment may reduce initial weed control
- Apply with manufacturer's specialist equipment [1]

Restrictions

- Maximum number of treatments 1 per yr
- Ornamental trees and shrubs should be established for at least 12 mth and be at least 50 mm tall
- Among ornamental trees and shrubs take care to avoid foliage or the stems of young plants
- Do not allow spray to contact desired plants or crops
- Do not use on tree and shrub nurseries or before planting
- Do not treat ornamental trees and shrubs on Sands or Very Light soils or those that are gravelly or with less than 1% organic matter [1]
- Do not decant; connect directly to Nomix applicator [1]

Crop-specific information

- Treat hard surfaces between Feb and end May

Environmental safety

- Dangerous for the environment
- Toxic to aquatic organisms
- Not to be used on food crops
- Do not apply over drains or in drainage channels, gullies or similar structures
- When treating paved or similar surfaces direct spray at joints and cracks using hand held equipment. If this is not possible ensure overall sprays do not lead to run-off

Hazard classification and safety precautions

Hazard H03, H11
Risk phrases R40, R48, R51, R53a
Operator protection A [2]; H, M; U02a [1]; U09a, U19a, U20b [1, 2]
Environmental protection E15a, E38 [1, 2]; E34 [2]
Consumer protection C01 [2]
Storage and disposal D01, D09a, D11a, D12a [1, 2]; D02 [2]; D05 [1]
Medical advice M03 [2]

191 dodine

A protectant and eradicant guanidine fungicide
FRAC mode of action code: M7

Products

1	Clayton Scabius	Clayton	450 g/l	SC	13339
2	Dodifun 400 SC	Hermoo	400 g/l	SC	11657
3	Greencrop Budburst	Greencrop	450 g/l	SC	11042
4	Radspor FL	Truchem	450 g/l	SC	01685

Uses

- Currant leaf spot in ***bilberries*** *(off-label)*, ***blueberries*** *(off-label)*, ***cranberries*** *(off-label)*, ***gooseberries*** *(off-label)*, ***redcurrants*** *(off-label)*, ***whitecurrants*** *(off-label)* [4]; ***blackcurrants*** [1-4]
- Scab in ***apples***, ***pears*** [1-4]; ***quinces*** *(off-label)* [4]

Specific Off-Label Approvals (SOLAs)

- ***bilberries*** *20060617* [4]
- ***blueberries*** *20060617* [4]
- ***cranberries*** *20060617* [4]
- ***gooseberries*** *20060617* [4]
- ***quinces*** *20060616* [4]
- ***redcurrants*** *20060617* [4]
- ***whitecurrants*** *20060617* [4]

Efficacy guidance

- Apply protective spray on apples and pears at bud-burst and at 10-14 d intervals until late Jun to early Jul
- Apply post-infection spray within 36 h of rain responsible for initiating infection. Where scab already present spray prevents production of spores
- On blackcurrants commence spraying at early grape stage and repeat at 2-3 wk intervals, and at least once after picking

Restrictions

- Do not apply in very cold weather (under 5 °C) or under slow drying conditions to pears or dessert apples during bloom or immediately after petal fall
- Do not mix with lime sulphur or tetradifon
- Consult processors before use on crops grown for processing

Crop-specific information

- Latest use: early Jul for culinary apples; pre-blossom for dessert apples and pears

Environmental safety

- Dangerous for the environment
- Very toxic to aquatic organisms

Hazard classification and safety precautions

Hazard H03, H11

Risk phrases R21 [2, 3]; R22a, R36, R38, R43, R50, R53a [1-4]; R37, R41 [1, 4]
Operator protection A, C; U05a, U08, U19a, U20b [1-4]; U11 [1, 4]
Environmental protection E15a, E34 [1-4]; E38 [4]
Storage and disposal D01, D02, D09a, D10b [1-4]; D05 [1-3]; D12a [1]
Medical advice M03 [1, 4]; M04a [2, 3]; M05a [1]

192 epoxiconazole

A systemic, protectant and curative triazole fungicide for use in cereals
FRAC mode of action code: 3

See also *boscalid + epoxiconazole*
carbendazim + epoxiconazole
dimoxystrobin + epoxiconazole

Products

1	Agriguard Epoxiconazole	AgriGuard	125 g/l	SC	09407
2	Clayton Oust	Clayton	125 g/l	SC	11588
3	Epic	BASF	125 g/l	SC	12136
4	Greencrop Martello	Greencrop	125 g/l	SC	12318
5	Landgold Epoxiconazole	Landgold	125 g/l	SC	09821
6	Landgold Epoxiconazole	Teliton	125 g/l	SC	12117
7	Opus	BASF	125 g/l	SC	12057
8	Overture	AgriGuard	125 g/l	SC	12587
9	Standon Epoxiconazole	Standon	125 g/l	SC	09517

Uses

- Brown rust in ***spring barley***, ***winter barley***, ***winter wheat*** [1-9]; ***spring rye***, ***spring wheat***, ***triticale***, ***winter rye*** [2-8]
- Eyespot in ***winter barley*** *(reduction)*, ***winter wheat*** *(reduction)* [1-9]
- Foliar disease control in ***durum wheat*** *(off-label)* [3, 7]; ***grass seed crops*** *(off-label)* [1, 2, 7, 9]
- Fusarium ear blight in ***spring wheat***, ***winter wheat*** [2-8]; ***winter wheat*** *(reduction)* [1, 9]
- Glume blotch in ***spring wheat***, ***triticale*** [2-8]; ***winter wheat*** [1-9]
- Net blotch in ***spring barley***, ***winter barley*** [1-9]
- Powdery mildew in ***spring barley***, ***winter barley***, ***winter wheat*** [1-9]; ***spring oats***, ***spring rye***, ***spring wheat***, ***triticale***, ***winter oats***, ***winter rye*** [2-8]
- Rhynchosporium in ***spring barley***, ***winter barley*** [1-9]; ***spring rye***, ***winter rye*** [2-8]
- Septoria leaf blotch in ***spring wheat***, ***triticale*** [2-8]; ***winter wheat*** [1-9]
- Sooty moulds in ***spring wheat*** *(reduction)* [2-8]; ***winter wheat*** *(reduction)* [1-9]
- Yellow rust in ***spring barley***, ***winter barley***, ***winter wheat*** [1-9]; ***spring rye***, ***spring wheat***, ***triticale***, ***winter rye*** [2-8]

Specific Off-Label Approvals (SOLAs)

- ***durum wheat*** *20061053* [3], *20061051* [7]
- ***grass seed crops*** *20061056* [1], *20061055* [2], *20061051* [7], *20061050* [9]

Approval information

- Accepted by BBPA for use on malting barley (before ear emergence only)
- Approval expiry 31 Dec 2008 [4]

Efficacy guidance

- Apply at the start of foliar disease attack
- Optimum effect against eyespot achieved by spraying between leaf-sheath erect and second node detectable stages (GS 30-32)
- Best control of ear diseases of wheat obtained by treatment during ear emergence
- Mildew control improved by use of tank mixtures. See label for details
- For Septoria spray after third node detectable stage (GS 33) when weather favouring disease development has occurred
- Epoxiconazole is a DMI fungicide. Resistance to some DMI fungicides has been identified in Septoria leaf blotch which may seriously affect performance of some products. For further advice contact a specialist advisor and visit the Fungicide Resistance Action Group (FRAG)-UK website

Restrictions

- Maximum total dose equivalent to two full dose treatments

- Product may cause damage to broad-leaved plant species
- Avoid spray drift onto neighbouring crops

Crop-specific information

- Latest use: up to and including flowering just complete (GS 69) in wheat, rye, triticale; up to and including emergence of ear just complete (GS 59) in barley, oats

Environmental safety

- Dangerous for the environment
- Very toxic to aquatic organisms
- LERAP Category B

Hazard classification and safety precautions

Hazard H03 [2-9]; H11 [1-9]
Risk phrases R38, R40, R50, R62, R63 [2-9]; R51 [1]; R53a [1-9]
Operator protection A; U05a [1-9]; U19a [4]; U20a [1]; U20b [2-9]
Environmental protection E15a, E16a [1-9]; E16b [1, 8, 9]; E38 [3, 7]
Storage and disposal D01, D02, D05, D09a [1-9]; D08 [5, 6]; D10b [1, 4-6, 9]; D10c [2, 3, 7, 8]; D12a [2, 3, 7]
Medical advice M05a [2, 3, 5-7]; M05b [9]

193 epoxiconazole + fenpropimorph

A systemic, protectant and curative fungicide mixture for cereals
FRAC mode of action code: 3 + 5

Products

1	Eclipse	BASF	84:250 g/l	SE	11731
2	Greencrop Galore	Greencrop	84:250 g/l	SE	09561
3	Opus Team	BASF	84:250 g/l	SE	11759
4	Standon Epoxifen	Standon	84:250 g/l	SE	08972

Uses

- Brown rust in ***spring barley, winter barley, winter wheat*** [1-4]; ***spring rye, spring wheat, triticale, winter rye*** [1, 3]
- Eyespot in ***winter barley*** *(reduction)*, ***winter wheat*** *(reduction)* [1-4]
- Foliar disease control in ***durum wheat*** *(off-label)* [1, 3]; ***grass seed crops*** *(off-label)* [1-4]
- Fusarium ear blight in ***spring wheat, winter wheat*** [1, 3]; ***winter wheat*** *(reduction)* [2, 4]
- Glume blotch in ***spring wheat, triticale*** [1, 3]; ***winter wheat*** [1-4]
- Net blotch in ***spring barley, winter barley*** [1-4]
- Powdery mildew in ***spring barley, winter barley, winter wheat*** [1-4]; ***spring oats, spring rye, spring wheat, triticale, winter oats, winter rye*** [1, 3]
- Rhynchosporium in ***spring barley, winter barley*** [1-4]; ***spring rye, winter rye*** [1, 3]
- Septoria leaf blotch in ***spring wheat, triticale*** [1, 3]; ***winter wheat*** [1-4]
- Sooty moulds in ***spring wheat*** *(reduction)* [1, 3]; ***winter wheat*** *(reduction)* [1-3]
- Yellow rust in ***spring barley, winter barley, winter wheat*** [1-4]; ***spring rye, spring wheat, triticale, winter rye*** [1, 3]

Specific Off-Label Approvals (SOLAs)

- ***durum wheat*** *20062644* [1], *20062645* [3]
- ***grass seed crops*** *20062644* [1], *20061077* [2], *20062645* [3], *20061081* [4]

Approval information

- Accepted by BBPA for use on malting barley

Efficacy guidance

- Apply at the start of foliar disease attack
- Optimum effect against eyespot achieved by spraying between leaf-sheath erect and second node detectable stages (GS 30-32)
- Best control of ear diseases obtained by treatment during ear emergence
- For Septoria spray after third node detectable stage (GS 33) when weather favouring disease development has occurred
- Epoxiconazole is a DMI fungicide. Resistance to some DMI fungicides has been identified in Septoria leaf blotch which may seriously affect performance of some products. For further advice contact a specialist advisor and visit the Fungicide Resistance Action Group (FRAG)-UK website

Restrictions

- Maximum total dose equivalent to two full dose treatments
- Product may cause damage to broad-leaved plant species

Crop-specific information

- Latest use: up to and including flowering just complete (GS 69) in wheat, rye, triticale; up to and including emergence of ear just complete (GS 59) in barley, oats

Environmental safety

- Dangerous for the environment
- Toxic to aquatic organisms
- LERAP Category B
- Avoid spray drift onto neighbouring crops

Hazard classification and safety precautions

Hazard H03, H11
Risk phrases R20, R43 [1-3]; R38 [4]; R40, R62, R63 [1, 3, 4]; R51, R53a [1-4]
Operator protection A [1-4]; H [1, 3]; U05a, U20b [1-4]; U14 [2]; U19a [2, 4]
Environmental protection E15a, E16a [1-4]; E16b [2, 4]; E38 [1, 3]
Storage and disposal D01, D02, D05, D09a [1-4]; D10a [2]; D10b [4]; D10c, D12a [1, 3]
Medical advice M05a [1, 3]; M05b [4]

194 epoxiconazole + fenpropimorph + kresoxim-methyl

A protectant, systemic and curative fungicide mixture for cereals
FRAC mode of action code: 3 + 5 + 11

Products

1	Mantra	BASF	125:150:125 g/l	SE	11728
2	Standon Kresoxim Super	Standon	125:150:125 g/l	SE	09794

Uses

- Brown rust in ***spring barley***, ***winter barley***, ***winter wheat*** [1, 2]; ***spring rye***, ***spring wheat***, ***triticale***, ***winter rye*** [1]
- Eyespot in ***spring oats*** *(reduction)*, ***spring rye*** *(reduction)*, ***triticale*** *(reduction)*, ***winter oats*** *(reduction)*, ***winter rye*** *(reduction)* [1]; ***winter barley*** *(reduction)*, ***winter wheat*** *(reduction)* [1, 2]
- Foliar disease control in ***durum wheat*** *(off-label)*, ***grass seed crops*** *(off-label)* [1]
- Fusarium ear blight in ***spring wheat*** *(reduction)* [1]; ***winter wheat*** *(reduction)* [1, 2]
- Glume blotch in ***spring wheat***, ***triticale*** [1]; ***winter wheat*** [1, 2]
- Net blotch in ***spring barley***, ***winter barley*** [1, 2]
- Powdery mildew in ***spring barley***, ***winter barley*** [1, 2]; ***spring oats***, ***spring rye***, ***triticale***, ***winter oats***, ***winter rye*** [1]
- Rhynchosporium in ***spring barley***, ***winter barley*** [1, 2]; ***spring rye***, ***winter rye*** [1]
- Septoria leaf blotch in ***spring wheat***, ***triticale*** [1]; ***winter wheat*** [1, 2]
- Sooty moulds in ***spring wheat*** *(reduction)* [1]; ***winter wheat*** *(reduction)* [1, 2]
- Yellow rust in ***spring barley***, ***winter barley***, ***winter wheat*** [1, 2]; ***spring rye***, ***spring wheat***, ***triticale***, ***winter rye*** [1]

Specific Off-Label Approvals (SOLAs)

- ***durum wheat*** *20061101* [1]
- ***grass seed crops*** *20061101* [1]

Approval information

- Kresoxim-methyl included in Annex I under EC Directive 91/414
- Accepted by BBPA for use on malting barley (before ear emergence only)

Efficacy guidance

- For best results spray at the start of foliar disease attack and repeat if infection conditions persist
- Optimum effect against eyespot obtained by treatment between leaf sheaths erect and first node detectable stages (GS 30-32)
- For protection against ear diseases apply during ear emergence
- Epoxiconazole is a DMI fungicide. Resistance to some DMI fungicides has been identified in Septoria leaf blotch which may seriously affect performance of some products. For further advice contact a specialist advisor and visit the Fungicide Resistance Action Group (FRAG)-UK website

- Kresoxim-methyl is a member of the QoI cross resistance group. Product should be used preventatively and not relied on for its curative potential
- Use product as part of an Integrated Crop Management strategy incorporating other methods of control, including where appropriate other fungicides with a different mode of action. Do not apply more than two foliar applications of QoI containing products to any cereal crop
- There is a significant risk of widespread resistance occurring in *Septoria tritici* populations in UK. Failure to follow resistance management action may result in reduced levels of disease control
- Strains of barley powdery mildew resistant to QoIs are common in the UK

Restrictions

- Maximum total dose equivalent to two full dose treatments

Crop-specific information

- Latest use: mid flowering (GS 65) for wheat, rye, triticale; completion of ear emergence (GS 59) for barley, oats

Environmental safety

- Dangerous for the environment
- Very toxic to aquatic organisms
- LERAP Category B
- Avoid spray drift onto neighbouring crops. Product may damage broad-leaved species

Hazard classification and safety precautions

Hazard H03, H11
Risk phrases R40, R50, R53a, R62, R63 [1, 2]; R43 [2]
Operator protection A [1, 2]; H [1]; U05a, U14, U20b
Environmental protection E15a, E16a [1, 2]; E16b [2]; E38 [1]
Storage and disposal D01, D02, D05, D09a [1, 2]; D08, D10b [2]; D10c, D12a [1]
Medical advice M05a [1]; M05b [2]

195 epoxiconazole + fenpropimorph + metrafenone

A broad spectrum fungicide mixture for cereals
FRAC mode of action code: 3 + 5 + U8

Products

Capalo	BASF	62.5:200:75 g/l	SE	13170

Uses

- Brown rust in ***spring barley***, ***spring rye***, ***spring wheat***, ***triticale***, ***winter barley***, ***winter rye***, ***winter wheat***
- Crown rust in ***spring oats***, ***winter oats***
- Fusarium ear blight in ***spring wheat*** *(reduction)*, ***winter wheat*** *(reduction)*
- Glume blotch in ***spring wheat***, ***triticale***, ***winter wheat***
- Net blotch in ***spring barley***, ***winter barley***
- Powdery mildew in ***spring barley***, ***spring oats***, ***spring rye***, ***spring wheat***, ***triticale***, ***winter barley***, ***winter oats***, ***winter rye***, ***winter wheat***
- Rhynchosporium in ***spring barley***, ***spring rye***, ***winter barley***, ***winter rye***
- Septoria leaf blotch in ***spring wheat***, ***triticale***, ***winter wheat***
- Sooty moulds in ***spring wheat*** *(reduction)*, ***winter wheat*** *(reduction)*
- Yellow rust in ***spring barley***, ***spring rye***, ***spring wheat***, ***triticale***, ***winter barley***, ***winter rye***, ***winter wheat***

Approval information

- Metrafenone included in Annex I under EC Directive 91/414

Efficacy guidance

- For best results spray at the start of foliar disease attack and repeat if infection conditions persist
- Optimum effect against eyespot obtained by treatment between leaf sheaths erect and first node detectable stages (GS 30-32)
- For protection against ear diseases apply during ear emergence
- Treatment for control of Septoria diseases should normally be made after third node detectable stage if the disease is present and favourable weather for disease development has occurred

- Epoxiconazole is a DMI fungicide. Resistance to some DMI fungicides has been identified in Septoria leaf blotch which may seriously affect performance of some products. For further advice contact a specialist advisor and visit the Fungicide Resistance Action Group (FRAG)-UK website
- Should be used as part of a resistance management strategy that includes mixtures or sequences effective against mildew and non-chemical methods
- Where mildew well established at time of treatment product should be mixed with fenpropimorph. See label

Restrictions

- Maximum number of treatments 2 per crop
- Do not apply more than two applications of any products containing metrafenone to any one crop

Crop-specific information

- Latest use: up to and including beginning of flowering (GS 61) for rye, wheat, triticale; up to and including ear emergence complete (GS 59) for barley, oats

Following crops guidance

- Cereals. oilseed rape, sugar beet, linseed, maize, clover, field beans, peas, turnips, carrots, cauliflowers, onions, lettuce or potatoes may be sown following normal harvest of a treated crop

Environmental safety

- Dangerous for the environment
- Very toxic to aquatic organisms
- LERAP Category B
- Avoid spray drift onto neighbouring crops. Broad-leaved species may be damaged

Hazard classification and safety precautions

Hazard H03, H11
Risk phrases R38, R40, R43, R50, R53a, R62, R63
Operator protection A, H; U05a, U08, U14, U15, U19a, U20b
Environmental protection E16a, E34, E38
Storage and disposal D01, D02, D05, D08, D09a, D10c, D12a
Medical advice M03

196 epoxiconazole + fenpropimorph + pyraclostrobin

A protectant and systemic fungicide mixture for cereals
FRAC mode of action code: 3 + 5 + 11

Products

Diamant	BASF	43:214:114 g/l	SE	13253

Uses

- Brown rust in ***spring barley***, ***spring wheat***, ***winter barley***, ***winter wheat***
- Crown rust in ***spring oats***, ***winter oats***
- Fusarium ear blight in ***spring wheat*** *(reduction)*, ***winter wheat*** *(reduction)*
- Glume blotch in ***spring wheat***, ***winter wheat***
- Net blotch in ***spring barley***, ***winter barley***
- Powdery mildew in ***spring barley***, ***spring oats***, ***winter barley***, ***winter oats***
- Rhynchosporium in ***spring barley***, ***winter barley***
- Septoria leaf blotch in ***spring wheat***, ***winter wheat***
- Yellow rust in ***spring barley***, ***spring wheat***, ***winter barley***, ***winter wheat***

Approval information

- Pyraclostrobin included in Annex I under EC Directive 91/414
- Accepted by BBPA for use on malting barley (before ear emergence only)

Efficacy guidance

- For best results spray at the start of foliar disease attack and repeat if infection conditions persist
- For protection against ear diseases apply during ear emergence
- Epoxiconazole is a DMI fungicide. Resistance to some DMI fungicides has been identified in Septoria leaf blotch which may seriously affect performance of some products. For further advice contact a specialist advisor and visit the Fungicide Resistance Action Group (FRAG)-UK website
- Pyraclostrobin is a member of the QoI cross resistance group. Product should be used preventatively and not relied on for its curative potential

- Use product as part of an Integrated Crop Management strategy incorporating other methods of control, including where appropriate other fungicides with a different mode of action. Do not apply more than two foliar applications of QoI containing products to any cereal crop
- There is a significant risk of widespread resistance occurring in *Septoria tritici* populations in UK. Failure to follow resistance management action may result in reduced levels of disease control
- Strains of barley powdery mildew resistant to QoIs are common in the UK

Restrictions

- Maximum number of treatments 2 per crop

Crop-specific information

- Latest use: up to and including ear emergence just complete (GS 59) for barley, oats; up to and including flowering just complete (GS 69) for wheat

Environmental safety

- Dangerous for the environment
- Very toxic to aquatic organisms
- Dangerous to fish or other aquatic life. Do not contaminate surface waters or ditches with chemical or used container
- LERAP Category B
- Avoid spray drift onto neighbouring crops. Product may damage broad-leaved species

Hazard classification and safety precautions

Hazard H03, H11
Risk phrases R20, R22a, R38, R40, R50, R53a, R63
Operator protection A; U05a, U14, U20b
Environmental protection E16a, E34, E38
Storage and disposal D01, D02, D05, D09a, D10c, D12a
Medical advice M03, M05a

SECTION 2

197 epoxiconazole + kresoxim-methyl

A protectant, systemic and curative fungicide mixture for cereals
FRAC mode of action code: 3 + 11

Products

1	Clayton Gantry	Clayton	125:125 g/l	SC	09482
2	Landmark	BASF	125:125 g/l	SC	11730
3	Standon Kresoxim-Epoxiconazole	Standon	125:125 g/l	SC	09281

Uses

- Brown rust in ***spring barley***, ***winter barley***, ***winter wheat*** [1-3]; ***spring rye***, ***spring wheat***, ***triticale***, ***winter rye*** [2]
- Eyespot in ***spring oats*** *(reduction)*, ***spring rye*** *(reduction)*, ***triticale*** *(reduction)*, ***winter oats*** *(reduction)*, ***winter rye*** *(reduction)* [2]; ***winter barley*** *(reduction)*, ***winter wheat*** *(reduction)* [1-3]
- Foliar disease control in ***durum wheat*** *(off-label)* [2]; ***grass seed crops*** *(off-label)* [1-3]
- Fusarium ear blight in ***spring wheat*** *(reduction)* [2]; ***winter wheat*** *(reduction)* [1-3]
- Glume blotch in ***spring wheat***, ***triticale*** [2]; ***winter wheat*** [1-3]
- Net blotch in ***spring barley***, ***winter barley*** [1-3]
- Powdery mildew in ***spring barley***, ***winter barley*** [1-3]; ***spring oats***, ***spring rye***, ***spring wheat***, ***triticale***, ***winter oats***, ***winter rye***, ***winter wheat*** [2]
- Rhynchosporium in ***spring barley***, ***winter barley*** [1-3]; ***spring rye***, ***winter rye*** [2]
- Septoria leaf blotch in ***spring wheat***, ***triticale*** [2]; ***winter wheat*** [1-3]
- Sooty moulds in ***spring wheat*** *(reduction)* [2]; ***winter wheat*** *(reduction)* [1-3]
- Yellow rust in ***spring barley***, ***winter barley***, ***winter wheat*** [1-3]; ***spring rye***, ***spring wheat***, ***triticale***, ***winter rye*** [2]

Specific Off-Label Approvals (SOLAs)

- ***durum wheat*** *20061084* [2]
- ***grass seed crops*** *20061047* [1], *20061084* [2], *20061085* [3]

Approval information

- Kresoxim-methyl included in Annex I under EC Directive 91/414
- Accepted by BBPA for use on malting barley (before ear emergence only)

SEE SECTION 3 FOR PRODUCTS ALSO REGISTERED

Efficacy guidance

- For best results spray at the start of foliar disease attack and repeat if infection conditions persist
- Optimum effect against eyespot obtained by treatment between leaf sheaths erect and first node detectable stages (GS 30-32)
- For protection against ear diseases apply during ear emergence
- Epoxiconazole is a DMI fungicide. Resistance to some DMI fungicides has been identified in Septoria leaf blotch which may seriously affect performance of some products. For further advice contact a specialist advisor and visit the Fungicide Resistance Action Group (FRAG)-UK website
- Kresoxim-methyl is a member of the QoI cross resistance group. Product should be used preventatively and not relied on for its curative potential
- Use product as part of an Integrated Crop Management strategy incorporating other methods of control, including where appropriate other fungicides with a different mode of action. Do not apply more than two foliar applications of QoI containing products to any cereal crop
- There is a significant risk of widespread resistance occurring in *Septoria tritici* populations in UK. Failure to follow resistance management action may result in reduced levels of disease control
- Strains of barley powdery mildew resistant to QoIs are common in the UK

Restrictions

- Maximum total dose equivalent to two full dose treatments

Crop-specific information

- Latest use: mid flowering (GS 65) for wheat, rye, triticale; completion of ear emergence (GS 59) for barley, oats

Environmental safety

- Dangerous for the environment
- Very toxic to aquatic organisms
- LERAP Category B
- Avoid spray drift onto neighbouring crops. Product may damage broad-leaved species

Hazard classification and safety precautions

Hazard H03, H11
Risk phrases R40, R50, R53a, R62, R63 [1-3]; R43 [1, 3]
Operator protection A; U05a, U14, U20b
Environmental protection E15a, E16a [1-3]; E16b [2, 3]; E38 [2]
Storage and disposal D01, D02, D09a [1-3]; D05, D10b [1, 3]; D08, D10c [2]; D12a [1, 2]
Medical advice M05a [2]; M05b [1, 3]

198 epoxiconazole + kresoxim-methyl + pyraclostrobin

A protectant, systemic and curative fungicide mixture for cereals
FRAC mode of action code: 3 + 11 + 11

Products

1	Covershield	BASF	50:67:133 g/l	SE	10900
2	Opponent	BASF	50:67:133 g/l	SE	10877

Uses

- Brown rust in ***spring barley***, ***spring wheat***, ***winter barley***, ***winter wheat***
- Crown rust in ***spring oats***, ***winter oats***
- Foliar disease control in ***durum wheat*** *(off-label)*, ***grass seed crops*** *(off-label)*, ***spring rye*** *(off-label)*, ***triticale*** *(off-label)*, ***winter rye*** *(off-label)*
- Glume blotch in ***spring wheat***, ***winter wheat***
- Net blotch in ***spring barley***, ***winter barley***
- Rhynchosporium in ***spring barley***, ***winter barley***
- Septoria leaf blotch in ***spring wheat***, ***winter wheat***
- Yellow rust in ***spring barley***, ***spring wheat***, ***winter barley***, ***winter wheat***

Specific Off-Label Approvals (SOLAs)

- ***durum wheat*** *20063644* [1], *20060539* [2]
- ***grass seed crops*** *20063644* [1], *20060539* [2]
- ***spring rye*** *20063644* [1], *20060539* [2]

- ***triticale*** *20063644* [1], *20060539* [2]
- ***winter rye*** *20063644* [1], *20060539* [2]

Approval information

- Kresoxim-methyl and pyraclostrobin included in Annex I under EC Directive 91/414
- Accepted by BBPA for use on malting barley (before ear emergence only)

Efficacy guidance

- For best results apply at the start of foliar disease attack
- Good reduction of Fusarium ear blight can be obtained from treatment during flowering
- Yield response may be obtained in the absence of visual disease symptoms
- Best results on Septoria leaf blotch when treated in the latent phase
- Epoxiconazole is a DMI fungicide. Resistance to some DMI fungicides has been identified in Septoria leaf blotch which may seriously affect performance of some products. For further advice contact a specialist advisor and visit the Fungicide Resistance Action Group (FRAG)-UK website
- Kresoxim-methyl and pyraclostrobin are members of the QoI cross resistance group. Product should be used preventatively and not relied on for its curative potential
- Use product as part of an Integrated Crop Management strategy incorporating other methods of control, including where appropriate other fungicides with a different mode of action. Do not apply more than two foliar applications of QoI containing products to any cereal crop
- There is a significant risk of widespread resistance occurring in *Septoria tritici* populations in UK. Failure to follow resistance management action may result in reduced levels of disease control

Restrictions

- Maximum total dose equivalent to two full dose treatments

Crop-specific information

- Latest use: before grain watery ripe (GS 71) for wheat; up to and including emergence of ear just complete (GS 59) for barley and oats

Environmental safety

- Dangerous for the environment
- Very toxic to aquatic organisms
- Dangerous to fish or other aquatic life. Do not contaminate surface waters or ditches with chemical or used container
- LERAP Category B

Hazard classification and safety precautions

Hazard H03, H11
Risk phrases R20, R22a, R40, R50, R53a
Operator protection A; U05a, U14, U20b
Environmental protection E13b [2]; E15a [1]; E16a, E16b, E34, E38 [1, 2]
Storage and disposal D01, D02, D09a, D10c, D12a [1, 2]; D05 [2]
Medical advice M05a

199 epoxiconazole + metrafenone

A broad spectrum fungicide mixture for cereals
FRAC mode of action code: 3 + U8

Products

Ceando	BASF	83:100 g/l	SC	13271

Uses

- Brown rust in ***spring barley***, ***spring rye***, ***spring wheat***, ***triticale***, ***winter barley***, ***winter rye***, ***winter wheat***
- Crown rust in ***spring oats***, ***winter oats***
- Fusarium ear blight in ***spring wheat*** *(reduction)*, ***winter wheat*** *(reduction)*
- Glume blotch in ***spring wheat***, ***triticale***, ***winter wheat***
- Net blotch in ***spring barley***, ***winter barley***
- Powdery mildew in ***spring barley***, ***spring oats***, ***spring rye***, ***spring wheat***, ***triticale***, ***winter barley***, ***winter oats***, ***winter rye***, ***winter wheat***
- Rhynchosporium in ***spring barley***, ***spring rye***, ***winter barley***, ***winter rye***
- Septoria leaf blotch in ***spring wheat***, ***triticale***, ***winter wheat***

- Sooty moulds in ***spring wheat*** *(reduction)*, ***winter wheat*** *(reduction)*
- Yellow rust in ***spring barley***, ***spring rye***, ***spring wheat***, ***triticale***, ***winter barley***, ***winter rye***, ***winter wheat***

Approval information

- Metrafenone included in Annex I under EC Directive 91/414

Efficacy guidance

- For best results spray at the start of foliar disease attack and repeat if infection conditions persist
- Optimum effect against eyespot obtained by treatment between leaf sheaths erect and first node detectable stages (GS 30-32)
- For protection against ear diseases apply during ear emergence
- Treatment for control of Septoria diseases should normally be made after third node detectable stage if the disease is present and favourable weather for disease development has occurred
- Epoxiconazole is a DMI fungicide. Resistance to some DMI fungicides has been identified in Septoria leaf blotch which may seriously affect performance of some products. For further advice contact a specialist advisor and visit the Fungicide Resistance Action Group (FRAG)-UK website
- Should be used as part of a resistance management strategy that includes mixtures or sequences effective against mildew and non-chemical methods
- Where mildew well established at time of treatment product should be mixed with fenpropimorph. See label

Restrictions

- Maximum number of treatments 2 per crop
- Do not apply more than two applications of any products containing metrafenone to any one crop

Crop-specific information

- Latest use: up to and including beginning of flowering (GS 61) for rye, wheat, triticale; up to and including ear emergence complete (GS 59) for barley, oats

Following crops guidance

- Cereals. oilseed rape, sugar beet, linseed, maize, clover, field beans, peas, turnips, carrots, cauliflowers, onions, lettuce or potatoes may be sown following normal harvest of a treated crop

Environmental safety

- Dangerous for the environment
- Very toxic to aquatic organisms
- LERAP Category B
- Avoid spray drift onto neighbouring crops. Broad-leaved species may be damaged

Hazard classification and safety precautions

Hazard H03, H11
Risk phrases R40, R50, R53a, R62, R63
Operator protection A, H; U05a, U20b
Environmental protection E15a, E16a, E34, E38
Storage and disposal D01, D02, D05, D08, D09a, D10c
Medical advice M03, M05a

200 epoxiconazole + pyraclostrobin

A protectant, systemic and curative fungicide mixture for cereals
FRAC mode of action code: 3 + 11

Products

1	Envoy	BASF	62.5:85 g/l	SE	12750
2	Ibex	BASF	50:133 g/l	SE	12168
3	Opera	BASF	50:133 g/l	SE	12167

Uses

- Brown rust in ***spring barley***, ***spring wheat***, ***winter barley***, ***winter wheat*** [1-3]
- Cercospora leaf spot in ***sugar beet*** [2, 3]
- Crown rust in ***spring oats***, ***winter oats*** [2, 3]; ***spring oats*** *(qualified minor use)*, ***winter oats*** *(qualified minor use)* [1]
- Foliar disease control in ***durum wheat*** *(off-label)*, ***fodder beet*** *(off-label)*, ***grass seed crops*** *(off-label)*, ***spring rye*** *(off-label)*, ***triticale*** *(off-label)*, ***winter rye*** *(off-label)* [3]

- Fusarium ear blight in ***spring wheat*** *(reduction)*, ***winter wheat*** *(reduction)* [1]
- Glume blotch in ***spring wheat***, ***winter wheat*** [1-3]
- Net blotch in ***spring barley***, ***winter barley*** [1-3]
- Powdery mildew in ***sugar beet*** [2, 3]
- Ramularia leaf spots in ***sugar beet*** [2, 3]
- Rhynchosporium in ***spring barley***, ***winter barley*** [1-3]
- Rust in ***sugar beet*** [2, 3]
- Septoria leaf blotch in ***spring wheat***, ***winter wheat*** [1-3]
- Tan spot in ***spring wheat***, ***winter wheat*** [1]
- Yellow rust in ***spring barley***, ***spring wheat***, ***winter barley***, ***winter wheat*** [1-3]

Specific Off-Label Approvals (SOLAs)

- ***durum wheat*** *20062653* [3]
- ***fodder beet*** *20062654* [3]
- ***grass seed crops*** *20062653* [3]
- ***spring rye*** *20062653* [3]
- ***triticale*** *20062653* [3]
- ***winter rye*** *20062653* [3]

Approval information

- Pyraclostrobin included in Annex I under EC Directive 91/414
- Accepted by BBPA for use on malting barley (before ear emergence only)

Efficacy guidance

- For best results apply at the start of foliar disease attack
- Good reduction of Fusarium ear blight can be obtained from treatment during flowering
- Yield response may be obtained in the absence of visual disease symptoms
- Epoxiconazole is a DMI fungicide. Resistance to some DMI fungicides has been identified in Septoria leaf blotch which may seriously affect performance of some products. For further advice contact a specialist advisor and visit the Fungicide Resistance Action Group (FRAG)-UK website
- Pyraclostrobin is a member of the QoI cross resistance group. Product should be used preventatively and not relied on for its curative potential
- Use product as part of an Integrated Crop Management strategy incorporating other methods of control, including where appropriate other fungicides with a different mode of action. Do not apply more than two foliar applications of QoI containing products to any cereal crop
- There is a significant risk of widespread resistance occurring in *Septoria tritici* populations in UK. Failure to follow resistance management action may result in reduced levels of disease control

Restrictions

- Maximum number of treatments 2 per crop
- Do not use on sugar beet crops grown for seed [3]

Crop-specific information

- Latest use: before grain watery ripe (GS 71) for wheat; up to and including emergence of ear just complete (GS 59) for barley and oats
- HI 6 wk for sugar beet [2, 3]

Environmental safety

- Dangerous for the environment
- Very toxic to aquatic organisms
- LERAP Category B

Hazard classification and safety precautions

Hazard H03, H11
Risk phrases R20, R22a, R38, R40, R43, R50, R53a [1-3]; R62, R63 [1]
Operator protection A; U05a, U14, U20b
Environmental protection E15a, E16a, E34, E38 [1-3]; E16b [2, 3]
Storage and disposal D01, D02, D09a, D10c, D12a [1-3]; D05 [2, 3]
Medical advice M03, M05a

201 esfenvalerate

A contact and ingested pyrethroid insecticide
IRAC mode of action code: 3

Products

1	Greencrop Cajole Ultra	Greencrop	25 g/l	EC	12967
2	Standon Hounddog	Standon	25 g/l	EC	13201
3	Sumi-Alpha	BASF	25 g/l	EC	10401

Uses

- Aphids in ***spring barley***, ***spring wheat***, ***winter barley***, ***winter wheat***

Approval information

- Esfenvalerate included in Annex I under EC Directive 91/414
- Accepted by BBPA for use on malting barley

Efficacy guidance

- For best reduction of spread of barley yellow dwarf virus winter sown crops at high risk (e.g. after grass or in areas with history of BYDV) should be treated when aphids first seen or by mid-Oct. Otherwise treat in late Oct-early Nov
- High risk winter sown crops will need a second treatment
- Spring sown crops should be treated from the 2-3 leaf stage if aphids are found colonising in the crop and a further application may be needed before the first node stage if they reinfest
- Product also recommended between onset of flowering and milky ripe stages (GS 61-73) for control of summer cereal aphids

Restrictions

- Maximum number of treatments 3 per crop of which 2 may be in autumn
- Do not use if another pyrethroid or dimethoate has been applied to crop after start of ear emergence (GS 51)
- Do not leave spray solution standing in spray tank

Crop-specific information

- Latest use: 31 Mar in yr of harvest (winter use dose); early milk stage for barley (summer use); late milk stage for wheat (summer use)

Environmental safety

- Dangerous for the environment
- Very toxic to aquatic organisms
- High risk to non-target insects or other arthropods. Do not spray within 6 m of the field boundary
- Flammable
- LERAP Category A
- Store product in dark away from direct sunlight

Hazard classification and safety precautions

Hazard H03, H08, H11
Risk phrases R20, R22a, R41, R43, R50, R53a
Operator protection A, C, H; U04a, U05a, U08, U11, U14, U19a, U20b
Environmental protection E15a, E22a [1, 3]; E15b, E22b [2]; E16c, E16d, E34 [1-3]; E38 [3]
Storage and disposal D01, D02, D09a [1-3]; D05 [1, 2]; D07, D10b [1]; D10c, D12a [2, 3]
Medical advice M03 [1-3]; M05b [2]

202 ethofumesate

A benzofuran herbicide for grass weed control in various crops
HRAC mode of action code: N

See also *bromoxynil + ethofumesate + ioxynil*
chloridazon + ethofumesate
desmedipham + ethofumesate + phenmedipham

Products

1	Alpha Ethofumesate	Makhteshim	500 g/l	SC	13055
2	Catalyst	AgriGuard	500 g/l	SC	13266
3	Ethosat 500	Makhteshim	500 g/l	SC	13050
4	Kubist Flo	Nufarm UK	500 g/l	SC	12987
5	Nortron Flo	Bayer CropScience	500 g/l	SC	12986

Uses

- Annual dicotyledons in ***amenity grassland*, *grass seed crops*** [1, 3]; ***fodder beet*, *mangels*, *red beet*, *sugar beet*** [1-5]
- Annual meadow grass in ***amenity grassland*, *grass seed crops*** [1, 3]; ***fodder beet*, *mangels*, *red beet*, *sugar beet*** [1-5]
- Blackgrass in ***amenity grassland*, *grass seed crops*** [1, 3]; ***fodder beet*, *mangels*, *red beet*, *sugar beet*** [1-5]
- Fat hen in ***poppies for morphine production*** *(off-label)* [5]
- Grass weeds in ***poppies for morphine production*** *(off-label)* [5]
- Mayweeds in ***poppies for morphine production*** *(off-label)* [5]

Specific Off-Label Approvals (SOLAs)

- ***poppies for morphine production*** *20070077* [5]

Approval information

- Ethofumesate included in Annex I under EC Directive 91/414

Efficacy guidance

- Most products may be applied pre- or post-emergence of crop or weeds but some restricted to pre-emergence or post-emergence use only. Check label
- Some products recommended for use only in mixtures. Check label
- Volunteer cereals not well controlled pre-emergence, weed grasses should be sprayed before fully tillered
- Grass crops may be sprayed during rain or when wet. Not recommended in very dry conditions or prolonged frost

Restrictions

- Maximum number of treatments 1 pre- plus 1 or 2 post-emergence per crop or yr for beet crops and grass seed crops; 2 per yr for amenity turf and established grassland. The maximum total dose must not exceed 1.0 kg ethofumesate per hectare in any three year period
- Do not use on Sands or Heavy soils, Very Light soils containing a high percentage of stones, or soils with more than 5-10% organic matter (percentage varies according to label)
- Do not use on swards reseeded without ploughing
- Clovers will be killed or severely checked
- Do not graze or cut grass for 14 d after, or roll less than 7 d before or after spraying

Crop-specific information

- Latest use: before crops meet across rows for beet crops and mangels; not specified for other crops
- Apply in beet crops in tank mixes with other pre- or post-emergence herbicides. Recommendations vary for different mixtures. See label for details
- Safe timing on beet crops varies with other ingredient of tank mix. See label for details
- In grass crops apply to moist soil as soon as possible after sowing or post-emergence when crop in active growth, normally mid-Oct to mid-Dec. See label for details
- May be used in Italian, hybrid and perennial ryegrass, timothy, cocksfoot, meadow fescue and tall fescue. Apply pre-emergence to autumn-sown leys, post-emergence after 2-3 leaf stage. See label for details

Following crops guidance

- In the event of failure of a treated crop only sugar beet, red beet, fodder beet and mangels may be redrilled within 3 mth of application
- Any crop may be sown 3 mth after application of mixtures in beet crops following ploughing, 5 mth after application in grass crops

Environmental safety

- Dangerous for the environment
- Toxic to aquatic organisms
- Do not empty into drains

Hazard classification and safety precautions

Hazard H11
Risk phrases R51, R53a
Operator protection U08 [1-3]; U09a [4, 5]; U19a, U20a [1-5]
Environmental protection E13c [2]; E15a [1-4]; E19b, E34 [1, 3]; E38 [4, 5]
Storage and disposal D05, D09a [1-5]; D10b [1-4]; D12a [1, 3-5]

203 ethofumesate + metamitron

A contact and residual herbicide mixture for beet crops
HRAC mode of action code: N + C1

Products

1	Galahad	Bayer CropScience	100:400 g/l	SC	10727
2	Goltix Super	Makhteshim	150:350 g/l	SC	12216
3	Torero	Makhteshim	150:350 g/l	SC	11158

Uses

- Annual dicotyledons in ***fodder beet***, ***mangels***, ***sugar beet***
- Annual meadow grass in ***fodder beet***, ***mangels***, ***sugar beet***

Approval information

- Ethofumesate included in Annex I under EC Directive 91/414

Efficacy guidance

- Best results obtained from a series of treatments applied as an overall fine spray commencing when earliest germinating weeds are no larger than fully expanded cotyledon and the majority of the crop at fully expanded cotyledon
- Apply subsequent sprays as each new flush of weeds reaches early cotyledon and continue until weed emergence ceases (maximum 3 sprays)
- Product may be used on all soil types but residual activity may be reduced on those with more than 5% organic matter

Restrictions

- Maximum total dose equivalent to three full dose treatments

Crop-specific information

- Latest use: before crop leaves meet between rows
- Crop tolerance may be reduced by stress caused by growing conditions, effects of pests, disease or other pesticides, nutrient deficiency etc

Following crops guidance

- Beet crops may be sown at any time after treatment. Any other crop may be sown after mouldboard ploughing to 15 cm and a minimum interval of 3 mth after treatment

Environmental safety

- Dangerous for the environment
- Very toxic to aquatic organisms
- Do not empty into drains

Hazard classification and safety precautions

Hazard H03 [1]; H11 [1-3]
Risk phrases R22a, R43, R50 [1]; R51 [2, 3]; R53a [1-3]
Operator protection A, H [1]; U05a, U08, U19a [1-3]; U14, U15, U20a [2, 3]; U20b [1]
Environmental protection E13c, E34 [1-3]; E19b [2, 3]; E38 [1]

Storage and disposal D01, D02, D05, D09a, D10b, D12a
Medical advice M03 [1]

204 ethofumesate + phenmedipham

A contact and residual herbicide for use in beet crops
HRAC mode of action code: N + C1

Products

1	Betosip Combi	Sipcam	190:200 g/l	SC	13254
2	Duo 400 SC	AgriGuard	200:200 g/l	SC	12561
3	Fenlander 2	Makhteshim	94:97 g/l	EC	13124
4	Powertwin	Makhteshim	200:200 g/l	SC	12373
5	Thunder	Makhteshim	200:200 g/l	SC	12464
6	Twin	Makhteshim	94:97 g/l	EC	13114

Uses
- Annual dicotyledons in ***fodder beet*, *mangels*, *sugar beet*** [1-6]; ***red beet*** [2]
- Annual meadow grass in ***fodder beet*, *mangels*, *sugar beet*** [1-6]; ***red beet*** [2]
- Blackgrass in ***fodder beet*, *mangels*, *sugar beet*** [1-6]; ***red beet*** [2]

Approval information
- Ethofumesate and phenmedipham included in Annex I under EC Directive 91/414

Efficacy guidance
- Best results achieved by repeat applications to cotyledon stage weeds. Larger susceptible weeds not killed by first treatment usually checked and controlled by second application
- Apply on all soil types at 5-10 d intervals
- On soils with more than 5-10% organic matter residual activity may be reduced

Restrictions
- Maximum number of treatments normally 3 per crop or maximum total dose equivalent to three full dose treatments, or less - see labels for details
- Must be used in conjunction with specified adjuvant. See label [1, 2, 4, 5]
- Do not spray wet foliage or if rain imminent
- Spray must be applied low volume to avoid crystallisation and nozzle blockage. See label for details
- Spray in evening if daytime temperatures above 21 °C expected
- Avoid or delay treatment if frost expected within 7 d
- Avoid or delay treating crops under stress from wind damage, manganese or lime deficiency, pest or disease attack etc

Crop-specific information
- Latest use: before crop foliage meets in the rows
- Check from which recovery may not be complete may occur if treatment made during conditions of sharp diurnal temperature fluctuation

Following crops guidance
- Beet crops may be sown at any time after treatment. Any other crop may be sown after mouldboard ploughing to 15 cm and a minimum interval of 3 mth after treatment

Environmental safety
- Dangerous for the environment [2-6]
- Toxic (harmful [1]) to aquatic organisms
- Do not empty into drains
- Extra care necessary to avoid drift because product is recommended for use as a fine spray

Hazard classification and safety precautions
Hazard H03 [2, 3, 6]; H04 [4, 5]; H11 [2-6]
Risk phrases R20 [3]; R37, R40 [6]; R43, R51 [2-6]; R52 [1]; R53a [1-6]
Operator protection A [2-6]; H [2, 4, 5]; U05a, U08, U19a, U20a [1-6]; U14 [2, 4, 5]
Environmental protection E15a, E34 [1-6]; E19b [2-6]; E38 [3]
Storage and disposal D01, D09a, D10b, D12a [1-6]; D02 [1, 3-6]; D05 [1, 2, 4]
Medical advice M05a [2-6]

SEE SECTION 3 FOR PRODUCTS ALSO REGISTERED

205 ethoprophos

An organophosphorus nematicide and insecticide
IRAC mode of action code: 1B

Products

Mocap 10G	Bayer CropScience	10% w/w	GR	10003

Uses

- Potato cyst nematode in ***potatoes***
- Wireworm in ***potatoes***

Approval information

- Ethoprophos included in Annex I under EC Directive 91/414
- In 2006 PSD required that all products containing this active ingredient should carry the following warning in the main area of the container label: "Ethoprofos is an anticholinesterase organophosphate. Handle with care"

Efficacy guidance

- Broadcast shortly before or during final soil preparation with suitable fertilizer spreader and incorporate immediately to 10-15 cm. Deeper incorporation may reduce efficacy. See label for details
- Treatment can be applied on all soil types. Control of pests reduced on organic soils
- Effectiveness dependent on soil moisture. Drought after application may reduce control
- Where pre-planting nematode populations are high, loss of crop vigour and subsequent loss of yield may occur in spite of treatment

Restrictions

- Contains an anticholinesterase organophosphorous compound. Do not use if under medical advice not to work with such compounds
- Maximum number of treatments 1 per crop
- Do not harvest crops for human or animal consumption for at least 8 wk after application
- Vehicles with a closed cab must be used when applying and incorporating
- Do not apply by hand or hand held equipment

Crop-specific information

- Latest use: pre-planting of crop
- Field scale application must only be through positive displacement type specialist granule applicators or dual purpose fertilizer applicators with microgranule setting

Environmental safety

- Dangerous for the environment
- Toxic to aquatic organisms
- Dangerous to livestock. Keep all livestock out of treated areas/away from treated water for at least 13 wk. Bury or remove spillages
- Dangerous to game, wild birds and animals
- Product supplied in returnable/refillable containers. Follow instructions on label

Hazard classification and safety precautions

Hazard H02, H11
Risk phrases R22a, R23, R43, R51, R53a
Operator protection A, D, H; U02a, U04a, U05a, U07, U08, U13, U14, U19a, U20a
Environmental protection E06b (13 wk); E10a, E13b, E34, E36a, E38
Consumer protection C02a (8 wk)
Storage and disposal D01, D02, D09a, D12a, D14
Medical advice M01, M03, M04a

206 ethylene (commodity substance)

A gas used for fruit ripening and potato storage

Products

ethylene	various	-	GA

Uses

- Fruit ripening in ***fruit crops*** *(in store)*
- Sprout suppression in ***potatoes*** *(in store)*

Approval information

- Ethylene is being supported for review in the fourth stage of the EC Review Programme under Directive 91/414. Whether or not it is included in Annex I, the existing commodity chemical approval will have to be revoked because substances listed in Annex I must be approved for marketing and use under the EC regime. If ethylene is included in Annex I, products containing it will need to gain approval in the normal way if they carry label claims for pesticidal activity

Efficacy guidance

- For use in stored fruit or potatoes after harvest

Restrictions

- Handling and release of ethylene must only be undertaken by operators suitably trained and competent to carry out the work
- Operators must vacate treated areas immediately after ethylene introduction
- Unprotected persons must be excluded from the treated areas until atmospheres have been thoroughly ventilated for 15 minutes minimum before re-entry
- Ambient atmospheric ethylene concentration must not exceed 1000 ppm. Suitable self-contained breathing apparatus must be worn in atmospheres containing ethylene in excess of 1000 ppm
- A minimum 3 d post treatment period is required before removal of treated crop from storage
- Ethylene treatment must only be undertaken in fully enclosed storage areas that are air tight with appropriate air circulation and venting facilities

207 etridiazole

A protective thiadiazole fungicide for soil or compost incorporation
FRAC mode of action code: 14

Products

1	Standon Etridiazole 35	Standon	35% w/w	WP	08778
2	Terraguard	AgriGuard	35% w/w	WP	12321

Uses

- Damping off and foot rot in ***cabbages***, ***cauliflowers*** *(seeds and seedlings)*, ***celery (outdoor)*** *(seeds and seedlings)*, ***cucumbers*** *(seeds and seedlings)*, ***mustard and cress***, ***ornamental plant production*** *(seeds and seedlings)*, ***tomatoes (outdoor)*** *(seeds and seedlings)* [1, 2]; ***protected tomatoes*** *(seeds and seedlings)* [1]
- Phytophthora root rot in ***cabbages***, ***cauliflowers*** *(transplants)*, ***celery (outdoor)*** *(transplants)*, ***cucumbers*** *(transplants)*, ***ornamental plant production*** *(rooted cuttings and transplants)*, ***tomatoes (outdoor)*** *(transplants)*, ***tulips*** [1, 2]; ***ornamental plant production***, ***protected tomatoes*** *(transplants)* [1]
- Phytophthora wilt in ***cabbages***, ***cauliflowers*** *(transplants)*, ***celery (outdoor)*** *(transplants)*, ***cucumbers*** *(transplants)*, ***ornamental plant production*** *(rooted cuttings and transplants)*, ***tomatoes (outdoor)*** *(transplants)*, ***tulips*** [1, 2]; ***ornamental plant production***, ***protected tomatoes*** *(transplants)* [1]

Approval information

- Approval expiry 31 Aug 2008 [1, 2]

Efficacy guidance

- Best results obtained when used as prophylactic treatment
- Compost/soil incorporation more effective than when applied as drench
- After drenching wash spray residue from crop foliage
- When treating compost for blocking reduce dose by 50% to allow for compaction
- Uniform distribution and thorough incorporation essential for best results. Dilute powder with dry sand to facilitate handling
- Treat compost as soon as possible before use
- Do not apply to wet soil

Restrictions

- Maximum number of treatments 1 per crop or yr for compost incorporation. No restrictions on drenches

SEE SECTION 3 FOR PRODUCTS ALSO REGISTERED

- Do not use on *Escallonia, Pyracantha, Gloxinia* spp., pansies or lettuces
- Do not drench seedlings until well established
- When using compost containing more than 20% inert material or a high proportion of young peat, check crop safety before widespread use
- Test on small numbers of plants in advance when treating subjects of unknown susceptibility or using compost with more than 20% inert material

Crop-specific information

- Latest use: normally before sowing or transplanting
- HI tomatoes, mustard and cress 3 d
- Germination of lettuce in previously treated soil may be impaired

Environmental safety

- Dangerous for the environment
- Very toxic to aquatic organisms

Hazard classification and safety precautions

Hazard H03, H11 [1]; H04 [2]
Risk phrases R36 [1, 2]; R38, R40, R50, R53a [1]
Operator protection A, C; U02a, U04a, U05a, U08 [1, 2]; U19a, U20a [2]; U20b [1]
Environmental protection E15a
Storage and disposal D01, D02, D09a, D11a [1, 2]; D07 [2]

208 famoxadone

A strobilurin fungicide available only in mixtures
FRAC mode of action code: 11

See also cymoxanil + famoxadone

209 famoxadone + flusilazole

A contact, preventative and curative fungicide mixture for cereals and oilseed rape
FRAC mode of action code: 11 + 3

Products

1	Charisma	DuPont	100:106.7 g/l	EC	10415
2	Medley	DuPont	100:106.7 g/l	EC	10933

Uses

- Brown rust in ***spring barley***, ***winter barley***, ***winter wheat*** [1, 2]
- Foliar disease control in ***durum wheat*** *(off-label)*, ***grass seed crops*** *(off-label)*, ***triticale*** *(off-label)* [1, 2]; ***spring rye*** *(off-label)*, ***winter rye*** *(off-label)* [1]
- Glume blotch in ***winter wheat*** [1, 2]
- Light leaf spot in ***spring oilseed rape***, ***winter oilseed rape*** [1, 2]
- Net blotch in ***spring barley***, ***winter barley*** [1, 2]
- Phoma in ***spring oilseed rape*** *(suppression)*, ***winter oilseed rape*** *(suppression)* [1, 2]
- Rhynchosporium in ***spring barley***, ***winter barley*** [1, 2]
- Septoria leaf blotch in ***winter wheat*** [1, 2]
- Yellow rust in ***spring barley***, ***winter barley***, ***winter wheat*** [1, 2]

Specific Off-Label Approvals (SOLAs)

- ***durum wheat*** *20060505* [1], *20060506* [2]
- ***grass seed crops*** *20060505* [1], *20060506* [2]
- ***spring rye*** *20060505* [1]
- ***triticale*** *20060505* [1], *20060506* [2]
- ***winter rye*** *20060505* [1]

Approval information

- Famoxadone included in Annex I under EC Directive 91/414
- Flusilazole has been included in Annex I under Directive 91/414 for 18 months from January 2007 and all approvals for products containing this active ingredient will expire in June 2008
- Accepted by BBPA for use on malting barley

Efficacy guidance

- Best results obtained from application at early stage of disease development before infection spreads to new growth
- Famoxadone is a member of the QoI cross resistance group. Product should be used preventatively and not relied on for its curative potential
- Use product as part of an Integrated Crop Management strategy incorporating other methods of control, including where appropriate other fungicides with a different mode of action. Do not apply more than two foliar applications of QoI containing products to any cereal crop
- There is a significant risk of widespread resistance occurring in *Septoria tritici* populations in UK. Failure to follow resistance management action may result in reduced levels of disease control
- Flusilazole is a DMI fungicide. Resistance to some DMI fungicides has been identified in Septoria leaf blotch which may seriously affect performance of some products. For further advice contact a specialist advisor and visit the Fungicide Resistance Action Group (FRAG)-UK website

Restrictions

- Maximum number of treatments 2 per crop for wheat and barley; 1 per crop for oilseed rape
- Do not apply to crops under stress
- Do not apply during frosty weather

Crop-specific information

- Latest use: before flowering (GS 60) for winter wheat; before quarter ear emergence (GS 53) for barley; up to decline of flowering (GS 67) for oilseed rape

Environmental safety

- Dangerous for the environment
- Very toxic to aquatic organisms
- Dangerous to fish or other aquatic life. Do not contaminate surface waters or ditches with chemical or used container
- LERAP Category B

Hazard classification and safety precautions

Hazard H02, H11
Risk phrases R36, R40, R51, R53a, R61
Operator protection A, C; U05a, U19a, U20b
Environmental protection E13b, E16a, E16b, E38
Storage and disposal D01, D02, D05, D09a, D10b, D12a
Medical advice M03, M04a

210 fatty acids

A soap concentrate insecticide and acaricide

Products

Savona	Koppert	49% w/w	SL	06057

Uses

- Aphids in ***broad beans, brussels sprouts, cabbages, cucumbers, fruit trees, lettuce, peas, peppers, protected tomatoes, pumpkins, runner beans, tomatoes (outdoor), woody ornamentals***
- Mealybugs in ***broad beans, brussels sprouts, cabbages, cucumbers, fruit trees, lettuce, peas, peppers, protected tomatoes, pumpkins, runner beans, tomatoes (outdoor), woody ornamentals***
- Scale insects in ***broad beans, brussels sprouts, cabbages, cucumbers, fruit trees, lettuce, peas, peppers, protected tomatoes, pumpkins, runner beans, tomatoes (outdoor), woody ornamentals***
- Spider mites in ***broad beans, brussels sprouts, cabbages, cucumbers, fruit trees, lettuce, peas, peppers, protected tomatoes, pumpkins, runner beans, tomatoes (outdoor), woody ornamentals***
- Whitefly in ***broad beans, brussels sprouts, cabbages, cucumbers, fruit trees, lettuce, peas, peppers, protected tomatoes, pumpkins, runner beans, tomatoes (outdoor), woody ornamentals***

Efficacy guidance

- Use only soft or rain water for diluting spray

SEE SECTION 3 FOR PRODUCTS ALSO REGISTERED

- Pests must be sprayed directly to achieve any control. Spray all plant parts thoroughly to run off
- For glasshouse use apply when insects first seen and repeat as necessary. For scale insects apply several applications at weekly intervals after egg hatch
- To control whitefly spray when required and use biological control after 12 h

Restrictions

- Do not use on new transplants, newly rooted cuttings or plants under stress
- Do not use on specified susceptible shrubs. See label for details

Crop-specific information

- HI zero

Environmental safety

- Harmful to fish or other aquatic life. Do not contaminate surface waters or ditches with chemical or used container

Hazard classification and safety precautions

Operator protection U20c
Environmental protection E13c
Storage and disposal D05, D09a, D10a

211 fenamidone

A strobilurin fungicide available only in mixtures
FRAC mode of action code: 11

212 fenamidone + propamocarb hydrochloride

A systemic fungicide mixture for potatoes
FRAC mode of action code: 11 + 28

Products

Consento	Bayer CropScience	75:375 g/l	SC	11889

Uses

- Blight in ***potatoes***

Approval information

- Fenamidone and propamocarb hydrochloride included in Annex I under EC Directive 91/414

Efficacy guidance

- Apply as soon as there is risk of blight infection or immediately after a blight warning
- Fenamidone is a member of the QoI cross resistance group. To minimise the likelihood of development of resistance these products should be used in a planned Resistance Management strategy. In addition before application consult and adhere to the latest FRAG-UK resistance guidance on application of QoI fungicides to potatoes. See Section 5 for more information

Restrictions

- Maximum number of treatments (including any other QoI containing fungicide) 6 per yr but no more than three should be applied consecutively. Use in alternation with fungicides from a different cross-resistance group
- Observe a 7 d interval between treatments
- Use only as a protective treatment. Do not use when blight has become readily visible (1% leaf area destroyed)

Crop-specific information

- HI 7 d for potatoes

Environmental safety

- Dangerous for the environment
- Very toxic to aquatic organisms
- Dangerous to fish or other aquatic life. Do not contaminate surface waters or ditches with chemical or used container
- LERAP Category B

Hazard classification and safety precautions

Hazard H04, H11

Risk phrases R36, R50, R53a
Operator protection A, C; U05a, U08, U11, U20a
Environmental protection E13b, E16a, E38
Storage and disposal D01, D02, D05, D09a, D10b, D12a

213 fenarimol

A systemic curative and protective pyrimidine fungicide
FRAC mode of action code: 3

Products

1	Power-Turf	AgriGuard	120 g/l	SC	12954
2	Rimidin	Rigby Taylor	120 g/l	SC	05907
3	Rubigan	Gowan	120 g/l	SC	12355

Uses

- Dollar spot in ***managed amenity turf*** [1, 2]
- Fusarium patch in ***managed amenity turf*** [1, 2]
- Powdery mildew in ***apples***, ***blackcurrants***, ***gooseberries***, ***marrows*** *(off-label)*, ***protected courgettes*** *(off-label)*, ***protected gherkins*** *(off-label)*, ***protected peppers*** *(off-label)*, ***protected tomatoes*** *(off-label)*, ***pumpkins*** *(off-label)*, ***raspberries***, ***roses***, ***squashes*** *(off-label)*, ***strawberries*** [3]
- Red thread in ***managed amenity turf*** [1, 2]
- Scab in ***apples*** [3]

Specific Off-Label Approvals (SOLAs)

- ***marrows*** *20050960* [3]
- ***protected courgettes*** *20050961* [3]
- ***protected gherkins*** *20050961* [3]
- ***protected peppers*** *20050962* [3]
- ***protected tomatoes*** *20050962* [3]
- ***pumpkins*** *20050960* [3]
- ***squashes*** *20050960* [3]

Approval information

- Fenarimol has been included in Annex I under Directive 91/414 for 18 months from January 2007 and all approvals for products containing this active ingredient will expire in June 2008
- Approval expiry 30 Jun 2008 [1, 2]

Efficacy guidance

- Best results obtained from treatment of actively growing turf when the leaves are dry
- For turf disease control spray as preventive treatment and repeat as necessary. See label for details

Restrictions

- Maximum number of treatments 15 per yr on apples; 4 per yr on managed amenity turf; 3 per yr on raspberries
- Do not mow within 24 h after treatment

Crop-specific information

- HI 14 d for apples, blackcurrants, gooseberries, raspberries, strawberries; 7 d for marrows, protected courgettes, protected gherkins, pumpkins, squashes; 2 d for protected peppers, protected tomatoes,

Environmental safety

- Dangerous for the environment
- Harmful to aquatic organisms

Hazard classification and safety precautions

Hazard H03 [1-3]; H11 [1, 2]
Risk phrases R51, R63, R64 [1, 2]; R52, R61 [3]; R53a, R62 [1-3]
Operator protection A, C [1-3]; H [1, 2]; U04a, U05a, U20c [1, 2]; U08 [1-3]; U19a [3]
Environmental protection E13c [1, 2]; E15a [1-3]
Consumer protection C02a [3] (14 d)

SEE SECTION 3 FOR PRODUCTS ALSO REGISTERED

Storage and disposal D01, D09a, D10b [1-3]; D02, D05, D07 [1, 2]
Medical advice M03 [1, 2]; M05b [3]

214 fenbuconazole

A systemic protectant and curative triazole fungicide for top fruit and grapevines
FRAC mode of action code: 3

Products

Indar 5 EW	Landseer	50 g/l	EW	09518

Uses

- Blossom wilt in ***cherries*** *(off-label)*, ***mirabelles*** *(off-label)*, ***plums*** *(off-label)*
- Brown rot in ***cherries*** *(off-label)*, ***mirabelles*** *(off-label)*, ***plums*** *(off-label)*
- Powdery mildew in ***apples*** *(reduction)*, ***pears*** *(reduction)*, ***wine grapes*** *(off-label)*
- Scab in ***apples***, ***pears***

Specific Off-Label Approvals (SOLAs)

- ***cherries*** *20031372*
- ***mirabelles*** *20031372*
- ***plums*** *20031372*
- ***wine grapes*** *20032081*

Efficacy guidance

- Most effective when used as part of a routine preventative programme from bud burst to onset of petal fall
- After petal fall, tank mix with other protectant fungicides to enhance scab control
- See label for recommended spray intervals. In periods of rapid growth or high disease pressure, a 7 d interval should be used

Restrictions

- Maximum total dose equivalent to ten full doses per yr on apples, pears
- Consult processors before using on pears for processing
- Do not harvest for human or animal consumption for at least 4 wk after last application

Crop-specific information

- HI 28 d for apples, pears; 21 d for grapevines; 3 d for cherries, mirabelles, plums
- Safe to use on all main commercial varieties of apples and pears in UK

Environmental safety

- Dangerous for the environment
- Toxic to aquatic organisms

Hazard classification and safety precautions

Hazard H04, H11
Risk phrases R36, R41, R51, R53a
Operator protection A, C; U05a, U08, U11, U20a
Environmental protection E15a, E38
Consumer protection C02a (4 wk)
Storage and disposal D01, D02, D05, D11a, D12a

215 fenhexamid

A protectant hydroxyanilide fungicide for soft fruit and a range of horticultural crops
FRAC mode of action code: 17

Products

Teldor	Bayer CropScience	50% w/w	WG	11229

Uses

- Botrytis in ***bilberries*** *(off-label)*, ***blackberries***, ***blackcurrants***, ***blueberries*** *(off-label)*, ***cherries*** *(off-label)*, ***chives*** *(off-label)*, ***cress*** *(off-label)*, ***gooseberries***, ***herbs (see appendix 6)*** *(off-label)*, ***lamb's lettuce*** *(off-label)*, ***leaf spinach*** *(off-label)*, ***lettuce*** *(off-label)*, ***loganberries***, ***parsley*** *(off-label)*, ***plums*** *(off-label)*, ***protected aubergines*** *(off-label)*, ***protected courgettes*** *(off-label)*, ***protected cucumbers*** *(off-label)*, ***protected gherkins*** *(off-label)*, ***protected peppers*** *(off-label)*,

protected squashes *(off-label)*, ***protected tomatoes*** *(off-label)*, ***raspberries***, ***redcurrants***, ***rubus hybrids***, ***salad brassicas*** *(off-label - for baby leaf production)*, ***scarole*** *(off-label)*, ***strawberries***, ***tomatoes (outdoor)*** *(off-label)*, ***whitecurrants***, ***wine grapes***

Specific Off-Label Approvals (SOLAs)

- ***bilberries*** *20061290*
- ***blueberries*** *20061290*
- ***cherries*** *20031866*
- ***chives*** *20050026*
- ***cress*** *20050026*
- ***herbs (see appendix 6)*** *20050026*
- ***lamb's lettuce*** *20050026*
- ***leaf spinach*** *20050026*
- ***lettuce*** *20050026*
- ***parsley*** *20050026*
- ***plums*** *20031866*
- ***protected aubergines*** *20042087*
- ***protected courgettes*** *20042085*
- ***protected cucumbers*** *20042085*
- ***protected gherkins*** *20042085*
- ***protected peppers*** *20042086*
- ***protected squashes*** *20042085*
- ***protected tomatoes*** *20042087*
- ***salad brassicas*** *(for baby leaf production) 20050026*
- ***scarole*** *20050026*
- ***tomatoes (outdoor)*** *20042399*

Approval information

- Fenhexamid included in Annex I under EC Directive 91/414

Efficacy guidance

- Use as part of a programme of sprays throughout the flowering period to achieve effective control of Botrytis
- To minimise possibility of development of resistance, no more than two sprays of the product may be applied consecutively. Other fungicides from a different chemical group should then be used for at least two consecutive sprays. If only two applications are made on grapevines, only one may include fenhexamid
- Complete spray cover of all flowers and fruitlets throughout the blossom period is essential for successful control of Botrytis
- Spray programmes should normally start at the start of flowering

Restrictions

- Maximum number of treatments 2 per yr on grapevines; 4 per yr on other listed crops but no more than 2 sprays may be applied consecutively

Crop-specific information

- HI 1 d for strawberries, raspberries, loganberries, blackberries, Rubus hybrids; 3 d for cherries; 7 d for blackcurrants, redcurrants, whitecurrants, gooseberries; 21d for outdoor grapes

Environmental safety

- Dangerous for the environment
- Harmful to fish or other aquatic life. Do not contaminate surface waters or ditches with chemical or used container

Hazard classification and safety precautions

Hazard H11
Risk phrases R53a
Operator protection U08, U19a, U20b
Environmental protection E13c, E38
Storage and disposal D05, D09a, D11a, D12a

216 fenoxaprop-P-ethyl

An aryloxyphenoxypropionate herbicide for use in wheat
HRAC mode of action code: A

Products

1	Cheetah Super	Bayer CropScience	55 g/l	EW	08723
2	Foxtrot EW	Headland	69 g/l	EW	13243
3	Oskar	Headland	69 g/l	EW	13344
4	Triumph	Bayer CropScience	120 g/l	EC	10902

Uses

- Blackgrass in ***spring barley***, ***winter barley*** [2, 3]; ***spring wheat***, ***winter wheat*** [1-4]
- Canary grass in ***spring barley***, ***winter barley*** [2, 3]; ***spring wheat***, ***winter wheat*** [1-4]
- Rough meadow grass in ***spring barley***, ***winter barley*** [2, 3]; ***spring wheat***, ***winter wheat*** [1-4]
- Wild oats in ***spring barley***, ***winter barley*** [2, 3]; ***spring wheat***, ***winter wheat*** [1-4]

Efficacy guidance

- Treat weeds from 2 fully expanded leaves up to flag leaf ligule just visible; for awned canary-grass and rough meadow-grass from 2 leaves to the end of tillering
- A second application may be made in spring where susceptible weeds emerge after an autumn application
- Spray is rainfast 1 h after application
- Dry conditions resulting in moisture stress may reduce effectiveness
- Fenoxaprop-P-ethyl is an ACCase inhibitor herbicide. To avoid the build up of resistance do not apply products containing an ACCase inhibitor herbicide more than twice to any crop. In addition do not use any product containing fenoxaprop-P-ethyl in mixture or sequence with any other product containing the same ingredient
- Use these products as part of a resistance management strategy that includes cultural methods of control and does not use ACCase inhibitors as the sole chemical method of grass weed control
- Applying a second product containing an ACCase inhibitor to a crop will increase the risk of resistance development; only use a second ACCase inhibitor to control different weeds at a different timing
- Always follow WRAG guidelines for preventing and managing herbicide resistant weeds. See Section 5

Restrictions

- Maximum total dose equivalent to one or two full dose treatments depending on product used
- Do not apply to barley, durum wheat, undersown crops or crops to be undersown
- Do not roll or harrow within 1 wk of spraying
- Do not spray crops under stress, suffering from drought, waterlogging or nutrient deficiency or those grazed or if soil compacted
- Avoid spraying immediately before or after a sudden drop in temperature or a period of warm days/cold nights
- Do not mix with hormone weedkillers

Crop-specific information

- Latest use: before flag leaf sheath extending (GS 41)
- Treat from crop emergence to flag leaf fully emerged (GS 41).
- Product may be sprayed in frosty weather provided crop hardened off but do not spray wet foliage or leaves covered with ice
- Broadcast crops should be sprayed post-emergence after plants have developed well-established root system

Environmental safety

- Dangerous for the environment
- Very toxic (toxic [1]) to aquatic organisms

Hazard classification and safety precautions

Hazard H04 [2-4]; H11 [1-4]
Risk phrases R36, R50 [4]; R38 [2-4]; R43 [2, 3]; R51 [1-3]; R53a [1-4]
Operator protection A, C, H; U05a, U20b [1-4]; U08 [2-4]; U09a [1]; U14, U19a [2, 3]
Environmental protection E13c [1, 4]; E15a, E34 [2, 3]; E38 [1-4]
Storage and disposal D01, D02, D09a, D10b [1-4]; D05 [2, 3]; D12a [1, 4]

217 fenoxycarb

An insect specific growth regulator for top fruit
IRAC mode of action code: 7B

Products

Insegar WG	Syngenta	25% w/w	WG	09789

Uses

- Summer-fruit tortrix moth in ***apples***, ***pears***

Efficacy guidance

- Best results from application at 5th instar stage before pupation. Product prevents transformation from larva to pupa
- Correct timing best identified from pest warnings
- Because of mode of action rapid knock-down of pest is not achieved and larvae continue to feed for a period after treatment
- Adequate water volume necessary to ensure complete coverage of leaves

Restrictions

- Maximum number of treatments 2 per crop
- Consult processors before use

Crop-specific information

- HI 42 d for apples, pears
- Use on all varieties of apples and pears

Environmental safety

- Dangerous for the environment
- Toxic to aquatic organisms
- High risk to bees. Do not apply to crops in flower or to those in which bees are actively foraging. Do not apply when flowering weeds are present
- Risk to certain non-target insects or other arthropods. See directions for use
- Broadcast air-assisted LERAP (8 m)
- Apply to minimise off-target drift to reduce effects on non-target organisms. Some margin of safety to beneficial arthropods is indicated.

Hazard classification and safety precautions

Hazard H11
Risk phrases R51, R53a
Operator protection A, C, H; U05a, U20a, U23a
Environmental protection E12a, E12e, E15a, E22b, E38; E17b (8 m)
Storage and disposal D01, D02, D07, D09a, D11a, D12a

218 fenpropidin

A systemic, curative and protective piperidine (morpholine) fungicide
FRAC mode of action code: 5

See also difenoconazole + fenpropidin

Products

1	Alpha Fenpropidin 750 EC	Makhteshim	750 g/l	EC	12843
2	Instinct	Headland	750 g/l	EC	12317
3	Tern	Syngenta	750 g/l	EC	08660

Uses

- Brown rust in ***spring barley***, ***winter barley*** [1-3]
- Foliar disease control in ***durum wheat*** *(off-label)* [1, 3]; ***spring rye*** *(off-label)*, ***triticale*** *(off-label)*, ***winter rye*** *(off-label)* [3]
- Powdery mildew in ***spring barley***, ***spring wheat***, ***winter barley***, ***winter wheat*** [1-3]
- Rhynchosporium in ***spring barley*** *(moderate control)*, ***winter barley*** *(moderate control)* [1-3]
- Yellow rust in ***spring barley*** *(moderate control)*, ***spring wheat*** *(moderate control)*, ***winter barley*** *(moderate control)*, ***winter wheat*** *(moderate control)* [1-3]

SEE SECTION 3 FOR PRODUCTS ALSO REGISTERED

Specific Off-Label Approvals (SOLAs)

- ***durum wheat*** *20070459* [1], *20060533* [3]
- ***spring rye*** *20060533* [3]
- ***triticale*** *20060533* [3]
- ***winter rye*** *20060533* [3]

Approval information

- Accepted by BBPA for use on malting barley

Efficacy guidance

- Best results obtained when applied at early stage of disease development. See label for details of recommended timing alone and in mixtures
- Disease control enhanced by vapour-phase activity. Control can persist for 4-6 wk
- Alternate with triazole fungicides to discourage build-up of resistance

Restrictions

- Maximum number of treatments 3 per crop (up to 2 in yr of harvest) for winter crops; 2 per crop for spring crops
- Treated crops must not be harvested for human or animal consumption for at least 5 wk after the last application

Crop-specific information

- Latest use: up to and including ear emergence complete (GS 59).
- HI 5 wk

Environmental safety

- Dangerous for the environment
- Very toxic to aquatic organisms

Hazard classification and safety precautions

Hazard H03, H11

Risk phrases R20, R21 [1]; R22a [1, 3]; R37 [3]; R38 [2]; R41 [1, 2]; R50, R53a [1-3]

Operator protection A, C, H; U02a, U05a [1-3]; U04a, U10 [1, 3]; U09a [2]; U11, U19a, U20b [1, 2]; U20a [3]

Environmental protection E15a, E34 [2, 3]; E19b [1]; E38 [1-3]

Consumer protection C02a [2, 3] (5 wk)

Storage and disposal D01, D02, D05, D09a, D12a [1-3]; D10c [2]

Medical advice M03

219 fenpropimorph

A contact and systemic morpholine fungicide
FRAC mode of action code: 5

See also *azoxystrobin + fenpropimorph*
epoxiconazole + fenpropimorph
epoxiconazole + fenpropimorph + kresoxim-methyl
epoxiconazole + fenpropimorph + metrafenone
epoxiconazole + fenpropimorph + pyraclostrobin

Products

1	Clayton Spigot	Clayton	750 g/l	EC	11560
2	Corbel	BASF	750 g/l	EC	00578
3	Landgold Fenpropimorph 750	Landgold	750 g/l	EC	10472
4	Landgold Fenpropimorph 750	Teliton	750 g/l	EC	12118
5	Standon Fenpropimorph 750	Standon	750 g/l	EC	08965

Uses

- Alternaria in ***carrots*** *(off-label)*, ***horseradish*** *(off-label)*, ***parsley root*** *(off-label)*, ***parsnips*** *(off-label)*, ***salsify*** *(off-label)* [2]
- Brown rust in ***spring barley***, ***spring wheat***, ***triticale***, ***winter barley***, ***winter wheat*** [1-5]
- Crown rot in ***carrots*** *(off-label)*, ***horseradish*** *(off-label)*, ***parsley root*** *(off-label)*, ***parsnips*** *(off-label)*, ***salsify*** *(off-label)* [2]
- Foliar disease control in ***durum wheat*** *(off-label)* [2]; ***grass seed crops*** *(off-label)* [1, 5]

- Powdery mildew in ***bilberries*** *(off-label)*, ***blackcurrants*** *(off-label)*, ***blueberries*** *(off-label)*, ***carrots*** *(off-label)*, ***cranberries*** *(off-label)*, ***dewberries*** *(off-label)*, ***gooseberries*** *(off-label)*, ***hops*** *(off-label)*, ***horseradish*** *(off-label)*, ***loganberries*** *(off-label)*, ***parsley root*** *(off-label)*, ***parsnips*** *(off-label)*, ***raspberries*** *(off-label)*, ***redcurrants*** *(off-label)*, ***salsify*** *(off-label)*, ***strawberries*** *(off-label)*, ***tayberries*** *(off-label)*, ***whitecurrants*** *(off-label)* [2]; ***spring barley***, ***spring oats***, ***spring wheat***, ***winter barley***, ***winter oats***, ***winter rye***, ***winter wheat*** [1-5]
- Rhynchosporium in ***spring barley***, ***winter barley*** [1-5]
- Rust in ***fodder beet*** *(off-label)*, ***red beet*** *(off-label)*, ***sugar beet seed crops*** *(off-label)* [2]; ***leeks*** [1]
- Yellow rust in ***spring barley***, ***spring wheat***, ***triticale***, ***winter barley***, ***winter wheat*** [1-5]

Specific Off-Label Approvals (SOLAs)

- ***bilberries*** *20040804* [2]
- ***blackcurrants*** *20040804* [2]
- ***blueberries*** *20040804* [2]
- ***carrots*** *20023753* [2]
- ***cranberries*** *20040804* [2]
- ***dewberries*** *20040804* [2]
- ***durum wheat*** *20061460* [2]
- ***fodder beet*** *20023757* [2]
- ***gooseberries*** *20040804* [2]
- ***grass seed crops*** *20061060* [1], *20061057* [5]
- ***hops*** *20023759* [2]
- ***horseradish*** *20023753* [2]
- ***loganberries*** *20040804* [2]
- ***parsley root*** *20023753* [2]
- ***parsnips*** *20023753* [2]
- ***raspberries*** *20040804* [2]
- ***red beet*** *20023751* [2]
- ***redcurrants*** *20040804* [2]
- ***salsify*** *20023753* [2]
- ***strawberries*** *20040804* [2]
- ***sugar beet seed crops*** *20023757* [2]
- ***tayberries*** *20040804* [2]
- ***whitecurrants*** *20040804* [2]

Approval information

- Accepted by BBPA for use on malting barley and hops

Efficacy guidance

- On all crops spray at start of disease attack. See labels for recommended tank mixes. Follow-up treatments may be needed if disease pressure remains high
- Product rainfast after 2 h

Restrictions

- Maximum number of treatments 2 per crop for spring cereals; 3 or 4 oer crop for winter cereals; 6 per crop for leeks
- Consult processors before using on crops for processing

Crop-specific information

- HI cereals 5 wk
- Scorch may occur on cereals if applied during frosty weather or in high temperatures
- Some leaf spotting may occur on undersown clovers
- Leeks should be treated every 2-3 wk as required, up to 6 applications

Environmental safety

- Dangerous for the environment
- Very toxic to aquatic organisms

Hazard classification and safety precautions

Hazard H03 [1-4]; H04 [5]; H11 [1-5]

Risk phrases R22a [1]; R36 [3, 4]; R38, R53a [1-5]; R50 [2]; R51 [1, 3-5]; R63 [1-4]

Operator protection A [1-5]; C [1, 3-5]; U05a, U08, U14, U15, U19a, U20b

Environmental protection E15a [1-5]; E38 [2]

SEE SECTION 3 FOR PRODUCTS ALSO REGISTERED

Storage and disposal D01, D02, D09a [1-5]; D05 [1, 3-5]; D08 [2]; D10b [5]; D10c [1-4]; D12a [1, 2]

Medical advice M05a

220 fenpropimorph + flusilazole

A broad-spectrum eradicant and protectant fungicide mixture for cereals

FRAC mode of action code: 5 + 3

Products

1	Colstar	DuPont	375:160 g/l	EC	12175
2	Pluton	DuPont	375:160 g/l	EC	12200

Uses

- Brown rust in ***spring barley***, ***winter barley***, ***winter wheat*** [1, 2]
- Foliar disease control in ***durum wheat*** *(off-label)*, ***grass seed crops*** *(off-label)* [1, 2]; ***spring rye*** *(off-label)*, ***triticale*** *(off-label)*, ***winter rye*** *(off-label)* [1]
- Glume blotch in ***winter wheat*** [1, 2]
- Net blotch in ***spring barley***, ***winter barley*** [1, 2]
- Powdery mildew in ***spring barley***, ***winter barley***, ***winter wheat*** [1, 2]
- Rhynchosporium in ***spring barley***, ***winter barley*** [1, 2]
- Septoria leaf blotch in ***winter wheat*** [1, 2]
- Yellow rust in ***spring barley***, ***winter barley***, ***winter wheat*** [1, 2]

Specific Off-Label Approvals (SOLAs)

- ***durum wheat*** *20060977* [1], *20060985* [2]
- ***grass seed crops*** *20060977* [1], *20060985* [2]
- ***spring rye*** *20060977* [1]
- ***triticale*** *20060977* [1]
- ***winter rye*** *20060977* [1]

Approval information

- Flusilazole has been included in Annex I under Directive 91/414 for 18 months from January 2007 and all approvals for products containing this active ingredient will expire in June 2008
- Accepted by BBPA for use on malting barley

Efficacy guidance

- Disease control is more effective if treatment made at an early stage of disease development
- Treat winter cereals in spring or early summer before diseases spread to new growth
- Spring barley should be treated when diseases are first evident
- Treatment may be repeated after 3-4 wk if necessary
- Flusilazole is a DMI fungicide. Resistance to some DMI fungicides has been identified in Septoria leaf blotch which may seriously affect performance of some products. For further advice contact a specialist advisor and visit the Fungicide Resistance Action Group (FRAG)-UK website

Restrictions

- Maximum number of treatments 2 per crop of barley or wheat
- Do not apply to crops under stress
- Do not apply during frosty weather

Crop-specific information

- Latest use: before beginning of anthesis (GS 60) for winter wheat; up to and including completion of ear emergence (GS 59) for barley

Environmental safety

- Dangerous for the environment
- Toxic to aquatic organisms
- Dangerous to fish or other aquatic life. Do not contaminate surface waters or ditches with chemical or used container

Hazard classification and safety precautions

Hazard H02, H11

Risk phrases R36, R40, R51, R53a, R61

Operator protection A, C; U05a, U11, U19a, U20b

Environmental protection E13b, E15a, E38

Storage and disposal D01, D02, D05, D09a, D10b, D12a
Medical advice M04a

221 fenpropimorph + kresoxim-methyl

A protectant and systemic fungicide mixture for cereals
FRAC mode of action code: 5 + 11

Products

1	Ensign	BASF	300:150 g/l	SE	11729
2	Greencrop Monsoon	Greencrop	300:150 g/l	SE	09573
3	Standon Kresoxim FM	Standon	300:150 g/l	SE	08922

Uses

- Foliar disease control in ***durum wheat*** *(off-label)* [1]; ***grass seed crops*** *(off-label)* [2, 3]
- Glume blotch in ***spring wheat*** *(reduction)*, ***triticale*** *(reduction)* [1]; ***winter wheat*** *(reduction)* [1-3]
- Powdery mildew in ***spring barley***, ***winter barley*** [1-3]; ***spring oats***, ***spring rye***, ***triticale***, ***winter oats***, ***winter rye*** [1]
- Rhynchosporium in ***spring barley***, ***winter barley*** [1-3]; ***spring rye***, ***triticale***, ***winter rye*** [1]
- Septoria leaf blotch in ***spring wheat*** *(reduction)*, ***triticale*** *(reduction)* [1]; ***winter wheat*** *(reduction)* [1-3]

Specific Off-Label Approvals (SOLAs)

- ***durum wheat*** *20062643* [1]
- ***grass seed crops*** *20060978* [2], *20060984* [3]

Approval information

- Kresoxim-methyl included in Annex I under EC Directive 91/414
- Accepted by BBPA for use on malting barley
- Approval expiry 31 Dec 2008 [11918]

Efficacy guidance

- For best results spray at the start of foliar disease attack and repeat if infection conditions persist
- Kresoxim-methyl is a member of the QoI cross resistance group. Product should be used preventatively and not relied on for its curative potential
- Use product as part of an Integrated Crop Management strategy incorporating other methods of control, including where appropriate other fungicides with a different mode of action. Do not apply more than two foliar applications of QoI containing products to any cereal crop
- There is a significant risk of widespread resistance occurring in *Septoria tritici* populations in UK. Strains of barley powdery mildew resistant to QoIs are common in UK. Failure to follow resistance management action may result in reduced levels of disease control
- Strains of barley powdery mildew resistant to QoIs are common in the UK

Restrictions

- Maximum total dose equivalent to two full dose treatments for all crops

Crop-specific information

- Latest use: completion of ear emergence (GS 59) for barley and oats; completion of flowering (GS 69) for wheat, rye and triticale

Environmental safety

- Dangerous for the environment
- Very toxic to aquatic organisms

Hazard classification and safety precautions

Hazard H03, H11
Risk phrases R40, R50, R53a [1-3]; R43 [2, 3]; R63 [1, 2]
Operator protection A [1-3]; H [1]; U05a, U14, U20b [1-3]; U19a [2]
Environmental protection E15a [1-3]; E38 [1]
Storage and disposal D01, D02, D05, D09a [1-3]; D10b [2, 3]; D10c, D12a [1]
Medical advice M05a [1, 3]

SEE SECTION 3 FOR PRODUCTS ALSO REGISTERED

222 fenpropimorph + pyraclostrobin

A protectant and curative fungicide mixture for cereals
FRAC mode of action code: 5 +11

Products

Jenton	BASF	375:100 g/l	EC	11898

Uses

- Brown rust in ***spring barley***, ***spring wheat***, ***winter barley***, ***winter wheat***
- Crown rust in ***spring oats***, ***winter oats***
- Drechslera leaf spot in ***spring wheat***, ***winter wheat***
- Foliar disease control in ***grass seed crops*** *(off-label)*, ***triticale*** *(off-label)*
- Glume blotch in ***spring wheat***, ***winter wheat***
- Net blotch in ***spring barley***, ***winter barley***
- Powdery mildew in ***spring barley***, ***spring oats***, ***winter barley***, ***winter oats***
- Rhynchosporium in ***spring barley***, ***winter barley***
- Septoria leaf blotch in ***spring wheat***, ***winter wheat***
- Yellow rust in ***spring barley***, ***spring wheat***, ***winter barley***, ***winter wheat***

Specific Off-Label Approvals (SOLAs)

- ***grass seed crops*** *20052650*
- ***triticale*** *20052650*

Approval information

- Pyraclostrobin included in Annex I under EC Directive 91/414
- Accepted by BBPA for use on malting barley

Efficacy guidance

- Best results obtained from treatment at the start of foliar disease attack
- Yield response may be obtained in the absence of visual disease
- Pyraclostrobin is a member of the QoI cross resistance group. Product should be used preventatively and not relied on for its curative potential
- Use product as part of an Integrated Crop Management strategy incorporating other methods of control, including where appropriate other fungicides with a different mode of action. Do not apply more than two foliar applications of QoI containing products to any cereal crop
- There is a significant risk of widespread resistance occurring in *Septoria tritici* populations in UK. Strains of barley powdery mildew resistant to QoIs are common in UK. Failure to follow resistance management action may result in reduced levels of disease control

Restrictions

- Maximum number of treatments 2 per crop

Crop-specific information

- Latest use: up to and including emergence of ear just complete (GS 59) for barley

Environmental safety

- Dangerous for the environment
- Very toxic to aquatic organisms
- LERAP Category B

Hazard classification and safety precautions

Hazard H03, H11
Risk phrases R20, R22a, R36, R38, R50, R53a, R63
Operator protection A, C; U05a, U14, U20b
Environmental protection E15a, E16a, E34, E38
Storage and disposal D01, D02, D05, D09a, D10c, D12a
Medical advice M03, M05a

223 fenpropimorph + quinoxyfen

A systemic fungicide mixture for cereals
FRAC mode of action code: 5 + 13

Products

Orka	Dow	250:66.7 g/l	EW	08879

Uses

- Powdery mildew in ***durum wheat***, ***grass seed crops*** *(off-label)*, ***spring barley***, ***spring oats***, ***spring rye***, ***spring wheat***, ***triticale***, ***winter barley***, ***winter oats***, ***winter rye***, ***winter wheat***

Specific Off-Label Approvals (SOLAs)

- ***grass seed crops*** *20060979*

Approval information

- Quinoxyfen included in Annex I under Directive 91/414
- Accepted by BBPA for use on malting barley

Efficacy guidance

- For best results treat at early stage of disease development before infection spreads to new crop growth. Further treatment may be necessary if disease pressure remains high
- For control of established infections and broad spectrum disease control use in tank mixtures. See label
- Product rainfast after 1 h
- Systemic activity may be reduced in severe drought

Restrictions

- Maximum total dose 3.0 l product per ha
- Apply only in the spring from mid-tillering stage (GS 25)

Crop-specific information

- Latest use: when first awns visible (GS 49)
- Crop scorch may occur when treatment made in high temperatures

Environmental safety

- Dangerous for the environment
- Very toxic to aquatic organisms
- LERAP Category B

Hazard classification and safety precautions

Hazard H03, H11
Risk phrases R43, R50, R53a, R61
Operator protection A, H; U05a, U14
Environmental protection E15a, E16a, E34, E38
Storage and disposal D01, D02, D12a

224 fenpyroximate

A mitochondrial electron transport inhibitor (METI) acaricide for apples
IRAC mode of action code: 21

Products

Sequel	Certis	51.3 g/l	SC	12657

Uses

- Fruit tree red spider mite in ***apples***, ***plums*** *(off-label)*
- Tarsonemid mites in ***protected strawberries*** *(off-label)*, ***strawberries*** *(off-label)*

Specific Off-Label Approvals (SOLAs)

- ***plums*** *20060444*
- ***protected strawberries*** *20070902*
- ***strawberries*** *20070902*

Efficacy guidance

- Kills motile stages of fruit tree red spider mite. Best results achieved if applied in warm weather

- Total spray cover of trees essential. Use higher volumes for large trees
- Apply when majority of winter eggs have hatched

Restrictions

- Maximum number of treatments 1 per yr or total dose equivalent to one full dose treatment
- Other mitochondrial electron transport inhibitor (METI) acaricides should not be applied to the same crop in the same calendar yr either separately or in mixture
- Do not apply when apple crops or pollinator are in flower
- Consult processor before use on crops for processing

Crop-specific information

- HI 2 wk

Environmental safety

- Dangerous for the environment
- Very toxic to aquatic organisms
- Risk to non-target insects or other arthropods
- Broadcast air-assisted LERAP (40 m)

Hazard classification and safety precautions

Hazard H03, H11
Risk phrases R20, R36, R43, R50, R53a
Operator protection A, C, H; U05a, U08, U14, U15, U20b
Environmental protection E13b, E22c; E17b (40 m)
Storage and disposal D01, D02, D05, D09a, D10c, D12a

225 ferric phosphate

A molluscicide bait for controlling slugs and snails

Products

1	Ferramol	Certis	1% w/w	GB 12274
2	Sluggo	Omex	1% w/w	GB 12529

Uses

- Slugs in ***all edible crops (outdoor)***, ***all non-edible crops (outdoor)***, ***protected crops***
- Snails in ***all edible crops (outdoor)***, ***all non-edible crops (outdoor)***, ***protected crops***

Efficacy guidance

- Treat as soon as damage first seen preferably in early evening. repeat as necessary to maintain control
- Best results obtained from moist soaked granules. This will occur naturally on moist soils or in humid conditions
- Ferric phosphate does not cause excessive slime secretion and has no requirement to collect moribund slugs from the soil surface
- Active ingredient is degraded by micro-organisms to beneficial plant nutrients

Restrictions

- Maximum total dose equivalent to four full dose treatments

Crop-specific information

- Latest use not specified for any crop

Hazard classification and safety precautions

Operator protection U05a, U20b
Environmental protection E15a, E34
Storage and disposal D01, D09a

226 ferrous sulphate

A herbicide/fertilizer combination for moss control in turf

See also *dicamba + dichlorprop-P + ferrous sulphate + MCPA*
dichlorprop-P + ferrous sulphate + MCPA

Products

1	Elliott's Lawn Sand	Elliott	9.3% w/w	SA	04860
2	Elliott's Mosskiller	Elliott	24.6% w/w	GR	04909
3	Ferromex Mosskiller Concentrate	Omex	16.4% w/w	SL	13180
4	Greenmaster Autumn	Scotts	18.2% w/w	GR	12196
5	Greentec Mosskiller	Headland Amenity	9% w/w	GR	12518
6	Scotts Moss Control	Scotts	18.2% w/w	GR	12928
7	SHL Lawn Sand	Sinclair	5.4% w/w	SA	05254
8	Taylors Lawn Sand	Rigby Taylor	4.1% w/w	SA	04451
9	Vitax Microgran 2	Vitax	10.9% w/w	MG	04541
10	Vitax Turf Tonic	Vitax	15% w/w	SA	04354

Uses

- Moss in ***amenity grassland*** [3, 5]; ***managed amenity turf*** [1-10]

Efficacy guidance

- For best results apply when turf is actively growing and the soil is moist
- Fertilizer component of most products encourages strong root growth and tillering
- Mow 3 d before treatment and do not mow for 3-4 d afterwards
- Water after 2 d if no rain
- Rake out dead moss thoroughly 7-14 d after treatment. Re-treatment may be necessary for heavy infestations

Restrictions

- Maximum number of treatments - see labels
- Do not apply during drought or when heavy rain expected
- Do not apply in frosty weather or when the ground is frozen
- Do not walk on treated areas until well watered
- Do not apply product undiluted [3]

Crop-specific information

- If spilt on paving, concrete, clothes etc brush off immediately to avoid discolouration
- Observe label restrictions for interval before cutting after treatment

Environmental safety

- Harmful to fish or other aquatic life. Do not contaminate surface waters or ditches with chemical or used container [4, 6]

Hazard classification and safety precautions

Operator protection A [3]; U20a [1, 2]; U20b [3, 5, 9, 10]; U20c [4, 6-8]
Environmental protection E13c [4, 6]; E15a [1, 2, 5, 8-10]; E15b [3]
Storage and disposal D01 [3, 5, 7, 10]; D09a [1-10]; D11a [1, 2, 4-10]; D12a [3]

227 ferrous sulphate + MCPA + mecoprop-P

A translocated herbicide and moss killer mixture
HRAC mode of action code: O + O

Products

Renovator Pro	Scotts	0.49:0.29:16.3% w/w	GR	12204

Uses

- Annual dicotyledons in ***managed amenity turf***
- Moss in ***managed amenity turf***
- Perennial dicotyledons in ***managed amenity turf***

Approval information

- MCPA and mecoprop-P included in Annex I under EC Directive 91/414

SEE SECTION 3 FOR PRODUCTS ALSO REGISTERED

Efficacy guidance

- Apply from Apr to Sep when weeds are growing
- For best results apply when light showers or heavy dews are expected
- Apply with a suitable calibrated fertilizer distributor
- Retreatment may be necessary after 6 wk if weeds or moss persist
- For best control of moss scarify vigorously after 2 wk to remove dead moss
- Where regrowth of moss or weeds occurs a repeat treatment may be made after 6 wk
- Avoid treatment of wet grass or during drought. If no rain falls within 48 h water in thoroughly

Restrictions

- Maximum number of treatments 3 per yr
- Do not treat new turf until established for 6 mth
- The first 4 mowings after treatment should not be used to mulch cultivated plants unless composted at least 6 mth
- Avoid walking on treated areas until it has rained or they have been watered
- Do not re-seed or turf within 8 wk of last treatment
- Do not cut grass for at least 3 d before and at least 4 d after treatment
- Do not apply during freezing conditions or when rain imminent

Environmental safety

- Keep livestock out of treated areas for up to two weeks following treatment and until poisonous weeds, such as ragwort, have died down and become unpalatable
- Harmful to fish or other aquatic life. Do not contaminate surface waters or ditches with chemical or used container
- Do not empty into drains

Hazard classification and safety precautions

Operator protection U20b
Environmental protection E07a, E13c, E19b
Storage and disposal D01, D09a, D12a
Medical advice M05a

228 fipronil

A phenylpyrazole insecticide for horticulture
IRAC mode of action code: 2B

Products

1	Vi-Nil	Certis	0.1% w/w	FG	11077
2	Vi-Nil	Certis	0.1% w/w	FG	13056

Uses

- Vine weevil in ***ornamental plant production***

Approval information

- Fipronil included in Annex I under EC Directive 91/414 as a seed treatment only

Efficacy guidance

- Product may be used at any stage of plant propagation to potting on
- Treated compost will give control of vine weevil larvae into second yr after treatment
- Incorporate into fresh compost each time the subject is re-potted
- Control not guaranteed if untreated liners or plugs are potted into treated compost

Restrictions

- Maximum number of treatments 1 per batch of compost
- Check tolerance on sample plants before large scale treatment
- Do not use in compost for aquatic plants or marginals

Crop-specific information

- Latest use: before planting in treated compost
- Product may be used in all compost types but dose should be reduced in proportion if more than 20% inert material included

Environmental safety

- Harmful to fish or other aquatic life. Do not contaminate surface waters or ditches with chemical or used container

Hazard classification and safety precautions

Operator protection A, D; U13, U20b
Environmental protection E13c
Consumer protection C01
Storage and disposal D01, D02, D11a

229 flonicamid

A selective feeding blocker aphicide
IRAC mode of action code: 9C

Products

Teppeki	Belchim	50% w/w	WG	12402

Uses

- Aphids in ***potatoes***, ***winter wheat***

Efficacy guidance

- Apply when warning systems forecast significant aphid infestations
- Persistence of action is 21 d
- Do not apply more than two consecutive treatments of flonicamid. If further treatment is needed use an insecticide with a different mode of action

Restrictions

- Maximum number of treatments 2 per crop for potatoes, winter wheat
- Must not be applied to winter wheat before 50% ear emerged stage (GS 53)

Crop-specific information

- HI: potatoes 14 d; winter wheat 28 d

Environmental safety

- Dangerous for the environment
- Harmful to aquatic organisms

Hazard classification and safety precautions

Hazard H11
Risk phrases R52, R53a
Operator protection A, H; U05a
Environmental protection E15a
Storage and disposal D01, D02, D07, D12b

230 florasulam

A triazolopyrimidine herbicide for cereals
HRAC mode of action code: B

See also *2,4-D + florasulam*
clopyralid + florasulam + fluroxypyr

Products

1	Barton WG	Dow	25% w/w	WG	13284
2	Boxer	Dow	50 g/l	SC	09819

Uses

- Annual dicotyledons in ***grass seed crops*** *(off-label)*, ***spring rye*** *(off-label)*, ***triticale*** *(off-label)*, ***winter rye*** *(off-label)* [2]; ***spring barley***, ***spring oats***, ***spring wheat***, ***winter barley***, ***winter oats***, ***winter wheat*** [1, 2]
- Chickweed in ***spring barley***, ***spring oats***, ***spring wheat***, ***winter barley***, ***winter oats***, ***winter wheat*** [1, 2]
- Cleavers in ***spring barley***, ***spring oats***, ***spring wheat***, ***winter barley***, ***winter oats***, ***winter wheat*** [1, 2]

SEE SECTION 3 FOR PRODUCTS ALSO REGISTERED

- Mayweeds in ***spring barley***, ***spring oats***, ***spring wheat***, ***winter barley***, ***winter oats***, ***winter wheat*** [1, 2]
- Volunteer oilseed rape in ***spring barley***, ***spring oats***, ***spring wheat***, ***winter barley***, ***winter oats***, ***winter wheat*** [1, 2]

Specific Off-Label Approvals (SOLAs)

- ***grass seed crops*** *20060997* [2]
- ***spring rye*** *20060997* [2]
- ***triticale*** *20060997* [2]
- ***winter rye*** *20060997* [2]

Approval information

- Florasulam included in Annex I under EC Directive 91/414
- Accepted by BBPA for use on malting barley

Efficacy guidance

- Best results obtained from treatment of small actively growing weeds in good conditions
- Apply in autumn or spring once crop has 3 leaves
- Product is mainly absorbed by leaves of weeds and is effective on all soil types
- Florasulam is a member of the ALS-inhibitor group of herbicides

Restrictions

- Maximum total dose on any crop equivalent to one full dose treatment
- Do not roll or harrow within 7 d before or after application
- Do not spray when crops under stress from cold, drought, pest damage, nutrient deficiency or any other cause
- Specific restrictions apply to use in sequence or tank mixture with other sulfonylurea or ALS-inhibiting herbicides. See label for details

Crop-specific information

- Latest use: up to and including flag leaf just visible stage for all crops

Following crops guidance

- Where the product has been used in mixture with certain named products (see label) only cereals or grass may be sown in the autumn following harvest [2]
- Unless otherwise restricted cereals, oilseed rape, field beans, grass or vegetable brassicas as transplants may be sown as a following crop in the same calendar yr as treatment. Oilseed rape may show some temporary reduction of vigour after a dry summer, but yields are not affected
- In addition to the above, linseed, peas, sugar beet, potatoes, maize, clover (for use in grass/clover mixtures) or carrots may be sown in the calendar yr following treatment
- In the event of failure of a treated crop in spring only spring wheat, spring barley, spring oats, maize or ryegrass may be sown

Environmental safety

- Dangerous for the environment
- Very toxic to aquatic organisms
- See label for detailed instructions on tank cleaning

Hazard classification and safety precautions

Hazard H11
Risk phrases R50, R53a
Operator protection U05a
Environmental protection E15a [2]; E15b [1]; E34, E38 [1, 2]
Storage and disposal D01, D02, D05, D09a, D10b, D12a

231 florasulam + fluroxypyr

A post-emergence herbicide mixture for cereals
HRAC mode of action code: B + O

Products

1	GF 184	Dow	2.5:100 g/l	SE	10878
2	Hiker	Dow	1.0:100 g/l	SE	11451
3	Hunter	Dow	2.5:100 g/l	SE	12836
4	Starane Gold	Dow	1.0:100 g/l	SE	10879
5	Starane Vantage	Dow	1.0:100 g/l	SE	10922
6	Starane XL	Dow	2.5:100 g/l	SE	10921

Uses

- Annual dicotyledons in ***durum wheat*** *(off-label)*, ***spring rye*** *(off-label)*, ***triticale*** *(off-label)*, ***winter rye*** *(off-label)* [6]; ***grass seed crops*** *(off-label)* [1, 6]; ***spring barley***, ***spring oats***, ***spring wheat***, ***winter barley***, ***winter oats***, ***winter wheat*** [1, 3, 6]
- Chickweed in ***durum wheat*** *(off-label)*, ***grass seed crops*** *(off-label)*, ***spring rye*** *(off-label)*, ***triticale*** *(off-label)*, ***winter rye*** *(off-label)* [2, 4, 5]; ***spring barley***, ***spring oats***, ***spring wheat***, ***winter barley***, ***winter oats***, ***winter wheat*** [1-6]
- Cleavers in ***durum wheat*** *(off-label)*, ***grass seed crops*** *(off-label)*, ***spring rye*** *(off-label)*, ***triticale*** *(off-label)*, ***winter rye*** *(off-label)* [2, 4, 5]; ***spring barley***, ***spring oats***, ***spring wheat***, ***winter barley***, ***winter oats***, ***winter wheat*** [1-6]
- Mayweeds in ***spring barley***, ***spring oats***, ***spring wheat***, ***winter barley***, ***winter oats***, ***winter wheat*** [1, 3, 6]

Specific Off-Label Approvals (SOLAs)

- ***durum wheat*** *20060038* [2], *20060039* [4], *20060040* [5], *20060041* [6]
- ***grass seed crops*** *20060037* [1], *20060038* [2], *20060039* [4], *20060040* [5], *20060041* [6]
- ***spring rye*** *20060038* [2], *20060039* [4], *20060040* [5], *20060041* [6]
- ***triticale*** *20060038* [2], *20060039* [4], *20060040* [5], *20060041* [6]
- ***winter rye*** *20060038* [2], *20060039* [4], *20060040* [5], *20060041* [6]

Approval information

- Florasulam and fluroxypyr included in Annex I under EC Directive 91/414
- Accepted by BBPA for use on malting barley

Efficacy guidance

- Best results obtained when weeds are small and growing actively
- Products are mainly absorbed through weed foliage. Cleavers emerging after application will not be controlled
- Florasulam is a member of the ALS-inhibitor group of herbicides

Restrictions

- Maximum total dose equivalent to one full dose treatment
- Do not roll or harrow 7 d before or after application
- Do not spray when crops are under stress from cold, drought, pest damage or nutrient deficiency
- Do not apply through CDA applicators
- Specific restrictions apply to use in sequence or tank mixture with other sulfonylurea or ALS-inhibiting herbicides. See label for details

Crop-specific information

- Latest use: before flag leaf sheath extended (before GS 41) for spring barley and spring wheat; before flag leaf sheath opening (before GS 47) for winter barley and winter wheat; before second node detectable (before GS 32) for winter oats

Following crops guidance

- Cereals, oilseed rape, field beans or grass may follow treated crops in the same yr. Oilseed rape may suffer temporary vigour reduction after a dry summer
- In addition to the above, linseed, peas, sugar beet, potatoes, maize or clover may be sown in the calendar yr following treatment
- In the event of failure of a treated crop in the spring, only spring cereals, maize or ryegrass may be planted

Environmental safety

- Dangerous for the environment

SEE SECTION 3 FOR PRODUCTS ALSO REGISTERED

- Toxic to aquatic organisms
- Take extreme care to avoid drift onto non-target crops or plants

Hazard classification and safety precautions

Hazard H04, H11

Risk phrases R36, R38, R51, R53a, R67

Operator protection A, C; U05a, U11, U19a

Environmental protection E15a, E34, E38

Storage and disposal D01, D02, D09a, D10c, D12a

232 fluazifop-P-butyl

A phenoxypropionic acid grass herbicide for broadleaved crops

HRAC mode of action code: A

Products

1	Clayton Maximus	Clayton	125 g/l	EC	12543
2	Fusilade Max	Syngenta	125 g/l	EC	11519
3	Greencrop Bantry	Greencrop	125 g/l	EC	12737

Uses

- Annual grasses in ***blackcurrants, bulb onions, carrots, combining peas, farm forestry, fodder beet, gooseberries, hops, kale*** *(stockfeed only)*, ***linseed, raspberries, spring field beans, spring oilseed rape, spring oilseed rape for industrial use, strawberries, sugar beet, swedes*** *(stockfeed only)*, ***turnips*** *(stockfeed only)*, ***vining peas, winter field beans, winter oilseed rape, winter oilseed rape for industrial use*** [1-3]; ***broad beans*** *(off-label)*, ***chinese cabbage*** *(off-label)*, ***collards*** *(off-label)*, ***field margins, flax, flax for industrial use, garlic*** *(off-label)*, ***haricot beans*** *(off-label)*, ***linseed for industrial use, navy beans*** *(off-label)*, ***ornamental plant production*** *(off-label)*, ***parsnips*** *(off-label)*, ***protected ornamentals*** *(off-label)*, ***red beet*** *(off-label)* [2]
- Annual meadow grass in ***lucerne*** *(off-label)* [2]
- Barren brome in ***field margins*** [2]
- Blackgrass in ***hops, spring field beans, winter field beans*** [1-3]; ***lucerne*** *(off-label)* [2]
- Green cover in ***land temporarily removed from production*** [1-3]
- Perennial grasses in ***blackcurrants, bulb onions, carrots, combining peas, farm forestry, fodder beet, gooseberries, hops, kale*** *(stockfeed only)*, ***linseed, raspberries, spring field beans, spring oilseed rape, spring oilseed rape for industrial use, strawberries, sugar beet, swedes*** *(stockfeed only)*, ***turnips*** *(stockfeed only)*, ***vining peas, winter field beans, winter oilseed rape, winter oilseed rape for industrial use*** [1-3]; ***broad beans*** *(off-label)*, ***chinese cabbage*** *(off-label)*, ***collards*** *(off-label)*, ***flax, flax for industrial use, garlic*** *(off-label)*, ***haricot beans*** *(off-label)*, ***linseed for industrial use, navy beans*** *(off-label)*, ***ornamental plant production*** *(off-label)*, ***parsnips*** *(off-label)*, ***protected ornamentals*** *(off-label)*, ***red beet*** *(off-label)* [2]
- Ryegrass in ***lucerne*** *(off-label)* [2]
- Volunteer cereals in ***blackcurrants, bulb onions, carrots, combining peas, farm forestry, fodder beet, gooseberries, hops, kale*** *(stockfeed only)*, ***linseed, raspberries, spring field beans, spring oilseed rape, spring oilseed rape for industrial use, strawberries, sugar beet, swedes*** *(stockfeed only)*, ***turnips*** *(stockfeed only)*, ***vining peas, winter field beans, winter oilseed rape, winter oilseed rape for industrial use*** [1-3]; ***field margins, flax, flax for industrial use, linseed for industrial use, lucerne*** *(off-label)* [2]
- Wild oats in ***blackcurrants, bulb onions, carrots, combining peas, farm forestry, fodder beet, gooseberries, hops, kale*** *(stockfeed only)*, ***linseed, raspberries, spring field beans, spring oilseed rape, spring oilseed rape for industrial use, strawberries, sugar beet, swedes*** *(stockfeed only)*, ***turnips*** *(stockfeed only)*, ***vining peas, winter field beans, winter oilseed rape, winter oilseed rape for industrial use*** [1-3]; ***field margins, flax, flax for industrial use, linseed for industrial use, lucerne*** *(off-label)* [2]

Specific Off-Label Approvals (SOLAs)

- ***broad beans*** *20032138* [2]
- ***chinese cabbage*** *20032138* [2]
- ***collards*** *20032138* [2]
- ***garlic*** *20032138* [2]
- ***haricot beans*** *20032138* [2]

- ***lucerne*** *20041321* [2]
- ***navy beans*** *20032138* [2]
- ***ornamental plant production*** *20032138* [2]
- ***parsnips*** *20032138* [2]
- ***protected ornamentals*** *20032138* [2]
- ***red beet*** *20032138* [2]

Approval information

- Accepted by BBPA for use on hops

Efficacy guidance

- Best results achieved by application when weed growth active under warm conditions with adequate soil moisture. Agral or other specified adjuvant must always be added to spray. See label
- Spray weeds from 2-expanded leaf stage to fully tillered, couch from 4 leaves when majority of shoots have emerged, with a second application if necessary
- Control may be reduced under dry conditions. Do not cultivate for 2 wk after spraying couch
- Annual meadow grass is not controlled
- May also be used to remove grass cover crops
- Fluazifop-p-butyl is an ACCase inhibitor herbicide. To avoid the build up of resistance do not apply products containing an ACCase inhibitor herbicide more than twice to any crop. In addition do not use any product containing fluazifop-p-butyl in mixture or sequence with any other product containing the same ingredient
- Use these products as part of a resistance management strategy that includes cultural methods of control and does not use ACCase inhibitors as the sole chemical method of grass weed control

Restrictions

- Maximum number of treatments 2 per crop for winter oilseed rape for industrial use; 2 per yr for farm forestry; 1 per crop or yr for all other crops
- Do not sow cereals for at least 8 wk after application of high rate or 2 wk after low rate
- Do not apply through CDA sprayer, with hand-held equipment or from air
- Avoid treatment before spring growth has hardened or when buds opening
- Do not treat bush and cane fruit or hops between flowering and harvest
- Consult processors before treating crops intended for processing
- Oilseed rape, linseed and flax for industrial use must not be harvested for human or animal consumption nor grazed
- Do not use for forestry establishment on land not previously under arable cultivation or improved grassland
- Treated vegetation in field margins, land temporarily removed from production etc, must not be grazed or harvested for human or animal consumption and unprotected persons must be kept out of treated areas for at least 24 h

Crop-specific information

- Latest use: before 50% ground cover for swedes, turnips; before 5 leaf stage for spring oilseed rape; before flowering for blackcurrants, gooseberries, hops, raspberries, strawberries; before flower buds visible for field beans, peas, linseed, flax, winter oilseed rape; 2 wk before sowing cereals or grass for field margins, land temporarily removed from production
- HI beet crops, kale, carrots 8 wk; onions 4 wk; oilseed rape for industrial use 2 wk
- Apply to sugar and fodder beet from 1-true leaf to 50% ground cover
- Apply to winter oilseed rape from 1-true leaf to established plant stage
- Apply to spring oilseed rape from 1-true leaf but before 5-true leaves
- Apply in fruit crops after harvest. See label for timing details on other crops
- Before using on onions or peas use crystal violet test to check that leaf wax is sufficient

Environmental safety

- Dangerous for the environment
- Very toxic to aquatic organisms

Hazard classification and safety precautions

Hazard H03, H11

Risk phrases R38, R50, R53a, R63

Operator protection A, C, H, M; U05a, U08, U20b

Environmental protection E15a [1-3]; E38 [2]

Storage and disposal D01, D02, D05, D09a, D12a [1-3]; D10b [1]; D10c [2, 3]

Medical advice M05b [1]

SEE SECTION 3 FOR PRODUCTS ALSO REGISTERED

233 fluazinam

A dinitroaniline fungicide for use in potatoes
FRAC mode of action code: 29

Products

1	Blizzard	AgriGuard	500 g/l	SC	12854
2	Greencrop Solanum	Greencrop	500 g/l	SC	11052
3	Landgold Fluazinam	Landgold	500 g/l	SC	08060
4	Landgold Fluazinam	Teliton	500 g/l	SC	12119
5	Shirlan	Syngenta	500 g/l	SC	10573
6	Volley	Makhteshim	500 g/l	SC	13591

Uses

- Blight in ***potatoes*** [1-6]
- Powdery scab in ***seed potatoes*** *(off-label)* [5]
- Root rot in ***blackberries*** *(off-label)*, ***protected blackberries*** *(off-label)*, ***protected raspberries*** *(off-label)*, ***protected rubus hybrids*** *(off-label)*, ***raspberries*** *(off-label)*, ***rubus hybrids*** *(off-label)* [5]

Specific Off-Label Approvals (SOLAs)

- ***blackberries*** *20032168* [5]
- ***protected blackberries*** *20032168* [5]
- ***protected raspberries*** *20032168* [5]
- ***protected rubus hybrids*** *20032168* [5]
- ***raspberries*** *20032168* [5]
- ***rubus hybrids*** *20032168* [5]
- ***seed potatoes*** *20060893* [5]

Efficacy guidance

- Commence treatment at the first blight risk warning (before blight enters the crop). Products are rainfast within 1 h
- In the absence of a warning, treatment should start before foliage of adjacent plants meets in the rows
- Spray at 7-14 d intervals (5-10 d intervals [5]) depending on severity of risk (see label)
- Ensure complete coverage of the foliage and stems, increasing volume as haulm growth progresses, in dense crops and if blight risk increases

Restrictions

- Maximum number of treatments 10 per crop on potatoes [1-4]
- Maximum total dose equivalent to 7.5 full dose treatments [5, 6]
- Do not use with hand-held sprayers

Crop-specific information

- HI 0 - 10 d for potatoes. Check label
- Ensure complete kill of potato haulm before lifting and do not lift crops for storage while there is any green tissue left on the leaves or stem bases

Environmental safety

- Dangerous for the environment
- Very toxic to aquatic organisms
- LERAP Category B

Hazard classification and safety precautions

Hazard H03 [1]; H04 [2-6]; H11 [1-6]

Risk phrases R36, R43, R50, R53a

Operator protection A, C, H; U02a, U04a [1-6]; U05a [2-6]; U08 [1-5]; U09a [6]; U14, U15 [3-6]; U20a [3-5]; U20b [1, 2, 6]

Environmental protection E15a, E16a [1-6]; E16b [1, 3-6]; E34 [1, 5, 6]; E38 [5, 6]

Storage and disposal D01, D02 [2-6]; D05 [1-5]; D09a [1-6]; D10b [2-4]; D10c [1, 5, 6]; D12a [5, 6]

Medical advice M03 [1-6]; M05a [5, 6]

234 fluazinam + metalaxyl-M

A mixture of contact and systemic fungicides for potatoes
FRAC mode of action code: 29 + 4

Products

Epok	Belchim	400:200 g/l	EC	11997

Uses
- Blight in ***potatoes***

Approval information
- Metalaxyl-M included in Annex I under EC Directive 91/414

Efficacy guidance
- Commence treatment at the first blight risk warning (before blight enters the crop). Products are rainfast within 1 h
- In the absence of a warning, treatment should start before foliage of adjacent plants meets in the rows
- Spray at 7-14 d intervals depending on severity of risk (see label)
- Ensure complete coverage of the foliage and stems, increasing volume as haulm growth progresses, in dense crops and if blight risk increases
- Metalaxyl-M works best on young actively growing foliage and efficacy declines with the onset of senescence. Therefore it is recommended that the product is only used for the first part of the blight control program, which should be completed with a reliable protectant fungicide

Restrictions
- Maximum number of treatments 3 per crop

Crop-specific information
- HI 7d for potatoes

Environmental safety
- Dangerous for the environment
- Very toxic to aquatic organisms
- LERAP Category B

Hazard classification and safety precautions

Hazard H03, H11
Risk phrases R20, R36, R38, R40, R43, R50, R53a
Operator protection A, C, H; U05a, U07, U11, U14, U15, U19a, U20a
Environmental protection E15a, E16a, E16b, E34, E38
Storage and disposal D01, D02, D09a, D10b, D12a
Medical advice M05a

235 fludioxonil

A phenylpyrrole fungicide seed treatment for wheat and barley
FRAC mode of action code: 12

See also *cymoxanil + fludioxonil + metalaxyl-M*
cyprodinil + fludioxonil

Products

Beret Gold	Syngenta	25 g/l	FS	12625

Uses
- Bunt in ***spring wheat*** *(seed treatment)*, ***winter wheat*** *(seed treatment)*
- Covered smut in ***spring barley*** *(seed treatment)*, ***winter barley*** *(seed treatment)*
- Fusarium foot rot and seedling blight in ***spring barley*** *(seed treatment)*, ***spring oats*** *(seed treatment)*, ***spring wheat*** *(seed treatment)*, ***winter barley*** *(seed treatment)*, ***winter oats*** *(seed treatment)*, ***winter wheat*** *(seed treatment)*
- Leaf stripe in ***spring barley*** *(seed treatment - reduction)*, ***winter barley*** *(seed treatment - reduction)*
- Pyrenophora leaf spot in ***spring oats*** *(seed treatment)*, ***winter oats*** *(seed treatment)*

- Septoria seedling blight in ***spring wheat*** *(seed treatment)*, ***winter wheat*** *(seed treatment)*
- Snow mould in ***spring barley*** *(seed treatment)*, ***spring oats*** *(seed treatment)*, ***spring wheat*** *(seed treatment)*, ***winter barley*** *(seed treatment)*, ***winter oats*** *(seed treatment)*, ***winter wheat*** *(seed treatment)*

Approval information

- Fludioxonil included in Annex I under EC Directive 91/414
- Accepted by BBPA for use on malting barley

Efficacy guidance

- Apply direct to seed using conventional seed treatment equipment. Continuous flow treaters should be calibrated using product before use
- Effective against benzimidazole-resistant strains of *Microdochium nivale*

Restrictions

- Maximum number of treatments 1 per seed batch
- Do not apply to cracked, split or sprouted seed
- Sow treated seed within 6 mth

Crop-specific information

- Latest use: before drilling
- Product may reduce flow rate of seed through drill. Recalibrate with treated seed before drilling

Environmental safety

- Dangerous for the environment
- Toxic to aquatic organisms
- Do not use treated seed as food or feed
- Treated seed harmful to game and wildlife

Hazard classification and safety precautions

Hazard H11
Risk phrases R51, R53a
Operator protection A, H; U05a, U20b
Environmental protection E03, E15a, E38
Storage and disposal D01, D02, D05, D09a, D11a, D12a
Treated seed S02, S05, S07

236 fludioxonil + flutriafol

A fungicide seed treatment mixture for cereals
FRAC mode of action code: 12 + 3

Products

Beret Multi	Syngenta	25:25 g/l	FS	13017

Uses

- Bunt in ***spring wheat*** *(seed treatment)*, ***winter wheat*** *(seed treatment)*
- Covered smut in ***spring barley*** *(seed treatment)*, ***winter barley*** *(seed treatment)*
- Fusarium foot rot and seedling blight in ***spring barley*** *(seed treatment)*, ***spring oats*** *(seed treatment)*, ***spring wheat*** *(seed treatment)*, ***winter barley*** *(seed treatment)*, ***winter oats*** *(seed treatment)*, ***winter wheat*** *(seed treatment)*
- Leaf stripe in ***spring barley*** *(seed treatment)*, ***winter barley*** *(seed treatment)*
- Loose smut in ***spring barley*** *(seed treatment)*, ***spring wheat*** *(seed treatment)*, ***winter barley*** *(seed treatment)*, ***winter wheat*** *(seed treatment)*
- Pyrenophora leaf spot in ***spring oats*** *(seed treatment)*, ***winter oats*** *(seed treatment)*
- Septoria seedling blight in ***spring wheat*** *(seed treatment)*, ***winter wheat*** *(seed treatment)*
- Snow mould in ***spring barley*** *(seed treatment)*, ***spring wheat*** *(seed treatment)*, ***winter barley*** *(seed treatment)*, ***winter wheat*** *(seed treatment)*

Approval information

- Fludioxonil included in Annex I under EC Directive 91/414

Efficacy guidance

- Apply directly to the seed using conventional seed treatment equipment
- Calibrate continuous flow treaters before use
- Effective against benzimidazole-resistant strains of *Microdochium nivale*

Restrictions

- Maximum number of treatments 1 per seed batch
- Do not apply to cracked, split or sprouted seed

Crop-specific information

- Latest use: before drilling for all cereals
- Treated seed may affect the flow rate through drills. Recalibrate with treated seed before drilling

Environmental safety

- Dangerous for the environment
- Toxic to aquatic organisms

Hazard classification and safety precautions

Hazard H11
Risk phrases R51, R53a
Operator protection A, H; U05a, U20c
Environmental protection E03, E15a, E38
Storage and disposal D01, D02, D05, D09a, D10c, D11a, D12a
Treated seed S02, S05, S07

237 fludioxonil + tefluthrin

A fungicide and insecticide seed treatment mixture for cereals
FRAC mode of action code: 12

Products

Austral Plus	Syngenta	10:40 g/l	FS	13314

Uses

- Bunt in ***spring wheat, winter wheat***
- Covered smut in ***spring barley, winter barley***
- Fusarium foot rot and seedling blight in ***spring barley, spring oats, spring wheat, winter barley, winter oats, winter wheat***
- Leaf stripe in ***spring barley*** *(partial control)*, ***winter barley*** *(partial control)*
- Pyrenophora leaf spot in ***spring oats, winter oats***
- Septoria seedling blight in ***spring wheat, winter wheat***
- Snow mould in ***spring barley, spring wheat, winter barley, winter wheat***
- Wheat bulb fly in ***spring barley, spring wheat, winter barley, winter wheat***
- Wireworm in ***spring barley, spring oats, spring wheat, winter barley, winter oats, winter wheat***

Approval information

- Fludioxonil included in Annex I under EC Directive 91/414

Efficacy guidance

- Apply direct to seed using conventional seed treatment equipment. Continuous flow treaters should be calibrated using product before use
- Best results obtained from seed drilled into a firm even seedbed
- Tefluthrin is released into soil after drilling and repels or kills larvae of wheat bulb fly and wireworm attacking below ground. Pest control may be reduced by deep or shallow drilling
- Where egg counts of wheat bulb fly or population counts of wireworm indicate a high risk of severe attack follow up spray treatments may be needed
- Control of leaf stripe in barley may not be sufficient in crops grown for seed certification
- Effective against benzimidazole-resistant strains of *Microdochium nivale*
- Under adverse soil or environmental conditions seed rates should be increased to compensate for possible reduced germination capacity

Restrictions

- Maximum number of treatments 1 per batch
- Store treated seed in cool dry conditions and drill within 3 mth
- Do not use on seed above 16% moisture content or on sprouted, cracked or damaged seed

Crop-specific information

- Latest use: before drilling

SEE SECTION 3 FOR PRODUCTS ALSO REGISTERED

Environmental safety

- Dangerous for the environment
- Very toxic to aquatic organisms

Hazard classification and safety precautions

Hazard H11
Risk phrases R50, R53a
Operator protection A, C, D, H; U02a, U04a, U05a, U07, U08, U14, U20b
Environmental protection E03, E15b, E34, E38
Storage and disposal D01, D02, D05, D09b, D10c, D11a, D12a
Treated seed S02, S04b, S05, S07, S08

238 flufenacet

A broad spectrum oxyacetamide herbicide available only in mixtures
HRAC mode of action code: K3

See also diflufenican + flufenacet

239 flufenacet + isoxaflutole

A residual herbicide mixture for maize
HRAC mode of action code: K3 + F2

Products

Cadou Star	Bayer CropScience	48:10 % w/w	WG	13242

Uses

- Annual dicotyledons in ***forage maize*, *grain maize***
- Grass weeds in ***forage maize*, *grain maize***

Approval information

- Flufenacet included in Annex I under EC Directive 91/414

Efficacy guidance

- Best results obtained from applications to a fine, firm seedbed in the presence of some soil moisture
- Efficacy may be reduced on cloddy seedbeds or under prolonged dry conditions
- Established perennial grasses and broad-leaved weeds growing from rootstocks will not be controlled
- Always follow WRAG guidelines for preventing and managing herbicide resistant weeds. See Section 5 for more information

Restrictions

- Maximum number of treatments 1 per crop
- Do not use on soils both above 70% sand and less than 2% organic matter
- Do not use on maize crops intended for seed production

Crop-specific information

- Latest use: before crop emergence
- Ideally treatment should be made within 4 d of sowing, before tha maize seeds have germinated

Following crops guidance

- In the event of failure of a treated crop maize, sweet corn or potatoes may be grown after surface cultivation or ploughing
- After normal harvest of a treated crop oats or barley may be sown after ploughing and wheat, rye, triticale, sugar beet, field beans, peas, soya, sorghum or sunflowers may be sown after surface cultivation or ploughing.

Environmental safety

- Dangerous for the environment
- Very toxic to aquatic organisms
- Take care to avoid drift over other crops. Beet crops and sunflowers are particularly sensitive
- LERAP Category B

Hazard classification and safety precautions

Hazard H03, H11

Risk phrases R22a, R36, R43, R48, R50, R53a, R63

Operator protection A, C, H; U05a, U11, U20b

Environmental protection E15b, E16a, E34, E38

Storage and disposal D01, D02, D05, D09a, D10a, D12a

Medical advice M03

240 flufenacet + metribuzin

A herbicide mixture for potatoes

HRAC mode of action code: K3 + C1

Products

Artist	Bayer CropScience	24:17.5% w/w	WP	11239

Uses

- Annual dicotyledons in ***early potatoes*, *maincrop potatoes***
- Annual meadow grass in ***early potatoes*, *maincrop potatoes***

Approval information

- Flufenacet and metribuzin included in Annex I under EC Directive 91/414

Efficacy guidance

- Product acts through root uptake and needs sufficient soil moisture at and shortly after application
- Effectiveness is reduced under dry soil conditions
- Residual activity is reduced on mineral soils with a high organic matter content and on peaty or organic soils
- Ensure application is made evenly to both sides of potato ridges
- Perennial weeds are not controlled

Restrictions

- Maximum total dose equivalent to one full dose treatment
- Potatoes must be sprayed before emergence of crop and weeds
- See label for list of tolerant varieties. Do not treat Maris Piper grown on Sands or Very Light soils
- Do not use on Sands
- On stony or gravelly soils there is risk of crop damage especially if heavy rain falls soon after application

Crop-specific information

- Latest use: before potato crop emergence
- Consult processor before use on crops for processing

Following crops guidance

- Before drilling or planting any succeeding crop soil must be mouldboard ploughed to at least 15 cm as soon as possible after lifting and no later than end Dec
- In W Cornwall on soils with more than 5% organic matter treated early potatoes may be followed by summer planted brassica crops 14 wk after treatment and after mouldboard ploughing. Elsewhere cereals or winter beans may be grown in the same year if at least 16 wk have elapsed since treatment
- In the yr following treatment any crop may be grown except lettuce or radish, or vegetable brassica crops on silt soils in Lincs

Environmental safety

- Dangerous for the environment
- Very toxic to aquatic organisms
- Take care to avoid spray drift onto neighbouring crops, especially lettuce or brassicas

Hazard classification and safety precautions

Hazard H03, H11

Risk phrases R22a, R43, R48, R50, R53a

Operator protection A, D, H; U05a, U08, U13, U14, U19a, U20b

Environmental protection E15a, E38

Storage and disposal D01, D02, D09a, D11a, D12a

Medical advice M03

SEE SECTION 3 FOR PRODUCTS ALSO REGISTERED

241 flufenacet + pendimethalin

A broad spectrum residual and contact herbicide mixture for winter cereals
HRAC mode of action code: K3 + K1

Products

1	Crystal	BASF	60:300 g/l	EC	10657
2	Greencrop Duet 603	Greencrop	60:300 g/l	EC	12171
3	Ice	BASF	60:300 g/l	EC	10681
4	Landgold Bedrock	Teliton	60:300 g/l	EC	12145
5	Standon FFA60 Plus	Standon	60:300 g/l	EC	11649
6	Trooper	BASF	60:300 g/l	EC	10682

Uses

- Annual dicotyledons in ***durum wheat*** *(off-label)*, ***spring rye*** *(off-label)*, ***triticale*** *(off-label)*, ***winter rye*** *(off-label)* [1, 3, 6]; ***grass seed crops*** *(off-label)* [1, 3-6]; ***winter barley***, ***winter wheat*** [1-6]
- Annual grasses in ***durum wheat*** *(off-label)*, ***spring rye*** *(off-label)*, ***triticale*** *(off-label)*, ***winter rye*** *(off-label)* [1, 3, 6]; ***grass seed crops*** *(off-label)* [1, 3-6]; ***winter barley***, ***winter wheat*** [1-6]
- Annual meadow grass in ***winter barley***, ***winter wheat*** [1-6]
- Blackgrass in ***winter barley***, ***winter wheat*** [1-6]
- Chickweed in ***winter barley***, ***winter wheat*** [1-6]
- Corn marigold in ***winter barley***, ***winter wheat*** [1-6]
- Field speedwell in ***winter barley***, ***winter wheat*** [1-6]
- Ivy-leaved speedwell in ***winter barley***, ***winter wheat*** [1-6]

Specific Off-Label Approvals (SOLAs)

- ***durum wheat*** *20063504* [1], *20063649* [3], *20063526* [6]
- ***grass seed crops*** *20063504* [1], *20063649* [3], *20063070* [4], *20063069* [5], *20063526* [6]
- ***spring rye*** *20063504* [1], *20063649* [3], *20063526* [6]
- ***triticale*** *20063504* [1], *20063649* [3], *20063526* [6]
- ***winter rye*** *20063504* [1], *20063649* [3], *20063526* [6]

Approval information

- Flufenacet and pendimethalin included in Annex I under EC Directive 91/414
- Accepted by BBPA for use on malting barley
- Approval expiry 31 Dec 2008 [1-6]

Efficacy guidance

- Best results achieved when applied from pre-emergence of weeds to 2-leaf stage but post emergence treatment is not recommended on clay soils
- Product requires some soil moisture to be activated ideally from rain within 7 d of application. Prolonged dry conditions may reduce residual control
- Product is slow acting and final level of weed control may take some time to appear
- For effective weed control seed bed preparations should ensure even incorporation of any trash, straw and ash to 15 cm
- Efficacy may be reduced on soils with more than 6% organic matter
- Always follow WRAG guidelines for preventing and managing herbicide resistant weeds. Section 5 for more information

Restrictions

- Maximum total dose equivalent to one full dose treatment
- For pre-emergence treatments seed should be covered with min 32 mm settled soil. Shallow drilled crops should be treated post-emergence only
- Do not treat undersown crops
- Avoid spraying during periods of prolonged or severe frosts
- Do not use on stony or gravelly soils or those with more than 10% organic matter
- Pre-emergence treatment may only be used on crops drilled before 30 Nov. All crops must be treated before 31 Dec in yr of planting.
- Concentrated or diluted product may stain clothing or skin

Crop-specific information

- Latest use: before third tiller stage (GS 23) and before 31 Dec in yr of planting
- Very wet weather before and after treatment may result in loss of crop vigour and reduced yield, particularly where soils become waterlogged

Following crops guidance

- Any crop may follow a failed or normally harvested treated crop provided ploughing to at least 15 cm is carried out beforehand

Environmental safety

- Dangerous for the environment
- Very toxic to aquatic organisms
- Risk to certain non-target insects or other arthropods
- LERAP Category B
- Some products supplied in small volume returnable packs. Follow instructions for use

Hazard classification and safety precautions

Hazard H03, H11
Risk phrases R22a, R22b, R38, R40, R50, R53a
Operator protection A, C, H; U02a, U05a, U20c
Environmental protection E15a, E16a, E16b, E34 [1-6]; E22b, E38 [1-4, 6]; E22c [5]
Storage and disposal D01, D02, D09a, D10a [1-6]; D12a [1-4, 6]
Medical advice M03, M05b

242 flumioxazin

A phenyphthalimide herbicide for winter wheat
HRAC mode of action code: E

Products

1	Digital	Interfarm	300 g/l	SC	13561
2	Guillotine	Interfarm	300 g/l	SC	13562
3	Sumimax	Interfarm	300 g/l	SC	13548

Uses

- Annual dicotyledons in ***winter wheat***
- Annual meadow grass in ***winter wheat***
- Chickweed in ***winter wheat***
- Cleavers in ***winter wheat***
- Loose silky bent in ***winter wheat***
- Mayweeds in ***winter wheat***
- Speedwells in ***winter wheat***
- Volunteer oilseed rape in ***winter wheat***

Efficacy guidance

- Best results obtained from applications to moist soil early post-emergence of the crop when weeds are germinating up to 1 leaf stage
- Flumioxazin is contact acting and relies on weeds germinating in moist soil and coming into sufficient contact with the herbicide
- Flumioxazin remains in the surface layer of the soil and does not migrate to lower layers
- Contact activity is long lasting during the autumn
- Seed beds should be firm, fine and free from clods. Crops should be drilled to 25 mm and well covered by soil
- Efficacy is reduced on soils with more than 10% organic matter
- Flumioxazin is a member of the protoporphyrin oxidase (PPO) inhibitor herbicides. To avoid the build up of resistance do not use PPO-inhibitor herbicides more than once on any crop
- Use as part of a resistance management strategy that includes cultural methods of control and does not use PPO-inhibitors as the sole chemical method of grass weed control

Restrictions

- Maximum number of treatments 1 per crop for winter wheat
- Only apply to crops that have been hardened by cool weather. Do not treat crops with lush or soft growth
- Do not apply to waterlogged soils or to soils with more than 10% organic matter
- Do not follow an application with another pesticide for 14 d
- Do not treat undersown crops
- Do not treat broadcast crops until they are past the 3 lf stage

- Do not treat crops under stress for any reason, or during prolonged frosty weather
- Do not roll or harrow for 2 wk before treatment or at any time afterwards

Crop-specific information

- Latest use: before 5th true leaf stage (GS 15) for winter wheat
- Light discolouration of leaf margins can occur after treatment and treatment of soft crops will cause transient leaf bleaching

Following crops guidance

- Any crop may be sown following normal harvest of a treated crop
- In the event of failure of a treated crop spring cereals, spring oilseed rape, sugar beet, maize or potatoes may be redrilled after ploughing

Environmental safety

- Dangerous for the environment
- Very toxic to aquatic organisms
- LERAP Category B
- Take extreme care to avoid drift onto plants outside the target area

Hazard classification and safety precautions

Hazard H02, H11
Risk phrases R50, R53a, R61
Operator protection A, H; U05a, U19a
Environmental protection E15b, E16a, E34, E38
Storage and disposal D01, D02, D05, D07, D09a, D12b
Medical advice M04a

243 fluopicolide

An benzamide fungicide available only in mixtures
FRAC mode of action code: 43

244 fluopicolide + propamocarb hydrochloride

A protectant and systemic fungicide mixture for potato blight
FRAC mode of action code: 43 + 28

Products

Infinito	Bayer CropScience	62.5:625 g/l	SC	12644

Uses

- Blight in ***potatoes***

Approval information

- Propamocarb hydrochloride included in Annex I under EC Directive 91/414

Efficacy guidance

- Commence spray programme before infection appears as soon as weather conditions favourable for disease development occur. At latest the first treatment should be made as the foliage meets along the rows
- Repeat treatments at 7-10 day intervals according to disease incidence and weather conditions
- Reduce spray interval if conditions are conducive to the spread of blight
- Spray as soon as possible after irrigation
- Increase water volume in dense crops
- When used from full canopy development to haulm desiccation as part of a full blight protection programme tubers will be protected from late blight after harvest and tuber blight incidence will be reduced
- To reduce the development of resistance product should be used in single or block applications with fungicides from a different cross-resistance group

Restrictions

- Maximum total dose on potatoes equivalent to four full dose treatments
- Do not apply if rainfall or irrigation is imminent. Product is rainfast in 1 h provided spray has dried on leaf

- Do not apply as a curative treatment when blight is present in the crop
- Do not apply more than 3 consecutive treatments of the product

Crop-specific information

- HI 7 d for potatoes
- All varieties of potatoes, including seed crops, may be treated

Environmental safety

- Dangerous for the environment
- Toxic to aquatic organisms

Hazard classification and safety precautions

Hazard H11
Risk phrases R51, R53a
Operator protection A, H; U05a, U08, U11, U19a, U20a
Environmental protection E15a, E38
Storage and disposal D01, D02, D09a, D10b, D12a

245 fluoxastrobin

A protectant stobilurin fungicide available in mixtures
FRAC mode of action code: 11

246 fluoxastrobin + prothioconazole

A strobilurin and triazole fungicide mixture for cereals
FRAC mode of action code: 11 + 3

Products

	Product	Company	Content	Formulation	Reg. no.
1	Clayton Edge	Clayton	100:100 g/l	EC	12911
2	Fandango	Bayer CropScience	100:100 g/l	EC	12276
3	Maestro	Bayer CropScience	100:100 g/l	EC	12307
4	Redigo Twin	Bayer CropScience	37.5:37.5 g/l	FS	12314

Uses

- Brown rust in ***spring barley***, ***winter barley***, ***winter rye***, ***winter wheat*** [1-3]
- Bunt in ***durum wheat*** *(seed treatment)*, ***spring rye*** *(seed treatment)*, ***triticale*** *(seed treatment)*, ***winter rye*** *(seed treatment)*, ***winter wheat*** *(seed treatment)* [4]
- Covered smut in ***winter oats*** *(seed treatment)* [4]
- Crown rust in ***spring oats***, ***winter oats*** [2, 3]
- Eyespot in ***spring barley*** *(reduction)*, ***winter barley*** *(reduction)*, ***winter rye*** *(reduction)*, ***winter wheat*** *(reduction)* [1-3]; ***spring oats***, ***winter oats*** [2, 3]
- Fusarium foot rot and seedling blight in ***durum wheat*** *(seed treatment)*, ***spring rye*** *(seed treatment)*, ***triticale*** *(seed treatment)*, ***winter oats*** *(seed treatment)*, ***winter rye*** *(seed treatment)*, ***winter wheat*** *(seed treatment)* [4]
- Glume blotch in ***winter wheat*** [1-3]
- Late ear diseases in ***spring barley***, ***winter barley***, ***winter wheat*** [1-3]
- Loose smut in ***durum wheat*** *(seed treatment)*, ***winter oats*** *(seed treatment)*, ***winter wheat*** *(seed treatment)* [4]
- Net blotch in ***spring barley***, ***winter barley*** [1-3]
- Powdery mildew in ***spring barley***, ***winter barley***, ***winter rye***, ***winter wheat*** [1-3]; ***spring oats***, ***winter oats*** [2, 3]
- Rhynchosporium in ***spring barley***, ***winter barley***, ***winter rye*** [1-3]
- Septoria leaf blotch in ***winter wheat*** [1-3]
- Take-all in ***winter barley*** *(reduction)*, ***winter wheat*** *(reduction)* [2, 3]
- Tan spot in ***winter wheat*** [1-3]
- Yellow rust in ***winter wheat*** [1-3]

Approval information

- Accepted by BBPA for use on malting barley

Efficacy guidance

- Seed treatments must be applied by manufacturer's recommended treatment application equipment [4]

SEE SECTION 3 FOR PRODUCTS ALSO REGISTERED

- Treated cereal seed should preferably be drilled in the same season [4]
- Follow-up treatments will be needed later in the season to give protection against air-borne and splash-borne diseases [4]
- Best results on foliar diseases obtained from treatment at early stages of disease development. Further treatment may be needed if disease attack is prolonged [1-3]
- Foliar applications to established infections of any disease are likely to be less effective [1-3]
- Best control of cereal ear diseases obtained by treatment during ear emergence [1-3]
- Fluoxastrobin is a member of the QoI cross resistance group. Foliar product should be used preventatively and not relied on for its curative potential [1-3]
- Use product as part of an Integrated Crop Management strategy incorporating other methods of control, including where appropriate other fungicides with a different mode of action. Do not apply more than two foliar applications of QoI containing products to any cereal crop
- There is a significant risk of widespread resistance occurring in *Septoria tritici* populations in UK. Failure to follow resistance management action may result in reduced levels of disease control
- Strains of wheat and barley powdery mildew resistant to QoIs are common in the UK. Control of wheat mildew can only be relied on from the triazole component
- Prothioconazole is a DMI fungicide. Resistance to some DMI fungicides has been identified in Septoria leaf blotch which may seriously affect performance of some products. For further advice contact a specialist advisor and visit the Fungicide Resistance Action Group (FRAG)-UK website
- Where specific control of wheat mildew is required this should be achieved through a programme of measures including products recommended for the control of mildew that contain a fungicide from a different cross-resistance group and applied at a dose that will give robust control

Restrictions

- Maximum number of seed treatments one per batch [4]
- Maximum total dose of foliar sprays equivalent to two full dose treatments for the crop [1-3]
- Seed treatment must be fully re-dispersed and homogeneous before use [4]
- Do not use on seed with more than 16% moisture content, or on sprouted, cracked or skinned seed [4]
- All seed batches should be tested to ensure they are suitable for treatment [4]

Crop-specific information

- Latest use: pre-drilling for seed treatment [4]; before grain milky ripe for spray treatments on wheat and rye; beginning of flowering for barley, oats [1-3]
- Some transient leaf chlorosis may occur after treatment of wheat or barley but this has not been found to affect yield [1-3]

Environmental safety

- Dangerous for the environment
- Toxic to aquatic organisms
- Risk to non-target insects or other arthropods. Avoid spraying within 6 m of the field boundary to reduce the effects on non-target insects or other arthropods [1-3]
- LERAP Category B [1-3]

Hazard classification and safety precautions

Hazard H04 [4]; H11 [1-4]
Risk phrases R43 [4]; R51, R53a [1-4]
Operator protection A [1-4]; H [4]; U05a [1-4]; U09b, U20a [1-3]; U14 [4]
Environmental protection E03 [4]; E15a, E34 [1-4]; E16a, E22c [1-3]; E38 [2-4]
Storage and disposal D01, D05, D09a, D12a [1-4]; D02 [1, 4]; D10b [1-3]; D11a [4]
Treated seed S01, S02, S03, S04a, S04b, S05, S06a, S07, S08 [4]
Medical advice M03 [1-3]

247 fluoxastrobin + prothioconazole + trifloxystrobin

A triazole and strobilurin fungicide mixture for cereals
FRAC mode of action code: 11 + 3 + 11

Products

Jaunt	Bayer CropScience	75:150:75 g/l	EC	12350

Uses

- Brown rust in ***spring barley***, ***winter barley***, ***winter wheat***

FOR FULL CONDITIONS OF USE ALWAYS READ THE PRODUCT LABEL

- Eyespot in ***spring barley*** *(reduction)*, ***winter barley*** *(reduction)*, ***winter wheat*** *(reduction)*
- Glume blotch in ***winter wheat***
- Late ear diseases in ***spring barley***, ***winter barley***, ***winter wheat***
- Net blotch in ***spring barley***, ***winter barley***
- Powdery mildew in ***spring barley***, ***winter barley***, ***winter wheat***
- Rhynchosporium in ***spring barley***, ***winter barley***
- Septoria leaf blotch in ***winter wheat***
- Tan spot in ***winter wheat***
- Yellow rust in ***winter wheat***

Approval information

- Trifloxystrobin included in Annex I under EC Directive 91/414
- Accepted by BBPA for use on malting barley

Efficacy guidance

- Best results obtained from treatment at early stages of disease development. Further treatment may be needed if disease attack is prolonged
- Applications to established infections of any disease are likely to be less effective
- Best control of cereal ear diseases obtained by treatment during ear emergence
- Fluoxastrobin and trifloxystrobin are members of the QoI cross resistance group. Product should be used preventatively and not relied on for its curative potential
- Use product as part of an Integrated Crop Management strategy incorporating other methods of control, including where appropriate other fungicides with a different mode of action. Do not apply more than two foliar applications of QoI containing products to any cereal crop
- There is a significant risk of widespread resistance occurring in *Septoria tritici* populations in UK. Failure to follow resistance management action may result in reduced levels of disease control
- Strains of wheat and barley powdery mildew resistant to QoIs are common in the UK. Control of wheat mildew can only be relied on from the triazole component
- Where specific control of wheat mildew is required this should be achieved through a programme of measures including products recommended for the control of mildew that contain a fungicide from a different cross-resistance group and applied at a dose that will give robust control
- Prothioconazole is a DMI fungicide. Resistance to some DMI fungicides has been identified in Septoria leaf blotch which may seriously affect performance of some products. For further advice contact a specialist advisor and visit the Fungicide Resistance Action Group (FRAG)-UK website

Restrictions

- Maximum total dose equivalent to two full dose treatments

Crop-specific information

- Latest use: before grain milky ripe for winter wheat; up to beginning of anthesis (GS 61) for barley

Environmental safety

- Dangerous for the environment
- Very toxic to aquatic organisms
- Risk to non-target insects or other arthropods. Avoid spraying within 6 m of the field boundary to reduce the effects on non-target insects or other arthropods
- LERAP Category B

Hazard classification and safety precautions

Hazard H04, H11
Risk phrases R37, R50, R53a
Operator protection A, C, H; U05a, U09b, U19a, U20b
Environmental protection E15a, E16a, E22c, E34, E38
Storage and disposal D01, D02, D05, D09a, D10b, D12a
Medical advice M03

SECTION 2

SEE SECTION 3 FOR PRODUCTS ALSO REGISTERED

248 flupyrsulfuron-methyl

A sulfonylurea herbicide for winter wheat and winter barley
HRAC mode of action code: B

See also *carfentrazone-ethyl + flupyrsulfuron-methyl*
diflufenican + flupyrsulfuron-methyl

Products

1	DP 459	DuPont	50% w/w	WG	12996
2	Lexus SX	DuPont	50% w/w	WG	12979

Uses
- Annual dicotyledons in ***winter barley***, ***winter linseed*** *(off-label)* [2]; ***winter wheat*** [1, 2]
- Blackgrass in ***winter barley***, ***winter linseed*** *(off-label)*, ***winter wheat*** [2]

Specific Off-Label Approvals (SOLAs)
- ***winter linseed*** *20063714* [2]

Approval information
- Flupyrsulfuron-methyl included in Annex I under EC Directive 91/414

Efficacy guidance
- Best results achieved from applications made in good growing conditions
- Good spray cover of weeds must be obtained
- For pre-emergence control of blackgrass application must be made in tank mixture with specified products (see label). Best control of blackgrass post-emergence in wheat obtained from application from the 1-leaf stage [2]
- Winter barley must be treated pre-emergence and used in mixture. See label [2]
- Growth of weeds is inhibited within hours of treatment but visible symptoms may not be apparent for up to 4 wk
- Flupyrsulfuron-methyl has moderate residual life in soil. Under normal moisture conditions susceptible weeds germinating soon after treatment will be controlled
- May be used on all soil types but residual activity and weed control is reduced on highly alkaline soils
- Flupyrsulfuron-methyl is a member of the ALS-inhibitor group of herbicides. To avoid the build up of resistance do not use any product containing an ALS-inhibitor herbicide with claims for control of grass weeds more than once on any crop
- Use these products as part of a resistance management strategy that includes cultural methods of control and does not use ALS inhibitors as the sole chemical method of grass weed control

Restrictions
- Maximum number of treatments 1 per crop [2]
- Maximum total dose equivalent to one full dose treatment [1]
- Do not use on wheat undersown with grasses, clover or other legumes, or any other broad-leaved crop
- Do not apply within 7 d of rolling
- Do not treat any crop suffering from drought, waterlogging, pest or disease attack, nutrient deficiency, or any other stress factors
- Specific restrictions apply to use in sequence or tank mixture with other sulfonylurea or ALS-inhibiting herbicides. See label for details

Crop-specific information
- Latest use: pre-emergence or before 1st node detectable (GS 31) for wheat; pre-emergence for barley [2]
- Under certain conditions chlorosis and stunting may occur, from which recovery is rapid

Following crops guidance
- Only cereals, oilseed rape, field beans, clover or grass may be sown in the yr of harvest of a treated crop
- In the event of crop failure only winter or spring wheat may be sown within 3 mth of treatment. Land should be ploughed and cultivated to 15 cm minimum before resowing. After 3 mth crops may be sown as shown above

Environmental safety
- Dangerous for the environment

- Very toxic to aquatic organisms
- Take extreme care to avoid drift onto broad-leaved plants outside the target area or onto ponds, waterways or ditches, or onto land intended for cropping
- Spraying equipment should not be drained or flushed onto land planted, or to be planted, with trees or crops other than cereals and should be thoroughly cleansed after use - see label for instructions

Hazard classification and safety precautions

Hazard H11
Risk phrases R50, R53a
Operator protection U08, U19a, U20b
Environmental protection E15a, E38
Storage and disposal D09a, D11a, D12a

249 flupyrsulfuron-methyl + thifensulfuron-methyl

A sulfonylurea herbicide mixture for winter wheat and winter oats
HRAC mode of action code: B + B

Products

1	Lancer	Headland	10:40% w/w	WG	13031
2	Lexus Millenium	DuPont	10:40% w/w	WG	09206

Uses

- Annual dicotyledons in ***winter oats***, ***winter wheat***
- Blackgrass in ***winter oats***, ***winter wheat***

Approval information

- Flupyrsulfuron-methyl and thifensulfuron-methyl included in Annex I under EC Directive 91/414

Efficacy guidance

- Best results obtained when applied to small actively growing weeds
- Good spray cover of weeds must be obtained
- Increased degradation of active ingredient in high soil temperatures reduces residual activity
- Product has moderate residual life in soil. Under normal moisture conditions susceptible weeds germinating soon after treatment will be controlled
- Product may be used on all soil types but residual activity and weed control is reduced on highly alkaline soils
- Blackgrass should be treated from 1 leaf up to first node stage
- Flupyrsulfuron-methyl and thifensulfuron-methyl are members of the ALS-inhibitor group of herbicides. To avoid the build up of resistance do not use any product containing an ALS-inhibitor herbicide with claims for control of grass weeds more than once on any crop
- Use these products as part of a resistance management strategy that includes cultural methods of control and does not use ALS inhibitors as the sole chemical method of grass weed control

Restrictions

- Maximum number of treatments 1 per crop
- Do not use on wheat undersown with grasses or legumes, or any other broad-leaved crop
- Do not apply within 7 d of rolling
- Do not treat any crop suffering from drought, waterlogging, pest or disease attack, nutrient deficiency, or any other stress factors
- Contact contract agents before use on crops grown for seed
- Specific restrictions apply to use in sequence or tank mixture with other sulfonylurea or ALS-inhibiting herbicides. See label for details

Crop-specific information

- Latest use: before 1st node detectable (GS 31) on wheat; before 31 Dec in yr of sowing on oats [2]
- Slight chlorosis, speckling and stunting may occur in certain conditions. Recovery is rapid and yield not affected

Following crops guidance

- Only cereals, oilseed rape, field beans or grass may be sown in the yr of harvest of a treated crop. Any crop may be sown the following spring
- In the event of crop failure only winter wheat may be sown before normal harvest date. Land should be ploughed and cultivated to 15 cm minimum before resowing

SEE SECTION 3 FOR PRODUCTS ALSO REGISTERED

Environmental safety

- Dangerous for the environment
- Very toxic to aquatic organisms
- LERAP Category B

Hazard classification and safety precautions

Hazard H04, H11
Risk phrases R43, R50, R53a
Operator protection A, H; U05a, U08, U14, U19a, U20b
Environmental protection E15a, E16a, E38
Storage and disposal D01, D02, D09a, D11a, D12a

250 flupyrsulfuron-methyl + tribenuron-methyl

A foliar applied sulfonylurea herbicide mixture
HRAC mode of action code: B + B

Products

Excalibur	DuPont	30:30 % w/w	WG	12047

Uses

- Annual dicotyledons in ***winter oats***, ***winter wheat***
- Blackgrass in ***winter oats***, ***winter wheat***

Approval information

- Flupyrsulfuron and tribenuron-methyl included in Annex I under EC Directive 91/414

Efficacy guidance

- Best results obtained when applied to small actively growing weeds
- Good spray cover of weeds must be obtained
- Increased degradation of active ingredient in high soil temperatures reduces residual activity
- Product has moderate residual life in soil. Under normal moisture conditions susceptible weeds germinating soon after treatment will be controlled
- Product may be used on all soil types but residual activity and weed control is reduced on highly alkaline soils
- Blackgrass should be treated from 2 leaf up to before first node detectable
- Flupyrsulfuron-methyl and tribenuron-methyl are members of the ALS-inhibitor group of herbicides. To avoid the build up of resistance do not use any product containing an ALS-inhibitor herbicide with claims for control of grass weeds more than once on any crop
- Use these products as part of a resistance management strategy that includes cultural methods of control and does not use ALS inhibitors as the sole chemical method of grass weed control

Restrictions

- Maximum number of treatments 1 per crop on oats and wheat
- Do not use on wheat undersown with grasses or legumes, or any other broad-leaved crop
- Do not apply within 7 d of rolling
- Do not treat any crop suffering from drought, waterlogging, pest or disease attack, nutrient deficiency, or any other stress factors
- Specific restrictions apply to use in sequence or tank mixture with other sulfonylurea or ALS-inhibiting herbicides. See label for details

Crop-specific information

- Latest use: 31 Dec for winter oats; before first node detectable stage (GS 31) for winter wheat
- Slight chlorosis and stunting may occur in certain conditions. Recovery is rapid and yield not affected

Following crops guidance

- Only cereals, oilseed rape or field beans may be sown in the yr of harvest of a treated crop
- In the event of crop failure only winter or spring wheat may be sown within 3 mth of treatment. Land should be ploughed and cultivated to 15 cm minimum before resowing. After 3 mth crops may be sown as indicated above

Environmental safety

- Dangerous for the environment

- Extremely dangerous to fish or other aquatic life. Do not contaminate surface waters or ditches with chemical or used container
- Take extreme care to avoid drift onto broad-leaved plants outside the target area or onto ponds, waterways or ditches, or onto land intended for cropping
- Spraying equipment should not be drained or flushed onto land planted, or to be planted, with trees or crops other than cereals and should be thoroughly cleansed after use - see label for instructions

Hazard classification and safety precautions

Hazard H04, H11
Risk phrases R43, R50, R53a
Operator protection A; U05a, U08, U20b
Environmental protection E13a, E38
Storage and disposal D01, D02, D09a, D11a, D12a

251 fluquinconazole

A protectant, eradicant and systematic triazole fungicide for winter cereals
FRAC mode of action code: 3

Products

1	Flamenco	BASF	100 g/l	SC	11699
2	Galmano	Bayer CropScience	167 g/l	FS	11650
3	Sahara	Bayer CropScience	100 g/l	SC	11905

Uses

- Brown rust in ***winter wheat*** [1, 3]
- Bunt in ***winter wheat*** *(seed treatment)* [2]
- Glume blotch in ***winter wheat*** [1, 3]
- Powdery mildew in ***winter wheat*** [1, 3]
- Septoria leaf blotch in ***winter wheat*** [1, 3]; ***winter wheat*** *(seed treatment)* [2]
- Take-all in ***winter wheat*** *(seed treatment - reduction)* [2]
- Yellow rust in ***winter wheat*** [1, 3]; ***winter wheat*** *(seed treatment)* [2]

Efficacy guidance

- Best results obtained from spray treatments when disease first becomes active in crop but before infection spreads to younger leaves [1, 3]
- Adequate disease protection throughout the season will usually require a programme of at least two fungicide treatments [1, 3]
- May also be applied as a protectant at end of ear emergence (but no later) if crop still disease free [1, 3]
- With seed treatments ensure good even coverage of seed to obtain reliable disease control [2]
- Seed treatment provides early control of Septoria leaf blotch and yellow rust but may require foliar treatment for later infection [2]
- Fluquinconazole is a DMI fungicide. Resistance to some DMI fungicides has been identified in Septoria leaf blotch which may seriously affect performance of some products. For further advice contact a specialist advisor and visit the Fungicide Resistance Action Group (FRAG)-UK website

Restrictions

- Maximum total spray dose equivalent to two full doses [1, 3]
- Maximum number of seed treatments 1 per batch [2]
- Do not apply spray treatments after the start of anthesis (GS 59) [1, 3]
- Do not mix with any other products [2]
- Do not treat cracked, split or sprouted seed [2]
- Do not use treated seed as food or feed [2]

Crop-specific information

- Latest use: before beginning of anthesis (GS 59) for spray treatment; before drilling for seed treatment
- Ensure good spray coverage and increase volume in dense crops [1, 3]
- Delayed emergence may result when treated seed is drilled into heavy or poorly drained soils which then become wet or waterlogged [2]

Environmental safety

- Dangerous for the environment
- Toxic to aquatic organisms
- Dangerous to fish or other aquatic life. Do not contaminate surface waters or ditches with chemical or used container
- Special PPE requirements and precautions apply where product supplied in returnable packs. Check label

Hazard classification and safety precautions

Hazard H02, H11
Risk phrases R22a, R48, R51, R53a [1-3]; R36, R43 [1, 3]
Operator protection A, H [1-3]; C [1, 3]; D [2]; M [1]; U05a [1-3]; U07, U20b [2]; U11, U14, U15 [1, 3]
Environmental protection E13b [2, 3]; E15a [1]; E34, E38 [1-3]; E36a [2]
Storage and disposal D01, D02, D05, D12a [1-3]; D08, D09a, D14 [2]; D10c [1, 3]
Treated seed S01, S02, S03, S04b, S05, S06a, S07, S08 [2]
Medical advice M03 [1-3]; M04a [1, 3]

252 fluquinconazole + prochloraz

A broad-spectrum triazole mixture for use as a spray and seed treatment in winter cereals
FRAC mode of action code: 3 + 3

Products

1	Epona	BASF	167:31 g/l	FS	12444
2	Foil	BASF	54:174 g/l	SE	11700
3	Galmano Plus	Bayer CropScience	167:31 g/l	FS	11645
4	Jockey	BASF	167:31 g/l	FS	11689

Uses

- Brown rust in ***winter barley***, ***winter wheat*** [2]; ***winter wheat*** *(seed treatment - moderate control)* [1]
- Bunt in ***winter wheat*** *(seed treatment)* [1, 3, 4]
- Covered smut in ***winter barley*** *(seed treatment)* [4]
- Fusarium foot rot and seedling blight in ***winter barley*** *(seed treatment)* [4]
- Fusarium root rot in ***winter wheat*** *(seed treatment)* [1, 3, 4]
- Loose smut in ***winter barley*** *(seed treatment)* [4]
- Powdery mildew in ***winter barley***, ***winter wheat*** *(moderate control)* [2]
- Rhynchosporium in ***winter barley*** [2]
- Septoria leaf blotch in ***winter wheat*** [2]; ***winter wheat*** *(seed treatment)* [3, 4]; ***winter wheat*** *(seed treatment - moderate control)* [1]
- Take-all in ***winter barley*** *(seed treatment - reduction)* [4]; ***winter wheat*** *(seed treatment - reduction)* [1, 3, 4]
- Yellow rust in ***winter barley*** *(seed treatment)* [4]; ***winter wheat*** [2]; ***winter wheat*** *(seed treatment)* [3, 4]; ***winter wheat*** *(seed treatment - moderate control)* [1]

Approval information

- Accepted by BBPA for use on malting barley (before ear emergence only)

Efficacy guidance

- Best results obtained from spray treatment when disease first becomes active in crop but before infection spreads to younger leaves [2]
- When treating seed ensure good even coverage of seed to obtain reliable disease control [1, 3, 4]
- Seed treatment provides early control of *Septoria* leaf spot and rust but may require foliar treatment for later infection [1, 3, 4]
- Fluquinconazole and prochloraz are DMI fungicides. Resistance to some DMI fungicides has been identified in Septoria leaf blotch which may seriously affect performance of some products. For further advice contact a specialist advisor and visit the Fungicide Resistance Action Group (FRAG)-UK website

Restrictions

- Maximum number of treatments 1 per seed batch (seed dressing)
- Maximum total dose equivalent to two full doses (spray)

- Do not treat cracked, split or sprouted seed [1, 3, 4]
- Do not use treated seed as food or feed [1, 3, 4]

Crop-specific information

- Latest use: before 1st awns visible (GS 47) for winter barley; before beginning of anthesis (GS 59) for winter wheat; before drilling (seed treatment)
- Ensure good spray coverage and increase volume in dense crops [2]
- Delayed emergence may result when treated seed is drilled into heavy or poorly drained soils which then become wet or waterlogged [1, 3, 4]

Environmental safety

- Dangerous for the environment
- Toxic to aquatic organisms
- Dangerous to fish or other aquatic life. Do not contaminate surface waters or ditches with chemical or used container
- Product supplied in returnable packs for which special PPE requirements and precautions apply. Check label

Hazard classification and safety precautions

Hazard H02 [1, 3, 4]; H03, H04 [2]; H11 [1-4]
Risk phrases R22a, R48, R51, R53a [1-4]; R36 [2, 3]; R43, R66 [2]
Operator protection A, H [1-4]; C [2]; D [3, 4]; G [1]; U05a [1-4]; U07 [3, 4]; U20b [1, 3, 4]
Environmental protection E13b [3]; E15a [1, 2, 4]; E34, E38 [1-4]; E36a [3, 4]
Storage and disposal D01, D02, D09a, D12a [1-4]; D05 [2, 3]; D08 [1, 3, 4]; D10c [2]; D11a [1, 4]; D14 [3, 4]
Treated seed S01, S03, S04b, S06a [3, 4]; S02, S05, S07, S08 [1, 3, 4]
Medical advice M03 [2, 3]; M04a [1, 4]; M05b [2]

253 fluroxypyr

A post-emergence pyridinecarboxylic acid herbicide
HRAC mode of action code: O

See also *aminopyralid + fluroxypyr*
clopyralid + florasulam + fluroxypyr
clopyralid + fluroxypyr + MCPA
clopyralid + fluroxypyr + triclopyr
florasulam + fluroxypyr

Products

1	Agriguard Fluroxypyr	AgriGuard	200 g/l	EC	12525
2	Agriguard Fluroxypyr	AgriGuard	200 g/l	EC	13335
3	Awac	Globachem	200 g/l	EC	13296
4	Crescent	AgriGuard	200 g/l	EC	12559
5	Crescent	AgriGuard	200 g/l	EC	13330
6	Flurostar 200	Globachem	200 g/l	EC	12903
7	Greencrop Reaper	Greencrop	200 g/l	EC	12261
8	Hatchet	AgriGuard	200 g/l	EC	12524
9	Hatchet Extra	AgriGuard	200 g/l	EC	13213
10	Klever	Globachem	200 g/l	EC	13295
11	Landgold Fluroxypyr	Teliton	200 g/l	EC	13356
12	Standon Homerun	Standon	200 g/l	EC	13177
13	Starane 2	Dow	200 g/l	EC	12018
14	Tandus	Nufarm UK	200 g/l	EC	13432
15	Tomahawk	Makhteshim	200 g/l	EC	09249

Uses

- Annual dicotyledons in ***bulb onions*** *(off-label)*, ***leeks*** *(off-label)* [13]; ***durum wheat***, ***triticale***, ***winter barley***, ***winter oats***, ***winter rye***, ***winter wheat*** [1-15]; ***forage maize***, ***permanent grassland***, ***rotational grass***, ***spring barley***, ***spring oats***, ***spring rye***, ***spring wheat*** [1-13, 15]
- Black bindweed in ***durum wheat***, ***triticale***, ***winter barley***, ***winter oats***, ***winter rye***, ***winter wheat*** [1-15]; ***forage maize***, ***permanent grassland***, ***rotational grass***, ***spring barley***, ***spring oats***, ***spring rye***, ***spring wheat*** [1-13, 15]

- Chickweed in ***durum wheat, triticale, winter barley, winter oats, winter rye, winter wheat*** [1-15]; ***forage maize, permanent grassland, rotational grass, spring barley, spring oats, spring rye, spring wheat*** [1-13, 15]
- Cleavers in ***apple orchards*** *(off-label)*, ***pear orchards*** *(off-label)*, ***poppies for morphine production*** *(off-label)* [13]; ***durum wheat, triticale, winter barley, winter oats, winter rye, winter wheat*** [1-15]; ***forage maize, permanent grassland, rotational grass, spring barley, spring oats, spring rye, spring wheat*** [1-13, 15]
- Docks in ***apple orchards*** *(off-label)*, ***pear orchards*** *(off-label)* [13]; ***durum wheat, triticale, winter barley, winter oats, winter rye, winter wheat*** [1-15]; ***forage maize, permanent grassland, rotational grass, spring barley, spring oats, spring rye, spring wheat*** [1-13, 15]
- Forget-me-not in ***durum wheat, triticale, winter barley, winter oats, winter rye, winter wheat*** [1-15]; ***forage maize, permanent grassland, rotational grass, spring barley, spring oats, spring rye, spring wheat*** [1-13, 15]
- Hemp-nettle in ***durum wheat, triticale, winter barley, winter oats, winter rye, winter wheat*** [1-15]; ***forage maize, permanent grassland, rotational grass, spring barley, spring oats, spring rye, spring wheat*** [1-13, 15]
- Poppies in ***poppies for morphine production*** *(off-label)* [13]
- Stinging nettle in ***apple orchards*** *(off-label)*, ***pear orchards*** *(off-label)* [13]
- Volunteer potatoes in ***bulb onions*** *(off-label)*, ***leeks*** *(off-label)*, ***poppies for morphine production*** *(off-label)*, ***sweetcorn*** *(off-label)* [13]; ***winter barley, winter wheat*** [1-15]

Specific Off-Label Approvals (SOLAs)

- ***apple orchards*** *20040988* [13]
- ***bulb onions*** *20040986* [13]
- ***leeks*** *20040987* [13]
- ***pear orchards*** *20040988* [13]
- ***poppies for morphine production*** *20052059* [13]
- ***sweetcorn*** *20051696* [13]

Approval information

- Fluroxypyr included in Annex I under EC Directive 91/414
- Accepted by BBPA for use on malting barley

Efficacy guidance

- Best results achieved under good growing conditions in a strongly competing crop
- A number of tank mixtures with HBN and other herbicides are recommended for use in autumn and spring to extend range of species controlled. See label for details
- Spray is rainfast in 1 h

Restrictions

- Maximum number of treatments 1 per crop or yr or maximum total dose equivalent to one full dose treatment
- Do not apply in any tank-mix on triticale or forage maize
- Do not use on crops undersown with clovers or other legumes
- Do not treat crops suffering stress caused by any factor
- Do not roll or harrow for 7 d before or after treatment
- Do not spray if frost imminent
- Do not apply before 1 Mar in yr of harvest [14]
- Do not use on crops grown for seed production [14]
- Straw from treated crops must not be returned directly to the soil but must be removed and used only for livestock bedding

Crop-specific information

- Latest use: before flag leaf sheath opening (GS 47) for winter wheat and barley; before flag leaf sheath extending (GS 41) for spring wheat and barley; before second node detectable (GS 32) for oats, rye, triticale and durum wheat; before 7 leaves unfolded and before buttress roots appear for maize
- Apply to new leys from 3 expanded leaf stage
- Timing varies in tank mixtures. See label for details
- Crops undersown with grass may be sprayed provided grasses are tillering

Following crops guidance

- Clovers, peas, beans and other legumes must not be sown for 12 mth following treatment at the highest dose

Environmental safety

- Dangerous for the environment
- Vey toxic to aquatic organisms
- Flammable
- Keep livestock out of treated areas for at least 3 d following treatment and until poisonous weeds, such as ragwort, have died down and become unpalatable
- Wash spray equipment thoroughly with water and detergent immediately after use. Traces of product can damage susceptible plants sprayed later

Hazard classification and safety precautions

Hazard H03, H11 [1-15]; H08 [1-13, 15]
Risk phrases R22b, R53a, R67 [1-15]; R36, R38 [1-6, 8-12]; R37 [1-6, 8-13, 15]; R43 [1-6, 8-12, 14]; R50 [2, 3, 5, 6, 9, 10, 14]; R51 [1, 4, 7, 8, 11-13, 15]
Operator protection A, H [1-6, 8-12, 14]; C [1-6, 8-12]; U05a [1-6, 8-10, 12]; U08, U19a [1-15]; U11 [1-6, 8-12]; U14 [1-6, 8-11, 14]; U20b [1-13, 15]
Environmental protection E06a [5] (3 d); E07a [1-4, 6-15]; E15a [7, 13, 15]; E15b [1-6, 8-12, 14]; E34 [1-6, 8-12, 15]; E38 [2, 3, 5, 6, 9, 10, 13-15]
Storage and disposal D01, D02 [1-6, 8-12]; D05 [1-3, 5-7, 9-12, 15]; D09a, D10b [1-15]; D12a [1-6, 8-15]
Treated seed S06a [14]
Medical advice M03 [1-6, 8-12]; M05b [1-15]

254 fluroxypyr + mecoprop-P

A post-emergence herbicide for broadleaved weeds in amenity turf
HRAC mode of action code: O + O

Products

Bastion T	Rigby Taylor	72:300 g/l	ME	06011

Uses

- Annual dicotyledons in ***managed amenity turf***
- Perennial dicotyledons in ***managed amenity turf***
- Slender speedwell in ***managed amenity turf***

Approval information

- Fluroxypyr and mecoprop-P included in Annex I under EC Directive 91/414

Efficacy guidance

- Best results achieved when soil moist and weeds in active growth, normally Apr-Sep

Restrictions

- Maximum number of treatments 2 per yr. The total amount of mecoprop-P applied in a single yr must not exceed the maximum total dose approved for any single product for the crop/situation
- Do not treat turf under stress from any cause, if night temperatures are low, if ground frost imminent or during prolonged cold weather
- Do not apply in drought period unless irrigation applied
- Do not spray if turf wet

Crop-specific information

- Avoid mowing for 3 d before or after spraying (5 d before for young turf)
- Young turf must only be treated in spring provided that at least 2 mth have elapsed between sowing and application

Environmental safety

- Dangerous for the environment
- Toxic to aquatic organisms
- Wash spray equipment thoroughly with water and detergent immediately after use. Traces of product can damage susceptible plants sprayed later
- Avoid drift onto all broad-leaved plants outside the target area

Hazard classification and safety precautions

Hazard H03, H11
Risk phrases R21, R22a, R22b, R41, R51, R53a
Operator protection A, C; U05a, U08, U11, U19a, U20b

Environmental protection E15a, E34, E38
Storage and disposal D01, D02, D05, D09a, D10b, D12a
Medical advice M03, M05a

255 fluroxypyr + thifensulfuron-methyl + tribenuron-methyl

A foliar applied herbicide mixture for cereals with some residual activity
HRAC mode of action code: O + B + B

Products

GEX 353	DuPont	200 g/l + 50.25% w/w	KK	12755

Uses
- Annual dicotyledons in ***spring barley***, ***spring wheat***, ***winter barley***, ***winter wheat***

Approval information
- Fluroxypyr, thifensulfuron-methyl and tribenuron-methyl included in Annex I under EC Directive 91/414
- Accepted by BBPA for use on malting barley

Efficacy guidance
- Best results obtained when weeds small and actively growing. This is particularly important for cleavers
- Ensure good spray cover
- Susceptible species cease growth immediately after application but may take 2 wk to show symptoms
- Weed control may be reduced in very dry conditions
- Rain within 4 hr may reduce effectiveness
- Thifensulfuron-methyl and tribenuron-methyl are members of the ALS-inhibitor group of herbicides and products should be used in a planned Resistance Management strategy. See Section 5 for more information

Restrictions
- Maximum number of treatments 1 per crop for all cereals
- Must only be applied after 1 Feb in yr of harvest and should not be applied more than once to any cereal crop
- Do not use on any crop suffering stress from drought, waterlogging, frost, deficiency, pest or disease attack or any other cause
- Do not use on crops undersown with grasses, clover or legumes, or any other broad leaved crop
- Specific restrictions apply to use in sequence or tank mixture with other sulfonylurea or ALS-inhibiting herbicides. See label for details
- Do not apply within 7 d of rolling

Crop-specific information
- Latest use: before flag leaf extending (GS 39) for all crops

Following crops guidance
- Only cereals, field beans or oilseed rape may be sown in the same calendar yr as harvest of a treated crop
- In the event of failure of a treated crop only a cereal crop may be sown within 3 mth
- Where imazamethabenz-methyl (MAPP 10218) or thifensulfuron-methyl (12285) are also applied to the same cereal crop, only wheat, barley, rye or triticale should be sown as a following crop [1]

Environmental safety
- Dangerous for the environment
- Very toxic to aquatic organisms
- Extremely dangerous to fish or other aquatic life. Do not contaminate surface waters or ditches with chemical or used container
- Keep livestock out of treated areas for at least 3 days and until foliage of any poisonous weeds such as ragwort has died and become unpalatable
- Take extreme care to avoid damage by drift onto broad-leaved plants outside the target area, or onto ponds, waterways or ditches

- Spraying equipment should not be drained or flushed onto land planted, or to be planted, with trees or crops other than cereals and should be thoroughly cleansed after use - see label for instructions

Hazard classification and safety precautions

Hazard H03, H08, H11
Risk phrases R22b, R37, R43, R50, R53a, R67
Operator protection A, C, H; U08, U19a, U20b
Environmental protection E07c (3 d); E13a, E16a, E16b, E38
Storage and disposal D05, D09a, D10b, D12a
Medical advice M05b

256 fluroxypyr + triclopyr

A foliar acting herbicide for docks in grassland
HRAC mode of action code: O + O

Products

Doxstar	Dow	100:100 g/l	EC	11063

Uses

- Chickweed in ***rotational grass***
- Docks in ***permanent grassland*, *rotational grass***

Approval information

- Fluroxypyr and triclopyr included in Annex I under EC Directive 91/414

Efficacy guidance

- Seedling docks in new leys are controlled up to 50 mm diameter. In established grass apply in spring or autumn or, at lower dose, in spring and autumn on docks up to 200 mm. A second application in the subsequent yr may be needed
- Allow 2-3 wk after cutting or grazing to allow sufficient regrowth of docks to occur before spraying
- Control may be reduced if rain falls within 2 h of application
- To allow maximum translocation to the roots of docks do not cut grass for 28 d after spraying

Restrictions

- Maximum total dose equivalent to one full dose treatment
- Do not roll or harrow for 10 d before or 7 d after spraying
- Do not spray in drought, very hot or very cold weather

Crop-specific information

- Latest use: 7 d before grazing or harvest of grass
- Grass less than one yr old may be treated at half dose from the third leaf visible stage
- Clover will be killed or severely checked by treatment

Following crops guidance

- Do not sow kale, turnips, swedes or grass mixtures containing clover by direct drilling or minimum cultivation techniques within 6 wk of application

Environmental safety

- Dangerous for the environment
- Toxic to aquatic organisms
- Keep livestock out of treated areas for at least 7 d following treatment and until poisonous weeds, such as ragwort, have died down and become unpalatable
- Do not allow drift to come into contact with crops, amenity plantings, gardens, ponds, lakes or watercourses
- Wash spray equipment thoroughly with water and detergent immediately after use. Traces of product can damage susceptible plants sprayed later

Hazard classification and safety precautions

Hazard H03, H08, H11
Risk phrases R22a, R22b, R37, R38, R43, R51, R53a, R67
Operator protection A, C; U02a, U05a, U08, U14, U19a, U20b
Environmental protection E07a, E15a, E34, E38
Consumer protection C01

SEE SECTION 3 FOR PRODUCTS ALSO REGISTERED

Storage and disposal D01, D02, D05, D09a, D10b, D12a
Medical advice M05b

257 flurtamone

A carotenoid synthesis inhibitor available only in mixtures
HRAC mode of action code: F1

See also *diflufenican + flurtamone*

258 flusilazole

A systemic, protective and curative triazole fungicide for cereals, oilseed rape and sugar beet
FRAC mode of action code: 3

See also *carbendazim + flusilazole*
chlorothalonil + flusilazole
famoxadone + flusilazole
fenpropimorph + flusilazole

Products

1	Capitan 25	DuPont	250 g/l	EW	13067
2	Genie 25	DuPont	250 g/l	EW	13065
3	Lyric	DuPont	250 g/l	OL	13022
4	Sanction 25	DuPont	250 g/l	EW	13066

Uses
- Brown rust in ***spring barley***, ***winter barley***, ***winter wheat***
- Canker in ***brussels sprouts*** *(off-label)*
- Eyespot in ***spring barley***, ***winter barley***, ***winter wheat***
- Glume blotch in ***spring barley***, ***winter barley***, ***winter wheat***
- Light leaf spot in ***spring oilseed rape***, ***winter oilseed rape***
- Net blotch in ***spring barley***, ***winter barley***
- Powdery mildew in ***spring barley***, ***sugar beet***, ***winter barley***, ***winter wheat***
- Rhynchosporium in ***spring barley***, ***winter barley***
- Rust in ***sugar beet***
- Septoria leaf blotch in ***spring barley***, ***winter barley***, ***winter wheat***
- Yellow rust in ***spring barley***, ***winter barley***, ***winter wheat***

Specific Off-Label Approvals (SOLAs)
- ***brussels sprouts*** *20070102* [1], *20070101* [2], *20063998* [3], *20070100* [4]

Approval information
- Flusilazole has been included in Annex I under Directive 91/414 for 18 months from January 2007 and all approvals for products containing this active ingredient will expire in June 2008
- Accepted by BBPA for use on malting barley

Efficacy guidance
- Best results obtained from treatment at onset of disease development
- Best control of eyespot in cereals achieved by spraying between leaf-sheath erect and second node detectable stages (GS 30-32)
- Product active against both MBC-sensitive and MBC-resistant strains of eyespot
- Rain occurring within 2 h after application may reduce effectiveness
- See label for recommended tank-mixes to give broader spectrum control
- Flusilazole is a DMI fungicide. Resistance to some DMI fungicides has been identified in Septoria leaf blotch which may seriously affect performance of some products. For further advice contact a specialist advisor and visit the Fungicide Resistance Action Group (FRAG)-UK website

Restrictions
- Maximum number of treatments (including other products containing flusilazole) on cereals and oilseed rape depends on dose and timing - see labels for details. High rate must not be used more than once in any crop
- Maximum total dose on sugar beet equivalent to one full dose
- Do not apply to crops under stress or during frosty weather

Crop-specific information

- Latest use: high dose before 3rd node detectable (GS 33) plus reduced dose before early milk stage (GS 72) on winter wheat and barley; before first flowers open (GS 4,0) on oilseed rape
- HI 7 wk for sugar beet
- Treat oilseed rape in autumn and/or spring at stem extension stage
- On sugar beet apply at an early stage of disease development, usually in early Aug

Environmental safety

- Dangerous for the environment
- Toxic to aquatic organisms
- Harmful to fish or other aquatic life. Do not contaminate surface waters or ditches with chemical or used container

Hazard classification and safety precautions

Hazard H02, H11
Risk phrases R22a, R38, R40, R51, R53a, R61
Operator protection A, C; U05a, U11, U19a, U20a
Environmental protection E13c, E34, E38
Storage and disposal D01, D02, D05, D10b, D12a
Medical advice M04a

259 flutolanil

An carboxamide fungicide for treatment of potato seed tubers
FRAC mode of action code: 7

Products

1	Rhino	Certis	460 g/l	FS	11802
2	Rhino	Certis	460 g/l	FS	13101
3	Rhino DS	Certis	6% w/w	DS	12763

Uses

- Black scurf in ***potatoes*** *(tuber treatment)*
- Stem canker in ***potatoes*** *(tuber treatment)*

Approval information

- Approval expiry 31 Aug 2008 [1]

Efficacy guidance

- Apply to clean tubers before chitting, prior to planting, or at planting
- Apply flowable concentrate through canopied, hydraulic or spinning disc equipment (with or without electrostatics) mounted on a rolling conveyor or table [1, 2]
- Flowable concentrate may be diluted with water up to 2.0 l per tonne to improve tuber coverage. Disease in areas not covered by spray will not be controlled [1, 2]
- Dry powder may be applied at hopper filling, at box filling, to the box surface, to chitting trays or via an on-planter applicator [3]

Restrictions

- Maximum number of treatments 1 per batch of seed tubers
- Check with processor before use on crops for processing
- Treated tubers must be planted within 3 mth
- Do not use on potatoes being grown for the seed market [3]

Crop-specific information

- Latest use: at planting
- HI 12 wk for potatoes [1, 2]
- Seed tubers should be of good quality and free from bacterial rots, physical damage or virus infection, and should not be sprouted to such an extent that mechanical damage to the shoots will occur during treatment or planting

Environmental safety

- Harmful to aquatic organisms

Hazard classification and safety precautions

Hazard H04
Risk phrases R36, R53a [3]; R43 [1, 2]; R51 [2]; R52 [1, 3]

SEE SECTION 3 FOR PRODUCTS ALSO REGISTERED

Operator protection A, H [1-3]; C [3]; U04a, U05a, U19a, U20a [1-3]; U14 [2]
Environmental protection E15a, E34 [1-3]; E38 [2, 3]
Storage and disposal D01, D02, D09a, D12a [1-3]; D05, D10a [1, 2]; D11a [3]
Treated seed S01, S03, S04a, S05 [1-3]; S02 [3]
Medical advice M03

260 flutriafol

A broad-spectrum triazole fungicide for cereals
FRAC mode of action code: 3

See also *chlorothalonil + flutriafol*
fludioxonil + flutriafol

Products

1	Consul	Headland	125 g/l	SC	12976
2	Pointer	Headland	125 g/l	SC	12975

Uses

- Brown rust in ***spring barley***, ***winter barley***, ***winter wheat***
- Powdery mildew in ***spring barley***, ***winter barley***, ***winter wheat***
- Rhynchosporium in ***spring barley***, ***winter barley***
- Septoria leaf blotch in ***winter wheat***
- Yellow rust in ***spring barley***, ***winter barley***, ***winter wheat***

Approval information

- Accepted by BBPA for use on malting barley

Efficacy guidance

- Best results obtained from treatment in early stages of disease development. See label for detailed guidance on spray timing for specific diseases and the need for repeat treatments
- Tank mix options available on the label to broaden activity spectrum
- Good spray coverage essential for optimum performance
- Flutriafol is a DMI fungicide. Resistance to some DMI fungicides has been identified in Septoria leaf blotch which may seriously affect performance of some products. For further advice contact a specialist advisor and visit the Fungicide Resistance Action Group (FRAG)-UK website

Restrictions

- Maximum number of treatments 2 per crop (including other products containing flutriafol)

Crop-specific information

- Latest use: before early grain milky ripe stage (GS 73)
- Flag leaf tip scorch on wheat caused by stress may be increased by fungicide treatment

Environmental safety

- Harmful to aquatic organisms
- Harmful to fish or other aquatic life. Do not contaminate surface waters or ditches with chemical or used container

Hazard classification and safety precautions

Hazard H03
Risk phrases R43 [1]; R48, R52 [1, 2]
Operator protection A, H; U05a, U09a, U19a, U20b
Environmental protection E13c, E38
Storage and disposal D01, D02, D09a, D10c

261 formaldehyde (commodity substance)

An agricultural/horticultural and animal husbandry fungicide

Products

1	formaldehyde	various	38-40%	SL
2	paraformaldehyde	various	-	ZZ

FOR FULL CONDITIONS OF USE ALWAYS READ THE PRODUCT LABEL

Uses

- Fungus diseases in ***flower bulbs*** *(dip)*, ***glasshouses*** *(spray, dip or fumigant)*, ***mushroom houses*** *(spray or fumigant)* [1]; ***livestock houses*** [2]
- Soil-borne diseases in ***soil and compost*** *(drench)* [1]

Approval information

- Approval for the use of formaldehyde as a commodity substance was granted on 1 March 1991 by Ministers under regulation 5 of the Control of Pesticides Regulations 1986.
- Formaldehyde will not be included in Annex I of EC Directive 91/414 and will not be granted Essential Uses. Formaldehyde will therefore be withdrawn and plant protection use must cease by 22 December 2008

Efficacy guidance

- Use as a dip to sterilize flower bulbs [1]
- Use as drench to sterilize soil and compost, indoors and outdoors [1]
- Use as spray or fumigant in mushroom houses [1]
- Use as spray, dip or fumigant for glasshouse hygiene [1]
- Use as fumigant to sterilize animal houses [2]

Restrictions

- Formaldehyde is subject to the Poisons Rules 1982 and the Poisons Act 1972. See notes in Section 5
- Operators must observe Occupational Exposure Standard as set out in HSE Guidance Note EH40/90 and ACOP 30 *Control of Substances Hazardous to Health in Fumigation*
- Operators must be supplied with a Section 6 (HSW) Safety Data Sheet before commencing work
- Observe maximum permitted concentrations. See PR 12, 1990, pp 9-10

Hazard classification and safety precautions

Operator protection A, D, P [1]

262 fosetyl-aluminium

A systemic phosphonic acid fungicide for various horticultural crops
FRAC mode of action code: 33

Products

1	Aliette 80 WG	Certis	80% w/w	WG	11213
2	Aliette 80 WG	Certis	80% w/w	WG	13130
3	Standon Fosetyl-AL 80 WG	Standon	80% w/w	WG	10667
4	Standon Fullstop	Standon	80% w/w	WG	13571

Uses

- Collar rot in ***apples*** [1-4]
- Crown rot in ***apples*** [1-4]; ***protected strawberries*** *(off-label)*, ***strawberries*** *(off-label)* [1, 2]
- Downy mildew in ***broad beans***, ***hops*** [1-4]; ***broccoli*** *(off-label - seedlings)*, ***brussels sprouts*** *(off-label - seedlings)*, ***cabbages*** *(off-label - seedlings)*, ***calabrese*** *(off-label - seedlings)*, ***cauliflowers*** *(off-label - seedlings)*, ***chinese cabbage*** *(off-label - seedlings)*, ***chives*** *(off-label)*, ***combining peas*** *(off-label - seed treatment)*, ***endives*** *(off-label)*, ***herbs (see appendix 6)*** *(off-label)*, ***kale*** *(off-label - seedlings)*, ***lamb's lettuce*** *(off-label)*, ***leaf spinach*** *(off-label)*, ***lupins*** *(off-label - seed treatment)*, ***parsley*** *(off-label)*, ***protected brassica seedlings*** *(off-label)*, ***protected chives*** *(off-label)*, ***protected endives*** *(off-label)*, ***protected herbs (see appendix 6)*** *(off-label)*, ***protected lamb's lettuce*** *(off-label)*, ***protected lettuce*** *(off-label)*, ***protected parsley*** *(off-label)*, ***protected salad brassicas*** *(off-label - for baby leaf production)*, ***protected scarole*** *(off-label)*, ***salad brassicas*** *(off-label - for baby leaf production)*, ***salad onions*** *(off-label)*, ***scarole*** *(off-label)*, ***spinach beet*** *(off-label)*, ***vining peas*** *(off-label - seed treatment)*, ***watercress*** *(off-label - during propagation)*, ***wine grapes*** *(off-label)* [1, 2]; ***protected lettuce*** [3, 4]
- Phytophthora in ***chicory*** *(off-label - for forcing)*, ***watercress*** *(off-label - during propagation)* [1, 2]
- Phytophthora root rot in ***capillary benches*** [1, 2]; ***protected pot plants*** [1-4]
- Phytophthora stem rot in ***capillary benches*** [1, 2]; ***protected pot plants*** [1-4]
- Phytophthora wilt in ***ornamental plant production*** [1-4]
- Pythium in ***managed amenity turf*** *(off-label)*, ***watercress*** *(off-label - during propagation)* [1, 2]
- Red core in ***protected strawberries*** *(off-label)*, ***strawberries*** *(off-label)* [1, 2]; ***strawberries*** [1-4]
- Root rot in ***capillary benches*** [1, 2]; ***protected pot plants*** [1-4]

Specific Off-Label Approvals (SOLAs)

- ***broccoli*** *(seedlings) 20063524* [2], *(seedlings) 20071902 expires Aug 2008* [1]
- ***brussels sprouts*** *(seedlings) 20063524* [2], *(seedlings) 20071902 expires Aug 2008* [1]
- ***cabbages*** *(seedlings) 20063524* [2], *(seedlings) 20071902 expires Aug 2008* [1]
- ***calabrese*** *(seedlings) 20063524* [2], *(seedlings) 20071902 expires Aug 2008* [1]
- ***cauliflowers*** *(seedlings) 20063524* [2], *(seedlings) 20071902 expires Aug 2008* [1]
- ***chicory*** *(for forcing) 20063521* [2], *(for forcing) 20071904 expires Aug 2008* [1]
- ***chinese cabbage*** *(seedlings) 20063524* [2], *(seedlings) 20071902 expires Aug 2008* [1]
- ***chives*** *20030868 expires Aug 2008* [1], *20063522* [2]
- ***combining peas*** *(seed treatment) 20030861 expires Aug 2008* [1], *(seed treatment) 20063518* [2]
- ***endives*** *20030868 expires Aug 2008* [1], *20063522* [2]
- ***herbs (see appendix 6)*** *20030868 expires Aug 2008* [1], *20063522* [2]
- ***kale*** *(seedlings) 20063524* [2], *(seedlings) 20071902 expires Aug 2008* [1]
- ***lamb's lettuce*** *20030868 expires Aug 2008* [1], *20063522* [2]
- ***leaf spinach*** *20063520* [2], *20071903 expires Aug 2008* [1]
- ***lupins*** *(seed treatment) 20030861 expires Aug 2008* [1, 2]
- ***managed amenity turf*** *20061446 expires Aug 2008* [1], *20063523* [2]
- ***parsley*** *20030868 expires Aug 2008* [1], *20063522* [2]
- ***protected brassica seedlings*** *20063524* [2], *20071902 expires Aug 2008* [1]
- ***protected chives*** *20030868 expires Aug 2008* [1], *20063522* [2]
- ***protected endives*** *20030868 expires Aug 2008* [1], *20063522* [2]
- ***protected herbs (see appendix 6)*** *20030868 expires Aug 2008* [1], *20063522* [2]
- ***protected lamb's lettuce*** *20030868 expires Aug 2008* [1], *20063522* [2]
- ***protected lettuce*** *20030868 expires Aug 2008* [1], *20063522* [2]
- ***protected parsley*** *20030868 expires Aug 2008* [1], *20063522* [2]
- ***protected salad brassicas*** *(for baby leaf production) 20030868 expires Aug 2008* [1], *(for baby leaf production) 20063522* [2]
- ***protected scarole*** *20030868 expires Aug 2008* [1], *20063522* [2]
- ***protected strawberries*** *20063517* [2], *20071900 expires Aug 2008* [1]
- ***salad brassicas*** *(for baby leaf production) 20030868 expires Aug 2008* [1], *(for baby leaf production) 20063522* [2]
- ***salad onions*** *20063515* [2], *20071906 expires Aug 2008* [1]
- ***scarole*** *20030868 expires Aug 2008* [1], *20063522* [2]
- ***spinach beet*** *20063520* [2], *20071903 expires Aug 2008* [1]
- ***strawberries*** *20030861 expires Aug 2008* [1], *20063518* [2]
- ***vining peas*** *(seed treatment) 20030861 expires Aug 2008* [1], *(seed treatment) 20063518* [2]
- ***watercress*** *(during propagation) 20071260* [2], *(during propagation) 20071905 expires Aug 2008* [1]
- ***wine grapes*** *20063519* [2], *20071901 expires Aug 2008* [1]

Approval information

- Fosetyl-aluminium included in Annex I under EC Directive 91/414
- Accepted by BBPA for use on hops
- Approval expiry 31 Aug 2008 [1, 3]

Restrictions

- Maximum number of treatments 1 per seed batch for peas; 1 per batch of compost for lettuce; 1 per yr for strawberries (root dip or foliar spray); 1 per yr for apples (bark paste), 2 per yr for apples (foliar spray); 2 per crop for broad beans, peas; 2 per yr for hops (basal spray); 6 per yr for hops (foliar spray); 1 per crop during propagation, 2 per crop after planting out for lettuce
- Check tolerance of ornamental species before large-scale treatment

Crop-specific information

- Latest use: pre-planting for strawberry runners; pre-sowing for protected lettuce
- HI apples 5 mth (bark paste) or 4 wk (spray); broad beans 17 d; hops 14 d
- Spray young orchards for crown rot protection after blossom when first leaves fully open and repeat after 4-6 wk. Apply as paste to bark of apples to control collar rot
- Apply to broad beans when infection appears (usually at flowering) and 14 d later. Consult before treating crops to be processed
- Use on strawberries only between harvest and 31 Dec

- Spray autumn-planted strawberry runners 2-3 wk after planting or use dip treatment at planting. Spray established crops in late summer/early autumn after picking and repeat annually
- Apply to hops as early season basal spray or as foliar spray every 10-14 d from when training is completed
- Use by wet incorporation in blocking compost for protected lettuce only from Sep to Apr and follow all directions carefully to avoid severe crop injury. Crop maturity may be delayed by a few days
- Apply as drench to rooted cuttings of hardy nursery stock after first potting and repeat mthly. Up to 6 applications may be needed
- See label for details of application to capillary benches and trays of young plants.

Environmental safety

- Harmful to aquatic organisms

Hazard classification and safety precautions

Hazard H04 [4]
Risk phrases R36 [4]; R52 [3, 4]
Operator protection A, C, H; U11 [4]; U20c [1-4]
Environmental protection E15a
Storage and disposal D01, D02, D12a [4]; D09a, D11a [1-4]

263 fosthiazate

An organophosphorus contact nematicide for potatoes
IRAC mode of action code: 1B

Products

Nemathorin 10G	Syngenta	10% w/w	FG	11003

Uses

- Potato cyst nematode in ***potatoes***
- Spraing vectors in ***potatoes*** *(reduction)*
- Wireworm in ***potatoes*** *(reduction)*

Approval information

- Fosthiazate included in Annex I under EC Directive 91/414
- In 2006 PSD required that all products containing this active ingredient should carry the following warning in the main area of the container label: "Fosthiazate is an anticholinesterase organophosphate. Handle with care"

Efficacy guidance

- Treatment may be in furrow or applied and incorporated in one operation to a uniform depth of 10-15 cm. Deeper incorporation will reduce control
- Application best achieved using equipment such as Horstine Farmery Microband Applicator, Matco or Stocks Micrometer applicators together with a rear mounted powered rotary cultivator
- Granules must not become wet or damp before use

Restrictions

- Contains an anticholinesterase organophosphorus compound. Do not use if under medical advice not to work with such compounds
- Maximum number of treatments 1 per crop
- Product must only be applied using tractor-mounted/drawn direct placement machinery. Do not use air assisted broadcast machinery other than that referenced on the product label
- Do not allow granules to stand overnight in the application hopper
- Do not apply more than once every four years on the same area of land
- Do not use on crops to be harvested less than 17 wk after treatment
- Consult before using on crops intended for processing

Crop-specific information

- Latest use: at planting
- HI 17 wk for potatoes

Environmental safety

- Dangerous for the environment
- Toxic to aquatic organisms

SEE SECTION 3 FOR PRODUCTS ALSO REGISTERED

- Dangerous to game, wild birds and animals
- Dangerous to livestock. Keep all livestock out of treated areas for at least 13 wk
- Incorporation to 10-15 cm and ridging up of treated soil must be carried out immediately after application. Powered rotary cultivators are preferred implements for incorporation but discs, power, spring tine or Dutch harrows may be used provided two passes are made at right angles
- To protect birds and wild mammals remove spillages
- Failure completely to bury granules immediately after application is hazardous to wildlife
- To protect groundwater do not apply any product containing fosthiazate more than once every four yr

Hazard classification and safety precautions

Hazard H03, H11
Risk phrases R22a, R43, R51, R53a
Operator protection A, E, G, H, K, M; U02a, U04a, U05a, U09a, U13, U14, U19a, U20a
Environmental protection E06b (13 wk); E15b, E34, E38
Storage and disposal D01, D02, D05, D09a, D11a, D12a
Medical advice M01, M03, M05a

264 fuberidazole

A benzimidazole (MBC) fungicide available only in mixtures
FRAC mode of action code: 1

265 fuberidazole + imidacloprid + triadimenol

A broad spectrum systemic fungicide and insecticide seed treatment for winter cereals
FRAC mode of action code: 1 + 3

Products

Tripod Plus	Makhteshim	15:117:125 g/l	LS	13168

Uses

- Blue mould in ***winter wheat*** *(seed treatment)*
- Brown foot rot in ***winter barley*** *(seed treatment)*
- Brown rust in ***winter barley*** *(seed treatment)*, ***winter wheat*** *(seed treatment)*
- Bunt in ***winter wheat*** *(seed treatment)*
- Covered smut in ***winter barley*** *(seed treatment)*
- Fusarium foot rot and seedling blight in ***winter barley*** *(seed treatment - reduction)*, ***winter oats*** *(seed treatment - reduction)*, ***winter wheat*** *(seed treatment - reduction)*
- Loose smut in ***winter barley*** *(seed treatment)*, ***winter oats*** *(seed treatment)*, ***winter wheat*** *(seed treatment)*
- Net blotch in ***winter barley*** *(seed treatment - seed-borne only)*
- Powdery mildew in ***winter barley*** *(seed treatment)*, ***winter oats*** *(seed treatment)*, ***winter wheat*** *(seed treatment)*
- Pyrenophora leaf spot in ***winter oats*** *(seed treatment)*
- Septoria leaf blotch in ***winter wheat*** *(seed treatment)*
- Septoria seedling blight in ***winter wheat*** *(seed treatment - reduction)*
- Snow rot in ***winter barley*** *(seed treatment - reduction)*
- Virus vectors in ***winter barley*** *(seed treatment)*, ***winter oats*** *(seed treatment)*, ***winter wheat*** *(seed treatment)*
- Wireworm in ***winter barley*** *(reduction of damage)*, ***winter oats*** *(reduction of damage)*, ***winter wheat*** *(reduction of damage)*
- Yellow rust in ***winter barley*** *(seed treatment)*, ***winter wheat*** *(seed treatment)*

Approval information

- Accepted by BBPA for use on malting barley

Efficacy guidance

- Apply through recommended seed treatment machinery
- Evenness of seed cover improved by simultaneous application of equal volumes of product and water
- Calibrate drill for treated seed and drill at 4 cm into firm, well prepared seedbed

- Use minimum 125 kg treated seed per ha and increase seed rate as drilling season progresses. Lower drilling rates and/or early drilling affect duration of BYDV protection needed and may require follow-up aphicide treatment
- When aphid activity unusually late, or is heavy and prolonged in areas of high risk, and mild weather predominates, follow-up treatment may be required
- In addition to seed-borne diseases early attacks of various foliar, air-borne diseases are controlled or suppressed. See label for details

Restrictions

- Maximum number of treatments 1 per batch of seed
- Do not use on naked oats
- Do not drill treated winter wheat seed after end of Nov
- Do not handle seed unnecessarily
- Do not use treated seed as food or feed
- Treated seed should not be broadcast, but drilled to a depth of 4 cm in a well prepared seedbed
- Germination tests should be done on all batches of seed to be treated to ensure seed viability and suitability for treatment
- Do not use on seed with moisture content above 16%, on sprouted, cracked or skinned seed or on seed already treated with another seed treatment

Crop-specific information

- Latest use: before drilling
- Store treated seed in cool, dry, well-ventilated store and drill as soon as possible, preferably in season of purchase
- Treatment may accentuate effects of adverse seedbed conditions on crop emergence

Environmental safety

- Dangerous to fish or other aquatic life. Do not contaminate surface waters or ditches with chemical or used container
- Dangerous to birds, game and other wildlife. Treated seed should not be left on the soil surface. Bury spillages
- Seed should be drilled to a depth of 4 cm into a well-prepared seed bed
- If seed is present on the soil surface, or if spills have occurred, the field should be harrowed and rolled if conditions are appropriate to ensure good incorporation

Hazard classification and safety precautions

Risk phrases R52, R53a
Operator protection A, H; U04a, U05a, U07, U13, U14, U20b
Environmental protection E03, E36a, E38
Storage and disposal D01, D02, D05, D09a, D12a
Treated seed S01, S02, S03, S04c, S05, S06a, S07, S08

266 fuberidazole + triadimenol

A broad spectrum systemic fungicide seed treatment for cereals
FRAC mode of action code: 1 + 3

Products

Tripod	Makhteshim	22.5:187.5 g/l	FS	13014

Uses

- Blue mould in ***spring wheat*** *(seed treatment)*, ***triticale*** *(seed treatment)*, ***winter wheat*** *(seed treatment)*
- Brown foot rot in ***spring barley*** *(seed treatment)*, ***spring oats*** *(seed treatment)*, ***spring rye*** *(seed treatment)*, ***spring wheat*** *(seed treatment)*, ***triticale*** *(seed treatment)*, ***winter barley*** *(seed treatment)*, ***winter oats*** *(seed treatment)*, ***winter rye*** *(seed treatment)*, ***winter wheat*** *(seed treatment)*
- Bunt in ***spring wheat*** *(seed treatment)*, ***winter wheat*** *(seed treatment)*
- Covered smut in ***spring barley*** *(seed treatment)*, ***winter barley*** *(seed treatment)*
- Leaf stripe in ***spring barley*** *(seed treatment)*

- Loose smut in ***spring barley*** *(seed treatment)*, ***spring oats*** *(seed treatment)*, ***spring wheat*** *(seed treatment)*, ***winter barley*** *(seed treatment)*, ***winter oats*** *(seed treatment)*, ***winter wheat*** *(seed treatment)*
- Pyrenophora leaf spot in ***spring oats*** *(seed treatment)*, ***winter oats*** *(seed treatment)*

Approval information

- Accepted by BBPA for use on malting barley

Efficacy guidance

- Apply through recommended seed treatment machinery
- Calibrate drill for treated seed and drill at 2.5-4 cm into firm, well prepared seedbed
- Product will give control of early attacks of foliar air-borne diseases such as powdery mildew, yellow rust, brown rust, crown rust, Septoria. On autumn-sown crops control of rust is prolonged and usually extends over winter, leading to a delayed build-up of the disease in the following spring, often resulting in fewer fungicide sprays being required to maintain control during spring/ summer
- Product will often delay and reduce the likelihood and extent of lodging in autumn sown wheat.

Restrictions

- Maximum number of treatments 1 per batch of seed
- Do not use on naked oats
- Do not drill treated winter wheat or rye seed after end of Nov. Seed rate should be increased as drilling season progresses
- Germination tests should be done on all batches of seed to be treated to ensure seed viability and suitability for treatment
- Do not use on seed with moisture content above 16%, on sprouted, cracked or skinned seed or on seed already treated with another seed treatment
- Do not use treated seed as food or feed

Crop-specific information

- Latest use: before drilling
- Treated spring wheat may be drilled in autumn up to end of Nov or from Feb onwards
- Store treated seed in cool, dry, well-ventilated store and drill as soon as possible, preferably in season of purchase
- Treatment may accentuate effects of adverse seedbed conditions on crop emergence

Environmental safety

- Dangerous for the environment
- Harmful to aquatic organisms

Hazard classification and safety precautions

Hazard H11
Risk phrases R52, R53a
Operator protection A; U20b
Environmental protection E03, E15a, E34
Storage and disposal D05, D09a, D11a
Treated seed S01, S02, S03, S04a, S05, S06a, S07

267 gibberellins

A plant growth regulator for use in top fruit

Products

1	GIBB 3	Globachem	10% w/w	TB	12290
2	GIBB Plus	Globachem	10 g/l	SL	11875
3	Gistar	Hermoo	10 g/l	SL	11397
4	Novagib	Fine	10 g/l	SL	08954

Uses

- Growth regulation in ***pears*** *(off-label)* [4]
- Increasing fruit set in ***pears*** [1]
- Increasing germination in ***nothofagus*** *(qualified minor use)* [2, 3]
- Increasing yield in ***celery (outdoor)***, ***protected rhubarb*** [1]
- Reducing fruit russeting in ***apples*** [2-4]; ***pears*** *(off-label)*, ***quinces*** *(off-label)* [2]

Specific Off-Label Approvals (SOLAs)

- ***pears*** *20061627* [2], *20012756* [4]
- ***quinces*** *20061627* [2]

Efficacy guidance

- For optimum results on apples spray under humid, slow drying conditions and ensure good spray cover [2-4]
- Treat apples and pears immediately after completion of petal fall and repeat as directed on the label [2-4]
- Nothofagus seeds should be stirred into the solution, soaked for 24 hr and then sown immediately [2, 3]
- Fruit set in pears can be improved when blossom is spare, setting is poor or where frost has killed many flowers [1]

Restrictions

- Maximum total dose varies with crop and product. Check label
- Prepared spray solutions are unstable. Do not leave in the sprayer during meal breaks or overnight
- Return bloom may be reduced in yr following treatment
- Consult before treating apple crops grown for processing [2-4]
- Avoid storage at temperatures above 32 °C [1]

Crop-specific information

- HI zero for apples
- Apply to apples at completion of petal fall and repeat 3 or 4 times at 7-10 d intervals. Number of sprays and spray interval depend on weather conditions and dose (see labels) [2, 4]
- Good results achieved on apple varieties Cox's Orange Pippin, Discovery, Golden Delicious and Karmijn. For other cultivars test on a small number of trees [2, 4]
- Split treatment on pears allows second spray to be omitted if pollinating conditions become very favourable [1]
- Pear variety Conference usually responds well to treatment; Doyenne du Comice can be variable [1]
- On celery the interval between application and harvest should never exceed 4 wk to avoid risk of premature seeding and root splitting [1]
- Treat washed rhubarb crowns on transfer to forcing shed [1]

Hazard classification and safety precautions

Operator protection U08 [1]; U20b [4]; U20c [1-3]
Environmental protection E15a
Storage and disposal D01, D03, D05, D07 [4]; D09a [1-4]; D10b [2, 4]; D10c [3]

268 glufosinate-ammonium

A non-selective, non-residual phosphinic acid contact herbicide
HRAC mode of action code: H

Products

1	Challenge 60	Fargro	60 g/l	SL	08236
2	Finale	Bayer Environ.	120 g/l	SL	10092
3	Harvest	Bayer CropScience	150 g/l	SL	07321
4	Kaspar	Certis	150 g/l	SL	11214

Uses

- Annual and perennial weeds in ***all cane fruit***, ***all currants***, ***all tree nuts***, ***apple orchards***, ***bilberries***, ***blueberries***, ***cherries***, ***cranberries***, ***cultivated land/soil***, ***damsons***, ***fallows***, ***forest***, ***non-crop farm areas***, ***pear orchards***, ***plums***, ***potatoes*** *(not for seed)*, ***strawberries***, ***stubbles***, ***sugar beet***, ***vegetables***, ***wine grapes*** [3]
- Annual dicotyledons in ***all bush fruit***, ***fruit trees***, ***ornamental trees***, ***shrubs*** [1]; ***all cane fruit***, ***cultivated land/soil*** [1, 4]; ***all currants***, ***all tree nuts***, ***apple orchards***, ***cherries***, ***damsons***, ***fallows***, ***forest***, ***headlands*** *(uncropped)*, ***non-crop farm areas***, ***pear orchards***, ***plums***, ***strawberries***, ***sugar beet***, ***vegetables***, ***ware potatoes***, ***wine grapes*** [4]; ***nursery stock***, ***woody ornamentals*** [2]
- Annual grasses in ***all bush fruit***, ***fruit trees***, ***ornamental trees***, ***shrubs*** [1]; ***all cane fruit***, ***cultivated land/soil*** [1, 4]; ***all currants***, ***all tree nuts***, ***apple orchards***, ***cherries***, ***damsons***,

fallows, forest, headlands** (uncropped), **non-crop farm areas, pear orchards, plums, strawberries, sugar beet, vegetables, ware potatoes, wine grapes [4]; ***nursery stock, woody ornamentals*** [2]
- Green cover in ***land temporarily removed from production*** [3, 4]
- Harvest management/desiccation in ***combining peas, ware potatoes*** [4]; ***combining peas** (not for seed), **potatoes*** [3]; ***linseed, spring field beans, spring oilseed rape, winter field beans, winter oilseed rape*** [3, 4]
- Line marking preparation in ***managed amenity turf*** [2]
- Perennial dicotyledons in ***all bush fruit, cultivated land/soil, fruit trees, ornamental trees, shrubs*** [1]; ***all cane fruit*** [1, 4]; ***all currants, all tree nuts, apple orchards, cherries, damsons, fallows, forest, headlands** (uncropped), **land temporarily removed from production, non-crop farm areas, pear orchards, plums, strawberries, wine grapes*** [4]; ***nursery stock, woody ornamentals*** [2]
- Perennial grasses in ***all bush fruit, cultivated land/soil, fruit trees, ornamental trees, shrubs*** [1]; ***all cane fruit*** [1, 4]; ***all currants, all tree nuts, apple orchards, cherries, damsons, fallows, forest, headlands** (uncropped), **land temporarily removed from production, non-crop farm areas, pear orchards, plums, strawberries, wine grapes*** [4]; ***nursery stock, woody ornamentals*** [2]
- Sward destruction in ***permanent grassland*** [3, 4]

Approval information
- Glufosinate-ammonium included in Annex I under EC Directive 91/414
- Accepted by BBPA for use on malting barley and hops

Efficacy guidance
- Activity quickest under warm, moist conditions. Light rainfall 3-4 h after application will not affect activity. Do not spray wet foliage or if rain likely within 6 h
- For weed control uses treat when weeds growing actively. Deep rooted weeds may require second treatment [1]
- On uncropped headlands apply in May/Jun to prevent weeds invading field
- Glufosinate kills all green tissue but does not harm mature bark

Restrictions
- Maximum number of treatments 4 per crop for potatoes (including 2 desiccant uses); 2 per crop (including 1 desiccant use) for oilseed rape, dried peas, field beans, linseed, wheat and barley; 3 (2 [1]) per yr for fruit and forestry; 2 per yr for strawberries, non-crop land and land temporarily removed from production, round fruit trees, canes or bushes; 1 per crop for sugar beet, vegetables and other crops; 1 per yr for grassland destruction, on cultivated land prior to planting edible crops
- Pre-harvest desiccation sprays should not be used on seed crops of wheat, barley, peas or potatoes but may be used on seed crops of oilseed rape, field beans and linseed [3, 4]
- Do not desiccate potatoes in exceptionally wet weather or in saturated soil. See label for details
- Do not spray potatoes after emergence if grown from small or diseased seed or under very dry conditions
- Application for weed control and line-marking uses in sports turf must be between 1 Mar and 30 Sep [2]
- Do not allow spray to contact dormant or green buds, suckers, damaged or green bark and foliage of wanted plants [1]
- Do not spray hedge bottoms or on broadcast crops [3]

Crop-specific information
- Latest use: 30 Sep for use on non-crop land, top fruit, soft fruit, cane fruit, bush fruit, forestry; pre-drilling, pre-planting or pre-emergence in sugar beet, vegetables and other crops; before winter dormancy for grassland destruction.
- HI potatoes, oilseed rape, combining peas, field beans, linseed 7 d; wheat, barley 14 d
- Ploughing or other cultivations can follow 4 h after spraying
- Crops can normally be sown/planted immediately after spraying or sprayed post-drilling. On sand, very light or immature peat soils allow at least 3 d before sowing/planting or expected emergence
- For weed control in potatoes apply pre-emergence or up to 10% emergence on earlies and seed crops, up to 40% on maincrop, on plants up to 15 cm high
- In sugar beet and vegetables apply just before crop emergence, using stale seedbed technique
- In top and soft fruit, grapevines and forestry apply between 1 Mar and 30 Sep as directed sprays

- For grass destruction apply before winter dormancy occurs. Heavily grazed fields should show active regrowth. Plough from the day after spraying
- Apply pre-harvest desiccation treatments 10-21 d before harvest (14-21 d for oilseed rape). See label for timing details on individual crops
- For potato haulm desiccation apply to listed varieties (not seed crops) at onset of senescence, 14-21 d before harvest

Environmental safety

- Harmful to fish or other aquatic life. Do not contaminate surface waters or ditches with chemical or used container
- Keep livestock out of treated areas until foliage of any poisonous weeds such as ragwort has died and become unpalatable
- Risk to certain non-target insects or other arthropods [3, 4]
- Treated pea haulm may be fed to livestock from 7 d after spraying, treated grain from 14 d [3]
- Product supplied in small volume returnable container - see label for filling and mixing instructions [3]

Hazard classification and safety precautions

Hazard H03 [3, 4]; H04 [2]
Risk phrases R21, R22a, R41 [3, 4]; R36 [2]
Operator protection A [1, 3, 4]; B [2]; C, H, M; U02a [1]; U04a, U11, U13, U14, U15, U19a [3, 4]; U05a [2-4]; U08 [1-4]; U20a [1, 3, 4]; U20b [2]
Environmental protection E07a [1-4]; E13c [1, 2, 4]; E15a, E22c [3]; E22b [4]; E34 [1, 3, 4]
Consumer protection C01 [1]
Storage and disposal D01, D02, D09a, D10b [1-4]; D05 [1, 2]; D12a [3, 4]
Medical advice M03 [3, 4]

269 glyphosate

A translocated non-residual glycine derivative herbicide
HRAC mode of action code: G

See also *diflufenican + glyphosate*
diuron + glyphosate

Products

1	Amega Duo	Nufarm UK	540 g/l	SL	13358
2	Amenity Glyphosate 360	Monsanto	360 g/l	SL	13000
3	Asteroid	Headland	360 g/l	SL	11118
4	Azural	Cardel	360 g/l	SL	11668
5	Azural	Cardel	360 g/l	SL	12666
6	Buggy XTG	Sipcam	36% w/w	SG	12951
7	CDA Vanquish Biactive	Bayer Environ.	120 g/l	RH	12586
8	Clinic Ace	Nufarm UK	360 g/l	SL	12980
9	Credit	Nufarm UK	540 g/l	SL	13032
10	Dow Agrosciences Glyphosate 360	Dow	360 g/l	SL	12720
11	Envision	Headland	450 g/l	SL	10569
12	Glyfos	Headland	360 g/l	SL	10995
13	Glyfos Dakar	Headland	68% w/w	SG	13054
14	Glyfos Dakar Pro	Headland Amenity	68% w/w	SG	13147
15	Glyfos Gold	Nomix Enviro	360 g/l	SL	10570
16	Glyfos Proactive	Nomix Enviro	360 g/l	SL	11976
17	Glyfos Supreme	Headland	450 g/l	SL	12371
18	Glyphogan	Makhteshim	360 g/l	SL	12668
19	Glyphosate 360	Monsanto	360 g/l	SL	11726
20	Glyphosate 360	Monsanto	360 g/l	SL	12669
21	Greenaway Gly-490	Greenaway	490 g/l	SL	12718
22	Hilite	Nomix Enviro	144 g/l	RH	12922
23	Kernel	Headland	480 g/l	SL	10993
24	Manifest	Headland	360 g/l	SL	11041
25	Roundup Ace	Monsanto	450 g/l	SL	12772
26	Roundup Biactive	Monsanto	360 g/l	SL	10320
27	Roundup Energy	Monsanto	450 g/l	SL	12945
28	Roundup Klik	Monsanto	450 g/l	SL	12866
29	Roundup Max	Monsanto	68% w/w	SG	10231

SEE SECTION 3 FOR PRODUCTS ALSO REGISTERED

Products – continued

30	Roundup Max	Monsanto	68% w/w	SG	12952
31	Roundup Pro Biactive	Monsanto	360 g/l	SL	10330
32	Roundup ProBiactive 450	Monsanto	450 g/l	SL	12778
33	Roundup Ultimate	Monsanto	540 g/l	SL	13081
34	Samurai	Monsanto	360 g/l	SL	10334
35	Samurai	Monsanto	360 g/l	SL	12674
36	Scorpion	Cardel	360 g/l	SL	12673
37	Stirrup	Nomix Enviro	144 g/l	RH	12923
38	Tangent	Headland Amenity	450 g/l	SL	11872
39	Touchdown Quattro	Syngenta	360 g/l	SL	10608
40	Tumbleweed Pro-Active	Scotts	450 g/l	SL	11963
41	Typhoon 360	Makhteshim	360 g/l	SL	12932

Uses

- Annual and perennial weeds in ***all edible crops (outdoor)*** *(before planting)*, ***all non-edible crops (outdoor)*** *(before planting)* [8, 13, 17, 39]; ***amenity grassland***, ***amenity grassland*** *(wiper application)*, ***amenity vegetation*** *(wiper application)*, ***broad-leaved trees***, ***conifers***, ***farm buildings/yards***, ***fencelines***, ***land clearance***, ***non-crop areas***, ***walls***, ***woody ornamentals*** [31]; ***amenity vegetation*** [7, 21, 22, 31, 32, 37]; ***apple orchards***, ***cherries***, ***damsons***, ***pear orchards***, ***plums*** [1, 3-6, 8-13, 17-21, 23-28, 33-36, 41]; ***asparagus*** *(off-label)*, ***bilberries*** *(off-label)*, ***blackcurrants*** *(off-label)*, ***blueberries*** *(off-label)*, ***chestnuts*** *(off-label)*, ***cob nuts*** *(off-label)*, ***cranberries*** *(off-label)*, ***gooseberries*** *(off-label)*, ***hazel nuts*** *(off-label)*, ***redcurrants*** *(off-label)*, ***rhubarb*** *(off-label)*, ***vaccinium spp.*** *(off-label)*, ***walnuts*** *(off-label)*, ***whitecurrants*** *(off-label)*, ***wine grapes*** *(off-label)* [26]; ***combining peas***, ***linseed***, ***spring barley***, ***spring field beans***, ***spring oilseed rape***, ***spring wheat***, ***winter barley***, ***winter field beans***, ***winter oilseed rape***, ***winter wheat*** [3, 39]; ***cultivated land/soil*** [3-5, 10, 19, 20, 34, 35]; ***durum wheat***, ***spring oats***, ***winter oats*** [3]; ***field crops*** *(wiper application)*, ***non-crop farm areas*** [4, 5, 10, 19, 20, 34, 35]; ***forest*** [1, 2, 6, 8-12, 15-18, 21-24, 31, 32, 37, 38, 40, 41]; ***hard surfaces*** [1, 2, 8, 9, 15, 17, 21, 22, 31, 32, 37, 38, 40]; ***industrial sites***, ***road verges*** [2, 15, 16, 31]; ***land immediately adjacent to aquatic areas*** [21, 32, 40]; ***land not intended to bear vegetation*** [4-7, 10-12, 16, 19-21, 23, 24, 26, 29-31, 34-36, 41]; ***land prior to cultivation*** [17, 38]; ***managed amenity turf*** *(pre-establishment)* [22, 37]; ***mustard*** [39]; ***natural surfaces not intended to bear vegetation***, ***permeable surfaces overlying soil*** [1, 2, 8, 9, 15, 17, 21, 22, 32, 37, 38, 40]; ***ornamental plant production*** [37]; ***paths and drives*** [2, 15, 31]; ***permanent grassland*** [4-6, 10, 18-20, 25-28, 34-36]; ***stubbles*** [1, 3-6, 8-10, 18-21, 25-30, 33-36, 39]
- Annual dicotyledons in ***all edible crops (outdoor)*** *(pre-planting or sowing)*, ***all non-edible crops (outdoor)*** *(pre-planting or sowing)* [1, 9]; ***bulb onions***, ***durum wheat***, ***leeks***, ***linseed***, ***mustard***, ***spring field beans***, ***spring oilseed rape***, ***sugar beet***, ***swedes***, ***turnips***, ***winter field beans*** [13, 25-30, 33]; ***combining peas*** [13, 29, 30]; ***cultivated land/soil*** [21]; ***peas*** [26]; ***permanent grassland*** [25-28, 33, 36]; ***spring barley***, ***spring oats***, ***spring wheat***, ***winter barley***, ***winter oats***, ***winter wheat*** [11-13, 17, 23-30, 33, 41]; ***vining peas*** [25, 27-30, 33]; ***winter oilseed rape*** [8, 13, 25-30, 33]
- Annual grasses in ***all edible crops (outdoor)*** *(pre-planting or sowing)*, ***all non-edible crops (outdoor)*** *(pre-planting or sowing)* [1, 9]; ***bulb onions***, ***durum wheat***, ***leeks***, ***linseed***, ***mustard***, ***spring barley***, ***spring field beans***, ***spring oats***, ***spring oilseed rape***, ***spring wheat***, ***sugar beet***, ***swedes***, ***turnips***, ***winter barley***, ***winter field beans***, ***winter oats***, ***winter oilseed rape***, ***winter wheat*** [13, 25-30, 33]; ***combining peas*** [13, 29, 30]; ***cultivated land/soil*** [11, 12, 21, 23, 24, 41]; ***peas*** [26]; ***permanent grassland*** [25-28, 33, 36]; ***vining peas*** [25, 27-30, 33]
- Aquatic weeds in ***open waters*** [21, 32, 40]
- Black bent in ***combining peas***, ***spring field beans***, ***spring oilseed rape***, ***stubbles***, ***winter field beans***, ***winter oilseed rape*** [11, 12, 17, 21, 23, 24, 41]; ***durum wheat***, ***spring barley***, ***spring oats***, ***spring wheat***, ***winter barley***, ***winter oats***, ***winter wheat*** [21]; ***linseed*** [11, 12, 17, 23, 24, 41]
- Bolters in ***sugar beet*** *(wiper application)* [4, 5, 10, 13, 19-21, 25-28, 33-36]
- Bracken in ***amenity vegetation***, ***fencelines***, ***land clearance***, ***non-crop areas***, ***paths and drives*** [31]; ***forest*** [1, 2, 6, 8-12, 14-18, 21, 23, 24, 31, 32, 38, 40, 41]; ***tolerant conifers*** [16, 31]
- Brambles in ***tolerant conifers*** [16, 31]
- Canary grass in ***aquatic areas*** [13]; ***enclosed waters***, ***land immediately adjacent to aquatic areas*** [14]

- Chemical thinning in ***forest*** [1, 2, 8, 9, 11, 12, 14, 15, 17, 18, 21, 23, 24, 31, 32, 38, 40, 41]; ***forest*** *(stump treatment)* [6]
- Couch in ***combining peas, spring field beans, spring oilseed rape, stubbles, winter field beans, winter oilseed rape*** [3-6, 10-13, 17-21, 23-28, 33-36, 39, 41]; ***durum wheat*** [3, 13, 21, 25-28, 33, 36]; ***forest*** [14, 18, 21, 31, 32, 40]; ***linseed*** [3, 6, 11-13, 17, 18, 23-28, 33, 36, 39, 41]; ***mustard*** [6, 18, 25-28, 33, 36, 39]; ***permanent grassland*** [6]; ***spring barley, spring wheat, winter barley, winter wheat*** [3-6, 10, 13, 18-21, 25-28, 33-36, 39]; ***spring oats, winter oats*** [3-6, 10, 13, 18-21, 25-28, 33-36]
- Creeping bent in ***aquatic areas*** [13]; ***combining peas, spring field beans, spring oilseed rape, stubbles, winter field beans, winter oilseed rape*** [11, 12, 17, 21, 23, 24, 41]; ***durum wheat, spring barley, spring oats, spring wheat, winter barley, winter oats, winter wheat*** [21]; ***enclosed waters, land immediately adjacent to aquatic areas*** [14]; ***linseed*** [11, 12, 17, 23, 24, 41]
- Desiccation in ***buckwheat*** *(off-label - for wild bird seed production),* ***canary seed*** *(off-label - for wild bird seed production),* ***hemp*** *(off-label - for wild bird seed production),* ***millet*** *(off-label - for wild bird seed production),* ***quinoa*** *(off-label - for wild bird seed production),* ***sorghum*** *(off-label - for wild bird seed production)* [25, 27, 28, 30]; ***lupins*** *(off-label)* [27]
- Green cover in ***land temporarily removed from production*** [1, 3, 6, 8, 9, 11-13, 17, 21, 23-30, 33, 39, 41]
- Harvest management/desiccation in ***combining peas, spring field beans, winter field beans*** [4-6, 8, 10, 18-20, 25-28, 33-36]; ***durum wheat*** [4, 5, 8, 10, 13, 19, 20, 25-28, 33-36]; ***linseed*** [1, 4-6, 8-10, 18-21, 25-28, 33-36]; ***mustard*** [1, 4-6, 8-10, 18-20, 25-28, 33-36, 39]; ***spring barley, winter barley*** [6, 8, 11-13, 17, 18, 23-28, 33, 36, 41]; ***spring oats, spring wheat, winter oats, winter wheat*** [4-6, 8, 10-13, 17-20, 23-28, 33-36, 41]; ***spring oilseed rape, winter oilseed rape*** [1, 3-6, 8-13, 17-21, 23-28, 33-36, 39, 41]
- Heather in ***forest*** [1, 2, 6, 8-12, 14-17, 21, 23, 24, 32, 38, 40, 41]
- Perennial dicotyledons in ***combining peas, linseed, spring field beans, spring oilseed rape, winter field beans, winter oilseed rape*** [11-13, 17, 23, 24, 41]; ***forest*** [2, 15, 38]; ***forest*** *(wiper application)* [31]; ***permanent grassland*** *(wiper application)* [25-28, 33, 36]
- Perennial grasses in ***aquatic areas*** [3, 6, 10-12, 15, 16, 21, 23, 24, 31, 38, 40]; ***combining peas, linseed, spring field beans, stubbles, winter field beans*** [13]; ***enclosed waters, land immediately adjacent to aquatic areas*** [15, 38]; ***tolerant conifers*** [16, 31]
- Pre-harvest desiccation in ***combining peas, linseed, spring barley, spring field beans, spring wheat, winter barley, winter field beans, winter wheat*** [39]; ***poppies for morphine production*** *(off-label)* [26]
- Reeds in ***aquatic areas*** [3, 6, 10-13, 15, 16, 21, 23, 24, 26, 31, 38, 40]; ***enclosed waters, land immediately adjacent to aquatic areas*** [14, 15, 21, 32, 38, 40]
- Rhododendrons in ***forest*** [2, 6, 8, 10-12, 14-17, 21, 23, 24, 31, 32, 38, 40, 41]
- Rushes in ***aquatic areas*** [3, 6, 10-13, 15, 16, 21, 23, 24, 26, 31, 38, 40]; ***enclosed waters, land immediately adjacent to aquatic areas*** [14, 15, 21, 32, 38, 40]; ***forest*** [18]
- Sedges in ***aquatic areas*** [3, 6, 10-13, 15, 16, 21, 23, 24, 26, 31, 38, 40]; ***enclosed waters, land immediately adjacent to aquatic areas*** [14, 15, 21, 32, 38, 40]
- Sucker control in ***apple orchards, cherries, damsons, pear orchards, plums*** [19, 20, 25-28, 34-36]
- Sward destruction in ***amenity grassland*** [31]; ***permanent grassland*** [1, 3-6, 8-13, 17-21, 23-28, 33-36, 39, 41]; ***rotational grass*** [1, 8, 9, 39]
- Total vegetation control in ***all edible crops (outdoor)*** *(pre-sowing/planting),* ***all non-edible crops (outdoor)*** *(pre-sowing/planting)* [25, 27, 28, 32]; ***amenity vegetation*** [7, 16, 31]; ***fencelines, industrial sites, road verges*** [16, 31]; ***hard surfaces*** [14, 16, 31]; ***land not intended to bear vegetation*** [7, 16, 31, 39]; ***natural surfaces not intended to bear vegetation, permeable surfaces overlying soil*** [14]
- Volunteer cereals in ***bulb onions, leeks, linseed, mustard, spring field beans, sugar beet, swedes, turnips, winter field beans*** [13, 25-30, 33]; ***combining peas*** [13, 29, 30]; ***cultivated land/soil*** [11, 12, 21, 23, 24, 41]; ***durum wheat, spring barley, spring oats, spring oilseed rape, spring wheat, winter barley, winter oats, winter oilseed rape, winter wheat*** [25-30, 33]; ***peas*** [26]; ***stubbles*** [3-6, 10-13, 17-21, 23-28, 33-36, 39, 41]; ***vining peas*** [25, 27-30, 33]
- Volunteer potatoes in ***stubbles*** [3-6, 10, 13, 18-21, 25-28, 33-36, 39]
- Waterlilies in ***aquatic areas*** [3, 6, 10-13, 15, 16, 21, 23, 24, 31, 38, 40]; ***enclosed waters*** [14, 21, 32, 40]

SEE SECTION 3 FOR PRODUCTS ALSO REGISTERED

- Weed beet in ***sugar beet*** *(wiper application)* [4, 5, 13, 19, 20, 34, 35]
- Wild oats in ***durum wheat***, ***spring barley***, ***spring oats***, ***spring wheat***, ***winter barley***, ***winter oats***, ***winter wheat*** [4, 5, 10, 19, 20, 34, 35]
- Woody weeds in ***forest*** [1, 2, 6, 8-12, 14-18, 21, 23, 24, 31, 32, 38, 40, 41]; ***tolerant conifers*** [16, 31]

Specific Off-Label Approvals (SOLAs)

- ***asparagus*** *20062036* [26]
- ***bilberries*** *20062035* [26]
- ***blackcurrants*** *20062035* [26]
- ***blueberries*** *20062035* [26]
- ***buckwheat*** *(for wild bird seed production) 20070584* [25], *(for wild bird seed production) 20071839* [27], *(for wild bird seed production) 20070580* [28], *(for wild bird seed production) 20070582* [30]
- ***canary seed*** *(for wild bird seed production) 20070584* [25], *(for wild bird seed production) 20071839* [27], *(for wild bird seed production) 20070580* [28], *(for wild bird seed production) 20070582* [30]
- ***chestnuts*** *20062028* [26]
- ***cob nuts*** *20062028* [26]
- ***cranberries*** *20062035* [26]
- ***gooseberries*** *20062035* [26]
- ***hazel nuts*** *20062028* [26]
- ***hemp*** *(for wild bird seed production) 20070583* [25], *(for wild bird seed production) 20071840* [27], *(for wild bird seed production) 20070579* [28], *(for wild bird seed production) 20070581* [30]
- ***lupins*** *20071279* [27]
- ***millet*** *(for wild bird seed production) 20070584* [25], *(for wild bird seed production) 20071839* [27], *(for wild bird seed production) 20070580* [28], *(for wild bird seed production) 20070582* [30]
- ***poppies for morphine production*** *20062030* [26]
- ***quinoa*** *(for wild bird seed production) 20070583* [25], *(for wild bird seed production) 20071840* [27], *(for wild bird seed production) 20070579* [28], *(for wild bird seed production) 20070581* [30]
- ***redcurrants*** *20062035* [26]
- ***rhubarb*** *20062029* [26]
- ***sorghum*** *(for wild bird seed production) 20070584* [25], *(for wild bird seed production) 20071839* [27], *(for wild bird seed production) 20070580* [28], *(for wild bird seed production) 20070582* [30]
- ***vaccinium spp.*** *20062035* [26]
- ***walnuts*** *20062028* [26]
- ***whitecurrants*** *20062035* [26]
- ***wine grapes*** *20062031* [26]

Approval information

- Glyphosate included in Annex I under EC Directive 91/414
- Accepted by BBPA for use on malting barley and hops
- Approval expiry 30 Jun 2008 [4, 29, 34, 40]
- Approval expiry 30 Nov 2008 [25, 28, 32]

Efficacy guidance

- For best results apply to actively growing weeds with enough leaf to absorb chemical
- For most products a rainfree period of at least 6 h (preferably 24 h) should follow spraying
- Adjuvants are obligatory for some products and recommended for some uses with others. See labels
- Mixtures with other pesticides or fertilizers may lead to reduced control.
- Products are formulated as isopropylamine, ammonium, potassium, or trimesium salts of glyphosate and may vary in the details of efficacy claims. See individual product labels
- Some products in ready-to-use formulations for use through hand-held applicators. See label for instructions [7, 22, 37]
- When applying products through rotary atomisers the spray droplet spectra must have a minimum Volume Median Diameter (VMD) of 200 microns
- With wiper application weeds should be at least 10 cm taller than crop
- Annual weed grasses should have at least 5 cm of leaf and annual broad-leaved weeds at least 2 expanded true leaves

- Perennial grass weeds should have 4-5 new leaves and be at least 10 cm long when treated. Perennial broad-leaved weeds should be treated at or near flowering but before onset of senescence
- Volunteer potatoes and polygonums are not controlled by harvest-aid rates
- Bracken must be treated at full frond expansion
- Fruit tree suckers best treated in late spring
- Chemical thinning treatment can be applied as stump spray or stem injection
- In order to allow translocation, do not cultivate before spraying and do not apply other pesticides, lime, fertilizer or farmyard manure within 5 d of treatment
- Recommended intervals after treatment and before cultivation vary. See labels

Restrictions

- Maximum total dose per crop or season normally equivalent to one full dose treatment on field and edible crops and no restriction for non-crop uses. However some older labels indicate a maximum number of treatments. Check for details
- Do not treat cereals grown for seed or undersown crops
- Consult grain merchant before treating crops grown on contract or intended for malting
- Do not use treated straw as a mulch or growing medium for horticultural crops
- For use in nursery stock, shrubberies, orchards, grapevines and tree nuts care must be taken to avoid contact with the trees. Do not use in orchards established less than 2 yr and keep off low-lying branches
- Certain conifers may be sprayed overall in dormant season. See label for details
- Use a tree guard when spraying in established forestry plantations
- Do not spray root suckers in orchards in late summer or autumn
- Do not use under glass or polythene as damage to crops may result
- Do not use trimesium salt products in aquatic situations or forestry
- Aquatic treatments must not be used in open water
- Do not mix, store or apply in galvanised or unlined mild steel containers or spray tanks
- Do not leave diluted chemical in spray tanks for long periods and make sure that tanks are well vented

Crop-specific information

- Harvest intervals: 4 wk for blackcurrants, blueberries; 14 d for linseed, oilseed rape; 8 d for mustard; 5-7 d for all other edible crops. Check label for exact details
- Latest use: for most products 2-14 d before cultivating, drilling or planting a crop in treated land; after harvest (post-leaf fall) but before bud formation in the following season for nuts and most fruit and vegetable crops; before fruit set for grapevines. See labels for details

Following crops guidance

- Decaying remains of plants killed by spraying must be dispersed before direct drilling
- Crops may be drilled 48 h after application. Trees and shrubs may be planted 7 d after application. Grass seed may be sown from 5 d after treatment

Environmental safety

- Products differ in their hazard and environmental safety classification. See labels
- Do not dump surplus herbicide in water or ditch bottoms or empty into drains
- Check label for maximum permitted concentration in treated water
- The Environment Agency or Local River Purification Authority must be consulted before use in or near water
- LERAP Category B [33]
- Take extreme care to avoid drift and possible damage to neighbouring crops or plants
- Treated poisonous plants must be removed before grazing or conserving
- Do not use in covered areas such as greenhouses or under polythene
- For field edge treatment direct spray away from hedge bottoms
- Do not dump surplus herbicide in water or ditch bottoms
- Some products require livestock to be excluded from treated areas and do not permit treated forage to be used for hay, silage or bedding. Check label for details

Hazard classification and safety precautions

Hazard H03 [2, 10]; H04 [4-6, 8, 15, 18-20, 29, 30, 33-37, 41]; H11 [2, 4, 5, 8, 10, 15, 18-20, 22-24, 29, 30, 33-37, 41]

Risk phrases R20 [2, 10]; R36 [4, 5, 10, 15, 18-20, 34-37]; R38 [33]; R41 [2, 6, 8, 29, 30, 41]; R50 [22, 33]; R51 [2, 4, 5, 8, 10, 15, 18-20, 23, 24, 29, 30, 34-37, 41]; R52 [13, 14, 17, 25, 27, 28, 32]; R53a [1, 2, 4, 5, 8-10, 13-15, 17-20, 22-25, 27-30, 32-37, 41]

Operator protection A [1-41]; C [2-6, 8, 10-12, 15-21, 23-32, 34-36, 38, 40, 41]; D [3, 11, 12, 15, 16, 23, 24, 38, 40]; F [4, 5, 18-20, 34-36]; H [1-7, 9, 11-24, 26, 29-41]; M [1-7, 9-41]; N [4, 5, 10, 18-20, 34-36, 41]; U02a [1, 8-10, 12-14, 17-24, 29, 30, 37, 40]; U05a [1, 2, 4, 5, 9, 10, 15, 18-20, 22, 29, 30, 33-37, 39, 41]; U08 [4, 5, 10, 12-15, 17-21, 23, 24, 29, 30, 34-36, 40]; U09a [22, 37, 41]; U11 [2, 4-6, 8, 10, 13-15, 18-20, 29, 30, 34-37, 41]; U15 [4, 5, 15, 18-20, 34-36]; U16b [37]; U19a [4, 5, 7, 10, 12-15, 17-24, 29, 30, 34-37, 40, 41]; U20a [2, 26, 33, 39]; U20b [1, 3-5, 7-25, 27-32, 34-38, 40, 41]

Environmental protection E06d, E08b [39]; E13c [12, 18, 21, 22]; E15a [3-5, 7, 10, 11, 15-20, 23, 24, 26, 29-31, 33-36, 38, 40]; E15b [1, 2, 6, 8, 9, 13, 14, 22, 25, 27, 28, 32, 37, 39]; E16a, E16b [33]; E19a [3, 10-12, 21, 23, 24, 38]; E19b [41]; E23 [40]; E34 [7, 17, 18]; E38 [2, 4-6, 8, 10, 13-15, 17-20, 29, 30, 33-37, 39]

Storage and disposal D01 [1-7, 9, 10, 12-20, 22-41]; D02 [1, 2, 4-7, 9, 10, 13-16, 18-20, 22, 25-37, 39-41]; D05 [1, 2, 4, 5, 7-9, 15-22, 25, 27-37, 39, 40]; D07 [29, 30]; D09a [2-8, 10-41]; D10a [7]; D10b [3, 10, 11, 18, 21, 23, 38, 41]; D10c [1, 2, 4, 5, 8, 9, 12, 15-17, 19, 20, 24-28, 31-36, 39, 40]; D11a [6, 13, 14, 22, 29, 30, 37]; D12a [2, 4, 5, 8, 10, 13-15, 17-20, 22, 29, 30, 33-37, 39, 41]; D12b [23, 24]

Medical advice M03 [39]; M05a [8, 41]

270 guazatine

A guanidine fungicide seed dressing for cereals
FRAC mode of action code: M7

Products

1	Panoctine	Makhteshim	300 g/l	LS	11404
2	Ravine	Makhteshim	300 g/l	LS	12211

Uses

- Brown foot rot in ***spring barley*** *(seed treatment - reduction)*, ***spring oats*** *(seed treatment - reduction)*, ***winter barley*** *(seed treatment - reduction)*, ***winter oats*** *(seed treatment - reduction)* [1]
- Foot rot in ***spring barley*** *(seed treatment)*, ***winter barley*** *(seed treatment)* [1]
- Fusarium foot rot and seedling blight in ***spring barley*** *(seed treatment - reduction)*, ***spring oats*** *(seed treatment - reduction)*, ***spring wheat*** *(seed treatment - reduction)*, ***winter barley*** *(seed treatment - reduction)*, ***winter oats*** *(seed treatment - reduction)*, ***winter wheat*** *(seed treatment - reduction)* [1, 2]
- Leaf stripe in ***spring barley*** *(seed treatment)*, ***winter barley*** *(seed treatment)* [1]
- Net blotch in ***spring barley*** *(seed treatment)*, ***winter barley*** *(seed treatment)* [1]
- Pyrenophora leaf spot in ***spring oats*** *(seed treatment)*, ***winter oats*** *(seed treatment)* [1]
- Septoria seedling blight in ***spring wheat*** *(seed treatment)*, ***winter wheat*** *(seed treatment)* [2]; ***spring wheat*** *(seed treatment - partial control)*, ***winter wheat*** *(seed treatment - partial control)* [1]

Approval information

- Accepted by BBPA for use on malting barley

Efficacy guidance

- Apply with conventional seed treatment machinery

Restrictions

- Maximum number of treatments 1 per batch
- Do not treat grain with moisture content above 16% and do not allow moisture content of treated seed to exceed 16%
- Do not apply to cracked, split or sprouted seed
- Do not use treated seed as food or feed

Crop-specific information

- Latest use: pre-drilling
- After treating, bag seed immediately and keep in dry, draught-free store
- Treatment may lower germination capacity, particularly if seed grown, harvested or stored under adverse conditions

Environmental safety

- Dangerous for the environment
- Very toxic to aquatic organisms
- Harmful to fish or other aquatic life. Do not contaminate surface waters or ditches with chemical or used container

Hazard classification and safety precautions

Hazard H03, H11
Risk phrases R20, R22a, R37, R41, R43, R50, R53a
Operator protection A, C, D, H; U05a, U20b [1, 2]; U09a, U14, U15 [1]
Environmental protection E13c [1]; E15a [2]; E34, E38 [1, 2]
Storage and disposal D01, D02, D05, D09a, D10a [1, 2]; D14 [1]
Treated seed S01, S02, S04b, S05, S06a, S07
Medical advice M03 [1]; M04a [1, 2]

271 guazatine + imazalil

A fungicide seed treatment for barley and oats
FRAC mode of action code: M7 + 3

Products

1	Panoctine Plus	Makhteshim	300:25 g/l	LS	11757
2	Ravine Plus	Makhteshim	300:25 g/l	LS	11832

Uses

- Brown foot rot in ***spring barley*** *(seed treatment)*, ***winter barley*** *(seed treatment)*
- Covered smut in ***spring barley*** *(seed treatment)*, ***winter barley*** *(seed treatment)*
- Leaf stripe in ***spring barley*** *(seed treatment)*, ***winter barley*** *(seed treatment)*
- Net blotch in ***spring barley*** *(seed treatment - moderate control)*, ***winter barley*** *(seed treatment - moderate control)*
- Pyrenophora leaf spot in ***spring oats*** *(seed treatment)*, ***winter oats*** *(seed treatment)*
- Seedling blight and foot rot in ***spring barley*** *(seed treatment - moderate control)*, ***winter barley*** *(seed treatment - moderate control)*
- Snow mould in ***spring barley*** *(seed treatment)*, ***winter barley*** *(seed treatment)*

Approval information

- Imazalil included in Annex I under EC Directive 91/414
- Accepted by BBPA for use on malting barley

Efficacy guidance

- Apply with conventional seed treatment machinery

Restrictions

- Maximum number of treatments 1 per batch
- Do not mix with other formulations
- Do not treat grain with moisture content above 16% and do not allow moisture content of treated seed to exceed 16%
- Do not apply to cracked, split or sprouted seed
- Do not store treated seed for more than 6 mth
- Treatment may lower germination capacity, particularly if seed grown, harvested or stored under adverse conditions

Crop-specific information

- Latest use: pre-drilling

Environmental safety

- Dangerous for the environment
- Very toxic to aquatic organisms
- Do not use treated seed as food or feed

Hazard classification and safety precautions

Hazard H02, H11
Risk phrases R22a, R23, R37, R41, R43, R50, R53a
Operator protection A, C, D, H; U05a, U10, U11, U14, U19a, U20b
Environmental protection E15a, E34, E38

SEE SECTION 3 FOR PRODUCTS ALSO REGISTERED

Storage and disposal D01, D02, D09a, D10b
Treated seed S02, S04b, S05, S07
Medical advice M04a

272 hymexazol

A systemic heteroaromatic fungicide for pelleting sugar beet seed
FRAC mode of action code: 32

Products

Tachigaren 70 WP	Summit Agro	70% w/w	WP	12568

Uses

- Aphanomyces cochlioides in ***red beet*** *(off-label - seed treatment)*
- Black leg in ***sugar beet*** *(seed treatment)*

Specific Off-Label Approvals (SOLAs)

- ***red beet*** *(seed treatment) 20062835*

Efficacy guidance

- Incorporate into pelleted seed using suitable seed pelleting machinery

Restrictions

- Maximum number of treatments 1 per batch of seed
- Do not use treated seed as food or feed

Crop-specific information

- Latest use: before planting sugar beet seed

Environmental safety

- Harmful to aquatic organisms
- Harmful to fish or other aquatic life. Do not contaminate surface waters or ditches with chemical or used container
- Treated seed harmful to game and wildlife

Hazard classification and safety precautions

Hazard H04, H07
Risk phrases R41, R52, R53a
Operator protection A, C, F; U05a, U11, U20b
Environmental protection E03, E13c
Storage and disposal D01, D02, D09a, D11a
Treated seed S01, S02, S03, S04a, S05

273 imazalil

A systemic and protectant imidazole fungicide
FRAC mode of action code: 3

See also guazatine + imazalil

Products

1	Fungaflor 100EC	Certis	100 g/l	EC	12978
2	Fungazil 100 SL	BASF	100 g/l	LS	11762
3	IMA 100 SL	Q-Chem	100 g/l	LS	13559
4	Magnate 100 SL	Makhteshim	100 g/l	SL	11705

Uses

- Dry rot in ***seed potatoes***, ***ware potatoes*** [2-4]
- Gangrene in ***seed potatoes***, ***ware potatoes*** [2-4]
- Powdery mildew in ***protected cucumbers***, ***protected tomatoes*** *(off-label - in synthetic/inert media)* [1]
- Silver scurf in ***seed potatoes***, ***ware potatoes*** [2-4]
- Skin spot in ***seed potatoes***, ***ware potatoes*** [2-4]

Specific Off-Label Approvals (SOLAs)

- ***protected tomatoes*** *(in synthetic/inert media) 20072799 expires Dec 2008* [1]

Approval information
- Imazalil included in Annex I under EC Directive 91/414
- Approval expiry 31 Dec 2008 [1]

Efficacy guidance
- For best control of skin and wound diseases of ware potatoes treat as soon as possible after harvest, preferably within 7-10 d, before any wounds have healed [2-4]
- Treat protected cucumbers as soon as mildew is present and repeat at 7-10 d intervals depending on disease pressure [1]
- Ensure spray has dried on treated cucumbers before nightfall [1]
- Treatment of protected cucumbers will also have activity against Botrytis and black stem rot [1]

Restrictions
- Maximum number of treatments 1 per batch of ware tubers [2-4]; 2 per batch of seed tubers [2-4]; 3 per crop of protected cucumbers [1]
- Consult processor before treating potatoes for processing [2-4]
- Do not treat protected cucumbers in bright sunny conditions [2-4]

Crop-specific information
- Latest use: during storage and before chitting for seed potatoes [2-4]
- HI 1 d for potatoes [2-4]; 72 h for protected cucumbers [1]
- Apply to clean soil-free potatoes post-harvest before putting into store, or at first grading. A further treatment may be applied in early spring before planting [2-4]
- Apply through canopied hydraulic or spinning disc equipment preferably diluted with up to two litres water per tonne of potatoes to obtain maximum skin cover and penetration [2-4]
- Use on ware potatoes subject to discharges of imazalil from potato washing plants being within emission limits set by the UK monitoring authority [2-4]

Environmental safety
- Dangerous for the environment
- Toxic to aquatic organisms
- Do not empty into drains [2-4]
- Personal protective equipment requirements may vary for each pack size. Check label

Hazard classification and safety precautions

Hazard H03 [1]; H04 [2-4]; H11 [1, 4]

Risk phrases R22a, R43, R51 [1]; R36, R50 [4]; R41, R52 [2, 3]; R53a [1-4]

Operator protection A [1-4]; C [2-4]; D [4]; H [1-3]; U02a [1]; U04a, U05a, U19a, U20a [1-4]; U11 [2, 3]; U14 [1-3]

Environmental protection E13c [4]; E15a, E19b [2, 3]; E15b, E38 [1]; E34 [2-4]

Storage and disposal D01, D02, D09a [1-4]; D05 [2-4]; D10a [4]; D10b [1]; D10c [2, 3]; D12a [1-3]

Treated seed S01, S03, S04a, S05, S06a [2-4]; S02 [2, 3]

Medical advice M03 [4]

274 imazalil + pencycuron

A fungicide mixture for treatment of seed potatoes
FRAC mode of action code: 3 + 20

Products

Monceren IM	Bayer CropScience	0.6:12.5% w/w	DS	11426

Uses
- Black scurf in ***potatoes*** *(tuber treatment)*
- Silver scurf in ***potatoes*** *(tuber treatment - reduction)*
- Stem canker in ***potatoes*** *(tuber treatment - reduction)*

Approval information
- Imazalil included in Annex I under EC Directive 91/414

Efficacy guidance
- Apply to clean seed tubers during planting (see label for suitable method) or sprinkle over tubers in chitting trays before loading into planter. It is essential to obtain an even distribution over tubers
- If seed tubers become damp from light rain distribution of product should not be affected. Tubers in the hopper should be covered if a shower interrupts planting

Restrictions

- Maximum number of treatments 1 per batch of tubers
- Operators must wear suitable respiratory equipment and gloves when handling product and when riding on planter. Wear gloves when handling treated tubers
- Do not use on tubers which have previously been treated with a dry powder seed treatment or hot water
- Do not use treated tubers for human or animal consumption

Crop-specific information

- Latest use: immediately before planting
- May be used on seed tubers previously treated with a liquid fungicide but not before 8 wk have elapsed if this contained imazalil

Environmental safety

- Harmful to aquatic organisms
- Harmful to fish or other aquatic life. Do not contaminate surface waters or ditches with chemical or used container
- Treated seed harmful to game and wildlife

Hazard classification and safety precautions

Risk phrases R52, R53a
Operator protection A, D; U20b
Environmental protection E03, E13c
Storage and disposal D09a, D11a
Treated seed S01, S02, S03, S04a, S05, S06a

275 imazalil + thiabendazole

A fungicide mixture for treatment of seed potatoes
FRAC mode of action code: 3 + 1

Products

Storite Super	Syngenta	125:250 g/l	FS	13135

Uses

- Dry rot in ***seed potatoes*** *(tuber treatment - post-harvest)*
- Gangrene in ***seed potatoes*** *(tuber treatment - post-harvest)*
- Silver scurf in ***seed potatoes*** *(tuber treatment - post-harvest)*

Approval information

- Imazalil and thiabendazole included in Annex I under EC Directive 91/414
- Approved for use through ULV equipment

Efficacy guidance

- Treat as soon after harvest as possible and preferably within 2 wk of lifting. Later applications will not be effective against established disease
- Apply undiluted with suitable spinning disc or diluted through ultra low volume hydraulic nozzles
- Avoid over-wetting tubers

Restrictions

- Maximum number of treatments 1 per batch of seed potatoes
- Only treat dormant tubers
- Only treat seed crops where there is a likely risk of disease development during storage
- Treated tubers must be planted using only automated equipment
- Do not use treated seed as food or feed

Crop-specific information

- Latest use: pre-planting

Environmental safety

- Dangerous for the environment
- Very toxic to aquatic organisms

Hazard classification and safety precautions

Hazard H03, H11
Risk phrases R22a, R41, R50, R53a

Operator protection A, C, H; U05a, U09a, U11, U16b, U19a, U20b
Environmental protection E15b, E38
Storage and disposal D01, D02, D05, D09a, D10c, D12a
Treated seed S01, S02, S03, S04a, S04b, S05
Medical advice M03

276 imazamox

An imidazolinone contact and residual herbicide available only in mixtures
HRAC mode of action code: B

277 imazamox + pendimethalin

A pre-emergence broad-spectrum herbicide mixture for legumes
HRAC mode of action code: B + K1

Products

Nirvana	BASF	16.7:250 g/l	EC	13220

Uses
- Annual dicotyledons in ***broad beans*** *(off-label)*, ***combining peas***, ***spring field beans***, ***vining peas***, ***winter field beans***

Specific Off-Label Approvals (SOLAs)
- ***broad beans*** *20070490*

Approval information
- Pendimethalin included in Annex I under EC Directive 91/414

Efficacy guidance
- Best results obtained from applications to fine firm seedbeds in the presence of adequate moisture
- Weed control may be reduced on cloddy seedbeds and on soils with over 6% organic matter
- Residual control may be reduced under prolonged dry conditions

Restrictions
- Maximum number of treatments 1 per crop
- Seed must be drilled to at least 2.5 cm of settled soil
- Do not use on soils containing more than 10% organic matter
- Do apply to soils that are waterlogged or are prone to waterlogging
- Do not apply if heavy rain is forecast
- Do not soil incorporate the product or disturb the soil after application
- Consult processors before use on crops destined for processing

Crop-specific information
- Latest use: pre-crop emergence for all crops
- Inadequately covered seed may result in cupping of the leaves after application from which recovery is normally complete
- Crop damage may occur on stony or gravelly soils especially if heavy rain follows treatment

Following crops guidance
- Winter wheat or winter barley may be drilled as a following crop provided 5 mth have elapsed since treatment and the land has been ploughed to minimum 150 mm and cultivated
- Winter oilseed rape or other brassica crops should not drilled as following crops
- A minimum of 12 mth must elapse between treatment and sowing red beet, sugar beet or spinach

Environmental safety
- Dangerous for the environment
- Very toxic to aquatic organisms
- LERAP Category B

Hazard classification and safety precautions
Hazard H04, H11
Risk phrases R38, R43, R50, R53a

Operator protection A, H; U05a, U14, U20b
Environmental protection E15b, E16a, E38
Storage and disposal D01, D02, D09a, D10c, D12a
Medical advice M03

278 imazaquin

An imidazolinone herbicide and plant growth regulator available only in mixtures
HRAC mode of action code: B

See also chlormequat + imazaquin

279 imidacloprid

A neonicotinoid insecticide for seed, soil, peat or foliar treatment
IRAC mode of action code: 4A

See also beta-cyfluthrin + imidacloprid
fuberidazole + imidacloprid + triadimenol

Products

1	Admire	Bayer CropScience	70% w/w	WG	11234
2	Gaucho	Bayer CropScience	70% w/w	WS	11281
3	Imidasect 5 GR	Fargro	5% w/w	GR	13581
4	Intercept 70WG	Scotts	70% w/w	WG	08585
5	Merit Turf	Bayer Environ.	0.5% w/w	GR	12415

Uses

- Aphids in ***bedding plants***, ***herbaceous perennials***, ***ornamental plant production***, ***pot plants*** [4]; ***ornamental plant production*** *(container grown)*, ***protected ornamentals*** [3]
- Beet virus yellows vectors in ***sugar beet*** *(seed treatment)* [2]
- Cabbage aphid in ***collards*** *(off-label - seed treatment)*, ***kale*** *(off-label - seed treatment)* [2]
- Chafer grubs in ***managed amenity turf*** [5]
- Damson-hop aphid in ***hops*** [1]
- Flea beetle in ***sugar beet*** *(seed treatment)* [2]
- Glasshouse whitefly in ***bedding plants***, ***herbaceous perennials***, ***ornamental plant production***, ***pot plants*** [4]; ***ornamental plant production*** *(container grown)*, ***protected ornamentals*** [3]
- Leatherjackets in ***managed amenity turf*** [5]
- Mangold fly in ***sugar beet*** *(seed treatment)* [2]
- Millipedes in ***sugar beet*** *(seed treatment)* [2]
- Peach-potato aphid in ***broccoli*** *(off-label - seed treatment)*, ***brussels sprouts*** *(off-label - seed treatment)*, ***cabbages*** *(off-label - seed treatment)*, ***calabrese*** *(off-label - seed treatment)*, ***cauliflowers*** *(off-label - seed treatment)*, ***collards*** *(off-label - seed treatment)*, ***kale*** *(off-label - seed treatment)* [2]
- Pygmy beetle in ***sugar beet*** *(seed treatment)* [2]
- Sciarid flies in ***bedding plants***, ***herbaceous perennials***, ***ornamental plant production***, ***pot plants*** [4]; ***ornamental plant production*** *(container grown)*, ***protected ornamentals*** [3]
- Springtails in ***sugar beet*** *(seed treatment)* [2]
- Symphylids in ***sugar beet*** *(seed treatment)* [2]
- Tobacco whitefly in ***bedding plants***, ***herbaceous perennials***, ***ornamental plant production***, ***pot plants*** [4]; ***ornamental plant production*** *(container grown)*, ***protected ornamentals*** [3]
- Vine weevil in ***bedding plants***, ***herbaceous perennials***, ***ornamental plant production***, ***pot plants*** [4]; ***ornamental plant production*** *(container grown)*, ***protected ornamentals*** [3]

Specific Off-Label Approvals (SOLAs)

- ***broccoli*** *(seed treatment) 20023927* [2]
- ***brussels sprouts*** *(seed treatment) 20023927* [2]
- ***cabbages*** *(seed treatment) 20023927* [2]
- ***calabrese*** *(seed treatment) 20023927* [2]
- ***cauliflowers*** *(seed treatment) 20023927* [2]

- ***collards*** *(seed treatment) 20023927* [2]
- ***kale*** *(seed treatment) 20023927* [2]

Approval information

- Accepted by BBPA for use on hops

Efficacy guidance

- Treated seed should be drilled within the season of purchase [2]
- Base of hop plants should be free of weeds and debris at application [1]
- Hop bines emerging away from the main stock or adjacent to poles may require a separate application [1]
- Uptake and movement within hops requires soil moisture and good growing conditions [1]
- Control may be impaired in plantations greater than 3640 plants/ha [1]
- To minimise likelihood of resistance in hops do not treat all the crop in any one yr [1]
- Best results on managed amenity turf achieved when treatment is followed by sufficient irrigation to move chemical through the thatch to wet the top 2.5 cm of soil [5]
- To avoid disturbing uniformity of treatment on turf do not mow until irrigation or rainfall has occurred [5]
- To minimise likelihood of resistance when using in compost adopt a planned programme to alternate with pesticides of different types or use other measures [3, 4]
- When applied as drench or incorporated as granules in moist compost imidacloprid is readily absorbed and translocated to aerial parts of plant [3, 4]

Restrictions

- Maximum number of treatments 1 per batch of seed [2] or growing medium [3, 4]; 1 per yr for hops and managed amenity turf [1, 5]
- Must not be used in compost that has already been treated with an imidacloprid-containing product, nor should compost be subsequently re-treated with imidacloprid [3, 4]
- Product formulated for use only as a compost incorporation treatment into peat-based growing media using suitable automated equipment [3]
- For use only on container grown ornamentals [3, 4]
- The safety of seeds sown into treated compost should not be assumed. Test treat before full-scale use [3]
- Product must not be used on crops for human or animal consumption and treated compost must not be re-used for this purpose [3, 4]
- Do not use treated seed as food or feed [2]
- Do not treat saturated or waterlogged turf, or when raining or in windy conditions [5]
- Avoid puddling or run-off of irrigation water after treatment [5]
- Do not use grass clippings from treated turf in situ, or use them for mulch [5]

Crop-specific information

- Latest use: before bines reach 2 m or before end 1st wk Jun for hops; before drilling for sugar beet; before sowing or planting for brassicas, bedding plants, hardy ornamental nursery stock, pot plants
- Apply to sugar beet seed as part of the normal commercial pelleting process using special treatment machinery [2]
- Apply to hops as a directed stem base spray before most bines reach a height of 2 m. If necessary treat both sides of the crown at half the normal concentration [1]
- The rate used per hop plant must be adjusted in accordance with the plant population if greater than 3640 per ha. See label for details [1]
- On managed amenity turf application should be made prior to egg-laying of the targets pests [5]
- Low levels of leaf scorch on some *Agrostis* turf species if application is made in mixture with fertilisers [5]

Environmental safety

- Harmful to aquatic organisms [1]
- High risk to bees. Do not apply to crops in flower or to those in which bees are actively foraging. Do not apply when flowering weeds are present [1, 4]
- Treated seed harmful to game and wildlife [2]

Hazard classification and safety precautions

Hazard H03 [1, 4]; H04 [2, 5]

Risk phrases R22a [1, 4]; R36 [5]; R43 [2]; R52 [1]

Operator protection A [1-3, 5]; C, D, H [2]; U05a [1, 3-5]; U20a [2]; U20b [1, 3, 4]; U20c [5]

Environmental protection E12a, E12e, E34 [1, 4]; E15a [1-5]
Storage and disposal D01, D02 [1, 3-5]; D09a, D11a [1-5]
Treated seed S02, S04b, S05, S06a, S07 [2]

280 indol-3-ylacetic acid

A plant growth regulator for promoting rooting of cuttings

Products

1	Rhizopon A Powder	Rhizopon	1% w/w	DP	09087
2	Rhizopon A Tablets	Rhizopon	50 mg a.i.	WT	09088

Uses
- Rooting of cuttings in ***ornamental plant production***

Efficacy guidance
- Apply by dipping end of prepared cuttings into powder or dissolved tablet solution
- Shake off excess powder and make planting holes to prevent powder stripping off [1]
- Consult manufacturer for details of application by spray or total immersion

Restrictions
- Maximum number of treatments 1 per cutting
- Store product in a cool, dark and dry place
- Use solutions once only. Discard after use [2]
- Use plastic, not metallic, container for solutions [2]

Crop-specific information
- Latest use: before cutting insertion

Hazard classification and safety precautions
Operator protection U19a [1]; U20a [1, 2]
Environmental protection E15a
Storage and disposal D09a, D11a

281 4-indol-3-ylbutyric acid

A plant growth regulator promoting the rooting of cuttings

Products

1	Chryzoplus Grey 0.8%	Rhizopon	0.8% w/w	DP	09094
2	Chryzopon Rose 0.1%	Rhizopon	0.1% w/w	DP	09092
3	Chryzosan White 0.6%	Rhizopon	0.6% w/w	DP	09093
4	Chryzotek Beige	Rhizopon	0.4% w/w	DP	09081
5	Chryzotop Green	Rhizopon	0.25% w/w	DP	09085
6	Rhizopon AA Powder (0.5%)	Rhizopon	0.5% w/w	DP	09082
7	Rhizopon AA Powder (1%)	Rhizopon	1% w/w	DP	09084
8	Rhizopon AA Powder (2%)	Rhizopon	2% w/w	DP	09083
9	Rhizopon AA Tablets	Rhizopon	50 mg a.i.	WT	09086
10	Seradix 1	Certis	0.1% w/w	DP	11330
11	Seradix 2	Certis	0.3% w/w	DP	11331
12	Seradix 3	Certis	0.8% w/w	DP	11332

Uses
- Rooting of cuttings in ***ornamental plant production***

Efficacy guidance
- Dip base of cuttings into powder immediately before planting
- Powders or solutions of different concentration are required for different types of cutting. Lowest concentration for softwood, intermediate for semi-ripe, highest for hardwood
- See label for details of concentration and timing recommended for different species
- Use of planting holes recommended for powder formulations to ensure product is not removed on insertion of cutting. Cuttings should be watered in if necessary

Restrictions
- Maximum number of treatments 1 per situation
- Use of too strong a powder or solution may cause injury to cuttings
- No unused moistened powder should be returned to container

Crop-specific information

- Latest use: before cutting insertion for ornamental specimens

Hazard classification and safety precautions

Operator protection A, C [10-12]; U14, U15 [10-12]; U19a, U20a [1-12]
Environmental protection E15a
Storage and disposal D01 [10-12]; D09a, D11a [1-12]

282 4-indol-3-ylbutyric acid + 1-naphthylacetic acid

A plant growth regulator for promoting rooting of cuttings

Products

Synergol	Certis	5.0:5.0 g/l	SL	07386

Uses

- Rooting of cuttings in ***ornamental plant production***

Efficacy guidance

- Dip base of cuttings into diluted concentrate immediately before planting
- Suitable for hardwood and softwood cuttings
- See label for details of concentration and timing for different species

Hazard classification and safety precautions

Hazard H03
Risk phrases R22a
Operator protection A, C; U05a, U13, U20b
Environmental protection E15a, E34
Storage and disposal D01, D02, D09a, D11a, D12a
Medical advice M03

283 indoxacarb

An oxadiazine insecticide for caterpillar control in a range of crops
IRAC mode of action code: 22A

Products

Steward	DuPont	30% w/w	WG	13149

Uses

- Caterpillars in ***apples***, ***broccoli***, ***cabbages***, ***cauliflowers***, ***pears***, ***protected aubergines***, ***protected courgettes***, ***protected cucumbers***, ***protected marrows***, ***protected melons***, ***protected ornamentals***, ***protected peppers***, ***protected pumpkins***, ***protected squashes***, ***protected tomatoes***

Approval information

- Indoxacarb included in Annex I under EC Directive 91/414

Efficacy guidance

- Best results in brassica and protected crops obtained from treatment when first caterpillars are detected, or when damage first seen, or 7-10 d after trapping first adults in pheromone traps
- In apples and pears apply at egg-hatch
- Subsequent treatments in all crops may be applied at 8-14 d intervals
- Indoxacarb acts by ingestion and contact. Only larval stages are controlled but there is some ovicidal action against some species

Restrictions

- Maximum number of treatments 3 per crop or yr for apples, pears, brassica crops; 6 per yr for protected crops
- Do not apply to any crop suffering from stress from any cause

Crop-specific information

- HI: brassica crops, protected crops 1 d; apples, pears 7 d

Environmental safety

- Dangerous for the environment
- Toxic to aquatic organisms

SEE SECTION 3 FOR PRODUCTS ALSO REGISTERED

- Broadcast air-assisted LERAP
- In accordance with good agricultural practice apply in early morning or late evening when bees are less active

Hazard classification and safety precautions

Hazard H03, H11
Risk phrases R22a, R51, R53a
Operator protection A, D; U04a, U05a
Environmental protection E15b, E17b, E34, E38
Storage and disposal D01, D02, D09a, D12a
Medical advice M03

284 iodosulfuron-methyl-sodium

A post-emergence sulfonylurea herbicide for winter sown cereals
HRAC mode of action code: B

See also *amidosulfuron + iodosulfuron-methyl-sodium*
diflufenican + iodosulfuron-methyl-sodium + mesosulfuron-methyl

Products

Hussar	Bayer CropScience	5% w/w	WG	12364

Uses

- Chickweed in ***spring barley***, ***triticale***, ***winter rye***, ***winter wheat***
- Cleavers in ***triticale***, ***winter rye***, ***winter wheat***
- Fat hen in ***spring barley***
- Field pansy in ***spring barley***
- Field speedwell in ***spring barley***, ***triticale***, ***winter rye***, ***winter wheat***
- Hemp-nettle in ***spring barley***
- Italian ryegrass in ***triticale*** *(from seed)*, ***winter rye*** *(from seed)*, ***winter wheat*** *(from seed)*
- Ivy-leaved speedwell in ***triticale***, ***winter rye***, ***winter wheat***
- Mayweeds in ***spring barley***, ***triticale***, ***winter rye***, ***winter wheat***
- Perennial ryegrass in ***triticale*** *(from seed)*, ***winter rye*** *(from seed)*, ***winter wheat*** *(from seed)*
- Red dead-nettle in ***spring barley***, ***triticale***, ***winter rye***, ***winter wheat***
- Volunteer oilseed rape in ***spring barley***

Approval information

- Iodosulfuron-methyl-sodium included in Annex I under EC Directive 91/414
- Accepted by BBPA for use on malting barley

Efficacy guidance

- Best results obtained from treatment in warm weather when soil is moist and the weeds are growing actively
- Weeds must be present at application to be controlled
- Weed control is slow especially under cool dry conditions
- Dry conditions resulting in moisture stress may reduce effectiveness
- Occasionally weeds may only be stunted but they will normally have little or no competitive effect on the crop
- Iodosulfuron is a member of the ALS-inhibitor group of herbicides. To avoid the build up of resistance do not use any product containing an ALS-inhibitor herbicide with claims for control of grass weeds more than once on any crop
- Use these products as part of a resistance management strategy that includes cultural methods of control and does not use ALS inhibitors as the sole chemical method of grass weed control

Restrictions

- Maximum number of treatments 1 per crop
- Must only be applied between 1 Feb in yr of harvest and specified latest time of application
- Do not apply to undersown crops, or crops to be undersown
- Do not roll or harrow within 1 wk of spraying
- Do not spray crops under stress from any cause or if the soil is compacted
- Treat broadcast crops after the plants have a well-established root system

- Do not spray if rain imminent or frost expected
- Do not apply in mixture or in sequence with any other ALS inhibitor

Crop-specific information

- Latest use: before third node detectable (GS 33)

Following crops guidance

- No restrictions apply to the sowing of cereal crops or sugar beet in the spring of the yr following treatment

Environmental safety

- Dangerous for the environment
- Very toxic to aquatic organisms
- Dangerous to fish or other aquatic life. Do not contaminate surface waters or ditches with chemical or used container
- LERAP Category B
- Take extreme care to avoid damage by drift onto broad-leaved plants outside the target area or onto ponds, waterways and ditches
- Observe carefully label instructions for sprayer cleaning

Hazard classification and safety precautions

Hazard H04, H11
Risk phrases R41, R50, R53a
Operator protection A, C, H; U05a, U08, U11, U14, U15, U20b
Environmental protection E15b, E16a, E16b, E38
Storage and disposal D01, D02, D10a, D12a

285 iodosulfuron-methyl-sodium + mesosulfuron-methyl

A sulfonyl urea herbicide mixture for winter wheat
HRAC mode of action code: B + B

Products

1	Atlantis WG	Bayer CropScience	0.6:3.0% w/w	WG	12478
2	Greencrop Doonbeg 2	Greencrop	0.6:3.0 % w/w	WG	13025
3	Pacifica	Bayer CropScience	1.0:3.0% w/w	WG	12049
4	Teliton Ocean	Teliton	0.6:3.0% w/w	WG	13163

Uses

- Annual meadow grass in ***winter wheat***
- Blackgrass in ***winter wheat***
- Chickweed in ***winter wheat***
- Italian ryegrass in ***winter wheat***
- Mayweeds in ***winter wheat***
- Perennial ryegrass in ***winter wheat***
- Rough meadow grass in ***winter wheat***
- Wild oats in ***winter wheat***

Approval information

- Iodosulfuron-methyl-sodium and mesosulfuron-methyl included in Annex I under EC Directive 91/414

Efficacy guidance

- Optimum grass weed control obtained when all grass weeds are emerged at spraying. Activity is primarily via foliar uptake and good spray coverage of the target weeds is essential
- Translocation occurs readily within the target weeds and growth is inhibited within hours of treatment but symptoms may not be apparent for up to 4 wk, depending on weed species, timing of treatment and weather conditions
- Residual activity is important for best results and is optimised by treatment on fine moist seedbeds. Avoid application under very dry conditions
- Residual efficacy may be reduced by high soil temperatures and cloddy seedbeds
- Iodosulfuron-methyl and mesosulfuron-methyl are both members of the ALS-inhibitor group of herbicides. To avoid the build up of resistance do not use any product containing an ALS-inhibitor herbicide with claims for control of grass weeds more than once on any crop

SEE SECTION 3 FOR PRODUCTS ALSO REGISTERED

- Use these products as part of a Resistance Management Strategy that includes cultural methods of control and does not use ALS inhibitors as the sole chemical method of grass weed control. See Section 5 for more information

Restrictions

- Maximum number of treatments 1 per crop with a maximum total dose equivalent to one full dose treatment
- Do not use on crops undersown with grasses, clover or other legumes or any other broad-leaved crop
- Do not use as a stand-alone treatment for blackgrass, ryegrass or chickweed control
- Do not use as the sole means of weed control in successive crops
- Do not use in mixture or in sequence with any other ALS-inhibitor herbicide except those (if any) specified on the label
- Do not mix with products containing isoproturon, chlorotoluron or linuron and allow at least 28 d after use of a residual urea herbicide before application, and at least 10 d after application before use of a residual urea herbicide [2]
- Do not apply earlier than 1 Feb in the yr of harvest [3]
- Do not apply to crops under stress from any cause
- Do not apply when rain is imminent or during periods of frosty weather
- Specified adjuvant must be used. See label

Crop-specific information

- Latest use: flag leaf just visible (GS 39)
- Winter wheat may be treated from the two-leaf stage of the crop
- Safety to crops grown for seed not established

Following crops guidance

- In the event of crop failure sow only winter wheat in the same cropping season
- Only winter wheat or winter barley (or winter oilseed rape [1]) may be sown in the year of harvest of a treated crop
- Spring wheat, spring barley, sugar beet or spring oilseed rape may be drilled in the following spring. Plough before drilling oilseed rape

Environmental safety

- Dangerous for the environment
- Very toxic to aquatic organisms
- Dangerous to fish or other aquatic life. Do not contaminate surface waters or ditches with chemical or used container
- LERAP Category B
- Take extreme care to avoid drift onto plants outside the target area or on to ponds, waterways or ditches

Hazard classification and safety precautions

Hazard H04, H11
Risk phrases R40 [3]; R41, R50, R53a [1-4]
Operator protection A, C, H; U05a, U11, U20b
Environmental protection E15a [1, 3]; E15b [2, 4]; E16a [1-4]; E38 [1-3]
Storage and disposal D01, D02, D05, D09a, D10a, D12a

286 ioxynil

A contact acting HBN herbicide for use in turf and onions
HRAC mode of action code: C3

See also *bromoxynil + diflufenican + ioxynil*
bromoxynil + ethofumesate + ioxynil
bromoxynil + ioxynil
bromoxynil + ioxynil + mecoprop-P

Products

Totril	Bayer CropScience	225 g/l	EC	13100

Uses

- Annual dicotyledons in ***bulb onions***, ***chives*** *(off-label)*, ***garlic***, ***leeks***, ***salad onions***, ***shallots***

Specific Off-Label Approvals (SOLAs)

- ***chives*** *20063162*

Approval information

- Ioxynil included in Annex I under EC Directive 91/414
- Ioxynil was reviewed in 1995 and approvals for home garden use, and most hand held applications revoked

Efficacy guidance

- Best results on seedling to 4-leaf stage weeds in active growth during mild weather

Restrictions

- Maximum number of treatments up to 4 per crop at split doses on onions; 1 per crop on carrots, garlic, leeks, shallots
- Do not apply by hand-held equipment or at concentrations higher than those recommended

Crop-specific information

- Latest use: pre-emergence for carrots, parsnips
- HI bulb onions, salad onions, shallots, garlic, leeks 14 d
- Apply to sown onion crops as soon as possible after plants have 3 true leaves or to transplanted crops when established

Environmental safety

- Dangerous for the environment
- Very toxic to aquatic organisms
- Flammable
- Harmful to bees. Do not apply to crops in flower or to those in which bees are actively foraging. Do not apply when flowering weeds are present
- Keep livestock out of treated areas for at least 6 wk after treatment and until foliage of any poisonous weeds such as ragwort has died and become unpalatable

Hazard classification and safety precautions

Hazard H03, H08, H11
Risk phrases R22a, R22b, R36, R37, R43, R50, R53a, R63, R66, R67
Operator protection A, C; U05a, U08, U19a, U20b, U23a
Environmental protection E06a (6 wk); E12d, E12e, E15a, E34, E38
Storage and disposal D01, D02, D05, D09a, D10b, D12a
Medical advice M03, M05b

287 iprodione

A protectant dicarboximide fungicide with some eradicant activity
FRAC mode of action code: 2

Products

1	Amenitywise Iprodione Green	Standon	250 g/l	SC	12237
2	Chipco Green	Bayer Environ.	250 g/l	SC	11211
3	Clayton Vert	Clayton	250 g/l	SC	12570
4	Governor	AgriGuard	255 g/l	SC	12736
5	Governor WP	AgriGuard	50% w/w	WP	12847
6	Rovral Flo	BASF	255 g/l	SC	11702
7	Rovral Liquid FS	BASF	500 g/l	FS	11703
8	Rovral WP	BASF	50% w/w	WP	13429
9	Standon Iprodione 50 WP	Standon	50% w/w	WP	12322
10	Super-Turf	AgriGuard	250 g/l	SC	12871
11	Surpass	Headland Amenity	250 g/l	SC	13590

Uses

- Alternaria in ***brassica seed crops***, ***mustard*** *(seed treatment)*, ***ornamental plant production*** *(seed treatment)*, ***stored cabbages***, ***stubble turnips*** *(seed treatment)*, ***swedes*** *(seed treatment)*, ***turnips*** *(seed treatment)* [5, 8, 9]; ***broccoli***, ***brussels sprouts***, ***cabbage seed crops***, ***calabrese***, ***cauliflower seed crops***, ***cauliflowers***, ***mustard seed crops***, ***spring oilseed rape***, ***stubble turnips***, ***winter oilseed rape*** [4, 6]; ***broccoli*** *(seed treatment)*, ***cabbages*** *(seed treatment)*, ***cauliflowers*** *(seed treatment)* [5]; ***chinese cabbage*** *(seed treatment)*, ***kohlrabi*** *(seed treatment)* [5, 8]; ***edible brassicas*** *(seed treatment)*, ***flower seeds*** *(seed treatment)* [8]; ***flax*** *(seed treatment)*

SEE SECTION 3 FOR PRODUCTS ALSO REGISTERED

[5, 7]; ***fodder rape*** *(seed treatment)*, ***leaf brassicas*** *(seed treatment)* [9]; ***spring oilseed rape*** *(seed treatment)*, ***winter oilseed rape*** *(seed treatment)* [5, 7-9]; ***swedes*** *(off-label)* [6]
- Black scurf in ***seed potatoes*** *(seed treatment)* [5, 7, 9]
- Black scurf and stem canker in ***potatoes*** *(seed treatment)* [8]
- Botrytis in ***cabbage seed crops***, ***cauliflower seed crops***, ***mustard seed crops***, ***spring oilseed rape***, ***winter oilseed rape*** [4, 6]; ***chicory*** *(off-label)*, ***chinese cabbage*** *(off-label)*, ***choi sum*** *(off-label)*, ***combining peas*** *(off-label)*, ***dwarf beans*** *(off-label)*, ***endives*** *(off-label)*, ***fennel*** *(off-label)*, ***french beans*** *(off-label)*, ***lettuce*** *(outdoor crops)*, ***mange-tout peas*** *(off-label)*, ***pak choi*** *(off-label)*, ***pears*** *(off-label)*, ***protected aubergines*** *(off-label)*, ***protected chives*** *(off-label)*, ***protected courgettes*** *(off-label)*, ***protected frise*** *(off-label)*, ***protected herbs (see appendix 6)*** *(off-label)*, ***protected parsley*** *(off-label)*, ***protected rhubarb*** *(off-label)*, ***protected salad brassicas*** *(off-label - for baby leaf production)*, ***protected scarole*** *(off-label)*, ***red beet*** *(off-label)*, ***runner beans*** *(off-label)*, ***vining peas*** *(off-label)* [8]; ***cress*** *(off-label)*, ***lamb's lettuce*** *(off-label)*, ***lettuce*** *(off-label)*, ***radicchio*** *(off-label)*, ***scarole*** *(off-label)*, ***wine grapes*** *(off-label)* [6]; ***lettuce*** [5, 9]; ***pot plants***, ***protected lettuce***, ***protected tomatoes***, ***raspberries***, ***stored cabbages***, ***tomatoes (outdoor)*** [5, 8, 9]; ***protected strawberries*** [5, 8]; ***strawberries*** [4-6, 8, 9]
- Botrytis bunch rot in ***wine grapes*** *(off-label)* [6]
- Brown patch in ***amenity grassland*** [2]; ***managed amenity turf*** [1-3, 10, 11]
- Canker in ***horseradish*** *(off-label)* [8]
- Chocolate spot in ***spring field beans***, ***winter field beans*** [4, 6]
- Collar rot in ***bulb onions***, ***salad onions*** [4, 6]
- Dollar spot in ***amenity grassland*** [2]; ***managed amenity turf*** [1-3, 10, 11]
- Fusarium patch in ***amenity grassland*** [2]; ***managed amenity turf*** [1-3, 10, 11]
- Glume blotch in ***winter wheat*** [4, 6]
- Grey mould in ***protected cucumbers*** *(off-label)* [8]
- Leaf rot in ***bulb onions***, ***salad onions*** [4, 6]; ***garlic*** *(off-label)*, ***shallots*** *(off-label)* [6]
- Melting out in ***amenity grassland*** [2]; ***managed amenity turf*** [1-3, 10, 11]
- Net blotch in ***spring barley***, ***winter barley*** [4, 6]
- Phoma in ***red beet*** *(off-label)* [8]
- Red thread in ***amenity grassland*** [2]; ***managed amenity turf*** [1-3, 10, 11]
- Sclerotinia in ***chicory*** *(off-label)*, ***combining peas*** *(off-label)*, ***dwarf beans*** *(off-label)*, ***french beans*** *(off-label)*, ***mange-tout peas*** *(off-label)*, ***runner beans*** *(off-label)*, ***vining peas*** *(off-label)* [8]; ***cress*** *(off-label)*, ***lamb's lettuce*** *(off-label)*, ***lettuce*** *(off-label)*, ***radicchio*** *(off-label)*, ***scarole*** *(off-label)* [6]
- Sclerotinia stem rot in ***spring oilseed rape***, ***winter oilseed rape*** [4, 6]
- Snow mould in ***amenity grassland*** [2]; ***managed amenity turf*** [1-3, 10, 11]
- Stemphylium in ***asparagus*** *(off-label)* [8]

Specific Off-Label Approvals (SOLAs)

- ***asparagus*** *20071796 expires Dec 2008* [8]
- ***chicory*** *20071790 expires Dec 2008* [8]
- ***chinese cabbage*** *20071794 expires Dec 2008* [8]
- ***choi sum*** *20071794 expires Dec 2008* [8]
- ***combining peas*** *20071797 expires Dec 2008* [8]
- ***cress*** *20040513* [6]
- ***dwarf beans*** *20071789 expires Dec 2008* [8]
- ***endives*** *20071798 expires Dec 2008* [8]
- ***fennel*** *20071798 expires Dec 2008* [8]
- ***french beans*** *20071789 expires Dec 2008* [8]
- ***garlic*** *20060536* [6]
- ***horseradish*** *20071788 expires Dec 2008* [8]
- ***lamb's lettuce*** *20040513* [6]
- ***lettuce*** *20040513* [6]
- ***mange-tout peas*** *20071789 expires Dec 2008* [8]
- ***pak choi*** *20071794 expires Dec 2008* [8]
- ***pears*** *20071792 expires Dec 2008* [8]
- ***protected aubergines*** *20072087 expires Dec 2008* [8]
- ***protected chives*** *20071800 expires Dec 2008* [8]
- ***protected courgettes*** *20071798 expires Dec 2008* [8]
- ***protected cucumbers*** *20071793 expires Dec 2008* [8]

- ***protected frise*** *20071800 expires Dec 2008* [8]
- ***protected herbs (see appendix 6)*** *20071800 expires Dec 2008* [8]
- ***protected parsley*** *20071800 expires Dec 2008* [8]
- ***protected rhubarb*** *20071795 expires Dec 2008* [8]
- ***protected salad brassicas*** *(for baby leaf production) 20071800 expires Dec 2008* [8]
- ***protected scarole*** *20071800 expires Dec 2008* [8]
- ***radicchio*** *20040513* [6]
- ***red beet*** *20071791 expires Dec 2008* [8]
- ***runner beans*** *20071789 expires Dec 2008* [8]
- ***scarole*** *20040513* [6]
- ***shallots*** *20060536* [6]
- ***swedes*** *20060534* [6]
- ***vining peas*** *20071797 expires Dec 2008* [8]
- ***wine grapes*** *20040514* [6]

Approval information

- Iprodione included in Annex I under EC Directive 91/414
- Accepted by BBPA for use on malting barley
- Approval expiry 31 Dec 2008 [1-4, 6, 8, 10, 11]
- Approval expiry 31 Aug 2008 [5, 7, 9]

Efficacy guidance

- Many diseases require a programme of 2 or more sprays at intervals of 2-4 wk. Recommendations vary with disease and crop - see label for details
- Use as a drench to control cabbage storage diseases. Spray ornamental pot plants and cucumbers to run-off
- Apply turf and amenity grass treatments after mowing to dry grass, free of dew. Do not mow again for at least 24 h
- Seed treatments on oilseed rape and linseed should be applied to dry seed prior to sowing
- Best results on seed potatoes achieved using hydraulic sprayer with solid or hollow cone jets with air or liquid pressure atomisation
- Apply to seed potatoes after harvest or after grading out and before traying out for chitting

Restrictions

- Maximum number of treatments or maximum total dose equivalent to 1 per batch for seed treatments; 1 per crop on cereals, cabbage (as drench), 2 per crop on field beans and stubble turnips; 3 per crop on brassicas (including seed crops), oilseed rape, protected winter lettuce (Oct-Feb); 4 per crop on strawberries, grapevines, salad onions, cucumbers; 5 per crop on raspberries; 6 per crop on bulb onions, tomatoes, turf; 7 per crop on lettuce (Mar-Sep)
- A minimum of 3 wk must elapse between treatments on leaf brassicas
- Treated brassica seed crops not to be used for human or animal consumption
- See label for pot plants showing good tolerance. Check other species before applying on a large scale
- Do not treat oilseed rape seed that is cracked or broken or of low viability
- Do not excessively wet seed potato skins
- Do not treat oats
- Do not use treated seed as food or feed [7]

Crop-specific information

- Latest use: pre-planting for seed treatments; at planting for potatoes; before grain watery ripe (GS 69) for wheat and barley; 21 d before consumption by livestock for stubble turnips
- HI strawberries, protected tomatoes 1 d; outdoor tomatoes 2 d; bulb onions, salad onions, raspberries, lettuce 7 d; brassicas, brassica seed crops, oilseed rape, mustard, field beans, stubble turnips 21 d
- Turf and amenity grass may become temporarily yellowed if frost or hot weather follows treatment
- Personal protective equipment requirements may vary for each pack size. Check label

Environmental safety

- Dangerous for the environment
- Very toxic to aquatic organisms
- See label for guidance on disposal of spent drench liquor
- Treatment harmless to *Encarsia* or *Phytoseiulus* being used for integrated pest control

SEE SECTION 3 FOR PRODUCTS ALSO REGISTERED

Hazard classification and safety precautions

Hazard H03, H11

Risk phrases R36, R38 [1-3, 11]; R40, R53a [1-11]; R50 [1-4, 6-8, 10, 11]; R51 [5, 9]

Operator protection A [1-4, 6, 7, 10, 11]; C [1-3, 10, 11]; H [7]; U04a [1-3, 10, 11]; U05a [1-4, 6, 7, 10, 11]; U08 [1-4, 6, 10, 11]; U19a [1, 5, 8, 9]; U20b [3, 4, 6]; U20c [1, 2, 5, 7-11]

Environmental protection E15a [1-11]; E34 [1, 2, 4, 5, 7-11]; E38 [2, 4-8]

Storage and disposal D01, D02 [1-4, 6, 7, 10, 11]; D05 [1-5, 7, 10, 11]; D07 [10]; D09a [1-11]; D10b [1-4, 6, 10, 11]; D11a [5, 7-9]; D12a [2-4, 6-8]

Treated seed S01, S05, S06a [5, 7-9]; S02 [5, 8, 9]; S03, S04a [5, 7]; S04b [5]; S07 [7-9]

Medical advice M03 [1-3, 10, 11]

288 iprodione + thiophanate-methyl

A protectant and systemic fungicide for oilseed rape

FRAC mode of action code: 2 + 1

Products

Compass	BASF	167:167 g/l	SC	11740

Uses

- Alternaria in ***carrots*** *(off-label)*, ***horseradish*** *(off-label)*, ***parsnips*** *(off-label)*, ***spring oilseed rape***, ***winter oilseed rape***
- Botrytis in ***meadowfoam (limnanthes alba)*** *(off-label)*
- Chocolate spot in ***spring field beans***, ***winter field beans***
- Crown rot in ***carrots*** *(off-label)*, ***horseradish*** *(off-label)*, ***parsnips*** *(off-label)*
- Grey mould in ***spring oilseed rape***, ***winter oilseed rape***
- Light leaf spot in ***spring oilseed rape***, ***winter oilseed rape***
- Sclerotinia stem rot in ***spring oilseed rape***, ***winter oilseed rape***
- Stem canker in ***spring oilseed rape***, ***winter oilseed rape***

Specific Off-Label Approvals (SOLAs)

- ***carrots*** *20040525*
- ***horseradish*** *20040525*
- ***meadowfoam (limnanthes alba)*** *20040507*
- ***parsnips*** *20040525*

Approval information

- Iprodione and thiophanate-methyl included in Annex I under EC Directive 91/414

Efficacy guidance

- Timing of sprays on oilseed rape varies with disease, see label for details
- For season-long control product should be applied as part of a disease control programme

Restrictions

- Maximum number of treatments (refers to total sprays containing benomyl, carbendazim and thiophanate methyl) 2 per crop for oilseed rape, field beans, meadowfoam; 3 per crop for carrots, horseradish, parsnips

Crop-specific information

- Latest use: before end of flowering for oilseed rape, field beans
- HI meadowfoam, field beans, oilseed rape 3 wk; carrots, horseradish, parsnips 28 d
- Treatment may extend duration of green leaf in winter oilseed rape

Environmental safety

- Dangerous for the environment
- Very toxic to aquatic organisms

Hazard classification and safety precautions

Hazard H03, H11

Risk phrases R40, R43, R50, R53a, R68

Operator protection A, C, H, M; U05a, U20c

Environmental protection E15a, E38

Storage and disposal D01, D02, D08, D09a, D10b, D12a

Medical advice M05a

289 isoproturon

A residual urea herbicide for use in cereals
HRAC mode of action code: C2

See also *bifenox + isoproturon*
chlorotoluron + isoproturon
diflufenican + isoproturon

Products

1	Alpha IPU 500	Makhteshim	500 g/l	SC	10733
2	Alpha Isoproturon 500	Makhteshim	500 g/l	SC	05882
3	IPU Minrinse	Makhteshim	80% w/w	WG	12457
4	Isotron 500	DAPT	500 g/l	SC	10668
5	Primer	AgriGuard	500 g/l	SC	11782
6	Primer	AgriGuard	500 g/l	SC	13226
7	Protugan	Makhteshim	500 g/l	SC	10043
8	Protugan 80 WDG	Makhteshim	80% w/w	WG	10430

Uses

- Annual dicotyledons in ***durum wheat*** *(off-label)* [2, 7, 8]; ***spring barley*** *(off-label)*, ***spring rye*** *(off-label)*, ***triticale*** *(off-label)*, ***winter rye*** *(off-label)* [2]; ***spring wheat*** *(autumn sown)* [3, 8]; ***triticale***, ***winter rye*** [1, 3, 7, 8]; ***winter barley***, ***winter wheat*** [1-8]
- Annual grasses in ***durum wheat*** *(off-label)* [2, 7, 8]; ***spring rye*** *(off-label)*, ***triticale*** *(off-label)*, ***winter rye*** *(off-label)* [2]; ***spring wheat*** *(autumn sown)* [3, 8]; ***triticale***, ***winter rye*** [1, 3, 7, 8]; ***winter barley***, ***winter wheat*** [1-8]
- Annual meadow grass in ***spring barley*** *(off-label)* [1, 2, 7, 8]
- Blackgrass in ***spring wheat*** *(autumn sown)* [3, 8]; ***triticale***, ***winter rye*** [1, 3, 7, 8]; ***winter barley***, ***winter wheat*** [1-8]
- Rough meadow grass in ***spring wheat*** *(autumn sown)* [3, 8]; ***triticale***, ***winter rye*** [1, 3, 7, 8]; ***winter barley***, ***winter wheat*** [1-8]
- Wild oats in ***spring wheat*** *(autumn sown)* [3, 8]; ***triticale***, ***winter rye*** [1, 3, 7, 8]; ***winter barley***, ***winter wheat*** [1-8]

Specific Off-Label Approvals (SOLAs)

- ***durum wheat*** *20052983 expires Jan 2008* [2], *20052991 expires Jan 2008* [7], *20052992 expires Jan 2008* [8]
- ***spring barley*** *20012166 expires Jan 2008* [1], *20011481 expires Jan 2008* [2], *20011488 expires Jan 2008* [7], *20023410 expires Jan 2008* [8]
- ***spring rye*** *20052984 expires Jan 2008* [2]
- ***triticale*** *20052984 expires Jan 2008* [2]
- ***winter rye*** *20052984 expires Jan 2008* [2]

Approval information

- Isoproturon is included in Annex I under EC Directive 91/414 but Member States have to 'pay particular attention to the protection of aquatic organisms...' before products containing isoproturon are re-registered. In UK the ACP have determined that there is an unacceptable risk to aquatic life arising from the movement of isoproturon into surface water courses. As a result all approvals for products containing isoproturon will be phased out by 30 June 2009. In the meantime the maximum total dose is restricted to 1500 g isoproturon per hectare
- Approval expiry 31 Aug 2008 [5]

Efficacy guidance

- Best control normally achieved by early post-emergence treatment
- See label for details of rates, timings and tank mixes for different weed problems
- Apply to moist soils. Effectiveness may be reduced in seasons of above average rainfall or by prolonged dry weather
- Residual activity reduced on soils with more than 10% organic matter. Only use on such soils in spring
- Always follow WRAG guidelines for preventing and managing herbicide resistant weeds. See Section 5 for more information

Restrictions

- Maximum number of treatments 1 per crop

- Maximum total dose 1.5 kg a.i./ha for any crop
- Do not apply pre-emergence to wheat or barley. On triticale and winter rye use pre-emergence only and on named varieties
- Do not use on durum wheats, oats, undersown cereals or crops to be undersown
- Do not apply to very wet or waterlogged soils, or when heavy rainfall is forecast
- Do not use on very cloddy soils or if frost is imminent or after onset of frosty weather
- Do not roll for 1 wk before or after treatment or harrow for 1 wk before or any time after treatment

Crop-specific information

- Latest use:not later than second node detectable stage for winter wheat, winter barley; pre-crop emergence for triticale, rye
- Recommended timing of treatment varies depending on crop to be treated, method of sowing, season of application, weeds to be controlled and product being used. See label for details
- Crop damage may occur on free draining, stony or gravelly soils if heavy rain falls soon after spraying. Early sown crops may be damaged if spraying precedes or coincides with a period of rapid growth

Environmental safety

- Dangerous for the environment
- Very toxic to aquatic organisms
- Do not apply to dry, cracked or waterlogged soils where heavy rain may lead to contamination of drains by isoproturon
- Do not empty into drains

Hazard classification and safety precautions

Hazard H03, H11 [1-3, 5-8]
Risk phrases R40, R50, R53a [1-3, 5-8]
Operator protection A, C, H [1-8]; D [3, 8]; U05a, U19a [1-4, 7, 8]; U08 [1-4, 6-8]; U14, U15 [3, 8]; U20a [2, 5, 6]; U20b [1, 3, 7, 8]
Environmental protection E13b [4]; E13c [2]; E15a [1, 3, 5-8]; E34 [1, 2, 7]
Storage and disposal D01, D02 [1-4, 7, 8]; D05, D09a [1-8]; D10b [1, 2, 4-7]; D11a [3, 8]; D12b [1, 2, 7]
Treated seed S06a [4]

290 isoxaben

A soil-acting benzamide herbicide for use in grass and fruit
HRAC mode of action code: L

Products

1	Agriguard Isoxaben	AgriGuard	125 g/l	SC	11652
2	Flexidor 125	Landseer	125 g/l	SC	10946

Uses

- Annual dicotyledons in ***amenity vegetation***, ***apple orchards***, ***blackberries***, ***blackcurrants***, ***cherries***, ***forestry transplants***, ***gooseberries***, ***hops***, ***ornamental plant production***, ***pear orchards***, ***plums***, ***raspberries***, ***strawberries***, ***wine grapes*** [1, 2]; ***canary grass*** *(off-label - grown for game cover)*, ***carrots*** *(off-label)*, ***clovers*** *(off-label - grown for game cover)*, ***courgettes*** *(off-label)*, ***forage maize*** *(off-label - grown for game cover)*, ***marrows*** *(off-label)*, ***millet*** *(off-label - grown for game cover)*, ***protected carrots*** *(off-label - temporary protection)*, ***protected courgettes*** *(off-label)*, ***protected marrows*** *(off-label)*, ***protected ornamentals*** *(off-label)*, ***protected parsnips*** *(off-label - temporary protection)*, ***protected pumpkins*** *(off-label)*, ***protected squashes*** *(off-label)*, ***pumpkins*** *(off-label)*, ***quinoa*** *(off-label - grown for game cover)*, ***squashes*** *(off-label)*, ***sunflowers*** *(off-label - grown for game cover)* [2]; ***table grapes*** [1]
- Cleavers in ***asparagus*** *(off-label)* [2]
- Fat hen in ***asparagus*** *(off-label)* [2]
- Groundsel in ***asparagus*** *(off-label)* [2]
- Knotgrass in ***asparagus*** *(off-label)* [2]

Specific Off-Label Approvals (SOLAs)

- ***asparagus*** *20050893* [2]
- ***canary grass*** *(grown for game cover) 20050896* [2]
- ***carrots*** *20050895* [2]

- ***clovers*** *(grown for game cover) 20050896* [2]
- ***courgettes*** *20050894* [2]
- ***forage maize*** *(grown for game cover) 20050896* [2]
- ***marrows*** *20050894* [2]
- ***millet*** *(grown for game cover) 20050896* [2]
- ***protected carrots*** *(temporary protection) 20050892* [2]
- ***protected courgettes*** *20050894* [2]
- ***protected marrows*** *20050894* [2]
- ***protected ornamentals*** *20050891* [2]
- ***protected parsnips*** *(temporary protection) 20050892* [2]
- ***protected pumpkins*** *20050894* [2]
- ***protected squashes*** *20050894* [2]
- ***pumpkins*** *20050894* [2]
- ***quinoa*** *(grown for game cover) 20050896* [2]
- ***squashes*** *20050894* [2]
- ***sunflowers*** *(grown for game cover) 20050896* [2]

Approval information
- Accepted by BBPA for use on hops

Efficacy guidance
- When used alone apply pre-weed emergence
- Effectiveness is reduced in dry conditions. Weed seeds germinating at depth are not controlled
- Activity reduced on soils with more than 10% organic matter. Do not use on peaty soils
- Various tank mixtures are recommended for early post-weed emergence treatment (especially for grass weeds). See label for details

Restrictions
- Maximum number of treatments 1 per crop for all edible crops; 2 per yr on amenity vegetation and non-edible crops

Crop-specific information
- Latest use: before 1 Apr in yr of harvest for other edible crops

Following crops guidance
- See label for details of crops which may be sown in the event of failure of a treated crop

Environmental safety
- Keep all livestock out of treated areas for at least 50 d

Hazard classification and safety precautions

Operator protection A, C; U05a, U20b
Environmental protection E06a (50 d); E15a
Storage and disposal D01, D09a, D11a [1, 2]; D05 [2]

291 isoxaben + terbuthylazine

A contact and residual herbicide for use in peas and spring field beans
HRAC mode of action code: L + C1

Products

Skirmish	Syngenta	75:420 g/l	SC	08444

Uses
- Annual dicotyledons in ***combining peas***, ***lupins*** *(off-label)*, ***spring field beans***, ***vining peas***

Specific Off-Label Approvals (SOLAs)
- ***lupins*** *20060540*

Efficacy guidance
- May be applied pre- or post-emergence of crop but before second node stage (GS 102)
- Product will give residual control of germinating weeds on mineral soils for up to 8 wk
- Product is slow acting and control may not be evident for 7-10 d or more after spraying
- Best results achieved when soil surface damp and with a fine, firm tilth. Do not use on very cloddy or stony soil
- Rain after spraying will normally improve weed control but excessive rainfall, very dry conditions or unusually low soil temperatures may lead to unsatisfactory control

Restrictions

- Maximum total dose equivalent to one full dose treatment on peas and spring field beans
- Do not use on forage peas
- Do not use on soils lighter than Coarse Sandy Loam, on very stony soils or soils with more than 10% organic matter
- Pea seed should be covered by at least 25 mm of soil
- Heavy rain after application may cause some crop damage, especially on light soils. Do not use on soils where surface water is likely to accumulate
- For post-emergence treatment a crystal violet test for cuticle wax is advised

Crop-specific information

- Latest use: before second node stage (GS 102) for peas; pre-emergence for spring field beans
- Product may be used on all varieties of spring sown vining and combining peas and spring field beans but pea varieties Vedette and Printana may be damaged from which recovery may not be complete

Environmental safety

- Dangerous for the environment
- Very toxic to aquatic organisms

Hazard classification and safety precautions

Hazard H03, H11
Risk phrases R22a, R50, R53a
Operator protection A; U05a, U20a
Environmental protection E15a, E38
Storage and disposal D01, D02, D05, D07, D09a, D10c, D12a

292 isoxaben + trifluralin

A residual herbicide mixture for amenity and horticultural use
HRAC mode of action code: L + K1

Products

1	Axit GR	Fargro	0.5:2.0% w/w	GR	08892
2	Premiere	Dow	0.5:2.0% w/w	GR	10866

Uses

- Annual dicotyledons in ***amenity vegetation*** [2]; ***ornamental plant production*** *(outdoor stock only)* [1]
- Annual meadow grass in ***amenity vegetation*** [2]; ***ornamental plant production*** *(outdoor stock only)* [1]
- Fat hen in ***amenity vegetation*** [2]; ***ornamental plant production*** *(outdoor stock only)* [1]
- Groundsel in ***amenity vegetation*** [2]
- Hairy bittercress in ***amenity vegetation*** [2]; ***ornamental plant production*** *(outdoor stock only)* [1]
- Shepherd's purse in ***ornamental plant production*** *(outdoor stock only)* [1]
- Sowthistle in ***amenity vegetation*** [2]; ***ornamental plant production*** *(outdoor stock only)* [1]

Approval information

- Trifluralin will not be included in Annex I of EC Directive 91/414. All approvals for the storage and use of all products listed in this profile will expire on 20 March 2009 unless an earlier date has been set

Efficacy guidance

- Best results obtained when granules applied to firm, moist soil, free from clods. 20-30 mm irrigation or rainfall required within 3 d after treatment
- Weed control reduced under dry soil conditions
- Hoeing after application will reduce weed control
- Treatment should be made before mulching and before weed emergence
- Uniform cover with granules optimises results. Improved uniformity of cover may be achieved by spreading the granules twice at half dose at right angles
- Control may be reduced on soils with high organic matter or where organic manure has been applied
- Perennial weeds regenerating from underground rootstocks or stolons are not controlled

Restrictions

- Maximum number of treatments 1 per yr
- Products contain up to 1% crystalline silica. The MEL for respirable silica dust is 0.4 mg/cu m
- Do not use on soils with more than 10% organic matter (except in container grown hardy nursery stock)
- Do not mix in the soil or compost
- Do not use on ornamentals grown under protection
- Do not apply to plants with wet foliage or where granules can lodge in the foliage
- Do not apply to heathers or *Potentilla spp.*

Crop-specific information

- Ensure soil has settled and no cracks are present before application to transplanted trees and shrubs
- May be used on Light, Medium and Heavy soils, but on Light soils early growth may be reduced under adverse conditions
- For newly planted containerised stock first irrigate and allow compost to settle and foliage and stems to dry before application [1]

Following crops guidance

- Land must be mouldboard ploughed to at least 20 cm before drilling or planting succeeding plants except well rooted forestry trees and ornamentals
- Do not use in the 12 mth prior to lifting field grown nursery stock if edible crops are to be grown as a following crop
- After an autumn application cereals or grass crops should not be sown until the following autumn
- Crops most sensitive to treatment residues are oilseed rape, stubble turnips, other brassicae, sugar beet, fodder beet, herbage seed, grass leys and amenity grass

Environmental safety

- Harmful to fish or other aquatic life. Do not contaminate surface waters or ditches with chemical or used container
- Do not allow direct applications of granules from ground based/vehicle mounted applicators to fall within 6 m, or from hand-held applicators to within 2 m, of surface waters or ditches. Direct applications away from water

Hazard classification and safety precautions

Operator protection A; U20b
Environmental protection E13c, E16f, E34
Storage and disposal D01, D09a, D11a

293 isoxaflutole

An isoxazole herbicide available only in mixtures
HRAC mode of action code: F2

See also *flufenacet + isoxaflutole*

294 kresoxim-methyl

A protectant strobilurin fungicide for apples
FRAC mode of action code: 11

See also *epoxiconazole + fenpropimorph + kresoxim-methyl*
epoxiconazole + kresoxim-methyl
epoxiconazole + kresoxim-methyl + pyraclostrobin
fenpropimorph + kresoxim-methyl

Products

1	Kresoxy 50 WG	AgriGuard	50% w/w	WG	12275
2	Stroby WG	BASF	50% w/w	WG	08653

Uses

- Black spot in ***protected roses*, *roses***

SEE SECTION 3 FOR PRODUCTS ALSO REGISTERED

- Powdery mildew in ***apples*** *(reduction)*, ***blackcurrants***, ***protected roses***, ***protected strawberries***, ***roses***, ***strawberries***
- Scab in ***apples***

Approval information

- Kresoxim-methyl included in Annex I under EC Directive 91/414
- Approved for use in ULV systems

Efficacy guidance

- Activity is protectant. Best results achieved from treatments prior to disease development. See label for timing details on each crop. Treatments should be repeated at 10-14 d intervals but note limitations below
- To minimise the likelihood of development of resistance to strobilurin fungicides these products should be used in a planned Resistance Management strategy. See Section 5 for more information
- Product may be applied in ultra low volumes (ULV) but disease control may be reduced

Restrictions

- Maximum number of treatments 4 per yr on apples; 3 per yr on other crops. See notes in Efficacy about limitations on consecutive treatments
- Consult before using on crops intended for processing

Crop-specific information

- HI 14 d for blackcurrants, protected strawberries, strawberries; 35 d for apples
- On apples do not spray product more than twice consecutively and separate each block of two consecutive treatments with at least two applications from a different cross-resistance group. For all other crops do not apply consecutively and use a maximum of once in every three fungicide sprays
- Product should not be used as final spray of the season on apples

Environmental safety

- Dangerous for the environment
- Very toxic to aquatic organisms
- Broadcast air-assisted LERAP (5 m)
- Harmless to ladybirds and predatory mites
- Harmless to honey bees and may be applied during flowering. Nevertheless local beekeepers should be notified when treatment of orchards in flower is to occur

Hazard classification and safety precautions

Hazard H03, H11
Risk phrases R40, R50, R53a
Operator protection A, H, J [1]; U05a, U20b
Environmental protection E15a, E38 [1, 2]; E16b [1]; E17b [1, 2] (5 m)
Storage and disposal D01, D02, D09a, D10c, D12a [1, 2]; D05 [1]; D08 [2]
Medical advice M05a

295 lambda-cyhalothrin

A quick-acting contact and ingested pyrethroid insecticide
IRAC mode of action code: 3

Products

1	Clayton Lambada	Clayton	100 g/l	CS	13231
2	Clayton Lanark	Clayton	100 g/l	CS	12942
3	Clayton Sparta	Clayton	50 g/l	EC	13457
4	Hallmark with Zeon Technology	Syngenta	100 g/l	CS	12629
5	Jackpot 2	AgriGuard	100 g/l	CS	12888
6	Landgold Lambda-Z	Teliton	100 g/l	CS	13533

Uses

- Aphids in ***borage*** *(off-label)*, ***canary flower (echium spp.)*** *(off-label)*, ***chives*** *(off-label)*, ***evening primrose*** *(off-label)*, ***frise*** *(off-label)*, ***grass seed crops*** *(off-label)*, ***herbs (see appendix 6)*** *(off-label)*, ***honesty*** *(off-label)*, ***lamb's lettuce*** *(off-label)*, ***mustard*** *(off-label)*, ***parsley*** *(off-label)*, ***salad brassicas*** *(off-label - for baby leaf production)*, ***scarole*** *(off-label)*, ***spring linseed*** *(off-label)*,

spring rye *(off-label)*, ***triticale*** *(off-label)*, ***winter linseed*** *(off-label)*, ***winter rye*** *(off-label)* [2, 4]; ***chestnuts*** *(off-label)*, ***cob nuts*** *(off-label)*, ***hazel nuts*** *(off-label)*, ***lupins*** *(off-label)*, ***walnuts*** *(off-label)* [4]; ***combining peas***, ***vining peas*** [3]; ***durum wheat***, ***spring barley***, ***spring oats***, ***winter barley***, ***winter oats*** [1, 2, 4-6]; ***potatoes***, ***spring wheat***, ***winter wheat*** [1-6]

- Barley yellow dwarf virus vectors in ***durum wheat***, ***winter barley***, ***winter oats***, ***winter wheat*** [1, 2, 4-6]; ***spring barley***, ***spring wheat*** [1, 5, 6]
- Bean beetle in ***broad beans*** *(off-label)* [2, 4]
- Beet leaf miner in ***fodder beet*** *(off-label)* [2, 4]; ***sugar beet*** [1, 2, 4-6]
- Beet virus yellows vectors in ***spring oilseed rape***, ***winter oilseed rape*** [1, 2, 4-6]
- Brassica pod midge in ***spring oilseed rape***, ***winter oilseed rape*** [3]
- Cabbage seed weevil in ***spring oilseed rape***, ***winter oilseed rape*** [1-6]
- Cabbage stem flea beetle in ***spring oilseed rape***, ***winter oilseed rape*** [1, 2, 4-6]
- Cabbage stem weevil in ***radishes*** *(off-label)* [2, 4]
- Capsids in ***blackberries*** *(off-label)*, ***dewberries*** *(off-label)*, ***raspberries*** *(off-label)*, ***rubus hybrids*** *(off-label)*, ***wine grapes*** *(off-label)* [2, 4]
- Carrot fly in ***carrots*** *(off-label)*, ***celeriac*** *(off-label)*, ***celery (outdoor)*** *(off-label)*, ***fennel*** *(off-label)*, ***horseradish*** *(off-label)*, ***mallow (althaea spp.)*** *(off-label)*, ***parsley root*** *(off-label)*, ***parsnips*** *(off-label)* [2, 4]; ***parsnips*** [6]
- Caterpillars in ***broccoli***, ***brussels sprouts***, ***cabbages***, ***calabrese***, ***cauliflowers*** [1, 2, 4-6]; ***chestnuts*** *(off-label)*, ***cob nuts*** *(off-label)*, ***hazel nuts*** *(off-label)*, ***walnuts*** *(off-label)* [4]; ***wine grapes*** *(off-label)* [2, 4]
- Clay-coloured weevil in ***blackberries*** *(off-label)*, ***dewberries*** *(off-label)*, ***raspberries*** *(off-label)*, ***rubus hybrids*** *(off-label)* [2, 4]
- Cutworms in ***carrots***, ***lettuce*** [1, 2, 4-6]; ***celeriac*** *(off-label)*, ***fennel*** *(off-label)*, ***fodder beet*** *(off-label)*, ***red beet*** *(off-label)* [2, 4]; ***chicory*** *(off-label - for forcing)* [4]; ***sugar beet*** [1-6]
- Flea beetle in ***fodder beet*** *(off-label)*, ***poppies for morphine production*** *(off-label)* [2, 4]; ***spring oilseed rape***, ***sugar beet***, ***winter oilseed rape*** [1-6]
- Frit fly in ***sweetcorn*** *(off-label)* [2, 4]
- Leaf miner in ***sugar beet*** [3]
- Midges in ***blackcurrants*** *(off-label)*, ***gooseberries*** *(off-label)*, ***redcurrants*** *(off-label)*, ***whitecurrants*** *(off-label)* [2, 4]
- Pea and bean weevil in ***combining peas***, ***vining peas*** [2-4]; ***edible podded peas*** [1, 2, 4]; ***peas*** [1, 5, 6]; ***spring field beans***, ***winter field beans*** [1-6]
- Pea aphid in ***combining peas***, ***vining peas*** [2, 4]; ***edible podded peas*** [1, 2, 4, 6]; ***peas*** [1, 5, 6]
- Pea midge in ***combining peas***, ***vining peas*** [2, 4]; ***edible podded peas*** [1, 2, 4, 6]; ***peas*** [1, 5, 6]
- Pea moth in ***combining peas***, ***vining peas*** [2-4]; ***edible podded peas*** [1, 2, 4, 6]; ***peas*** [1, 5, 6]
- Pear sucker in ***pears*** [1-6]; ***winter wheat*** [3]
- Pod midge in ***spring oilseed rape***, ***winter oilseed rape*** [1, 2, 4-6]
- Pollen beetle in ***poppies for morphine production*** *(off-label)* [2, 4]; ***spring oilseed rape***, ***winter oilseed rape*** [1-6]
- Sawflies in ***blackcurrants*** *(off-label)*, ***gooseberries*** *(off-label)*, ***redcurrants*** *(off-label)*, ***whitecurrants*** *(off-label)* [2, 4]
- Silver Y moth in ***celery (outdoor)*** *(off-label)*, ***dwarf beans*** *(off-label)*, ***french beans*** *(off-label)*, ***navy beans*** *(off-label)*, ***red beet*** *(off-label)*, ***runner beans*** *(off-label)* [2, 4]
- Thrips in ***broad beans*** *(off-label)*, ***bulb onions*** *(off-label)*, ***garlic*** *(off-label)*, ***leeks*** *(off-label)*, ***salad onions*** *(off-label)*, ***shallots*** *(off-label)* [2, 4]
- Wasps in ***wine grapes*** *(off-label)* [2, 4]
- Wheat-blossom midge in ***winter wheat*** [2, 4]
- Whitefly in ***broccoli***, ***brussels sprouts***, ***cabbages***, ***calabrese***, ***cauliflowers*** [1, 2, 4-6]
- Willow aphid in ***short rotation coppice willow*** *(off-label)* [4]
- Willow beetle in ***short rotation coppice willow*** *(off-label)* [4]
- Willow sawfly in ***short rotation coppice willow*** *(off-label)* [4]
- Yellow cereal fly in ***winter wheat*** [1-6]

Specific Off-Label Approvals (SOLAs)

- ***blackberries*** *20063755* [2], *20060728 expires Nov 2008* [4]
- ***blackcurrants*** *20063752* [2], *20060727 expires Nov 2008* [4]
- ***borage*** *20063748* [2], *20060634 expires Nov 2008* [4]
- ***broad beans*** *20063764* [2], *20060753 expires Nov 2008* [4]
- ***bulb onions*** *20063756* [2], *20060730 expires Nov 2008* [4]

- ***canary flower (echium spp.)*** *20063748* [2], *20060634 expires Nov 2008* [4]
- ***carrots*** *20063750* [2], *20060729 expires Nov 2008* [4]
- ***celeriac*** *20063757* [2], *20060731 expires Nov 2008* [4]
- ***celery (outdoor)*** *20063762* [2], *20060744 expires Nov 2008* [4]
- ***chestnuts*** *20060742 expires Nov 2008* [4]
- ***chicory*** *(for forcing) 20060740 expires Nov 2008* [4]
- ***chives*** *20063751* [2], *20060636 expires Nov 2008* [4]
- ***cob nuts*** *20060742 expires Nov 2008* [4]
- ***dewberries*** *20063755* [2], *20060728 expires Nov 2008* [4]
- ***dwarf beans*** *20063758* [2], *20060739 expires Nov 2008* [4]
- ***evening primrose*** *20063664* [2], *20060634 expires Nov 2008* [4]
- ***fennel*** *20060733 expires Nov 2008* [2], *20063751* [2], *20060733 expires Nov 2008* [4]
- ***fodder beet*** *20063665* [2], *20060637 expires Nov 2008* [4]
- ***french beans*** *20063758* [2], *20060739 expires Nov 2008* [4]
- ***frise*** *20063751* [2], *20060636 expires Nov 2008* [4]
- ***garlic*** *20063756* [2], *20060730 expires Nov 2008* [4]
- ***gooseberries*** *20063752* [2], *20060727 expires Nov 2008* [4]
- ***grass seed crops*** *20063749* [2], *20060624 expires Nov 2008* [4]
- ***hazel nuts*** *20060742 expires Nov 2008* [4]
- ***herbs (see appendix 6)*** *20063751* [2], *20060636 expires Nov 2008* [4]
- ***honesty*** *20063748* [2], *20060634 expires Nov 2008* [4]
- ***horseradish*** *20071307* [2], *20060729 expires Nov 2008* [4]
- ***lamb's lettuce*** *20063751* [2], *20060636 expires Nov 2008* [4]
- ***leeks*** *20063756* [2], *20060730 expires Nov 2008* [4]
- ***lupins*** *20060635 expires Nov 2008* [4]
- ***mallow (althaea spp.)*** *20071307* [2], *20060729 expires Nov 2008* [4]
- ***mustard*** *20063664* [2], *20060634 expires Nov 2008* [4]
- ***navy beans*** *20063758* [2], *20060739 expires Nov 2008* [4]
- ***parsley*** *20063751* [2], *20060636 expires Nov 2008* [4]
- ***parsley root*** *20071307* [2], *20060729 expires Nov 2008* [4]
- ***parsnips*** *20063750* [2], *20060729 expires Nov 2008* [4]
- ***poppies for morphine production*** *20063763* [2], *20060749 expires Nov 2008* [4]
- ***radishes*** *20063757* [2], *20060731 expires Nov 2008* [4]
- ***raspberries*** *20063755* [2], *20060728 expires Nov 2008* [4]
- ***red beet*** *20063761* [2], *20060743 expires Nov 2008* [4]
- ***redcurrants*** *20063752* [2], *20060727 expires Nov 2008* [4]
- ***rubus hybrids*** *20063755* [2], *20060728 expires Nov 2008* [4]
- ***runner beans*** *20063758* [2], *20060739 expires Nov 2008* [4]
- ***salad brassicas*** *(for baby leaf production) 20063751* [2], *(for baby leaf production) 20060636 expires Nov 2008* [4]
- ***salad onions*** *20060730 expires Nov 2008* [2, 4]
- ***scarole*** *20063751* [2], *20060636 expires Nov 2008* [4]
- ***shallots*** *20063756* [2], *20060730 expires Nov 2008* [4]
- ***short rotation coppice willow*** *20060748 expires Nov 2008* [4]
- ***spring linseed*** *20063748* [2], *20060634 expires Nov 2008* [4]
- ***spring rye*** *20063749* [2], *20060624 expires Nov 2008* [4]
- ***sweetcorn*** *20063760* [2], *20060732 expires Nov 2008* [4]
- ***triticale*** *20063749* [2], *20060624 expires Nov 2008* [4]
- ***walnuts*** *20060742 expires Nov 2008* [4]
- ***whitecurrants*** *20063752* [2], *20060727 expires Nov 2008* [4]
- ***wine grapes*** *20060266 expires Nov 2008* [2], *20063747* [2], *20060266 expires Nov 2008* [4]
- ***winter linseed*** *20063748* [2], *20060634 expires Nov 2008* [4]
- ***winter rye*** *20063749* [2], *20060624 expires Nov 2008* [4]

Approval information

- Lambda-cyhalothrin included in Annex I under EC Directive 91/414
- Accepted by BBPA for use on malting barley
- Approval expiry 13 Nov 2008 [5]

Efficacy guidance

- Best results normally obtained from treatment when pest attack first seen. See label for detailed recommendations on each crop
- Timing for control of barley yellow dwarf virus vectors depends on specialist assessment of the level of risk in the area
- Repeat applications recommended in some crops where prolonged attack occurs, up to maximum total dose. See label for details
- Where strains of aphids resistant to lambda-cyhalothrin occur control is unlikely to be satisfactory
- Addition of wetter recommended for control of certain pests in brassicas and oilseed rape
- Use of sufficient water volume to ensure thorough crop penetration recommended for optimum results
- Use of drop-legged sprayer gives improved results in crops such as Brussels sprouts

Restrictions

- Maximum number of applications or maximum total dose per crop varies - see labels
- Do not apply to a cereal crop if any product containing a pyrethroid insecticide or dimethoate has been applied to the crop after the start of ear emergence (GS 51)
- Do not spray cereals in the spring/summer (ie after 1 Apr) within 6 m of edge of crop

Crop-specific information

- Latest use before late milk stage on cereals; before end of flowering for winter oilseed rape
- HI 3 d for lettuce [1, 5], radishes, red beet; 7 d for lettuce [2, 4]; 14 d for carrots; 25 d for peas, field beans; 6 wk for spring oilseed rape; 8 wk for sugar beet

Environmental safety

- Dangerous for the environment
- Very toxic to aquatic organisms
- Flammable [3]
- Risk to certain non-target insects or other arthropods [4]
- To protect non-target arthropods respect an untreated buffer zone of 5 m to non-crop land
- LERAP Category B
- Broadcast air-assisted LERAP (25 m) [1, 2, 4-6]; 38 m [3]

Hazard classification and safety precautions

Hazard H03, H11 [1-6]; H08 [3]

Risk phrases R20, R22a, R50, R53a [1-6]; R21, R22b, R36, R37, R38, R67 [3]; R43 [1, 2, 4-6]

Operator protection A, H [1-6]; C [3, 5]; J, K, L, M [5]; U02a, U14, U20b [1, 2, 4-6]; U05a, U08 [1-6]; U19a, U20a [3]

Environmental protection E15a [1]; E15b [3, 6]; E16a, E16b [1, 2, 4-6]; E16c [3]; E17b [1, 2] (25 m); E17b [3] (38 m); E17b [4-6] (25 m); E22b [4]; E34 [1-6]; E38 [2, 4, 5]

Storage and disposal D01, D02, D09a, D12a [1-6]; D05 [1, 3, 6]; D07, D10b [3]; D10c [1, 2, 4-6]

Medical advice M03 [1, 2, 4-6]; M05a [1, 6]; M05b [3]

296 lambda-cyhalothrin + pirimicarb

An insecticide mixture combining translaminar, contact, fumigant and stomach activity

IRAC mode of action code: 3 + 1A

Products

1	Clayton Groove	Clayton	5:100 g/l	EC	13246
2	Dovetail	Syngenta	5:100 g/l	EC	12550
3	Mortice	AgriGuard	5:100 g/l	EC	12961

Uses

- Aphids in ***borage*** *(off-label)*, ***canary flower (echium spp.)*** *(off-label)*, ***canary seed*** *(off-label - for wild bird seed production)*, ***evening primrose*** *(off-label)*, ***fodder beet*** *(off-label)*, ***grass seed crops*** *(off-label)*, ***honesty*** *(off-label)*, ***horseradish*** *(off-label)*, ***lupins*** *(off-label)*, ***mallow (althaea spp.)*** *(off-label)*, ***mustard*** *(off-label)*, ***parsnips*** *(off-label)*, ***protected chives*** *(off-label)*, ***protected frise*** *(off-label)*, ***protected herbs (see appendix 6)*** *(off-label)*, ***protected parsley*** *(off-label)*, ***protected radicchio*** *(off-label)*, ***protected salad brassicas*** *(off-label - for baby leaf production)*, ***protected scarole*** *(off-label)*, ***spring linseed*** *(off-label)*, ***winter linseed*** *(off-label)* [2]; ***broccoli***, ***brussels sprouts***, ***cabbages***, ***calabrese***, ***carrots***, ***cauliflowers***, ***lettuce***, ***peas***, ***potatoes***, ***spring***

SEE SECTION 3 FOR PRODUCTS ALSO REGISTERED

barley, spring oats, spring rye, spring wheat, sugar beet, triticale, winter barley, winter oats, winter rye, winter wheat [1-3]; ***durum wheat*** [2, 3]
- Beet virus yellows vectors in ***spring oilseed rape, winter oilseed rape*** [1-3]
- Cabbage seed weevil in ***spring oilseed rape, winter oilseed rape*** [1-3]
- Cabbage stem flea beetle in ***spring oilseed rape, winter oilseed rape*** [1-3]
- Caterpillars in ***broccoli, brussels sprouts, cabbages, calabrese, cauliflowers*** [1-3]
- Cutworms in ***carrots, lettuce, potatoes, sugar beet*** [1-3]; ***fodder beet*** *(off-label)*, ***horseradish*** *(off-label)*, ***mallow (althaea spp.)*** *(off-label)*, ***parsnips*** *(off-label)* [2]
- Flea beetle in ***fodder beet*** *(off-label)* [2]; ***spring oilseed rape, sugar beet, winter oilseed rape*** [1-3]
- Leaf miner in ***fodder beet*** *(off-label)* [2]; ***sugar beet*** [1-3]
- Mealy aphid in ***broccoli, brussels sprouts, cabbages, calabrese, cauliflowers, spring oilseed rape, winter oilseed rape*** [1-3]
- Pea and bean weevil in ***peas, spring field beans, winter field beans*** [1-3]
- Pea midge in ***peas*** [1-3]
- Pea moth in ***peas*** [1-3]
- Pod midge in ***spring oilseed rape, winter oilseed rape*** [1-3]
- Pollen beetle in ***broccoli, brussels sprouts, cabbages, calabrese, cauliflowers, spring oilseed rape, winter oilseed rape*** [1-3]
- Whitefly in ***broccoli, brussels sprouts, cabbages, calabrese, cauliflowers*** [1-3]

Specific Off-Label Approvals (SOLAs)

- ***borage*** *20060641* [2]
- ***canary flower (echium spp.)*** *20060641* [2]
- ***canary seed*** *(for wild bird seed production) 20070870* [2]
- ***evening primrose*** *20060641* [2]
- ***fodder beet*** *20060642* [2]
- ***grass seed crops*** *20060640* [2]
- ***honesty*** *20060641* [2]
- ***horseradish*** *20060639* [2]
- ***lupins*** *20060638* [2]
- ***mallow (althaea spp.)*** *20060639* [2]
- ***mustard*** *20060641* [2]
- ***parsnips*** *20060639* [2]
- ***protected chives*** *20060643* [2]
- ***protected frise*** *20060643* [2]
- ***protected herbs (see appendix 6)*** *20060643* [2]
- ***protected parsley*** *20060643* [2]
- ***protected radicchio*** *20060643* [2]
- ***protected salad brassicas*** *(for baby leaf production) 20060643* [2]
- ***protected scarole*** *20060643* [2]
- ***spring linseed*** *20060641* [2]
- ***winter linseed*** *20060641* [2]

Approval information

- Lambda-cyhalothrin and pirimicarb included in Annex I under EC Directive 91/414
- Accepted by BBPA for use on malting barley
- In 2006 PSD required that all products containing pirimicarb should carry the following warning in the main area of the container label: "Pirimicarb is an anticholinesterase carbamate. Handle with care"

Efficacy guidance

- Best results obtained from treatment when pest attack first seen or after warning issued
- Repeat applications recommended in some crops where prolonged attacks occur, up to maximum total dose
- Addition of Agral recommended for certain uses in brassicas and oilseed rape. See label
- Use of drop-leg sprayers improves efficacy in crops such as Brussels sprouts
- Control unlikely to be satisfactory if aphids resistant to lambda-cyhalothrin or pirimicarb present

Restrictions

- Maximum total dose equivalent to two full dose treatments on carrots, lettuce, peas, field beans; three full dose treatments on brassicas, oilseed rape, cereals; four full dose treatments on sugar beet; eight full dose treatments on potatoes
- Must not be applied to cereals if any product containing a pyrethroid insecticide or dimethoate has been sprayed after the start of ear emergence (GS 51)

Crop-specific information

- Latest use: before late milky ripe (GS 77) for cereals; before end of flowering for oilseed rape
- HI sugar beet 8 wk; carrots 14 d; brassicas, lettuce, potatoes, peas, field beans 3 d

Environmental safety

- Dangerous for the environment
- Very toxic to aquatic organisms
- Do not spray cereals after 1 Apr within 6 m of the edge of the crop
- Harmful to livestock. Keep all livestock out of treated areas for at least 7 d following treatment
- Keep in original container, tightly closed, in a safe place, under lock and key
- LERAP Category A

Hazard classification and safety precautions

Hazard H03, H11 [1-3]; H08 [3]

Risk phrases R20, R22a, R38, R50, R53a [1-3]; R22b, R37, R41, R43 [3]

Operator protection A, C, D, E, H, J, K, M; U02a, U14 [1, 2]; U05a, U08, U19a, U20b [1-3]

Environmental protection E06a [1, 3] (7 d); E06c [2] (7 d); E15a, E16c, E16d, E34 [1-3]; E22c [1]; E38 [2]

Storage and disposal D01, D02, D09b [1-3]; D05, D10c, D12a [1, 2]; D11a [3]

Medical advice M02, M03 [1-3]; M05b [1, 2]

297 lenacil

A residual, soil-acting uracil herbicide for beet crops
HRAC mode of action code: C1

See also chloridazon + lenacil

Products

1	Agriguard Lenacil	AgriGuard	80% w/w	WP	10488
2	Clayton Lenacil 80 W	Clayton	80% w/w	WP	09488
3	Clayton Lenaflo	Clayton	440 g/l	SC	11031
4	Fernpath Lenzo Flo	AgriGuard	440 g/l	SC	11919
5	Lenacil FL	United Phosphorus	440 g/l	SC	12183
6	Lenazar Flo	Hermoo	440 g/l	SC	12286
7	Lenazar Flowable	Hermoo	440 g/l	SC	11068
8	Venzar Flowable	DuPont	440 g/l	SC	06907
9	Volcano	United Phosphorus	440 g/l	SC	12301

Uses

- Annual dicotyledons in ***blackberries*** *(off-label)*, ***blackcurrants***, ***established woody ornamentals***, ***gooseberries***, ***raspberries***, ***roses***, ***rubus hybrids*** *(off-label)*, ***strawberries*** [2]; ***farm woodland*** *(off-label)* [8]; ***fodder beet***, ***mangels***, ***sugar beet*** [1-9]; ***red beet*** [1-4, 6, 8]
- Annual meadow grass in ***blackberries*** *(off-label)*, ***blackcurrants***, ***established woody ornamentals***, ***gooseberries***, ***raspberries***, ***roses***, ***rubus hybrids*** *(off-label)*, ***strawberries*** [2]; ***farm woodland*** *(off-label)* [8]; ***fodder beet***, ***mangels***, ***sugar beet*** [1-9]; ***red beet*** [1-4, 6, 8]

Specific Off-Label Approvals (SOLAs)

- ***blackberries*** *20060884* [2]
- ***farm woodland*** *971282* [8]
- ***rubus hybrids*** *20060884* [2]

Efficacy guidance

- Best results, especially from pre-emergence treatments, achieved on fine, even, firm and moist soils free from clods. Continuing presence of moisture from rain or irrigation gives improved residual control of later germinating weeds. Effectiveness may be reduced by dry conditions
- On beet crops may be used pre- or post-emergence, alone or in mixture to broaden weed spectrum

- Apply overall or as band spray to beet crops pre-drilling incorporated, pre- or post-emergence
- All labels have limitations on soil types that may be treated. Residual activity reduced on soils with high OM content

Restrictions

- Maximum number of treatments in beet crops 1 pre-emergence + 3 post-emergence per crop; 2 per yr on established woody ornamentals and roses
- See label for soil type restrictions
- Do not use any other residual herbicide within 3 mth of the initial application to fruit or ornamental crops
- Do not treat crops under stress from drought, low temperatures, nutrient deficiency, pest or disease attack, or waterlogging

Crop-specific information

- Latest use: pre-emergence for red beet, fodder beet, spinach, spinach beet, mangels; before leaves meet over rows when used on these crops post-emergence; 24 h after planting new strawberry runners or before flowering for established strawberry crops, blackcurrants, gooseberries, raspberries [2]
- Heavy rain after application to beet crops may cause damage especially if followed by very hot weather
- Reduction in beet stand may occur where crop emergence or vigour is impaired by soil capping or pest attack
- Strawberry runner beds to be treated should be level without depressions around the roots
- New soft fruit cuttings should be planted at least 15 cm deep and firmed before treatment
- Check varietal tolerance of ornamentals before large scale treatment

Following crops guidance

- Succeeding crops should not be planted or sown for at least 4 mth (6 mth on organic soils) after treatment following ploughing to at least 150 mm. Only beet crops may be sown within 4 mth of treatment

Environmental safety

- Dangerous for the environment
- Very toxic to aquatic organisms

Hazard classification and safety precautions

Hazard H04 [1-5, 7, 9]; H11 [1-9]
Risk phrases R36, R38 [1-5, 7, 9]; R37 [1, 2, 4]; R50, R53a [2-9]
Operator protection A [1-9]; C [1-7, 9]; U05a, U08 [1-9]; U11 [5]; U14, U15 [1, 4]; U19a [1-4, 6-8]; U20a [1, 3-9]; U20b [2]
Environmental protection E13b [5, 9]; E13c, E38 [4, 6-8]; E15a [1-3, 5]; E34 [1]
Storage and disposal D01, D02, D09a [1-9]; D05 [3-9]; D10a [4, 6-8]; D10b [1, 3]; D10c [5, 9]; D11a [2]; D12a [3-8]; D12b [9]
Medical advice M05b [5, 9]

298 lenacil + triflusulfuron-methyl

A foliar and residual herbicide mixture for sugar beet
HRAC mode of action code: C1 + B

Products

Safari Lite WSB	DuPont	71.4:5.4% w/w	WG	12169

Uses

- Annual dicotyledons in ***sugar beet***
- Volunteer oilseed rape in ***sugar beet***

Efficacy guidance

- Best results obtained when weeds are small and growing actively
- Product recommended for use in a programme of treatments in tank mixture with a suitable herbicide partner to broaden the weed spectrum
- Ensure good spray cover of weeds. Apply when first weeds have emerged provided crop has reached cotyledon stage

- Susceptible plants cease to grow almost immediately after treatment and symptoms can be seen 5-10 d later
- Weed control may be reduced in very dry soil conditions
- Triflusulfuron-methyl is a member of the ALS-inhibitor group of herbicides

Restrictions

- Maximum number of treatments on sugar beet 3 per crop and do not apply more than 4 applications of any product containing triflusulfuron-methyl
- Do not apply to crops suffering from stress caused by drought, water-logging, low temperatures, pest or disease attack, nutrient deficiency or any other factors affecting crop growth
- Do not use on Sands, stony or gravelly soils or on soils with more than 10% organic matter
- Do not apply when temperature above or likely to exceed 21 °C on day of spraying or under conditions of high light intensity

Crop-specific information

- Latest use: before crop leaves meet between rows for sugar beet

Following crops guidance

- Only cereals may be sown in the same calendar yr as a treated sugar beet crop. Any crop may be sown in the following spring
- In the event of crop failure sow only sugar beet within 4 mth of treatment

Environmental safety

- Dangerous for the environment
- Very toxic to aquatic organisms
- LERAP Category B
- Take extreme care to avoid drift onto broad-leaved plants outside the target area or onto surface waters or ditches, or land intended for cropping
- Spraying equipment should not be drained or flushed onto land planted, or to be planted, with trees or crops other than cereals and should be thoroughly cleansed after use - see label for instructions

Hazard classification and safety precautions

Hazard H04, H11
Risk phrases R36, R37, R38, R43, R50, R53a
Operator protection A, C; U05a, U08, U11, U19a, U20b, U22a
Environmental protection E15a, E16a, E38
Storage and disposal D01, D02, D09a, D12a

299 lindane

An organochlorine insecticide all approvals for which were revoked in 2001 following non-inclusion of the substance in Annex I of EC Directive 91/414
IRAC mode of action code: 2A

300 linuron

A contact and residual urea herbicide for various field crops
HRAC mode of action code: C2

See also 2,4-DB + linuron + MCPA

Products

1	Afalon	Makhteshim	450 g/l	SC	11665
2	Alpha Linuron 50 SC	Makhteshim	500 g/l	SC	06967
3	Lincon 50	DAPT	500 g/l	SC	10639
4	Linuron 500	Nufarm UK	480 g/l	SC	11590
5	Luas	AgriGuard	500 g/l	SC	12357
6	UPL Linuron 45% Flowable	United Phosphorus	450 g/l	SC	07435

Uses

- Annual dicotyledons in ***asparagus*** *(off-label)*, ***celery (outdoor)*** [1]; ***bulb onions*** *(off-label)*, ***celeriac*** *(off-label)*, ***chives*** *(off-label)*, ***dwarf beans*** *(off-label)*, ***garlic*** *(off-label)*, ***herbs (see appendix 6)*** *(off-label)*, ***herbs for medicinal uses (see appendix 6)*** *(off-label)*, ***horseradish*** *(off-label)*, ***leeks*** *(off-label)*, ***parsley*** *(off-label)*, ***parsley root*** *(off-label)*, ***runner beans*** *(off-label)* [2]; ***carrots***, ***potatoes*** [1-6]; ***parsnips*** [1-3, 5, 6]; ***spring barley***, ***spring wheat*** [3-6]; ***spring oats*** [2, 3, 5]

- Annual meadow grass in ***asparagus*** *(off-label)*, ***celery (outdoor)*** [1]; ***carrots***, ***potatoes*** [1-6]; ***parsnips*** [1-3, 5, 6]; ***spring barley***, ***spring wheat*** [2-6]; ***spring oats*** [3, 5]
- Black bindweed in ***carrots***, ***potatoes*** [1-6]; ***celery (outdoor)*** [1]; ***parsnips*** [1-3, 5, 6]; ***spring barley***, ***spring wheat*** [2-6]; ***spring oats*** [2, 3, 5]
- Chickweed in ***carrots***, ***potatoes*** [1-6]; ***celery (outdoor)*** [1]; ***parsnips*** [1-3, 5, 6]; ***spring barley***, ***spring wheat*** [2-6]; ***spring oats*** [2, 3, 5]
- Corn marigold in ***carrots***, ***potatoes***, ***spring barley***, ***spring wheat*** [2-5]; ***parsnips***, ***spring oats*** [2, 3, 5]
- Fat hen in ***carrots***, ***potatoes*** [1-6]; ***celery (outdoor)*** [1]; ***parsnips*** [1-3, 5, 6]; ***spring barley***, ***spring wheat*** [2-6]; ***spring oats*** [2, 3, 5]
- General weed control in ***bulb onions*** *(off-label)*, ***celeriac*** *(off-label)*, ***garlic*** *(off-label)*, ***horseradish*** *(off-label)*, ***leeks*** *(off-label)*, ***mallow (althaea spp.)*** *(off-label)*, ***parsley root*** *(off-label)* [1]
- Redshank in ***carrots***, ***potatoes*** [1-6]; ***celery (outdoor)*** [1]; ***parsnips*** [1-3, 5, 6]; ***spring barley***, ***spring wheat*** [2-6]; ***spring oats*** [2, 3, 5]

Specific Off-Label Approvals (SOLAs)

- ***asparagus*** *20060651* [1]
- ***bulb onions*** *20041755* [1], *20021748* [2]
- ***celeriac*** *20041755* [1], *20021748* [2]
- ***chives*** *20061251* [2]
- ***dwarf beans*** *20051698* [2]
- ***garlic*** *20041755* [1], *20021748* [2]
- ***herbs (see appendix 6)*** *20061251* [2]
- ***herbs for medicinal uses (see appendix 6)*** *20050154* [2]
- ***horseradish*** *20053017* [1], *20053020* [2]
- ***leeks*** *20041755* [1], *20021748* [2]
- ***mallow (althaea spp.)*** *20053017* [1]
- ***parsley*** *20061251* [2]
- ***parsley root*** *20053017* [1], *20053020* [2]
- ***runner beans*** *20051698* [2]

Approval information

- Linuron included in Annex I under EC Directive 91/414
- All approvals for use of linuron on parsley were revoked in June 2005
- Accepted by BBPA for use on malting barley
- Approval expiry 31 Dec 2008 [1-6]

Efficacy guidance

- Many weeds controlled pre-emergence or post-emergence to 2-3 leaf stage, some (annual meadow grass, mayweed) only susceptible pre-emergence. See label for details
- Best results achieved by application to firm, moist soil of fine tilth
- Little residual effect on soil with more than 10% organic matter

Restrictions

- Maximum number of treatments 1 per crop on most crops or 1 pre-em + 1 post-em for carrots, parsley, parsnips
- Maximum total dose equivalent to one full dose treatment
- Do not use on undersown cereals or crops grown on Sands or Very Light soils or soils heavier than Sandy Clay Loam or with more than 10% organic matter
- Do not apply to emerged crops of carrots, parsnips or parsley under stress
- Do not apply by hand-held sprayers

Crop-specific information

- Latest use: pre-emergence for most crops; pre- or up to 40% emergence for potatoes (products differ, see label for details).
- HI onions, garlic 8 wk; celeriac 12 wk; leeks 16 wk
- Drill spring cereals at least 3 cm deep and apply pre-emergence of crop or weeds
- Apply to potatoes well earthed up to a rounded ridge pre-crop emergence and do not cultivate after spraying

- Apply to carrots at any time after drilling on organic soils, and within 4 d of drilling on other soils. Apply post-emergence as soon as weeds appear but after first rough leaf stage.
- Recommendations for parsnips, parsley and celery vary. See label for details

Following crops guidance

- Potatoes, carrots and parsnips may be planted at any time after application. Lettuce should not be grown within 12 mth of treatment. Transplanted brassicas may be grown from 3 mth after treatment

Environmental safety

- Dangerous for the environment
- Very toxic to aquatic organisms
- Keep livestock out of treated areas for at least 5 mth [4]
- LERAP Category B

Hazard classification and safety precautions

Hazard H03, H11

Risk phrases R22a [1, 2, 6]; R40, R48, R50, R53a [1-6]; R61, R62, R68 [4]

Operator protection A, H [1-6]; C [2-6]; U05a [1-6]; U08, U20b [1, 2, 4, 6]; U09a, U20a [3, 5]; U19a [2-6]; U23a [1]

Environmental protection E07b [4] (5 mth); E15a, E16a [1-6]; E34 [3, 5]; E38 [4]

Storage and disposal D01, D05, D09a [1-6]; D02 [1-3, 5, 6]; D08 [1]; D10b, D12a [1, 4]; D10c [3, 5, 6]; D11a [2]; D12b [1, 2, 6]

301 linuron + trifluralin

A residual pre-emergence herbicide for use in winter cereals
HRAC mode of action code: C2 + K1

Products

1	Arizona	Makhteshim	120:240 g/l	EC	12615
2	Blois	Makhteshim	128:256 g/l	EC	12376
3	Uranus	Makhteshim	120:240 g/l	EC	12506

Uses

- Annual dicotyledons in ***broad beans***, ***combining peas***, ***lupins*** *(off-label)*, ***spring field beans***, ***vining peas***, ***winter field beans*** [2]; ***durum wheat*** *(off-label)*, ***grass seed crops*** *(off-label)*, ***spring rye*** *(off-label)*, ***triticale***, ***winter barley***, ***winter rye*** *(off-label)*, ***winter wheat*** [1, 3]
- Annual grasses in ***broad beans***, ***combining peas***, ***lupins*** *(off-label)*, ***spring field beans***, ***vining peas***, ***winter field beans*** [2]; ***durum wheat*** *(off-label)*, ***grass seed crops*** *(off-label)*, ***spring rye*** *(off-label)*, ***triticale***, ***winter barley***, ***winter rye*** *(off-label)*, ***winter wheat*** [1, 3]
- Annual meadow grass in ***winter oats*** *(off-label)* [3]
- Blackgrass in ***spring oats*** *(off-label)* [3]

Specific Off-Label Approvals (SOLAs)

- ***durum wheat*** *20053000* [1], *20053015* [3]
- ***grass seed crops*** *20053000* [1], *20053015* [3]
- ***lupins*** *20053014* [2]
- ***spring oats*** *20052106* [3]
- ***spring rye*** *20053000* [1], *20053015* [3]
- ***winter oats*** *20052106* [3]
- ***winter rye*** *20053000* [1], *20053015* [3]

Approval information

- Linuron included in Annex I under EC Directive 91/414
- Trifluralin will not be included in Annex I of EC Directive 91/414. All approvals for the storage and use of all products listed in this profile will expire on 20 March 2009 unless an earlier date has been set
- Accepted by BBPA for use on malting barley

Efficacy guidance

- Effective against weeds germinating near soil surface. Best results achieved by application to fine, firm, moist seedbed free of clods, crop residues or established weeds
- Effectiveness reduced by long dry periods after application or on waterlogged soil

SEE SECTION 3 FOR PRODUCTS ALSO REGISTERED

- Autumn application residual effects normally last until spring but further herbicide treatment may be needed on thin or backward crops
- Results in loose seedbeds on lighter soils improved by rolling after drilling

Restrictions

- Maximum number of treatments 1 per crop
- Do not treat durum wheat or undersown crops
- Do not use on soils classed as Sands. Do not harrow after treatment
- Do not use on peaty soils or where organic matter exceeds 10%

Crop-specific information

- Latest use: pre-emergence of crop
- Apply without incorporation as soon as possible after drilling and before crop emergence, within 3 d on early drilled crops
- Crop seed must be well covered to a minimum depth of 30 mm

Following crops guidance

- In the event of failure of a treated crop, only peas, carrots or sunflowers may be sown within 5 mth of application. Grass and cereals should not be planted until the following autumn, and sugar beet not for one yr
- Before drilling or planting a following crop mouldboard plough to at least 150 mm

Environmental safety

- Dangerous for the environment
- Very toxic to aquatic organisms
- Flammable
- LERAP Category B
- Do not store near heat or flame

Hazard classification and safety precautions

Hazard H03, H08, H11
Risk phrases R20, R21 [3]; R22a, R36, R40, R43, R48, R50, R53a [1-3]
Operator protection A, C, H; U05a, U08, U11, U13, U19a, U20b [1-3]; U14, U15 [3]
Environmental protection E15a, E16a, E16b
Storage and disposal D01, D02, D05, D09a, D10b, D12b
Medical advice M05b

302 magnesium phosphide

A phosphine generating compound used to control insect pests in stored commodities

Products

Degesch Plates	Rentokil	56% w/w	GE	07603

Uses

- Insect pests in ***stored grain***

Approval information

- Accepted by BBPA for use in stores for malting barley

Efficacy guidance

- Product acts as fumigant by releasing poisonous hydrogen phosphide gas on contact with moisture in the air
- Place plates on the floor or wall of the building or on the surface of the commodity. Exposure time varies depending on temperature and pest. See label

Restrictions

- Magnesium phosphide is subject to the Poisons Rules 1982 and the Poisons Act 1972. See notes in Section 5
- Only to be used by professional operators trained in the use of magnesium phosphide and familiar with the precautionary measures to be observed. See label for full precautions

Environmental safety

- Highly flammable
- Prevent access to buildings under fumigation by livestock, pets and other non-target mammals and birds

- Dangerous to fish or other aquatic life. Do not contaminate surface waters or ditches with chemical or used container
- Keep in original container, tightly closed, in a safe place, under lock and key
- Do not allow plates or their spent residues to come into contact with food other than raw cereal grains
- Remove used plates after treatment. Do not bulk spent plates and residues: spontaneous ignition could result
- Keep livestock out of treated areas

Hazard classification and safety precautions

Hazard H01, H07
Risk phrases R21, R26, R28
Operator protection A, D, H; U01, U05b, U07, U13, U19a, U20a
Environmental protection E02a (4 h min); E02b, E13b, E34
Storage and disposal D01, D02, D05, D07, D09b, D11b
Medical advice M04a

303 malathion

A broad-spectrum contact organophosphorus insecticide and acaricide
IRAC mode of action code: 1B

See also bifenthrin + malathion

Products

Fyfanon 440	Headland	440 g/l	EW	11134

Uses

- Aphids in ***carnations***, ***chrysanthemums***, ***dahlias***, ***gladioli***, ***outdoor cyclamen***, ***protected carnations***, ***protected chrysanthemums***, ***protected cyclamen***, ***protected dahlias***, ***protected gladioli***, ***protected roses***, ***protected tulips***, ***roses***, ***tulips***

Approval information

- Malathion will not be included in Annex I of EC Directive 91/414. To comply with this EC decision approval for all products listed in this profile finally expire on 6 December 2008
- Approval expiry 06 Dec 2008 [1]

Efficacy guidance

- Spray when pest first seen and repeat as necessary, usually at 7-14 d intervals
- Number and timing of sprays vary with crop and pest. See label for details
- Repeat spray routinely for scale insect and whitefly control in glasshouses
- Where aphids resistant to malathion occur control is unlikely to be satisfactory

Restrictions

- Malathion is an anticholinesterase organophosphorus compound. Do not use if under medical advice not to work with such compounds
- Do not use on antirrhinums, crassula, ferns, fuchsias, gerberas, petunias, pileas, sweet peas or zinnias as damage may occur
- Do not apply by tractor mounted/drawn or broadcast air assisted sprayers

Environmental safety

- Dangerous for the environment
- Very toxic to aquatic organisms
- Dangerous to bees. Do not apply to crops in flower or to those in which bees are actively foraging. Do not apply when flowering weeds are present
- Risk to certain non-target insects or other arthropods. For advice on risk management and use in Integrated Pest Management (IPM) see directions for use

Hazard classification and safety precautions

Hazard H11
Risk phrases R50, R53a
Operator protection A, H, M, P; U08, U19a, U20b
Environmental protection E12c, E12e, E15a, E22b, E34, E38
Storage and disposal D05, D09a, D10a
Medical advice M01

304 maleic hydrazide

A pyridazine plant growth regulator suppressing sprout and bud growth

Products

1	Fazor	Dow	60% w/w	SG	05558
2	Regulox K	Bayer Environ.	250 g/l	SL	09937
3	Royal MH 180	Nufarm UK	180 g/l	SL	10982
4	Source II	Chiltern	60% w/w	WB	08314

Uses

- Growth suppression in ***amenity grassland***, ***grass near water*** [2, 3]; ***areas around farm buildings***, ***golf courses***, ***road verges*** [3]; ***carrots*** *(off-label)*, ***parsnips*** *(off-label)* [1]; ***industrial sites***, ***roadside grass***, ***waste ground*** [2]
- Re-growth suppression in ***carrots*** *(off-label)*, ***parsnips*** *(off-label)* [4]
- Sprout suppression in ***bulb onions***, ***garlic*** *(off-label)*, ***shallots*** *(off-label)*, ***ware potatoes*** [1, 4]
- Sucker inhibition in ***amenity vegetation*** [2]
- Volunteer suppression in ***ware potatoes*** [1, 4]

Specific Off-Label Approvals (SOLAs)

- ***carrots*** *20012159* [1], *20064032* [4]
- ***garlic*** *20052437* [1], *20052436* [4]
- ***parsnips*** *20012159* [1], *20064032* [4]
- ***shallots*** *20052437* [1], *20052436* [4]

Approval information

- Maleic hydrazide included in Annex I under EC Directive 91/414
- Approved for use on grass near water [3] . See notes in Section 5 on use of herbicides in or near water
- Approval expiry 31 Dec 2008 [1, 2, 4]
- Approval expiry 31 Aug 2008 [3]

Efficacy guidance

- Apply to grass at any time of yr when growth active, best when growth starting in Apr-May and repeated when growth recommences
- Uniform coverage and dry weather necessary for effective results
- Accurate timing essential for good results on potatoes but rain or irrigation within 24 h may reduce effectiveness on onions and potatoes
- Mow 2-3 d before and 5-10 d after spraying for best results. Need for mowing reduced for up to 6 wk
- When used for suppression of volunteer potatoes treatment may also give some suppression of sprouting in store but separate treatment will be necessary if sprouting occurs

Restrictions

- Maximum number of treatments 2 per yr on amenity grass, land not intended for cropping and land adjacent to aquatic areas; 1 per crop on onions, potatoes and on or around tree trunks
- Do not apply in drought or when crops are suffering from pest, disease or herbicide damage. Do not treat fine turf or grass seeded less than 8 mth previously
- Do not treat potatoes within 3 wk of applying a haulm desiccant or if temperatures above 26 °C
- Consult processor before use on potato crops for processing
- Do not apply by knapsack sprayer [1]

Crop-specific information

- Latest use: 3 wk before haulm destruction for potatoes; before 50% necking for onions
- HI onions 1 wk; potatoes 3 wk
- Apply to onions at 10% necking and not later than 50% necking stage when the tops are still green
- Only treat onions in good condition and properly cured, and do not treat more than 2 wk before maturing. Treated onions may be stored until Mar but must then be removed to avoid browning
- Apply to second early or maincrop potatoes at least 3 wk before haulm destruction
- Only treat potatoes of good keeping quality; not on seed, first earlies or crops grown under polythene
- May be applied to grass along water courses but not to water surface [2, 3]

Environmental safety

- Only apply to grass not to be used for grazing

- Do not use treated water for irrigation purposes within 3 wk of treatment or until concentration in water falls below 0.02 ppm [2, 3]
- Maximum permitted concentration in water 2 ppm
- Do not dump surplus product in water or ditch bottoms
- Avoid drift onto nearby vegetables, flowers or other garden plants

Hazard classification and safety precautions

Hazard H04 [4]

Risk phrases R41 [4]

Operator protection A [1, 4]; C [4]; H [1]; U02a, U05a, U13, U14, U15, U19a, U20a, U22a [4]; U08 [1]; U20b [1, 3]; U20c [2]

Environmental protection E15a [1-4]; E19a [2, 3]; E21 [2, 3] (3 wk)

Storage and disposal D01, D02 [4]; D05 [2, 4]; D07, D10c [2]; D09a [1-4]; D10a [3]; D11a [1]

305 mancozeb

A protective dithiocarbamate fungicide for potatoes and other crops
FRAC mode of action code: M3

See also *benalaxyl + mancozeb*
benthiavalicarb-isopropyl + mancozeb
chlorothalonil + mancozeb
cymoxanil + mancozeb
dimethomorph + mancozeb

Products

1	Dithane 945	Dow	80% w/w	WP	12545
2	Dithane 945	Interfarm	80% w/w	WP	12585
3	Dithane NT Dry Flowable	Interfarm	75% w/w	WG	12565
4	Hotspur	AgriGuard	75% w/w	WG	13449
5	Karamate Dry Flo Newtec	Landseer	75% w/w	WG	12691
6	Manzate 75 WG	Headland	75% w/w	WG	12070
7	Micene 80	Sipcam	80% w/w	WP	09112
8	Micene DF	Sipcam	77% w/w	WG	09957
9	Penncozeb WDG	Nufarm UK	77% w/w	WG	12156
10	Quell Flo	Interfarm	455 g/l	SC	09894
11	SpudGun	AgriGuard	80% w/w	WP	13447

Uses

- Anthracnose in ***courgettes*** *(off-label)*, ***marrows*** *(off-label)*, ***pumpkins*** *(off-label)*, ***squashes*** *(off-label)* [1]
- Black spot in ***roses*** [5]
- Blight in ***potatoes*** [2-4, 6-11]
- Brown rust in ***spring barley***, ***winter barley*** [7-9]; ***spring wheat***, ***winter wheat*** [2, 3, 8-11]
- Celery leaf spot in ***celeriac*** *(off-label)*, ***celery (outdoor)*** *(off-label)* [1, 2]
- Currant leaf spot in ***blackcurrants***, ***gooseberries*** [5]
- Downy mildew in ***bulb onions*** *(off-label)*, ***garlic*** *(off-label)*, ***shallots*** *(off-label)* [2, 3, 9]; ***chicory*** *(off-label)*, ***courgettes*** *(off-label)*, ***marrows*** *(off-label)*, ***pumpkins*** *(off-label)*, ***red beet*** *(off-label)*, ***squashes*** *(off-label)*, ***witloof*** *(off-label)* [1, 2]; ***lettuce***, ***protected lettuce*** [5, 11]; ***poppies for morphine production*** *(off-label)* [3, 9]; ***winter oilseed rape*** [2, 3, 10, 11]
- Fire in ***tulips*** [5]
- Foliar disease control in ***chives*** *(off-label)*, ***frise*** *(off-label)*, ***herbs (see appendix 6)*** *(off-label)*, ***lamb's lettuce*** *(off-label)*, ***parsley*** *(off-label)*, ***protected chives*** *(off-label)*, ***protected herbs (see appendix 6)*** *(off-label)*, ***protected parsley*** *(off-label)*, ***radicchio*** *(off-label)*, ***redcurrants*** *(off-label)*, ***whitecurrants*** *(off-label)*, ***wine grapes*** *(off-label)* [5]; ***durum wheat*** *(off-label)* [3]; ***grass seed crops*** *(off-label)* [3, 7-10]; ***spring rye*** *(off-label)*, ***winter rye*** *(off-label)* [10]
- Fungus diseases in ***wine grapes*** *(off-label)* [9]
- Glume blotch in ***spring wheat***, ***winter wheat*** [8, 9]
- Gummosis in ***courgettes*** *(off-label)*, ***marrows*** *(off-label)*, ***pumpkins*** *(off-label)*, ***squashes*** *(off-label)* [1]
- Net blotch in ***spring barley***, ***winter barley*** [7-9]
- Ray blight in ***chrysanthemums*** [5]

SEE SECTION 3 FOR PRODUCTS ALSO REGISTERED

- Rhynchosporium in ***spring barley*** [8, 9]; ***winter barley*** [2, 8, 9, 11]
- Root malformation disorder in ***red beet*** *(off-label)* [1, 2]
- Rust in ***asparagus*** *(off-label)*, ***chicory*** *(off-label)*, ***witloof*** *(off-label)* [1, 2]; ***carnations***, ***geraniums***, ***roses*** [5]
- Scab in ***apples*** [5, 7-9]; ***pears*** [5]
- Septoria leaf blotch in ***spring wheat***, ***winter wheat*** [2, 3, 7-11]
- Sooty moulds in ***spring barley***, ***winter barley*** [8, 9]; ***spring wheat***, ***winter wheat*** [2, 3, 8-11]
- Stemphylium in ***asparagus*** *(off-label)* [2]
- White mould in ***daffodils*** *(off-label - for galanthamine production)* [1, 9]
- White tip in ***leeks*** *(off-label)* [9]
- Yellow rust in ***spring barley***, ***spring wheat***, ***winter barley*** [8, 9]; ***winter wheat*** [2, 3, 8-11]

Specific Off-Label Approvals (SOLAs)

- ***asparagus*** *20060381* [2], *20060390* [1]
- ***bulb onions*** *20053038* [2], *20053040* [3], *20062543* [9]
- ***celeriac*** *20060384* [2], *20060393* [1]
- ***celery (outdoor)*** *20060383* [2], *20060391* [1]
- ***chicory*** *20060382* [2], *20060392* [1]
- ***chives*** *20061978* [5]
- ***courgettes*** *20060380* [2], *20060394* [1]
- ***daffodils*** *(for galanthamine production) 20060386* [1], *(for galanthamine production) 20051476* [9]
- ***durum wheat*** *20052532* [3]
- ***frise*** *20061978* [5]
- ***garlic*** *20053038* [2], *20053040* [3], *20062543* [9]
- ***grass seed crops*** *20052532* [3], *20052492* [7], *20052487* [8], *20052485* [9], *20052496* [10]
- ***herbs (see appendix 6)*** *20061978* [5]
- ***lamb's lettuce*** *20061978* [5]
- ***leeks*** *20051479* [9]
- ***marrows*** *20060380* [2], *20060394* [1]
- ***parsley*** *20061978* [5]
- ***poppies for morphine production*** *20060458* [3], *20051475* [9]
- ***protected chives*** *20061978* [5]
- ***protected herbs (see appendix 6)*** *20061978* [5]
- ***protected parsley*** *20061978* [5]
- ***pumpkins*** *20060380* [2], *20060394* [1]
- ***radicchio*** *20061978* [5]
- ***red beet*** *20060385* [2], *20060387* [1]
- ***redcurrants*** *20061979* [5]
- ***shallots*** *20053038* [2], *20053040* [3], *20062543* [9]
- ***spring rye*** *20052496* [10]
- ***squashes*** *20060380* [2], *20060394* [1]
- ***whitecurrants*** *20061979* [5]
- ***wine grapes*** *20063434* [5], *20052053* [9]
- ***winter rye*** *20052496* [10]
- ***witloof*** *20060382* [2], *20060392* [1]

Approval information

- Mancozeb included in Annex I under EC Directive 91/414
- Approved for aerial application on potatoes [1, 3, 7-10]. See notes in Section 5
- Accepted by BBPA for use on malting barley

Efficacy guidance

- Mancozeb is a protectant fungicide and will give moderate control, suppression or reduction of the cereal diseases listed if treated before they are established but in many cases mixture with carbendazim is essential to achieve satisfactory results. See labels for details
- May be recommended for suppression or control of mildew in cereals depending on product and tank mix. See label for details

Restrictions

- Maximum number of treatments varies with crop and product used - check labels for details
- Check labels for minimum interval that must elapse between treatments
- On protected lettuce only 2 post-planting applications of mancozeb or of any combination of products containing EBDC fungicide (mancozeb, maneb, thiram, zineb) either as a spray or a dust are permitted within 2 wk of planting out and none thereafter.
- Avoid treating wet cereal crops or those suffering from drought or other stress
- Keep dry formulations away from fire and sparks
- Use dry formulations immediately. Do not store

Crop-specific information

- Latest use: before early milk stage (GS 73) for cereals; before 6 true leaf stage and before 31 Dec for winter oilseed rape.
- HI potatoes 7 d; outdoor lettuce 14 d; protected lettuce 21 d; apples, blackcurrants 28 d
- Apply to potatoes before haulm meets across rows (usually mid-Jun) or at earlier blight warning, and repeat every 7-14 d depending on conditions and product used (see label)
- May be used on potatoes up to desiccation of haulm
- On oilseed rape apply as soon as disease develops between cotyledon and 5-leaf stage (GS 1,0-1,5)
- Apply to cereals from 4-leaf stage to before early milk stage (GS 71). Recommendations vary, see labels for details
- Treat winter oilseed rape before 6 true leaf stage (GS 1,6) and before 31 Dec

Environmental safety

- Dangerous for the environment
- Very toxic to aquatic organisms
- Harmful to fish or other aquatic life. Do not contaminate surface waters or ditches with chemical or used container
- Do not empty into drains

Hazard classification and safety precautions

Hazard H03 [9]; H04 [1-8, 10, 11]; H11 [1-6, 9-11]

Risk phrases R36 [6-8]; R37 [1-3, 5, 7-11]; R38 [7, 8]; R42 [9]; R43 [1-5, 9-11]; R50, R53a [1-6, 9-11]

Operator protection A [1-11]; C [4, 6-8]; D [1-3, 5, 7, 8, 11]; U05a [1-11]; U08 [6-8, 10]; U11 [6]; U13 [7, 8]; U14 [1-3, 5, 10]; U19a [6-9]; U20a [7]; U20b [1-6, 8, 10, 11]

Environmental protection E13c [7, 8]; E15a [1-6, 9-11]; E19b [9]; E34 [6, 10]; E38 [1-6, 10, 11]

Storage and disposal D01 [1-3, 5-10]; D02, D09a [1-11]; D05 [3-8, 10, 11]; D07, D10a [6]; D11a [1-5, 7-11]; D12a [1-6, 10, 11]

306 mancozeb + metalaxyl-M

A systemic and protectant fungicide mixture

FRAC mode of action code: M3 + 4

Products

Fubol Gold WG	Syngenta	64:4% w/w	WG	10184

Uses

- Blight in ***potatoes***
- Downy mildew in ***salad onions*** *(off-label)*
- Fungus diseases in ***chives*** *(off-label)*, ***herbs (see appendix 6)*** *(off-label)*, ***lettuce*** *(off-label)*, ***parsley*** *(off-label)*, ***protected chives*** *(off-label)*, ***protected herbs (see appendix 6)*** *(off-label)*, ***protected lettuce*** *(off-label)*, ***protected parsley*** *(off-label)*
- Phytophthora fruit rot in ***apples*** *(off-label - applied to orchard floor)*
- White blister in ***cabbages*** *(off-label)*, ***chinese cabbage*** *(off-label)*, ***choi sum*** *(off-label)*, ***collards*** *(off-label)*, ***kale*** *(off-label)*, ***pak choi*** *(off-label)*, ***protected radishes*** *(off-label)*, ***radishes*** *(off-label)*

Specific Off-Label Approvals (SOLAs)

- ***apples*** *(applied to orchard floor) 20011610*
- ***cabbages*** *20011610*
- ***chinese cabbage*** *20063643*
- ***chives*** *20032142*

- ***choi sum*** *20063643*
- ***collards*** *20063643*
- ***herbs (see appendix 6)*** *20032142*
- ***kale*** *20063643*
- ***lettuce*** *20032142*
- ***pak choi*** *20063643*
- ***parsley*** *20032142*
- ***protected chives*** *20032142*
- ***protected herbs (see appendix 6)*** *20032142*
- ***protected lettuce*** *20032142*
- ***protected parsley*** *20032142*
- ***protected radishes*** *20011610*
- ***radishes*** *20011610*
- ***salad onions*** *20032324*

Approval information

- Mancozeb and metalaxyl-M included in Annex I under EC Directive 91/414

Efficacy guidance

- Commence potato blight programme before risk of infection occurs as crops begin to meet along the rows and repeat every 10-14 d according to blight risk. Do not exceed a 14 d interval between sprays
- If infection risk conditions occur earlier than the above growth stage commence spraying potatoes immediately
- Complete the potato blight programme using a protectant fungicide starting no later than 10 d after the last phenylamide spray. At least 2 such sprays should be applied
- To minimise the likelihood of development of resistance these products should be used in a planned Resistance Management strategy. See Section 5 for more information

Restrictions

- Maximum number of treatments 5 per crop

Crop-specific information

- Latest use: before end of active potato haulm growth or before end Aug, whichever is earlier
- HI 7 d for potatoes
- After treating early potatoes destroy and remove any remaining haulm after harvest to minimise blight pressure on neighbouring maincrop potatoes

Environmental safety

- Dangerous for the environment
- Very toxic to aquatic organisms
- Do not harvest crops for human consumption for at least 7 d after final application

Hazard classification and safety precautions

Hazard H03, H11
Risk phrases R37, R43, R50, R53a, R63
Operator protection A; U05a, U08, U20b
Environmental protection E15a, E34, E38
Consumer protection C02a (7 d)
Storage and disposal D01, D02, D05, D09a, D11a, D12a

307 mancozeb + propamocarb hydrochloride

A systemic and contact protectant fungicide for potato blight control
FRAC mode of action code: M3 + 28

Products

Tattoo	Bayer CropScience	301.6:248 g/l	SC	07293

Uses

- Blight in ***potatoes***

Approval information

- Mancozeb and propamocarb hydrochloride included in Annex I under EC Directive 91/414

Efficacy guidance

- Commence treatment as soon as there is risk of blight infection
- In the absence of a warning treatment should start before the crop meets within the row
- Repeat treatments every 10-14 d depending on degree of infection risk
- Irrigated crops should be regarded as at high risk and treated every 10 d
- Do not spray when rainfall is imminent and apply only to dry foliage
- To complete the spray programme after the end of Aug, make subsequent treatments up to haulm destruction with a protectant fungicide, preferably fentin based
- To minimise the likelihood of development of resistance these products should be used in a planned Resistance Management strategy. See Section 5 for more information

Restrictions

- Maximum number of treatments 5 per crop
- Do not use once blight has become readily visible

Crop-specific information

- HI potatoes 14 d

Environmental safety

- Dangerous for the environment
- Very toxic to aquatic organisms
- Harmful to fish or other aquatic life. Do not contaminate surface waters or ditches with chemical or used container

Hazard classification and safety precautions

Hazard H04, H11
Risk phrases R43, R50, R53a
Operator protection A, H; U05a, U08, U14, U19a, U20a
Environmental protection E13c, E38
Consumer protection C02a (14 d)
Storage and disposal D01, D02, D05, D09a, D10b, D12a

308 mancozeb + zoxamide

A protectant fungicide mixture for potatoes
FRAC mode of action code: M3 + 22

Products

1	Electis 75 WG	Dow	66.7:8.3% w/w	WG	11013
2	Roxam 75 WG	Dow	66.7:8.3% w/w	WG	11017

Uses

- Blight in ***potatoes***

Approval information

- Mancozeb and zoxamide included in Annex I under EC Directive 91/414

Efficacy guidance

- Apply as protectant spray on potatoes immediately risk of blight in district or as crops begin to meet along the rows and repeat every 7-14 d according to blight risk
- Do not use if potato blight present in crop. Products are not curative
- Spray irrigated potato crops as soon as possible after irrigation once the crop leaves are dry

Restrictions

- Maximum number of treatments 10 per crop for potatoes

Crop-specific information

- HI 7 d for potatoes

Environmental safety

- Dangerous for the environment
- Very toxic to aquatic organisms

- LERAP Category B
- Keep away from fire and sparks

Hazard classification and safety precautions

Hazard H04, H11
Risk phrases R37, R43, R50, R53a
Operator protection A, H; U05a, U14, U20b
Environmental protection E15a, E16a, E16b, E38
Storage and disposal D01, D02, D05, D09a, D11a, D12a

309 mandipropamid

A mandelamide fungicide for the control of potato blight
FRAC mode of action code: 40

Products

Revus	Syngenta	250 g/l	SC	13484

Uses

- Blight in ***potatoes***

Efficacy guidance

- Mandipropamid acts preventatively by preventing spore germination and inhibiting mycelial growth during incubation. Apply immediately after blight warning or as soon as local conditions favour disease development but before blight enters the crop
- Spray at 7-10 d intervals reducing the interval as blight risk increases
- Spray programme should include a complete haulm desiccant to prevent tuber infection at harvest
- Eliminate other potential infection sources
- To minimise the likelihood of development of resistance this product should be used in a planned Resistance Management strategy. See Section 5 for more information
- See label for details of tank mixtures that may be used as part of a resistance management strategy

Restrictions

- Maximum number of treatments 4 per crop. Do not apply more than 3 treatments of this, or any other fungicide in the same resistance category, consecutively

Crop-specific information

- HI 3 d for potatoes

Hazard classification and safety precautions

Risk phrases R52, R53a
Operator protection A, H; U05a, U20b
Environmental protection E15b, E34, E38
Storage and disposal D01, D02, D05, D09a, D10c, D12a
Medical advice M03, M05a

310 MCPA

A translocated phenoxycarboxylic acid herbicide for cereals and grassland
HRAC mode of action code: O

See also *2,4-D + dichlorprop-P + MCPA + mecoprop-P*
2,4-D + MCPA
2,4-DB + linuron + MCPA
2,4-DB + MCPA
bentazone + MCPA + MCPB
clopyralid + 2,4-D + MCPA
clopyralid + diflufenican + MCPA
clopyralid + fluroxypyr + MCPA
dicamba + dichlorprop-P + ferrous sulphate + MCPA
dicamba + dichlorprop-P + MCPA
dicamba + MCPA + mecoprop-P
dichlorprop-P + ferrous sulphate + MCPA
dichlorprop-P + MCPA
dichlorprop-P + MCPA + mecoprop-P
ferrous sulphate + MCPA + mecoprop-P

Products

1	Agritox 50	Nufarm UK	500 g/l	SL	07400
2	Agroxone	Headland	500 g/l	SL	09947
3	Campbell's MCPA 50	United Phosphorus	500 g/l	SL	00381
4	Headland Spear	Headland	500 g/l	SL	07115
5	HY-MCPA	Agrichem	500 g/l	SL	06293
6	Tasker 75	Headland	750 g/l	SL	10544

Uses

- Annual dicotyledons in ***amenity grassland, managed amenity turf*** [1, 2, 4]; ***grass seed crops*** [1, 2, 6]; ***land not intended to bear vegetation*** [1, 2]; ***linseed*** [1, 2, 5, 6]; ***permanent grassland, spring barley, spring oats, spring wheat, winter barley, winter oats, winter wheat*** [1-6]; ***rotational grass*** [2, 5, 6]; ***spring rye*** [1]; ***undersown barley*** *(red clover or grass)*, ***undersown wheat*** *(red clover or grass)* [2, 6]; ***undersown spring cereals, undersown winter cereals*** [5]; ***winter rye*** [1, 3, 4]
- Buttercups in ***amenity grassland, managed amenity turf*** [4, 5]; ***permanent grassland*** [3-5]
- Charlock in ***grass seed crops*** [6]; ***linseed*** [5, 6]; ***permanent grassland*** [1, 2, 6]; ***rotational grass*** [2, 5, 6]; ***spring barley, spring oats, spring wheat, winter barley, winter oats, winter wheat*** [1-6]; ***spring rye*** [1]; ***undersown barley*** *(red clover or grass)*, ***undersown wheat*** *(red clover or grass)* [2, 6]; ***winter rye*** [1, 3, 4]
- Dandelions in ***permanent grassland*** [3, 5]
- Docks in ***permanent grassland*** [3-5]
- Fat hen in ***amenity grassland, hard surfaces, managed amenity turf, natural surfaces not intended to bear vegetation, permeable surfaces overlying soil*** [5]; ***grass seed crops*** [6]; ***linseed*** [5, 6]; ***permanent grassland*** [1, 2, 6]; ***rotational grass*** [2, 5, 6]; ***spring barley, spring oats, spring wheat, winter barley, winter oats, winter wheat*** [1-6]; ***spring rye*** [1]; ***undersown barley*** *(red clover or grass)*, ***undersown wheat*** *(red clover or grass)* [2, 6]; ***winter rye*** [1, 3, 4]
- Hemp-nettle in ***grass seed crops*** [6]; ***linseed*** [5, 6]; ***permanent grassland*** [1, 2, 6]; ***rotational grass, undersown barley*** *(red clover or grass)*, ***undersown wheat*** *(red clover or grass)* [2, 6]; ***spring barley, spring oats, spring wheat, winter barley, winter oats, winter wheat*** [1-6]; ***spring rye*** [1]; ***winter rye*** [1, 3, 4]
- Perennial dicotyledons in ***amenity grassland, managed amenity turf*** [1, 2, 4]; ***grass seed crops, linseed*** [6]; ***land not intended to bear vegetation*** [1, 2]; ***permanent grassland, spring barley, spring oats, spring wheat, winter barley, winter oats, winter wheat*** [1-6]; ***rotational grass*** [5, 6]; ***spring rye*** [1]; ***undersown barley*** *(red clover or grass)*, ***undersown wheat*** *(red clover or grass)* [2, 6]; ***winter rye*** [1, 3, 4]
- Plantains in ***amenity grassland, grass seed crops, managed amenity turf, permanent grassland*** [5]
- Ragwort in ***permanent grassland*** [4]
- Rushes in ***permanent grassland*** [4]

SEE SECTION 3 FOR PRODUCTS ALSO REGISTERED

- Stinging nettle in ***hard surfaces, natural surfaces not intended to bear vegetation, permeable surfaces overlying soil*** [5]; ***industrial sites, land not intended to bear vegetation*** [4]
- Thistles in ***hard surfaces, natural surfaces not intended to bear vegetation, permeable surfaces overlying soil*** [5]; ***industrial sites, land not intended to bear vegetation*** [4]; ***permanent grassland*** [3-5]
- Wild radish in ***grass seed crops*** [6]; ***linseed*** [5, 6]; ***permanent grassland*** [1, 2, 6]; ***rotational grass, undersown barley*** *(red clover or grass)*, ***undersown wheat*** *(red clover or grass)* [2, 6]; ***spring barley, spring oats, spring wheat, winter barley, winter oats, winter wheat*** [1-6]; ***spring rye*** [1]; ***winter rye*** [1, 3, 4]

Approval information

- MCPA included in Annex I under EC Directive 91/414
- Accepted by BBPA for use on malting barley

Efficacy guidance

- Best results achieved by application to weeds in seedling to young plant stage under good growing conditions when crop growing actively
- Spray perennial weeds in grassland before flowering. Most susceptible growth stage varies between species. See label for details
- Do not spray during cold weather, drought, if rain or frost expected or if crop wet

Restrictions

- Maximum number of treatments normally 1 per crop or yr except grass (2 per yr) for some products. See label
- Do not treat grass within 3 mth of germination and preferably not in the first yr of a direct sown ley or after reseeding
- Do not use on cereals before undersowing
- Do not roll, harrow or graze for a few days before or after spraying; see label
- Do not use on grassland where clovers are an important part of the sward
- Do not use on any crop suffering from stress or herbicide damage
- Avoid spray drift onto nearby susceptible crops

Crop-specific information

- Latest use: before 1st node detectable (GS 31) for cereals; 4-6 wk before heading for grass seed crops; before crop 15-25 cm high for linseed
- Apply to winter cereals in spring from fully tillered, leaf sheath erect stage to before first node detectable (GS 31)
- Apply to spring barley and wheat from 5-leaves unfolded (GS 15), to oats from 1-leaf unfolded (GS 11) to before first node detectable (GS 31)
- Apply to cereals undersown with grass after grass has 2-3 leaves unfolded
- Recommendations for crops undersown with legumes vary. Red clover may withstand low doses after 2-trifoliate leaf stage, especially if shielded by taller weeds but white clover is more sensitive. See label for details
- Apply to grass seed crops from 2-3 leaf stage to 5 wk before head emergence
- Temporary wilting may occur on linseed but without long term effects

Following crops guidance

- Do not direct drill brassicas or legumes within 6 wk of spraying grassland

Environmental safety

- Harmful to aquatic organisms
- MCPA is active at low concentrations. Take extreme care to avoid drift onto neighbouring crops, especially beet crops, brassicas, most market garden crops including lettuce and tomatoes under glass, pears and vines
- Keep livestock out of treated areas for at least 2 wk and until foliage of poisonous weeds such as ragwort has died and become unpalatable

Hazard classification and safety precautions

Hazard H03

Risk phrases R20, R21 [2, 3]; R22a [2-6]; R41 [1, 3-6]; R52 [4, 5]; R53a [5]

Operator protection A, C [1-5]; U05a, U08, U11 [1-6]; U15 [4]; U20a [3]; U20b [1, 2, 4-6]

Environmental protection E06a [5] (2 wk); E07a [1-4, 6]; E15a, E34 [1-6]; E38 [4]

Storage and disposal D01, D02, D05, D09a [1-6]; D10a [1, 2, 5, 6]; D10b [4]; D10c [3]

Medical advice M03 [1-6]; M05a [3, 5]

311 MCPA + MCPB

A translocated herbicide for undersown cereals, grassland and various legumes
HRAC mode of action code: O + O

Products

1	Bellmac Plus	United Phosphorus	38:262 g/l	SL	07521
2	Impetus	Headland	25:275 g/l	SL	11021
3	Tropotox Plus	Nufarm UK	37.5:262.5 g/l	SL	11142

Uses

- Annual dicotyledons in ***all cereals***, ***clover seed crops***, ***direct-sown seedling clovers***, ***dredge corn containing peas*** [3]; ***combining peas***, ***red clover***, ***spring barley***, ***spring oats***, ***spring wheat***, ***vining peas***, ***white clover***, ***winter barley***, ***winter oats***, ***winter wheat*** [2]; ***permanent grassland*** [2, 3]; ***rotational grass***, ***undersown spring cereals***, ***undersown winter cereals*** [1-3]; ***sainfoin*** [1, 3]
- Perennial dicotyledons in ***all cereals***, ***clover seed crops***, ***direct-sown seedling clovers***, ***dredge corn containing peas*** [3]; ***combining peas***, ***red clover***, ***spring barley***, ***spring oats***, ***spring wheat***, ***vining peas***, ***white clover***, ***winter barley***, ***winter oats***, ***winter wheat*** [2]; ***permanent grassland*** [2, 3]; ***rotational grass***, ***undersown spring cereals***, ***undersown winter cereals*** [1-3]; ***sainfoin*** [1, 3]

Approval information

- MCPA and MCPB included in Annex I under EC Directive 91/414
- Accepted by BBPA for use on malting barley

Efficacy guidance

- Best results achieved by application to weeds in seedling to young plant stage under good growing conditions when crop growing actively. Spray perennials when adequate leaf surface before flowering. Retreatment often needed in following year
- Spray leys and sainfoin before crop provides cover for weeds
- Rain, cold or drought may reduce effectiveness

Restrictions

- Maximum number of treatments 1 per crop or yr
- Do not spray clovers for seed
- Do not spray peas [1]
- Do not roll or harrow for a few days before or after spraying

Crop-specific information

- Latest use: before first node detectable (GS 31) for cereals; before flower buds visible for peas
- Apply to cereals from 2-expanded leaf stage to before jointing (GS 12-30) and, where undersown, after 1-trifoliate leaf stage of clover
- Apply to direct-sown seedling clover after 1-trifoliate leaf stage
- Apply to mature white clover for fodder at any stage. Do not spray red clover after flower stalk has begun to form
- Apply to sainfoin after first trifoliate leaf stage

Environmental safety

- Harmful to aquatic organisms
- Harmful to fish or other aquatic life. Do not contaminate surface waters or ditches with chemical or used container
- Keep livestock out of treated areas until foliage of any poisonous weeds such as ragwort has died and become unpalatable
- Take extreme care to avoid drift onto neighbouring crops, especially beet crops, brassicas, most market garden crops including lettuce and tomatoes under glass, pears and vines

Hazard classification and safety precautions

Hazard H03
Risk phrases R20, R21 [2]; R22a [1-3]; R38, R53a [1]; R41, R52 [1, 2]
Operator protection A, C [2]; U05a [2, 3]; U08, U20b [1-3]; U11, U14, U15 [1, 2]
Environmental protection E07a, E34 [1-3]; E13c [1]; E15a [2, 3]
Storage and disposal D01, D10b [2, 3]; D02 [2]; D05, D09a [1-3]; D10c [1]
Medical advice M03 [2]; M05a [1-3]

SEE SECTION 3 FOR PRODUCTS ALSO REGISTERED

312 MCPA + mecoprop-P

A translocated selective herbicide for amenity grass
HRAC mode of action code: O + O

Products

1	Cleanrun Pro	Scotts	0.49:0.29% w/w	GR	12083
2	Greenmaster Extra	Scotts	0.49:0.29 % w/w	GR	11563

Uses

- Annual dicotyledons in ***managed amenity turf***
- Perennial dicotyledons in ***managed amenity turf***

Approval information

- MCPA and mecoprop-P included in Annex I under EC Directive 91/414

Efficacy guidance

- Apply from Apr to Sep, when weeds growing actively and have large leaf area available for chemical absorption

Restrictions

- Maximum total dose 105 g/sq m per yr. The total amount of mecoprop-P applied in a single yr must not exceed the maximum total dose approved for any single product for use on turf
- Avoid contact with cultivated plants
- Do not use first 4 mowings as compost or mulch unless composted for 6 mth
- Do not treat newly sown or turfed areas for at least 6 mth
- Do not reseed bare patches for 8 wk after treatment
- Do not apply when heavy rain expected or during prolonged drought. Irrigate after 1-2 d unless rain has fallen
- Do not mow within 2-3 d of treatment
- Treat areas planted with bulbs only after the foliage has died down
- Avoid walking on treated areas until it has rained or irrigation has been applied

Crop-specific information

- Granules contain NPK fertilizer to encourage grass growth

Environmental safety

- Take extreme care to avoid drift onto neighbouring crops, especially beet crops, brassicas, most market garden crops including lettuce and tomatoes under glass, pears and vines
- Harmful to fish or other aquatic life. Do not contaminate surface waters or ditches with chemical or used container
- Keep livestock out of treated areas for at least 2 wk and until foliage of any poisonous weeds such as ragwort has died and become unpalatable

Hazard classification and safety precautions

Operator protection A, C, H, M; U20b
Environmental protection E07a, E13c, E19b
Storage and disposal D01, D09a, D12a
Medical advice M05a

313 MCPB

A translocated phenoxycarboxylic acid herbicide
HRAC mode of action code: O

See also *bentazone + MCPA + MCPB*
MCPA + MCPB

Products

1	Bellmac Straight	United Phosphorus	400 g/l	SL	07522
2	Butoxone	Headland	400 g/l	SL	10501
3	Tropotox	Nufarm UK	400 g/l	SL	11141

Uses

- Annual dicotyledons in ***blackberries***, ***combining peas***, ***gooseberries***, ***loganberries***, ***permanent grassland***, ***raspberries***, ***spring barley***, ***spring oats***, ***spring wheat***, ***vining peas***, ***white clover*** *(seed crops)*, ***winter barley, winter oats, winter wheat*** [2]; ***blackcurrants*** [1-3]; ***clover seed crops, peas*** [1, 3]; ***rotational grass, undersown spring cereals, undersown winter cereals*** [1, 2]

- Perennial dicotyledons in ***blackberries, combining peas, gooseberries, loganberries, permanent grassland, raspberries, spring barley, spring oats, spring wheat, vining peas, white clover*** *(seed crops)*, ***winter barley, winter oats, winter wheat*** [2]; ***blackcurrants*** [1-3]; ***clover seed crops, peas*** [1, 3]; ***rotational grass, undersown spring cereals, undersown winter cereals*** [1, 2]

Approval information

- MCPB included in Annex I under EC Directive 91/414
- Accepted by BBPA for use on malting barley

Efficacy guidance

- Best results achieved by spraying young seedling weeds in good growing conditions
- Best results on perennials by spraying before flowering
- Effectiveness may be reduced by rain within 12 h, by very cold or dry conditions

Restrictions

- Maximum number of treatments 1 per crop or yr. One label allows two treatments on blackcurrants [3]
- Do not roll or harrow for 7-10 d before or after treatment (check label)

Crop-specific information

- Latest use: first node detectable stage (GS 31) for cereals; before flower buds appear in terminal leaf (GS 201) for peas; before flower buds form for clover; before weeds damaged by frost for cane and bush fruit
- Apply to undersown cereals from 2-leaves unfolded to first node detectable (GS 12-31), and after first trifoliate leaf stage of clover
- Red clover seedlings may be temporarily damaged but later growth is normal
- Apply to white clover seed crops in Mar to early Apr, not after mid-May, and allow 3 wk before cutting and closing up for seed
- Apply to peas from 3-6 leaf stage but before flower bud detectable (GS 103-201). Consult PGRO (see Appendix 2) or label for information on susceptibility of cultivars. Do not treat peas grown for seed [2]
- Do not use on leguminous crops not mentioned on the label
- Apply to cane and bush fruit after harvest and after shoot growth ceased but before weeds are damaged by frost, usually in late Aug or Sep; direct spray onto weeds as far as possible

Environmental safety

- Harmful to aquatic organisms
- Harmful to fish or other aquatic life. Do not contaminate surface waters or ditches with chemical or used container
- Keep livestock out of treated areas until foliage of any poisonous weeds such as ragwort has died and become unpalatable
- Take extreme care to avoid drift onto neighbouring sensitive crops

Hazard classification and safety precautions

Hazard H03

Risk phrases R22a [1-3]; R38, R41 [1, 2]; R52, R53a [1]

Operator protection A, C; U05a, U08, U20b [1-3]; U11, U14, U15 [1, 2]; U19a [3]

Environmental protection E07a, E34 [1-3]; E13c [1, 2]; E15a [3]

Storage and disposal D01, D02, D09a [1-3]; D05 [1, 2]; D10b [2, 3]; D10c [1]

Medical advice M03, M05a

314 mecoprop-P

A translocated phenoxycarboxylic acid herbicide for cereals and grassland
HRAC mode of action code: O

See also *2,4-D + dichlorprop-P + MCPA + mecoprop-P*
2,4-D + mecoprop-P
bromoxynil + ioxynil + mecoprop-P
carfentrazone-ethyl + mecoprop-P
dicamba + MCPA + mecoprop-P
dicamba + mecoprop-P
dichlorprop-P + MCPA + mecoprop-P
ferrous sulphate + MCPA + mecoprop-P
fluroxypyr + mecoprop-P
MCPA + mecoprop-P

Products

1	Clenecorn Super	Nufarm UK	600 g/l	SL	09818
2	Clovotox	Bayer Environ.	142.5 g/l	SL	09928
3	Compitox Plus	Nufarm UK	600 g/l	SL	10077
4	Duplosan KV	Nufarm UK	600 g/l	SL	12073
5	Isomec	Nufarm UK	600 g/l	SL	11156
6	Landgold Mecoprop-P 600	Landgold	600 g/l	SL	09014
7	Landgold Mecoprop-P 600	Teliton	600 g/l	SL	12122
8	Optica	Headland	600 g/l	SL	09963

Uses

- Annual dicotyledons in ***durum wheat*** *(off-label)* [8]; ***grass seed crops*** [1, 3-5, 8]; ***managed amenity turf*** [2, 3, 5, 8]; ***permanent grassland***, ***rotational grass***, ***spring barley***, ***spring oats***, ***spring wheat***, ***winter barley***, ***winter oats***, ***winter wheat*** [1, 3-8]; ***undersown spring cereals***, ***undersown winter cereals*** [1]
- Chickweed in ***durum wheat*** *(off-label)* [8]; ***grass seed crops*** [1, 3-5, 8]; ***managed amenity turf*** [2, 3, 5, 8]; ***permanent grassland***, ***rotational grass***, ***spring barley***, ***spring oats***, ***spring wheat***, ***winter barley***, ***winter oats***, ***winter wheat*** [1, 3-8]; ***undersown spring cereals***, ***undersown winter cereals*** [1]
- Cleavers in ***durum wheat*** *(off-label)* [8]; ***grass seed crops*** [1, 3-5, 8]; ***managed amenity turf*** [3, 5, 8]; ***permanent grassland***, ***rotational grass***, ***spring barley***, ***spring oats***, ***spring wheat***, ***winter barley***, ***winter oats***, ***winter wheat*** [1, 3-8]; ***undersown spring cereals***, ***undersown winter cereals*** [1]
- Clovers in ***managed amenity turf*** [2]
- Perennial dicotyledons in ***grass seed crops*** [3-5]; ***managed amenity turf*** [2, 3, 5]; ***permanent grassland***, ***rotational grass***, ***spring barley***, ***spring oats***, ***spring wheat***, ***winter barley***, ***winter oats***, ***winter wheat*** [3-7]

Specific Off-Label Approvals (SOLAs)

- ***durum wheat*** *20070455* [8]

Approval information

- Mecoprop-P included in Annex I under EC Directive 91/414
- Accepted by BBPA for use on malting barley

Efficacy guidance

- Best results achieved by application to seedling weeds which have not been frost hardened, when soil warm and moist and expected to remain so for several days

Restrictions

- Maximum number of treatments normally 1 per crop for spring cereals and 1 per yr for newly sown grass; 2 per crop or yr for winter cereals and grass crops. Check labels for details
- The total amount of mecoprop-P applied in a single yr must not exceed the maximum total dose approved for any single product for the crop/situation
- Do not spray cereals undersown with clovers or legumes or to be undersown with legumes or grasses
- Do not spray grass seed crops within 5 wk of seed head emergence
- Do not spray crops suffering from herbicide damage or physical stress

- Do not spray during cold weather, periods of drought, if rain or frost expected or if crop wet
- Do not roll or harrow for 7 d before or after treatment

Crop-specific information

- Latest use: generally before 1st node detectable (GS 31) for spring cereals and before 3rd node detectable (GS 33) for winter cereals, but individual labels vary; 5 wk before emergence of seed head for grass seed crops
- Spray winter cereals from 1 leaf stage in autumn up to and including first node detectable in spring (GS 10-31) or up to second node detectable (GS 32) if necessary. Apply to spring cereals from first fully expanded leaf stage (GS 11) but before first node detectable (GS 31)
- Spray cereals undersown with grass after grass starts to tiller
- Spray newly sown grass leys when grasses have at least 3 fully expanded leaves and have begun to tiller. Any clovers will be damaged

Environmental safety

- Harmful to aquatic organisms
- Harmful to fish or other aquatic life. Do not contaminate surface waters or ditches with chemical or used container
- Keep livestock out of treated areas for at least 2 wk and until foliage of any poisonous weeds, such as ragwort, has died and become unpalatable
- Take extreme care to avoid drift onto neighbouring crops, especially beet crops, brassicas, most market garden crops including lettuce and tomatoes under glass, pears and vines

Hazard classification and safety precautions

Hazard H03 [1, 3-8]; H04 [2]

Risk phrases R21 [6, 7]; R22a [1, 3-8]; R36 [2, 6, 7]; R38 [4, 6-8]; R41, R52 [1, 3-5, 8]; R43 [1, 3, 5]; R53a [1, 3-5]

Operator protection A, C [1-8]; B [1]; H [2, 4, 8]; M [1, 4, 8]; U05a, U20b [1-8]; U08 [1, 2, 6-8]; U09a [3-5]; U11 [1, 3-5]; U14 [1, 3, 5]; U15 [8]

Environmental protection E07a [1-8]; E13c [1-3, 5-8]; E15a [4]; E34 [1, 3-8]; E38 [8]

Storage and disposal D01, D02, D05 [1-8]; D08 [1, 3-5]; D09a [1-7]; D10a [6, 7]; D10b [1, 3, 5, 8]; D10c [2, 4]

Medical advice M03 [1, 3-8]; M05a [4]

315 mepanipyrim

An anilinopyrimidine fungicide for use in horticulture

FRAC mode of action code: 9

Products

Frupica SC	Certis	450 g/l	SC	12067

Uses

- Botrytis in ***courgettes*** *(off-label)*, ***protected strawberries***, ***strawberries***

Specific Off-Label Approvals (SOLAs)

- ***courgettes*** *20051383*

Approval information

- Mepanipyrim included in Annex I under Directive 91/414

Efficacy guidance

- Product is protectant and should be applied as a preventative spray when conditions favourable for Botrytis development occur
- To maintain Botrytis control use as part of a programme with other fungicides that control the disease
- To minimise the possibility of development of resistance adopt resistance management procedures by using products from different chemical groups as part of a mixed spray programme

Restrictions

- Maximum number of treatments 2 per crop (including other anilinopyrimidine products)
- Consult processor before use on crops for processing
- Use spray mixture immediately after preparation

Crop-specific information

- HI 3 d

Environmental safety

- Dangerous for the environment
- Very toxic to aquatic organisms
- LERAP Category B

Hazard classification and safety precautions

Hazard H11
Risk phrases R50, R53a
Operator protection A, H; U20c
Environmental protection E16a, E16b, E34
Storage and disposal D01, D02, D05, D10c, D11a, D12b

316 mepiquat chloride

A quaternary ammonium plant growth regulator available only in mixtures

See also *2-chloroethylphosphonic acid + mepiquat chloride*
chlormequat + 2-chloroethylphosphonic acid + mepiquat chloride
chlormequat + mepiquat chloride

317 mepiquat chloride + prohexadione-calcium

A growth regulator mixture for cereals

Products

Canopy	BASF	500:50 g/l	SE	13181

Uses

- Increasing yield in ***durum wheat*** *(off-label)*, ***winter barley***, ***winter wheat***
- Lodging control in ***durum wheat*** *(off-label)*, ***winter barley***, ***winter wheat***

Specific Off-Label Approvals (SOLAs)

- ***durum wheat*** *20072403*

Approval information

- Prohexadione-calcium included in Annex I under EC Directive 91/414

Efficacy guidance

- Best results obtained from treatments applied to healthy crops from the beginning of stem extension

Restrictions

- Maximum totoal dose equivalent to one full dose treatment on all crops
- Do not apply to any crop suffering from physical stress caused by waterlogging, drought or other conditions
- Do not treat on soils with a substantial moisture deficit
- Consult grain merchant or processor before use on crops for bread making or brewing. Effects on these processes have not been tested

Crop-specific information

- Latest use: before flag leaf fully emerged on wheat and barley

Following crops guidance

- Any crop may follow a normally harvested treated crop. Ploughing is not essential.

Environmental safety

- Harmful to aquatic organisms
- Avoid spray drift onto neighbouring crops

Hazard classification and safety precautions

Hazard H03
Risk phrases R22a, R52, R53a
Operator protection A; U05a, U20b
Environmental protection E15a, E34, E38
Storage and disposal D01, D02, D05, D09a, D10c
Medical advice M05a

318 mesosulfuron-methyl

A sulfonyl urea herbicide for cereals available only in mixtures
HRAC mode of action code: B

See also *diflufenican + iodosulfuron-methyl-sodium + mesosulfuron-methyl*
iodosulfuron-methyl-sodium + mesosulfuron-methyl

319 mesotrione

A foliar applied triketone herbicide for maize
HRAC mode of action code: F2

Products

1	Callisto	Syngenta	100 g/l	SC	12323
2	Greencrop Goldcob	Greencrop	100 g/l	SC	12863

Uses

- Annual dicotyledons in ***forage maize*** [1, 2]; ***grain maize*** *(off-label)*, ***sweetcorn*** *(off-label)* [1]
- Volunteer oilseed rape in ***forage maize*** [1, 2]; ***grain maize*** *(off-label)*, ***sweetcorn*** *(off-label)* [1]

Specific Off-Label Approvals (SOLAs)

- ***grain maize*** *20064210 expires Feb 2008* [1]
- ***sweetcorn*** *20051893 expires Feb 2008* [1]

Approval information

- Mesotrione included in Annex I under EC Directive 91/414
- Approval expiry 17 Feb 2008 [2]

Efficacy guidance

- Best results obtained from treatment of young actively growing weed seedlings in the presence of adequate soil moisture
- Treatment in poor growing conditions or in dry soil may give less reliable control
- Activity is mostly by foliar uptake with some soil uptake
- To minimise the possible development of resistance where continuous maize is grown the product should not be used for more than two consecutive seasons

Restrictions

- Maximum number of treatments 1 per crop of forage maize
- Do not use on seed crops or on sweetcorn varieties
- Do not spray when crop foliage wet or when excessive rainfall is expected to follow application
- Do not treat crops suffering from stress from cold or drought conditions, or when wide temperature fluctuations are anticipated

Crop-specific information

- Latest use: 8 leaves unfolded stage (GS 18) for forage maize
- Treatment under adverse conditions may cause mild to moderate chlorosis. The effect is transient and does not affect yield

Following crops guidance

- Winter wheat, durum wheat, winter barley or ryegrass may follow a normally harvested treated crop of maize. Oilseed rape may be sown provided it is preceded by deep ploughing to more than 15 cm
- In the spring following application only forage maize, ryegrass, spring wheat or spring barley may be sown
- In the event of crop failure maize may be re-seeded immediately. Some slight crop effects may be seen soon after emergence but these are normally transient

Environmental safety

- Dangerous for the environment
- Very toxic to aquatic organisms
- LERAP Category B
- Take extreme care to avoid drift onto all plants outside the target area

Hazard classification and safety precautions

Hazard H04, H11

Risk phrases R36, R50, R53a
Operator protection A, C; U05a, U20b [1, 2]; U08, U19a [2]; U09a [1]
Environmental protection E15b, E16a, E16b [1, 2]; E38 [1]
Storage and disposal D01, D02, D05, D09a, D12a [1, 2]; D10b [2]; D10c, D11a [1]
Medical advice M05a [2]

320 mesotrione + terbuthylazine

A foliar and soil acting herbicide mixture for maize
HRAC mode of action code: F2 + C1

Products

Calaris	Syngenta	70:330 g/l	SC	12405

Uses

- Annual dicotyledons in ***forage maize***, ***forage maize (under plastic mulches)*** *(off-label)*, ***grain maize*** *(off-label)*, ***sweetcorn*** *(off-label)*
- Annual meadow grass in ***forage maize***, ***forage maize (under plastic mulches)*** *(off-label)*, ***grain maize*** *(off-label)*, ***sweetcorn*** *(off-label)*

Specific Off-Label Approvals (SOLAs)

- ***forage maize (under plastic mulches)*** *20061439*
- ***grain maize*** *20064209*
- ***sweetcorn*** *20051892*

Approval information

- Mesotrione included in Annex I under EC Directive 91/414

Efficacy guidance

- Best results obtained from treatment of young actively growing weed seedlings in the presence of adequate soil moisture
- Treatment in poor growing conditions or in dry soil may give less reliable control
- Residual weed control is reduced on soils with more than 10% organic matter
- To minimise the possible development of resistance where continuous maize is grown the product should not be used for more than two consecutive seasons

Restrictions

- Maximum number of treatments 1 per crop of forage maize
- Do not use on seed crops or on sweetcorn varieties
- Do not spray when crop foliage wet or when excessive rainfall is expected to follow application
- Do not treat crops suffering from stress from cold or drought conditions, or when wide temperature fluctuations are anticipated
- Do not apply on Sands or Very Light soils

Crop-specific information

- Latest use: 8 leaves unfolded stage (GS 18) for forage maize
- Treatment under adverse conditions may cause mild to moderate chlorosis. The effect is transient and does not affect yield

Following crops guidance

- Winter wheat, durum wheat, winter barley or ryegrass may follow a normally harvested treated crop of maize. Oilseed rape may be sown provided it is preceded by deep ploughing to more than 15 cm
- In the spring following application forage maize, ryegrass, spring wheat or spring barley may be sown
- Spinach, beet crops, peas, beans, lettuce and cabbages must not be sown in the yr following application
- In the event of crop failure maize may be re-seeded immediately. Some slight crop effects may be seen soon after emergence but these are normally transient

Environmental safety

- Dangerous for the environment
- Very toxic to aquatic organisms
- LERAP Category B
- Take extreme care to avoid drift onto all plants outside the target area

Hazard classification and safety precautions

Hazard H03, H11
Risk phrases R22a, R50, R53a
Operator protection A; U05a
Environmental protection E15a, E16a, E16b, E34, E38
Storage and disposal D01, D02, D05, D09a, D10c, D12a

321 metalaxyl-M

A phenylamide systemic fungicide
FRAC mode of action code: 4

See also *chlorothalonil + metalaxyl-M*
cymoxanil + fludioxonil + metalaxyl-M
fluazinam + metalaxyl-M
mancozeb + metalaxyl-M

Products

1	SL 567A	Syngenta	465.2 g/l	SL	12380
2	Subdue	Fargro	465 g/l	SL	12503

Uses

- Cavity spot in ***carrots, parsnips*** *(off-label)* [1]
- Crown rot in ***water lilies*** *(off-label)* [1]
- Downy mildew in ***hops*** *(off-label)*, ***horseradish*** *(off-label)*, ***leaf spinach*** *(off-label)*, ***protected cucumbers*** *(off-label)*, ***protected spinach*** *(off-label)*, ***spinach beet*** *(off-label)*, ***wine grapes*** *(off-label)* [1]
- Phytophthora in ***asparagus*** *(off-label)* [1]
- Phytophthora root rot in ***ornamental plant production***, ***protected ornamentals*** [2]
- Pythium in ***ornamental plant production***, ***protected ornamentals*** [2]
- Root malformation disorder in ***red beet*** *(off-label)* [1]
- Storage rots in ***cabbages*** *(off-label)* [1]
- White blister in ***horseradish*** *(off-label)* [1]

Specific Off-Label Approvals (SOLAs)

- ***asparagus*** *20051502 expires Apr 2008* [1]
- ***cabbages*** *20062117* [1]
- ***hops*** *20051500 expires Apr 2008* [1]
- ***horseradish*** *20061040* [1]
- ***leaf spinach*** *20051507* [1]
- ***parsnips*** *20051508 expires Apr 2008* [1]
- ***protected cucumbers*** *20051503 expires Apr 2008* [1]
- ***protected spinach*** *20051507* [1]
- ***red beet*** *20051307 expires Apr 2008* [1]
- ***spinach beet*** *20051507* [1]
- ***water lilies*** *20051501 expires Apr 2008* [1]
- ***wine grapes*** *20051504 expires Apr 2008* [1]

Approval information

- Metalaxyl-M included in Annex I under EC Directive 91/414
- Accepted by BBPA for use on hops

Efficacy guidance

- Best results achieved when applied to damp soil or potting media
- Treatments to ornamentals should be followed immediately by irrigation to wash any residues from the leaves and allow penetration to the rooting zone [2]
- Efficacy may be reduced in prolonged dry weather [1]
- Results may not be satisfactory on soils with high organic matter content [1]
- Control of cavity spot on carrots overwintered in the ground or lifted in winter may be lower than expected [1]
- Use in an integrated pest management strategy and, where appropriate, alternate with products from different chemical groups

- Product should ideally be used preventatively and the number of phenylamide applications should be limited to 1-2 consecutive treatments
- Always follow FRAG guidelines for preventing and managing fungicide resistance. See Section 5 for more information

Restrictions

- Maximum number of treatments on ornamentals 1 per situation for media treatment. See label for details of drench treatment of protected ornamentals [2]
- Maximum total dose on carrots equivalent to one full dose treatment [1]
- Do not use where carrots have been grown on the same site within the previous eight yrs [1]
- Consult before use on crops intended for processing [1]
- Do not re-use potting media from treated plants for subsequent crops [2]
- Disinfect pots thoroughly prior to re-use [2]

Crop-specific information

- Latest use: 6 wk after drilling for Carrots [1]
- Because of the large number of species and ornamental cultivars susceptibility should be checked before large scale treatment [2]
- Treatment of *Viburnum* and *Prunus* species not recommended [2]

Environmental safety

- Harmful to aquatic organisms
- Limited evidence suggests that metalaxyl-M is not harmful to soil dwelling predatory mites

Hazard classification and safety precautions

Hazard H03
Risk phrases R22a, R37, R52, R53a
Operator protection A [1, 2]; C, H [2]; U02a, U04a, U05a, U10, U19a, U20b
Environmental protection E15b, E34, E38
Storage and disposal D01, D02, D05, D07, D09a, D10c, D12a
Medical advice M03

322 metaldehyde

A molluscicide bait for controlling slugs and snails

Products

1	Allure	Chiltern	1.5% w/w	RB	11089
2	Appeal	Chiltern	1.5% w/w	RB	12022
3	Attract	Chiltern	1.5% w/w	RB	12023
4	Bristol Blues	Doff Portland	6% w/w	PT	11798
5	CDP Minis	De Sangosse	5% w/w	PT	12955
6	Certis Deal 5	Certis	5% w/w	PT	13488
7	Certis Metaldehyde 5	Certis	5% w/w	RB	13490
8	Certis Red 5	Certis	5% w/w	RB	13491
9	Certis Trigger 5	Certis	5% w/w	RB	13493
10	Chiltern Blues	Chiltern	6% w/w	RB	10071
11	Chiltern Hardy	Chiltern	6% w/w	RB	06948
12	Chiltern Hundreds	Chiltern	3% w/w	RB	10072
13	Condor	Doff Portland	6% w/w	PT	11791
14	Condor 5	Doff Portland	5% w/w	PT	12484
15	Dixie 6	Greencrop	6% w/w	PT	11790
16	Doff Horticultural Slug Killer Blue Mini Pellets	Doff Portland	3% w/w	RB	11463
17	Entice	De Sangosse	4% w/w	RB	12933
18	Escar-Go 6	Chiltern	6% w/w	RB	06076
19	ESP	De Sangosse	5% w/w	PT	12999
20	Gusto	Makhteshim	5% w/w	RB	13448
21	Helimax S	De Sangosse	5% w/w	RB	13109
22	Keel Over	De Sangosse	5% w/w	RB	12966
23	Lincoln VI	Doff Portland	6% w/w	PT	11793
24	Lynx S	De Sangosse	5% w/w	PT	13276
25	Metarex Amba	De Sangosse	5% w/w	RB	11157
26	Metarex Green	De Sangosse	5% w/w	RB	10113
27	Metarex RG	De Sangosse	5% w/w	RH	10115
28	Molotov	Chiltern	3% w/w	RB	08295
29	Optimol	De Sangosse	4% w/w	RB	12997

Products – continued

30	Pathfinder Excel	Barclay	5% w/w	PT	12501
31	Pesta M	De Sangosse	5% w/w	PT	13275
32	Polymet 5	Unicrop	5% w/w	RB	13480
33	Prego	Makhteshim	5% w/w	RB	13582
34	Regel	De Sangosse	5% w/w	RB	10114
35	Rowent	Chiltern	6% w/w	PT	12024
36	Slug-Lentils	Luxan	3% w/w	PT	12533
37	Steam	Chiltern	3% w/w	RB	12473
38	Super 5	Unicrop	5% w/w	RB	12535
39	TDS Major	De Sangosse	4% w/w	CB	13462
40	TDS Metarex Amba	De Sangosse	4% w/w	CB	13461

Uses

- Slugs in ***all edible crops (outdoor)***, ***all non-edible crops (outdoor)*** [1-37, 39, 40]; ***all edible crops (outdoor)*** *(around)*, ***all non-edible crops (outdoor)*** *(around)* [38]; ***amenity grassland***, ***managed amenity turf*** [6-9]; ***cultivated land/soil*** [1-3, 10-12, 16, 18, 28, 30, 35, 37]; ***natural surfaces not intended to bear vegetation*** [4-9, 13-15, 17, 21-24, 29, 31, 32, 36, 38-40]; ***ornamental plant production*** [20, 25-27, 33, 34]; ***protected crops*** [1-3, 10-12, 18, 20, 25-28, 30, 33-35, 37, 39, 40]
- Snails in ***all edible crops (outdoor)***, ***all non-edible crops (outdoor)*** [1-37, 39, 40]; ***all edible crops (outdoor)*** *(around)*, ***all non-edible crops (outdoor)*** *(around)* [38]; ***amenity grassland***, ***managed amenity turf*** [6-9]; ***cultivated land/soil*** [1-3, 10-12, 16, 18, 28, 30, 35, 37]; ***natural surfaces not intended to bear vegetation*** [4-9, 13-15, 17, 21-24, 29, 31, 32, 36, 38-40]; ***ornamental plant production*** [20, 25-27, 33, 34]; ***protected crops*** [1-3, 10-12, 18, 20, 25-28, 30, 33-35, 37, 39, 40]

Approval information

- Approved for aerial application on all edible crops, all non-edible crops [11101, 14-16, 24, 31, 38]; on natural surfaces not intended to be cultivated [14, 15, 38]; on bare soil [16]. See notes in Section 5
- Conditions of approval for all metaldehyde products were strengthened in 2006
- Accepted by BBPA for use on malting barley and hops
- Approval expiry 31 Mar 2008 [22]
- Approval expiry 31 Oct 2008 [25-27, 34]

Efficacy guidance

- Apply pellets by hand, fiddle drill, fertilizer distributor, by air (check label) or in admixture with seed. See labels for rates and timing.
- Best results achieved from an even spread of granules applied during mild, damp weather when slugs and snails most active. May be applied in standing crops
- To establish the need for pellet application on winter wheat or winter oilseed rape, monitor for slug activity. Where bait traps are used, use a foodstuff attractive to slugs e.g. chicken layer's mash
- Varieties of oilseed rape low in glucosinolates can be more acceptable to slugs than "single low" varieties and control may not be as good
- To prevent slug build up apply at end of season to brassicas and other leafy crops
- To reduce tuber damage in potatoes apply twice in Jul and Aug
- For information on slug trapping and damage risk assessment refer to HGCA Topic Sheets No. 84 (winter wheat) and 85 (winter oilseed rape), available from the HGCA website (www.hgca.com)

Restrictions

- Do not apply when rain imminent or water glasshouse crops within 4 d of application
- Take care to avoid lodging of pellets in the foliage when making late applications to edible crops

Crop-specific information

- Put slug traps out before cultivation, when the soil surface is visibly moist and the weather mild (5-25 °C) (see label for guidance)
- For winter wheat, a catch of 4 or more slugs/trap indicates a possible risk, where soil and weather conditions favour slug activity
- For winter oilseed rape a catch of 4 or more slugs in standing cereals, or 1 or more in cereal stubble, if other conditions were met, would indicate possible risk of damage

Environmental safety

- Dangerous to game, wild birds and animals
- Some products contain proprietary cat and dog deterrent

SEE SECTION 3 FOR PRODUCTS ALSO REGISTERED

- Keep poultry out of treated areas for at least 7 d
- Do not use slug pellets in traps in winter wheat or winter oilseed rape since they are a potential hazard to wildlife and pets

Hazard classification and safety precautions

Operator protection A, H [1-9, 12-15, 17, 19-40]; J [4, 5, 13-15, 17, 19, 23, 24, 29, 31, 32, 38]; U05a [1-3, 5-12, 17, 18, 24, 28, 30, 31, 35, 36, 39, 40]; U15, U20a [20, 25-27, 33, 34]; U20b [10-12, 18, 28, 32, 35]; U20c [1-9, 13, 15-17, 19, 21-24, 29-31, 37, 39, 40]

Environmental protection E05b [1-13] (7 d); E05b [14]; E05b [15-23] (7 d); E05b [24]; E05b [25-30] (7 d); E05b [31, 32]; E05b [33-40] (7 d); E10a, E15a [1-40]

Storage and disposal D01 [1-13, 17-31, 33-40]; D02 [1-3, 5-12, 17, 18, 20-22, 24-28, 30, 31, 33-36, 39, 40]; D05 [5-9, 17]; D07 [1-3, 10-12, 14, 16, 18, 20-22, 25-28, 32-35, 37-40]; D09a [1-15, 17-35, 37-40]; D11a [1-20, 23, 25-30, 32-35, 37, 38]; D11b [21, 22, 24, 31, 39, 40]; D12a [20-22, 25-27, 33, 34]

Treated seed S04a [4, 13, 19-27, 29-31, 33, 34, 36, 38-40]

Medical advice M04a [5-9, 17]; M05a [20-22, 25-27, 33, 34, 36, 39, 40]

323 metamitron

A contact and residual triazinone herbicide for use in beet crops
HRAC mode of action code: C1

See also *chloridazon + chlorpropham + metamitron*
chloridazon + metamitron
ethofumesate + metamitron

Products

1	Agriguard Metamitron	AgriGuard	70% w/w	WG	09859
2	Alpha Metamitron	Makhteshim	70% w/w	WG	11081
3	Alpha Metamitron 70 SC	Makhteshim	700 g/l	SC	12868
4	Barclay Seismic	Barclay	70% w/w	ZZ	11377
5	Bettix 70 WG	United Phosphorus	70% w/w	WG	11154
6	Bettix Flo	United Phosphorus	700 g/l	SC	11959
7	Clayton Mitrex	Clayton	70% w/w	WG	11921
8	Defiant SC	United Phosphorus	700 g/l	SC	12302
9	Defiant WG	United Phosphorus	70% w/w	WG	12300
10	Fernpath Haptol	AgriGuard	70% w/w	WG	11951
11	Fernpath Haptol Flo	AgriGuard	700 g/l	SC	11481
12	Goldbeet	Makhteshim	90% w/w	WG	11538
13	Goltix 90	Makhteshim	90% w/w	WG	11578
14	Goltix Flowable	Makhteshim	700 g/l	SC	12851
15	Goltix WG	Makhteshim	70% w/w	WG	11539
16	Landgold Metamitron	Landgold	70% w/w	WG	06287
17	Landgold Metamitron	Teliton	70% w/w	WG	12123
18	Marquise	Makhteshim	70% w/w	WG	08738
19	Mitron 70 WG	Hermoo	70% w/w	WG	11516
20	Mitron 90 WG	Hermoo	90% w/w	WG	10888
21	Mitron SC	Hermoo	700 g/l	SC	11643
22	MM 70 Flo	Nufarm UK	700 g/l	SC	13318
23	Predator WG	AgriGuard	70% w/w	WG	12630
24	Skater	Makhteshim	700 g/l	SC	12857
25	Volcan	Sipcam	70% w/w	WG	09295

Uses

- Annual dicotyledons in ***asparagus*** *(off-label)*, ***forest*** *(off-label)* [18]; ***cardoons*** *(off-label)*, ***celery (outdoor)*** *(off-label)*, ***rhubarb*** *(off-label)* [24]; ***fodder beet***, ***sugar beet*** [1-25]; ***horseradish*** *(off-label)*, ***parsnips*** *(off-label)*, ***strawberries*** *(off-label)* [14]; ***mangels*** [1-18, 20-25]; ***red beet*** [1-18, 20-23, 25]
- Annual grasses in ***asparagus*** *(off-label)*, ***forest*** *(off-label)* [18]; ***fodder beet***, ***sugar beet*** [1-25]; ***horseradish*** *(off-label)*, ***parsnips*** *(off-label)*, ***strawberries*** *(off-label)* [14]; ***mangels*** [1-18, 20-25]; ***red beet*** [1-18, 20-23, 25]
- Annual meadow grass in ***fodder beet***, ***sugar beet*** [1-25]; ***mangels*** [1-18, 20-25]; ***red beet*** [1-18, 20-23, 25]
- Fat hen in ***fodder beet***, ***sugar beet*** [1-25]; ***mangels*** [1-18, 20-25]; ***red beet*** [1-18, 20-23, 25]

- General weed control in ***asparagus*** *(off-label)*, ***forestry transplants*** *(off-label)*, ***horseradish*** *(off-label)*, ***parsnips*** *(off-label)* [15]
- Groundsel in ***chives*** *(off-label)*, ***herbs (see appendix 6)*** *(off-label)*, ***parsley*** *(off-label)* [15, 18]
- Polygonums in ***cardoons*** *(off-label)*, ***celery (outdoor)*** *(off-label)*, ***rhubarb*** *(off-label)* [24]

Specific Off-Label Approvals (SOLAs)

- ***asparagus*** *20041758* [15], *20030737* [18]
- ***cardoons*** *20070844* [24]
- ***celery (outdoor)*** *20070844* [24]
- ***chives*** *20041756* [15], *20030737* [18]
- ***forest*** *20030737* [18]
- ***forestry transplants*** *20041757* [15]
- ***herbs (see appendix 6)*** *20041756* [15], *20030737* [18]
- ***horseradish*** *20070513* [14], *20061637* [15]
- ***parsley*** *20041756* [15], *20030737* [18]
- ***parsnips*** *20070513* [14], *20061637* [15]
- ***rhubarb*** *20070844* [24]
- ***strawberries*** *20070856* [14]

Approval information

- Approval expiry 31 Aug 2008 [11]

Efficacy guidance

- May be used pre-emergence alone or post-emergence in tank mixture or with an authorised adjuvant oil
- Low dose programme (LDP). Apply a series of low-dose post-weed emergence sprays, including adjuvant oil, timing each treatment according to weed emergence and size. See label for details and for recommended tank mixes and sequential treatments. On mineral soils the LDP should be preceded by pre-drilling or pre-emergence treatment
- Traditional application. Apply either pre-drilling before final cultivation with incorporation to 8-10 cm, or pre-crop emergence at or soon after drilling into firm, moist seedbed to emerged weeds from cotyledon to first true leaf stage
- On emerged weeds at or beyond 2-leaf stage addition of adjuvant oil advised
- Up to 3 post-emergence sprays may be used on soils with over 10% organic matter
- For control of wild oats and certain other weeds, tank mixes with other herbicides or sequential treatments are recommended. See label for details

Restrictions

- Maximum total dose equivalent to three full dose treatments for most products. Check label
- Using traditional method post-crop emergence on mineral soils do not apply before first true leaves have reached 1 cm long

Crop-specific information

- Latest use: before crop foliage meets across rows for beet crops
- HI herbs 6 wk
- Crop tolerance may be reduced by stress caused by growing conditions, effects of pests, disease or other pesticides, nutrient deficiency etc

Following crops guidance

- Only sugar beet, fodder beet or mangels may be drilled within 4 mth after treatment. Winter cereals may be sown in same season after ploughing, provided 16 wk passed since last treatment

Environmental safety

- Dangerous for the environment
- Very toxic to aquatic organisms
- Dangerous to fish or other aquatic life. Do not contaminate surface waters or ditches with chemical or used container
- Do not empty into drains

Hazard classification and safety precautions

Hazard H03 [1-3, 5, 6, 8, 9, 11-18, 20-23]; H04 [4, 7, 10]; H11 [1-3, 5-24]

Risk phrases R22a [1-3, 5, 6, 8, 9, 11-18, 20-23]; R41 [1, 4, 7, 10, 11]; R43 [1, 2, 4, 6-8, 10, 11, 18]; R50 [2, 3, 5-9, 12-14, 16-22]; R51 [1, 10, 15, 23, 24]; R53a [1-3, 5-10, 12-24]

Operator protection A [1-4, 6-8, 10, 11, 14, 15, 18, 21-23]; C [1, 4, 7, 10]; H [1, 2, 4, 7, 10, 18, 23]; U05a [1, 2, 4, 5, 7, 9-11, 18, 19, 22]; U08 [2, 11, 18, 19, 22]; U09a [6, 8]; U11, U15 [7]; U14 [2, 6-8,

11, 18, 19, 22]; U19a [1, 2, 5, 7, 9, 11, 15, 18, 19, 22]; U20a [1, 2, 7, 10-13, 15-20, 22]; U20b [3-5, 9, 14, 21]; U20c [6, 8, 24, 25]
Environmental protection E13b [5, 6, 8, 9]; E15a [1-4, 7, 10-25]; E19b [3, 12-14, 20, 21, 23, 24]; E34 [3, 11, 14, 21, 22]; E38 [1, 15, 19]
Storage and disposal D01 [1-14, 18-23]; D02 [1-11, 14, 18, 19, 21, 22]; D05 [6-8, 11, 22]; D09a [1-25]; D10c [6, 8, 24]; D11a [1-5, 7, 9-22, 25]; D12a [1, 3, 5-9, 12-15, 20, 21, 23, 24]; D12b [2, 18, 22]
Medical advice M03 [3, 5, 6, 8, 9, 11, 14, 21]; M04a [20]; M05a [3, 5, 6, 8, 9, 12-14, 16, 17, 21, 23]

324 metam-sodium

A methyl isothiocyanate producing sterilant for glasshouse, nursery and outdoor soils
HRAC mode of action code: Z

Products

1	Discovery	United Phosphorus	510 g/l	SL	10416
2	Sistan 51	Unicrop	510 g/l	SL	10046

Uses

- Nematodes in ***glasshouse soils, nursery soils, outdoor soils, potting soils***
- Soil pests in ***glasshouse soils, nursery soils, outdoor soils, potting soils***
- Soil-borne diseases in ***glasshouse soils, nursery soils, outdoor soils, potting soils***
- Weed seeds in ***glasshouse soils, nursery soils, outdoor soils, potting soils***

Efficacy guidance

- Metam-sodium is a partial soil sterilant and acts by breaking down in contact with soil to release methyl isothiocyanate (MIT)
- Apply to glasshouse soils as a drench, or inject undiluted to 20 cm at 30 cm intervals and seal immediately, or apply to surface and rotavate
- May also be used by mixing into potting soils
- Apply when soil temperatures exceed 7 °C, preferably above 10 °C, between 1 Apr and 31 Oct. Soil must be of fine tilth, free from debris and with 'potting moisture' content. If soil is too dry postpone treatment and water soil

Restrictions

- No plants must be present during treatment
- Crops must not be planted until a cress germination test has been completed satisfactorily
- Do not treat glasshouses within 2 m of growing crops. Fumes are damaging to all plants
- Avoid using in equipment incorporating natural rubber parts

Crop-specific information

- Latest use: pre-planting of crop

Following crops guidance

- Do not plant until soil is entirely free of fumes

Environmental safety

- Dangerous for the environment
- Very toxic to aquatic organisms
- Keep unprotected persons, livestock and pets out of treated areas for at least 24 h following treatment
- When diluted breakdown commences almost immediately. Only quantities for immediate use should be made up
- Divert or block drains which could carry solution under untreated glasshouses
- After treatment allow sufficient time (several weeks) for residues to dissipate and aerate soil by forking. Time varies with soil and season. Soils with high clay or organic matter content will retain gas longer than lighter soils

Hazard classification and safety precautions

Hazard H03, H05, H11
Risk phrases R22a, R34, R43, R50, R53a
Operator protection A, C, H, M; U02a, U07, U08, U11, U20b
Environmental protection E15a, E36a
Storage and disposal D01, D02, D12b, D14
Medical advice M04a

325 metazachlor

A residual anilide herbicide for use in brassicas, nurseries and forestry
HRAC mode of action code: K3

See also *dimethenamid-p + metazachlor*

Products

1	Agriguard Metazachlor	AgriGuard	500 g/l	SC	10417
2	Alpha Metazachlor 50 SC	Makhteshim	500 g/l	SC	10669
3	Butisan S	BASF	500 g/l	SC	11733
4	Clayton Buzz	Clayton	500 g/l	SC	12509
5	Clayton Metazachlor	Clayton	500 g/l	SC	09688
6	Clayton Metazachlor 50 SC	Clayton	500 g/l	SC	11719
7	Greencrop Monogram	Greencrop	500 g/l	SC	12048
8	Landgold Metazachlor 50	Landgold	500 g/l	SC	09726
9	Landgold Metazachlor 50	Teliton	500 g/l	SC	12124
10	Landgold Metazachlor 50 SC	Teliton	500 g/l	SC	12133
11	Marksman	AgriGuard	500 g/l	SC	12456
12	Rapsan 500 SC	Nufarm UK	500 g/l	SC	12365
13	Standon Metazachlor 50	Standon	500 g/l	SC	05581
14	Standon Metazachlor 500	Standon	500 g/l	SC	12012
15	Sultan 50 SC	Makhteshim	500 g/l	SC	10418

Uses

- Annual dicotyledons in ***amenity vegetation*** [4]; ***borage*** *(off-label)* [2, 3, 5-7, 12-15]; ***broccoli*** [2-10, 13-15]; ***brussels sprouts***, ***cabbages***, ***cauliflowers***, ***spring oilseed rape***, ***swedes***, ***turnips***, ***winter oilseed rape*** [1-15]; ***calabrese*** [1-6, 8-15]; ***canary flower (echium spp.)*** *(off-label)* [2, 3, 5-7, 12-14]; ***evening primrose*** *(off-label)* [1-3, 5, 6, 10, 12-15]; ***farm forestry*** [6, 10]; ***farm woodland*** [1-3, 11, 12, 14, 15]; ***forest*** [1-7, 10-12, 14, 15]; ***honesty*** *(off-label)*, ***mustard*** *(off-label)* [1-3, 5-7, 10, 12-15]; ***nursery fruit trees and bushes***, ***ornamental trees***, ***shrubs*** [1-7, 11, 12, 14, 15]; ***spring linseed*** *(off-label)* [2, 3]; ***transplanted leeks*** *(off-label)* [3]; ***winter linseed*** *(off-label)* [2, 3, 15]
- Annual grasses in ***amenity vegetation*** [4]; ***borage*** *(off-label)* [2, 5-7, 12-15]; ***canary flower (echium spp.)*** *(off-label)* [2, 5-7, 12-14]; ***evening primrose*** *(off-label)* [1, 2, 5, 6, 10, 12-15]; ***farm forestry*** [10]; ***farm woodland*** [1-3, 11, 12, 14, 15]; ***forest*** [1-7, 10-12, 14, 15]; ***honesty*** *(off-label)*, ***mustard*** *(off-label)* [1, 2, 5-7, 10, 12-15]; ***ornamental trees***, ***shrubs*** [1-7, 11, 12, 14, 15]; ***spring linseed*** *(off-label)* [2]; ***transplanted leeks*** *(off-label)* [3]; ***winter linseed*** *(off-label)* [2, 15]
- Annual meadow grass in ***amenity vegetation*** [4]; ***broccoli***, ***brussels sprouts***, ***cabbages***, ***cauliflowers***, ***spring oilseed rape***, ***swedes***, ***turnips***, ***winter oilseed rape*** [1-15]; ***calabrese*** [1-6, 8-15]; ***farm forestry*** [6]; ***nursery fruit trees and bushes***, ***ornamental trees***, ***shrubs*** [1-7, 11, 12, 14, 15]
- Blackgrass in ***amenity vegetation*** [4]; ***broccoli*** [2-10, 13-15]; ***brussels sprouts***, ***cabbages***, ***cauliflowers***, ***spring oilseed rape***, ***swedes***, ***turnips***, ***winter oilseed rape*** [1-15]; ***calabrese*** [1-6, 8-15]; ***farm forestry*** [6]; ***nursery fruit trees and bushes***, ***ornamental trees***, ***shrubs*** [1-7, 11, 12, 14, 15]
- Groundsel in ***chinese cabbage*** *(off-label)*, ***choi sum*** *(off-label)*, ***collards*** *(off-label)*, ***kale*** *(off-label)*, ***pak choi*** *(off-label)* [3]
- Mayweeds in ***chinese cabbage*** *(off-label)*, ***choi sum*** *(off-label)*, ***collards*** *(off-label)*, ***kale*** *(off-label)*, ***pak choi*** *(off-label)* [3]
- Shepherd's purse in ***chinese cabbage*** *(off-label)*, ***choi sum*** *(off-label)*, ***collards*** *(off-label)*, ***kale*** *(off-label)*, ***pak choi*** *(off-label)* [3]

Specific Off-Label Approvals (SOLAs)

- ***borage*** *20060901* [2], *20060912* [3], *20060905* [5], *20060902* [6], *20060898* [7], *20060892* [12], *20060891* [13], *20060890* [14], *20060889* [15]
- ***canary flower (echium spp.)*** *20060901* [2], *20060912* [3], *20060905* [5], *20060902* [6], *20060898* [7], *20060892* [12], *20060891* [13], *20060890* [14]
- ***chinese cabbage*** *20050344* [3]
- ***choi sum*** *20050344* [3]
- ***collards*** *20050344* [3]
- ***evening primrose*** *20060911* [1], *20060901* [2], *20060912* [3], *20060905* [5], *20060902* [6], *20060897* [10], *20060892* [12], *20060891* [13], *20060890* [14], *20060889* [15]

- ***honesty*** *20060911* [1], *20060901* [2], *20060912* [3], *20060905* [5], *20060902* [6], *20060898* [7], *20060897* [10], *20060892* [12], *20060891* [13], *20060890* [14], *20060889* [15]
- ***kale*** *20050344* [3]
- ***mustard*** *20060911* [1], *20060901* [2], *20060912* [3], *20060905* [5], *20060902* [6], *20060898* [7], *20060897* [10], *20060892* [12], *20060891* [13], *20060890* [14], *20060889* [15]
- ***pak choi*** *20050344* [3]
- ***spring linseed*** *20060901* [2], *20060912* [3]
- ***transplanted leeks*** *20063698* [3]
- ***winter linseed*** *20060901* [2], *20060912* [3], *20060889* [15]

Efficacy guidance

- Activity is dependent on root uptake. For pre-emergence use apply to firm, moist, clod-free seedbed
- Some weeds (chickweed, mayweed, blackgrass etc) susceptible up to 2- or 4-leaf stage. Moderate control of cleavers achieved provided weeds not emerged and adequate soil moisture present
- Split pre- and post-emergence treatments recommended for certain weeds in winter oilseed rape on light and/or stony soils
- Effectiveness is reduced on soils with more than 10% organic matter
- Various tank-mixtures recommended to broaden spectrum. See label for details
- Always follow WRAG guidelines for preventing and managing herbicide resistant weeds. See Section 5 for more information

Restrictions

- Maximum number of treatments 1 per crop for spring oilseed rape, swedes, turnips and brassicas; 2 per crop for winter oilseed rape (split dose treatment); 3 per yr for ornamentals, nursery stock, nursery fruit trees, forestry and farm forestry
- Do not use on sand, very light or poorly drained soils
- Do not treat protected crops or spray overall on ornamentals with soft foliage
- Do not spray crops suffering from wilting, pest or disease
- Do not spray broadcast crops or if a period of heavy rain forecast
- When used on nursery fruit trees any fruit harvested within 1 yr of treatment must be destroyed

Crop-specific information

- Latest use: pre-emergence for swedes and turnips; before 10 leaf stage for spring oilseed rape; before end of Jan for winter oilseed rape
- HI brassicas 6 wk
- On winter oilseed rape may be applied pre-emergence from drilling until seed chits, post-emergence after fully expanded cotyledon stage (GS 1,0) or by split dose technique depending on soil and weeds. See label for details
- On spring oilseed rape, swedes, turnips metazachlor recommended as a pre-emergence sequential treatment following trifluralin
- On spring oilseed rape may also be used pre-weed-emergence from cotyledon to 10-leaf stage of crop (GS 1,0-1,10)
- With pre-emergence treatment ensure seed covered by 15 mm of well consolidated soil. Harrow across slits of direct-drilled crops
- Ensure brassica transplants have roots well covered and are well established. Direct drilled brassicas should not be treated before 3 leaf stage
- In ornamentals and hardy nursery stock apply after plants established and hardened off as a directed spray or, on some subjects, as an overall spray. See label for list of tolerant subjects. Do not treat plants in containers

Following crops guidance

- Any crop can follow normally harvested treated winter oilseed rape. See label for details of crops which may be planted after spring treatment and in event of crop failure

Environmental safety

- Dangerous for the environment
- Very toxic to aquatic organisms
- Keep livestock out of treated areas until foliage of any poisonous weeds such as ragwort has died and become unpalatable
- Keep livestock out of treated areas of swede and turnip for at least 5 wk following treatment

Hazard classification and safety precautions

Hazard H03, H11

Risk phrases R22a, R43, R50 [1-15]; R38 [2-10, 13, 15]; R41 [12]; R53a [3-6, 8-13]

Operator protection A, C, H, M; U05a, U19a [1-15]; U08, U20b [1-11, 13-15]; U09a, U20a [12]; U14 [2, 3, 15]; U15 [2, 15]

Environmental protection E06a [1-15] (5 wk for swedes, turnips); E07a, E15a, E34 [1-15]; E38 [3, 4]

Consumer protection C02a [12]

Storage and disposal D01, D02, D09a [1-15]; D05 [1, 2, 4-15]; D10b [1, 5-10, 13, 14]; D10c [2-4, 11, 12, 15]; D12a [3-6]; D12b [2, 15]

Medical advice M03 [1-15]; M05a [3-6]

326 metazachlor + quinmerac

A residual herbicide mixture for oilseed rape

HRAC mode of action code: K3 + O

Products

1	Clayton Mazarac	Clayton	375:125 g/l	SC	11161
2	Katamaran	BASF	375:125 g/l	SC	11732
3	Novall	BASF	400:100 g/l	SC	12031
4	Oryx	BASF	333:83 g/l	SC	13527
5	Standon Metazachlor-Q	Standon	375:125 g/l	SC	09676
6	Syndicate	AgriGuard	400:100 g/l	SC	12262

Uses

- Annual dicotyledons in ***borage*** *(off-label)* [1-3, 5]; ***canary flower (echium spp.)*** *(off-label)* [1-3]; ***evening primrose*** *(off-label)*, ***honesty*** *(off-label)*, ***linseed*** *(off-label)*, ***mustard*** *(off-label)* [2, 3]; ***winter oilseed rape*** [1-6]
- Annual grasses in ***borage*** *(off-label)* [1-3, 5]; ***canary flower (echium spp.)*** *(off-label)* [1-3]; ***evening primrose*** *(off-label)*, ***honesty*** *(off-label)*, ***linseed*** *(off-label)*, ***mustard*** *(off-label)* [2, 3]
- Annual meadow grass in ***winter oilseed rape*** [1-6]
- Blackgrass in ***winter oilseed rape*** [1-6]
- Cleavers in ***winter oilseed rape*** [1-6]
- Poppies in ***winter oilseed rape*** [1-6]

Specific Off-Label Approvals (SOLAs)

- ***borage*** *20060503* [1], *20060547* [2], *20060548* [3], *20060502* [5]
- ***canary flower (echium spp.)*** *20060503* [1], *20060547* [2], *20060548* [3]
- ***evening primrose*** *20060547* [2], *20060548* [3]
- ***honesty*** *20060547* [2], *20060548* [3]
- ***linseed*** *20060547* [2], *20060548* [3]
- ***mustard*** *20060547* [2], *20060548* [3]

Efficacy guidance

- Activity is dependent on root uptake. Pre-emergence treatments should be applied to firm moist seedbeds. Applications to dry soil do not become effective until after rain has fallen
- Maximum activity achieved from treatment before weed emergence for some species
- Weed control may be reduced if excessive rain falls shortly after application especially on light soils
- May be used on all soil types except Sands, Very Light Soils, and soils containing more than 10% organic matter. Crop vigour and/or plant stand may be reduced on brashy and stony soils

Restrictions

- Maximum total dose equivalent to one full dose treatment
- Damage may occur in waterlogged conditions. Do not use on poorly drained soils
- Do not treat stressed crops. In frosty conditions transient scorch may occur

Crop-specific information

- Latest use: end Jan in yr of harvest
- To ensure crop safety it is essential that crop seed is well covered with soil to 15 mm. Loose or puffy seedbeds must be consolidated before treatment. Do not use on broadcast crops
- Crop vigour and possibly plant stand may be reduced if excessive rain falls shortly after treatment especially on light soils

SEE SECTION 3 FOR PRODUCTS ALSO REGISTERED

Following crops guidance

- In the event of crop failure after use, wheat or barley may be sown in the autumn after ploughing to 15 cm. Spring cereals or brassicas may be planted after ploughing in the spring

Environmental safety

- Dangerous for the environment
- Very toxic to aquatic organisms
- Keep livestock out of treated areas until foliage of any poisonous weeds such as ragwort has died and become unpalatable
- To reduce risk of movement to water do not apply to dry soil or if heavy rain is forecast. On clay soils create a fine consolidated seedbed
- LERAP Category B

Hazard classification and safety precautions

Hazard H04, H11

Risk phrases R43, R50, R53a

Operator protection A [1-6]; C [3, 4, 6]; H [3, 4]; U04a [3, 4, 6]; U05a, U08, U19a, U20b [1-6]; U14 [1-5]

Environmental protection E07a, E15a, E16a [1-6]; E16b [5]; E34 [1, 6]; E38 [2-4]

Storage and disposal D01, D02, D09a [1-6]; D05 [1, 5]; D08 [2-4]; D10b [1, 5, 6]; D10c, D12a [1-4]

Medical advice M05a [2-4]

327 metconazole

A triazole fungicide for cereals and oilseed rape

FRAC mode of action code: 3

Products

1	Caramba	BASF	60 g/l	SL	10213
2	Caramba 90	BASF	90 g/l	SL	12655
3	Clayton Tunik	Clayton	60 g/l	SL	10545
4	Juventus	BASF	90 g/l	SL	13340
5	Sunorg Pro	BASF	90 g/l	SL	11112

Uses

- Alternaria in ***spring oilseed rape***, ***winter oilseed rape*** [1-5]
- Ascochyta in ***combining peas*** *(reduction)*, ***lupins*** *(qualified minor use)*, ***vining peas*** *(reduction)* [1, 2, 4, 5]
- Botrytis in ***combining peas*** *(reduction)*, ***lupins*** *(qualified minor use)*, ***vining peas*** *(reduction)* [1, 2, 4, 5]; ***spring linseed*** *(off-label)*, ***winter linseed*** *(off-label)* [1, 5]
- Brown rust in ***spring barley***, ***winter barley***, ***winter wheat*** [1-5]
- Foliar disease control in ***durum wheat*** *(off-label)*, ***spring rye*** *(off-label)*, ***triticale*** *(off-label)*, ***winter rye*** *(off-label)* [1]; ***grass seed crops*** *(off-label)* [1, 3]
- Fusarium ear blight in ***winter wheat*** *(reduction)* [1-5]
- Light leaf spot in ***spring oilseed rape***, ***winter oilseed rape*** [1-5]
- Mycosphaerella in ***combining peas*** *(reduction)*, ***vining peas*** *(reduction)* [1, 2, 4, 5]
- Net blotch in ***spring barley*** *(reduction)*, ***winter barley*** *(reduction)* [1-5]
- Phoma in ***spring oilseed rape*** *(reduction)*, ***winter oilseed rape*** *(reduction)* [1-5]
- Powdery mildew in ***spring barley*** *(moderate control)*, ***winter barley***, ***winter wheat*** *(moderate control)* [1-5]
- Rhynchosporium in ***spring barley***, ***winter barley*** [1-5]
- Rust in ***broad beans*** *(off-label)* [1, 5]; ***combining peas***, ***lupins*** *(qualified minor use)*, ***spring field beans***, ***vining peas***, ***winter field beans*** [1, 2, 4, 5]
- Septoria leaf blotch in ***winter wheat*** [1-5]
- Yellow rust in ***winter wheat*** [1-5]

Specific Off-Label Approvals (SOLAs)

- ***broad beans*** *20063078* [1], *20063080* [5]
- ***durum wheat*** *20063077* [1]
- ***grass seed crops*** *20063077* [1], *20061092* [3]
- ***spring linseed*** *20063075* [1], *20063079* [5]
- ***spring rye*** *20063077* [1]

- ***triticale*** *20063077* [1]
- ***winter linseed*** *20063075* [1], *20063079* [5]
- ***winter rye*** *20063077* [1]

Approval information

- Metconazole included in Annex I under EC Directive 91/414
- Accepted by BBPA for use on malting barley

Efficacy guidance

- Best results from application to healthy, vigorous crops when disease starts to develop
- Good spray cover of the target is essential for best results. Spray volume should be increased to improve spray penetration into dense crops
- Metconazole is a DMI fungicide. Resistance to some DMI fungicides has been identified in Septoria leaf blotch which may seriously affect performance of some products. For further advice contact a specialist advisor and visit the Fungicide Resistance Action Group (FRAG)-UK website

Restrictions

- Maximum total dose equivalent to two full dose treatments on all crops
- Do not apply to oilseed rape crops that are damaged or stressed from previous treatments, adverse weather, nutrient deficiency or pest attack. Spring application may lead to reduction of crop height
- The addition of adjuvants is neither advised nor necessary and can lead to enhanced growth regulatory effects on stressed crops of oilseed rape
- Ensure sprayer is free from residues of previous treatments that may harm the crop, especially oilseed rape. Use of a detergent cleaner is advised before and after use
- Do not apply with pyrethroid insecticides on oilseed rape at flowering [2, 4, 5]

Crop-specific information

- Latest use: up to and including milky ripe stage (GS 71) for cereals; 10% pods at final size for oilseed rape
- HI: 14 d for peas, field beans, lupins
- Spring treatments on oilseed rape can reduce the height of the crop
- Treatment for Septoria leaf spot in wheat should be made before second node detectable stage and when weather favouring development of the disease has occurred. If conditions continue to favour disease development a follow-up treatment may be needed
- Treat mildew infections in cereals before 3% infection on any green leaf. A specific mildewicide will improve control of established infections
- Treat yellow rust on wheat before 1% infection on any leaf or as preventive treatment after GS 39
- For brown rust in cereals spray susceptible varieties before any of top 3 leaves has more than 2% infection
- Peas should be treated at the start of flowering and repeat 3-4 wk later if required
- Field beans and lupins should be treated at petal fall and repeat 3-4 wk later if required

Following crops guidance

- Only cereals, oilseed rape, sugar beet, linseed, maize, clover, beans, peas, carrots, potatoes or onions may be sown as following crops after treatment

Environmental safety

- Dangerous for the environment
- Very toxic to aquatic organisms
- Flammable [1, 3]
- LERAP Category B but avoid treatment close to field boundary, even if permitted by LERAP assessment, to reduce effects on non-target insects or other arthropods [1, 3]

Hazard classification and safety precautions

Hazard H03, H08 [1, 3]; H04 [2, 4, 5]; H11 [1-5]
Risk phrases R22b, R38, R41, R43, R50 [1, 3]; R36, R51 [2, 4, 5]; R37 [3]; R53a, R63 [1-5]
Operator protection A, C [1-5]; H [1, 3]; U02a, U05a, U20c [1-5]; U11 [1, 3]; U14 [1]; U19a [3]
Environmental protection E15a, E34 [1-5]; E16a [1, 3]; E16b [1]; E38 [1, 2, 4, 5]
Storage and disposal D01, D02, D05, D09a, D10a, D12a [1-5]; D06c [2, 4, 5]
Medical advice M05b [1, 3]

SEE SECTION 3 FOR PRODUCTS ALSO REGISTERED

328 methiocarb

A stomach acting carbamate molluscicide and insecticide
IRAC mode of action code: 1A

Products

1	Decoy Wetex	Bayer CropScience	2% w/w	PT	11266
2	Draza Forte	Bayer CropScience	4% w/w	PT	13306
3	Exit Wetex	Bayer CropScience	3% w/w	PT	11276
4	Huron	Bayer CropScience	3% w/w	PT	11288
5	Karan	Bayer CropScience	3% w/w	PT	11289
6	Rivet	Bayer CropScience	3% w/w	PT	11300

Uses

- Cutworms in ***sugar beet*** *(reduction)* [1-6]
- Leatherjackets in ***all cereals*** *(reduction)*, ***potatoes*** *(reduction)*, ***sugar beet*** *(reduction)* [1-6]; ***rotational grass*** *(seed admixture)* [1, 3-6]
- Millipedes in ***sugar beet*** *(reduction)* [1-6]
- Slugs in ***all cereals***, ***all non-edible crops (outdoor)*** *(outdoor only)*, ***brussels sprouts***, ***cabbages***, ***cauliflowers***, ***forage maize***, ***leaf spinach***, ***lettuce***, ***potatoes***, ***spring oilseed rape***, ***strawberries***, ***sugar beet***, ***sunflowers***, ***winter oilseed rape*** [1-6]; ***rotational grass*** *(seed admixture)* [1, 3-6]
- Strawberry seed beetle in ***strawberries*** [1-6]

Approval information

- Methiocarb included in Annex I under EC Directive 91/414
- In 2006 PSD required that all products containing this active ingredient should carry the following warning in the main area of the container label: "Methiocarb is an anticholinesterase carbamate. Handle with care"
- Accepted by BBPA for use on malting barley

Efficacy guidance

- Use as a surface, overall application when pests active (normally mild, damp weather), pre-drilling or post-emergence. May also be used on cereals or ryegrass in admixture with seed at time of drilling
- Also reduces populations of cutworms and millipedes
- Best on potatoes in late Jul to Aug
- Apply to strawberries before strawing down to prevent seed beetles contaminating crop
- See label for details of suitable application equipment

Restrictions

- This product contains an anticholinesterase carbamate compound. Do not use if under medical advice not to work with such compounds
- Maximum number of treatments 3 per crop for potatoes; 2 per crop for cereals, cauliflowers, maize, oilseed rape, sunflowers; 1 per crop for cabbages, lettuce, leaf spinach, sugar beet; 1 per yr for strawberries. Other crops vary according to product - see labels
- Do not treat any protected crops
- Do not allow pellets to lodge in edible crops

Crop-specific information

- Latest use: before first node detectable for cereals, maize; before three visibly extended internodes for oilseed rape, sunflowers
- HI 7 d for spinach, strawberries; 14 d for Brussels sprouts, cabbages, cauliflowers, lettuce; 18 d for potatoes; 6 mth for sugar beet

Environmental safety

- Dangerous for the environment [2]
- Very toxic to aquatic organisms [2]
- Harmful to aquatic organisms [1, 3-6]
- Dangerous to game, wild birds and animals
- Risk to certain non-target insects or other arthropods
- Avoid surface broadcasting application within 6 m of field boundary to reduce effects on non-target species
- Admixed seed should be drilled and not broadcast, and may not be applied from the air

Hazard classification and safety precautions

Hazard H03 [1-6]; H11 [2]

Risk phrases R22a, R53a [1-6]; R50 [2]; R52 [1, 3-6]

Operator protection A, H [1-6]; J [1, 3-6]; U05a, U20b

Environmental protection E05b [1-6] (7 d); E10a, E22b, E34 [1-6]; E13b [1, 3-6]

Storage and disposal D01, D02, D09a, D11a

Medical advice M02, M03

329 methoxyfenozide

A moulting accelerating diacylhydrazine insecticide
IRAC mode of action code: 18A

Products

Runner	Bayer CropScience	240 g/l	SC	12940

Uses

- Codling moth in ***apples***, ***pears***
- Tortrix moths in ***apples***, ***pears***
- Winter moth in ***apples***, ***pears***

Approval information

- Approval expiry 30 Sep 2007 [11470]

Efficacy guidance

- To achieve best results uniform coverage of the foliage and full spray penetration of the leaf canopy is important, particularly when spraying post-blossom
- For maximum effectiveness on winter moth and tortrix spray pre-blossom when first signs of active larvae are seen, followed by a further spray in June if larvae of the summer generation are present
- For codling moth spray post-blossom to coincide with early to peak egg deposition. Follow-up treatments will normally be needed
- Methoxyfenozide is a moulting accelerating compound (MAC) and may be used in an anti-resistance strategy with other top fruit insecticides (including chitin biosynthesis inhibitors and juvenile hormones) which have a different mode of action
- To reduce further the likelihood of resistance development use at full recommended dose in sufficient water volume to achieve required spray penetration

Restrictions

- Maximum number of treatments 3 per yr but no more than two should be sprayed consecutively

Crop-specific information

- HI 14 d for apples, pears

Environmental safety

- Risk to non-target insects or other arthropods
- LERAP Category B
- Broadcast air-assisted LERAP (5 m)

Hazard classification and safety precautions

Operator protection U20b

Environmental protection E15b, E16b, E22c; E17b (5 m)

Storage and disposal D05, D09a, D11a

330 methyl bromide

The EC Ozone Depleting Substances Regulation (2037/2000) has banned the use of methyl bromide within the UK with certain exceptions, including where a 'Critical Use Exemption' (CUE) has been granted. All other uses ceased in December 2005
IRAC mode of action code: 8A

331 1-methylcyclopropene

An inhibitor of ethylene production for use in stored apples

Products

SmartFresh	Landseer	3.3% w/w	SP	11799

Uses

- Ethylene inhibition in ***apples*** *(post-harvest use)*
- Scald in ***apples*** *(post-harvest use)*

Efficacy guidance

- Best results obtained from treatment of fruit in good condition and of proper quality for long-term storage
- Effects may be reduced in fruit that is in poor condition or ripe prior to storage or harvested late
- Product acts by releasing vapour into store when mixed with water
- Apply as soon as possible after harvest
- Treatment controls superficial scald and maintains fruit firmness and acid content for 3-6 mth in normal air, and 6-9 mth in controlled atmosphere storage
- Ethylene production recommences after removal from storage

Restrictions

- Maximum number of treatments 1 per batch of apples
- Must only be used by suitably trained and competent persons in fumigation operations
- Consult processors before treatment of fruit destined for processing or cider making
- Do not apply in mixture with other products
- Ventilate all areas thoroughly with all refrigeration fans operating at maximum power for at least 15 min before re-entry

Crop-specific information

- Latest use: 7d after harvest of apples
- Product tested on Granny Smith, Gala, Jonagold, Bramley and Cox. Consult distributor or supplier before treating other varieties

Environmental safety

- Unprotected persons must be kept out of treated stores during the 24 h treatment period
- Prior to application ensure that the store can be properly and promptly sealed

Hazard classification and safety precautions

Operator protection U05a
Environmental protection E02a (24 h); E15a, E34
Consumer protection C12
Storage and disposal D01, D02, D07, D14

332 metrafenone

A benzophenone protectant and curative fungicide for cereals
FRAC mode of action code: U8

See also *epoxiconazole + fenpropimorph + metrafenone*
epoxiconazole + metrafenone

Products

1	Attenzo	BASF	300 g/l	SC	11917
2	Flexity	BASF	300 g/l	SC	11775

Uses

- Eyespot in ***spring wheat*** *(reduction)*, ***winter wheat*** *(reduction)* [1, 2]
- Foliar disease control in ***durum wheat*** *(off-label)*, ***grass seed crops*** *(off-label)*, ***spring rye*** *(off-label)*, ***triticale*** *(off-label)*, ***winter rye*** *(off-label)* [2]
- Powdery mildew in ***spring barley***, ***spring oats*** *(qualified minor use)*, ***spring wheat***, ***winter barley***, ***winter oats*** *(qualified minor use)*, ***winter wheat*** [1, 2]

Specific Off-Label Approvals (SOLAs)

- ***durum wheat*** *20070146 expires Dec 2008* [2]
- ***grass seed crops*** *20070146 expires Dec 2008* [2]

- ***spring rye*** *20070146 expires Dec 2008* [2]
- ***triticale*** *20070146 expires Dec 2008* [2]
- ***winter rye*** *20070146 expires Dec 2008* [2]

Approval information
- Metrafenone included in Annex I under EC Directive 91/414
- Accepted by BBPA for use on malting barley
- Approval expiry 31 Dec 2008 [1, 2]

Efficacy guidance
- Best results obtained from treatment at the start of foliar disease attack
- Activity against mildew in wheat is mainly protectant with moderate curative control in the latent phase; activity is entirely protectant in barley
- Useful reduction of eyespot in wheat is obtained if treatment applied at GS 30-32
- Should be used as part of a resistance management strategy that includes mixtures or sequences effective against mildew and non-chemical methods

Restrictions
- Maximum number of treatments 2 per crop
- Do not use sequential treatments containing metrafenone

Crop-specific information
- Latest use: beginning of flowering (GS 31) for wheat, barley, oats

Following crops guidance
- Cereals, oilseed rape, sugar beet, linseed, maize, clover, field beans, peas, turnips, carrots, cauliflowers, onions, lettuce or potatoes may follow a treated cereal crop

Environmental safety
- Dangerous for the environment
- Toxic to aquatic organisms

Hazard classification and safety precautions

Hazard H04, H11
Risk phrases R43, R51, R53a
Operator protection A, H; U05a, U20b
Environmental protection E15a, E34
Storage and disposal D01, D02, D05, D09a, D10c
Medical advice M03

333 metribuzin

A contact and residual triazinone herbicide for use in potatoes
HRAC mode of action code: C1

See also flufenacet + metribuzin

Products

1	Agriguard Metribuzin	AgriGuard	70% w/w	WG	09853
2	Citation WG	United Phosphorus	70% w/w	WG	11929
3	Greencrop Majestic	Greencrop	70% w/w	WG	12422
4	Greencrop Majestic 2	Greencrop	70% w/w	WG	12781
5	Landgold Metribuzin WG	Teliton	70% w/w	WG	12468
6	Lexone 2	DuPont	70% w/w	WG	13804
7	Python	Makhteshim	70% w/w	WG	11166
8	Sencorex WG	Bayer CropScience	70% w/w	WG	11304
9	Shotput	Makhteshim	70% w/w	WG	11960

Uses
- Annual and perennial weeds in ***asparagus*** *(off-label)* [8]
- Annual dicotyledons in ***asparagus*** *(off-label)* [7, 9]; ***carrots*** *(off-label)*, ***parsnips*** *(off-label)* [7]; ***early potatoes***, ***maincrop potatoes*** [1-9]; ***sweet potato*** *(off-label)* [8]
- Annual grasses in ***asparagus*** *(off-label)*, ***carrots*** *(off-label)*, ***parsnips*** *(off-label)* [7]; ***early potatoes***, ***maincrop potatoes*** [1-9]; ***sweet potato*** *(off-label)* [8]
- Fool's parsley in ***carrots*** *(off-label)*, ***parsnips*** *(off-label)* [9]
- Perennial dicotyledons in ***asparagus*** *(off-label)*, ***carrots*** *(off-label)*, ***parsnips*** *(off-label)* [7]

SEE SECTION 3 FOR PRODUCTS ALSO REGISTERED

- Volunteer oilseed rape in ***early potatoes*, *maincrop potatoes*** [1-9]
- Wild mignonette in ***carrots*** *(off-label)*, ***parsnips*** *(off-label)* [8, 9]; ***mallow (althaea spp.)*** *(off-label)* [8]

Specific Off-Label Approvals (SOLAs)

- ***asparagus*** *20030368* [7], *20050102* [8], *20064276* [9]
- ***carrots*** *20030368* [7], *20031887* [8], *20064278* [9]
- ***mallow (althaea spp.)*** *20031887* [8]
- ***parsnips*** *20030368* [7], *20031887* [8], *20064278* [9]
- ***sweet potato*** *20070733* [8]

Approval information

- Metribuzin included in Annex I under EC Directive 91/414

Efficacy guidance

- Best results achieved on weeds at cotyledon to 1-leaf stage
- May be applied pre- or post-emergence of crop on named maincrop varieties and cv. Marfona; pre-emergence only on named early varieties
- Apply to moist soil with well-rounded ridges and few clods
- Activity reduced by dry conditions and on soils with high organic matter content
- On fen and moss soils pre-planting incorporation to 10-15 cm gives increased activity. Incorporate thoroughly and evenly
- With named maincrop and second early potato varieties on soils with more than 10% organic matter shallow pre- or post-planting incorporation may be used. See label for details
- Effective control using a programme of reduced doses is made possible by using a spray of smaller droplets, thus improving retention

Restrictions

- Maximum total dose equivalent to one full dose treatment on early potato varieties and one and a third full doses on maincrop varieties
- Only certain varieties may be treated. Apply pre-emergence only on named first earlies, pre- or post-emergence on named second earlies. On named maincrop varieties apply pre-emergence (except for certain varieties on Sands or Very Light soils) or post-emergence. See label
- All post-emergence treatments must be carried out before longest shoots reach 15 cm
- Do not cultivate after treatment
- Some recommended varieties may be sensitive to post-emergence treatment if crop under stress

Crop-specific information

- Latest use: pre-crop emergence for named early potato varieties; before most advanced shoots have reached 15 cm for post-emergence treatment of potatoes
- On stony or gravelly soils there is risk of crop damage, especially if heavy rain falls soon after application
- When days are hot and sunny delay spraying until evening

Following crops guidance

- Ryegrass, cereals or winter beans may be sown in same season provided at least 16 wk elapsed after treatment and ground ploughed to 15 cm and thoroughly cultivated as soon as possible after harvest and no later than end Dec
- In W Cornwall on soil with more than 5% organic matter early potatoes treated as recommended may be followed by summer planted brassica crops provided the soil has been ploughed, spring rainfall has been normal and at least 14 wk have elapsed since treatment
- Do not grow any vegetable brassicas, lettuces or radishes on land treated the previous yr. Other crops may be sown normally in spring of next yr

Environmental safety

- Dangerous for the environment
- Very toxic to aquatic organisms
- Do not empty into drains
- LERAP Category B

Hazard classification and safety precautions

Hazard H03, H11

Risk phrases R22a [1-9]; R43 [6, 7, 9]; R50, R53a [3-9]

Operator protection A, H [7, 9]; U05a [3, 6, 7, 9]; U08, U13 [1-9]; U14 [6, 7, 9]; U19a [1, 2, 4-9]; U20a [3-6, 8]; U20b [1, 2, 7, 9]

Environmental protection E13b [5]; E15a [1, 2, 6-9]; E16a [1-9]; E16b [1, 3-5, 7-9]; E19b [6, 7, 9]; E34 [2, 3]; E38 [5, 8]

Storage and disposal D01, D02 [2-4, 6, 7, 9]; D05 [1, 2]; D07 [1, 2, 4]; D09a [1-9]; D10c [2]; D11a [1, 3-9]; D12a [3, 5-9]

Medical advice M03 [3]; M05a [6, 7, 9]

334 metsulfuron-methyl

A contact and residual sulfonylurea herbicide used in cereals, linseed and set-aside
HRAC mode of action code: B

See also carfentrazone-ethyl + metsulfuron-methyl

Products

1	Avail XT	AgriGuard	20% w/w	SG	13194
2	Clayton Peak	Clayton	20% w/w	WG	13547
3	Deft	Sipcam	20% w/w	WG	13263
4	Jubilee SX	DuPont	20% w/w	SG	12203
5	Landgold Metsulfuron	Teliton	20% w/w	SG	13421
6	Minx PX	Nufarm UK	20% w/w	WG	13337
7	Pike	Nufarm UK	20% w/w	WG	12746

Uses

- Annual dicotyledons in ***durum wheat*** [6, 7]; ***linseed*** [1, 2, 4-7]; ***spring barley***, ***spring oats***, ***spring wheat***, ***triticale***, ***winter barley***, ***winter oats***, ***winter wheat*** [1-7]
- Chickweed in ***durum wheat*** [6, 7]; ***linseed*** [1, 2, 4-7]; ***spring barley***, ***spring oats***, ***spring wheat***, ***triticale***, ***winter barley***, ***winter oats***, ***winter wheat*** [1-7]
- Green cover in ***land temporarily removed from production*** [1, 4, 5]
- Mayweeds in ***durum wheat*** [6, 7]; ***linseed*** [1, 2, 4-7]; ***spring barley***, ***spring oats***, ***spring wheat***, ***triticale***, ***winter barley***, ***winter oats***, ***winter wheat*** [1-7]

Approval information

- Metsulfuron-methyl included in Annex I under EC Directive 91/414
- Accepted by BBPA for use on malting barley

Efficacy guidance

- Best results achieved on small, actively growing weeds up to 6-true leaf stage. Good spray cover is important
- Weed control may be reduced in dry soil conditions
- Commonly used in tank-mixture on wheat and barley with other cereal herbicides to improve control of resistant dicotyledons (cleavers, fumitory, ivy-leaved speedwell), larger weeds and grasses. See label for recommended mixtures
- Metsulfuron-methyl is a member of the ALS-inhibitor group of herbicides and products should be used in a planned Resistance Management strategy. See Section 5 for more information

Restrictions

- Maximum number of treatments 1 per crop or per yr (for set-aside)
- Product must only be used after 1 Feb
- Do not apply within 7 d of rolling
- Do not use on cereal crops undersown with grass or legumes
- Do not use in tank mixture on oats, triticale or linseed. On linseed allow at least 7 d before or after other treatments
- Do not tank mix with chlorpyrifos. Allow at least 14 d before or after chlorpyrifos treatments
- Consult contract agents before use on a cereal crop for seed
- Do not use on any crop suffering stress from drought, waterlogging, frost, deficiency, pest or disease attack or apply within 7 d of rolling
- Specific restrictions apply to use in sequence or tank mixture with other sulfonylurea or ALS-inhibiting herbicides. See label for details
- Spraying equipment should not be drained or flushed onto land planted, or to be planted, with trees or crops other than cereals and should be thoroughly cleansed after use - see label for instructions

Crop-specific information

- Latest use: before flag leaf sheath extending stage for cereals (GS 41); before flower buds visible or crop 30 cm tall for linseed [1, 2, 4-7]; before 1 Aug in yr of treatment for land not being used for crop production [1, 4, 5]
- Apply after 1 Feb to wheat, oats and triticale from 2-leaf (GS 12), and to barley from 3-leaf (GS 13) until flag-leaf sheath extending (GS 41)
- On linseed allow at least 7 d (10 d if crop is growing poorly or under stress) before or after other treatments
- Use in set-aside when a full green cover is established and made up predominantly of grassland, wheat, barley, oats or triticale. Do not use on seedling grasses

Following crops guidance

- Only cereals, oilseed rape, field beans or grass may be sown in same calendar year after treating cereals with the product alone. Other restrictions apply to tank mixtures. See label for details
- Only cereals should be planted within 16 mth of applying to a linseed crop or set-aside
- In the event of failure of a treated crop sow only wheat within 3 mth after treatment

Environmental safety

- Dangerous for the environment
- Very toxic to aquatic organisms
- Take extreme care to avoid damage by drift onto broad-leaved plants outside the target area, onto surface waters or ditches or onto land intended for cropping
- LERAP Category B
- A range of broad leaved species will be fully or partially controlled when used in land temporarily removed from production, hence product may not be suitable where wild flower borders or other forms of conservation headland are being developed
- Before use on land temporarily removed from production as part of grant-aided scheme, ensure compliance with the management rules
- Green cover on land temporarily removed from production must not be grazed by livestock or harvested for human or animal consumption or used for animal bedding [1, 4, 5]

Hazard classification and safety precautions

Hazard H11
Risk phrases R50, R53a
Operator protection U08, U20b [1]; U19a [1, 3]; U20a [3]
Environmental protection E15b, E16a [1-7]; E16b [6, 7]; E34 [1]; E38 [1, 3-7]
Storage and disposal D01 [1]; D09a [1-3, 6, 7]; D10b [3]; D11a [2, 6, 7]; D12a [1-7]
Medical advice M05a [1]

335 metsulfuron-methyl + thifensulfuron-methyl

A contact residual and translocated sulfonylurea herbicide mixture for use in cereals
HRAC mode of action code: B + B

Products

1	Chimera SX	DuPont	2.9:42.9% w/w	SG	13171
2	Concert SX	DuPont	4:40% w/w	SG	12288
3	Harmony M SX	DuPont	4:40% w/w	SG	12258
4	Pennant	Headland	4:40% w/w	SG	13350

Uses

- Annual dicotyledons in ***spring barley***, ***spring wheat***, ***winter wheat*** [1-4]; ***winter barley*** [1]
- Chickweed in ***spring barley***, ***spring wheat***, ***winter wheat*** [1-4]; ***winter barley*** [1]
- Creeping thistle in ***spring barley***, ***spring wheat***, ***winter barley***, ***winter wheat*** [1]
- Mayweeds in ***spring barley***, ***spring wheat***, ***winter wheat*** [1-4]; ***winter barley*** [1]
- Polygonums in ***spring barley***, ***spring wheat***, ***winter wheat*** [1-4]; ***winter barley*** [1]
- Red dead-nettle in ***spring barley***, ***spring wheat***, ***winter wheat*** [1-4]; ***winter barley*** [1]
- Shepherd's purse in ***spring barley***, ***spring wheat***, ***winter wheat*** [1-4]; ***winter barley*** [1]

Approval information

- Metsulfuron-methyl and thifensulfuron-methyl included in Annex I under EC Directive 91/414
- Accepted by BBPA for use on malting barley

Efficacy guidance

- Best results by application to small, actively growing weeds up to 6-true leaf stage
- Ensure good spray cover
- Susceptible weeds stop growing almost immediately but symptoms may not be visible for about 2 wk
- Effectiveness may be reduced by heavy rain or if soil conditions very dry
- Metsulfuron-methyl and thifensulfuron-methyl are members of the ALS-inhibitor group of herbicides and products should be used in a planned Resistance Management strategy. See Section 5 for more information

Restrictions

- Maximum number of treatments 1 per crop
- Products may only be used after 1 Feb
- Do not use on any crop suffering stress from drought, waterlogging, frost, deficiency, pest or disease attack or any other cause
- Do not use on crops undersown with grasses, clover or legumes, or any other broad leaved crop
- Specific restrictions apply to use in sequence or tank mixture with other sulfonylurea or ALS-inhibiting herbicides. See label for details
- Do not apply within 7 d of rolling
- Consult contract agents before use on a cereal crop grown for seed

Crop-specific information

- Latest use: before flag leaf sheath extending (GS 39) for barley and wheat

Following crops guidance

- Only cereals, oilseed rape, field beans or grass may be sown in same calendar year after treatment
- Additional constraints apply after use of certain tank mixtures. See label
- In the event of crop failure sow only winter wheat within 3 mth after treatment and after ploughing and cultivating to a depth of at least 15 cm

Environmental safety

- Dangerous for the environment
- Very toxic to aquatic organisms
- LERAP Category B
- Take extreme care to avoid damage by drift onto broad-leaved plants outside the target area, or onto ponds, waterways or ditches
- Spraying equipment should not be drained or flushed onto land planted, or to be planted, with trees or crops other than cereals and should be thoroughly cleansed after use - see label for instructions

Hazard classification and safety precautions

Hazard H11
Risk phrases R50, R53a
Operator protection U08, U19a, U20b
Environmental protection E15b [2-4]; E16a, E38 [1-4]
Storage and disposal D09a, D12a [1-4]; D11a [2-4]

336 metsulfuron-methyl + tribenuron-methyl

A sulfonylurea herbicide mixture for cereals
HRAC mode of action code: B + B

Products

1	Ally Max SX	DuPont	14.3:14.3% w/w	SG	12631
2	Biplay SX	DuPont	11.1:22.2% w/w	SG	12246
3	Traton SX	DuPont	11.1:22.2% w/w	SG	12270

Uses

- Annual dicotyledons in ***durum wheat*** *(off-label)*, ***spring oats*** [1]; ***spring barley***, ***spring wheat***, ***triticale***, ***winter barley***, ***winter oats***, ***winter wheat*** [1-3]
- Chickweed in ***durum wheat*** *(off-label)*, ***spring oats*** [1]; ***spring barley***, ***spring wheat***, ***triticale***, ***winter barley***, ***winter oats***, ***winter wheat*** [1-3]
- Field pansy in ***durum wheat*** *(off-label)* [1]; ***spring barley***, ***spring wheat***, ***triticale***, ***winter barley***, ***winter oats***, ***winter wheat*** [1-3]

SEE SECTION 3 FOR PRODUCTS ALSO REGISTERED

- Field speedwell in ***spring oats*** [1]
- Hemp-nettle in ***durum wheat*** *(off-label)*, ***spring oats*** [1]; ***spring barley***, ***spring wheat***, ***triticale***, ***winter barley***, ***winter oats***, ***winter wheat*** [1-3]
- Mayweeds in ***durum wheat*** *(off-label)*, ***spring oats*** [1]; ***spring barley***, ***spring wheat***, ***triticale***, ***winter barley***, ***winter oats***, ***winter wheat*** [1-3]
- Red dead-nettle in ***durum wheat*** *(off-label)*, ***spring oats*** [1]; ***spring barley***, ***spring wheat***, ***triticale***, ***winter barley***, ***winter oats***, ***winter wheat*** [1-3]
- Volunteer oilseed rape in ***durum wheat*** *(off-label)*, ***spring barley***, ***spring oats***, ***spring wheat***, ***triticale***, ***winter barley***, ***winter oats***, ***winter wheat*** [1]
- Volunteer sugar beet in ***durum wheat*** *(off-label)*, ***spring barley***, ***spring oats***, ***spring wheat***, ***triticale***, ***winter barley***, ***winter oats***, ***winter wheat*** [1]

Specific Off-Label Approvals (SOLAs)

- ***durum wheat*** *20070453* [1]

Approval information

- Metsulfuron-methyl and tribenuron-methyl included in Annex I under EC Directive 91/414
- Accepted by BBPA for use on malting barley

Efficacy guidance

- Best results obtained when applied to small actively growing weeds
- Product acts by foliar and root uptake. Good spray cover essential but performance may be reduced when soil conditions are very dry and residual effects may be reduced by heavy rain
- Weed growth inhibited within hours of treatment and many show marked colour changes as they die back. Full effects may not be apparent for up to 4 wk
- Metsulfuron-methyl and tribenuron-methyl are members of the ALS-inhibitor group of herbicides and products should be used in a planned Resistance Management strategy. See Section 5 for more information

Restrictions

- Maximum number of treatments 1 per crop
- Product must only be used after 1 Feb and after crop has three leaves
- Do not apply to a crop suffering from drought, water-logging, low temperatures, pest or disease attack, nutrient deficiency, soil compaction or any other stress
- Do not use on crops undersown with grasses, clover or other legumes
- Do not apply within 7 d of rolling
- Specific restrictions apply to use in sequence or tank mixture with other sulfonylurea or ALS-inhibiting herbicides. See label for details

Crop-specific information

- Latest use: before flag leaf sheath extending

Following crops guidance

- Only cereals, field beans or oilseed rape may be sown in the same calendar yr as harvest of a treated crop [1]
- Only cereals, field beans, grass or oilseed rape may be sown in the same calendar yr as harvest of a treated crop [2, 3]
- In the event of failure of a treated crop only winter wheat may sown within 3 mth of treatment, and only after ploughing and cultivation to 15 cm min

Environmental safety

- Dangerous for the environment
- Very toxic to aquatic organisms
- Some non-target crops are highly sensitive. Take extreme care to avoid drift outside the target area, or onto ponds, waterways or ditches
- Spraying equipment should be thoroughly cleaned in accordance with manufacturer's instructions
- LERAP Category B [1]

Hazard classification and safety precautions

Hazard H04, H11
Risk phrases R43, R50, R53a
Operator protection A, H; U08, U19a, U20b
Environmental protection E15a, E38 [1-3]; E16a [1]
Storage and disposal D01, D02, D09a, D11a, D12a

337 myclobutanil

A systemic, protectant and curative triazole fungicide
FRAC mode of action code: 3

Products

1	Aristocrat	AgriGuard	200 g/l	EW	12343
2	Masalon	Rigby Taylor	45 g/l	EW	12385
3	Systhane 20EW	Landseer	200 g/l	EW	09396

Uses

- American gooseberry mildew in ***blackcurrants***, ***gooseberries*** [1, 3]
- Black spot in ***ornamental plant production***, ***roses*** [1, 3]
- Blossom wilt in ***cherries*** *(off-label)*, ***mirabelles*** *(off-label)* [3]
- Brown rot in ***apricots*** *(off-label)*, ***peaches*** *(off-label)* [3]
- Fusarium patch in ***managed amenity turf*** [2]
- Plum rust in ***plums*** *(off-label)* [3]
- Powdery mildew in ***apples***, ***ornamental plant production***, ***pears***, ***roses***, ***strawberries*** [1, 3]; ***apricots*** *(off-label)*, ***artichokes*** *(off-label)*, ***courgettes*** *(off-label)*, ***gherkins*** *(off-label)*, ***hops*** *(off-label)*, ***peaches*** *(off-label)*, ***protected aubergines*** *(off-label)*, ***protected blackberries*** *(off-label)*, ***protected courgettes*** *(off-label)*, ***protected cucumbers*** *(off-label)*, ***protected gherkins*** *(off-label)*, ***protected raspberries*** *(off-label)*, ***protected rubus hybrids*** *(off-label)*, ***protected tomatoes*** *(off-label)*, ***wine grapes*** *(off-label)* [3]
- Rust in ***ornamental plant production***, ***roses*** [1, 3]
- Scab in ***apples***, ***pears*** [1, 3]

Specific Off-Label Approvals (SOLAs)

- ***apricots*** *20070626* [3]
- ***artichokes*** *20070624* [3]
- ***cherries*** *991535* [3]
- ***courgettes*** *20070627* [3]
- ***gherkins*** *20070627* [3]
- ***hops*** *20021412* [3]
- ***mirabelles*** *991535* [3]
- ***peaches*** *20070626* [3]
- ***plums*** *20012459* [3]
- ***protected aubergines*** *20070625* [3]
- ***protected blackberries*** *20051189* [3]
- ***protected courgettes*** *20070627* [3]
- ***protected cucumbers*** *20070627* [3]
- ***protected gherkins*** *20070627* [3]
- ***protected raspberries*** *20051189* [3]
- ***protected rubus hybrids*** *20051189* [3]
- ***protected tomatoes*** *20070625* [3]
- ***wine grapes*** *20040116* [3]

Approval information

- Accepted by BBPA for use on hops

Efficacy guidance

- Best results achieved when used as part of routine preventive spray programme from bud burst to end of flowering in apples and pears and from just before the signs of mildew infection in blackcurrants and gooseberries [3]
- In strawberries commence spraying at, or just prior to, first flower. Post-harvest sprays may be required on mildew-susceptible varieties where mildew is present and likely to be damaging [3]
- Spray at 7-14 d intervals depending on disease pressure and dose applied [3]
- For improved scab control on apples in post-blossom period tank-mix with mancozeb or captan [3]
- Apply alone from mid-Jun for control of secondary mildew on apples and pears
- On roses spray at first signs of disease and repeat every 2 wk. In high risk areas spray when leaves emerge in spring, repeat 1 wk later and then continue normal programme [3]

SEE SECTION 3 FOR PRODUCTS ALSO REGISTERED

- Treatment of managed amenity turf may be carried out at any time of year before, or at, the first sign of disease [2]
- Myclobutanil products should be used in conjunction with other fungicides with a different mode of action to reduce the possibility of resistance developing

Restrictions

- Maximum total dose equivalent to ten full dose treatments in apples and pears; six full dose treatments in blackcurrants, gooseberries, strawberries
- Maximum number of treatments on managed amenity turf: two per yr [2]
- Do not mow turf within 24 hr after treatment [2]

Crop-specific information

- HI 28 d (grapevines); 21 d (cherries, mirabelles); 14 d (apples, pears, blackcurrants, gooseberries, hops); 3 d (plums, protected blackberries, protected raspberries, protected Rubus hybrids, strawberries)
- Product may be applied to newly sown turf after the two-leaf stage but in view of the large number of turf grass cultivars a safety test on a small area is recommended before large scale treatment [2]

Environmental safety

- Dangerous for the environment
- Toxic (harmful [2]) to aquatic organisms
- LERAP Category B [2]

Hazard classification and safety precautions

Hazard H03, H11 [1, 3]
Risk phrases R22b, R51, R63 [1, 3]; R52 [2]; R53a [1-3]
Operator protection A, H [1-3]; J [1, 3]; U08, U20a [1-3]; U15 [1, 3]
Environmental protection E15a [1-3]; E16a [2]
Storage and disposal D01, D02 [1, 3]; D05, D10c, D12a [1-3]
Medical advice M05b [1, 3]

338 1-naphthylacetic acid

A plant growth regulator to promote rooting of cuttings

See also 4-indol-3-ylbutyric acid + 1-naphthylacetic acid

Products

1	Rhizopon B Powder (0.1%)	Rhizopon	0.1% w/w	DP	09089
2	Rhizopon B Powder (0.2%)	Rhizopon	0.2% w/w	DP	09090
3	Rhizopon B Tablets	Rhizopon	25 mg a.i.	WT	09091

Uses

- Rooting of cuttings in ***ornamental plant production***

Efficacy guidance

- Dip moistened base of cuttings into powder immediately before planting [1, 2]
- See label for details of concentrations recommended for promotion of rooting in cuttings of different species [3]
- Dip prepared cuttings in solution for 4-24 h depending on species [3]

Restrictions

- Maximum number of treatments 1 per cutting

Crop-specific information

- Latest use: before cutting insertion for ornamental specimens

Hazard classification and safety precautions

Operator protection U08, U19a, U20a
Environmental protection E15a, E34
Storage and disposal D09a, D11a

339 napropamide

A soil applied alkanamide herbicide for oilseed rape, fruit and woody ornamentals
HRAC mode of action code: K3

Products

1	AC 650	United Phosphorus	450 g/l	SC	11102
2	Devrinol	United Phosphorus	450 g/l	SC	09374
3	Jouster	AgriGuard	450 g/l	SC	12741
4	Naprop	Globachem	450 g/l	SC	13402

Uses

- Annual dicotyledons in ***blackcurrants, forest nurseries, gooseberries, ornamental plant production, raspberries, strawberries*** [2, 3]; ***broccoli, brussels sprouts, cabbages, calabrese, cauliflowers, kale*** [2-4]; ***winter oilseed rape*** [1, 3, 4]; ***woody ornamentals*** [2]
- Annual grasses in ***blackcurrants, forest nurseries, gooseberries, ornamental plant production, raspberries, strawberries*** [2, 3]; ***broccoli, brussels sprouts, cabbages, calabrese, cauliflowers, kale*** [2-4]; ***winter oilseed rape*** [1, 3, 4]; ***woody ornamentals*** [2]
- Cleavers in ***blackcurrants, forest nurseries, gooseberries, ornamental plant production, raspberries, strawberries*** [2, 3]; ***broccoli, brussels sprouts, cabbages, calabrese, cauliflowers, kale*** [2-4]; ***winter oilseed rape*** [3, 4]; ***woody ornamentals*** [2]
- Groundsel in ***blackcurrants, forest nurseries, gooseberries, ornamental plant production, raspberries, strawberries*** [2, 3]; ***broccoli, brussels sprouts, cabbages, calabrese, cauliflowers, kale*** [2-4]; ***winter oilseed rape*** [3, 4]; ***woody ornamentals*** [2]

Efficacy guidance

- Best results obtained from treatment pre-emergence of weeds but product may be used in conjunction with contact herbicide such as paraquat for control of emerged weeds. Otherwise remove existing weeds before application
- Weed control may be reduced where spray is mixed too deeply in the soil
- Manure, crop debris or other organic matter may reduce weed control
- Napropamide broken down by sunlight, so application during such conditions not recommended. Most crops recommended for treatment between Nov and end-Feb
- Apply to winter oilseed rape pre-emergence or as pre-drilling treatment in tank-mixture with trifluralin and incorporate within 30 min [1, 3, 4]
- Post-emergence use of specific grass weedkilller recommended where volunteer cereals are a serious problem
- Increase water volume to ensure adequate dampening of compost when treating containerised nursery stock with dense leaf canopy [2]

Restrictions

- Maximum number of treatments 1 per crop or yr
- Do not use on Sands
- Do not use on soils with more than 10% organic matter
- Strawberries, blackcurrants, gooseberries and raspberries must be treated between 1 Nov and end Feb [2, 3]
- Applications to strawberries must only be made where the mature foliage has been previously removed [2, 3]
- Applications to ornamentals and forest nurseries must be made after 1 Nov and before end Apr
- Do not treat ornamentals in containers of less than 1 litre [2]
- Some phytotoxicity seen on yellow and golden varieties of conifers and container grown alpines. On any ornamental variety treat only a small number of plants in first season [2]
- Consult processors before use on any crop for processing

Crop-specific information

- Latest use: before transplanting for brassicas; pre-emergence for winter oilseed rape; before end Feb for strawberries, bush and cane fruit; before end of Apr for field and container grown ornamental trees and shrubs
- Where minimal cultivation used to establish oilseed rape, tank-mixture may be applied directly to stubble and mixed into top 25 mm as part of surface cultivations [1]
- Apply up to 14 d prior to drilling winter oilseed rape [1, 3, 4]

- Apply to strawberries established for at least one season or to maiden crops as long as planted carefully and no roots exposed, between 1 Nov and end Feb. Do not treat runners of poor vigour or with shallow roots, or runner beds [2, 3]
- Newly planted ornamentals should have no roots exposed. Do not treat stock of poor vigour or with shallow roots. Treatments made in Mar and Apr must be followed by 25 mm irrigation within 24 h [2]
- Bush and cane fruit must be established for at least 10 mth before spraying and treated between 1 Nov and end Feb [2, 3]

Following crops guidance

- After use in fruit or ornamentals no crop can be drilled within 7 mth of treatment. Leaf, flowerhead, root and fodder brassica crops may be drilled after 7 mth; potatoes, maize, peas or dwarf beans after 9 mth; autumn sown wheat or grass after 18 mth; any crop after 2 yr
- After use in oilseed rape only oilseed rape, swedes, fodder turnips, brassicas or potatoes should be sown within 12 mth of application
- Soil should be mould-board ploughed to a depth of at least 200 mm before drilling or planting any following crop

Environmental safety

- Dangerous for the environment
- Toxic to aquatic organisms

Hazard classification and safety precautions

Hazard H04, H11
Risk phrases R36, R38, R51, R53a
Operator protection A [1-4]; C [2-4]; U05a, U09a, U11, U20b
Environmental protection E15a [2-4]; E34, E38 [1-4]
Storage and disposal D01, D02, D05, D09a, D10c [1-4]; D12a [1]; D12b [2-4]

340 nicosulfuron

A sulfonylurea herbicide for maize
HRAC mode of action code: B

Products

1	Clayton Delilah	Clayton	40 g/l	SC	12610
2	Greencrop Folklore	Greencrop	40 g/l	SC	12867
3	Landgold Nicosulfuron	Teliton	40 g/l	SC	13464
4	Samson	Syngenta	40 g/l	SC	12141
5	Standon Frontrunner	Standon	40 g/l	SC	13308

Uses

- Annual dicotyledons in ***forage maize***
- Annual meadow grass in ***forage maize***
- Chickweed in ***forage maize***
- Couch in ***forage maize***
- Mayweeds in ***forage maize***
- Ryegrass in ***forage maize***
- Shepherd's purse in ***forage maize***
- Volunteer oilseed rape in ***forage maize***

Efficacy guidance

- Product should be applied post-emergence between 2 and 8 crop leaf stage to emerged weeds from the 2-leaf stage
- Product acts mainly by foliar activity. Ensure good spray cover of the weeds
- Nicosulfuron is a member of the ALS-inhibitor group of herbicides. To avoid the build up of resistance do not use any product containing an ALS-inhibitor herbicide with claims for control of grass weeds more than once on any crop
- Use these products as part of a resistance management strategy that includes cultural methods of control and does not use ALS inhibitors as the sole chemical method of grass weed control

Restrictions

- Maximum number of treatments 1 per crop
- Do not use if an organophosphorus soil insecticide has been used on the same crop

- Do not mix with foliar or liquid fertilisers or specified herbicides. See label for details
- Do not apply in mixture, or sequence, with any other sulfonyl-urea containing product
- Do not treat crops under stress
- Do not apply if rainfall is forecast to occur within 6 h of application

Crop-specific information

- Latest use: up to and including 8 true leaves for maize
- Some transient yellowing may be seen from 1-2 wk after treatment
- Only healthy maize crops growing in good field conditions should be treated

Following crops guidance

- Winter wheat (not undersown) may be sown 4 mth after treatment; maize may be sown in the spring following treatment; all other crops may be sown from the next autumn

Environmental safety

- Dangerous for the environment [2-5]
- Very toxic to aquatic organisms [2-5]
- LERAP Category B

Hazard classification and safety precautions

Hazard H04 [1-5]; H11 [2-5]
Risk phrases R38 [1-5]; R50, R53a [2-5]
Operator protection A, H; U02a, U05a, U14, U20b [1-5]; U10 [2-5]
Environmental protection E15a [1, 2]; E15b [3-5]; E16a, E16b [1-5]; E38 [2, 4, 5]
Storage and disposal D01, D02 [1-5]; D05, D10c, D12a [2-5]; D07 [3]; D09a [1, 3, 5]; D10b [1]
Medical advice M05a

SECTION 2

341 nicotine

A general purpose, non-persistent, contact, alkaloid insecticide
IRAC mode of action code: 4B

Products

1	Nicotine 40% Shreds	Dow	40% w/w	FU	05725
2	No-Fid	Certis	75 g/l	LI	11183
3	Stalwart	United Phosphorus	75 g/l	SL	11877
4	XL-All Nicotine 95%	Vitax	950 g/l	SL	07402

Uses

- Aphids in ***all bush fruit, all top fruit, flower bulbs, leaf brassicas, navy beans, nursery stock, parsley, peas, protected carnations, protected chrysanthemums, protected crops, protected figs, protected flowers, protected melons, protected peaches, protected roses, protected wine grapes*** [4]; ***apples, artichokes, asparagus, broccoli, brussels sprouts, bulb onions, cabbages, calabrese, cauliflowers, celeriac, chicory, chinese cabbage, chives, combining peas, courgettes, cress, cucumbers, dwarf beans, edible podded peas, endives, fennel, frise, garlic, kale, lamb's lettuce, leeks, marrows, peppers, potatoes, pumpkins, radicchio, radishes, shallots, spring field beans, squashes, sweetcorn, tomatoes (outdoor), vining peas, winter field beans*** [2, 3]; ***aubergines, protected courgettes, protected herbs (see appendix 6), protected peppers, protected potatoes, protected salad onions*** [1]; ***broad beans, carrots, celery (outdoor), leaf spinach, lettuce, ornamental plant production, parsnips, red beet, runner beans, strawberries, swedes, turnips*** [2-4]; ***protected cucumbers, protected lettuce*** [1, 4]; ***protected ornamentals, protected tomatoes*** [1-4]
- Capsids in ***all bush fruit, all top fruit, flower bulbs, protected crops*** [4]; ***apples*** [2, 3]
- Caterpillars in ***artichokes, asparagus, broad beans, broccoli, brussels sprouts, bulb onions, cabbages, calabrese, carrots, cauliflowers, celeriac, celery (outdoor), chicory, chinese cabbage, chives, combining peas, courgettes, cress, cucumbers, dwarf beans, edible podded peas, endives, fennel, frise, garlic, kale, lamb's lettuce, leaf spinach, leeks, lettuce, marrows, ornamental plant production, parsnips, peppers, potatoes, protected ornamentals, protected tomatoes, pumpkins, radicchio, radishes, red beet, runner beans, shallots, spring field beans, squashes, swedes, sweetcorn, tomatoes (outdoor), turnips, vining peas, winter field beans*** [2, 3]

- Glasshouse whitefly in ***aubergines, protected courgettes, protected cucumbers, protected herbs (see appendix 6), protected lettuce, protected ornamentals, protected peppers, protected potatoes, protected salad onions, protected tomatoes*** [1]
- Green leafhopper in ***aubergines, protected courgettes, protected cucumbers, protected herbs (see appendix 6), protected lettuce, protected ornamentals, protected peppers, protected potatoes, protected salad onions, protected tomatoes*** [1]
- Leaf miner in ***artichokes, asparagus, broad beans, broccoli, brussels sprouts, bulb onions, cabbages, calabrese, carrots, cauliflowers, celeriac, celery (outdoor), chicory, chinese cabbage, chives, combining peas, courgettes, cress, cucumbers, dwarf beans, edible podded peas, endives, fennel, frise, garlic, kale, lamb's lettuce, leaf spinach, leeks, lettuce, marrows, ornamental plant production, parsnips, peppers, potatoes, protected ornamentals, protected tomatoes, pumpkins, radicchio, radishes, red beet, runner beans, shallots, spring field beans, squashes, swedes, sweetcorn, tomatoes (outdoor), turnips, vining peas, winter field beans*** [2, 3]; ***flower bulbs, leaf brassicas, protected crops*** [4]
- Leafhoppers in ***all bush fruit, flower bulbs, protected crops*** [4]
- Potato virus vectors in ***chitting potatoes*** [1]
- Sawflies in ***all bush fruit, all top fruit, flower bulbs, protected crops*** [4]; ***gooseberries*** [2, 3]
- Thrips in ***aubergines, protected courgettes, protected cucumbers, protected herbs (see appendix 6), protected lettuce, protected ornamentals, protected peppers, protected potatoes, protected salad onions, protected tomatoes*** [1]; ***flower bulbs, protected crops*** [4]

Efficacy guidance

- Apply as foliar spray, taking care to cover undersides of leaves and repeat as necessary or dip young plants, cuttings or strawberry runners before planting out
- Best results achieved by spraying at air temperatures above 16 °C but do not treat in bright sunlight or windy weather [1]
- Fumigate glasshouse crops at temperatures of at least 16 °C. See label for details of recommended fumigation procedure [1]
- Potatoes may be fumigated in chitting houses to control virus-spreading aphids [1]
- May be used in integrated control systems to give partial control of organophosphorus-, organochlorine- and pyrethroid-resistant whitefly

Restrictions

- Nicotine is subject to the Poisons Rules 1982 and the Poisons Act 1972. See notes in Section 5 [1, 4]
- Maximum number of treatments not specified
- On plants of unknown sensitivity test first on a small scale
- Keep unprotected persons out of treated glasshouses for at least 12 h
- Wear overall, hood, rubber gloves and respirator if entering glasshouse within 12 h of fumigating [1]

Crop-specific information

- HI 2 d for most crops - check labels

Environmental safety

- Dangerous for the environment [4]
- Toxic to aquatic organisms [4]
- Dangerous to fish or other aquatic life. Do not contaminate surface waters or ditches with chemical or used container
- Dangerous to livestock. Keep all livestock out of treated areas/away from treated water for at least 12 h
- Dangerous to game, wild birds and animals
- Harmful to bees. Do not apply to crops in flower or to those in which bees are actively foraging. Do not apply when flowering weeds are present
- Keep in original container, tightly closed, in a safe place, under lock and key [1, 4]

Hazard classification and safety precautions

Hazard H01 [2-4]; H03, H07 [1]; H11 [4]

Risk phrases R20, R21 [1]; R22a [1, 2]; R25, R27, R53a [2-4]; R36, R52 [2, 3]; R37 [2]; R38 [3]; R51 [4]

Operator protection A [1-4]; C [2-4]; D, J [1]; H [1, 4]; K, M [4]; U02a, U04a, U05a, U10, U13 [2-4]; U05b [1]; U08 [4]; U09a [2, 3]; U19a, U20a [1-4]

Environmental protection E02a, E06b [1-4] (12 h); E10a, E12d, E12e [1-4]; E13b [1-3]; E15a [4]; E34 [1, 3, 4]

Consumer protection C02a [1] (24 h); C02a [4] (2 d)

Storage and disposal D01, D02 [1, 3, 4]; D05, D09a [2, 3]; D09b [1, 4]; D10a [4]; D10b [2]; D10c [3]; D11a [1]

Medical advice M03 [1]; M04a [2-4]

342 oxadiazon

A residual and contact oxadiazole herbicide for fruit and ornamentals

HRAC mode of action code: E

Products

1	Clayton Oxen FL	Clayton	250 g/l	EC	12861
2	Clayton Oxen G	Clayton	2% w/w	GR	12770
3	Ronstar 2G	Certis	2% w/w	GR	12965
4	Ronstar Liquid	Certis	250 g/l	EC	11215
5	Standon Roxx L	Standon	250 g/l	EC	13309

Uses

- Annual dicotyledons in ***almonds*** *(off-label)*, ***bilberries*** *(off-label)*, ***blackberries*** *(off-label)*, ***blueberries*** *(off-label)*, ***chestnuts*** *(off-label)*, ***cranberries*** *(off-label)*, ***hazel nuts*** *(off-label)*, ***redcurrants*** *(off-label)*, ***rubus hybrids*** *(off-label)*, ***walnuts*** *(off-label)*, ***whitecurrants*** *(off-label)* [4]; ***apple orchards***, ***blackcurrants***, ***gooseberries***, ***hops***, ***pear orchards***, ***raspberries***, ***wine grapes***, ***woody ornamentals*** [1, 4, 5]; ***ornamental plant production*** [2, 3]
- Annual grasses in ***almonds*** *(off-label)*, ***bilberries*** *(off-label)*, ***blackberries*** *(off-label)*, ***blueberries*** *(off-label)*, ***chestnuts*** *(off-label)*, ***cranberries*** *(off-label)*, ***hazel nuts*** *(off-label)*, ***redcurrants*** *(off-label)*, ***rubus hybrids*** *(off-label)*, ***walnuts*** *(off-label)*, ***whitecurrants*** *(off-label)* [4]; ***apple orchards***, ***blackcurrants***, ***gooseberries***, ***hops***, ***pear orchards***, ***raspberries***, ***wine grapes***, ***woody ornamentals*** [1, 4, 5]; ***ornamental plant production*** [2, 3]
- Bindweeds in ***almonds*** *(off-label)*, ***bilberries*** *(off-label)*, ***blackberries*** *(off-label)*, ***blueberries*** *(off-label)*, ***chestnuts*** *(off-label)*, ***cranberries*** *(off-label)*, ***hazel nuts*** *(off-label)*, ***redcurrants*** *(off-label)*, ***rubus hybrids*** *(off-label)*, ***walnuts*** *(off-label)*, ***whitecurrants*** *(off-label)* [4]; ***apple orchards***, ***blackcurrants***, ***gooseberries***, ***hops***, ***pear orchards***, ***raspberries***, ***wine grapes***, ***woody ornamentals*** [1, 4, 5]
- Cleavers in ***almonds*** *(off-label)*, ***bilberries*** *(off-label)*, ***blackberries*** *(off-label)*, ***blueberries*** *(off-label)*, ***chestnuts*** *(off-label)*, ***cranberries*** *(off-label)*, ***hazel nuts*** *(off-label)*, ***redcurrants*** *(off-label)*, ***rubus hybrids*** *(off-label)*, ***walnuts*** *(off-label)*, ***whitecurrants*** *(off-label)* [4]; ***apple orchards***, ***blackcurrants***, ***gooseberries***, ***hops***, ***pear orchards***, ***raspberries***, ***wine grapes***, ***woody ornamentals*** [1, 4, 5]
- Knotgrass in ***almonds*** *(off-label)*, ***bilberries*** *(off-label)*, ***blackberries*** *(off-label)*, ***blueberries*** *(off-label)*, ***chestnuts*** *(off-label)*, ***cranberries*** *(off-label)*, ***hazel nuts*** *(off-label)*, ***redcurrants*** *(off-label)*, ***rubus hybrids*** *(off-label)*, ***walnuts*** *(off-label)*, ***whitecurrants*** *(off-label)* [4]; ***apple orchards***, ***blackcurrants***, ***gooseberries***, ***hops***, ***pear orchards***, ***raspberries***, ***wine grapes***, ***woody ornamentals*** [1, 4, 5]

Specific Off-Label Approvals (SOLAs)

- ***almonds*** *20063295* [4]
- ***bilberries*** *20063296* [4]
- ***blackberries*** *20063297* [4]
- ***blueberries*** *20063296* [4]
- ***chestnuts*** *20063295* [4]
- ***cranberries*** *20063296* [4]
- ***hazel nuts*** *20063295* [4]
- ***redcurrants*** *20063296* [4]
- ***rubus hybrids*** *20063297* [4]
- ***walnuts*** *20063295* [4]
- ***whitecurrants*** *20063296* [4]

Approval information

- Accepted by BBPA for use on hops

SEE SECTION 3 FOR PRODUCTS ALSO REGISTERED

Efficacy guidance

- Apply as a directed spray or as a spot treatment as directed. Avoid direct contact with the crop. See label [1, 4, 5]
- Apply granules as soon after potting as possible and before crop plant is making soft growth [2, 3]
- Best results from spray treatments obtained when weeds are dry at application and for a period afterwards. Later rain or overhead watering is needed for effective results [1, 4, 5]
- Residual activity reduced on soils with more than 10% organic matter and, in these conditions, post-emergence treatment is more effective [1, 4, 5]
- Best results on bindweed when first shoots are 10-15 cm long [1, 4, 5]
- Adjust water volume to achieve good coverage when weed density is high [1, 4, 5]

Restrictions

- Maximum number of treatments 2 per yr for ornamental plant production [2, 3]
- Maximum total dose per yr equivalent to one full dose treatment on edible crops [1, 4, 5]
- See label for list of ornamental species which may be treated with granules. Treat small numbers of other species to check safety. Do not treat Hydrangea, Spiraea or Genista [2, 3]
- Do not cultivate after treatment [1, 4, 5]
- Do not treat container stock under glass or use on plants rooted in media with high sand or non-organic content. Do not apply to plants with wet foliage [2, 3]
- Consult processors before treatment of crops intended for processing [1, 4, 5]

Crop-specific information

- Latest use: Jul for apples, grapevines, hops, pears; Jun for raspberries, woody ornamentals
- Apply spray to apples and pears from Jan to Jul, avoiding young growth [1, 4, 5]
- Treat bush fruit from Jan to bud-break, avoiding bushes, grapevines in Feb/Mar before start of new growth or in Jun/Jul avoiding foliage [1, 4, 5]
- Treat hops cropped for at least 2 yr in Feb or in Jun/Jul after deleafing [1, 4, 5]
- Treat woody ornamentals from Jan to Jun, avoiding young growth. Do not spray container stock overall [1, 4, 5]

Following crops guidance

- A period of at least 6 mth must elapse between treatment at half the recommended maximum dose and planting a following crop. 12 mth must elapse after use of the maximum recommended dose. In either case soil must be ploughed to min 15 cm and cultivated before sowing or planting a following crop [1, 4, 5]

Environmental safety

- Flammable [1, 4, 5]
- Dangerous for the environment [1, 4, 5]
- Very toxic to aquatic organisms [1, 4, 5]
- Harmful to aquatic organisms [2, 3]

Hazard classification and safety precautions

Hazard H03, H08, H11 [1, 4, 5]
Risk phrases R22b, R36, R38, R50, R67 [1, 4, 5]; R41 [4]; R52 [2, 3]; R53a [1-5]
Operator protection A, C [1-5]; H [5]; U02a [1, 5]; U05a, U20b [1-5]; U08 [1, 4, 5]; U09a, U19a [2, 3]; U14, U15 [3-5]
Environmental protection E15a [1-5]; E38 [4]
Storage and disposal D01, D09a, D12a [1-5]; D02 [1, 3-5]; D05 [5]; D10a [1, 4, 5]; D11a [2, 3]
Medical advice M05b [1, 4, 5]

343 oxamyl

A soil-applied, systemic carbamate nematicide and insecticide
IRAC mode of action code: 1A

Products

Vydate 10G	DuPont	10% w/w	GR	02322

Uses

- American serpentine leaf miner in ***aubergines*** *(off-label)*, ***ornamental plant production*** *(off-label)*, ***protected broad beans*** *(off-label)*, ***protected ornamentals*** *(off-label)*, ***protected peppers*** *(off-label)*, ***protected soya beans*** *(off-label)*, ***protected tomatoes*** *(off-label)*

- Aphids in ***fodder beet*** *(off-label)*, ***potatoes***, ***sugar beet***
- Docking disorder vectors in ***fodder beet*** *(off-label)*, ***sugar beet***
- Eelworm in ***bulb onions*** *(off-label)*, ***garlic*** *(off-label)*, ***shallots*** *(off-label)*
- Free-living nematodes in ***potatoes***
- Leaf miner in ***aubergines*** *(off-label)*, ***ornamental plant production*** *(off-label)*, ***protected broad beans*** *(off-label)*, ***protected ornamentals*** *(off-label)*, ***protected peppers*** *(off-label)*, ***protected soya beans*** *(off-label)*, ***protected tomatoes*** *(off-label)*
- Mangold fly in ***fodder beet*** *(off-label)*, ***sugar beet***
- Millipedes in ***fodder beet*** *(off-label)*, ***sugar beet***
- Potato cyst nematode in ***potatoes***
- Pygmy beetle in ***fodder beet*** *(off-label)*, ***sugar beet***
- South American leaf miner in ***aubergines*** *(off-label)*, ***ornamental plant production*** *(off-label)*, ***protected broad beans*** *(off-label)*, ***protected ornamentals*** *(off-label)*, ***protected peppers*** *(off-label)*, ***protected soya beans*** *(off-label)*, ***protected tomatoes*** *(off-label)*
- Spraing vectors in ***potatoes***
- Stem nematodes in ***bulb onions*** *(off-label)*, ***carrots***, ***garlic*** *(off-label)*, ***parsnips***, ***protected garlic*** *(off-label)*, ***protected onions*** *(off-label)*, ***shallots*** *(off-label)*

Specific Off-Label Approvals (SOLAs)

- ***aubergines*** *930020*
- ***bulb onions*** *20061890*
- ***fodder beet*** *20061107*
- ***garlic*** *20061890*
- ***ornamental plant production*** *930020*
- ***protected broad beans*** *930020*
- ***protected garlic*** *940925*
- ***protected onions*** *940925*
- ***protected ornamentals*** *930020*
- ***protected peppers*** *930020*
- ***protected soya beans*** *930020*
- ***protected tomatoes*** *930020*
- ***shallots*** *20061890*

Approval information

- Oxamyl included in Annex I under EC Directive 91/414
- In 2006 PSD required that all products containing this active ingredient should carry the following warning in the main area of the container label: "Oxamyl is an anticholinesterase carbamate. Handle with care"

Efficacy guidance

- Apply granules with suitable applicator before drilling or planting. See label for details of recommended machines
- In potatoes incorporate thoroughly to 10 cm and plant within 3-4 d
- In sugar beet apply in seed furrow at drilling

Restrictions

- Oxamyl is subject to the Poisons Rules 1982 and the Poisons Act 1972. See notes in Section 5
- Contains an anticholinesterase carbamate compound. Do not use if under medical advice not to work with such compounds
- Maximum number of treatments 1 per crop or yr
- Keep in original container, tightly closed, in a safe place, under lock and key
- Wear protective gloves if handling treated compost or soil within 2 wk after treatment
- Allow at least 12 h, followed by at least 1 h ventilation, before entry of unprotected persons into treated glasshouses

Crop-specific information

- Latest use: at drilling/planting for vegetables; before drilling/planting for potatoes and peas.
- HI:potatoes, sugar beet, carrots, parsnips 12 wk

Environmental safety

- Dangerous for the environment
- Toxic to aquatic organisms

SEE SECTION 3 FOR PRODUCTS ALSO REGISTERED

- Dangerous to fish or other aquatic life. Do not contaminate surface waters or ditches with chemical or used container
- Dangerous to game, wild birds and animals. Bury spillages

Hazard classification and safety precautions

Hazard H02, H11

Risk phrases R23, R25, R51, R53a

Operator protection A, B, H, K, M [1]; C [1] (or D); U02a, U04a, U05a, U07, U09a, U13, U19a, U20a

Environmental protection E10a, E13b, E34, E36a, E38

Storage and disposal D01, D02, D09a, D11a, D14

Medical advice M04a

344 paclobutrazol

A triazole plant growth regulator for ornamentals and fruit

Products

1	Agrovista Paclo	Agrovista	250 g/l	SC	13424
2	Bonzi	Syngenta	4 g/l	SC	10517
3	Carousel	Fine	250 g/l	SC	13225
4	Cultar	Syngenta	250 g/l	SC	10523
5	Pirouette	Fine	4 g/l	SC	13073

Uses

- Control of shoot growth in ***apples***, ***pears*** [1, 3, 4]
- Growth regulation in ***cherries*** *(off-label)*, ***plums*** *(off-label)* [4]
- Improving colour in ***poinsettias*** [2, 5]
- Increasing flowering in ***azaleas***, ***bedding plants***, ***begonias***, ***kalanchoes***, ***lilies***, ***roses***, ***tulips*** [2, 5]
- Increasing fruit set in ***apples***, ***pears*** [1, 3, 4]
- Stem shortening in ***azaleas***, ***bedding plants***, ***begonias***, ***kalanchoes***, ***lilies***, ***poinsettias***, ***roses***, ***tulips*** [2, 5]

Specific Off-Label Approvals (SOLAs)

- ***cherries*** *20031235* [4]
- ***plums*** *20031235* [4]

Efficacy guidance

- Chemical is active via both foliage and root uptake. For best results apply in dull weather when relative humidity not high
- Apply as spray to produce compact pot plants and to improve bract colour of poinsettias [2]
- Apply as compost drench to reduce flower stem length of potted tulips and Mid-Century hybrid lilies [2]
- Timing is critical and varies with species. See label for details [2]

Restrictions

- Maximum number of treatments 1 per specimen for some species [2, 5]
- Maximum total dose equivalent to 3 full dose treatments for pears and 4 full dose treatments for apples [1, 3, 4]
- Some varietal restrictions apply in top fruit (see label for details) [1, 3, 4]
- Do not use on trees of low vigour or under stress [1, 3, 4]
- Do not use on trees from green cluster to 2 wk after full petal fall [1, 3, 4]
- Do not use in underplanted orchards or those recently interplanted [1, 3, 4]

Crop-specific information

- HI apples, pears 14 d [1, 3, 4]
- Apply to apple and pear trees under good growing conditions as pre-blossom spray (apples only) and post-blossom at 7-14 d intervals [1, 3, 4]
- Timing and dose of orchard treatments vary with species and cultivar. See label [1, 3, 4]

Following crops guidance

- Chemical has residual soil activity which can affect growth of following crops. Withhold treatment from orchards due for grubbing to allow the following intervals between the last treatment and planting the next crop: apples, pears 1 yr; beans, peas, onions 2 yr; stone fruit 3 yr; cereals, grass, oilseed rape, carrots 4 yr; market brassicas 6 yr; potatoes and other crops 7 yr [1, 3, 4]

Environmental safety
- Harmful to aquatic organisms
- Do not use on food crops [2, 5]
- Keep livestock out of treated areas for at least 2 years after treatment [1, 3, 4]

Hazard classification and safety precautions

Hazard H04 [1, 3, 4]
Risk phrases R36, R52, R53a [1, 3, 4]
Operator protection A, H, M [1, 3, 4]; U05a, U20a [1-5]; U08 [2, 5]
Environmental protection E06a [1, 3, 4] (2 yr); E13c [5]; E15a [2, 4, 5]; E15b [1, 3]; E34, E38 [1, 3, 4]
Consumer protection C01 [2, 5]; C02a [1, 3, 4] (2 wk)
Storage and disposal D01, D02, D09a [1-5]; D05 [2, 5]; D10b [5]; D10c [1-4]; D12a [1, 3, 4]
Medical advice M03 [1, 3, 4]

345 paraffin oil (commodity substance)

An agent for the control of birds by egg treatment

Products

paraffin oil	various	OL

Uses
- Birds in ***miscellaneous pest control situations*** *(egg treatment)*

Approval information
- Approval for the use of paraffin oil as a commodity substance was granted on 25 April 1995 by Ministers under regulation 5 of the Control of Pesticides Regulations 1986
- Only to be used where a licence has been approved in accordance with Section 16(1) of the Wildlife and Countryside Act (1981)

Efficacy guidance
- Egg treatment should be undertaken as soon as clutch is complete
- Eggs should be treated by complete immersion in liquid paraffin

Restrictions
- Use to control eggs of birds covered by licences issued by the Agriculture and Environment Departments under Section 16(1) of the Wildlife and Countryside Act (1981)
- Treat eggs once only

Hazard classification and safety precautions

Operator protection A, C

346 paraquat

A non-selective, non-residual contact bipyridilium herbicide
HRAC mode of action code: D

See also diquat + paraquat

Products

Agriguard Paraquat	AgriGuard	200 g/l	SL	09951

Uses
- Annual dicotyledons in ***blackcurrants***, ***cultivated land/soil*** *(minimum cultivation)*, ***flower bulbs***, ***forest***, ***forest nursery beds*** *(stale seedbed)*, ***forestry transplants***, ***gooseberries***, ***hops***, ***non-crop areas***, ***orchards***, ***potatoes***, ***raspberries***, ***redcurrants***, ***row crops***, ***stubbles***, ***sugar beet***, ***whitecurrants***
- Annual grasses in ***blackcurrants***, ***cultivated land/soil*** *(minimum cultivation)*, ***flower bulbs***, ***forest***, ***forest nursery beds*** *(stale seedbed)*, ***forestry transplants***, ***gooseberries***, ***hops***, ***non-crop areas***, ***orchards***, ***permanent grassland*** *(sward destruction/direct drilling)*, ***potatoes***, ***raspberries***, ***redcurrants***, ***row crops***, ***stubbles***, ***sugar beet***, ***whitecurrants***

- Barren brome in ***field crops*** *(stubble treatment)*
- Chemical stripping in ***hops***
- Creeping bent in ***blackcurrants***, ***cultivated land/soil*** *(minimum cultivation)*, ***flower bulbs***, ***forest***, ***forestry transplants***, ***gooseberries***, ***hops***, ***non-crop areas***, ***orchards***, ***permanent grassland*** *(sward destruction/direct drilling)*, ***raspberries***, ***redcurrants***, ***stubbles***, ***whitecurrants***
- Firebreak desiccation in ***forest***
- Green cover in ***land temporarily removed from production***
- Perennial ryegrass in ***blackcurrants***, ***gooseberries***, ***hops***, ***non-crop areas***, ***orchards***, ***permanent grassland*** *(sward destruction/direct drilling)*, ***raspberries***, ***redcurrants***, ***whitecurrants***
- Rough meadow grass in ***blackcurrants***, ***gooseberries***, ***hops***, ***non-crop areas***, ***orchards***, ***permanent grassland*** *(sward destruction/direct drilling)*, ***raspberries***, ***redcurrants***, ***whitecurrants***
- Runner desiccation in ***strawberries***
- Volunteer cereals in ***cultivated land/soil*** *(minimum cultivation)*, ***potatoes***, ***row crops***, ***stubbles***, ***sugar beet***
- Wild oats in ***stubbles***

Approval information

- Paraquat was originally included in Annex I under EC Directive 91/414 but this decision was annulled in July 2007. As a result all remaining approvals for use of products containing paraquat will expire in July 2008
- Accepted by BBPA for use on hops

Efficacy guidance

- Apply to green weeds preferably less than 15 cm high. Spray is rainfast after 10 min
- Addition of wetter recommended with lower application rates
- Spray in autumn to suppress couch when shoots have 2 leaves, repeat as necessary and plough after last treatment
- For direct-drilling land should be free of perennial weeds and protection against slugs should be provided. Allow 7-10 d before drilling into sprayed grass
- Best time for use in top fruit is Nov-Apr
- For forestry fire-break use apply in Jul-Aug and fire 7-10 d later
- For interrow use in row crops apply with guarded, no-drift sprayer
- Use as a carefully directed spray in blackcurrants, gooseberries, grapevines and other fruit crops, in raspberries only when dormant
- For stawberry runner control use guarded sprayer, not when flowers or fruit present
- Apply to bulbs pre-emergence (at least 3 d pre-emergence on sandy soils) or at end of season, provided no attached foliage and bulbs well covered (not on sandy soils)
- In forestry seedbed apply up to 3 d before seedling emergence

Restrictions

- Paraquat is subject to the Poisons Rules 1982 and the Poisons Act 1972. See notes in Section 5
- Maximum number of treatments not stated for many situations but normally 1 per crop or yr; 3 per yr on leafy herbs
- Products in this profile are for professional use only
- Do not put in a food or drinks container
- Do not use on straw or other artificial growing media
- Do not use on potatoes under hot, dry conditions or on hops under drought conditions
- If using around glasshouses ensure vents and doors closed
- Only use clean water for mixing up spray
- Allow at least 4 h before cultivating, leave overnight if possible
- Observe label restrictions for maximum permitted concentration when spraying at low volume

Crop-specific information

- Latest use: before 10% (earlies) or 40% (maincrop) emergence and before 15 cm high for potatoes; before active crop growth starts for fruit crops, bulbs and hops or after bine reaches highest wire (stripping) for hops; before drilling or crop emergence for field crops, row crops, sugar beet; before first flower for strawberries; before planting or crop seedling emergence for uses in forestry

Following crops guidance

- Chemical is inactivated on contact with soil. Crops can be sown or planted soon after spraying on most soils but 3 d should be allowed on sandy or immature peat soils

Environmental safety

- Dangerous for the environment
- Very toxic to aquatic organisms
- Paraquat can be harmful to hares; cereal stubbles must be sprayed early in the day
- Keep in original container, tightly closed, in a safe place, under lock and key
- Harmful to livestock
- Keep all livestock out of treated areas for at least 24 h
- Do not contaminate surface waters or ditches with chemical or used container

Hazard classification and safety precautions

Hazard H02, H11
Risk phrases R20, R21, R25, R36, R37, R38
Operator protection A, C; U02a, U04a, U05a, U09a, U19a, U20a
Environmental protection E06c (24 h); E11, E15a, E34
Storage and disposal D01, D02, D05, D09b, D10a
Medical advice M04a

347 penconazole

A protectant triazole fungicide with antisporulant activity
FRAC mode of action code: 3

Products

1	Agrovista Penco	Agrovista	100 g/l	EC	13425
2	Topas	Syngenta	100 g/l	EC	09717
3	Topenco 100 EC	Globachem	100 g/l	EC	11972

Uses

- Powdery mildew in ***apples*** [1-3]; ***bilberries*** *(off-label)*, ***blueberries*** *(off-label)*, ***cranberries*** *(off-label)*, ***gooseberries*** *(off-label)*, ***redcurrants*** *(off-label)*, ***whitecurrants*** *(off-label)* [2]; ***blackcurrants***, ***hops***, ***ornamental trees*** [1, 2]
- Rust in ***roses*** [1, 2]
- Scab in ***ornamental trees*** [1, 2]

Specific Off-Label Approvals (SOLAs)

- ***bilberries*** *20061106* [2]
- ***blueberries*** *20061106* [2]
- ***cranberries*** *20061106* [2]
- ***gooseberries*** *20061106* [2]
- ***redcurrants*** *20061106* [2]
- ***whitecurrants*** *20061106* [2]

Approval information

- Accepted by BBPA for use on hops

Efficacy guidance

- Use as a protectant fungicide by treating at the earliest signs of disease
- Treat crops every 10-14 d (every 7-10 d in warm, humid weather) at first sign of infection or as a protective spray ensuring complete coverage. See label for details of timing
- Increase dose and volume with growth of hops but do not exceed 2000 l/ha. Little or no activity will be seen on established powdery mildew
- Antisporulant activity reduces development of secondary mildew in apples

Restrictions

- Maximum number of treatments 10 per yr for apples, 6 per yr for hops, 4 per yr for blackcurrants
- Check for varietal susceptibility in roses. Some defoliation may occur after repeat applications on Dearest

Crop-specific information

- HI apples, hops 14 d; currants, bilberries, blueberries, cranberries, gooseberries 4 wk

SEE SECTION 3 FOR PRODUCTS ALSO REGISTERED

Environmental safety

- Dangerous for the environment
- Toxic to aquatic organisms

Hazard classification and safety precautions

Hazard H04, H11
Risk phrases R36, R51, R53a [1-3]; R38 [3]
Operator protection A, C; U02a, U05a, U20b [1-3]; U08 [1, 2]; U11 [3]
Environmental protection E15a [1-3]; E22c [3]; E38 [1, 2]
Storage and disposal D01, D02, D05, D09a [1-3]; D07 [1, 3]; D10b [3]; D10c, D12a [1, 2]
Medical advice M05b

348 pencycuron

A non-systemic urea fungicide for use on seed potatoes
FRAC mode of action code: 20

See also imazalil + pencycuron

Products

1	Agriguard Pencycuron	AgriGuard	12.5% w/w	DS	10445
2	Monceren DS	Bayer CropScience	12.5% w/w	DS	11292
3	Pency Flowable	Globachem	250 g/l	FS	12637
4	Standon Pencycuron DP	Standon	12.5% w/w	DS	08774

Uses

- Black scurf in ***potatoes*** *(tuber treatment)*
- Stem canker in ***potatoes*** *(tuber treatment)*

Efficacy guidance

- Provides control of tuber-borne disease and gives some reduction of stem canker
- Seed tubers should be of good quality and vigour, and free from soil deposits when treated
- Dust formulations should be applied immediately before, or during, the planting process by treating seed tubers in chitting trays, in bulk bins immediately before planting or in hopper at planting. Liquid formulations may be applied by misting equipment at any time, into or out of store but treatment over a roller table at the end of grading out, or at planting, is usually most convenient
- Whichever method is chosen, complete and even distribution on the tubers is essential for optimum efficacy
- Apply in accordance with detailed guidelines in manufacturer's literature
- If rain interrupts planting, cover dust treated tubers in hopper

Restrictions

- Maximum number of treatments 1 per batch of seed potatoes
- Treated tubers must be used only as seed and not for human or animal consumption
- Do not use on tubers previously treated with a dry powder seed treatment or hot water
- To prevent rapid absorption through damaged 'eyes' seed tubers should only be treated before the 'eyes' begin to open or immediately before or during planting [3]
- Tubers removed from cold storage must be allowed to attain a temperature of at least 8 °C before treatment [3]
- Use of suitable dust mask is mandatory when applying dust, filling the hopper or riding on planter [1, 2, 4]
- Treated tubers in bulk bins must be allowed to dry before being stored, especially when intended for cold storage [3]
- Some internal staining of boxes used to store treated tubers may occur. Ware tubers should not be stored or supplied to packers in such stained boxes [3]

Crop-specific information

- Latest use: at planting

Environmental safety

- Do not use treated seed as food or feed
- Treated seed harmful to game and wildlife

Hazard classification and safety precautions

Hazard H04 [3]

Risk phrases R43 [3]
Operator protection A [1-4]; F [1, 2, 4]; U14, U20a [3]; U20b [1, 2, 4]
Environmental protection E03, E15a
Storage and disposal D05 [3]; D07 [1, 4]; D09a, D11a [1-4]
Treated seed S01, S02, S03, S04a, S05, S06a

349 pendimethalin

A residual dinitroaniline herbicide for cereals and other crops
HRAC mode of action code: K1

See also *flufenacet + pendimethalin*
imazamox + pendimethalin

Products

1	Agriguard Pendimethalin	AgriGuard	400 g/l	SC	10864
2	Alpha Pendimethalin 330 EC	Makhteshim	330 g/l	EC	12214
3	Blazer	Headland	330 g/l	EC	11217
4	Bunker	Makhteshim	40% w/w	WG	12864
5	Claymore	BASF	400 g/l	SC	13441
6	Greencrop Estuary	Greencrop	400 g/l	SC	12095
7	Landgold Pendimethalin 400	Teliton	400 g/l	SC	12125
8	PDM 330 EC	BASF	330 g/l	EC	13406
9	Stomp 400 SC	BASF	400 g/l	SC	13405

Uses

- Annual dicotyledons in ***almonds*** *(off-label)*, ***chestnuts*** *(off-label)*, ***grass seed crops*** *(off-label)*, ***hazel nuts*** *(off-label)*, ***lupins*** *(off-label)*, ***walnuts*** *(off-label)* [2, 3, 6, 9]; ***apple orchards***, ***blackberries***, ***blackcurrants***, ***broccoli***, ***brussels sprouts***, ***bulb onions***, ***cabbages***, ***calabrese***, ***carrots***, ***cauliflowers***, ***cherries***, ***forage maize***, ***gooseberries***, ***leeks***, ***loganberries***, ***parsnips***, ***pear orchards***, ***plums***, ***raspberries***, ***strawberries*** [1-6, 9]; ***asparagus*** *(off-label)*, ***bilberries*** *(off-label)*, ***blueberries*** *(off-label)*, ***cranberries*** *(off-label)*, ***fennel*** *(off-label)*, ***quinces*** *(off-label)*, ***redcurrants*** *(off-label)*, ***runner beans*** *(off-label)*, ***salad onions*** *(off-label)*, ***whitecurrants*** *(off-label)* [2, 4, 9]; ***broad beans*** *(off-label)*, ***bulb onions*** *(off-label - pre + post emergence treatment)*, ***celery (outdoor)*** *(off-label)*, ***forest*** *(off-label)*, ***leeks*** *(off-label)*, ***rhubarb*** *(off-label)*, ***sweetcorn*** *(off-label - under covers)* [9]; ***bulb onions*** *(off-label)*, ***chervil*** *(off-label)*, ***coriander*** *(off-label)*, ***daffodils*** *(off-label - for galanthamine production)*, ***dill*** *(off-label)*, ***edible podded peas*** *(off-label)*, ***mallow (althaea spp.)*** *(off-label)*, ***vining peas*** *(off-label)* [2, 4]; ***chives*** *(off-label)*, ***herbs (see appendix 6)*** *(off-label)* [2-4, 9]; ***combining peas***, ***durum wheat***, ***potatoes***, ***spring barley***, ***triticale***, ***winter barley***, ***winter rye***, ***winter wheat*** [1-9]; ***cress*** *(off-label)*, ***evening primrose*** *(off-label)*, ***farm forestry*** *(off-label)*, ***frise*** *(off-label)*, ***lamb's lettuce*** *(off-label)*, ***lettuce*** *(off-label)*, ***radicchio*** *(off-label)*, ***salad brassicas*** *(off-label - for baby leaf production)*, ***scarole*** *(off-label)* [2, 9]; ***garlic*** *(off-label)*, ***shallots*** *(off-label)* [2-4, 6, 9]; ***hops***, ***parsley*** [1-4, 6, 9]; ***parsley root*** *(off-label)* [1-3, 6, 9]; ***protected herbs (see appendix 6)*** *(off-label)*, ***protected salad brassicas*** *(off-label)*, ***salad brassicas*** *(off-label)*, ***sweetcorn*** *(off-label - under covers and mulches)* [4]; ***rubus hybrids*** [1, 4-6, 9]; ***spring field beans*** *(off-label)* [2, 4, 8, 9]; ***sunflowers*** [1-6, 8, 9]; ***sweetcorn*** *(off-label - under crop covers or mulches)* [2]; ***tayberries*** [2, 3]; ***winter field beans*** *(off-label)* [4, 8]
- Annual grasses in ***almonds*** *(off-label)*, ***chestnuts*** *(off-label)*, ***garlic*** *(off-label)*, ***grass seed crops*** *(off-label)*, ***hazel nuts*** *(off-label)*, ***lupins*** *(off-label)*, ***shallots*** *(off-label)*, ***walnuts*** *(off-label)* [2, 3, 6]; ***asparagus*** *(off-label)*, ***bilberries*** *(off-label)*, ***blueberries*** *(off-label)*, ***chervil*** *(off-label)*, ***coriander*** *(off-label)*, ***cranberries*** *(off-label)*, ***cress*** *(off-label)*, ***dill*** *(off-label)*, ***evening primrose*** *(off-label)*, ***farm forestry*** *(off-label)*, ***fennel*** *(off-label)*, ***frise*** *(off-label)*, ***lamb's lettuce*** *(off-label)*, ***lettuce*** *(off-label)*, ***mallow (althaea spp.)*** *(off-label)*, ***quinces*** *(off-label)*, ***radicchio*** *(off-label)*, ***redcurrants*** *(off-label)*, ***runner beans*** *(off-label)*, ***salad brassicas*** *(off-label - for baby leaf production)*, ***scarole*** *(off-label)*, ***sweetcorn*** *(off-label - under crop covers or mulches)*, ***whitecurrants*** *(off-label)* [2]; ***chives*** *(off-label)*, ***herbs (see appendix 6)*** *(off-label)* [2, 3]; ***combining peas***, ***durum wheat***, ***potatoes***, ***spring barley***, ***triticale***, ***winter barley***, ***winter rye***, ***winter wheat*** [7, 8]; ***parsley root*** *(off-label)* [1-3, 6]; ***spring field beans*** *(off-label)* [2, 8]; ***sunflowers***, ***winter field beans*** *(off-label)* [8]
- Annual meadow grass in ***apple orchards***, ***blackberries***, ***blackcurrants***, ***broccoli***, ***brussels sprouts***, ***bulb onions***, ***cabbages***, ***calabrese***, ***carrots***, ***cauliflowers***, ***cherries***, ***combining peas***,

durum wheat, ***forage maize***, ***gooseberries***, ***leeks***, ***loganberries***, ***parsnips***, ***pear orchards***, ***plums***, ***potatoes***, ***raspberries***, ***spring barley***, ***strawberries***, ***sunflowers***, ***triticale***, ***winter barley***, ***winter rye***, ***winter wheat*** [1-6, 9]; ***hops***, ***parsley*** [1-4, 6, 9]; ***rubus hybrids*** [1, 4-6, 9]; ***tayberries*** [2, 3]

- Blackgrass in ***apple orchards***, ***blackberries***, ***blackcurrants***, ***broccoli***, ***brussels sprouts***, ***bulb onions***, ***cabbages***, ***calabrese***, ***carrots***, ***cauliflowers***, ***cherries***, ***combining peas***, ***durum wheat***, ***gooseberries***, ***leeks***, ***loganberries***, ***parsnips***, ***pear orchards***, ***plums***, ***raspberries***, ***spring barley***, ***strawberries***, ***sunflowers***, ***triticale***, ***winter barley***, ***winter rye***, ***winter wheat*** [1-6, 9]; ***hops***, ***parsley*** [1-4, 6, 9]; ***rubus hybrids*** [1, 4-6, 9]; ***tayberries*** [2, 3]
- Cleavers in ***winter field beans*** *(off-label)* [2, 9]
- Field pansy in ***daffodils*** *(off-label - for galanthamine production)* [2]
- Field speedwell in ***daffodils*** *(off-label - for galanthamine production)* [2]
- Ivy-leaved speedwell in ***daffodils*** *(off-label - for galanthamine production)* [2]
- Knotgrass in ***daffodils*** *(off-label - for galanthamine production)* [2]; ***edible podded peas*** *(off-label)*, ***vining peas*** *(off-label)* [9]
- Polygonums in ***edible podded peas*** *(off-label)*, ***vining peas*** *(off-label)* [2]
- Rough meadow grass in ***apple orchards***, ***blackberries***, ***blackcurrants***, ***broccoli***, ***brussels sprouts***, ***bulb onions***, ***cabbages***, ***calabrese***, ***carrots***, ***cauliflowers***, ***cherries***, ***combining peas***, ***durum wheat***, ***forage maize***, ***gooseberries***, ***leeks***, ***loganberries***, ***parsnips***, ***pear orchards***, ***plums***, ***potatoes***, ***raspberries***, ***spring barley***, ***strawberries***, ***sunflowers***, ***triticale***, ***winter barley***, ***winter rye***, ***winter wheat*** [1-6, 9]; ***hops***, ***parsley*** [1-4, 6, 9]; ***rubus hybrids*** [1, 4-6, 9]; ***tayberries*** [2, 3]
- Volunteer oilseed rape in ***apple orchards***, ***blackberries***, ***blackcurrants***, ***broccoli***, ***brussels sprouts***, ***bulb onions***, ***cabbages***, ***calabrese***, ***carrots***, ***cauliflowers***, ***cherries***, ***forage maize***, ***gooseberries***, ***leeks***, ***loganberries***, ***parsnips***, ***pear orchards***, ***plums***, ***raspberries***, ***spring barley***, ***strawberries***, ***sunflowers*** [1-6, 9]; ***combining peas***, ***durum wheat***, ***potatoes***, ***triticale***, ***winter barley***, ***winter rye***, ***winter wheat*** [1-6, 8, 9]; ***daffodils*** *(off-label - for galanthamine production)* [2]; ***hops***, ***parsley*** [1-4, 6, 9]; ***rubus hybrids*** [1, 4-6, 9]; ***spring field beans*** *(off-label)*, ***winter field beans*** *(off-label)* [8]; ***tayberries*** [2, 3]
- Wild oats in ***durum wheat***, ***triticale***, ***winter barley***, ***winter rye***, ***winter wheat*** [1-6, 9]

Specific Off-Label Approvals (SOLAs)

- ***almonds*** *20062367 expires Dec 2008* [2], *20052639* [3], *20052609* [6], *20071442* [9]
- ***asparagus*** *20062364 expires Dec 2008* [2], *20062113 expires Dec 2008* [4], *20071431* [9]
- ***bilberries*** *20062370 expires Dec 2008* [2], *20062112 expires Dec 2008* [4], *20071444* [9]
- ***blueberries*** *20062370 expires Dec 2008* [2], *20062112 expires Dec 2008* [4], *20071444* [9]
- ***broad beans*** *20071017* [9]
- ***bulb onions*** *20062375 expires Dec 2008* [2], *20062106 expires Dec 2008* [4], *(pre + post emergence treatment) 20071438* [9]
- ***celery (outdoor)*** *20071430* [9]
- ***chervil*** *20062373 expires Sep 2008* [2], *20062109 expires Sep 2008* [4]
- ***chestnuts*** *20062367 expires Dec 2008* [2], *20052639* [3], *20052609* [6], *20071442* [9]
- ***chives*** *20062373 expires Sep 2008* [2], *20052645* [3], *20062109 expires Sep 2008* [4], *20071432* [9]
- ***coriander*** *20062373 expires Sep 2008* [2], *20062109 expires Sep 2008* [4]
- ***cranberries*** *20062370 expires Dec 2008* [2], *20062112 expires Dec 2008* [4], *20071444* [9]
- ***cress*** *20062373 expires Sep 2008* [2], *20071432* [9]
- ***daffodils*** *(for galanthamine production) 20062362 expires Dec 2008* [2], *(for galanthamine production) 20062111 expires Dec 2008* [4]
- ***dill*** *20062373 expires Sep 2008* [2], *20062109 expires Sep 2008* [4]
- ***edible podded peas*** *20062368 expires Dec 2008* [2], *20062108 expires Dec 2008* [4], *20071437* [9]
- ***evening primrose*** *20062359 expires Dec 2008* [2], *20071434* [9]
- ***farm forestry*** *20062361 expires Dec 2008* [2], *20071436* [9]
- ***fennel*** *20062356 expires Dec 2008* [2], *20062114 expires Dec 2008* [4], *20071430* [9]
- ***forest*** *20071436* [9]
- ***frise*** *20062373 expires Sep 2008* [2], *20071432* [9]
- ***garlic*** *20062365 expires Dec 2008* [2], *20052640* [3], *20062106 expires Dec 2008* [4], *20052610* [6], *20071438* [9]
- ***grass seed crops*** *20062366 expires Dec 2008* [2], *20052642* [3], *20052612* [6], *20071429* [9]

- ***hazel nuts*** *20062367 expires Dec 2008* [2], *20052639* [3], *20052609* [6], *20071442* [9]
- ***herbs (see appendix 6)*** *20062373 expires Sep 2008* [2], *20052645* [3], *20062109 expires Sep 2008* [4], *20071432* [9]
- ***lamb's lettuce*** *20062373 expires Sep 2008* [2], *20071432* [9]
- ***leeks*** *20071438* [9]
- ***lettuce*** *20062373 expires Sep 2008* [2], *20071432* [9]
- ***lupins*** *20062372 expires Dec 2008* [2], *20052630* [3], *20052608* [6], *20071437* [9]
- ***mallow (althaea spp.)*** *20062373 expires Sep 2008* [2], *20062109 expires Sep 2008* [4]
- ***parsley root*** *20052556* [1], *20062358 expires Dec 2008* [2], *20052641* [3], *20052611* [6], *20071433* [9]
- ***protected herbs (see appendix 6)*** *20062116 expires Dec 2008* [4]
- ***protected salad brassicas*** *20062116 expires Dec 2008* [4]
- ***quinces*** *20062369 expires Dec 2008* [2], *20062107 expires Dec 2008* [4], *20071443* [9]
- ***radicchio*** *20062373 expires Sep 2008* [2], *20071432* [9]
- ***redcurrants*** *20062370 expires Dec 2008* [2], *20062112 expires Dec 2008* [4], *20071444* [9]
- ***rhubarb*** *20071430* [9]
- ***runner beans*** *20062360 expires Dec 2008* [2], *20062115 expires Dec 2008* [4], *20071437* [9]
- ***salad brassicas*** *(for baby leaf production) 20062373 expires Sep 2008* [2], *20062109 expires Sep 2008* [4], *(for baby leaf production) 20071432* [9]
- ***salad onions*** *20062375 expires Dec 2008* [2], *20062106 expires Dec 2008* [4], *20071438* [9]
- ***scarole*** *20062373 expires Sep 2008* [2], *20071432* [9]
- ***shallots*** *20062365 expires Dec 2008* [2], *20052640* [3], *20062106 expires Dec 2008* [4], *20052610* [6], *20071438* [9]
- ***spring field beans*** *20070928 expires Dec 2008* [2], *20062017 expires Dec 2008* [4], *20070876* [8], *20071437* [9]
- ***sweetcorn*** *(under crop covers or mulches) 20062357 expires Dec 2008* [2], *(under covers and mulches) 20062105 expires Dec 2008* [4], *(under covers) 20071435* [9]
- ***vining peas*** *20062368 expires Dec 2008* [2], *20062108 expires Dec 2008* [4], *20071437* [9]
- ***walnuts*** *20062367 expires Dec 2008* [2], *20052639* [3], *20052609* [6], *20071442* [9]
- ***whitecurrants*** *20062370 expires Dec 2008* [2], *20062112 expires Dec 2008* [4], *20071444* [9]
- ***winter field beans*** *20062363 expires Dec 2008* [2], *20062110 expires Dec 2008* [4], *20070876* [8], *20071437* [9]

Approval information

- Pendimethalin included in Annex I under EC Directive 91/414
- Accepted by BBPA for use on malting barley and hops
- Approval expiry 31 Aug 2008 [1, 7]
- Approval expiry 31 Dec 2008 [2-4, 6]

Efficacy guidance

- Apply as soon as possible after drilling. Weeds are controlled as they germinate and emerged weeds will not be controlled by use of the product alone
- For effective blackgrass control apply not more than 2 d after final cultivation and before weed seeds germinate
- Best results by application to fine firm, moist, clod-free seedbeds when rain follows treatment. Effectiveness reduced by prolonged dry weather after treatment
- Effectiveness reduced on soils with more than 6% organic matter. Do not use where organic matter exceeds 10%
- Any trash, ash or straw should be incorporated evenly during seedbed preparation
- Do not disturb soil after treatment
- Apply to potatoes as soon as possible after planting and ridging in tank-mix with metribuzin but note that this will restrict the varieties that may be treated
- Always follow WRAG guidelines for preventing and managing herbicide resistant weeds. Section 5 for more information

Restrictions

- Maximum number of treatments 1 per crop or yr
- Maximum total dose equivalent to one full dose treatment on most crops; 2 full dose treatments on leeks
- May be applied pre-emergence of cereal crops sown before 30 Nov provided seed covered by at least 32 mm soil, or post-emergence to early tillering stage (GS 23)
- Do not undersow treated crops

- Do not use on crops suffering stress due to disease, drought, waterlogging, poor seedbed conditions or chemical treatment or on soils where water may accumulate

Crop-specific information

- Latest use: pre-emergence for spring barley, carrots, lettuce, fodder maize, parsnips, parsley, combining peas, potatoes, onions and leeks; before transplanting for brassicas; before leaf sheaths erect for winter cereals; before bud burst for blackcurrants, gooseberries, cane fruit, hops; before flower trusses emerge for strawberries; 14 d after transplanting for leaf herbs
- HI: parsley 8 wk [4]
- Do not use on spring barley after end Mar (mid-Apr in Scotland on some labels) because dry conditions likely. Do not apply to dry seedbeds in spring unless rain imminent
- Apply to combining peas as soon as possible after sowing. Do not spray if plumule less than 13 mm below soil surface
- Apply to potatoes up to 7 d before first shoot emerges
- Apply to drilled crops as soon as possible after drilling but before crop and weed emergence
- Apply in top fruit, bush fruit and hops from autumn to early spring when crop dormant
- In cane fruit apply to weed free soil from autumn to early spring, immediately after planting new crops and after cutting out canes in established crops
- Apply in strawberries from autumn to early spring (not before Oct on newly planted bed). Do not apply pre-planting or during flower initiation period (post-harvest to mid-Sep)
- Apply pre-emergence in drilled onions or leeks, not on Sands, Very Light, organic or peaty soils or when heavy rain forecast
- Apply to brassicas after final plant-bed cultivation but before transplanting. Avoid unnecessary soil disturbance after application and take care not to introduce treated soil into the root zone when transplanting. Follow transplanting with specified post-planting treatments - see label
- Do not use on protected crops or in greenhouses

Following crops guidance

- Before ryegrass is drilled after a very dry season plough or cultivate to at least 15 cm. If treated spring crops are to be followed by crops other than cereals, plough or cultivate to at least 15 cm
- In the event of crop failure land must be ploughed or thoroughly cultivated to at least 15 cm. See label for minimum intervals that should elapse between treatment and sowing a range of replacement crops

Environmental safety

- Dangerous for the environment
- Very toxic to aquatic organisms
- Dangerous to fish or other aquatic life. Do not contaminate surface waters or ditches with chemical or used container
- Flammable [2, 3]
- Do not empty into drains [4]
- LERAP Category B [2, 3, 5, 8, 9]

Hazard classification and safety precautions

Hazard H03 [2, 3, 8]; H04 [1, 4]; H08 [2, 3]; H11 [1-9]

Risk phrases R22a, R36, R38 [8]; R22b, R51 [2, 3]; R43 [4]; R50 [1, 4-9]; R53a [1-9]

Operator protection A [4-6, 8, 9]; C [8]; H [4]; U02a, U14 [5, 9]; U05a [1-3, 5, 8, 9]; U08, U19a [1-9]; U11 [8]; U13, U20b [1, 4-9]

Environmental protection E15a [1-4, 6, 7]; E15b [5, 8, 9]; E16a [2, 3, 5, 8, 9]; E19b [4]; E34 [2, 3, 8]; E38 [1-3, 5, 8, 9]

Storage and disposal D01, D12a [1-5, 8, 9]; D02 [1-3, 5, 8, 9]; D05 [1-3, 6, 7]; D08 [5, 8, 9]; D09a [1-9]; D10b [1-3, 5-9]; D11a [4]

Medical advice M03, M05b [2, 3, 8]; M05a [4, 5, 9]

350 pendimethalin + picolinafen

A post-emergence broad-spectrum herbicide mixture for winter cereals
HRAC mode of action code: K1 + F1

Products

1	Flight	BASF	330:7.5 g/l	SC	12534
2	Orient	BASF	330:7.5 g/l	SC	12541
3	PicoMax	BASF	320:16 g/l	SC	13456
4	Picona	BASF	320:16 g/l	SC	13428
5	PicoPro	BASF	320:16 g/l	SC	13454
6	PicoStomp	BASF	320:16 g/l	SC	13455

Uses

- Annual dicotyledons in ***winter barley***, ***winter wheat*** [1-6]
- Annual meadow grass in ***winter barley***, ***winter wheat*** [1-6]
- Chickweed in ***winter barley***, ***winter wheat*** [1-6]
- Cleavers in ***winter barley***, ***winter wheat*** [3-6]
- Field speedwell in ***winter barley***, ***winter wheat*** [1-6]
- Ivy-leaved speedwell in ***winter barley***, ***winter wheat*** [1-6]
- Rough meadow grass in ***winter barley***, ***winter wheat*** [1-6]

Approval information

- Pendimethalin and picolinafen included in Annex I under EC Directive 91/414

Efficacy guidance

- Best results obtained on crops growing in a fine, firm tilth and when rain falls within 7 d of application
- Loose or cloddy seed beds must be consolidated prior to application otherwise reduced weed control may occur
- Residual weed control may be reduced on soils with more than 6% organic matter or under prolonged dry conditions
- Always follow WRAG guidelines for preventing and managing herbicide resistant weeds. Section 5 for more information

Restrictions

- Maximum number of treatments 1 per crop
- Do not treat undersown cereals or those to be undersown
- Do not roll emerged crops before treatment nor autumn treated crops until the following spring
- Do not use on stony or gravelly soils, on soils that are waterlogged or prone to waterlogging, or on soils with more than 10% organic matter
- Do not treat crops under stress from any cause
- Do not apply pre-emergence to crops drilled after 30 Nov [1, 2]
- Consult processor before treating crops for processing [1, 2]

Crop-specific information

- Latest use: before pseudo-stem erect stage (GS 30)
- Transient bleaching may occur after treatment but it does not lead to yield loss

Following crops guidance

- In the event of failure of a treated crop plough to at least 15 cm to ensure any residues are evenly dispersed
- In the event of failure of a treated crop allow at least 8 wk from the time of treatment before re-drilling either spring wheat or spring barley. After a normally harvested crop there are no restrictions on following crops apart from rye-grass, as indicated below [3-6]
- In a very dry season, plough or cultivate to at least 15 cm before drilling rye-grass
- A minimum of 5 mth must follow autumn applications before sowing spring barley, spring wheat, maize, peas, potatoes, spring field beans, dwarf beans, turnips, Brussels sprouts, cabbage, calabrese, linseed, carrots, parsnips, parsley, cauliflower [1, 2]
- After spring or summer applications a minimum of 2 mth must follow before sowing spring field beans, dwarf beans, broad beans, peas, turnips, Brussels sprouts, cabbage, calabrese, cauliflower, carrots, parsnips, parsley, linseed and any crop may be sown after 5 mth except red beet, sugar beet and spinach, for which an interval of 12 mth must be allowed [1, 2]

Environmental safety
- Dangerous for the environment
- Very toxic to aquatic organisms
- Product binds strongly to soil minimising likelihood of movement into groundwater
- LERAP Category B

Hazard classification and safety precautions

Hazard H11
Risk phrases R50, R53a
Operator protection A, H [1-3]; U02a, U05a, U20c
Environmental protection E15a [1, 2]; E15b [3-6]; E16a, E34, E38 [1-6]
Storage and disposal D01, D02, D05, D09a, D10c, D12a
Medical advice M03

351 pendimethalin + trifluralin

A dinitroaniline herbicide mixture for winter cereals
HRAC mode of action code: K1 + K1

Products

1	O-Tan	DuPont	261:287 g/l	EC	12924
2	Zimbali	Makhteshim	290:261 g/l	EC	12748

Uses
- Annual dicotyledons in ***spring field beans*** *(off-label)*, ***winter field beans*** *(off-label)* [2]; ***winter barley***, ***winter wheat*** [1, 2]
- Annual meadow grass in ***spring field beans*** *(off-label)*, ***winter field beans*** *(off-label)* [2]; ***winter barley***, ***winter wheat*** [1, 2]

Specific Off-Label Approvals (SOLAs)
- ***spring field beans*** *20070533* [2]
- ***winter field beans*** *20070533* [2]

Approval information
- Pendimethalin included in Annex I under EC Directive 91/414
- Trifluralin will not be included in Annex I of EC Directive 91/414. All approvals for the storage and use of all products listed in this profile will expire on 20 March 2009 unless an earlier date has been set

Efficacy guidance
- Best results obtained from applications made under good growing conditions before weeds emerge or soon after emergence. Weeds are controlled throughout the season as they germinate
- Apply to drilled crops on firm, fine seedbeds, free from clods
- Weed control may be reduced if rainfall occurs within 6 h of application
- Residual control may be reduced under prolonged dry conditions
- Residual control may be reduced on soils with a high Kd factor, where organic matter exceeds 6% or where ash content is high
- Always follow WRAG guidelines for preventing and managing herbicide resistant weeds. Section 5 for more information

Restrictions
- Maximum number of treatments 1 per crop
- Do not apply pre-emergence to winter crops drilled after 30th Nov
- Do not treat shallow drilled crops. Seed must be covered by a minimum of 32 mm settled soil
- Do not treat undersown crops
- Do not roll after treatment until the following spring
- Do not incorporate or disturb soil after application
- Do not use on stony or gravelly soils or those which are waterlogged or prone to waterlogging
- Do not use on soils with more than 10% organic matter
- Do not treat stressed crops

Crop-specific information
- Latest use: before 2-tiller stage of wheat and barley
- If germination is impaired by adverse conditions such as waterlogging or compaction, crop damage may occur

Following crops guidance
- In the event of failure of an autumn-treated crop, spring sown field, broad or dwarf beans, peas, maize, brassica crops, carrots, parsnips, parsley or linseed may be drilled after ploughing to at least 15 cm and provided a minimum of 5 mth have elapsed since treatment
- Following spring or early summer treatment, the above crops, except forage maize, may be drilled after a minimum of 2 mth have elapsed
- After 5 mth from application any crop except beet crops or spinach may be drilled
- Sugar beet, red beet or spinach may be drilled 12 mth after application

Environmental safety
- Dangerous for the environment
- Very toxic to aquatic organisms

Hazard classification and safety precautions
Hazard H03, H08, H11
Risk phrases R22b, R36, R50, R53a [1, 2]; R43 [1]
Operator protection A, C, H; U05a, U11 [1, 2]; U08, U13, U20b [2]; U14 [1]
Environmental protection E15a, E34
Storage and disposal D01, D02, D06a, D08, D09a, D10b, D12b
Medical advice M05b

352 permethrin

A broad spectrum, contact and ingested pyrethroid insecticide
IRAC mode of action code: 3

Products

Fortefog 'P' Fumer	Agropharm	13.5% w/w	FU	H7187

Uses
- Flies in ***agricultural premises***
- Grain storage pests in ***grain stores***

Approval information
- All approvals for plant protection uses of permethrin were revoked in Jun 2001. Off-label uses on young plants in forestry remained valid until 31 Dec 2003. Several grain store and non-agricultural uses remain approved

Efficacy guidance
- For best results in farm buildings fumigate in the late afternoon and leave closed overnight
- Make areas to be fumigated as smoke tight as possible
- Place smoke generators as low as possible
- Space treatment controls insects present at the time by direct contact action

Restrictions
- For use only by professional operators
- Maximum number of treatments up to 3 fumigations at 5-7 d intervals

Environmental safety
- Extremely dangerous to fish or other aquatic life. Do not contaminate surface waters or ditches with chemical or used container
- Domestic animals, birds and fish should be removed from the vicinity of buildings to be treated. Foodstuffs should be protected from smoke
- Before treating any structure used by bats consult English Nature, Scottish Natural Heritage or the Countryside Council for Wales. All bats are protected under the Wildlife and Countryside Act 1981

Hazard classification and safety precautions
Operator protection A, E, H; U02b, U14, U18, U19a, U20b
Environmental protection E02c, E13a

Consumer protection C04, C09, C11
Storage and disposal D01, D09a, D11a

353 peroxyacetic acid (commodity substance)

An inorganic fungicide for use on potato tubers

Products

peroxyacetic acid	various	5% w/w	SL

Uses

- Fungus diseases in ***flower bulbs*** *(dip)*, ***potatoes*** *(tuber treatment)*
- Nematodes in ***flower bulbs*** *(dip)*, ***potatoes*** *(tuber treatment)*

Approval information

- Approval for the use of peroxyacetic acid as a commodity substance was granted on 18 January 2007 by Ministers under regulation 5 of the Control of Pesticides Regulations 1986
- Peroxyacetic acid will not be included in Annex I of EC Directive 91/414 but Essential Uses on flower bulbs, potato tubers and disinfection of glasshouses, warehouses and agricultural tools and equipment will be permitted in UK until December 2010.

Efficacy guidance

- Flower bulbs and seed potato tubers must be immersed for one minute in a 50:1 dilution of proprietary 5% w/w peroxyacetic acid formulation

Environmental safety

- Used dip must be disposed of in accordance with Part Five of the Code of Practice for Using Plant Protection Products

Hazard classification and safety precautions

Operator protection A, C, D, H, M

354 petroleum oil

An insecticidal and acaricidal hydrocarbon oil

Products

Certis Spraying Oil	Certis	710 g/l	EC	-

Uses

- Mealybugs in ***all bush fruit, all cane fruit, fruit trees, hops, nursery stock, protected cucumbers, protected pot plants, protected tomatoes, wine grapes***
- Red spider mites in ***all bush fruit, all cane fruit, fruit trees, hops, nursery stock, protected cucumbers, protected pot plants, protected tomatoes, wine grapes***
- Scale insects in ***all bush fruit, all cane fruit, fruit trees, hops, nursery stock, protected cucumbers, protected pot plants, protected tomatoes, wine grapes***

Approval information

- Product not controlled by Control of Pesticides Regulations because it acts by physical means only

Efficacy guidance

- Spray at 1% (0.5% on tender foliage) to wet plants thoroughly, particularly the underside of leaves, and repeat as necessary
- Petroleum oil acts by blocking pest breathing pores and making leaf surfaces inhospitable to pests seeking to attack or attach to the leaf surface
- On outdoor crops apply when dormant or on plants with a known tolerance
- On plants of unknown sensitivity test first on a small scale
- Mixtures with certain pesticides may damage crop plants. If mixing, spray a few plants to test for tolerance before treating larger areas

Restrictions

- Do not mix with sulphur or iprodione products or use such mixtures within 28 d of treatment
- Do not treat protected crops in bright sunshine unless the glass is well shaded
- Test species tolerance before large scale treatment

Crop-specific information

- HI zero
- Treat grapevines before flowering

Hazard classification and safety precautions

Hazard H03
Risk phrases R22a, R36, R37, R38
Operator protection A, C; U05a, U08, U11, U19a
Environmental protection E34
Storage and disposal D01, D02, D09a, D10a, D12a
Medical advice M03

355 phenmedipham

A contact phenyl carbamate herbicide for beet crops and strawberries
HRAC mode of action code: C1

See also desmedipham + ethofumesate + phenmedipham
desmedipham + phenmedipham
ethofumesate + phenmedipham

Products

1	Alpha Phenmedipham 320 SC	Makhteshim	320 g/l	SC	12320
2	Beetup	United Phosphorus	114 g/l	EC	07520
3	Beetup Flo	United Phosphorus	160 g/l	SC	11916
4	Betamax	AgriGuard	160 g/l	SC	12885
5	Betanal Flow	Bayer CropScience	160 g/l	SE	10997
6	Betasana SC	United Phosphorus	160 g/l	SC	12353
7	Cleancrop Phenmedipham	United Phosphorus	114 g/l	EC	11586
8	Crotale	Sipcam	471 g/l	SC	11593
9	Dancer Flow	Sipcam	160 g/l	SE	11405
10	Herbasan Flow	Nufarm UK	160 g/l	SC	11448
11	Mandolin Flow	Nufarm UK	160 g/l	SE	11162

Uses

- Amaranthus in ***chives*** *(off-label)*, ***herbs (see appendix 6)*** *(off-label)*, ***lettuce*** *(off-label)*, ***parsley*** *(off-label)*, ***salad brassicas*** *(off-label - for baby leaf production)* [5]
- Annual dicotyledons in ***fodder beet***, ***mangels***, ***red beet***, ***sugar beet*** [1-11]; ***leaf spinach*** *(off-label)*, ***spinach beet*** *(off-label)* [5]; ***strawberries*** [2, 7]
- Fat hen in ***chives*** *(off-label)*, ***herbs (see appendix 6)*** *(off-label)*, ***lettuce*** *(off-label)*, ***parsley*** *(off-label)*, ***salad brassicas*** *(off-label - for baby leaf production)* [5]
- Groundsel in ***chives*** *(off-label)*, ***herbs (see appendix 6)*** *(off-label)*, ***lettuce*** *(off-label)*, ***parsley*** *(off-label)*, ***salad brassicas*** *(off-label - for baby leaf production)* [5]
- Mayweeds in ***chives*** *(off-label)*, ***herbs (see appendix 6)*** *(off-label)*, ***lettuce*** *(off-label)*, ***parsley*** *(off-label)*, ***salad brassicas*** *(off-label - for baby leaf production)* [5]

Specific Off-Label Approvals (SOLAs)

- ***chives*** *20062914* [5]
- ***herbs (see appendix 6)*** *20062914* [5]
- ***leaf spinach*** *20050633 expires Dec 2008* [5]
- ***lettuce*** *20062914* [5]
- ***parsley*** *20062914* [5]
- ***salad brassicas*** *(for baby leaf production) 20062914* [5]
- ***spinach beet*** *20050633 expires Dec 2008* [5]

Approval information

- Phenmedipham included in Annex I under EC Directive 91/414
- Approval expiry 31 Aug 2008 [2, 7]

Efficacy guidance

- Best results achieved by application to young seedling weeds, preferably cotyledon stage, under good growing conditions when low doses are effective
- If using a low-dose programme 2-3 repeat applications at 7-10 d intervals are recommended on mineral soils, 3-5 applications may be needed on organic soils

- Addition of adjuvant oil may improve effectiveness on some weeds
- Various tank-mixtures with other beet herbicides recommended. See label for details
- Use of certain pre-emergence herbicides is recommended in combination with post-emergence treatment. See label for details

Restrictions

- Maximum number of treatments and maximum total dose varies with crop and product used. See label for details
- At high temperatures (above 21 °C) reduce rate and spray after 5 pm
- Do not apply immediately after frost or if frost expected
- Do not spray wet foliage or if rain imminent
- Do not spray crops stressed by wind damage, nutrient deficiency, pest or disease attack etc. Do not roll or harrow for 7 d before or after treatment
- Do not use on strawberries under cloches or polythene tunnels
- Do not use after 31 Jul in yr of harvest [8, 10, 11]
- Consult processor before use on crops for processing

Crop-specific information

- Latest use: before crop leaves meet between rows for beet crops; before flowering for strawberries
- Apply to beet crops at any stage as low dose/low volume spray or from fully developed cotyledon stage with full rate. Apply to red beet after fully developed cotyledon stage
- Apply to strawberries at any time when weeds in susceptible stage, except in period from start of flowering to picking

Following crops guidance

- Beet crops may follow at any time after a treated crop. 3 mth must elapse from treatment before any other crop is sown and must be preceded by mould-board ploughing to 15 cm

Environmental safety

- Dangerous for the environment
- Very toxic to aquatic organisms
- Harmful to fish or other aquatic life. Do not contaminate surface waters or ditches with chemical or used container

Hazard classification and safety precautions

Hazard H03 [2, 7]; H04 [1, 3, 4, 6, 8]; H11 [1-7, 9-11]
Risk phrases R21, R22a, R37, R40 [2, 7]; R36, R51 [2-4, 6, 7]; R38 [1]; R43 [1, 3, 4, 6, 8]; R50 [1, 5, 9-11]; R52 [8]; R53a [1-11]
Operator protection A [1-6, 8, 9, 11]; C [2-4, 6]; H [1, 3, 4, 6, 8]; U05a [1-4, 7, 8]; U08 [1-4, 6, 7]; U11, U20c [3, 6]; U13 [2, 7]; U14 [1, 3, 6, 8]; U19a [1, 2, 6, 7]; U20a [1, 10, 11]; U20b [2, 4, 5, 7-9]
Environmental protection E13c [5, 8, 9]; E15a [1, 2, 4, 7, 10, 11]; E15b [3, 6]; E19b [1]; E34 [1, 2, 7]; E38 [1, 4, 5, 9-11]
Storage and disposal D01 [1-4, 6-8]; D02 [2-4, 6-8]; D05 [1-7, 11]; D08 [8, 9]; D09a [1-11]; D10b [1, 2, 5, 7-11]; D10c [3, 4, 6]; D12a [1, 3-6, 9-11]
Medical advice M03 [2, 7]; M05a [1]

356 d-phenothrin

A non-systemic pyrethroid insecticide available only in mixtures

357 d-phenothrin + tetramethrin

A pyrethroid insecticide mixture for control of flying insects
IRAC mode of action code: 3 + 3

Products

Killgerm ULV 500	Killgerm	36.8:18.4 g/l	UL	H4647

Uses

- Flies in ***agricultural premises***
- Grain storage mite in ***grain stores***

- Mosquitoes in ***agricultural premises***
- Wasps in ***agricultural premises***

Approval information
- Product approved for ULV application. See label for details [1]

Efficacy guidance
- Close doors and windows and spray in all directions for 3-5 sec. Keep room closed for at least 10 min

Restrictions
- For use only by professional operators
- Do not use space sprays containing pyrethrins or pyrethroid more than once per week in intensive or controlled environment animal houses in order to avoid development of resistance. If necessary, use a different control method or product

Crop-specific information
- May be used in the presence of poultry and livestock

Environmental safety
- Dangerous for the environment
- Toxic to aquatic organisms
- Do not apply directly to livestock/poultry
- Remove exposed milk and collect eggs before application. Protect milk machinery and containers from contamination

Hazard classification and safety precautions

Hazard H03, H11
Risk phrases R22b, R51, R53a
Operator protection A, C, D, E, H; U02b, U09b, U14, U19a, U20a, U20b
Environmental protection E05a, E15a, E38
Consumer protection C06, C07, C08, C09, C11, C12
Storage and disposal D01, D05, D06a, D09a, D10a, D12a
Medical advice M05b

358 physical pest control

Products that work by physical action only

Products

1	Majestik	Certis	SL	-
2	SB Plant	Invigorator Fargro	SL	-

Uses
- Aphids in ***all edible crops (outdoor)*** [2]; ***fruit crops, protected fruit, protected vegetables, vegetables*** [1]; ***ornamental plant production, protected ornamentals*** [1, 2]
- Leafhoppers in ***fruit crops, ornamental plant production, protected fruit, protected ornamentals, protected vegetables, vegetables*** [1]
- Mealybugs in ***all edible crops (outdoor), ornamental plant production, protected ornamentals*** [2]
- Mites in ***fruit crops, ornamental plant production, protected fruit, protected ornamentals, protected vegetables, vegetables*** [1]
- Spider mites in ***all edible crops (outdoor), ornamental plant production, protected ornamentals*** [2]
- Thrips in ***fruit crops, ornamental plant production, protected fruit, protected ornamentals, protected vegetables, vegetables*** [1]
- Whitefly in ***all edible crops (outdoor)*** [2]; ***fruit crops, protected fruit, protected vegetables, vegetables*** [1]; ***ornamental plant production, protected ornamentals*** [1, 2]

Approval information
- Products included in this profile are not subject to the Control of Pesticides Regulations/Plant Protection Products Regulations because they act by physical means only

Efficacy guidance
- Products act by physical means following direct contact with spray. They may therefore be used at any time of year on all pest growth stages

- Ensure thorough spray coverage of plant, paying special attention to growing points and the underside of leaves
- Treat as soon as target pests are seen and repeat as often as necessary
- Heavy infestations will require 2-3 rapid applications before extending the spray interval [2]
- Consult manufacturer for guidance on efficacy before treating outdoor crops for the first time [1]
- Frequent applications may result in temporary brown staining at petal and leaf edges or at the bottom of fruits [2]
- Helps prevent chlorosis, improves leaf colour and can offer some control of powdery mildew [2]

Restrictions

- No limit on the number of treatments and no minimum interval between applications
- Before large scale use on a new crop, treat a few plants to check for crop safety
- Do not treat ornamental crops when in flower
- Avoid treating fruit or vegetables close to harvest as product may leave a sticky deposit [1]
- Do not mix with any other product [2]

Crop-specific information

- HI zero

Environmental safety

- May be used in conjunction with biological control agents. Spray 24 h before they are introduced
- Where bees are being used for pollination close hives before application and ensure spray has completely dried before release [1]
- Ensure spray equipment is thoroughly washed out before use. Product can release chemical residues from equipment and cause scorch [2]

Hazard classification and safety precautions

Operator protection U05a, U09a, U19a, U20c [1]; U10, U14, U15 [2]
Storage and disposal D01, D02, D05, D06d [2]

359 picloram

A persistent, translocated pyridine carboxylic acid herbicide for non-crop areas
HRAC mode of action code: O

See also *2,4-D + picloram*
clopyralid + picloram

Products

1	Pantheon	Pan Agriculture	240 g/l	SC	13695
2	Tordon 22K	Nomix Enviro	240 g/l	SL	05790

Uses

- Annual dicotyledons in ***land not intended for cropping***
- Bracken in ***land not intended for cropping***
- Japanese knotweed in ***land not intended for cropping***
- Perennial dicotyledons in ***land not intended for cropping***
- Woody weeds in ***land not intended for cropping***

Efficacy guidance

- May be applied at any time of year. Best results achieved by application as foliage spray in late winter to early spring
- For bracken control apply 2-4 wk before frond emergence
- Clovers are highly sensitive and eliminated at very low doses
- Persists in soil for up to 2 yr

Restrictions

- Maximum number of treatments 1 per yr on land not intended for cropping
- Do not apply around desirable trees or shrubs where roots may absorb chemical
- Do not apply on slopes where chemical may be leached onto areas of desirable plants

Environmental safety

- Harmful to fish or other aquatic life. Do not contaminate surface waters or ditches with chemical or used container
- Keep livestock out of treated areas for at least 2 wk and until foliage of any poisonous weeds such as ragwort has died and become unpalatable

Hazard classification and safety precautions

Hazard H04

Risk phrases R43

Operator protection A, H; U05a, U08, U14, U20b

Environmental protection E07a, E13c

Storage and disposal D01, D02, D09a, D10b

360 picolinafen

A pyridinecarboxamide herbicide for cereals

HRAC mode of action code: F1

See also *pendimethalin + picolinafen*

Products

Vixen	BASF	75% w/w	WG	13621

Uses
- Chickweed in ***winter barley, winter wheat***
- Cleavers in ***winter barley, winter wheat***
- Field pansy in ***winter barley, winter wheat***
- Field speedwell in ***winter barley, winter wheat***
- Ivy-leaved speedwell in ***winter barley, winter wheat***
- Shepherd's purse in ***winter barley, winter wheat***

Approval information
- Picolinafen included in Annex I under EC Directive 91/414

Efficacy guidance
- Picolinafen is contact acting. Ensure good spray cover of the target weeds
- Best results obtained from treatment of small actively growing weeds
- Weeds emerging after application will not be controlled
- Always follow WRAG guidelines for preventing and managing herbicide resistant weeds. Section 5 for more information

Restrictions
- Maximum number of treatments 1 per crop
- Do not disturb soil surface after application
- Do not roll emerged crops prior to application
- Do not treat undersown crops or those to be undersown
- Do not spray during periods of prolonged or severe frost
- Do not apply to crops under stress from any cause

Crop-specific information
- Latest Use: before pseudo-stem erect stage (GS 30) for wheat and barley
- Treatment may cause transient bleaching in some crops

Following crops guidance
- No restrictions on following crops after normal harvest of a treated crop
- In the event of failure of a treated crop further crops of wheat or barley may be drilled after a minimum interval of 8 wk from treatment and after ploughing to at least 15 cm

Environmental safety
- Dangerous for the environment
- Very toxic to aquatic organisms
- LERAP Category B
- Take extreme care to avoid drift onto plants outside the target area

Hazard classification and safety precautions

Hazard H11

Risk phrases R50, R53a

Operator protection A, D; U05a, U08, U13, U19a, U20b

Environmental protection E15a, E16a, E38

Storage and disposal D01, D02, D09a, D10b, D12a

361 picoxystrobin

A broad spectrum strobilurin fungicide for cereals
FRAC mode of action code: 11

See also *chlorothalonil + picoxystrobin*
cyprodinil + picoxystrobin

Products

Galileo	DuPont	250 g/l	SC	13252

Uses

- Brown rust in ***spring barley, spring wheat, winter barley, winter wheat***
- Crown rust in ***spring oats, winter oats***
- Eyespot in ***spring wheat*** *(reduction)*, ***winter wheat*** *(reduction)*
- Glume blotch in ***spring wheat, winter wheat***
- Late ear diseases in ***spring wheat, winter wheat***
- Net blotch in ***spring barley, winter barley***
- Powdery mildew in ***spring barley, spring oats, winter barley, winter oats***
- Rhynchosporium in ***spring barley, winter barley***
- Septoria leaf blotch in ***spring wheat, winter wheat***
- Yellow rust in ***spring wheat, winter wheat***

Approval information

- Picoxystrobin included in Annex I under EC Directive 91/414
- Accepted by BBPA for use on malting barley

Efficacy guidance

- Best results achieved from protectant treatments made before disease establishes in the crop
- Use of tank mixtures is recommended where disease has become established at the time of treatment
- Persistence of yellow rust control is less than for other foliar diseases but can be improved by use of an appropriate tank mixture
- Eyespot control is not reliable and an appropriate tank mix should be used where crops are at risk
- Picoxystrobin is a member of the QoI cross resistance group. Product should be used preventatively and not relied on for its curative potential
- Use product as part of an Integrated Crop Management strategy incorporating other methods of control, including where appropriate other fungicides with a different mode of action. Do not apply more than two foliar applications of QoI containing products to any cereal crop
- There is a significant risk of widespread resistance occurring in *Septoria tritici* populations in UK. Failure to follow resistance management action may result in reduced levels of disease control
- On cereal crops product must always be used in mixture with another product, recommended for control of the same target disease, that contains a fungicide from a different cross resistance group and is applied at a dose that will give robust control
- Strains of barley powdery mildew resistant to QoIs are common in the UK

Restrictions

- Maximum number of treatments 2 per crop per yr

Crop-specific information

- Latest use: grain watery ripe (GS 71)

Environmental safety

- Dangerous for the environment
- Very toxic to aquatic organisms

Hazard classification and safety precautions

Hazard H11
Risk phrases R50, R53a
Operator protection U05a, U09a, U14, U15, U20a
Environmental protection E15a, E38
Storage and disposal D01, D02, D05, D09a, D10c, D12a

362 pinoxaden

A phenylpyrazoline grass weed herbicide for cereals
HRAC mode of action code: A

See also *clodinafop-propargyl + pinoxaden*

Products

1	Axial	Syngenta	100 g/l	EC	12521
2	Greencrop Helvick	Greencrop	100 g/l	EC	13052

Uses

- Italian ryegrass in ***spring barley, winter barley, winter wheat***
- Perennial ryegrass in ***spring barley, winter barley, winter wheat***
- Wild oats in ***spring barley, winter barley, winter wheat***

Efficacy guidance

- Products contain a herbicide safener (cloquintocet-mexyl)
- Best results obtained from treatment when all grass weeds have emerged. There is no residual activity
- Broad-leaved weeds are not controlled
- Treat before emerged weed competition reduces yield
- Blackgrass in winter and spring barley is also controlled when used as part of an integrated control strategy. Product is not recommended for blackgrass control in winter wheat
- Grass weed control may be reduced if rain falls within 1 hr of application
- Pinoxaden is an ACCase inhibitor herbicide. To avoid the build up of resistance do not apply products containing an ACCase inhibitor herbicide more than twice to any crop. In addition do not use any product containing pinoxaden in mixture or sequence with any other product containing the same ingredient
- Use these products as part of a resistance management strategy that includes cultural methods of control and does not use ACCase inhibitors as the sole chemical method of grass weed control
- Applying a second product containing an ACCase inhibitor to a crop will increase the risk of resistance development; only use a second ACCase inhibitor to control different weeds at a different timing
- Always follow WRAG guidelines for preventing and managing herbicide resistant weeds. See Section 5

Restrictions

- Maximum number of treatments 1 per crop
- Do not spray crops under stress or suffering from waterlogging, pest attack, disease or frost damage
- Do not spray crops undersown with grass mixtures
- Avoid the use of hormone-containing herbicides in mixture or in sequence. Allow 21 d after, or 7 d before, hormone application
- Product must be used with prescribed adjuvant. See label

Crop-specific information

- Latest use: before flag leaf extending stage (GS 41)
- Spray cereals in autumn, winter or spring from the 2-leaf stage (GS 12)

Following crops guidance

- There are no restrictions on succeeding crops in a normal rotation
- In the event of failure of a treated crop ryegrass, maize, oats or any broad-leaved crop may be planted after a minimum interval of 4 wk from application

Environmental safety

- Dangerous for the environment
- Toxic to aquatic organisms
- Do not allow spray to drift onto neighbouring crops of oats, ryegrass or maize

Hazard classification and safety precautions

Hazard H04, H11
Risk phrases R36, R38, R51, R53a
Operator protection A, C, H; U05a, U09a, U20b

Environmental protection E15b, E38
Storage and disposal D01, D02, D05, D09a, D10c, D11a, D12a

363 pirimicarb

A carbamate insecticide for aphid control
IRAC mode of action code: 1A

See also lambda-cyhalothrin + pirimicarb

Products

1	Agriguard Pirimicarb	AgriGuard	50% w/w	WG	09620
2	Aphox	Syngenta	50% w/w	WG	10515
3	Clayton Pirimicarb 50	Clayton	50% w/w	WG	12910
4	Greencrop Glenroe	Greencrop	50% w/w	WG	09903
5	Landgold Pirimicarb 50	Landgold	50% w/w	WG	09018
6	Landgold Pirimicarb 50	Teliton	50% w/w	WG	12126
7	Phantom	Syngenta	50% w/w	WG	11954
8	Standon Pirimicarb 50	Standon	50% w/w	WG	08878
9	Standon Pirimicarb 50	Standon	50% w/w	WG	13290

Uses

- Aphids in ***apples***, ***broad beans***, ***broccoli***, ***brussels sprouts***, ***cabbages***, ***calabrese***, ***cauliflowers***, ***durum wheat***, ***pears***, ***peas***, ***potatoes***, ***spring barley***, ***spring field beans***, ***spring oats***, ***spring rye***, ***spring wheat***, ***strawberries***, ***sugar beet***, ***swedes***, ***triticale***, ***turnips***, ***winter barley***, ***winter field beans***, ***winter oats***, ***winter rye***, ***winter wheat*** [1-9]; ***bilberries*** *(off-label)*, ***blueberries*** *(off-label)*, ***cranberries*** *(off-label)*, ***evening primrose*** *(off-label)*, ***mallow (althaea spp.)*** *(off-label)*, ***mustard*** *(off-label)*, ***protected chives*** *(off-label)*, ***protected herbs (see appendix 6)*** *(off-label)*, ***protected lamb's lettuce*** *(off-label)*, ***protected spinach*** *(off-label)*, ***radicchio*** *(off-label)*, ***scarole*** *(off-label)*, ***spring linseed*** *(off-label)*, ***sweetcorn*** *(off-label)*, ***winter linseed*** *(off-label)* [2]; ***blackberries*** *(off-label)*, ***choi sum*** *(off-label)*, ***pak choi*** *(off-label)*, ***rubus hybrids*** *(off-label)* [1, 2, 7, 8]; ***blackcurrants***, ***carrots***, ***chinese cabbage***, ***collards***, ***dwarf beans***, ***forage maize***, ***gooseberries***, ***parsnips***, ***raspberries***, ***redcurrants***, ***runner beans***, ***spring oilseed rape***, ***sweetcorn***, ***winter oilseed rape*** [1-4, 7-9]; ***celeriac*** *(off-label)*, ***celery (outdoor)*** *(off-label)*, ***chicory*** *(off-label - for forcing)*, ***chives*** *(off-label)*, ***courgettes*** *(off-label)*, ***endives*** *(off-label)*, ***fennel*** *(off-label)*, ***fodder beet*** *(off-label)*, ***frise*** *(off-label)*, ***gherkins*** *(off-label)*, ***herbs (see appendix 6)*** *(off-label)*, ***honesty*** *(off-label)*, ***leaf spinach*** *(off-label)*, ***marrows*** *(off-label)*, ***parsley*** *(off-label)*, ***parsley root*** *(off-label)*, ***poppies for morphine production*** *(off-label)*, ***protected courgettes*** *(off-label)*, ***protected endives*** *(off-label)*, ***protected frise*** *(off-label)*, ***protected gherkins*** *(off-label)*, ***protected parsley*** *(off-label)*, ***protected radicchio*** *(off-label)*, ***protected radishes*** *(off-label)*, ***protected scarole*** *(off-label)*, ***protected spinach beet*** *(off-label)*, ***radishes*** *(off-label)*, ***red beet*** *(off-label)*, ***spinach beet*** *(off-label)* [2, 7]; ***cherries*** [1-4, 7, 9]; ***cucumbers***, ***forest nurseries***, ***ornamental plant production***, ***peppers***, ***protected carnations***, ***protected chrysanthemums***, ***protected cinerarias***, ***protected cyclamen***, ***protected lettuce***, ***protected roses***, ***protected tomatoes***, ***tomatoes (outdoor)*** [2-4, 7, 9]; ***horseradish*** *(off-label)* [1, 2, 4, 7, 8]; ***kale*** [1-3, 5-9]; ***lamb's lettuce*** *(off-label)*, ***protected horseradish*** *(off-label)*, ***protected leaf spinach*** *(off-label)*, ***rhubarb*** *(off-label)*, ***salsify*** *(off-label)* [7]; ***lettuce*** [2, 9]; ***lettuce*** *(outdoor crops)* [3, 4, 7]; ***permanent grassland*** [1]; ***protected aubergines*** *(off-label)*, ***protected cayenne peppers*** *(off-label)*, ***salad brassicas*** *(off-label - for baby leaf production)* [2, 4, 7]; ***protected salad brassicas*** *(off-label - for baby leaf production)* [2, 4]
- Blackfly in ***cherries*** [8]
- Carrot willow aphid in ***celery (outdoor)*** [1]
- Leaf curling plum aphid in ***plums*** *(off-label)* [2, 7]
- Mealy plum aphid in ***plums*** *(off-label)* [2, 7]
- Thrips in ***endives*** *(off-label)*, ***protected endives*** *(off-label)* [2, 7]; ***protected parsley*** *(off-label)* [7]

Specific Off-Label Approvals (SOLAs)

- ***bilberries*** *20063151* [2]
- ***blackberries*** *20063004* [1], *20063110* [2], *20063445* [7], *20062998* [8]
- ***blueberries*** *20063151* [2]
- ***celeriac*** *20011292* [2], *20051747* [7]
- ***celery (outdoor)*** *20040680* [2], *20051744* [7]

- ***chicory*** *(for forcing) 20011283* [2], *(for forcing) 20051749* [7]
- ***chives*** *20063150* [2], *20063448* [7]
- ***choi sum*** *20063000* [1], *20063109* [2], *20063449* [7], *20062997* [8]
- ***courgettes*** *20011298* [2], *20051746* [7]
- ***cranberries*** *20063151* [2]
- ***endives*** *20011286* [2], *20051741* [7]
- ***evening primrose*** *20063155* [2]
- ***fennel*** *20011292* [2], *20051747* [7]
- ***fodder beet*** *20063152* [2], *20063446* [7]
- ***frise*** *20063150* [2], *20063448* [7]
- ***gherkins*** *20011298* [2], *20051746* [7]
- ***herbs (see appendix 6)*** *20063150* [2], *20063448* [7]
- ***honesty*** *20063155* [2], *20051742* [7]
- ***horseradish*** *20063005* [1], *20063154* [2], *20063469* [4], *20063450* [7], *20062996* [8]
- ***lamb's lettuce*** *20063448* [7]
- ***leaf spinach*** *20011298* [2], *20051746* [7]
- ***mallow (althaea spp.)*** *20063154* [2]
- ***marrows*** *20011298* [2], *20051746* [7]
- ***mustard*** *20063155* [2]
- ***pak choi*** *20063000* [1], *20063109* [2], *20063449* [7], *20062997* [8]
- ***parsley*** *20063150* [2], *20063448* [7]
- ***parsley root*** *20011298* [2], *20051746* [7]
- ***plums*** *20012687* [2], *20051738* [7]
- ***poppies for morphine production*** *20030767* [2], *20051748* [7]
- ***protected aubergines*** *20063153* [2], *20063470* [4], *20063447* [7]
- ***protected cayenne peppers*** *20063153* [2], *20063470* [4], *20063447* [7]
- ***protected chives*** *20063150* [2]
- ***protected courgettes*** *20011301* [2], *20051745* [7]
- ***protected endives*** *20011286* [2], *20051741* [7]
- ***protected frise*** *20063150* [2], *20063448* [7]
- ***protected gherkins*** *20011301* [2], *20051745* [7]
- ***protected herbs (see appendix 6)*** *20063150* [2]
- ***protected horseradish*** *20051746* [7]
- ***protected lamb's lettuce*** *20063150* [2]
- ***protected leaf spinach*** *20051746* [7]
- ***protected parsley*** *20063150* [2], *20051741* [7]
- ***protected radicchio*** *20063150* [2], *20063448* [7]
- ***protected radishes*** *20011298* [2], *20051746* [7]
- ***protected salad brassicas*** *(for baby leaf production) 20063150* [2], *(for baby leaf production) 20063468* [4]
- ***protected scarole*** *20063150* [2], *20063448* [7]
- ***protected spinach*** *20011298* [2]
- ***protected spinach beet*** *20011298* [2], *20051746* [7]
- ***radicchio*** *20063150* [2]
- ***radishes*** *20011298* [2], *20051746* [7]
- ***red beet*** *20011292* [2], *20051747* [7]
- ***rhubarb*** *20051744* [7]
- ***rubus hybrids*** *20063004* [1], *20063110* [2], *20063445* [7], *20062998* [8]
- ***salad brassicas*** *(for baby leaf production) 20011295* [2], *(for baby leaf production) 20063468* [4], *(for baby leaf production) 20063448* [7]
- ***salsify*** *20051747* [7]
- ***scarole*** *20063150* [2]
- ***spinach beet*** *20011298* [2], *20051746* [7]
- ***spring linseed*** *20063155* [2]
- ***sweetcorn*** *20011298* [2]
- ***winter linseed*** *20063155* [2]

Approval information

- Pirimicarb included in Annex I under EC Directive 91/414
- Approved for aerial application on cereals [1, 2, 4, 7]. See notes in Section 5

- Accepted by BBPA for use on malting barley
- In 2006 PSD required that all products containing this active ingredient should carry the following warning in the main area of the container label: "Pirimicarb is an anticholinesterase carbamate. Handle with care"

Efficacy guidance

- Chemical has contact, fumigant and translaminar activity
- Best results achieved under warm, calm conditions when plants not wilting and spray does not dry too rapidly. Little vapour activity at temperatures below 15 °C
- Apply as soon as aphids seen or warning issued and repeat as necessary
- Addition of non-ionic wetter recommended for use on brassicas
- On cucumbers and tomatoes a root drench is preferable to spraying when using predators in an integrated control programme
- Where aphids resistant to pirimicarb occur control is unlikely to be satisfactory

Restrictions

- Contains an anticholinesterase carbamate compound. Do not use if under medical advice not to work with such compounds
- Maximum number of treatments not specified in some cases but normally 2-6 depending on crop
- When treating ornamentals check safety by treating a small number of plants first
- Spray equipment must only be used where the operator's normal working position is within a closed cab on a tractor or self-propelled sprayer when making air-assisted applications to apples or other top fruit

Crop-specific information

- Latest use in accordance with harvest intervals below
- HI oilseed rape, cereals, maize, sweetcorn, lettuce under glass 14 d (sweetcorn off-label 3 d); grassland, chicory, plums 7 d; cucumbers, tomatoes and peppers under glass 2 d; protected courgettes and gherkins 24 h; other edible crops 3 d; flowers and ornamentals zero

Environmental safety

- Dangerous for the environment
- Very toxic to aquatic organisms
- Dangerous to fish or other aquatic life. Do not contaminate surface waters or ditches with chemical or used container
- Chemical has little effect on bees, ladybirds and other insects and is suitable for use in integrated control programmes on apples and pears
- Keep all livestock out of treated areas for at least 7 d. Bury or remove spillages

Hazard classification and safety precautions

Hazard H02, H11 [2-9]; H03 [1]
Risk phrases R20 [2, 3, 7, 9]; R22a [1]; R25, R36, R50, R53a [2-9]
Operator protection A, C, D, H, J, M; U05a, U08, U19a [1-9]; U20a [3, 4, 7]; U20b [1, 2, 5, 6, 8, 9]
Environmental protection E06c [1-9] (7 d); E13b [1]; E15a [3-6, 8]; E15b, E38 [2, 7, 9]; E34 [1-9]
Storage and disposal D01, D02, D09a, D11a [1-9]; D05 [4]; D07 [3, 4]; D12a [2, 3, 7, 9]
Medical advice M02 [1-9]; M03 [1, 2, 7, 9]; M04a [2-9]; M04b [3]

364 pirimiphos-methyl

A contact, fumigant and translaminar organophosphorus insecticide
IRAC mode of action code: 1B

Products

1	Actellic 50 EC	Syngenta	500 g/l	EC	12726
2	Actellic Smoke Generator No 20	Syngenta	22.5% w/w	FU	10540
3	Fumite Pirimiphos Methyl Smoke	Certis	22.5% w/w	FU	00941

Uses

- Ants in ***protected cucumbers, protected ornamentals, protected peppers, protected tomatoes*** [3]
- Aphids in ***protected cucumbers, protected ornamentals, protected peppers, protected tomatoes*** [3]

- Capsids in ***protected cucumbers, protected ornamentals, protected peppers, protected tomatoes*** [3]
- Earwigs in ***protected cucumbers, protected ornamentals, protected peppers, protected tomatoes*** [3]
- Flour beetle in ***stored grain*** [1]
- Flour moth in ***stored grain*** [1]
- Grain beetle in ***stored grain*** [1]
- Grain storage mite in ***stored grain*** [1]
- Grain storage pests in ***grain stores*** [1, 2]
- Grain weevil in ***stored grain*** [1]
- Leaf miner in ***protected cucumbers, protected ornamentals, protected peppers, protected tomatoes*** [3]
- Red spider mites in ***protected cucumbers, protected ornamentals, protected peppers, protected tomatoes*** [3]
- Sawflies in ***protected cucumbers, protected ornamentals, protected peppers, protected tomatoes*** [3]
- Thrips in ***protected cucumbers, protected ornamentals, protected peppers, protected tomatoes*** [3]
- Warehouse moth in ***stored grain*** [1]
- Whitefly in ***protected cucumbers, protected ornamentals, protected peppers, protected tomatoes*** [3]

Approval information

- Pirimiphos-methyl included in Annex I under EC Directive 91/414
- In 2006 PSD required that all products containing this active ingredient should carry the following warning in the main area of the container label: "Pirimiphos-methyl is an anticholinesterase organophosphate. Handle with care"
- Accepted by BBPA for use in stores for malting barley

Efficacy guidance

- Chemical acts rapidly and has short persistence in plants, but persists for long periods on inert surfaces
- Best results for protection of stored grain achieved by cleaning store thoroughly before use and employing a combination of pre-harvest and grain/seed treatments [1, 2]
- Best results for admixture treatment obtained when grain stored at 15% moisture or less. Dry and cool moist grain coming into store but then treat as soon as possible, ideally as it is loaded [1]
- Surface admixture can be highly effective on localised surface infestations but should not be relied on for long-term control unless application can be made to the full depth of the infestation [1]
- For control of whitefly and other glasshouse pests apply as smoke or by thermal fogging. See label for details of techniques and suitable machines [3]
- Where insect pests resistant to pirimiphos-methyl occur control is unlikely to be satisfactory

Restrictions

- Contains an organophosphorus anticholinesterase compound. Do not use if under medical advice not to work with such compounds
- Maximum number of treatments 2 per grain store [1, 2]; 1 per batch of stored grain [1]
- Do not apply surface admixture or complete admixture using hand held equipment [1]
- Surface admixture recommended only where the conditions for store preparation, treatment and storage detailed in the label can be met [1]
- Do not fumigate glasshouses in bright sunshine or when foliage is wet or roots dry [3]
- Do not fumigate young seedlings or plants being hardened off [3]
- Do not fog open flowers of ornamentals without first consulting firm [3]
- Do not fog mushrooms when wet as slight spotting may occur [3]

Crop-specific information

- Latest use: before storing grain [1, 2]
- HI for edible crops zero [3]
- Disinfect empty grain stores by spraying surfaces and/or fumigation and treat grain by full or surface admixture. Treat well before harvest in late spring or early summer and repeat 6 wk later or just before harvest if heavily infested. See label for details of treatment and suitable application machinery

- Treatment volumes on structural surfaces should be adjusted according to surface porosity. See label for guidelines
- Treat inaccessible areas with smoke generating product used in conjunction with spray treatment of remainder of store
- Predatory mites (*Cheyletus*) found in stored grain feeding on infestations of grain mites may survive treatment and can lead to rejection by buyers. They do not remain for long but grain should be inspected carefully before selling [1, 2]

Environmental safety

- Dangerous for the environment
- Very toxic to aquatic organisms [1]
- Toxic to aquatic organisms [2, 3]
- Highly flammable [2, 3]
- Flammable [1]
- Ventilate fumigated or fogged spaces thoroughly before re-entry
- Unprotected persons must be kept out of fumigated areas within 3 h of ignition and for 4 h after treatment
- Do not harvest treated crops for human or animal consumption for at least 5 d after last application [3]
- Extremely dangerous to bees. Do not apply to crops in flower or to those in which bees are actively foraging. Do not apply when flowering weeds are present [3]
- Wildlife must be excluded from buildings during treatment [1]
- Keep away from combustible materials [2]

Hazard classification and safety precautions

Hazard H03, H11 [1-3]; H07 [2, 3]; H08 [1]
Risk phrases R20, R53a [1-3]; R22a, R22b, R50 [1]; R36, R37 [1, 3]; R51 [2, 3]
Operator protection A, C, J, M [1]; D, H [1, 2]; U04a, U09a, U16a, U24 [1]; U05a, U19a [1-3]; U14 [2, 3]; U20a [3]; U20b [1, 2]
Environmental protection E02a [2, 3] (4 h); E12b, E12e, E13a [3]; E15a, E34 [1-3]; E38 [1, 2]
Consumer protection C02a [3] (5 d); C09 [1]; C12 [1-3]
Storage and disposal D01, D02, D09a [1-3]; D05, D10c [1]; D07 [2]; D11a [2, 3]; D12a [1, 2]
Treated seed S06a [1]
Medical advice M01 [1-3]; M03 [2]; M05b [1]

365 potassium bicarbonate (commodity substance)

An inorganic fungicide for use in horticulture

Products

potassium bicarbonate	various	>99% w/w	SL

Uses

- Fungus diseases in ***outdoor crops, protected crops***

Approval information

- Approval for the use of potassium bicarbonate as a commodity substance was granted on 26 July 2005 by Ministers under regulation 5 of the Control of Pesticides Regulations 1986
- Potassium bicarbonate is being supported for review in the fourth stage of the EC Review Programme under Directive 91/414. Whether or not it is included in Annex I, the existing commodity chemical approval will have to be revoked because substances listed in Annex I must be approved for marketing and use under the EC regime. If potassium bicarbonate is included in Annex I, products containing it will need to gain approval in the normal way if they carry label claims for pesticidal activity
- Approval expiry 31 Dec 2008 [1]

Efficacy guidance

- Adjuvants authorised by PSD may be used in conjunction with potassium bicarbonate
- Crop phytotoxicity can occur

Restrictions

- Food grade potassium bicarbonate must only be used

Hazard classification and safety precautions

Operator protection A, C, G, H

366 prochloraz

A broad-spectrum protectant and eradicant imidazole fungicide
FRAC mode of action code: 3

See also fluquinconazole + prochloraz

Products

1	Alpha Prochloraz 40 EC	Makhteshim	400 g/l	EC	11002
2	Barclay Eyetak 40	Barclay	400 g/l	EC	11352
3	Mirage 40 EC	Makhteshim	400 g/l	EC	06770
4	Panache 40	DAPT	400 g/l	EC	10632
5	Poraz	BASF	450 g/l	EC	11701
6	Prelude 20LF	Agrichem	200 g/l	LS	04371
7	Prospero	Makhteshim	400 g/l	EC	11931
8	Scotts Octave	Scotts	46% w/w	WP	09275
9	Warbler	Nufarm UK	400 g/l	EC	13389

Uses

- Alternaria in ***spring oilseed rape*** [1, 3, 9]; ***winter oilseed rape*** [1, 3, 5, 7, 9]
- Disease control in ***durum wheat*** *(off-label)*, ***grass seed crops*** *(off-label)* [1]
- Eyespot in ***spring wheat*** [1-4]; ***winter barley***, ***winter wheat*** [1-5, 7, 9]; ***winter rye*** [1, 3, 5, 7, 9]
- Foliar disease control in ***grass seed crops*** *(off-label)* [7]; ***triticale*** *(off-label)*, ***winter linseed*** *(off-label)* [3]
- Fungus diseases in ***hardy ornamentals***, ***ornamental plant production***, ***woody ornamentals*** [8]
- Glume blotch in ***spring wheat*** [1, 3]; ***winter wheat*** [1, 3, 5, 7, 9]
- Grey mould in ***spring oilseed rape*** [1, 3, 9]; ***winter oilseed rape*** [1, 3, 5, 7, 9]
- Light leaf spot in ***spring oilseed rape*** [1, 3, 4, 9]; ***winter oilseed rape*** [1, 3-5, 7, 9]
- Net blotch in ***spring barley***, ***winter barley*** [1, 3, 5, 7, 9]
- Phoma in ***spring oilseed rape*** [1, 3, 4, 9]; ***winter oilseed rape*** [1, 3-5, 7, 9]
- Powdery mildew in ***spring barley***, ***spring wheat***, ***winter barley***, ***winter wheat*** [5, 7, 9]; ***winter rye*** [1, 3, 5, 7, 9]
- Rhynchosporium in ***spring barley***, ***winter barley***, ***winter rye*** [1, 3, 5, 7, 9]
- Ring spot in ***chives*** *(off-label)*, ***herbs (see appendix 6)*** *(off-label)*, ***lettuce*** *(off-label)*, ***parsley*** *(off-label)*, ***protected chives*** *(off-label)*, ***protected herbs (see appendix 6)*** *(off-label)*, ***protected lettuce*** *(off-label)*, ***protected parsley*** *(off-label)* [3]
- Sclerotinia stem rot in ***spring oilseed rape*** [1, 3, 9]; ***winter oilseed rape*** [1, 3, 5, 7, 9]
- Seed-borne diseases in ***flax*** *(seed treatment)*, ***hemp for oilseed production*** *(off-label - seed treatment)*, ***linseed*** *(seed treatment)* [6]
- Septoria leaf blotch in ***spring wheat*** [1, 3]; ***winter rye*** [5, 7, 9]; ***winter wheat*** [1, 3, 5, 7, 9]
- White leaf spot in ***spring oilseed rape*** [1, 3, 9]; ***winter oilseed rape*** [1, 3, 5, 7, 9]

Specific Off-Label Approvals (SOLAs)

- ***chives*** *20032623* [3]
- ***durum wheat*** *20061102* [1]
- ***grass seed crops*** *20061102* [1], *20062920* [7]
- ***hemp for oilseed production*** *(seed treatment) 20052562* [6]
- ***herbs (see appendix 6)*** *20032623* [3]
- ***lettuce*** *20032623* [3]
- ***parsley*** *20032623* [3]
- ***protected chives*** *20032623* [3]
- ***protected herbs (see appendix 6)*** *20032623* [3]
- ***protected lettuce*** *20032623* [3]
- ***protected parsley*** *20032623* [3]
- ***triticale*** *20062686* [3]
- ***winter linseed*** *20062686* [3]

Approval information

- Accepted by BBPA for use on malting barley

SEE SECTION 3 FOR PRODUCTS ALSO REGISTERED

Efficacy guidance

- Spray cereals at first signs of disease. Protection of winter crops through season usually requires at least 2 treatments. See label for details of rates and timing. Treatment active against strains of eyespot resistant to benzimidazole fungicides [1-5, 7, 9]
- Tank mixes with other fungicides recommended to improve control of rusts in wheat and barley. See label for details [1-5, 7, 9]
- A period of at least 3 h without rain should follow spraying [1-5, 7, 9]
- Can be used through most seed treatment machines if good even seed coverage is obtained. Check drill calibration before drilling treated seed. Use inert seed flow agent supplied by manufacturer [6]
- Use methylated spirits, rather than water, to clean residual material from seed treatment machinery, then carry out final rinse with water and detergent [6]
- Apply as drench against soil diseases, as a spray against aerial diseases, as a dip for cuttings, or as a drench at propagation. Spray applications may be repeated at 10-14 d intervals. Under mist propagation use 7 d intervals [8]
- Prochloraz is a DMI fungicide. Resistance to some DMI fungicides has been identified in Septoria leaf blotch which may seriously affect performance of some products. For further advice contact a specialist advisor and visit the Fungicide Resistance Action Group (FRAG)-UK website

Restrictions

- Maximum number of treatments 1 per batch of seed for flax, linseed [6]; 2 per crop for foliar spray uses
- Do not treat linseed varieties Linda, Bolas, Karen, Laura, Mikael, Norlin, Moonraker, Abbey, Agriace [6]
- Do not use treated seed as food or feed [6]

Crop-specific information

- Latest use: before drilling for seed treatments on flax or linseed [6]; watery ripe to early milk stage (GS 71-73) for cereals
- HI 6 wk for cereals
- Application with other fungicides may cause cereal crop scorch [1-5, 7, 9]

Environmental safety

- Dangerous for the environment
- Very toxic to aquatic organisms
- Treated seed harmful to game and wildlife. Product contains red dye for easy identification of treated seed [6]
- Flammable [2, 4, 9]

Hazard classification and safety precautions

Hazard H03 [5, 6, 9]; H04 [1-4, 7, 8]; H08 [2, 4, 9]; H11 [1-3, 5-9]

Risk phrases R22a [5, 6, 9]; R36 [1-8]; R38 [1-5, 7-9]; R41, R67 [9]; R43 [2, 4]; R50 [2, 5, 8, 9]; R51 [1, 3, 6, 7]; R53a [1-3, 5-9]

Operator protection A [1-9]; C [1-6, 8]; U05a, U20b [1-9]; U08 [1-7, 9]; U09a [8]; U11 [1, 3, 6, 7, 9]; U14 [1, 3, 5, 7]; U15 [1, 3, 7]; U19a [2, 8]

Environmental protection E13b [4]; E15a [1-3, 5-9]; E34 [5, 9]; E38 [5, 6, 9]

Storage and disposal D01, D09a [1-9]; D02, D10b [1-5, 7-9]; D05 [1-4, 6, 7, 9]; D10a [6]; D12a [5, 6, 8, 9]

Treated seed S02, S03, S04b, S05, S06a, S07 [6]

Medical advice M03, M05a [5]; M05b [1, 3, 7]

367 prochloraz + propiconazole

A broad spectrum fungicide mixture for wheat, barley and oilseed rape

FRAC mode of action code: 3 + 3

Products

1	Bumper Excell	Makhteshim	400:90 g/l	EC	11015
2	Bumper P	Makhteshim	400:90 g/l	EC	08548
3	Greencrop Twinstar	Greencrop	400:90 g/l	EC	09516

Uses

- Disease control in ***grass seed crops*** *(off-label)* [1]

- Eyespot in ***spring barley*, *spring wheat*** [1, 2]; ***winter barley*, *winter wheat*** [1-3]
- Light leaf spot in ***spring oilseed rape*, *winter oilseed rape*** [1, 2]
- Phoma in ***spring oilseed rape*, *winter oilseed rape*** [1, 2]
- Rhynchosporium in ***spring barley*** [1, 2]; ***winter barley*** [1-3]
- Septoria leaf blotch in ***spring wheat*** [1, 2]; ***winter wheat*** [1-3]

Specific Off-Label Approvals (SOLAs)
- ***grass seed crops*** *20052289* [1]

Approval information
- Propiconazole included in Annex I under EC Directive 91/414
- Accepted by BBPA for use on malting barley

Efficacy guidance
- Best results obtained from treatment when disease is active but not well established
- Treat wheat normally from flag leaf ligule just visible stage (GS 39) but earlier if there is a high risk of Septoria
- Treat barley from the first node detectable stage (GS 31)
- If disease pressure persists a second application may be necessary
- Prochloraz and propiconazole are DMI fungicides. Resistance to some DMI fungicides has been identified in Septoria leaf blotch which may seriously affect performance of some products. For further advice contact a specialist advisor and visit the Fungicide Resistance Action Group (FRAG)-UK website

Restrictions
- Maximum number of treatments 2 per crop

Crop-specific information
- Latest use: before grain watery ripe (GS 71) for cereals; before most seeds green for winter oilseed rape

Environmental safety
- Dangerous for the environment
- Very toxic to aquatic organisms

Hazard classification and safety precautions

Hazard H04, H11
Risk phrases R36, R53a [1-3]; R50 [3]; R51 [1, 2]
Operator protection A, C; U05a, U08, U20b [1-3]; U11, U15 [1, 2]
Environmental protection E15a
Storage and disposal D01, D02, D05, D09a, D10b
Medical advice M05a [1, 2]

368 prochloraz + tebuconazole

A broad spectrum systemic fungicide mixture for cereals, oilseed rape and amenity use
FRAC mode of action code: 3 + 3

Products

1	Agate	Makhteshim	267:133 g/l	EC	12016
2	Agate	Makhteshim	267:133 g/l	EC	13024
3	Grail	Nufarm UK	300:200 g/l	EC	13385
4	Lunar	Sherriff Amenity	267:132 g/l	EC	12366
5	Monkey	Makhteshim	267:133 g/l	EC	12906
6	Pan Proteb	Pan Agriculture	267:132 g/l	EC	13427

Uses
- Brown rust in ***spring barley*, *spring rye*, *spring wheat*, *winter barley*, *winter rye*, *winter wheat*** [1, 2, 5]
- Disease control in ***grass seed crops*** *(off-label)*, ***linseed*** *(off-label)* [1]
- Eyespot in ***spring barley*, *spring rye*, *spring wheat*, *winter barley*, *winter rye*, *winter wheat*** [1, 2, 5]
- Fusarium patch in ***managed amenity turf*** [4, 6]
- Glume blotch in ***spring wheat*, *winter wheat*** [1-3, 5]
- Late ear diseases in ***spring wheat*, *winter wheat*** [1, 2, 5]
- Net blotch in ***spring barley*, *winter barley*** [1, 2, 5]

SEE SECTION 3 FOR PRODUCTS ALSO REGISTERED

- Powdery mildew in ***spring barley***, ***spring rye***, ***spring wheat***, ***winter barley***, ***winter rye***, ***winter wheat*** [1, 2, 5]
- Rhynchosporium in ***spring barley***, ***spring rye***, ***winter barley***, ***winter rye*** [1, 2, 5]
- Sclerotinia stem rot in ***spring oilseed rape***, ***winter oilseed rape*** [1, 2, 5]
- Septoria leaf blotch in ***spring wheat***, ***winter wheat*** [1-3, 5]
- Yellow rust in ***spring barley***, ***spring rye***, ***spring wheat***, ***winter barley***, ***winter rye***, ***winter wheat*** [1, 2, 5]

Specific Off-Label Approvals (SOLAs)

- ***grass seed crops*** *20062917* [1]
- ***linseed*** *20062918* [1]

Approval information

- Accepted by BBPA for use on malting barley

Efficacy guidance

- Best results achieved from applications at an early stage of disease development
- Complete protection of winter cereals will usually require a programme of at least two treatments [1-3, 5]
- Protection of flag leaf and ear from Septoria diseases requires treatment at flag leaf emergence and repeated when ear fully emerged [1-3, 5]
- Two treatments separated by a 2 wk gap may be needed for difficult Fusarium patch attacks [4, 6]
- Optimum application timing is normally when disease first seen but varies with main target disease - see labels
- Prochloraz and tebuconazole are DMI fungicides. Resistance to some DMI fungicides has been identified in Septoria leaf blotch which may seriously affect performance of some products. For further advice contact a specialist advisor and visit the Fungicide Resistance Action Group (FRAG)-UK website

Restrictions

- Maximum total dose on cereals equivalent to two full dose treatments
- Maximum number of treatments on managed amenity turf 2 per yr
- Do not apply during drought or to frozen turf [4, 6]

Crop-specific information

- Latest use: before beginning of anthesis (GS 61) for cereals [3]; before grain milky ripe (GS 73) for cereals [1, 2, 5]; most seeds green for oilseed rape [1, 2, 5]
- HI 6 wk for cereals [1, 2, 5]
- Occasionally transient leaf speckling may occur after treating wheat. Yield responses should not be affected [1, 2]

Environmental safety

- Dangerous for the environment
- Very toxic to aquatic organisms

Hazard classification and safety precautions

Hazard H03, H11
Risk phrases R22a, R36, R50 [1, 2, 4-6]; R37, R51 [3]; R43, R53a [1-6]; R63 [3, 4, 6]
Operator protection A, H [1-6]; C [1, 2, 4-6]; M [4, 6]; U05a, U14 [1-6]; U09b [1, 2, 5]; U11 [3, 4, 6]; U19a [3]; U20a [1-3, 5]
Environmental protection E13b [1, 2, 5]; E15a [3, 4, 6]; E34, E38 [1-6]
Storage and disposal D01 [1, 2, 4-6]; D02, D05, D09a, D12a [1-6]; D10b [3, 4, 6]; D10c [1, 2, 5]
Medical advice M03 [1-6]; M05a [4, 6]

369 prochloraz + thiram

A seed treatment fungicide mixture for oilseed rape
FRAC mode of action code: 3 + M3

Products

Agrichem Hy-Pro Duet	Agrichem	150:333 g/l	FS	12469

Uses

- Phoma in ***winter oilseed rape*** *(seed treatment)*

Approval information

- Thiram included in Annex I under EC Directive 91/414

Efficacy guidance

- Can be applied through most types of seed treatment machinery
- Good coverage of the seed surface essential for best results
- Product is active against seed-borne disease. It will not control leaf spotting from leaf spores on trash on the seed bed

Restrictions

- Maximum number of treatments 1 per crop
- Seed must be of satisfactory quality and moisture content
- Sow seed as soon as possible after treatment
- Do not store treated seed from one season to the next
- Do not co-apply product with water or mix with any other seed treatment product

Crop-specific information

- Latest use: pre-sowing
- Product will also control soil-borne fungi that cause damping-off of rape seedlings

Following crops guidance

- Any crop may be grown following a treated winter oilseed rape crop

Hazard classification and safety precautions

Hazard H03, H11
Risk phrases R20, R22a, R36, R38, R43, R50, R53a
Operator protection A, C, H; U02a, U08, U14, U15, U20a
Environmental protection E15a, E36b
Storage and disposal D01, D02, D05, D08, D09a, D12b, D13, D15
Treated seed S01, S02, S03, S04d, S05, S08
Medical advice M04a

370 prochloraz + triticonazole

A broad spectrum fungicide seed treatment mixture for winter cereals
FRAC mode of action code: 3 + 3

Products

Kinto	BASF	60:20 g/l	FS	12038

Uses

- Bunt in ***durum wheat*** *(off-label - seed treatment)*, ***winter wheat*** *(seed treatment)*
- Covered smut in ***winter barley*** *(seed treatment)*, ***winter oats*** *(seed treatment)*
- Fusarium foot rot and seedling blight in ***durum wheat*** *(off-label - seed treatment)*, ***triticale*** *(seed treatment)*, ***winter barley*** *(seed treatment)*, ***winter oats*** *(seed treatment)*, ***winter rye*** *(seed treatment)*, ***winter wheat*** *(seed treatment)*
- Leaf stripe in ***winter barley*** *(seed treatment)*
- Loose smut in ***durum wheat*** *(off-label - seed treatment)*, ***winter barley*** *(seed treatment)*, ***winter oats*** *(seed treatment)*, ***winter wheat*** *(seed treatment)*
- Septoria seedling blight in ***durum wheat*** *(off-label - seed treatment)*, ***winter wheat*** *(seed treatment)*

Specific Off-Label Approvals (SOLAs)

- ***durum wheat*** *(seed treatment) 20061094*

Approval information

- Triticonazole included in Annex I under EC Directive 91/414
- Accepted by BBPA for use on malting barley

Efficacy guidance

- Apply undiluted through conventional seed treatment machine
- Calibrate seed drill with treated seed before sowing
- Bag treated seed immediately and keep in a dry, draught free store
- Ensure good even coverage of seed to achieve best results
- Treatment may cause some slight delay in crop emergence

Restrictions

- Maximum number of treatments 1 per batch of seed
- Do not treat seed with moisture content above 16% and do not allow moisture content of treated seed to exceed 16%
- Do not treat cracked, split or sprouted seed
- Do not handle treated seed unnecessarily
- Do not use treated seed as food or feed

Crop-specific information

- Latest use: before drilling wheat, barley, oats, rye or triticale

Environmental safety

- Dangerous for the environment
- Toxic to aquatic organisms

Hazard classification and safety precautions

Hazard H11
Risk phrases R51, R53a
Operator protection A, H; U05a, U07, U20b
Environmental protection E15a, E34, E38
Storage and disposal D01, D02, D08, D09a, D11a, D12a
Treated seed S01, S02, S03, S04a, S04b, S05, S07

371 prohexadione-calcium

A cyclohexanecarboxylate growth regulator for use in apples

See also mepiquat chloride + prohexadione-calcium

Products

Regalis	BASF	10% w/w	WG	12414

Uses

- Control of shoot growth in ***apples***

Approval information

- Prohexadione-calcium included in Annex I under EC Directive 91/414

Efficacy guidance

- For a standard orchard treat at the start of active growth and again after 3-5 wk depending on growth conditions
- Alternatively follow the first application with four reduced rate applications at two wk intervals

Restrictions

- Maximum total dose equivalent to two full dose treatments
- Do not apply in conjunction with calcium based foliar fertilisers

Crop-specific information

- HI 55 d for apples

Environmental safety

- Harmful to aquatic organisms
- Avoid spray drift onto neighbouring crops and other non-target plants
- Product does not harm natural insect predators when used as directed

Hazard classification and safety precautions

Risk phrases R52, R53a
Operator protection U05a
Environmental protection E38
Storage and disposal D01, D02, D09a, D10c, D12a

372 propachlor

A pre-emergence chloroacetanilide herbicide for various horticultural crops
HRAC mode of action code: K3

See also *chloridazon + propachlor*
chlorthal-dimethyl + propachlor

Products

1	Alpha Propachlor 50 SC	Makhteshim	500 g/l	SC	04873
2	Alpha Propachlor 50 SC	Makhteshim	500 g/l	SC	13585
3	Ramrod Flowable	Monsanto	480 g/l	SC	10314
4	Titan 480	AgriGuard	480 g/l	SC	12596

Uses

- Annual dicotyledons in ***blackcurrants*** *(off-label)*, ***blueberries*** *(off-label)*, ***celery (outdoor)*** *(off-label)*, ***chinese cabbage*** *(off-label)*, ***chives*** *(off-label)*, ***gooseberries*** *(off-label)*, ***herbs (see appendix 6)*** *(off-label)*, ***kohlrabi*** *(off-label)*, ***lettuce*** *(off-label)*, ***parsley*** *(off-label)*, ***radicchio*** *(off-label)*, ***redcurrants*** *(off-label)*, ***rhubarb*** *(off-label)*, ***salad brassicas*** *(off-label - for baby leaf production)*, ***strawberries*** *(off-label)*, ***whitecurrants*** *(off-label)* [1, 3]; ***broccoli***, ***brussels sprouts***, ***bulb onions***, ***cabbages***, ***calabrese***, ***cauliflowers***, ***fodder rape***, ***kale***, ***leeks***, ***ornamental plant production***, ***spring oilseed rape***, ***swedes***, ***turnips*** [1-4]; ***brown mustard***, ***sage***, ***white mustard*** [3, 4]; ***cress*** *(off-label)*, ***frise*** *(off-label - under crop covers)*, ***lamb's lettuce*** *(off-label - under crop covers)*, ***lettuce*** *(off-label - under crop covers)*, ***radicchio*** *(off-label - under crop covers)*, ***ribes hybrids*** *(off-label)*, ***scarole*** *(off-label - under crop covers)* [3]; ***mustard***, ***salad onions***, ***strawberries***, ***winter oilseed rape*** [1, 2]; ***rubus hybrids*** *(off-label)* [1]
- Annual grasses in ***blackcurrants*** *(off-label)*, ***blueberries*** *(off-label)*, ***chinese cabbage*** *(off-label)*, ***chives*** *(off-label)*, ***gooseberries*** *(off-label)*, ***herbs (see appendix 6)*** *(off-label)*, ***kohlrabi*** *(off-label)*, ***lettuce*** *(off-label)*, ***parsley*** *(off-label)*, ***radicchio*** *(off-label)*, ***redcurrants*** *(off-label)*, ***strawberries*** *(off-label)*, ***whitecurrants*** *(off-label)* [1, 3]; ***broccoli***, ***brussels sprouts***, ***bulb onions***, ***cabbages***, ***calabrese***, ***cauliflowers***, ***fodder rape***, ***kale***, ***leeks***, ***ornamental plant production***, ***spring oilseed rape***, ***swedes***, ***turnips*** [1-4]; ***brown mustard***, ***sage***, ***white mustard*** [3, 4]; ***celery (outdoor)*** *(off-label)*, ***rhubarb*** *(off-label)*, ***rubus hybrids*** *(off-label)*, ***salad brassicas*** *(off-label - for baby leaf production)* [1]; ***cress*** *(off-label)*, ***frise*** *(off-label - under crop covers)*, ***lamb's lettuce*** *(off-label - under crop covers)*, ***lettuce*** *(off-label - under crop covers)*, ***radicchio*** *(off-label - under crop covers)*, ***ribes hybrids*** *(off-label)*, ***scarole*** *(off-label - under crop covers)* [3]; ***mustard***, ***salad onions***, ***strawberries***, ***winter oilseed rape*** [1, 2]

Specific Off-Label Approvals (SOLAs)

- ***blackcurrants*** *20002424* [1], *20021159* [3]
- ***blueberries*** *20002424* [1], *20021159* [3]
- ***celery (outdoor)*** *20051699* [1], *20051070* [3]
- ***chinese cabbage*** *20002424* [1], *20021159* [3]
- ***chives*** *20002424* [1], *20021159* [3]
- ***cress*** *20021159* [3]
- ***frise*** *(under crop covers) 20021159* [3]
- ***gooseberries*** *20002424* [1], *20021159* [3]
- ***herbs (see appendix 6)*** *20002424* [1], *20021159* [3]
- ***kohlrabi*** *20002424* [1], *20021159* [3]
- ***lamb's lettuce*** *(under crop covers) 20021159* [3]
- ***lettuce*** *20002424* [1], *(under crop covers) 20021159* [3], *20021159* [3]
- ***parsley*** *20002424* [1], *20021159* [3]
- ***radicchio*** *20002424* [1], *(under crop covers) 20021159* [3], *20021159* [3]
- ***redcurrants*** *20002424* [1], *20021159* [3]
- ***rhubarb*** *20051699* [1], *20051070* [3]
- ***ribes hybrids*** *20021159* [3]
- ***rubus hybrids*** *20002424* [1]
- ***salad brassicas*** *(for baby leaf production) 20051700* [1], *(for baby leaf production) 20030436* [3]
- ***scarole*** *(under crop covers) 20021159* [3]
- ***strawberries*** *20002424* [1], *20021159* [3]
- ***whitecurrants*** *20002424* [1], *20021159* [3]

SEE SECTION 3 FOR PRODUCTS ALSO REGISTERED

Efficacy guidance

- Controls germinating (not emerged) weeds for 6-8 wk
- Best results achieved by application to fine, firm, moist seedbed free of established weeds in spring, summer and early autumn
- Use higher rate on soils with more than 10% organic matter
- Recommended as tank-mix with glyphosate or chlorthal-dimethyl on onions, leeks, swedes, turnips. See label for details

Restrictions

- Maximum number of treatments 1 or 2 per crop, varies with product and crop - see label for details
- Maximum total dose equivalent to two full dose treatments on brassicas and leeks; one full dose treatment on most other crops [2]
- Do not use after products that de-wax the leaves of plants, especially brassica crops
- Do not use on crops under glass or polythene
- Do not use under extremely wet, dry or other adverse growth conditions

Crop-specific information

- Latest use normally pre-emergence; young plant (post crook stage) for onions, leeks [3]
- Apply in brassicas from drilling to time seed chits or after 3-4 true leaf stage but before weed emergence, in swedes and turnips pre-emergence only
- Apply in transplanted brassicas within 48 h of planting in warm weather. Plants must be hardened off and special care needed with block sown or modular propagated plants
- Apply in onions and leeks pre-emergence or from post-crook to young plant stage
- Apply to newly planted strawberries soon after transplanting, to weed-free soil in established crops in early spring before new weeds emerge
- Granules may be applied to onion, leek and brassica nurseries and to most flower crops after bedding out and hardening off
- Do not use pre-emergence of drilled wallflower seed

Following crops guidance

- In the event of crop failure only replant recommended crops in treated soil

Environmental safety

- Dangerous for the environment
- Very toxic to aquatic organisms

Hazard classification and safety precautions

Hazard H03 [1-4]; H11 [1-3]

Risk phrases R21, R36 [1, 2]; R22a, R38, R43 [1-4]; R50, R53a [1-3]

Operator protection A, C; U02a, U04a, U05a, U08, U13, U19a, U20b [1-4]; U10, U11 [1, 2]; U14 [3]

Environmental protection E15a, E34

Storage and disposal D01, D02, D05, D09a [1-4]; D10b [1, 2, 4]; D10c [3]; D12a [3, 4]; D12b [1, 2]

Medical advice M03 [1-4]; M04a [1, 2]

373 propamocarb hydrochloride

A translocated protectant carbamate fungicide
FRAC mode of action code: 28

See also *chlorothalonil + propamocarb hydrochloride*
fenamidone + propamocarb hydrochloride
fluopicolide + propamocarb hydrochloride
mancozeb + propamocarb hydrochloride

Products

1	Filex	Scotts	722 g/l	SL	07631
2	Flash	AgriGuard	722 g/l	SL	11989
3	Pan PCH	Pan Agriculture	722 g/l	SL	12842
4	Proplant	Fargro	722 g/l	SL	08572
5	Proplant	Fargro	722 g/l	SL	13359

Uses

- Blight in ***protected tomatoes, tomatoes (outdoor)*** [2]
- Botrytis in ***chives*** *(off-label)*, ***herbs (see appendix 6)*** *(off-label)*, ***parsley*** *(off-label)*, ***protected herbs (see appendix 6)*** *(off-label)* [1, 4]; ***cress*** *(off-label)* [1]; ***lettuce, protected lettuce*** [3-5]; ***protected chives*** *(off-label)*, ***protected parsley*** *(off-label)*, ***protected radishes*** *(off-label)* [4]
- Damping off in ***inert substrate cucumbers*** *(off-label)*, ***inert substrate peppers*** *(off-label)*, ***inert substrate tomatoes*** *(off-label)*, ***nft peppers*** *(off-label)*, ***nft tomatoes*** *(off-label)* [1, 4]; ***protected peppers*** *(off-label)*, ***protected tomatoes*** *(off-label)* [1]
- Downy mildew in ***broccoli, brussels sprouts, cabbages, calabrese, cauliflowers, chinese cabbage*** [1-5]; ***chives*** *(off-label)*, ***herbs (see appendix 6)*** *(off-label)*, ***parsley*** *(off-label)*, ***protected herbs (see appendix 6)*** *(off-label)*, ***protected radishes*** *(off-label)*, ***radishes*** *(off-label)* [1, 4]; ***cress*** *(off-label)*, ***watercress*** *(off-label - under protection)* [1]; ***lettuce, protected lettuce*** [3-5]; ***protected chives*** *(off-label)*, ***protected parsley*** *(off-label)*, ***watercress*** *(off-label - during propagation)* [4]
- Phytophthora in ***aubergines, bedding plants, broccoli, brussels sprouts, bulb onions, cabbages, calabrese, cauliflowers, chinese cabbage, cucumbers, leeks, nursery stock, peppers, pot plants*** [1-5]; ***flower bulbs, tomatoes (outdoor)*** [2-5]; ***ornamental plant production*** [1, 3-5]; ***protected tomatoes, rockwool aubergines, rockwool cucumbers, rockwool peppers, rockwool tomatoes*** [2]; ***tulips*** [1, 2]; ***watercress*** *(off-label - during propagation)* [4]; ***watercress*** *(off-label - under protection)* [1]
- Pythium in ***aubergines, bedding plants, broccoli, brussels sprouts, bulb onions, cabbages, calabrese, cauliflowers, chinese cabbage, cucumbers, flower bulbs, leeks, nursery stock, peppers, pot plants, tomatoes (outdoor)*** [1-5]; ***ornamental plant production*** [1, 3-5]; ***protected tomatoes, rockwool aubergines, rockwool cucumbers, rockwool peppers, rockwool tomatoes, tulips*** [1, 2]; ***watercress*** *(off-label - during propagation)* [4]; ***watercress*** *(off-label - under protection)* [1]
- Root malformation disorder in ***red beet*** *(off-label)* [1, 4]
- Root rot in ***inert substrate cucumbers*** *(off-label)*, ***inert substrate peppers*** *(off-label)*, ***inert substrate tomatoes*** *(off-label)*, ***nft peppers*** *(off-label)*, ***nft tomatoes*** *(off-label)* [4]
- White blister in ***horseradish*** *(off-label)*, ***protected horseradish*** *(off-label)* [1, 4]

Specific Off-Label Approvals (SOLAs)

- ***chives*** *20040626* [1], *20040625* [4]
- ***cress*** *20040626* [1]
- ***herbs (see appendix 6)*** *20040626* [1], *20040625* [4]
- ***horseradish*** *992036* [1], *20040625* [4]
- ***inert substrate cucumbers*** *992032* [1], *20002527* [4]
- ***inert substrate peppers*** *992032* [1], *20002527* [4]
- ***inert substrate tomatoes*** *992032* [1], *20002527* [4]
- ***nft peppers*** *992032* [1], *20002527* [4]
- ***nft tomatoes*** *992032* [1], *20002527* [4]
- ***parsley*** *20040626* [1], *20040625* [4]
- ***protected chives*** *20040625* [4]
- ***protected herbs (see appendix 6)*** *20040626* [1], *20040625* [4]
- ***protected horseradish*** *992036* [1], *20040625* [4]
- ***protected parsley*** *20040625* [4]
- ***protected peppers*** *992032* [1]
- ***protected radishes*** *992030* [1], *20040625* [4]
- ***protected tomatoes*** *992032* [1]
- ***radishes*** *992030* [1], *20040625* [4]
- ***red beet*** *031247 expires Jun 2008* [1], *20031453 expires Jul 2008* [4]
- ***watercress*** *(under protection) 20010439* [1], *(during propagation) 20040625* [4]

Approval information

- Propamocarb hydrochloride included in Annex I under EC Directive 91/414

Efficacy guidance

- Chemical is absorbed through roots and translocated throughout plant
- Incorporate in compost before use or drench moist compost or soil before sowing, pricking out, striking cuttings or potting up
- Drench treatment can be repeated at 3-6 wk intervals
- Concentrated solution is corrosive to all metals other than stainless steel

SEE SECTION 3 FOR PRODUCTS ALSO REGISTERED

- May also be applied in trickle irrigation systems
- To prevent root rot in tulip bulbs apply as dip for 20 min but full protection only achieved by soil drench treatment as well

Restrictions

- Maximum number of treatments 4 per crop for cucumbers, tomatoes, peppers, aubergines; 3 per crop for outdoor and protected lettuce; 1 per crop for listed brassicas; 1 compost incorporation and/or 1 drench treatment for leeks, onions; 1 dip and 1 soil drench for flower bulbs
- Do not apply to lettuce crops grown in non-soil systems
- When applied over established seedlings rinse off foliage with water (except lettuce) and do not apply under hot, dry conditions
- On plants of unknown tolerance test first on a small scale
- Do not apply in a recirculating irrigation/drip system
- Store away from seeds and fertilizers

Crop-specific information

- Latest use: before transplanting for brassicas
- HI 14 d for cucumbers, tomatoes, peppers, aubergines; 4 wk for brassicas; 19 wk for leeks, onions

Hazard classification and safety precautions

Operator protection A, C, H, K, M [1-5]; B [1, 2]; U02a, U19a [3-5]; U05a [2-5]; U08, U20b [1-5]
Environmental protection E15a [1-5]; E34 [3-5]
Storage and disposal D01, D02, D05 [2-5]; D07 [2]; D09a [1-5]; D10b [3-5]; D11a [1, 2]

374 propaquizafop

A phenoxy alkanoic acid foliar acting herbicide for grass weeds in a range of crops
HRAC mode of action code: A

Products

1	Bulldog	Makhteshim	100 g/l	EC	11723
2	Cleancrop GYR	Makhteshim	100 g/l	EC	10646
3	Falcon	Makhteshim	100 g/l	EC	10585
4	Greencrop Satchmo	Greencrop	100 g/l	EC	10748
5	Landgold PQF 100	Landgold	100 g/l	EC	10763
6	Landgold PQF 100	Teliton	100 g/l	EC	12127
7	Raptor	Makhteshim	100 g/l	EC	11092
8	Shogun	Makhteshim	100 g/l	EC	10584
9	Standon Propaquizafop	Standon	100 g/l	EC	11536
10	Zealot	AgriGuard	100 g/l	EC	12548

Uses

- Annual grasses in ***borage*** *(off-label)*, ***mallow (althaea spp.)*** *(off-label)* [3]; ***borage for oilseed production*** *(off-label)* [7, 8]; ***bulb onions***, ***carrots***, ***combining peas***, ***early potatoes***, ***farm forestry***, ***fodder beet***, ***linseed***, ***maincrop potatoes***, ***parsnips***, ***spring field beans***, ***spring oilseed rape***, ***sugar beet***, ***swedes***, ***turnips***, ***winter field beans***, ***winter oilseed rape*** [1-10]; ***canary flower (echium spp.)*** *(off-label)*, ***honesty*** *(off-label)* [3, 7, 8]; ***cut logs***, ***forest***, ***forest nurseries*** [1-3, 7, 8]; ***evening primrose*** *(off-label)*, ***garlic*** *(off-label)*, ***lupins*** *(off-label)*, ***mustard*** *(off-label)*, ***shallots*** *(off-label)* [1, 3, 7, 8]; ***horseradish*** *(off-label)* [8]; ***poppies for morphine production*** *(off-label)* [7]; ***spring linseed*** *(off-label)* [3, 8]
- Annual meadow grass in ***chives*** *(off-label)*, ***parsley*** *(off-label)* [2, 3, 7, 8]; ***herbs (see appendix 6)*** *(off-label)*, ***leaf spinach*** *(off-label)*, ***leeks*** *(off-label)*, ***red beet*** *(off-label)* [1-3, 7, 8]
- Couch in ***chives*** *(off-label)*, ***parsley*** *(off-label)* [2, 3, 7, 8]; ***herbs (see appendix 6)*** *(off-label)*, ***leaf spinach*** *(off-label)*, ***leeks*** *(off-label)*, ***red beet*** *(off-label)* [1-3, 7, 8]
- Perennial grasses in ***borage*** *(off-label)*, ***mallow (althaea spp.)*** *(off-label)* [3]; ***borage for oilseed production*** *(off-label)* [7, 8]; ***bulb onions***, ***carrots***, ***combining peas***, ***early potatoes***, ***farm forestry***, ***fodder beet***, ***linseed***, ***maincrop potatoes***, ***parsnips***, ***spring field beans***, ***spring oilseed rape***, ***sugar beet***, ***swedes***, ***turnips***, ***winter field beans***, ***winter oilseed rape*** [1-10]; ***canary flower (echium spp.)*** *(off-label)*, ***honesty*** *(off-label)* [3, 7, 8]; ***celery (outdoor)*** *(off-label)* [2, 3, 7, 8]; ***cut logs***, ***forest***, ***forest nurseries*** [1-3, 7, 8]; ***evening primrose*** *(off-label)*, ***garlic*** *(off-label)*, ***lupins*** *(off-label)*, ***mustard*** *(off-label)*, ***shallots*** *(off-label)* [1, 3, 7, 8]; ***horseradish*** *(off-label)* [8]; ***spring linseed*** *(off-label)* [3, 8]

- Volunteer cereals in ***celery (outdoor)*** *(off-label)*, ***chives*** *(off-label)*, ***parsley*** *(off-label)* [2, 3, 7, 8]; ***herbs (see appendix 6)*** *(off-label)*, ***leaf spinach*** *(off-label)*, ***leeks*** *(off-label)*, ***red beet*** *(off-label)* [1-3, 7, 8]; ***poppies for morphine production*** *(off-label)* [7]
- Wild oats in ***celery (outdoor)*** *(off-label)* [2, 3, 7, 8]

Specific Off-Label Approvals (SOLAs)

- ***borage*** *20061711* [3]
- ***borage for oilseed production*** *20061688* [7], *20061726* [8]
- ***canary flower (echium spp.)*** *20061711* [3], *20061688* [7], *20061726* [8]
- ***celery (outdoor)*** *20021634* [2], *20021502* [3], *20021632* [7], *20021629* [8]
- ***chives*** *20021634* [2], *20020148* [3], *20021632* [7], *20021629* [8]
- ***evening primrose*** *20061569* [1], *20061711* [3], *20061688* [7], *20061726* [8]
- ***garlic*** *20061568* [1], *20061713* [3], *20061689* [7], *20061729* [8]
- ***herbs (see appendix 6)*** *20032022* [1], *20021634* [2], *20020148* [3], *20021632* [7], *20021629* [8]
- ***honesty*** *20061711* [3], *20061688* [7], *20061726* [8]
- ***horseradish*** *20061728* [8]
- ***leaf spinach*** *20032022* [1], *20021634* [2], *20020148* [3], *20021632* [7], *20021629* [8]
- ***leeks*** *20032022* [1], *20021634* [2], *20020148* [3], *20021632* [7], *20021629* [8]
- ***lupins*** *20061570* [1], *20061714* [3], *20061687* [7], *20061725* [8]
- ***mallow (althaea spp.)*** *20061710* [3]
- ***mustard*** *20061569* [1], *20061711* [3], *20061688* [7], *20061726* [8]
- ***parsley*** *20021634* [2], *20020148* [3], *20021632* [7], *20021629* [8]
- ***poppies for morphine production*** *20030770* [7]
- ***red beet*** *20032022* [1], *20021634* [2], *20020148* [3], *20021632* [7], *20021629* [8]
- ***shallots*** *20061568* [1], *20061713* [3], *20061689* [7], *20061729* [8]
- ***spring linseed*** *20061711* [3], *20061726* [8]

Efficacy guidance

- Apply to emerged weeds when they are growing actively with adequate soil moisture
- Activity is slower under cool conditions
- Broad-leaved weeds and any weeds germinating after treatment are not controlled
- Annual meadow grass up to 3 leaves checked at low doses and severely checked at highest dose
- Spray barley cover crops when risk of wind blow has passed and before there is serious competition with the crop
- Various tank mixtures and sequences recommended for broader spectrum weed control in oilseed rape, peas and sugar beet. See label for details
- Severe couch infestations may require a second application at reduced dose when regrowth has 3-4 leaves unfolded
- Products contain surfactants. Tank mixing with adjuvants not required or recommended
- Propaquizafop is an ACCase inhibitor herbicide. To avoid the build up of resistance do not apply products containing an ACCase inhibitor herbicide more than twice to any crop. In addition do not use any product containing propaquizafop in mixture or sequence with any other product containing the same ingredient
- Use these products as part of a resistance management strategy that includes cultural methods of control and does not use ACCase inhibitors as the sole chemical method of grass weed control
- Applying a second product containing an ACCase inhibitor to a crop will increase the risk of resistance development; only use a second ACCase inhibitor to control different weeds at a different timing
- Always follow WRAG guidelines for preventing and managing herbicide resistant weeds. See Section 5

Restrictions

- Maximum number of treatments 1 or 2 per crop or yr. See label for details
- See label for list of tolerant tree species
- Do not treat seed potatoes

Crop-specific information

- Latest use: before crop flower buds visible for winter oilseed rape, linseed, field beans; before 8 fully expanded leaf stage for spring oilseed rape; before weeds are covered by the crop for potatoes, sugar beet, fodder beet, swedes, turnips; when flower buds visible for peas
- HI onions, carrots, early potatoes, parsnips 4 wk; combining peas, early potatoes 7 wk; sugar beet, fodder beet, maincrop potatoes, swedes, turnips 8 wk; field beans 14 wk

- Application in high temperatures and/or low soil moisture content may cause chlorotic spotting especially on combining peas and field beans
- Overlaps at the highest dose can cause damage from early applications to carrots and parsnips

Following crops guidance

- In the event of a failed treated crop an interval of 2 wk must elapse between the last application and redrilling with winter wheat or winter barley. 4 wk must elapse before sowing oilseed rape, peas or field beans, and 16 wk before sowing ryegrass or oats

Environmental safety

- Dangerous for the environment
- Toxic to aquatic organisms
- Risk to certain non-target insects or other arthropods. See directions for use

Hazard classification and safety precautions

Hazard H04, H11
Risk phrases R36, R38, R51, R53a
Operator protection A, C [3-6, 9, 10]; U02a, U05a, U08, U19a, U20b [1-10]; U11 [3, 5, 6, 9]; U13 [1-3, 7, 8, 10]; U14, U15 [1-3, 5-10]; U23a [4]
Environmental protection E15a, E22b [1-10]; E34 [1-3, 7, 8, 10]
Storage and disposal D01, D02, D05, D09a, D10b [1-10]; D12b [1-3, 7, 8]; D14 [1, 3, 7, 8]
Medical advice M05b

375 propiconazole

A systemic, curative and protectant triazole fungicide
FRAC mode of action code: 3

See also *chlorothalonil + cyproconazole + propiconazole*
cyproconazole + propiconazole
prochloraz + propiconazole

Products

1	Banner Maxx	Syngenta	156 g/l	EC	13167
2	Bumper 250 EC	Makhteshim	250 g/l	EC	09039
3	Tonik	AgriGuard	250 g/l	EC	12247

Uses

- Anthracnose in ***amenity grassland*** *(qualified minor use)*, ***managed amenity turf*** *(qualified minor use)* [1]
- Brown patch in ***amenity grassland*** *(qualified minor use)*, ***managed amenity turf*** *(qualified minor use)* [1]
- Brown rust in ***spring barley***, ***spring wheat***, ***winter barley***, ***winter rye***, ***winter wheat*** [2, 3]
- Crown rust in ***grass seed crops***, ***spring oats***, ***winter oats*** [2, 3]; ***permanent grassland*** [2]; ***permanent grassland*** *(for animal fodder)*, ***rotational grass*** *(for animal fodder)* [3]
- Dollar spot in ***amenity grassland***, ***managed amenity turf*** [1]
- Drechslera leaf spot in ***grass seed crops*** [2, 3]; ***permanent grassland*** [2]; ***permanent grassland*** *(for animal fodder)*, ***rotational grass*** *(for animal fodder)* [3]
- Fusarium patch in ***amenity grassland***, ***managed amenity turf*** [1]
- Glume blotch in ***spring wheat***, ***winter rye***, ***winter wheat*** [2, 3]
- Leaf blotch in ***bulb onions*** *(off-label)*, ***garlic*** *(off-label)*, ***shallots*** *(off-label)* [2]
- Light leaf spot in ***spring oilseed rape*** *(reduction)*, ***winter oilseed rape*** *(reduction)* [2, 3]
- Mildew in ***grass seed crops*** [2, 3]; ***permanent grassland*** [2]; ***permanent grassland*** *(for animal fodder)*, ***rotational grass*** *(for animal fodder)* [3]
- Powdery mildew in ***spring barley***, ***spring oats***, ***spring wheat***, ***winter barley***, ***winter oats***, ***winter rye***, ***winter wheat*** [2, 3]
- Ramularia leaf spots in ***sugar beet*** *(reduction)* [2, 3]
- Rhynchosporium in ***grass seed crops***, ***spring barley***, ***winter barley***, ***winter rye*** [2, 3]; ***permanent grassland*** [2]; ***permanent grassland*** *(for animal fodder)*, ***rotational grass*** *(for animal fodder)* [3]
- Rust in ***garlic*** *(off-label)* [2]; ***sugar beet*** [2, 3]
- Septoria leaf blotch in ***spring wheat***, ***winter rye***, ***winter wheat*** [2, 3]
- Sooty moulds in ***spring wheat***, ***winter wheat*** [2, 3]

- White blister in ***honesty*** *(off-label)* [2]
- White rust in ***chrysanthemums*** *(off-label)*, ***protected chrysanthemums*** *(off-label)* [2]
- Yellow rust in ***spring barley***, ***spring wheat***, ***winter barley***, ***winter wheat*** [2, 3]

Specific Off-Label Approvals (SOLAs)

- ***bulb onions*** *20012142* [2]
- ***chrysanthemums*** *20012142* [2]
- ***garlic*** *20012142* [2]
- ***honesty*** *20012142* [2]
- ***protected chrysanthemums*** *20012142* [2]
- ***shallots*** *20012142* [2]

Approval information

- Propiconazole included in Annex I under EC Directive 91/414
- Accepted by BBPA for use on malting barley

Efficacy guidance

- Best results achieved by applying at early stage of disease development. Recommended spray programmes vary with crop, disease, season, soil type and product. See label for details
- For optimum turf quality and control use in conjunction with turf management practices that promote good plant health [1]
- Propiconazole is a DMI fungicide. Resistance to some DMI fungicides has been identified in Septoria leaf blotch which may seriously affect performance of some products. For further advice contact a specialist advisor and visit the Fungicide Resistance Action Group (FRAG)-UK website

Restrictions

- Maximum number of treatments 4 per crop for wheat (up to 3 in yr of harvest), amenity grassland, managed amenity turf; 3 per crop for barley, oats, rye, triticale (up to 2 in yr of harvest); 2 per crop or yr for oilseed rape, sugar beet, grass seed crops, honesty; 1 per yr on grass for ensiling
- On oilseed rape do not apply during flowering. Apply first treatment before flowering, the second afterwards
- Grass seed crops must be treated in yr of harvest
- A minimum interval of 14 d must elapse between treatments on leeks and 21 d on sugar beet
- Avoid spraying crops or turf under stress, eg during cold weather or periods of frost

Crop-specific information

- Latest use: before 4 pairs of true leaves for honesty
- HI cereals, grass seed crops 35 d; oilseed rape, sugar beet, garlic, onions, shallots, grass for ensiling 28 d
- May be used on all common turf grass species [1]

Environmental safety

- Dangerous for the environment
- Toxic to aquatic organisms
- Risk to non-target insects or other arthropods [1]
- When treating turf with vehicle mounted or drawn hydraulic boom sprayers avoid spraying within 5 m of unmanaged land to reduce effects on non-target arthropods [1]

Hazard classification and safety precautions

Hazard H03 [1]; H11 [1-3]
Risk phrases R20, R43, R52 [1]; R51 [2, 3]; R53a [1-3]
Operator protection A [1-3]; C [2, 3]; H [1]; U02a, U05a, U19a, U20b [1-3]; U08 [2, 3]; U09a, U14 [1]
Environmental protection E15a, E34 [2, 3]; E15b, E22c, E38 [1]
Storage and disposal D01, D02, D05, D09a [1-3]; D10b [2, 3]; D10c, D12a [1]
Treated seed S06a [2, 3]
Medical advice M05a [2, 3]

376 propoxycarbazone-sodium

A sulfonylaminocarbonyltriazolinone residual grass weed herbicide for winter wheat
HRAC mode of action code: B

Products

Attribut	Bayer CropScience	70% w/w	SG	12730

SEE SECTION 3 FOR PRODUCTS ALSO REGISTERED

Uses

- Blackgrass in ***winter wheat***
- Couch in ***winter wheat***

Approval information

- Propoxycarbazone-sodium included in Annex I under EC Directive 91/414

Efficacy guidance

- Activity by root and foliar absorption but depends on presence of sufficient soil moisture to ensure root uptake by weeds. Control is enhanced by use of an adjuvant to encourage foliar uptake
- Best results obtained from treatments applied when weed grasses are growing actively. Symptoms may not become apparent for 3-4 wk after application
- Ensure good even spray coverage
- To achieve best control of couch and reduction of infestation in subsequent crop treat between 2 true leaves and first node stage of the weed
- Blackgrass should be treated after using specialist blackgrass herbicides with a different mode of action
- Propoxycarbazone-sodium is a member of the ALS-inhibitor group of herbicides. To avoid the build up of resistance do not use any product containing an ALS-inhibitor herbicide with claims for control of grass weeds more than once on any crop
- Use these products as part of a resistance management strategy that includes cultural methods of control and does not use ALS inhibitors as the sole chemical method of grass weed control

Restrictions

- Maximum total dose equivalent to one full dose treatment
- Do not use in a programme with other aceto-lactase synthesis (ALS) inhibitors
- Do not apply when temperature near or below freezing
- Avoid treatment under dry soil conditions

Crop-specific information

- Latest use: third node detectable in wheat (GS 33)

Following crops guidance

- Only winter wheat, field beans or winter barley may be sown in the autumn following spring treatment
- Any crop may be grown in the spring on land treated during the previous calendar yr

Environmental safety

- Dangerous for the environment
- Very toxic to aquatic organisms
- LERAP Category B
- Take extreme care to avoid drift onto adjacent plants or land as this could result in severe damage

Hazard classification and safety precautions

Hazard H11
Risk phrases R50, R53a
Operator protection A; U05a, U20b
Environmental protection E15b, E16a, E16b, E38
Storage and disposal D01, D02, D09a, D11a, D12a

377 propyzamide

A residual benzamide herbicide for use in a wide range of crops
HRAC mode of action code: K1

Products

1	Agriguard Propyzamide 50 WP	AgriGuard	50% w/w	WP	10187
2	Bulwark Flo	Interfarm	400 g/l	SC	10161
3	Clayton Propel 80 WG	Clayton	80% w/w	WG	11216
4	Engage	Interfarm	50% w/w	WP	13690
5	Flomide	Interfarm	400 g/l	SC	11171
6	Greencrop Saffron FL	Greencrop	400 g/l	SC	10244
7	Kerb 50 W	Dow	50% w/w	WP	12296
8	Kerb Flo	Dow	400 g/l	SC	10768
9	Kerb Pro Flo	Dow	400 g/l	SC	10760

Products – continued

10	Landgold Propyzamide 400 SC	Teliton	400 g/l	SC	13614
11	Menace 80 EDF	Dow	80% w/w	WG	10766
12	Precis	Dow	400 g/l	SC	10762
13	Quaver Flo	Dow	400 g/l	SC	10769
14	Setanta 50 WP	AgriGuard	50% w/w	WP	12475
15	Setanta Flo	AgriGuard	400 g/l	SC	12450
16	Standon Propyzamide 400 SC	Standon	400 g/l	SC	10255
17	Standon Propyzamide 50	Standon	50% w/w	WP	10814

Uses

- Annual dicotyledons in ***almonds*** *(off-label)*, ***chestnuts*** *(off-label)*, ***hazel nuts*** *(off-label)*, ***lupins*** *(off-label)*, ***walnuts*** *(off-label)* [3, 7, 8, 11-13]; ***apple orchards***, ***blackberries***, ***blackcurrants***, ***clover seed crops***, ***gooseberries***, ***loganberries***, ***lucerne***, ***pear orchards***, ***plums***, ***redcurrants***, ***strawberries***, ***sugar beet seed crops***, ***winter field beans***, ***winter oilseed rape*** [1, 3-8, 10-17]; ***bilberries*** *(off-label)*, ***blueberries*** *(off-label)*, ***cranberries*** *(off-label)* [7, 8]; ***brassica seed crops***, ***lettuce*** *(outdoor crops)*, ***rhubarb*** *(outdoor)* [3, 11]; ***cherries*** *(off-label)*, ***chicory*** *(off-label)*, ***chives*** *(off-label - outdoor or under crop covers)*, ***courgettes*** *(off-label - outdoor or under crop covers)*, ***evening primrose*** *(off-label)*, ***frise*** *(off-label - outdoor or under crop covers)*, ***herbs (see appendix 6)*** *(off-label - outdoor or under crop covers)*, ***honesty*** *(off-label)*, ***hops*** *(off-label)*, ***lamb's lettuce*** *(off-label - outdoor or under crop covers)*, ***marrows*** *(off-label - outdoor or under crop covers)*, ***mirabelles*** *(off-label)*, ***parsley*** *(off-label - outdoor or under crop covers)*, ***protected chives*** *(off-label)*, ***protected frise*** *(off-label)*, ***protected herbs (see appendix 6)*** *(off-label)*, ***protected lettuce*** *(off-label)*, ***protected parsley*** *(off-label)*, ***protected radicchio*** *(off-label)*, ***protected salad brassicas*** *(off-label - for baby leaf production)*, ***protected scarole*** *(off-label)*, ***pumpkins*** *(off-label - outdoor or under crop covers)*, ***quinces*** *(off-label)*, ***salad brassicas*** *(off-label - for baby leaf production outdoors or under crop covers)*, ***scarole*** *(off-label - outdoor or under crop covers)*, ***squashes*** *(off-label - outdoor or under crop covers)*, ***wine grapes*** *(off-label)* [7]; ***chives*** *(off-label)*, ***herbs (see appendix 6)*** *(off-label)*, ***parsley*** *(off-label)* [8]; ***fodder rape seed crops***, ***kale seed crops***, ***lettuce***, ***rhubarb***, ***turnip seed crops*** [1, 4-8, 10, 12-17]; ***forest*** [1, 2, 4, 7, 9, 14]; ***forest nurseries*** [14]; ***frise*** *(off-label)*, ***scarole*** *(off-label)* [8, 13]; ***lamb's lettuce*** *(off-label)* [8, 12, 13]; ***ornamental plant production*** [7, 14, 17]; ***radicchio*** *(off-label)* [7, 8, 13]; ***raspberries*** [17]; ***raspberries*** *(England only)* [1, 3-8, 10-16]; ***roses*** [1, 4, 7, 14, 17]; ***trees and shrubs*** [1, 2, 4, 9]; ***woody ornamentals*** [1-4, 7, 9, 11]
- Annual grasses in ***almonds*** *(off-label)*, ***chestnuts*** *(off-label)*, ***hazel nuts*** *(off-label)*, ***lupins*** *(off-label)*, ***walnuts*** *(off-label)* [3, 7, 8, 11-13]; ***apple orchards***, ***blackberries***, ***blackcurrants***, ***clover seed crops***, ***gooseberries***, ***loganberries***, ***lucerne***, ***pear orchards***, ***plums***, ***redcurrants***, ***strawberries***, ***sugar beet seed crops***, ***winter field beans***, ***winter oilseed rape*** [1, 3-8, 10-17]; ***bilberries*** *(off-label)*, ***blueberries*** *(off-label)*, ***cranberries*** *(off-label)* [7, 8]; ***brassica seed crops***, ***lettuce*** *(outdoor crops)*, ***rhubarb*** *(outdoor)* [3, 11]; ***cherries*** *(off-label)*, ***chicory*** *(off-label)*, ***chives*** *(off-label - outdoor or under crop covers)*, ***courgettes*** *(off-label - outdoor or under crop covers)*, ***evening primrose*** *(off-label)*, ***frise*** *(off-label - outdoor or under crop covers)*, ***herbs (see appendix 6)*** *(off-label - outdoor or under crop covers)*, ***honesty*** *(off-label)*, ***hops*** *(off-label)*, ***lamb's lettuce*** *(off-label - outdoor or under crop covers)*, ***marrows*** *(off-label - outdoor or under crop covers)*, ***mirabelles*** *(off-label)*, ***parsley*** *(off-label - outdoor or under crop covers)*, ***protected chives*** *(off-label)*, ***protected frise*** *(off-label)*, ***protected herbs (see appendix 6)*** *(off-label)*, ***protected lettuce*** *(off-label)*, ***protected parsley*** *(off-label)*, ***protected radicchio*** *(off-label)*, ***protected salad brassicas*** *(off-label - for baby leaf production)*, ***protected scarole*** *(off-label)*, ***pumpkins*** *(off-label - outdoor or under crop covers)*, ***quinces*** *(off-label)*, ***salad brassicas*** *(off-label - for baby leaf production outdoors or under crop covers)*, ***scarole*** *(off-label - outdoor or under crop covers)*, ***squashes*** *(off-label - outdoor or under crop covers)*, ***wine grapes*** *(off-label)* [7]; ***chives*** *(off-label)*, ***herbs (see appendix 6)*** *(off-label)*, ***parsley*** *(off-label)* [8]; ***fodder rape seed crops***, ***kale seed crops***, ***lettuce***, ***rhubarb***, ***turnip seed crops*** [1, 4-8, 10, 12-17]; ***forest*** [1, 2, 4, 7, 9, 14]; ***forest nurseries*** [14]; ***frise*** *(off-label)*, ***scarole*** *(off-label)* [8, 13]; ***lamb's lettuce*** *(off-label)* [8, 12, 13]; ***ornamental plant production*** [7, 14, 17]; ***radicchio*** *(off-label)* [7, 8, 13]; ***raspberries*** [17]; ***raspberries*** *(England only)* [1, 3-8, 10-16]; ***roses*** [1, 4, 7, 14, 17]; ***trees and shrubs*** [1, 2, 4, 9]; ***woody ornamentals*** [1-4, 7, 9, 11]
- Horsetails in ***forest*** [2, 3, 9, 11]; ***woody ornamentals*** [2, 9]
- Perennial grasses in ***almonds*** *(off-label)*, ***chestnuts*** *(off-label)*, ***hazel nuts*** *(off-label)*, ***walnuts*** *(off-label)* [3, 7, 11-13]; ***apple orchards***, ***blackberries***, ***blackcurrants***, ***gooseberries***,

loganberries, ***pear orchards***, ***plums***, ***redcurrants*** [1, 3-8, 10-17]; ***bilberries*** *(off-label)*, ***blueberries*** *(off-label)*, ***cranberries*** *(off-label)* [7, 8]; ***cherries*** *(off-label)*, ***chicory*** *(off-label)*, ***chives*** *(off-label - outdoor or under crop covers)*, ***courgettes*** *(off-label - outdoor or under crop covers)*, ***evening primrose*** *(off-label)*, ***frise*** *(off-label - outdoor or under crop covers)*, ***herbs (see appendix 6)*** *(off-label - outdoor or under crop covers)*, ***honesty*** *(off-label)*, ***hops*** *(off-label)*, ***lamb's lettuce*** *(off-label - outdoor or under crop covers)*, ***marrows*** *(off-label - outdoor or under crop covers)*, ***mirabelles*** *(off-label)*, ***parsley*** *(off-label - outdoor or under crop covers)*, ***protected chives*** *(off-label)*, ***protected frise*** *(off-label)*, ***protected herbs (see appendix 6)*** *(off-label)*, ***protected lettuce*** *(off-label)*, ***protected parsley*** *(off-label)*, ***protected radicchio*** *(off-label)*, ***protected salad brassicas*** *(off-label - for baby leaf production)*, ***protected scarole*** *(off-label)*, ***pumpkins*** *(off-label - outdoor or under crop covers)*, ***quinces*** *(off-label)*, ***salad brassicas*** *(off-label - for baby leaf production outdoors or under crop covers)*, ***scarole*** *(off-label - outdoor or under crop covers)*, ***squashes*** *(off-label - outdoor or under crop covers)*, ***wine grapes*** *(off-label)* [7]; ***chives*** *(off-label)*, ***herbs (see appendix 6)*** *(off-label)*, ***parsley*** *(off-label)* [8]; ***clover seed crops***, ***fodder rape seed crops***, ***kale seed crops***, ***lettuce***, ***lucerne***, ***rhubarb***, ***strawberries***, ***sugar beet seed crops***, ***turnip seed crops***, ***winter field beans***, ***winter oilseed rape*** [1, 4-8, 10, 12-17]; ***forest*** [1-4, 7, 9, 11, 14]; ***forest nurseries*** [14]; ***frise*** *(off-label)*, ***scarole*** *(off-label)* [8, 13]; ***lamb's lettuce*** *(off-label)* [8, 12, 13]; ***lupins*** *(off-label)* [3, 7, 8, 11-13]; ***ornamental plant production*** [7, 14, 17]; ***radicchio*** *(off-label)* [7, 8, 13]; ***raspberries*** [17]; ***raspberries*** *(England only)* [1, 3-8, 10-16]; ***rhubarb*** *(outdoor)* [3, 11]; ***roses*** [1, 4, 7, 14, 17]; ***trees and shrubs*** [1, 2, 4, 9]; ***woody ornamentals*** [1-4, 7, 9, 11]

- Perennial weeds in ***almonds*** *(off-label)*, ***chestnuts*** *(off-label)*, ***hazel nuts*** *(off-label)*, ***walnuts*** *(off-label)* [8]
- Sedges in ***forest*** [2, 3, 9, 11]; ***woody ornamentals*** [2, 9]
- Volunteer cereals in ***sugar beet seed crops***, ***winter field beans***, ***winter oilseed rape*** [3, 11]
- Wild oats in ***sugar beet seed crops***, ***winter field beans***, ***winter oilseed rape*** [3, 11]

Specific Off-Label Approvals (SOLAs)

- ***almonds*** *20052664* [3], *20060211* [7], *20052680* [8], *20052681* [11], *20052692* [12], *20052697* [13]
- ***bilberries*** *20060213* [7], *20052679* [8]
- ***blueberries*** *20060213* [7], *20052679* [8]
- ***cherries*** *20052650* [7]
- ***chestnuts*** *20052664* [3], *20060211* [7], *20052680* [8], *20052681* [11], *20052692* [12], *20052697* [13]
- ***chicory*** *20052650* [7]
- ***chives*** *(outdoor or under crop covers) 20052650* [7], *20052675* [8]
- ***courgettes*** *(outdoor or under crop covers) 20052650* [7]
- ***cranberries*** *20060213* [7], *20052679* [8]
- ***evening primrose*** *20052650* [7]
- ***frise*** *(outdoor or under crop covers) 20052650* [7], *20052675* [8], *20052699* [13]
- ***hazel nuts*** *20052664* [3], *20060211* [7], *20052680* [8], *20052681* [11], *20052692* [12], *20052697* [13]
- ***herbs (see appendix 6)*** *(outdoor or under crop covers) 20052650* [7], *20052675* [8]
- ***honesty*** *20052650* [7]
- ***hops*** *20052650* [7]
- ***lamb's lettuce*** *(outdoor or under crop covers) 20052650* [7], *20052675* [8], *20052693* [12], *20052699* [13]
- ***lupins*** *20052665* [3], *20060210* [7], *20052678* [8], *20052682* [11], *20052691* [12], *20052698* [13]
- ***marrows*** *(outdoor or under crop covers) 20052650* [7]
- ***mirabelles*** *20052650* [7]
- ***parsley*** *(outdoor or under crop covers) 20052650* [7], *20052675* [8]
- ***protected chives*** *20052650* [7]
- ***protected frise*** *20052650* [7]
- ***protected herbs (see appendix 6)*** *20052650* [7]
- ***protected lettuce*** *20052650* [7]
- ***protected parsley*** *20052650* [7]
- ***protected radicchio*** *20052650* [7]
- ***protected salad brassicas*** *(for baby leaf production) 20052650* [7]
- ***protected scarole*** *20052650* [7]
- ***pumpkins*** *(outdoor or under crop covers) 20052650* [7]

- ***quinces*** *20060212* [7]
- ***radicchio*** *20052650* [7], *20052675* [8], *20052699* [13]
- ***salad brassicas*** *(for baby leaf production outdoors or under crop covers) 20052650* [7]
- ***scarole*** *(outdoor or under crop covers) 20052650* [7], *20052675* [8], *20052699* [13]
- ***squashes*** *(outdoor or under crop covers) 20052650* [7]
- ***walnuts*** *20052664* [3], *20060211* [7], *20052680* [8], *20052681* [11], *20052692* [12], *20052697* [13]
- ***wine grapes*** *20052650* [7]

Approval information

- Propyzamide included in Annex I under EC Directive 91/414
- Some products may be applied through CDA equipment. See labels for details
- Accepted by BBPA for use on hops
- Approval expiry 31 Aug 2008 [1, 17]

Efficacy guidance

- Active via root uptake. Weeds controlled from germination to young seedling stage, some species (including many grasses) also when established
- Best results achieved by winter application to fine, firm, moist soil. Rain is required after application if soil dry
- Uptake is slow and and may take up to 12 wk
- Excessive organic debris or ploughed-up turf may reduce efficacy
- For heavy couch infestations a repeat application may be needed in following winter
- Always follow WRAG guidelines for preventing and managing herbicide resistant weeds. Section 5 for more information

Restrictions

- Maximum number of treatments 1 per crop or yr
- Maximum total dose equivalent to one full dose treatment for all crops
- Do not treat protected crops
- Apply to listed edible crops only between 1 Oct and the date specified as the latest time of application (except lettuce)
- Do not apply in windy weather and avoid drift onto non-target crops
- Do not use on soils with more than 10% organic matter except in forestry

Crop-specific information

- Latest use: labels vary but normally before 31 Dec for rhubarb, lucerne, strawberries and winter field beans; before 31 Jan for other crops
- HI: 6 wk for edible crops
- Apply as soon as possible after 3-true leaf stage of oilseed rape (GS 1,3) and seed brassicas, after 4-leaf stage of sugar beet for seed, within 7 d after sowing but before emergence for field beans, after perennial crops established for at least 1 season, strawberries after 1 yr
- Only apply to strawberries on heavy soils. Do not use on matted row crops
- Only apply to field beans on medium and heavy soils
- Only apply to established lucerne not less than 7 d after last cut
- In lettuce lightly incorporate in top 25 mm pre-drilling or irrigate on dry soil
- See label for lists of ornamental and forest species which may be treated

Following crops guidance

- Following an application between 1 Apr and 31 Jul at any dose the following minimum intervals must be observed before sowing the next crop: lettuce 0 wk; broad beans, chicory, clover, field beans, lucerne, radishes, peas 5 wk; brassicas, celery, leeks, oilseed rape, onions, parsley, parsnips 10 wk
- Following an application between 1 Aug and 31 Mar at any dose the following minimum intervals must be observed before sowing the next crop: lettuce 0 wk; broad beans, chicory, clover, field beans, lucerne, radishes, peas 10 wk; brassicas, celery, leeks, oilseed rape, onions, parsley, parsnips 25 wk or after 15 Jun, whichever occurs sooner
- Cereals or grasses or other crops not listed may be sown 30 wk after treatment up to 840 g ai/ha between 1 Aug and 31 Mar or 40 wk after treatment at higher doses at any time and after mouldboard ploughing to at least 15 cm
- A period at least 9 mth must elapse between applications of propyzamide to the same land

Environmental safety

- Dangerous for the environment
- Very toxic to aquatic organisms

Hazard classification and safety precautions

Hazard H03, H11 [1-8, 10-12, 14-17]

Risk phrases R40, R53a [1-8, 10-12, 14-17]; R50 [2, 3, 5, 6, 8, 10-12, 15-17]; R51 [1, 4, 7, 14]

Operator protection A, H [2, 4]; U20c

Environmental protection E15a [1-17]; E34 [1, 14, 15]; E38 [2, 4, 5, 7, 8, 11, 12, 15]

Consumer protection C02a [1, 3-8, 10-17] (6 wk)

Storage and disposal D01 [3, 6, 10, 16, 17]; D05 [1, 2, 4-17]; D07 [15]; D09a [1-17]; D10b [6, 10, 16]; D11a [1-5, 7-9, 11-15, 17]; D12a [2, 4, 5, 7, 8, 11, 12, 15]

378 proquinazid

A quinazolinone fungicide for powdery mildew control in cereals
FRAC mode of action code: U7

Products

1	Justice	DuPont	200 g/l	EC	12835
2	Talius	DuPont	200 g/l	EC	12752

Uses

- Powdery mildew in ***durum wheat*** *(off-label)*, ***spring barley***, ***spring oats***, ***spring rye***, ***spring wheat***, ***triticale***, ***winter barley***, ***winter oats***, ***winter rye***, ***winter wheat***

Specific Off-Label Approvals (SOLAs)

- ***durum wheat*** *20070451* [1], *20070450* [2]

Efficacy guidance

- Best results obtained from preventive treatment before disease is established in the crop
- Where mildew has already spread to new growth a tank mix with a curative fungicide with an alternative mode of action should be used
- Use as part of an integrated crop management (ICM) strategy incorporating other methods of control or fungicides with different modes of action

Restrictions

- Maximum number of treatments 2 per crop
- Do not apply to any crop suffering from stress from any cause
- Avoid application in either frosty or hot, sunny conditions

Crop-specific information

- Latest use: before beginning of heading (GS 49) for barley, oats, rye, triticale; before full flowering (GS 65) for wheat

Environmental safety

- Dangerous for the environment
- Toxic to aquatic organisms
- Dangerous to fish or other aquatic life. Do not contaminate surface waters or ditches with chemical or used container
- LERAP Category B

Hazard classification and safety precautions

Hazard H03, H11

Risk phrases R38, R40, R41, R51, R53a

Operator protection A, C, H; U05a, U11

Environmental protection E13b, E16a, E16b, E34, E38

Storage and disposal D01, D02, D05, D09a, D10b, D11a, D12a

Medical advice M03, M05a

379 prosulfocarb

A thiocarbamate herbicide for grass and broad-leaved weed control in cereals
HRAC mode of action code: N

Products

Defy	Syngenta	800 g/l	EC	12606

Uses

- Annual dicotyledons in ***potatoes***, ***winter barley***, ***winter wheat***
- Annual meadow grass in ***potatoes***, ***winter barley***, ***winter wheat***
- Chickweed in ***celeriac*** *(off-label)*, ***potatoes***, ***winter barley***, ***winter wheat***
- Cleavers in ***potatoes***, ***winter barley***, ***winter wheat***
- Fat hen in ***celeriac*** *(off-label)*
- Field speedwell in ***potatoes***, ***winter barley***, ***winter wheat***
- Fumitory in ***bulb onions*** *(off-label)*, ***garlic*** *(off-label)*, ***leeks*** *(off-label)*, ***onion sets*** *(off-label)*, ***shallots*** *(off-label)*, ***spring onions*** *(off-label)*
- Ivy-leaved speedwell in ***potatoes***, ***winter barley***, ***winter wheat***
- Loose silky bent in ***potatoes***, ***winter barley***, ***winter wheat***
- Mayweeds in ***celeriac*** *(off-label)*
- Polygonums in ***bulb onions*** *(off-label)*, ***celeriac*** *(off-label)*, ***garlic*** *(off-label)*, ***leeks*** *(off-label)*, ***onion sets*** *(off-label)*, ***shallots*** *(off-label)*
- Rough meadow grass in ***potatoes***, ***winter barley***, ***winter wheat***
- Stinging nettle in ***celeriac*** *(off-label)*
- Thistles in ***celeriac*** *(off-label)*
- Volunteer oilseed rape in ***spring onions*** *(off-label)*

Specific Off-Label Approvals (SOLAs)

- ***bulb onions*** *20062435*
- ***celeriac*** *20062437*
- ***garlic*** *20062435*
- ***leeks*** *20062436*
- ***onion sets*** *20062435*
- ***shallots*** *20062435*
- ***spring onions*** *20062434*

Approval information

- Prosulfocarb included in Annex I under EC Directive 91/414

Efficacy guidance

- Best results obtained in cereals from treatment of crops in a firm moist seedbed, free from clods
- Pre-emergence use will reduce blackgrass populations but the product should only be used against this weed as part of a management strategy involving sequences with products of alternative modes of action
- Always follow WRAG guidelines for preventing and managing herbicide resistant weeds. Section 5 for more information

Restrictions

- Maximum number of treatments 1 per crop for winter barley, winter wheat and potatoes
- Do not apply to crops under stress from any cause. Transient yellowing can occur from which recovery is complete
- Winter cereals must be covered by 3 cm of settled soil

Crop-specific information

- Latest use: at emergence (soil rising over emerging potato shoots) for potatoes, up to and including early tillering (GS 21) for winter barley, winter wheat
- When applied pre-emergence to cereals crop emergence may occasionally be slowed down but yield is not affected
- For potatoes, complete ridge formation before application and do not disturb treated soil afterwards

Following crops guidance

- Do not sow field or broad beans within 12 mth of treatment
- In the event of failure of a treated cereal crop, winter wheat or winter barley may be re-sown immediately. In the following spring sunflowers, maize, flax, spring cereals, peas, oilseed rape or soya beans may be sown without ploughing, and carrots, lettuce, onions, sugar beet or potatoes may be sown or planted after ploughing

Environmental safety

- Dangerous for the environment
- Very toxic to aquatic organisms
- LERAP Category B

Hazard classification and safety precautions

Hazard H04, H11

Risk phrases R38, R43, R50, R53a

Operator protection A, C, H; U02a, U05a, U08, U14, U15, U20b

Environmental protection E15b, E16a, E16b, E38

Storage and disposal D01, D02, D05, D09a, D10c, D12a

Medical advice M05a

380 prosulfuron

A contact and residual sulfonyl urea herbicide available only in mixture

See also *bromoxynil + prosulfuron*

381 prothioconazole

A systemic, protectant and curative triazole fungicide
FRAC mode of action code: 3

See also *clothianidin + prothioconazole*
clothianidin + prothioconazole + tebuconazole + triazoxide
fluoxastrobin + prothioconazole
fluoxastrobin + prothioconazole + trifloxystrobin

Products

1	Clayton Impress	Clayton	250 g/l	EC	12912
2	Proline	Bayer CropScience	250 g/l	EC	12084
3	Redigo	Bayer CropScience	100 g/l	FS	12085
4	Standon Win-Win	Standon	250 g/l	EC	12386

Uses

- Brown rust in ***spring barley***, ***winter barley***, ***winter rye***, ***winter wheat*** [1, 2, 4]
- Bunt in ***durum wheat*** *(seed treatment)*, ***spring rye*** *(seed treatment)*, ***spring wheat*** *(seed treatment)*, ***triticale*** *(seed treatment)*, ***winter rye*** *(seed treatment)*, ***winter wheat*** *(seed treatment)* [3]
- Covered smut in ***winter barley*** *(seed treatment)* [3]
- Crown rust in ***spring oats***, ***winter oats*** [2]
- Eyespot in ***spring barley***, ***winter barley***, ***winter rye***, ***winter wheat*** [1, 2, 4]; ***spring oats***, ***winter oats*** [2]
- Fusarium foot rot and seedling blight in ***durum wheat*** *(seed treatment)*, ***spring oats*** *(seed treatment)*, ***spring rye*** *(seed treatment)*, ***spring wheat*** *(seed treatment)*, ***triticale*** *(seed treatment)*, ***winter barley*** *(seed treatment)*, ***winter oats*** *(seed treatment)*, ***winter rye*** *(seed treatment)*, ***winter wheat*** *(seed treatment)* [3]
- Glume blotch in ***winter wheat*** [1, 2, 4]
- Late ear diseases in ***spring barley***, ***winter barley***, ***winter wheat*** [1, 2, 4]
- Leaf stripe in ***winter barley*** *(seed treatment)* [3]
- Loose smut in ***durum wheat*** *(seed treatment)*, ***spring oats*** *(seed treatment)*, ***spring wheat*** *(seed treatment)*, ***winter barley*** *(seed treatment)*, ***winter oats*** *(seed treatment)*, ***winter wheat*** *(seed treatment)* [3]
- Net blotch in ***spring barley***, ***winter barley*** [1, 2, 4]
- Powdery mildew in ***spring barley***, ***winter barley***, ***winter rye***, ***winter wheat*** [1, 2, 4]; ***spring oats***, ***winter oats*** [2]
- Rhynchosporium in ***spring barley***, ***winter barley***, ***winter rye*** [1, 2, 4]
- Sclerotinia stem rot in ***winter oilseed rape*** [1, 2, 4]
- Septoria leaf blotch in ***winter wheat*** [1, 2, 4]
- Tan spot in ***winter wheat*** [1, 2, 4]
- Yellow rust in ***winter barley***, ***winter wheat*** [1, 2, 4]

Approval information

- Accepted by BBPA for use on malting barley

Efficacy guidance

- Seed treatments must be applied by manufacturer's recommended treatment application equipment [3]
- Treated cereal seed should preferably be drilled in the same season [3]
- Follow-up treatments will be needed later in the season to give protection against air-borne and splash-borne diseases [3]
- Best results on cereal foliar diseases obtained from treatment at early stages of disease development. Further treatment may be needed if disease attack is prolonged [2, 4]
- Foliar applications to established infections of any disease are likely to be less effective [2, 4]
- Best control of cereal ear diseases obtained by treatment during ear emergence [2, 4]
- Treat oilseed rape at early to full flower [2, 4]
- Prothioconazole is a DMI fungicide. Resistance to some DMI fungicides has been identified in Septoria leaf blotch which may seriously affect performance of some products. For further advice contact a specialist advisor and visit the Fungicide Resistance Action Group (FRAG)-UK website

Restrictions

- Maximum number of seed treatments one per batch [3]
- Maximum total dose equivalent to one full dose treatment on oilseed rape, two full dose treatments on barley, oats; three full dose treatments on wheat, rye [2, 4]
- Seed treatment must be fully re-dispersed and homogeneous before use [3]
- Do not use on seed with more than 16% moisture content, or on sprouted, cracked or skinned seed [3]
- All seed batches should be tested to ensure they are suitable for treatment [3]
- Treated winter barley seed must be used within the season of treatment; treated winter wheat seed should preferably be drilled in the same season [3]
- Do not make repeated treatments of the product alone to the same crop against pathogens such as powdery mildew. Use tank mixtures or alternate with fungicides having a different mode of action [3]
- Do not treat oilseed rape crops to be used for seed production [2, 4]

Crop-specific information

- Latest use: pre-drilling for seed treatments [3]; before grain milky ripe for foliar sprays on winter rye, winter wheat; beginning of flowering for barley, oats [2, 4]
- HI 56 d [2] or 65 d [4] for winter oilseed rape

Environmental safety

- Dangerous for the environment [2, 4]
- Toxic to aquatic organisms [2, 4]
- Harmful to aquatic organisms [3]
- LERAP Category B [2, 4]

Hazard classification and safety precautions

Hazard H04 [1-4]; H11 [1, 2, 4]
Risk phrases R36, R51 [1, 2, 4]; R43, R52 [3]; R53a [1-4]
Operator protection A, H [1-4]; C [1, 2, 4]; U05a, U09b, U20b [1, 2, 4]; U07, U14, U20a [3]
Environmental protection E15a, E34 [1-4]; E16a, E38 [1, 2, 4]; E36a [3]
Storage and disposal D01, D02, D05, D09a, D12a [1-4]; D10b [1, 2, 4]; D14 [3]
Treated seed S01, S02, S03, S04a, S04b, S05, S06b, S07, S08 [3]
Medical advice M03 [1, 2, 4]

382 prothioconazole + spiroxamine

A broad spectrum fungicide mixture for cereals
FRAC mode of action code: 3 + 5

Products

Helix	Bayer CropScience	160:300 g/l	EC	12264

Uses

- Brown rust in ***spring barley***, ***winter barley***, ***winter rye***, ***winter wheat***
- Crown rust in ***spring oats***, ***winter oats***
- Eyespot in ***spring barley***, ***spring oats***, ***winter barley***, ***winter oats***, ***winter rye***, ***winter wheat***
- Glume blotch in ***winter wheat***

SEE SECTION 3 FOR PRODUCTS ALSO REGISTERED

- Late ear diseases in ***winter rye***, ***winter wheat***
- Net blotch in ***spring barley***, ***winter barley***
- Powdery mildew in ***spring barley***, ***spring oats***, ***winter barley***, ***winter oats***, ***winter rye***, ***winter wheat***
- Rhynchosporium in ***spring barley***, ***winter barley***, ***winter rye***
- Septoria leaf blotch in ***winter wheat***
- Tan spot in ***winter wheat***
- Yellow rust in ***spring barley***, ***winter barley***, ***winter wheat***

Approval information

- Spiroxamine included in Annex I under EC Directive 91/414
- Accepted by BBPA for use on malting barley

Efficacy guidance

- Best results obtained from treatment at early stages of disease development. Further treatment may be needed if disease attack is prolonged
- Applications to established infections of any disease are likely to be less effective
- Best control of cereal ear diseases obtained by treatment during ear emergence
- Prothioconazole is a DMI fungicide. Resistance to some DMI fungicides has been identified in Septoria leaf blotch which may seriously affect performance of some products. For further advice contact a specialist advisor and visit the Fungicide Resistance Action Group (FRAG)-UK website

Restrictions

- Maximum total dose equivalent to two full dose treatments on barley and oats and three full dose treatments on wheat and rye

Crop-specific information

- Latest use: before grain watery ripe for rye, oats, winter wheat; up to beginning of anthesis for barley

Environmental safety

- Dangerous for the environment
- Very toxic to aquatic organisms
- LERAP Category B

Hazard classification and safety precautions

Hazard H03, H11
Risk phrases R20, R22a, R36, R38, R50, R53a
Operator protection A, C, H; U05a, U09b, U11, U19a, U20b
Environmental protection E15a, E16a, E34, E38
Storage and disposal D01, D02, D05, D09a, D10b, D12a
Medical advice M03

383 prothioconazole + spiroxamine + tebuconazole

A broad spectrum fungicide mixture for cereals
FRAC mode of action code: 3 + 5 + 3

Products

Cello	Bayer CropScience	100:250:100 g/l	EC	13178

Uses

- Brown rust in ***spring barley***, ***winter barley***, ***winter rye***, ***winter wheat***
- Crown rust in ***spring oats***, ***winter oats***
- Eyespot in ***spring barley*** *(reduction)*, ***spring oats***, ***winter barley*** *(reducton)*, ***winter oats***, ***winter rye*** *(reducton)*, ***winter wheat*** *(reduction)*
- Glume blotch in ***winter wheat***
- Late ear diseases in ***spring barley***, ***winter barley***, ***winter wheat***
- Net blotch in ***spring barley***, ***winter barley***
- Powdery mildew in ***spring barley***, ***spring oats***, ***winter barley***, ***winter oats***, ***winter rye***, ***winter wheat***
- Rhynchosporium in ***spring barley***, ***winter barley***, ***winter rye***
- Septoria leaf blotch in ***winter wheat***
- Yellow rust in ***spring barley***, ***winter barley***, ***winter wheat***

Approval information

- Spiroxamine included in Annex I under EC Directive 91/414

Efficacy guidance

- Best results obtained from treatment at early stages of disease development. Further treatment may be needed if disease attack is prolonged
- Applications to established infections of any disease are likely to be less effective
- Best control of cereal ear diseases obtained by treatment during ear emergence
- Prothioconazole and tebuconazole are DMI fungicides. Resistance to some DMI fungicides has been identified in Septoria leaf blotch which may seriously affect performance of some products. For further advice contact a specialist advisor and visit the Fungicide Resistance Action Group (FRAG)-UK website

Restrictions

- Maximum total dose equivalent to two full dose treatments

Crop-specific information

- Latest use: before grain milky ripe stage for rye, wheat; up to beginning of anthesis barley, oats

Environmental safety

- Dangerous for the environment
- Very toxic to aquatic organisms
- LERAP Category B

Hazard classification and safety precautions

Hazard H03, H11
Risk phrases R20, R36, R37, R38, R43, R50, R53a, R63
Operator protection A, C, H; U05a, U09b, U11, U14, U19a, U20b
Environmental protection E15a, E16a, E34, E38
Storage and disposal D01, D02, D05, D09a, D10b, D12a
Medical advice M03

384 prothioconazole + tebuconazole

A triazole fungicide mixture for cereals
FRAC mode of action code: 3 + 3

Products

Prosaro	Bayer CropScience	125:125 g/l	EC	12263

Uses

- Brown rust in ***spring barley, winter barley, winter rye, winter wheat***
- Crown rust in ***spring oats, winter oats***
- Eyespot in ***spring barley*** *(reduction)*, ***spring oats, winter barley*** *(reduction)*, ***winter oats, winter rye*** *(reduction)*, ***winter wheat*** *(reduction)*
- Glume blotch in ***winter wheat***
- Late ear diseases in ***spring barley, winter barley, winter wheat***
- Light leaf spot in ***spring oilseed rape, winter oilseed rape***
- Net blotch in ***spring barley, winter barley***
- Phoma in ***spring oilseed rape, winter oilseed rape***
- Powdery mildew in ***spring barley, spring oats, winter barley, winter oats, winter rye, winter wheat***
- Rhynchosporium in ***spring barley, winter barley, winter rye***
- Sclerotinia stem rot in ***spring oilseed rape, winter oilseed rape***
- Septoria leaf blotch in ***winter wheat***
- Tan spot in ***winter wheat***
- Yellow rust in ***spring barley, winter barley, winter wheat***

Approval information

- Accepted by BBPA for use on malting barley

Efficacy guidance

- Best results on cereal foliar diseases obtained from treatment at early stages of disease development. Further treatment may be needed if disease attack is prolonged

SEE SECTION 3 FOR PRODUCTS ALSO REGISTERED

- On oilseed rape apply a protective treatment in autumn/winter for Phoma followed by a further spray in early spring from the onset of stem elongation, if necessary. For control of Sclerotinia apply at early to full flower
- Applications to established infections of any disease are likely to be less effective
- Best control of cereal ear diseases obtained by treatment during ear emergence
- Prothioconazole and tebuconazole are DMI fungicides. Resistance to some DMI fungicides has been identified in Septoria leaf blotch which may seriously affect performance of some products. For further advice contact a specialist advisor and visit the Fungicide Resistance Action Group (FRAG)-UK website

Restrictions

- Maximum total dose equivalent to two full dose treatments on barley, oats, oilseed rape; three full dose treatments on wheat and rye

Crop-specific information

- Latest use: before grain milky ripe for rye, wheat; beginning of flowering for barley, oats
- HI 56 d for oilseed rape

Environmental safety

- Dangerous for the environment
- Toxic to aquatic organisms
- Risk to non-target insects or other arthropods. Avoid spraying within 6 m of the field boundary to reduce the effects on non-target insects or other arthropods
- LERAP Category B

Hazard classification and safety precautions

Hazard H04, H11
Risk phrases R38, R43, R51, R53a, R63
Operator protection A, H; U05a, U09b, U19a, U20b
Environmental protection E15a, E16a, E22c, E34, E38
Storage and disposal D01, D02, D05, D09a, D10b, D12a
Medical advice M03

385 prothioconazole + tebuconazole + triazoxide

A broad spectrum fungicide seed treatment for cereals
FRAC mode of action code: 3 + 3 + 35

Products

Raxil Pro	Bayer CropScience	25:15:10 g/l	FS	12432

Uses

- Covered smut in ***spring barley*** *(seed treatment)*, ***winter barley*** *(seed treatment)*
- Fusarium foot rot and seedling blight in ***spring barley*** *(seed treatment)*, ***winter barley*** *(seed treatment)*
- Leaf stripe in ***spring barley*** *(seed treatment)*, ***winter barley*** *(seed treatment)*
- Loose smut in ***spring barley*** *(seed treatment)*, ***winter barley*** *(seed treatment)*
- Net blotch in ***spring barley*** *(seed treatment)*, ***winter barley*** *(seed treatment)*

Approval information

- Accepted by BBPA for use on malting barley

Efficacy guidance

- Treatments must be applied by manufacturer's recommended treatment application equipment
- Treated seed should preferably be drilled in the same season
- Follow-up treatments will be needed later in the season to give protection against air-borne and splash-borne diseases

Restrictions

- Maximum number of seed treatments one per batch

Crop-specific information

- Latest use for seed treatment: pre-drilling

Environmental safety

- Harmful to aquatic organisms

Hazard classification and safety precautions

Hazard H04
Risk phrases R37, R43, R52, R53a
Operator protection A, H; U07, U14, U19a, U20a
Environmental protection E15a, E34, E36a, E38
Storage and disposal D05, D09a, D14
Treated seed S01, S02, S03, S04b, S05, S06b, S07, S08

386 prothioconazole + trifloxystrobin

A triazole and strobilurin fungicide mixture for cereals
FRAC mode of action code: 3 + 11

Products

Mobius	Bayer CropScience	175:150 g/l	SC	12324

Uses

- Brown rust in ***spring barley***, ***winter barley***, ***winter wheat***
- Eyespot in ***spring barley*** *(reduction)*, ***winter barley*** *(reduction)*, ***winter wheat***
- Glume blotch in ***winter wheat***
- Late ear diseases in ***spring barley***, ***winter barley***, ***winter wheat***
- Net blotch in ***spring barley***, ***winter barley***
- Powdery mildew in ***spring barley***, ***winter barley***, ***winter wheat***
- Rhynchosporium in ***spring barley***, ***winter barley***
- Septoria leaf blotch in ***winter wheat***
- Yellow rust in ***winter barley***, ***winter wheat***

Approval information

- Trifloxystrobin included in Annex I under EC Directive 91/414
- Accepted by BBPA for use on malting barley
- Approval expiry 30 Sep 2008 [1]

Efficacy guidance

- Best results obtained from treatment at early stages of disease development. Further treatment may be needed if disease attack is prolonged
- Applications to established infections of any disease are likely to be less effective
- Best control of cereal ear diseases obtained by treatment during ear emergence
- Prothioconazole is a DMI fungicide. Resistance to some DMI fungicides has been identified in Septoria leaf blotch which may seriously affect performance of some products. For further advice contact a specialist advisor and visit the Fungicide Resistance Action Group (FRAG)-UK website
- Trifloxystrobin is a member of the QoI cross resistance group. Product should be used preventatively and not relied on for its curative potential
- Use product as part of an Integrated Crop Management strategy incorporating other methods of control, including where appropriate other fungicides with a different mode of action. Do not apply more than two foliar applications of QoI containing products to any cereal crop
- There is a significant risk of widespread resistance occurring in *Septoria tritici* populations in UK. Failure to follow resistance management action may result in reduced levels of disease control
- Strains of wheat and barley powdery mildew resistant to QoIs are common in the UK. Control of wheat mildew can only be relied on from the triazole component
- Where specific control of wheat mildew is required this should be achieved through a programme of measures including products recommended for the control of mildew that contain a fungicide from a different cross-resistance group and applied at a dose that will give robust control

Restrictions

- Maximum total dose equivalent to two full dose treatments

Crop-specific information

- Latest use: before grain milky ripe for wheat; beginning of flowering for barley

Environmental safety

- Dangerous for the environment
- Very toxic to aquatic organisms
- LERAP Category B

SEE SECTION 3 FOR PRODUCTS ALSO REGISTERED

Hazard classification and safety precautions

Hazard H04, H11

Risk phrases R37, R50, R53a

Operator protection A, C, H; U05a, U09b, U19a, U20b

Environmental protection E15a, E16a, E34, E38

Storage and disposal D01, D02, D05, D09a, D10c, D12a

Medical advice M03

387 pymetrozine

A novel azomethine insecticide

IRAC mode of action code: 9B

Products

1	Chess WG	Syngenta Bioline	50% w/w	WG	13310
2	Plenum WG	Syngenta	50% w/w	WG	10652

Uses

- Aphids in ***bilberries*** *(off-label)*, ***blackberries*** *(off-label)*, ***blackcurrants*** *(off-label)*, ***blueberries*** *(off-label)*, ***celeriac*** *(off-label)*, ***celery (outdoor)*** *(off-label)*, ***cranberries*** *(off-label)*, ***frise*** *(off-label)*, ***gooseberries*** *(off-label)*, ***lamb's lettuce*** *(off-label)*, ***leaf spinach***, ***lettuce*** *(off-label)*, ***loganberries*** *(off-label)*, ***radicchio*** *(off-label)*, ***raspberries*** *(off-label)*, ***redcurrants*** *(off-label)*, ***rubus hybrids*** *(off-label)*, ***salad brassicas*** *(off-label - for baby leaf production)*, ***seed potatoes***, ***strawberries*** *(off-label)*, ***sweetcorn*** *(off-label)*, ***vaccinium spp.*** *(off-label)*, ***ware potatoes***, ***whitecurrants*** *(off-label)* [2]; ***brassica seed beds*** *(off-label)*, ***ornamental plant production***, ***protected blackberries*** *(off-label)*, ***protected blackcurrants*** *(off-label)*, ***protected celery*** *(off-label)*, ***protected chinese cabbage*** *(off-label)*, ***protected chives*** *(off-label)*, ***protected choi sum*** *(off-label)*, ***protected cucumbers***, ***protected gooseberries*** *(off-label)*, ***protected herbs (see appendix 6)*** *(off-label)*, ***protected lettuce*** *(off-label)*, ***protected loganberries*** *(off-label)*, ***protected ornamentals***, ***protected pak choi*** *(off-label)*, ***protected parsley*** *(off-label)*, ***protected raspberries*** *(off-label)*, ***protected redcurrants*** *(off-label)*, ***protected ribes hybrids*** *(off-label)*, ***protected rubus hybrids*** *(off-label)*, ***protected salad brassicas*** *(off-label - for baby leaf production)*, ***protected strawberries*** *(off-label)* [1]
- Damson-hop aphid in ***hops*** *(off-label)* [2]
- Glasshouse whitefly in ***protected tomatoes*** *(off-label)* [1]
- Peach-potato aphid in ***broccoli*** *(off-label)*, ***brussels sprouts*** *(off-label)*, ***cabbages*** *(off-label)*, ***calabrese*** *(off-label)*, ***cauliflowers*** *(off-label)*, ***collards*** *(off-label)*, ***kale*** *(off-label)* [2]; ***protected aubergines*** *(off-label)*, ***protected peppers*** *(off-label)* [1]
- Tobacco whitefly in ***protected tomatoes*** *(off-label)* [1]

Specific Off-Label Approvals (SOLAs)

- ***bilberries*** *20061702* [2]
- ***blackberries*** *20061633* [2]
- ***blackcurrants*** *20060946* [2]
- ***blueberries*** *20061702* [2]
- ***brassica seed beds*** *20070788* [1]
- ***broccoli*** *20031478* [2]
- ***brussels sprouts*** *20031478* [2]
- ***cabbages*** *20031478* [2]
- ***calabrese*** *20031478* [2]
- ***cauliflowers*** *20031478* [2]
- ***celeriac*** *20051062* [2]
- ***celery (outdoor)*** *20051062* [2]
- ***collards*** *20050385* [2]
- ***cranberries*** *20061702* [2]
- ***frise*** *20070060* [2]
- ***gooseberries*** *20061702* [2]
- ***hops*** *20031423* [2]
- ***kale*** *20050385* [2]
- ***lamb's lettuce*** *20070060* [2]
- ***lettuce*** *20070060* [2]

- ***loganberries*** *20061702* [2]
- ***protected aubergines*** *20070501* [1]
- ***protected blackberries*** *20070498* [1]
- ***protected blackcurrants*** *20070504* [1]
- ***protected celery*** *20070503* [1]
- ***protected chinese cabbage*** *20070500* [1]
- ***protected chives*** *20070502* [1]
- ***protected choi sum*** *20070500* [1]
- ***protected gooseberries*** *20070504* [1]
- ***protected herbs (see appendix 6)*** *20070502* [1]
- ***protected lettuce*** *20070502* [1]
- ***protected loganberries*** *20070504* [1]
- ***protected pak choi*** *20070500* [1]
- ***protected parsley*** *20070502* [1]
- ***protected peppers*** *20070501* [1]
- ***protected raspberries*** *20070498* [1]
- ***protected redcurrants*** *20070504* [1]
- ***protected ribes hybrids*** *20070504* [1]
- ***protected rubus hybrids*** *20070504* [1]
- ***protected salad brassicas*** *(for baby leaf production) 20070502* [1]
- ***protected strawberries*** *20070499* [1]
- ***protected tomatoes*** *20070501* [1]
- ***radicchio*** *20070060* [2]
- ***raspberries*** *20061633* [2]
- ***redcurrants*** *20061702* [2]
- ***rubus hybrids*** *20061702* [2]
- ***salad brassicas*** *(for baby leaf production) 20070060* [2]
- ***strawberries*** *20060461* [2]
- ***sweetcorn*** *20041318* [2]
- ***vaccinium spp.*** *20061702* [2]
- ***whitecurrants*** *20061702* [2]

Approval information

- Pymetrozine included in Annex I under EC Directive 91/414
- Accepted by BBPA for use on hops

Efficacy guidance

- Pymetrozine moves systemically in the plant and acts by preventing feeding leading to death by starvation in 1-4 d. There is no immediate knockdown
- Aphids controlled include those resistant to organophosphorus and carbamate insecticides
- To prevent development of resistance do not use continuously or as the sole method of control
- Best results achieved by starting spraying as soon as aphids seen in crop and repeating as necessary.
- To limit spread of persistent viruses such as potato leaf roll virus, apply from 90% crop emergence

Restrictions

- Maximum total dose equivalent to two full dose treatments on ware potatoes; three full dose treatments on seed potatoes, leaf spinach [2]; four full dose treatments on cucumbers, ornamentals [1]
- Consult processors before use on potatoes for processing
- Check tolerance of ornamental species before large scale use. See label for list of species known to have been treated without damage. Visible spray deposits may be seen on leaves of some species [1]

Crop-specific information

- HI cucumbers 3 d; potatoes, leaf spinach 7 d

Environmental safety

- High risk to bees (outdoor use only). Do not apply to crops in flower or to those in which bees are actively foraging. Do not apply when flowering weeds are present
- Avoid spraying within 6 m of field boundaries to reduce effects on non-target insects or arthropods

Hazard classification and safety precautions

Hazard H03

Risk phrases R40

Operator protection A [1, 2]; H [2]; U05a, U20c

Environmental protection E12a, E12e, E15a

Storage and disposal D01, D02, D09a, D11a

388 pyraclostrobin

A protectant and curative strobilurin fungicide for cereals
FRAC mode of action code: 11

See also *boscalid + pyraclostrobin*
dithianon + pyraclostrobin
epoxiconazole + fenpropimorph + pyraclostrobin
epoxiconazole + kresoxim-methyl + pyraclostrobin
epoxiconazole + pyraclostrobin
fenpropimorph + pyraclostrobin

Products

1	Comet 200	BASF	200 g/l	EC	12639
2	Flyer	BASF	250 g/l	EC	12654
3	Insignia	Vitax	20% w/w	WG	11865
4	Platoon 250	BASF	250 g/l	EC	12640
5	Tucana	BASF	250 g/l	EC	10899
6	Vivid	BASF	250 g/l	EC	10898

Uses

- Brown rust in ***spring barley***, ***spring wheat***, ***winter barley***, ***winter wheat*** [1, 2, 4-6]
- Crown rust in ***spring oats***, ***winter oats*** [1, 2, 4-6]
- Dollar spot in ***managed amenity turf*** *(reduction)* [3]
- Fusarium patch in ***managed amenity turf*** [3]
- Glume blotch in ***spring wheat***, ***winter wheat*** [1, 2, 4-6]
- Net blotch in ***spring barley***, ***winter barley*** [1, 2, 4-6]
- Red thread in ***managed amenity turf*** [3]
- Rhynchosporium in ***spring barley*** *(moderate)*, ***winter barley*** *(moderate)* [1, 2, 4-6]
- Septoria leaf blotch in ***spring wheat***, ***winter wheat*** [1, 2, 4-6]
- Yellow rust in ***spring barley***, ***spring wheat***, ***winter barley***, ***winter wheat*** [1, 2, 4-6]

Approval information

- Pyraclostrobin included in Annex I under EC Directive 91/414
- Accepted by BBPA for use on malting barley (before ear emergence only)

Efficacy guidance

- For best results apply at the start of disease attack on cereals [1-6]
- For Fusarium Patch treat early as severe damage to turf can occur once the disease is established [3]
- Regular turf aeration, appropriate scarification and judicious use of nitrogenous fertiliser will assist the control of Fusarium Patch [3]
- Best results on Septoria glume blotch achieved when used as a protective treatment and against Septoria leaf blotch when treated in the latent phase [1-6]
- Yield response may be obtained in the absence of visual disease symptoms [1-6]
- Pyraclostrobin is a member of the QoI cross resistance group. Product should be used preventatively and not relied on for its curative potential
- Use product as part of an Integrated Crop Management strategy incorporating other methods of control, including where appropriate other fungicides with a different mode of action. Do not apply more than two foliar applications of QoI containing products to any cereal crop or to grass
- There is a significant risk of widespread resistance occurring in *Septoria tritici* populations in UK. Failure to follow resistance management action may result in reduced levels of disease control [1-6]
- On cereal crops product must always be used in mixture with another product, recommended for control of the same target disease, that contains a fungicide from a different cross resistance group and is applied at a dose that will give robust control [1-6]

Restrictions

- Maximum number of treatments 2 per crop on cereals [1, 2, 4-6]
- Maximum total dose on turf equivalent to two full dose treatments [3]
- Do not apply during drought conditions or to frozen turf [3]

Crop-specific information

- Latest use: before grain watery ripe (GS 71) for wheat; up to and including emergence of ear just complete (GS 59) for barley and oats [1, 2, 4-6]
- Avoid applying to turf immediately after cutting or 48 h before mowing [3]

Environmental safety

- Dangerous for the environment
- Very toxic to aquatic organisms
- LERAP Category B

Hazard classification and safety precautions

Hazard H03, H11
Risk phrases R20, R50 [1-6]; R22a, R38, R53a [1, 2, 4-6]
Operator protection A [1-6]; D [3]; U05a, U14, U20b [1, 2, 4-6]
Environmental protection E15a [3]; E15b, E34 [1, 2, 4-6]; E16a, E16b, E38 [1-6]
Storage and disposal D01, D02 [1, 2, 4-6]; D08, D09a, D10c, D12a [1-6]
Medical advice M03, M05a [1, 2, 4-6]

389 pyrethrins

A non-persistent, contact acting insecticide extracted from Pyrethrum
IRAC mode of action code: 3

Products

1	Dairy Fly Spray	B H & B	0.75 g/l	AL	H5579
2	Killgerm ULV 400	Killgerm	30 g/l	UL	H4838
3	Pyblast	Agropharm	30 g/l	UL	H7485
4	Pyrethrum 5 EC	Agropharm	50 g/l	EC	12685
5	Pyrethrum Spray	Killgerm	3.3 g/l	UL	H4636
6	Spruzit	Certis	4.59 g/l	EC	13438
7	Turbair Super Flydown	Summit Agro	2.5 g/l	RH	H7225

Uses

- Aphids in ***all bush fruit, all cane fruit, all protected bush fruit, all protected cane fruit, broccoli, brussels sprouts, cabbages, calabrese, cauliflowers, lettuce, ornamental plant production, protected broccoli, protected brussels sprouts, protected cabbages, protected calabrese, protected cauliflowers, protected lettuce, protected ornamentals, protected tomatoes, tomatoes (outdoor)*** [4]; ***all edible crops (outdoor), all non-edible crops (outdoor), protected crops*** [6]
- Caterpillars in ***all bush fruit, all cane fruit, all protected bush fruit, all protected cane fruit, broccoli, brussels sprouts, cabbages, calabrese, cauliflowers, lettuce, ornamental plant production, protected broccoli, protected brussels sprouts, protected cabbages, protected calabrese, protected cauliflowers, protected lettuce, protected ornamentals, protected tomatoes, tomatoes (outdoor)*** [4]; ***all edible crops (outdoor), all non-edible crops (outdoor), protected crops*** [6]
- Flea beetle in ***broccoli, brussels sprouts, cabbages, calabrese, cauliflowers, protected broccoli, protected brussels sprouts, protected cabbages, protected calabrese, protected cauliflowers, protected tomatoes, tomatoes (outdoor)*** [4]; ***protected crops*** [6]
- Insect pests in ***dairies, farm buildings*** [1-3, 5]; ***glasshouses*** [5]; ***livestock houses, poultry houses*** [1-3, 5, 7]
- Macrolophus caliginosus in ***protected tomatoes*** *(off-label)* [4]
- Mealybugs in ***protected crops*** [6]; ***protected tomatoes*** *(off-label)* [4]
- Scale insects in ***protected crops*** [6]
- Spider mites in ***all edible crops (outdoor), all non-edible crops (outdoor), protected crops*** [6]
- Thrips in ***all edible crops (outdoor), all non-edible crops (outdoor), protected crops*** [6]
- Whitefly in ***broccoli, brussels sprouts, cabbages, calabrese, cauliflowers, protected broccoli, protected brussels sprouts, protected cabbages, protected calabrese, protected cauliflowers, protected tomatoes, tomatoes (outdoor)*** [4]; ***protected crops*** [6]

Approval information

- Products formulated for ULV application [1-3]. May be applied through fogging machine or sprayer [1, 3, 5]. See label for details

Efficacy guidance

- For indoor fly control close doors and windows and spray or apply fog as appropriate [1-3, 7]
- For best fly control outdoors spray during early morning or late afternoon and evening when conditions are still [1-3, 5, 7]
- For all uses ensure good spray coverage of the target area or plants by increasing spray volume where necessary
- Best results on outdoor or protected crops achieved from treatment at first signs of pest attack in early morning or evening [6]
- Product formulated for use through Turbair Flydowner machines [7]
- To avoid possibility of development of resistance do not spray more frequently than once per week

Restrictions

- Maximum number of treatments on edible crops 3 per crop or season [4]
- For use only by professional operators [1-3, 5, 7]
- Do not allow spray to contact open food products or food preparing equipment or utensils [1-3, 5, 7]
- Remove exposed milk and collect eggs before application [1-3, 5, 7]
- Do not treat plants [1-3, 5, 7]
- Avoid direct application to open flowers [6]
- Do not mix with, or apply closely before or after, products containing dithianon or tolylfluanid [6]
- Do not use space sprays containing pyrethrins or pyrethroid more than once per week in intensive or controlled environment animal houses in order to avoid development of resistance. If necessary, use a different control method or product [1-3, 5, 7]
- Store away from strong sunlight [4]

Crop-specific information

- HI: 24 h for edible crops [4]
- Some plant species including *Ageratum*, ferns, *Ficus, Lantana, Poinsettia, Petroselium crispum*, and some strawberries may be sensitive especially where more than one application is made. Test before large scale treatment [6]

Environmental safety

- Dangerous for the environment [2-6]
- Very toxic (toxic [2, 3], [6]) to aquatic organisms
- High risk to bees. Do not apply to crops in flower or to those in which bees are actively foraging. Do not apply when flowering weeds are present [4]
- Risk to non-target insects or other arthropods [4]
- Do not apply directly to livestock and exclude all persons and animals during treatment [1-3, 5, 7]
- Wash spray equipment thoroughly after use to avoid traces of pyrethrum causing damage to susceptible crops sprayed later [4]
- LERAP Category A [4]

Hazard classification and safety precautions

Hazard H03 [1, 2, 5, 7]; H04 [1, 3, 4]; H11 [2-6]

Risk phrases R22a, R36, R38 [1]; R22b [2, 5, 7]; R41 [3, 4]; R50 [4]; R51 [2, 3, 5, 6]; R53a [2-7]; R66 [7]

Operator protection A [1-7]; B [1]; C [2, 3, 5, 7]; D [2, 3]; E [1-3]; H [2-6]; P [7]; U02b, U14 [2, 7]; U05a [1, 4, 5, 7]; U09a [1, 5]; U09b, U20a [2]; U11 [3, 4, 7]; U15 [3, 7]; U19a [1-4, 7]; U20b [1-3, 5, 7]; U20c [4]; U20d [6]

Environmental protection E02a [7] (until surfaces dry); E02c [3]; E02d, E05c [5]; E05a [1-3, 5]; E12a, E12e, E16c, E16d, E22c [4]; E13c [1]; E15a [2-5]; E15b [6]; E19b [3, 6]; E34 [7]; E38 [2-4, 6]

Consumer protection C04 [1]; C05 [5]; C06 [1, 2, 5, 7]; C07, C09 [1-3, 5]; C08 [1-3]; C10 [1, 3]; C11 [1-3, 5, 7]; C12 [2, 3, 5]

Storage and disposal D01, D09a [1-7]; D02 [1, 3-7]; D05 [2, 4]; D06a [2, 7]; D10a [2]; D11a [1, 4, 6]; D12a [2-6]; D12b [7]

Medical advice M05a [3]; M05b [2, 5, 7]

390 pyrimethanil

An anilinopyrimidine fungicide for apples and strawberries
FRAC mode of action code: 9

See also *chlorothalonil + pyrimethanil*

Products

Scala	BASF	400 g/l	SC	11695

Uses

- Botrytis in ***aubergines*** *(off-label)*, ***bilberries*** *(off-label)*, ***blackberries*** *(off-label)*, ***blackcurrants*** *(off-label)*, ***cranberries*** *(off-label)*, ***gooseberries*** *(off-label)*, ***protected blackberries*** *(off-label)*, ***protected boysenberries*** *(off-label)*, ***protected lettuce*** *(off-label)*, ***protected loganberries*** *(off-label)*, ***protected marionberries*** *(off-label)*, ***protected raspberries*** *(off-label)*, ***protected sunberries*** *(off-label)*, ***protected tayberries*** *(off-label)*, ***protected tomatoes*** *(off-label)*, ***quinces*** *(off-label)*, ***raspberries*** *(off-label)*, ***redcurrants*** *(off-label)*, ***strawberries***, ***tomatoes (outdoor)*** *(off-label)*, ***whitecurrants*** *(off-label)*, ***wine grapes*** *(off-label)*
- Scab in ***apples***, ***pears*** *(off-label)*

Specific Off-Label Approvals (SOLAs)

- ***aubergines*** *20040516*
- ***bilberries*** *20040519*
- ***blackberries*** *20051737*
- ***blackcurrants*** *20040519*
- ***cranberries*** *20040519*
- ***gooseberries*** *20040519*
- ***pears*** *20070200*
- ***protected blackberries*** *20061379*
- ***protected boysenberries*** *20061379*
- ***protected lettuce*** *20040518*
- ***protected loganberries*** *20061379*
- ***protected marionberries*** *20061379*
- ***protected raspberries*** *20061379*
- ***protected sunberries*** *20061379*
- ***protected tayberries*** *20061379*
- ***protected tomatoes*** *20040516*
- ***quinces*** *20061379*
- ***raspberries*** *20051737*
- ***redcurrants*** *20040519*
- ***tomatoes (outdoor)*** *20040516*
- ***whitecurrants*** *20040519*
- ***wine grapes*** *20040517*

Approval information

- Pyrimethanil included in Annex I under EC Directive 91/414. However a Standing Committee decision in July 2007 to reduce to the limit of determination the MRL set for pyrimenthanil in cane fruit (other than blackberries and raspberries) and the MRL for brassicas grown for baby leaf production led to immediate revocation of SOLAs for these crops

Efficacy guidance

- On apples a programme of sprays will give early season control of scab. Season long control can be achieved by continuing programme with other approved fungicides
- In strawberries product should be used as part of a programme of disease control treatments which should alternate with other materials to prevent or limit development of less sensitive strains of grey mould

Restrictions

- Maximum number of treatments 5 per yr for apples; 4 per yr for protected crops; 3 per yr for grapevines; 2 per yr for cane fruit, bush fruit, strawberries, tomatoes
- Product does not taint apples. Processors should be consulted before use on strawberries

Crop-specific information

- Latest use: before end of flowering for apples

SEE SECTION 3 FOR PRODUCTS ALSO REGISTERED

- HI protected cane fruit, strawberries 1 d; aubergines, tomatoes 3 d; protected lettuce 14 d; bush fruit, grapevines 21 d
- All varieties of apples and strawberries may be treated
- Treat apples from bud burst at 10-14 d intervals
- In strawberries start treatments at white bud to give maximum protection of flowers against grey mould and treat every 7-10 d. Product should not be used more than once in a 3 or 4 spray programme

Environmental safety

- Harmful to aquatic organisms
- Product has negligible effect on hoverflies and lacewings. Limited evidence indicates some margin of safety to *Typhlodromus pyri*
- Broadcast air-assisted LERAP (20 m)

Hazard classification and safety precautions

Risk phrases R52, R53a
Operator protection U08, U20b
Environmental protection E15a; E17b (20 m)
Storage and disposal D01, D02, D05, D08, D09a, D10b, D12a

391 quinmerac

A residual herbicide available only in mixtures
HRAC mode of action code: O

See also *chloridazon + quinmerac*
metazachlor + quinmerac

392 quinoxyfen

A systemic protectant quinoline fungicide for cereals
FRAC mode of action code: 13

See also *cyproconazole + quinoxyfen*
fenpropimorph + quinoxyfen

Products

1	Apres	Dow	500 g/l	SC	08881
2	Erysto	Dow	250 g/l	SC	08697
3	Fortress	Dow	500 g/l	SC	08279

Uses

- Powdery mildew in ***durum wheat***, ***spring barley***, ***spring oats***, ***spring rye***, ***spring wheat***, ***sugar beet***, ***triticale***, ***winter barley***, ***winter oats***, ***winter rye***, ***winter wheat*** [1-3]; ***protected strawberries*** *(off-label)*, ***strawberries*** *(off-label)* [3]

Specific Off-Label Approvals (SOLAs)

- ***protected strawberries*** *20041923* [3]
- ***strawberries*** *20041923* [3]

Approval information

- Quinoxyfen included in Annex I under Directive 91/414
- Accepted by BBPA for use on malting barley (before ear emergence only)

Efficacy guidance

- For best results treat at early stage of disease development before infection spreads to new crop growth. Further treatment may be necessary if disease pressure remains high
- Product not curative and will not control latent or established disease infections
- For broad spectrum control in cereals use in tank mixtures - see label
- Product rainfast after 1 h
- Systemic activity may be reduced in severe drought
- Use as part of an integrated crop management (ICM) strategy incorporating other methods of control or fungicides with different modes of action

Restrictions

- Maximum total dose equivalent to two full dose treatments in cereals; see labels for split dose recommendation in sugar beet
- On cereals apply only in the spring from mid-tillering stage (GS 25)

Crop-specific information

- Latest use: first awns visible stage (GS 49) for cereals
- HI sugar beet 28 d

Environmental safety

- Dangerous for the environment
- Very toxic to aquatic organisms

Hazard classification and safety precautions

Hazard H04, H11
Risk phrases R43, R50, R53a
Operator protection A, C, H; U05a, U14
Environmental protection E15a, E16a, E16b, E34, E38
Storage and disposal D01, D02, D12a

393 quizalofop-P-ethyl

An aryl phenoxypropionic acid post-emergence herbicide for grass weed control
HRAC mode of action code: A

Products

1	Leopard 5 EC	Makhteshim	50 g/l	EC	13720
2	Pilot Ultra	Nissan	50 g/l	SC	12929

Uses

- Annual grasses in ***combining peas, fodder beet, linseed, mangels, red beet, spring field beans, spring oilseed rape, sugar beet, vining peas, winter field beans, winter oilseed rape***
- Couch in ***combining peas, fodder beet, linseed, mangels, red beet, spring field beans, spring oilseed rape, sugar beet, vining peas, winter field beans, winter oilseed rape***
- Perennial grasses in ***combining peas, fodder beet, linseed, mangels, red beet, spring field beans, spring oilseed rape, sugar beet, vining peas, winter field beans, winter oilseed rape***
- Volunteer cereals in ***combining peas, fodder beet, linseed, mangels, red beet, spring field beans, spring oilseed rape, sugar beet, vining peas, winter field beans, winter oilseed rape***

Efficacy guidance

- Best results achieved by application to emerged weeds growing actively in warm conditions with adequate soil moisture
- Weed control may be reduced under conditions such as drought that limit uptake and translocation
- Annual meadow-grass is not controlled
- For effective couch control do not hoe beet crops within 21 d after spraying
- At least 2 h without rain should follow application otherwise results may be reduced
- Quizalofop-p-ethyl is an ACCase inhibitor herbicide. To avoid the build up of resistance do not apply products containing an ACCase inhibitor herbicide more than twice to any crop. In addition do not use any product containing quizalofop-p-ethyl in mixture or sequence with any other product containing the same ingredient
- Use these products as part of a resistance management strategy that includes cultural methods of control and does not use ACCase inhibitors as the sole chemical method of grass weed control
- Applying a second product containing an ACCase inhibitor to a crop will increase the risk of resistance development; only use a second ACCase inhibitor to control different weeds at a different timing
- Always follow WRAG guidelines for preventing and managing herbicide resistant weeds. See Section 5

Restrictions

- Maximum number of treatments 1 on all recommended crops
- Do not spray crops under stress from any cause or in frosty weather
- Consult processor before use on peas or beans for processing

- An interval of at least 3 d must elapse between treatment and use of another herbicide on beet crops, 14 d on oilseed rape, 21 d on linseed and other recommended crops

Crop-specific information

- HI 16 wk for beet crops; 11 wk for oilseed rape, linseed; 8 wk for field beans; 5 wk for peas
- In some situations treatment can cause yellow patches on foliage of peas, especially vining varieties. Symptoms usually rapidly and completely outgrown
- May cause taint in peas

Following crops guidance

- In the event of failure of a treated crop broad-leaved crops may be resown after a minimum interval of 2 wk, and cereals after 2-6 wk depending on dose applied
- Onions, leeks and maize are not recommended to follow a failed treated crop

Environmental safety

- Dangerous for the environment
- Toxic to aquatic organisms
- Flammable

Hazard classification and safety precautions

Hazard H03, H08 [1]; H11 [1, 2]
Risk phrases R22b, R37, R38, R41, R67 [1]; R51, R53a [1, 2]
Operator protection A, C [1, 2]; H [2]; U05a, U11, U19a [1]; U09a [2]; U20b [1, 2]
Environmental protection E13b [2]; E15a [1]; E34, E38 [1, 2]
Storage and disposal D01, D02, D05, D10b [1]; D09a, D12a [1, 2]; D10c [2]
Medical advice M03, M05b [1]

394 quizalofop-P-tefuryl

An aryloxyphenoxypropionate herbicide for grass weed control
HRAC mode of action code: A

Products

Panarex	Certis	40 g/l	EC	12532

Uses

- Blackgrass in ***amenity vegetation*** *(off-label)*, ***cereal cover crops, combining peas, fodder beet, natural surfaces not intended to bear vegetation*** *(off-label)*, ***ornamental plant production*** *(off-label)*, ***permeable surfaces overlying soil*** *(off-label)*, ***potatoes, spring field beans, spring linseed, spring oilseed rape, sugar beet, winter field beans, winter linseed, winter oilseed rape***
- Couch in ***amenity vegetation*** *(off-label)*, ***cereal cover crops, combining peas, fodder beet, natural surfaces not intended to bear vegetation*** *(off-label)*, ***ornamental plant production*** *(off-label)*, ***permeable surfaces overlying soil*** *(off-label)*, ***potatoes, spring field beans, spring linseed, spring oilseed rape, sugar beet, winter field beans, winter linseed, winter oilseed rape***
- Italian ryegrass in ***cereal cover crops, combining peas, fodder beet, potatoes, spring field beans, spring linseed, spring oilseed rape, sugar beet, winter field beans, winter linseed, winter oilseed rape***
- Perennial ryegrass in ***amenity vegetation*** *(off-label)*, ***cereal cover crops, combining peas, fodder beet, natural surfaces not intended to bear vegetation*** *(off-label)*, ***ornamental plant production*** *(off-label)*, ***permeable surfaces overlying soil*** *(off-label)*, ***potatoes, spring field beans, spring linseed, spring oilseed rape, sugar beet, winter field beans, winter linseed, winter oilseed rape***
- Volunteer cereals in ***cereal cover crops, combining peas, fodder beet, potatoes, spring field beans, spring linseed, spring oilseed rape, sugar beet, winter field beans, winter linseed, winter oilseed rape***
- Wild oats in ***cereal cover crops, combining peas, fodder beet, potatoes, spring field beans, spring linseed, spring oilseed rape, sugar beet, winter field beans, winter linseed, winter oilseed rape***

Specific Off-Label Approvals (SOLAs)

- ***amenity vegetation*** *20060364*

- ***natural surfaces not intended to bear vegetation*** *20060364*
- ***ornamental plant production*** *20060364*
- ***permeable surfaces overlying soil*** *20060364*

Efficacy guidance

- Best results on annual grass weeds when growing actively and treated from 2 leaves to the start of tillering
- Treat cover crops when they have served their purpose and the threat of wind blow has passed
- Best results on couch achieved when the weed is growing actively and commencing new rhizome growth
- Grass weeds germinating after treatment will not be controlled
- Treatment quickly stops growth and visible colour changes to the leaf tips appear after about 7 d. Complete kill takes 3-4 wk under good growing conditions
- Quizalofop-p-tefuryl is an ACCase inhibitor herbicide. To avoid the build up of resistance do not apply products containing an ACCase inhibitor herbicide more than twice to any crop. In addition do not use any product containing quizalofop-p-tefuryl in mixture or sequence with any other product containing the same ingredient
- Use these products as part of a resistance management strategy that includes cultural methods of control and does not use ACCase inhibitors as the sole chemical method of grass weed control
- Applying a second product containing an ACCase inhibitor to a crop will increase the risk of resistance development; only use a second ACCase inhibitor to control different weeds at a different timing
- Always follow WRAG guidelines for preventing and managing herbicide resistant weeds. See Section 5

Restrictions

- Maximum number of treatments 1 per crop
- Consult processors before use on peas or potatoes for processing
- Do not treat crops and weeds growing under stress from any cause

Crop-specific information

- HI 60 d for all crops
- Treat oilseed rape from the fully expanded cotyledon stage
- Treat linseed, peas and field beans from 2-3 unfolded leaves
- Treat sugar beet from 2 unfolded leaves

Following crops guidance

- In the event of failure of a treated crop any broad-leaved crop may be planted at any time. Cereals may be drilled from 4 wk after treatment

Environmental safety

- Dangerous for the environment
- Very toxic to aquatic organisms
- Risk to certain non-target insects or other arthropods. For advice on risk management and use in Integrated Pest Management (IPM) see directions for use
- Avoid spraying within 6 m of the field boundary to reduce effects on non-target insects and other arthropods

Hazard classification and safety precautions

Hazard H04, H11
Risk phrases R41, R43, R50, R53a
Operator protection A, C, H; U02a, U04a, U05a, U10, U11, U14, U15, U19a, U20b
Environmental protection E15a, E22b, E38
Storage and disposal D01, D02, D09a, D10b, D12a

395 rimsulfuron

A selective systemic sulfonylurea herbicide
HRAC mode of action code: B

Products

1	Rigid	AgriGuard	25% w/w	SG	12319
2	Standon Rimsulfuron	Standon	25% w/w	SG	09955
3	Tarot	Makhteshim	25% w/w	SG	11896
4	Titus	Makhteshim	25% w/w	SG	11895

Uses

- Annual dicotyledons in ***forage maize, potatoes***
- Volunteer oilseed rape in ***forage maize, potatoes***

Approval information

- Rimsulfuron included in Annex I under EC Directive 91/414
- Approval expiry 31 Jul 2008 [2]

Efficacy guidance

- Product should be used with a suitable adjuvant or a suitable herbicide tank-mix partner. See label for details
- Product acts by foliar action. Best results obtained from good spray cover of small actively growing weeds. Effectiveness is reduced in very dry conditions
- Split application provides control over a longer period and improves control of fat hen and polygonums
- Weed spectrum can be broadened by tank mixture with other herbicides. See label for details
- Susceptible weeds cease growth immediately and symptoms can be seen 10 d later
- Rimsulfuron is a member of the ALS-inhibitor group of herbicides

Restrictions

- Maximum number of treatments 1 per crop
- Do not treat maize previously treated with organophosphorus insecticides
- Do not apply to potatoes grown for certified seed
- Consult processor before use on crops grown for processing
- Avoid high light intensity (full sunlight) and high temperatures on the day of spraying
- Do not treat during periods of substantial diurnal temperature fluctuation or when frost anticipated
- Do not apply to any crop stressed by drought, water-logging, low temperatures, pest or disease attack, nutrient or lime deficiency
- Do not apply to forage maize treated with organophosphate insecticides
- Do not apply to forage maize undersown with grass or clover

Crop-specific information

- Latest use: before most advanced potato plants are 25 cm high; before 4-collar stage of fodder maize
- All varieties of ware potatoes may be treated, but variety restrictions of any tank-mix partner must be observed
- Only certain named varieties of forage maize may be treated. See label

Following crops guidance

- Only winter wheat should follow a treated crop in the same calendar yr
- Only barley, wheat or maize should be sown in the spring of the yr following treatment
- In the second autumn after treatment any crop except brassicas or oilseed rape may be drilled

Environmental safety

- Dangerous for the environment
- Toxic to aquatic organisms
- Extremely dangerous to fish or other aquatic life. Do not contaminate surface waters or ditches with chemical or used container
- Herbicide is very active. Take particular care to avoid drift onto plants outside the target area
- Spraying equipment should not be drained or flushed onto land planted, or to be planted, with trees or crops other than potatoes or forage maize and should be thoroughly cleansed after use - see label for instructions

Hazard classification and safety precautions

Hazard H11 [2-4]

Risk phrases R51, R53a [2-4]
Operator protection U08, U19a [1-4]; U20a [1]; U20b [2-4]
Environmental protection E13a [1, 3, 4]; E15a [2]; E34 [1-4]; E38 [3, 4]
Storage and disposal D01, D09a, D11a [1-4]; D05 [1]; D12a [3, 4]

396 rotenone

A natural, contact insecticide of low persistence
IRAC mode of action code: 21

Products

Liquid Derris	Law Fertilisers	50 g/l	EC	13370

Uses

- Aphids in ***all bush fruit***, ***all cane fruit***, ***all top fruit***, ***miscellaneous flowers***, ***ornamental plant production***, ***protected crops***, ***vegetables***
- Raspberry beetle in ***all cane fruit***
- Sawflies in ***gooseberries***
- Slug sawfly in ***pears***, ***roses***

Approval information

- Rotenone will not be included in Annex I of EC Directive 91/414. All approvals for the storage and use of products in this profile will expire in about September 2009. The UK has been granted Essential Uses for professional users on apples, pears, peaches, cherries, ornamentals and potatoes

Efficacy guidance

- Apply as high volume spray when pest first seen and repeat as necessary
- Spray raspberries at first pink fruit, loganberries when most of blossom over, blackberries as first blossoms open. Repeat 2 wk later if necessary
- Spray to obtain thorough coverage, especially on undersurfaces of leaves

Crop-specific information

- HI 1 d for all crops

Environmental safety

- Flammable
- Dangerous to fish or other aquatic life. Do not contaminate surface waters or ditches with chemical or used container

Hazard classification and safety precautions

Hazard H08
Operator protection U20a
Environmental protection E13b, E34
Consumer protection C02a (1 d)
Storage and disposal D09a, D10a

397 silthiofam

A thiophene carboxamide fungicide seed dressing for cereals
FRAC mode of action code: 38

Products

Latitude	Monsanto	125 g/l	FS	10695

Uses

- Take-all in ***spring wheat*** *(seed treatment)*, ***winter barley*** *(seed treatment)*, ***winter wheat*** *(seed treatment)*

Approval information

- Silthiofam included in Annex I under EC Directive 91/414
- Accepted by BBPA for use on malting barley

Efficacy guidance

- Apply using approved seed treatment equipment which has been accurately calibrated
- Apply simultaneously (i.e. not in mixture) with a standard seed treatment

- Drill treated seed in the season of purchase. The viability of treated seed and fungicide activity may be reduced by physical storage
- Drill treated seed at 2.5-4 cm into a well prepared firm seedbed
- As precaution against possible development of disease resistance do not treat more than three consecutive susceptible cereal crops in any one rotation

Restrictions

- Maximum number of treatments one per batch of seed
- Do not use on seed with more than 16% moisture content, or on sprouted, cracked or skinned seed
- Test germination of all seed batches before treatment

Crop-specific information

- Latest use: immediately prior to drilling

Hazard classification and safety precautions

Operator protection A, H; U20b
Environmental protection E03, E15a, E34
Storage and disposal D05, D09a, D10a
Treated seed S01, S02, S04b, S05, S06a, S07

398 sodium chlorate

A non-selective inorganic herbicide for total vegetation control

Products

1	Atlacide Soluble Powder Weedkiller	Nomix Enviro	58% w/w	SP	00125
2	Doff Sodium Chlorate Weedkiller	Doff Portland	53% w/w	SP	06049

Uses

- Total vegetation control in ***land not intended to bear vegetation*** [1]; ***non-crop areas, paths and drives*** [2]

Approval information

- Sodium chlorate will not be included in Annex I of EC Directive 91/414 and will not be granted Essential Uses. Sodium chlorate will therefore be withdrawn and plant protection use must cease by 22 December 2008

Efficacy guidance

- Active through foliar and root uptake
- Apply as overall spray at any time during growing season. Best results obtained from application to moist soil in spring or early summer

Restrictions

- Maximum number of treatments 2 per yr for non-crop areas, paths and drives
- Treated ground must not be replanted for at least 6 mth after treatment
- Do not apply before heavy rain

Environmental safety

- Dangerous for the environment
- Contact with combustible material may cause fire
- Harmful to aquatic organisms
- Keep livestock out of treated areas for at least two weeks following treatment and until poisonous weeds, such as ragwort, have died down and become unpalatable
- Clothing, paper, plant debris etc become highly inflammable when dry if contaminated with sodium chlorate
- Fire risk has been reduced by inclusion of fire depressant but product should not be used in areas of exceptionally high fire risk eg oil installations, timber yards
- Wash clothing thoroughly after use
- If clothes become contaminated do not stand near an open fire
- Must not be sold to persons under 18

Hazard classification and safety precautions

Hazard H03, H09 [1, 2]; H11 [1]

Risk phrases R08, R22a [1, 2]; R52 [1]
Operator protection A, H [2]; L [1]; M; U02a [1]; U05a, U20b [1, 2]
Environmental protection E07a [1]; E13c [2]; E15a, E34 [1, 2]
Storage and disposal D01, D02, D09a, D11a [1, 2]; D07 [2]; D12a [1]
Medical advice M03 [1, 2]; M05a [2]

399 sodium chloride (commodity substance)

An inorganic salt for use as a sugar beet herbicide

Products

sodium chloride	various	-	SL

Uses

- Polygonums in ***sugar beet***
- Volunteer potatoes in ***sugar beet***

Approval information

- Sodium chloride will not be included in Annex I of EC Directive 91/414 and will not be granted Essential Uses. Sodium chloride will therefore be withdrawn and plant protection use must cease by 22 December 2008

Efficacy guidance

- Best results obtained from good spray cover treatment in hot, humid weather
- Use as a follow-up treatment where earlier sprays have failed to control large volunteer potatoes or polygonums
- Acts by contact scorch and has no direct effect on daughter potato tubers
- Saturated spray solution should contain 0.1% w/w non-ionic wetter and be applied at 1000 l/ha
- Spray between three true leaf stage of crops and end Jul
- Crop scorch may occur after treatment

400 sodium hypochlorite (commodity substance)

An inorganic horticultural bactericide for use in mushrooms

Products

sodium hypochlorite	various	100% w/v	ZZ

Uses

- Bacterial blotch in ***mushrooms***

Approval information

- Approval for the use of sodium hypochlorite as a commodity substance was granted on 5 December 1996 by Ministers under regulation 5 of the Control of Pesticides Regulations 1986
- Sodium hypochlorite is being supported for review in the fourth stage of the EC Review Programme under Directive 91/414. Whether or not it is included in Annex I, the existing commodity chemical approval will have to be revoked because substances listed in Annex I must be approved for marketing and use under the EC regime. If sodium hypochlorite is included in Annex I, products containing it will need to gain approval in the normal way if they carry label claims for pesticidal activity

Restrictions

- Maximum concentration 315 mg/litre of water
- Mixing and loading must only take place in a ventilated area
- Must only be used by suitably trained and competent operators

Crop-specific information

- HI 1 d

Environmental safety

- Harmful to fish or other aquatic life. Do not contaminate surface waters or ditches with chemical or used container

Hazard classification and safety precautions

Operator protection A, C, H
Environmental protection E13c

401 spinosad

A selective insecticide derived from naturally occurring soil fungi (naturalyte)
IRAC mode of action code: 5

Products

1	Conserve	Fargro	120 g/l	SC	12058
2	Tracer	Landseer	480 g/l	SC	12438

Uses

- Cabbage moth in ***brussels sprouts***, ***cabbages***, ***cauliflowers*** [2]
- Cabbage root fly in ***broccoli*** *(off-label)*, ***brussels sprouts*** *(off-label)*, ***cabbages*** *(off-label)*, ***calabrese*** *(off-label)*, ***cauliflowers*** *(off-label)* [2]
- Cabbage white butterfly in ***brussels sprouts***, ***cabbages***, ***cauliflowers*** [2]
- Caterpillars in ***broccoli*** *(off-label)*, ***calabrese*** *(off-label)*, ***protected aubergines*** *(off-label)*, ***protected peppers*** *(off-label)*, ***protected tomatoes*** *(off-label)* [2]
- Codling moth in ***apples***, ***crab apples*** *(off-label)*, ***pears***, ***quinces*** *(off-label)* [2]
- Diamond-back moth in ***brussels sprouts***, ***cabbages***, ***cauliflowers*** [2]
- Small white butterfly in ***brussels sprouts***, ***cabbages***, ***cauliflowers*** [2]
- Thrips in ***bulb onions***, ***garlic*** *(off-label)*, ***leeks***, ***protected aubergines*** *(off-label)*, ***protected peppers*** *(off-label)*, ***protected tomatoes*** *(off-label)*, ***salad onions***, ***shallots*** *(off-label)* [2]
- Tortrix moths in ***apples***, ***crab apples*** *(off-label)*, ***pears***, ***quinces*** *(off-label)* [2]
- Western flower thrips in ***protected cucumbers***, ***protected ornamentals*** [1]

Specific Off-Label Approvals (SOLAs)

- ***broccoli*** *20062086* [2]
- ***calabrese*** *20062086* [2]
- ***crab apples*** *20063647* [2]
- ***garlic*** *20063646* [2]
- ***protected aubergines*** *20062085* [2]
- ***protected peppers*** *20062085* [2]
- ***protected tomatoes*** *20062085* [2]
- ***quinces*** *20063647* [2]
- ***shallots*** *20063646* [2]

Approval information

- Spinosad included in Annex I under EC Directive 91/414

Efficacy guidance

- Product enters insects by contact from a treated surface or ingestion of treated plant material therefore good spray coverage is essential
- Some plants, for example Fuchsia flowers, can provide effective refuges from spray deposits and control of western flower thrips may be reduced [1]
- Apply to protected crops when western flower thrip nymphs or adults are first seen [1]
- Monitor western flower thrip development carefully to see whether further applications are necessary. A 2 or 3 spray programme at 5-7 d intervals may be needed when conditions favour rapid pest development [1]
- Treat top fruit and field crops when pests are first seen or at very first signs of crop damage [2]
- Ensure a rain-free period of 12 h after treatment before applying irrigation [2]
- To reduce possibility of development of resistance, adopt resistance management measures. See label and Section 5

Restrictions

- Maximum number of treatments 3 per crop for protected cucumbers or 6 per structure per yr for protected ornamentals; 4 per crop for brassicas, onions, leeks; 1 pre-blossom and 3 post blossom for apples, pears
- Apply no more than 3 consecutive sprays. Rotate with another insecticide with a different mode of action or use no further treatment after applying the maximum number of treatments
- Establish whether any incoming plants have been treated and apply no more than 3 consecutive sprays. Maximum number of treatments 6 per structure per yr [1]
- Avoid application in bright sunlight or into open flowers [1]
- Contact processor before use on a crop for processing

Crop-specific information

- HI 3 d for protected cucumbers, brassicas; 7 d for apples, pears, onions, leeks
- Test for tolerance on a small number of ornamentals or cucumbers before large scale treatment
- Some spotting of african violet flowers may occur

Environmental safety

- Dangerous for the environment
- Very toxic (toxic [1]) to aquatic organisms
- Whenever possible use an Integrated Pest Management system. Spinosad presents low risk to beneficial arthropods
- Product has low impact on many insect and mite predators but is harmful to adults of most parasitic wasps. Most beneficials may be introduced to treated plants when spray deposits are dry but an interval of 2 wk should elapse before introduction of parasitic wasps. See label for details
- Treatment may cause temporary reduction in abundance of insect and mite predators if present at application
- Exposure to direct spray is harmful to bees but dry spray deposits are harmless. Treatment of field crops should not be made in the heat of the day when bees are actively foraging
- LERAP Category B and 40 m broadcast air-assisted LERAP [2]

Hazard classification and safety precautions

Hazard H11
Risk phrases R50 [2]; R51 [1]; R53a [1, 2]
Operator protection A, H; U05a [1]; U08, U20b [2]
Environmental protection E15a, E34, E38 [1, 2]; E16a [2]; E17b [2] (40 m)
Storage and disposal D01, D05 [1]; D02, D10b, D12a [1, 2]

402 spiromesifen

A tetronic acid insecticide for protected tomatoes
IRAC mode of action code: 23

Products

Oberon	Certis	240 g/l	SC	11819

Uses

- Red spider mites in ***protected aubergines*** *(off-label - on inert media or by NFT)*, ***protected cayenne peppers*** *(off-label - on inert media or by NFT)*, ***protected cucumbers*** *(off-label - on inert media or by NFT)*, ***protected gherkins*** *(off-label - on inert media or by NFT)*, ***protected peppers*** *(off-label - on inert media or by NFT)*, ***protected strawberries*** *(off-label - on inert media or by NFT)*
- Whitefly in ***inert substrate tomatoes***, ***nft tomatoes***, ***protected aubergines*** *(off-label - on inert media or by NFT)*, ***protected cayenne peppers*** *(off-label - on inert media or by NFT)*, ***protected cucumbers*** *(off-label - on inert media or by NFT)*, ***protected gherkins*** *(off-label - on inert media or by NFT)*, ***protected peppers*** *(off-label - on inert media or by NFT)*, ***protected strawberries*** *(off-label - on inert media or by NFT)*

Specific Off-Label Approvals (SOLAs)

- ***protected aubergines*** *(on inert media or by NFT) 20063645*
- ***protected cayenne peppers*** *(on inert media or by NFT) 20062149*
- ***protected cucumbers*** *(on inert media or by NFT) 20050958*
- ***protected gherkins*** *(on inert media or by NFT) 20050958*
- ***protected peppers*** *(on inert media or by NFT) 20062149*
- ***protected strawberries*** *(on inert media or by NFT) 20050957*

Efficacy guidance

- Apply to run off and ensure that all sides of the crop are covered
- Apply when infestation first observed and repeat 7-10 d later if necessary
- Always follow guidelines for preventing and managing insect resistance. Section 5 for more information

Restrictions

- Maximum number of treatments 2 per cropping cycle
- Do not use on cherry tomatoes or on any outdoor crops
- Product should be used in rotation with compounds with different modes of action

Crop-specific information

- HI 3 d for tomatoes

Environmental safety

- Dangerous for the environment
- Very toxic to aquatic organisms

Hazard classification and safety precautions

Hazard H04, H11
Risk phrases R43, R50, R53a
Operator protection A, H; U05a, U14, U20b
Environmental protection E15a, E34, E38
Storage and disposal D01, D02, D05, D09a, D11a, D12a

403 spiroxamine

A spiroketal amine fungicide for cereals
FRAC mode of action code: 5

See also *prothioconazole + spiroxamine*
prothioconazole + spiroxamine + tebuconazole

Products

Torch Extra	Bayer CropScience	800 g/l	EC	12140

Uses

- Brown rust in ***spring barley***, ***spring rye***, ***spring wheat***, ***winter barley***, ***winter rye***, ***winter wheat***
- Powdery mildew in ***spring barley***, ***spring rye***, ***spring wheat***, ***winter barley***, ***winter rye***, ***winter wheat***
- Rhynchosporium in ***spring barley*** *(reduction)*, ***spring rye*** *(reduction)*, ***winter barley*** *(reduction)*, ***winter rye*** *(reduction)*
- Yellow rust in ***spring barley***, ***spring rye***, ***spring wheat***, ***winter barley***, ***winter rye***, ***winter wheat***

Approval information

- Spiroxamine included in Annex I under EC Directive 91/414
- Accepted by BBPA for use on malting barley

Efficacy guidance

- For best results treat at an early stage of disease development before infection spreads to new growth
- To reduce the risk of development of resistance avoid repeat treatments on diseases such as powdery mildew. If necessary tank mix or alternate with other non-morpholine fungicides

Restrictions

- Maximum total dose equivalent to two full dose treatments

Crop-specific information

- Latest use: before caryopsis watery ripe (GS 71) for wheat, rye; ear emergence complete (GS 59) for barley

Environmental safety

- Dangerous for the environment
- Very toxic to aquatic organisms
- LERAP Category B

Hazard classification and safety precautions

Hazard H03, H11
Risk phrases R20, R21, R22a, R41, R43, R50, R53a
Operator protection A, C, H; U05a, U11, U13, U14, U15, U20a
Environmental protection E15a, E16a, E34, E38
Storage and disposal D01, D02, D09a, D10b, D12a
Medical advice M03

SECTION 2

404 spiroxamine + tebuconazole

A broad spectrum systemic fungicide mixture for cereals
FRAC mode of action code: 5 + 3

Products

Sage	Bayer CropScience	250:133 g/l	EW	11303

Uses

- Brown rust in ***spring barley, spring rye, spring wheat, winter barley, winter rye, winter wheat***
- Foliar disease control in ***grass seed crops*** *(off-label)*
- Fusarium ear blight in ***spring wheat, winter wheat***
- Glume blotch in ***spring wheat, winter wheat***
- Net blotch in ***spring barley, winter barley***
- Powdery mildew in ***spring barley, spring rye, spring wheat, winter barley, winter rye, winter wheat***
- Rhynchosporium in ***spring barley, winter barley***
- Septoria leaf blotch in ***spring wheat, winter wheat***
- Sooty moulds in ***spring wheat, winter wheat***
- Yellow rust in ***spring barley, spring rye, spring wheat, winter barley, winter rye, winter wheat***

Specific Off-Label Approvals (SOLAs)

- ***grass seed crops*** *20060983*

Approval information

- Spiroxamine included in Annex I under EC Directive 91/414
- Accepted by BBPA for use on malting barley

Efficacy guidance

- Best results achieved from treatment at early stage of disease development before infection spreads to new growth
- To protect flag leaf and ear from Septoria apply from flag leaf emergence to ear fully emerged (GS 37-59). Earlier treatment may be necessary where there is high disease risk
- Control of rusts, powdery mildew, leaf blotch and net blotch may require second treatment 2-3 wk later
- Tebuconazole is a DMI fungicide. Resistance to some DMI fungicides has been identified in Septoria leaf blotch which may seriously affect performance of some products. For further advice contact a specialist advisor and visit the Fungicide Resistance Action Group (FRAG)-UK website

Restrictions

- Maximum total dose equivalent to two full dose treatments
- Do not use on durum wheat

Crop-specific information

- Latest use: before caryopsis watery ripe (GS 71) for rye, wheat; ear emergence complete (GS 59) for barley
- Some transient leaf speckling may occur on wheat but this has not been shown to reduce yield reponse or disease control

Environmental safety

- Dangerous for the environment
- Very toxic to aquatic organisms
- Dangerous to fish or other aquatic life. Do not contaminate surface waters or ditches with chemical or used container
- LERAP Category B

Hazard classification and safety precautions

Hazard H03, H11
Risk phrases R20, R22a, R38, R41, R43, R50, R53a
Operator protection A, C, H; U05a, U11, U13, U14, U15, U20b
Environmental protection E13b, E16a, E16b, E34, E38
Storage and disposal D01, D02, D05, D09a, D10b, D12a
Medical advice M03

SEE SECTION 3 FOR PRODUCTS ALSO REGISTERED

405 strychnine hydrochloride (commodity substance)

A vertebrate control agent for destruction of moles underground

Products

strychnine	various	-	ZZ

Uses

- Moles in ***miscellaneous pest control situations*** *(areas of restricted public access)*, ***permanent grassland*** *(areas of restricted public access)*

Approval information

- Approval for the use of strychnine hydrochloride as a commodity substance was granted on 19 June 1997 by Ministers under regulation 5 of the Control of Pesticides Regulations 1986
- Strychnine has not been supported in the EC Review Programme (see Section 5) and use for mole control ceased to be authorised from September 2006. It has not been permitted as a plant protection product (e.g. to prevent crop damage) since 1 January 2005

Efficacy guidance

- Only for use as poison bait against moles on commercial agricultural/horticultural land where public access restricted, on grassland associated with aircraft landing strips, horse paddocks, race and golf courses and other areas specifically approved by Defra in England and the appropriate authorities in Wales and Scotland

Restrictions

- Strychnine is subject to the Poisons Rules 1982 and the Poisons Act 1972. See notes in Section 5
- Must only be supplied to holders of a permit issued by Defra in England or the appropriate authorities in Wales and Scotland. Permits may only be issued to persons who satisfy the appropriate authority that they are trained and competent in its use
- Only to be supplied in original sealed pack in units up to 2 g. Quantities of more than 8 g may be supplied only to providers of a commercial service
- Other restrictions apply, see the PSD website at: www.pesticides.gov.uk/approvals.asp?id=310

Environmental safety

- Store in original container under lock and key and only on premises under control of a permit holder or of a named individual who satisfies the designated requirements of competence
- A written COSHH assessment must be made before use
- Must be prepared for application with great care so that there is no contamination of the ground surface
- Any prepared bait remaining at the end of the day must be buried
- Operators must be provided with a Chemicals (Hazard Information and Supply) (CHIP) Safety Data Sheet, obtainable from the supplier, before commencing work

Hazard classification and safety precautions

Operator protection A

406 sulfosulfuron

A sulfonylurea herbicide for grass and broad-leaved weed control in winter wheat
HRAC mode of action code: B

Products

Safeguard	AgriGuard	80% w/w	WG	12020

Uses

- Brome grasses in ***winter wheat***
- Chickweed in ***winter wheat***
- Cleavers in ***winter wheat***
- Loose silky bent in ***winter wheat***
- Mayweeds in ***winter wheat***
- Onion couch in ***winter wheat***

Approval information

- Sulfosulfuron included in Annex I under EC Directive 91/414
- Approval expiry 31 Aug 2008 [1]

Efficacy guidance

- For best results treat in early spring when annual weeds are small and growing actively. Avoid treatment when weeds are dormant for any reason
- Best control of onion couch is achieved when the weed has more than two leaves. Effects on bulbils or on growth in the following yr have not been examined
- An extended period of dry weather before or after treatment may result in reduced control
- The addition of a recommended surfactant is essential for full activity
- Specific follow-up treatments may be needed for complete control of some weeds
- Use only where competitively damaging weed populations have emerged otherwise yield may be reduced
- Sulfosulfuron is a member of the ALS-inhibitor group of herbicides. To avoid the build up of resistance do not use any product containing an ALS-inhibitor herbicide with claims for control of grass weeds more than once on any crop
- Use these products as part of a resistance management strategy that includes cultural methods of control and does not use ALS inhibitors as the sole chemical method of grass weed control

Restrictions

- Maximum total dose equivalent to one treatment at full dose
- Do not treat crops under stress
- Apply only after 1 Feb and from 3 expanded leaf stage
- Do not treat durum wheat or any undersown wheat crop
- Do not use in mixture or in sequence with any other sulfonyl urea herbicide on the same crop

Crop-specific information

- Latest use: flag leaf ligule just visible for winter wheat (GS 39)

Following crops guidance

- In the autumn following treatment winter wheat, winter rye, winter oats, triticale, winter oilseed rape, winter peas or winter field beans may be sown on any soil, and winter barley on soils with less than 60% sand
- In the next spring following application crops of wheat, barley, oats, maize, peas, beans, linseed, oilseed rape, potatoes or grass may be sown
- In the second autumn following application winter linseed may be sown, and winter barley on soils with more than 60% sand
- Sugar beet or any other crop not mentioned above must not be drilled until the second spring following application
- Where winter oilseed rape is to be sown in the autumn following treatment soil cultivation to a minimum of 10 cm is recommended

Environmental safety

- Dangerous for the environment
- Very toxic to aquatic organisms
- Extremely dangerous to fish or other aquatic life. Do not contaminate surface waters or ditches with chemical or used container
- LERAP Category B
- Take extreme care to avoid drift onto broad leaved plants or other crops, or onto ponds, waterways or ditches.
- Follow detailed label instructions for cleaning the sprayer to avoid damage to sensitive crops during subsequent use

Hazard classification and safety precautions

Hazard H11

Risk phrases R50, R53a

Operator protection U20b

Environmental protection E13a, E16a, E16b, E34

Storage and disposal D02, D09a, D10b

SEE SECTION 3 FOR PRODUCTS ALSO REGISTERED

407 sulphur

A broad-spectrum inorganic protectant fungicide, foliar feed and acaricide
FRAC mode of action code: M2

Products

1	Cosavet DF	Headland	80% w/w	WG	11477
2	Headland Sulphur	Headland	800 g/l	SC	12879
3	Kumulus DF	BASF	80% w/w	SG	04707
4	Luxan Micro-Sulphur	Luxan	80% w/w	WG	06565
5	Solfa WG	Nufarm UK	80% w/w	WG	11602
6	Sulphur Flowable	United Phosphorus	800 g/l	SC	07526
7	Thiovit Jet	Syngenta	80% w/w	WG	10928
8	Venus	Headland	80% w/w	WG	11856

Uses

- American gooseberry mildew in ***gooseberries*** [5]
- Foliar feed in ***permanent grassland*** [4, 5, 7]; ***spring barley***, ***spring wheat***, ***winter barley***, ***winter wheat*** [1, 3, 4, 7, 8]; ***spring oilseed rape*** [1, 4, 5, 7, 8]; ***sugar beet***, ***swedes*** [4]; ***winter oilseed rape*** [1, 3-5, 7, 8]
- Gall mite in ***blackcurrants*** [1, 3, 5, 8]
- Powdery mildew in ***apples***, ***strawberries*** [1-3, 5, 6, 8]; ***aubergines*** *(off-label)*, ***borage for oilseed production*** *(off-label)*, ***parsnips*** *(off-label)*, ***protected chives*** *(off-label)*, ***protected cucumbers*** *(off-label)*, ***protected herbs (see appendix 6)*** *(off-label)*, ***protected parsley*** *(off-label)*, ***protected peppers*** *(off-label)*, ***protected tomatoes*** *(off-label)* [7]; ***fodder beet*** *(off-label)*, ***turnips*** *(off-label)* [2, 7]; ***gooseberries*** [1, 3, 8]; ***grass seed crops*** *(off-label)*, ***spring oilseed rape***, ***spring rye***, ***triticale*** *(off-label)*, ***winter oilseed rape***, ***winter rye*** [2]; ***hops*** [1-3, 5-8]; ***pears*** [2, 6]; ***spring barley***, ***spring wheat***, ***winter barley***, ***winter wheat*** [2, 4-6]; ***spring oats***, ***winter oats*** [2, 5]; ***sugar beet*** [1-8]; ***swedes*** [2, 5-7]; ***turnips*** [5]; ***wine grapes*** [1, 3, 5, 8]
- Scab in ***apples*** [1-3, 6, 8]; ***pears*** [2, 6]; ***quinces*** *(off-label)* [2]

Specific Off-Label Approvals (SOLAs)

- ***aubergines*** *20023652* [7]
- ***borage for oilseed production*** *20061719* [7]
- ***fodder beet*** *20062690* [2], *20061383* [7]
- ***grass seed crops*** *20062691* [2]
- ***parsnips*** *20023654* [7]
- ***protected chives*** *20023652* [7]
- ***protected cucumbers*** *20023652* [7]
- ***protected herbs (see appendix 6)*** *20023652* [7]
- ***protected parsley*** *20023652* [7]
- ***protected peppers*** *20023652* [7]
- ***protected tomatoes*** *20023652* [7]
- ***quinces*** *20062688* [2]
- ***triticale*** *20062691* [2]
- ***turnips*** *20062689* [2], *20061384* [7]

Approval information

- Accepted by BBPA for use on malting barley and hops (before burr)

Efficacy guidance

- Apply when disease first appears and repeat 2-3 wk later. Details of application rates and timing vary with crop, disease and product. See label for information
- Sulphur acts as foliar feed as well as fungicide and with some crops product labels vary in whether treatment recommended for disease control or growth promotion
- In grassland best results obtained at least 2 wk before cutting for hay or silage, 3 wk before grazing
- Treatment unlikely to be effective if disease already established in crop

Restrictions

- Maximum number of treatments normally 2 per crop for grassland, sugar beet, parsnips, swedes, hops, protected herbs; 3 per yr for blackcurrants, gooseberries; 4 per crop on apples, pears but labels vary

- Do not use on sulphur-shy apples (Beauty of Bath, Belle de Boskoop, Cox's Orange Pippin, Lanes Prince Albert, Lord Derby, Newton Wonder, Rival, Stirling Castle) or pears (Doyenne du Comice)
- Do not use on gooseberry cultivars Careless, Early Sulphur, Golden Drop, Leveller, Lord Derby, Roaring Lion, or Yellow Rough
- Do not use on apples or gooseberries when young, under stress or if frost imminent
- Do not use on fruit for processing, on grapevines during flowering or near harvest on grapes for wine-making
- Do not use on hops at or after burr stage
- Do not spray top or soft fruit with oil or within 30 d of an oil-containing spray

Crop-specific information

- Latest use: before burr stage for hops; before end Sep for parsnips, swedes, sugar beet; fruit swell for gooseberries; milky ripe stage for cereals
- HI cutting grass for hay or silage 2 wk; grazing grassland 3 wk

Environmental safety

- Sulphur products are attractive to livestock and must be kept out of their reach
- Do not empty into drains

Hazard classification and safety precautions

Hazard H04 [5]
Risk phrases R37 [5]
Operator protection A, H [7]; U02a, U08, U14 [1, 7, 8]; U05a [5, 7]; U19a [5]; U20a [2]; U20b [6]; U20c [1, 3-5, 7, 8]
Environmental protection E15a [1-8]; E19b [5]; E34 [2, 4]
Storage and disposal D01 [4, 5, 7]; D02 [5, 7]; D05 [6, 7]; D09a [1-8]; D10b [2]; D10c [6]; D11a [1, 3, 4, 7, 8]

408 sulphuric acid (commodity substance)

A strong acid used as an agricultural desiccant

Products

sulphuric acid	various	77% w/w	SL

Uses

- Haulm destruction in ***potatoes***
- Pre-harvest desiccation in ***bulbs/corms***, ***peas***

Approval information

- Approval for the use of sulphuric acid as a commodity substance was granted on 23 November 1995 by Ministers under regulation 5 of the Control of Pesticides Regulations 1986
- Sulphuric acid is being supported for review in the fourth stage of the EC Review Programme under Directive 91/414. Whether or not it is included in Annex I, the existing commodity chemical approval will have to be revoked because substances listed in Annex I must be approved for marketing and use under the EC regime. If sulphuric acid is included in Annex I, products containing it will need to gain approval in the normal way if they carry label claims for pesticidal activity

Efficacy guidance

- Apply with suitable equipment between 1 Mar and 15 Nov

Restrictions

- Sulphuric acid is subject to the Poisons Rules 1982 and the Poisons Act 1972. See notes in Section 5
- Must only be used by suitably trained operators competent in use of equipment for applying sulphuric acid
- Not to be applied using hand-held or pedestrian controlled applicators
- A written COSHH assessment must be made before use. Operators should observe OES set out in HSE guidance note EH40/90 or subsequent issues
- Operators must have liquid suitable for eye irrigation immediately available at all times throughout spraying operation

- Maximum number of treatments 3 per crop for potatoes; 1 per crop for peas; 1 per yr for bulbs and corms
- Only 'sulphur burnt' sulphuric acid to be used

Crop-specific information

- Latest use: 15 Nov for bulbs, corms, peas, potatoes

Environmental safety

- Spray must not be deposited within 1 m of public footpaths
- Written notice of any intended spraying must be given to owners of neighbouring land and readable warning notices posted beforehand and left in place for 96 h afterwards
- Unprotected persons must be kept out of treated areas for at least 96 h after treatment
- Do not apply to crops in which bees are actively foraging. Do not apply when flowering weeds are present

Hazard classification and safety precautions

Hazard H05
Risk phrases R34
Operator protection B, C, D, H, K, M

409 tau-fluvalinate

A contact pyrethroid insecticide for cereals and oilseed rape
IRAC mode of action code: 3

Products

1	Greencrop Malin	Greencrop	240 g/l	EW	11787
2	Klartan	Makhteshim	240 g/l	EW	11074
3	Mavrik	Makhteshim	240 g/l	EW	10612

Uses

- Aphids in ***borage for oilseed production*** *(off-label)* [3]; ***durum wheat*** *(off-label)*, ***evening primrose*** *(off-label)*, ***grass seed crops*** *(off-label)*, ***honesty*** *(off-label)*, ***mustard*** *(off-label)*, ***spring linseed*** *(off-label)*, ***spring rye*** *(off-label)*, ***triticale*** *(off-label)*, ***winter rye*** *(off-label)* [2, 3]; ***spring barley***, ***spring oilseed rape***, ***spring wheat***, ***winter barley***, ***winter oilseed rape***, ***winter wheat*** [1-3]
- Barley yellow dwarf virus vectors in ***winter barley***, ***winter wheat*** [1-3]
- Cabbage stem flea beetle in ***winter oilseed rape*** [2, 3]
- Pollen beetle in ***spring oilseed rape***, ***winter oilseed rape*** [1-3]

Specific Off-Label Approvals (SOLAs)

- ***borage for oilseed production*** *20061366* [3]
- ***durum wheat*** *20061369* [2], *20061365* [3]
- ***evening primrose*** *20061370* [2], *20061366* [3]
- ***grass seed crops*** *20061369* [2], *20061365* [3]
- ***honesty*** *20061370* [2], *20061366* [3]
- ***mustard*** *20061370* [2], *20061366* [3]
- ***spring linseed*** *20061370* [2], *20061366* [3]
- ***spring rye*** *20061369* [2], *20061365* [3]
- ***triticale*** *20061369* [2], *20061365* [3]
- ***winter rye*** *20061369* [2], *20061365* [3]

Approval information

- Accepted by BBPA for use on malting barley

Efficacy guidance

- For BYDV control on winter cereals follow local warnings or spray high risk crops in mid-Oct and make repeat application in late autumn/early winter if aphid activity persists
- For summer aphid control on cereals spray once when aphids present on two thirds of ears and increasing
- On oilseed rape treat peach potato aphids in autumn in response to local warning and repeat if necessary
- Best control of pollen beetle in oilseed rape obtained from treatment at green to yellow bud stage and repeat if necessary
- Good spray cover of target essential for best results

Restrictions

- Maximum total dose equivalent to two full dose treatments on oilseed rape. See label for dose rates on cereals
- A minimum of 14 d must elapse between applications to cereals

Crop-specific information

- Latest use: before caryopsis watery ripe (GS 71) for barley; before flowering for oilseed rape; before kernel medium milk (GS 75) for wheat

Environmental safety

- Dangerous for the environment
- Very toxic to aquatic organisms
- High risk to non-target insects or other arthropods. Do not spray within 6 m of the field boundary [1-3]
- LERAP Category A
- Avoid spraying oilseed rape within 6 m of field boundary to reduce effects on certain non-target species or other arthropods
- Must not be applied to cereals if any product containing a pyrethroid insecticide or dimethoate has been sprayed after the start of ear emergence (GS 51)

Hazard classification and safety precautions

Hazard H04 [1]; H11 [1-3]
Risk phrases R36, R38, R51 [1]; R50 [2, 3]; R53a [1-3]
Operator protection A, C, H; U05a, U10, U11, U19a, U20a
Environmental protection E15a, E16c, E16d, E22a
Storage and disposal D01, D02, D05, D10c [1-3]; D09a [2, 3]; D09b [1]

410 tebuconazole

A systemic triazole fungicide for cereals and other field crops
FRAC mode of action code: 3

See also *clothianidin + prothioconazole + tebuconazole + triazoxide*
prochloraz + tebuconazole
prothioconazole + spiroxamine + tebuconazole
prothioconazole + tebuconazole
prothioconazole + tebuconazole + triazoxide
spiroxamine + tebuconazole

Products

1	Alpha Tebuconazole 20 EW	Makhteshim	200 g/l	EW	12893
2	Bezel	Bayer CropScience	2% w/w	PA	12753
3	Clayton Tebucon	Clayton	250 g/l	EW	08707
4	Folicur	Bayer CropScience	250 g/l	EW	11278
5	Greencrop Tabloid	Greencrop	250 g/l	EW	11969
6	Grisou	Sipcam	250 g/l	EC	13166
7	Icon	AgriGuard	250 g/l	EW	12202
8	Mitre	Makhteshim	200 g/l	EW	12901
9	Mystique	Nufarm UK	250 g/l	EC	12348
10	Orius	Makhteshim	250 g/l	EW	12105
11	Orius 20 EW	Makhteshim	200 g/l	EW	12311
12	Riza	Headland	250 g/l	EW	12696
13	Standon Tebuconazole	Standon	250 g/l	EC	09056

Uses

- Alternaria in ***brussels sprouts*** [11]; ***cabbages*** [1, 4-12]; ***carrots***, ***horseradish*** [1, 4, 5, 7, 10, 12]; ***parsnips*** [12]; ***spring oilseed rape***, ***winter oilseed rape*** [1, 3-13]
- Botrytis in ***daffodils*** *(off-label - for galanthamine production)* [1, 4, 8, 10]; ***linseed*** *(reduction)* [1, 3-5, 7, 10, 13]
- Brown rust in ***spring barley***, ***spring wheat***, ***winter barley***, ***winter wheat*** [1, 3-13]; ***spring rye***, ***winter rye*** [1, 3-5, 7, 8, 10-13]
- Cane blight in ***blackberries*** *(off-label)*, ***raspberries*** *(off-label)*, ***rubus hybrids*** *(off-label)* [1, 4, 8, 12]

- Canker in ***all pome fruit***, ***all stone fruit***, ***all tree nuts***, ***farm forestry***, ***forest***, ***forest nurseries***, ***ornamental plant production*** [2]; ***apples*** *(off-label)*, ***chestnuts*** *(off-label)*, ***cob nuts*** *(off-label)*, ***crab apples*** *(off-label)*, ***pears*** *(off-label)*, ***quinces*** *(off-label)*, ***walnuts*** *(off-label)* [1, 4, 8, 12]
- Chocolate spot in ***spring field beans***, ***winter field beans*** [1, 3-13]
- Crown rust in ***spring oats***, ***winter oats*** [9, 12]; ***spring oats*** *(reduction)*, ***winter oats*** *(reduction)* [4, 5, 7, 10]
- Foliar disease control in ***borage*** *(off-label)* [1, 3, 4, 8, 13]; ***borage for oilseed production*** *(off-label)*, ***horseradish*** *(off-label)* [12]; ***canary flower (echium spp.)*** *(off-label)*, ***evening primrose*** *(off-label)*, ***honesty*** *(off-label)*, ***mustard*** *(off-label)* [1, 3, 4, 8, 12, 13]; ***grass seed crops*** *(off-label)* [1, 3, 8, 13]; ***mallow (althaea spp.)*** *(off-label)*, ***parsley root*** *(off-label)* [1, 4, 8, 12]; ***triticale*** *(off-label)* [4, 8, 12]
- Fungus diseases in ***kohlrabi*** *(off-label)* [1, 4, 8, 10]
- Fusarium ear blight in ***spring wheat***, ***winter wheat*** [1, 4, 5, 7-12]
- Glume blotch in ***spring wheat***, ***winter wheat*** [1, 3-13]
- Light leaf spot in ***brussels sprouts*** [11]; ***cabbages*** [1, 3-5, 7-13]; ***spring oilseed rape***, ***winter oilseed rape*** [1, 3-13]
- Lodging control in ***spring oilseed rape***, ***winter oilseed rape*** [1, 4, 5, 7, 8, 10, 11]
- Net blotch in ***spring barley***, ***winter barley*** [1, 3-13]
- Phoma in ***spring oilseed rape***, ***winter oilseed rape*** [1, 4, 5, 7-12]
- Powdery mildew in ***brussels sprouts*** [11]; ***cabbages***, ***spring barley***, ***spring wheat***, ***swedes***, ***turnips***, ***winter barley***, ***winter wheat*** [1, 3-13]; ***carrots***, ***linseed*** [1, 4, 5, 7, 10, 12]; ***chives*** *(off-label)*, ***herbs (see appendix 6)*** *(off-label)*, ***parsley*** *(off-label)* [1, 4, 8, 10, 12]; ***horseradish*** [12]; ***parsnips*** [1, 4, 10, 12]; ***spring oats***, ***winter oats*** [4-7, 9, 10, 12, 13]; ***spring rye***, ***winter rye*** [1, 3-5, 7, 8, 10-13]
- Pruning wounds in ***all pome fruit***, ***all stone fruit***, ***all tree nuts***, ***farm forestry***, ***forest***, ***forest nurseries***, ***ornamental plant production*** [2]
- Rhynchosporium in ***spring barley***, ***winter barley*** [1, 3-13]; ***spring rye***, ***winter rye*** [1, 3-5, 7, 8, 10-13]
- Ring spot in ***broccoli*** *(off-label)*, ***calabrese*** *(off-label)*, ***cauliflowers*** *(off-label)*, ***chinese cabbage*** *(off-label)*, ***collards*** *(off-label)*, ***kale*** *(off-label)*, ***salad brassicas*** *(off-label - for baby leaf production)* [1, 4, 8, 10, 12]; ***brussels sprouts*** [11]; ***cabbages*** [1, 3-13]; ***choi sum*** *(off-label)*, ***kohlrabi*** *(off-label)*, ***pak choi*** *(off-label)* [12]; ***spring oilseed rape***, ***winter oilseed rape*** [9]; ***spring oilseed rape*** *(reduction)*, ***winter oilseed rape*** *(reduction)* [1, 4, 5, 7, 8, 10, 11]
- Rust in ***broad beans*** *(off-label)*, ***chives*** *(off-label)*, ***dwarf beans*** *(off-label)*, ***french beans*** *(off-label)*, ***herbs (see appendix 6)*** *(off-label)*, ***parsley*** *(off-label)*, ***runner beans*** *(off-label)* [1, 4, 8, 10, 12]; ***leeks*** [1, 4, 5, 7, 8, 10-12]; ***spring field beans***, ***winter field beans*** [1, 3-13]
- Sclerotinia in ***carrots*** [1, 4, 5, 7, 10, 12]; ***horseradish***, ***parsnips*** [12]
- Sclerotinia stem rot in ***spring oilseed rape***, ***winter oilseed rape*** [1, 3-13]
- Septoria leaf blotch in ***spring wheat***, ***winter wheat*** [1, 3-13]
- Sooty moulds in ***spring wheat***, ***winter wheat*** [1, 4, 5, 7, 8, 10-12]
- Stem canker in ***spring oilseed rape***, ***winter oilseed rape*** [1, 4, 5, 7, 8, 10-12]
- White rot in ***bulb onions*** *(off-label)*, ***garlic*** *(off-label)*, ***salad onions*** *(off-label)*, ***shallots*** *(off-label)* [1, 4, 8, 10, 12]; ***onion sets*** *(off-label)* [12]
- Yellow rust in ***spring barley***, ***spring wheat***, ***winter barley***, ***winter wheat*** [1, 3-13]; ***spring rye***, ***winter rye*** [1, 3-5, 7, 8, 10-13]

Specific Off-Label Approvals (SOLAs)

- ***apples*** *20071380* [1], *20051734* [4], *20071368* [8], *20070554* [12]
- ***blackberries*** *20071376* [1], *20050897* [4], *20071358* [8], *20070543* [12]
- ***borage*** *20071389* [1], *20061371* [3], *20062694* [4], *20071359* [8], *20061399* [13]
- ***borage for oilseed production*** *20070558* [12]
- ***broad beans*** *20071390* [1], *20031878* [4], *20071364* [8], *20051393* [10], *20070549* [12]
- ***broccoli*** *20071375* [1], *20031874* [4], *20071371* [8], *20051397* [10], *20070546* [12]
- ***bulb onions*** *20071378* [1], *20071384* [1], *20031877* [4], *20031879* [4], *20071365* [8], *20071369* [8], *20051394* [10], *20051403* [10], *20070550* [12]
- ***calabrese*** *20071375* [1], *20031874* [4], *20071371* [8], *20051397* [10], *20070546* [12]
- ***canary flower (echium spp.)*** *20071389* [1], *20061371* [3], *20062694* [4], *20071359* [8], *20070558* [12], *20061399* [13]
- ***cauliflowers*** *20071375* [1], *20031874* [4], *20071371* [8], *20051397* [10], *20070546* [12]
- ***chestnuts*** *20071393* [1], *20051735* [4], *20071372* [8], *20070555* [12]

- ***chinese cabbage*** *20071375* [1], *20031874* [4], *20071371* [8], *20051397* [10], *20070546* [12]
- ***chives*** *20071383* [1], *20031873* [4], *20071374* [8], *20051396* [10], *20070544* [12]
- ***choi sum*** *20070545* [12]
- ***cob nuts*** *20071393* [1], *20051735* [4], *20071372* [8], *20070555* [12]
- ***collards*** *20071381* [1], *20031872* [4], *20071367* [8], *20051395* [10], *20070545* [12]
- ***crab apples*** *20071380* [1], *20051734* [4], *20071368* [8], *20070554* [12]
- ***daffodils*** *(for galanthamine production) 20071379* [1], *(for galanthamine production) 20041516* [4], *(for galanthamine production) 20071370* [8], *(for galanthamine production) 20051402* [10]
- ***dwarf beans*** *20071385* [1], *20031880* [4], *20071357* [8], *20051401* [10], *20070551* [12]
- ***evening primrose*** *20071389* [1], *20061371* [3], *20062694* [4], *20071359* [8], *20070558* [12], *20061399* [13]
- ***french beans*** *20071385* [1], *20031880* [4], *20071357* [8], *20051401* [10], *20070551* [12]
- ***garlic*** *20071378* [1], *20031879* [4], *20071369* [8], *20051394* [10], *20070550* [12]
- ***grass seed crops*** *20071381* [1], *20061372* [3], *20071362* [8], *20061400* [13]
- ***herbs (see appendix 6)*** *20071383* [1], *20031873* [4], *20071374* [8], *20051396* [10], *20070544* [12]
- ***honesty*** *20071389* [1], *20061371* [3], *20062694* [4], *20071359* [8], *20070558* [12], *20061399* [13]
- ***horseradish*** *20070556* [12]
- ***kale*** *20071381* [1], *20031872* [4], *20071365* [8], *20051395* [10], *20070545* [12]
- ***kohlrabi*** *20071377* [1], *20031881* [4], *20071360* [8], *20051400* [10], *20070552* [12]
- ***mallow (althaea spp.)*** *20071391* [1], *20062692* [4], *20071363* [8], *20070556* [12]
- ***mustard*** *20071389* [1], *20061371* [3], *20062694* [4], *20071359* [8], *20070558* [12], *20061399* [13]
- ***onion sets*** *20070548* [12]
- ***pak choi*** *20070545* [12]
- ***parsley*** *20071383* [1], *20031873* [4], *20071374* [8], *20051396* [10], *20070544* [12]
- ***parsley root*** *20071391* [1], *20062692* [4], *20071363* [8], *20070556* [12]
- ***pears*** *20071380* [1], *20051734* [4], *20071368* [8], *20070554* [12]
- ***quinces*** *20071380* [1], *20051734* [4], *20071368* [8], *20070554* [12]
- ***raspberries*** *20071376* [1], *20050897* [4], *20071358* [8], *20070543* [12]
- ***rubus hybrids*** *20071376* [1], *20050897* [4], *20071358* [8], *20070543* [12]
- ***runner beans*** *20071385* [1], *20031880* [4], *20071357* [8], *20051401* [10], *20070551* [12]
- ***salad brassicas*** *(for baby leaf production) 20071382* [1], *(for baby leaf production) 20051398* [4], *(for baby leaf production) 20071373* [8], *(for baby leaf production) 20051398* [10], *(for baby leaf production) 20070547* [12]
- ***salad onions*** *20071387* [1], *20042408* [4], *20071361* [8], *20051399* [10], *20070553* [12]
- ***shallots*** *20071378* [1], *20031879* [4], *20071369* [8], *20051394* [10], *20070550* [12]
- ***triticale*** *20062693* [4], *20071366* [8], *20070557* [12]
- ***walnuts*** *20071393* [1], *20051735* [4], *20071372* [8], *20070555* [12]

Approval information

- Accepted by BBPA for use on malting barley

Efficacy guidance

- For best results apply at an early stage of disease development before infection spreads to new crop growth
- To protect flag leaf and ear from Septoria diseases apply from flag leaf emergence to ear fully emerged (GS 37-59). Earlier application may be necessary where there is a high risk of infection
- Improved control of established cereal mildew can be obtained by tank mixture with fenpropimorph
- For light leaf spot control in oilseed rape apply in autumn/winter with a follow-up spray in spring/ summer if required
- For control of most other diseases spray at first signs of infection with a follow-up spray 2-4 wk later if necessary. See label for details
- For disease control in cabbages a 3-spray programme at 21-28 d intervals will give good control
- Tebuconazole is a DMI fungicide. Resistance to some DMI fungicides has been identified in Septoria leaf blotch which may seriously affect performance of some products. For further advice contact a specialist advisor and visit the Fungicide Resistance Action Group (FRAG)-UK website

Restrictions

- Maximum number of treatments 1 per wound for all pome fruit, all stone fruit, all tree nuts, forestry, ornamental plant production [2]
- Maximum total dose equivalent to 1 full dose treatment on linseed; 2 full dose treatments on wheat, barley, rye, oats, field beans, swedes, turnips, onions; 2.25 full dose treatments on

cabbages; 2.5 full dose treatments on oilseed rape; 3 full dose treatments on leeks, parsnips, carrots
- Do not treat durum wheat
- Apply only to listed oat varieties (see label) and do not apply to oats in tank mixture
- Do not apply before swedes and turnips have a root diameter of 2.5 cm, or before heart formation in cabbages, or before button formation in Brussels sprouts
- Consult processor before use on crops for processing

Crop-specific information
- Latest use: before grain milky-ripe for cereals (GS 71); when most seed green-brown mottled for oilseed rape (GS 6,3); before brown capsule for linseed
- HI field beans, swedes, turnips, linseed, winter oats 35 d; market brassicas, carrots, horseradish, parsnips 21 d; leeks 14 d
- Some transient leaf speckling on wheat or leaf reddening/scorch on oats may occur but this has not been shown to reduce yield response to disease control

Environmental safety
- Dangerous for the environment
- Toxic to aquatic organisms

Hazard classification and safety precautions

Hazard H03, H11 [3-10, 12, 13]; H04 [1, 11]
Risk phrases R20 [3, 4, 8, 10]; R22a [3-5, 7, 8, 10, 12, 13]; R36, R52 [1, 11]; R38 [1, 5-7, 9, 11-13]; R41, R51 [3-10, 12, 13]; R53a [1, 3-13]; R63 [3, 6, 9]
Operator protection A, H [1-13]; C [1, 3-13]; U05a, U11, U20a [1, 3-13]; U14, U15 [1, 11]; U19a [3, 5, 7, 12, 13]; U20b [2]
Environmental protection E15a [1, 3-13]; E34 [1, 4-12]; E38 [4, 6, 8-10]
Storage and disposal D01, D02 [1, 3-13]; D05 [2-10, 12, 13]; D09a [1-13]; D10b [3-10, 12, 13]; D11a [2]; D12a [4, 6, 8-10]
Medical advice M03 [3-10, 12, 13]

411 tebuconazole + triadimenol

A broad spectrum systemic fungicide for cereals
FRAC mode of action code: 3 + 3

Products

1	Silvacur	Bayer CropScience	250:125 g/l	EC	11309
2	Veto F	Bayer CropScience	225:75 g/l	EC	11317

Uses
- Brown rust in ***spring barley***, ***spring rye***, ***spring wheat***, ***winter barley***, ***winter rye***, ***winter wheat*** [1, 2]
- Crown rust in ***spring oats***, ***winter oats*** [1, 2]
- Foliar disease control in ***grass seed crops*** *(off-label)* [1, 2]; ***triticale*** *(off-label)* [2]
- Fusarium ear blight in ***spring wheat***, ***winter wheat*** [1, 2]
- Glume blotch in ***spring wheat***, ***winter wheat*** [1, 2]
- Net blotch in ***spring barley***, ***winter barley*** [1, 2]
- Powdery mildew in ***spring barley***, ***spring oats***, ***spring rye***, ***spring wheat***, ***winter barley***, ***winter oats***, ***winter rye***, ***winter wheat*** [1, 2]
- Rhynchosporium in ***spring barley***, ***winter barley*** [1, 2]
- Septoria leaf blotch in ***spring wheat***, ***winter wheat*** [1, 2]
- Sooty moulds in ***spring wheat***, ***winter wheat*** [1, 2]
- Yellow rust in ***spring barley***, ***spring rye***, ***spring wheat***, ***winter barley***, ***winter rye***, ***winter wheat*** [1, 2]

Specific Off-Label Approvals (SOLAs)
- ***grass seed crops*** *20061023* [1], *20061024* [2]
- ***triticale*** *20061024* [2]

Approval information
- Accepted by BBPA for use on malting barley

Efficacy guidance

- For best results apply at an early stage of disease development before infection spreads to new crop growth
- To protect flag leaf and ear from Septoria diseases apply from flag leaf emergence to ear fully emerged (GS 37-59). Earlier application may be necessary where there is a high risk of infection
- For control of rust, powdery mildew, leaf and net blotch apply at first signs of disease with a second application 2-3 wk later if necessary
- Tebuconazole is a DMI fungicide. Resistance to some DMI fungicides has been identified in Septoria leaf blotch which may seriously affect performance of some products. For further advice contact a specialist advisor and visit the Fungicide Resistance Action Group (FRAG)-UK website

Restrictions

- Maximum total dose equivalent to two full dose treatments
- Do not use on durum wheat
- Use only on listed varieties of oats. See label [1]

Crop-specific information

- Latest use: before grain milky-ripe (GS 71)
- Some transient leaf speckling may occur on wheat or leaf reddening/scorch on oats but this has not been shown to reduce yield response or disease control

Environmental safety

- Harmful to aquatic organisms
- Harmful to fish or other aquatic life. Do not contaminate surface waters or ditches with chemical or used container

Hazard classification and safety precautions

Hazard H03 [1]
Risk phrases R20, R36 [1]; R52, R53a [1, 2]
Operator protection A, C, H; U05a, U09b, U20a
Environmental protection E13c [1, 2]; E34 [1]
Storage and disposal D01, D02, D05, D09a [1, 2]; D10b [1]; D10c [2]
Medical advice M03 [2]

412 tebuconazole + trifloxystrobin

A conazole and stobilurin fungicide mixture for wheat and vegetable crops
FRAC mode of action code: 3 + 11

Products

1	Coronet	Bayer CropScience	200:100 g/l	SC	12267
2	Nativo 75 WG	Bayer CropScience	50:25% w/w	WG	13057

Uses

- Alternaria in ***broccoli***, ***brussels sprouts***, ***cabbages***, ***calabrese***, ***carrots***, ***cauliflowers*** [2]
- Brown rust in ***winter wheat*** [1]
- Foliar disease control in ***horseradish*** *(off-label)*, ***parsley root*** *(off-label)*, ***parsnips*** *(off-label)*, ***salsify*** *(off-label)* [2]
- Glume blotch in ***winter wheat*** [1]
- Late ear diseases in ***winter wheat*** [1]
- Light leaf spot in ***broccoli***, ***brussels sprouts***, ***cabbages***, ***calabrese***, ***cauliflowers*** [2]
- Phoma leaf spot in ***broccoli***, ***brussels sprouts***, ***cabbages***, ***calabrese***, ***cauliflowers*** [2]
- Powdery mildew in ***broccoli***, ***brussels sprouts***, ***cabbages***, ***calabrese***, ***carrots***, ***cauliflowers*** [2]; ***winter wheat*** [1]
- Purple blotch in ***leeks*** [2]
- Ring spot in ***broccoli***, ***brussels sprouts***, ***cabbages***, ***calabrese***, ***cauliflowers*** [2]
- Rust in ***leeks*** [2]
- Sclerotinia rot in ***carrots*** [2]
- Septoria leaf blotch in ***winter wheat*** [1]
- White blister in ***broccoli***, ***brussels sprouts***, ***cabbages***, ***calabrese***, ***cauliflowers*** [2]
- White tip in ***leeks*** [2]
- Yellow rust in ***winter wheat*** [1]

Specific Off-Label Approvals (SOLAs)

- ***horseradish*** *20064163* [2]
- ***parsley root*** *20064163* [2]
- ***parsnips*** *20064163* [2]
- ***salsify*** *20064163* [2]

Approval information

- Trifloxystrobin included in Annex I under EC Directive 91/414

Efficacy guidance

- Best results obtained from treatment at early stages of disease development. Further treatment may be needed if disease attack is prolonged
- Applications to established infections of any disease are likely to be less effective
- Best control of cereal ear diseases obtained by treatment during ear emergence [1]
- Treatment may give some control of *Stemphyllium botryosum* and leaf blotch on leeks [2]
- Tebuconazole is a DMI fungicide. Resistance to some DMI fungicides has been identified in Septoria leaf blotch which may seriously affect performance of some products. For further advice contact a specialist advisor and visit the Fungicide Resistance Action Group (FRAG)-UK website
- Trifloxystrobin is a member of the QoI cross resistance group. Product should be used preventatively and not relied on for its curative potential
- Use product as part of an Integrated Crop Management strategy incorporating other methods of control, including where appropriate other fungicides with a different mode of action. Do not apply more than two foliar applications of QoI containing products to any cereal crop, broccoli, calabrese or cauliflower. Do not apply more than three applications to Brussels sprouts, cabbage, carrots or leeks
- There is a significant risk of widespread resistance occurring in *Septoria tritici* populations in UK. Failure to follow resistance management action may result in reduced levels of disease control
- Strains of wheat powdery mildew resistant to QoIs are common in the UK. Control of wheat mildew can only be relied on from the triazole component
- Where specific control of wheat mildew is required this should be achieved through a programme of measures including products recommended for the control of mildew that contain a fungicide from a different cross-resistance group and applied at a dose that will give robust control

Restrictions

- Maximum total dose on wheat equivalent to two full dose treatments [1]
- Maximum number of treatments 2 per crop for broccoli, calabrese, cauliflowers; 3 per crop for Brussels sprouts, cabbages, carrots, leeks [2]
- In addition to the maximum number of treatments per crop a maximum of 3 applications may be applied to the same ground in one calendar year [2]
- Consult processor before use on vegetable crops for processing [2]

Crop-specific information

- Latest use: before milky ripe stage on winter wheat
- HI 35 d for wheat; 21 d for all other crops
- Performance against leaf spot diseases of brassicas, *Alternaria* leaf blight of carrots, and rust in leeks may be improved by mixing with an approved sticker/wetter [2]

Environmental safety

- Dangerous for the environment
- Very toxic to aquatic organisms

Hazard classification and safety precautions

Hazard H03, H11
Risk phrases R37 [1]; R50, R53a, R63 [1, 2]
Operator protection A, H; U05a [2]; U09b, U20b [1, 2]; U19a [1]
Environmental protection E15a, E34, E38
Storage and disposal D01, D02, D09a, D10b, D12a [1, 2]; D05 [1]
Medical advice M03

413 tebufenpyrad

A pyrazole mitochondrial electron transport inhibitor (METI) aphicide and acaricide
IRAC mode of action code: 21

Products

Masai	BASF	20% w/w	WB	13082

Uses

- Damson-hop aphid in ***hops***
- Red spider mites in ***apples***, ***pears***
- Two-spotted spider mite in ***hops***, ***protected roses***, ***strawberries***

Approval information

- Accepted by BBPA for use on hops

Efficacy guidance

- Acts on eggs (except winter eggs) and all motile stages of spider mites up to adults
- Treat spider mites from 80% egg hatch but before mites become established
- For effective control total spray cover of the crop is required
- Product can be used in a programme to give season-long control of damson-hop aphids coupled with mite control
- Where aphids resistant to tebufenpyrad occur in hops control is unlikely to be satisfactory and repeat treatments may result in lower levels of control. Where possible use different active ingredients in a programme

Restrictions

- Maximum total dose equivalent to one full dose treatment on apples, pears, strawberries; 3 full dose treatments on hops
- Other mitochondrial electron transport inhibitor (METI) acaricides should not be applied to the same crop in the same calendar yr either separately or in mixture
- Do not treat apples before 90% petal fall
- Small-scale testing of rose varieties to establish tolerance recommended before use
- Inner liner of container must not be removed

Crop-specific information

- Latest use: end of burr stage for hops
- HI strawberries 3 d; apples, bilberries, blackcurrants, blueberries, cranberries, gooseberries, pears, redcurrants, whitecurrants 7d; blackberries, plums, raspberries 21 d
- Product has no effect on fruit quality or finish

Environmental safety

- Dangerous for the environment
- Very toxic to aquatic organisms
- High risk to bees. Do not apply to crops in flower or to those in which bees are actively foraging. Do not apply when flowering weeds are present
- LERAP Category B
- Broadcast air-assisted LERAP (18 m)

Hazard classification and safety precautions

Hazard H03, H11
Risk phrases R20, R22a, R50, R53a
Operator protection A; U02a, U05a, U09a, U13, U14, U20b
Environmental protection E12a, E12e, E15a, E16a, E16b, E38; E17b (18 m)
Storage and disposal D01, D02, D09a, D11a, D12a
Medical advice M05a

414 teflubenzuron

A benzoylurea insecticide for use on ornamentals
IRAC mode of action code: 15

Products

Nemolt	Fargro	150 g/l	SC	10226

SEE SECTION 3 FOR PRODUCTS ALSO REGISTERED

Uses

- Browntail moth in ***ornamental plant production***
- Caterpillars in ***ornamental plant production***
- Whitefly in ***ornamental plant production***

Efficacy guidance

- Product acts as larval stomach poison interfering with moulting process leading to cessation of feeding and larval death
- Apply as soon as first stage larvae seen. This will often coincide with the peak of moth flight

Restrictions

- Maximum number of treatments 3 per yr
- Test specific varieties before carrying out extensive treatments

Environmental safety

- Dangerous for the environment
- Very toxic to aquatic organisms
- Limited evidence suggests some margin of safety to *Encarsia formosa*. Effects on other parasites and predators not fully tested

Hazard classification and safety precautions

Hazard H11
Risk phrases R50, R53a
Operator protection A, C; U05a, U20c
Environmental protection E15a, E38
Storage and disposal D01, D02, D05, D09a, D11a, D12a

415 tefluthrin

A soil acting pyrethroid insecticide seed treatment
IRAC mode of action code: 3

See also fludioxonil + tefluthrin

Products

Force ST	Syngenta	200 g/l	CF	11752

Uses

- Bean seed fly in ***bulb onions*** *(off-label - seed treatment)*, ***chard*** *(off-label - seed treatment)*, ***chives*** *(off-label - seed treatment)*, ***green mustard*** *(off-label - seed treatment)*, ***herbs (see appendix 6)*** *(off-label - seed treatment)*, ***leaf spinach*** *(off-label - seed treatment)*, ***leeks*** *(off-label - seed treatment)*, ***miduna*** *(off-label - seed treatment)*, ***mizuna*** *(off-label - seed treatment)*, ***pak choi*** *(off-label - seed treatment)*, ***parsley*** *(off-label - seed treatment)*, ***red mustard*** *(off-label - seed treatment)*, ***salad onions*** *(off-label - seed treatment)*, ***spinach beet*** *(off-label - seed treatment)*, ***tatsoi*** *(off-label - seed treatment)*
- Carrot fly in ***carrots*** *(off-label - seed treatment)*, ***parsnips*** *(off-label - seed treatment)*
- Millipedes in ***fodder beet*** *(seed treatment)*, ***sugar beet*** *(seed treatment)*
- Onion fly in ***bulb onions*** *(off-label - seed treatment)*, ***leeks*** *(off-label - seed treatment)*, ***salad onions*** *(off-label - seed treatment)*
- Pygmy beetle in ***fodder beet*** *(seed treatment)*, ***sugar beet*** *(seed treatment)*
- Springtails in ***fodder beet*** *(seed treatment)*, ***sugar beet*** *(seed treatment)*
- Symphylids in ***fodder beet*** *(seed treatment)*, ***sugar beet*** *(seed treatment)*

Specific Off-Label Approvals (SOLAs)

- ***bulb onions*** *(seed treatment) 20050546*
- ***carrots*** *(seed treatment) 20050547*
- ***chard*** *(seed treatment) 20050545*
- ***chives*** *(seed treatment) 20050545*
- ***green mustard*** *(seed treatment) 20050545*
- ***herbs (see appendix 6)*** *(seed treatment) 20050545*
- ***leaf spinach*** *(seed treatment) 20050545*
- ***leeks*** *(seed treatment) 20050546*
- ***miduna*** *(seed treatment) 20050545*
- ***mizuna*** *(seed treatment) 20050545*

- ***pak choi*** *(seed treatment) 20050545*
- ***parsley*** *(seed treatment) 20050545*
- ***parsnips*** *(seed treatment) 20050547*
- ***red mustard*** *(seed treatment) 20050545*
- ***salad onions*** *(seed treatment) 20050546*
- ***spinach beet*** *(seed treatment) 20050545*
- ***tatsoi*** *(seed treatment) 20050545*

Approval information
- Accepted by BBPA for use on malting barley

Efficacy guidance
- Apply during process of pelleting beet seed. Consult manufacturer for details of specialist equipment required
- Micro-capsule formulation allows slow release to provide a protection zone around treated seed during establishment

Restrictions
- Maximum number of treatments 1 per batch of seed
- Sow treated seed as soon as possible. Do not store treated seed from one drilling season to next
- If used in areas where soil erosion by wind or water likely, measures must be taken to prevent this happening
- Can cause a transient tingling or numbing sensation to exposed skin. Avoid skin contact with product, treated seed and dust throughout all operations in the seed treatment plant and at drilling

Crop-specific information
- Latest use: before drilling seed
- Treated seed must be drilled within the season of treatment

Environmental safety
- Dangerous for the environment
- Very toxic to aquatic organisms
- Keep treated seed secure from people, domestic stock/pets and wildlife at all times during storage and use
- Treated seed harmful to game and wild life. Bury spillages
- In the event of seed spillage clean up as much as possible into the related seed sack and bury the remainder completely
- Do not apply treated seed from the air
- Keep livestock out of areas drilled with treated seed for at least 80 d

Hazard classification and safety precautions

Hazard H03, H11
Risk phrases R20, R22b, R43, R50, R53a
Operator protection A, D, E, H; U02a, U04a, U05a, U08, U20b
Environmental protection E06a (80 d); E15a, E34, E38
Storage and disposal D01, D02, D05, D09a, D11a, D12a
Treated seed S02, S03, S04b, S05, S07
Medical advice M05b

416 tepraloxydim

A systemic post-emergence herbicide for control of annual grass weeds
HRAC mode of action code: A

Products

1	Aramo	BASF	50 g/l	EC	10280
2	Clayton Rally	Clayton	50 g/l	EC	13415
3	Landgold Tepraloxydim	Teliton	50 g/l	EC	13379
4	Standon Tonga	Standon	50 g/l	EC	13394

Uses
- Annual grasses in ***borage*** *(off-label)*, ***canary flower (echium spp.)*** *(off-label)*, ***evening primrose*** *(off-label)*, ***flax***, ***garlic*** *(off-label)*, ***honesty*** *(off-label)*, ***linseed for industrial use***, ***lupins*** *(off-label)*, ***mallow (althaea spp.)*** *(off-label)*, ***mustard*** *(off-label)*, ***parsnips*** *(off-label)*, ***shallots*** *(off-label)*, ***winter oilseed rape for industrial use*** [1]; ***bulb onions***, ***cabbages***, ***carrots***, ***cauliflowers***,

SEE SECTION 3 FOR PRODUCTS ALSO REGISTERED

combining peas, fodder beet, land temporarily removed from production, leeks, linseed, spring field beans, sugar beet, vining peas, winter field beans, winter oilseed rape [1, 2]

- Annual meadow grass in ***bulb onions, cabbages, carrots, cauliflowers, combining peas, fodder beet, leeks, linseed, spring field beans, sugar beet, vining peas, winter field beans, winter oilseed rape*** [3, 4]; ***golf courses*** *(off-label)*, ***poppies for morphine production*** *(off-label)*, ***salad onions*** *(off-label)* [1]
- Blackgrass in ***bulb onions, cabbages, carrots, cauliflowers, combining peas, fodder beet, leeks, linseed, spring field beans, sugar beet, vining peas, winter field beans, winter oilseed rape*** [3, 4]
- Green cover in ***land temporarily removed from production*** [3, 4]
- Perennial grasses in ***borage*** *(off-label)*, ***canary flower (echium spp.)*** *(off-label)*, ***evening primrose*** *(off-label)*, ***flax***, ***garlic*** *(off-label)*, ***honesty*** *(off-label)*, ***linseed for industrial use***, ***lupins*** *(off-label)*, ***mallow (althaea spp.)*** *(off-label)*, ***mustard*** *(off-label)*, ***parsnips*** *(off-label)*, ***shallots*** *(off-label)*, ***winter oilseed rape for industrial use*** [1]; ***bulb onions, cabbages, carrots, cauliflowers, combining peas, fodder beet, land temporarily removed from production, leeks, linseed, spring field beans, sugar beet, vining peas, winter field beans, winter oilseed rape*** [1, 2]
- Volunteer barley in ***bulb onions, cabbages, carrots, cauliflowers, combining peas, fodder beet, leeks, linseed, spring field beans, sugar beet, vining peas, winter field beans, winter oilseed rape*** [3, 4]
- Volunteer cereals in ***bulb onions, cabbages, carrots, cauliflowers, combining peas, fodder beet, land temporarily removed from production, leeks, linseed, spring field beans, sugar beet, vining peas, winter field beans, winter oilseed rape*** [1, 2]; ***flax, linseed for industrial use, winter oilseed rape for industrial use*** [1]
- Volunteer wheat in ***bulb onions, cabbages, carrots, cauliflowers, combining peas, fodder beet, leeks, linseed, spring field beans, sugar beet, vining peas, winter field beans, winter oilseed rape*** [3, 4]

Specific Off-Label Approvals (SOLAs)

- ***borage*** *20064019* [1]
- ***canary flower (echium spp.)*** *20064019* [1]
- ***evening primrose*** *20064019* [1]
- ***garlic*** *20064021* [1]
- ***golf courses*** *20064025* [1]
- ***honesty*** *20064019* [1]
- ***lupins*** *20064022* [1]
- ***mallow (althaea spp.)*** *20064023* [1]
- ***mustard*** *20064019* [1]
- ***parsnips*** *20064023* [1]
- ***poppies for morphine production*** *20064026* [1]
- ***salad onions*** *20064024* [1]
- ***shallots*** *20064021* [1]

Approval information

- Tepraloxydim included in Annex I under EC Directive 91/414

Efficacy guidance

- Best results obtained from applications when weeds small and have not begun to compete with crop
- Only emerged weeds are controlled
- Cool conditions slow down activity and very dry conditions reduce activity by interfering with uptake and translocation
- Foliar death of susceptible weeds evident after 3-4 wks in warm conditions
- Reduced doses must not be used on resistant grass weed populations
- Tepraloxydim is an ACCase inhibitor herbicide. To avoid the build up of resistance do not apply products containing an ACCase inhibitor herbicide more than twice to any crop. In addition do not use any product containing tepraloxydim in mixture or sequence with any other product containing the same ingredient
- Use these products as part of a resistance management strategy that includes cultural methods of control and does not use ACCase inhibitors as the sole chemical method of grass weed control

- Applying a second product containing an ACCase inhibitor to a crop will increase the risk of resistance development; only use a second ACCase inhibitor to control different weeds at a different timing
- Always follow WRAG guidelines for preventing and managing herbicide resistant weeds. See Section 5

Restrictions

- Maximum number of treatments 1 per crop or yr
- Consult processors or contract agents before treatment of crops intended for processing or for seed
- Do not apply to crops damaged or stressed by factors such as previous herbicide treatments, pest or disease attack
- Do not spray if rain or frost expected, or if foliage is wet
- Do not treat oilseed rape with very low vigour and poor yield potential. Overlapping on oilseed rape may cause damage and reduce yields
- On peas a satisfactory crystal violet leaf wax test must be carried out before treatment. Winter varieties may be treated only in the spring
- For sugar beet, fodder beet, linseed and green cover on land temporarily removed from production, applications are prohibited between 1 Nov and 31 Mar
- For field beans, peas, leeks, bulb onions, carrots, cabbages and cauliflowers applications are prohibited between 1 Nov and 1 Mar

Crop-specific information

- Latest use : before end Nov or before crop has 9 true leaves, whichever is first for winter oilseed rape; before flower buds visible for flax [1], linseed; before head formation for cabbages, cauliflowers
- HI 3 wk for carrots; 4 wk for bulb onions, leeks; 5 wk for peas; 8 wk for fodder beet, field beans, sugar beet
- Treatment prohibited between 1 Nov and 31 Mar on golf courses (off-label use), sugar beet, fodder beet, linseed and green cover on land temporarily removed from production
- Green cover must be fully established on land temporarily removed from production before treatment, and treated plants must not be grazed by livestock or harvested for human or animal consumption

Following crops guidance

- In the event of failure of a treated crop, wheat or barley may be drilled after 2 wk following normal seedbed cultivations, maize or Italian rye-grass may be planted after 8 wk following cultivation to 20 cm
- Graminaceous crops other than those mentioned above should not follow a treated crop in the rotation.
- Broad leaved crops may be planted at any time following a failed or normally harvested treated crop

Environmental safety

- Dangerous for the environment
- Toxic to aquatic organisms

Hazard classification and safety precautions

Hazard H03, H11
Risk phrases R22b, R38, R51, R53a, R63, R66 [1-4]; R40, R62 [2-4]
Operator protection A; U05a, U08, U20b, U23a [1-4]; U19a [3, 4]
Environmental protection E15a [1, 2]; E15b [3, 4]; E38 [1]
Storage and disposal D01, D02, D09a, D10c, D12a [1-4]; D05 [2-4]; D07 [2]; D08 [1, 3, 4]
Medical advice M04a [3, 4]; M05b [1-4]

417 terbuthylazine

A triazine herbicide available only in mixtures
HRAC mode of action code: C1

See also *bromoxynil + terbuthylazine*
isoxaben + terbuthylazine
mesotrione + terbuthylazine

418 tetraconazole

A systemic, protectant and curative triazole fungicide for cereals
FRAC mode of action code: 3

Products

Trident	AgriGuard	125 g/l	ME	12982

Uses

- Brown rust in ***spring barley***, ***spring wheat***, ***winter barley***, ***winter wheat***
- Crown rust in ***spring oats***, ***winter oats***
- Powdery mildew in ***spring barley***, ***spring wheat***, ***winter barley***, ***winter wheat***
- Septoria leaf blotch in ***spring wheat***, ***winter wheat***
- Sooty moulds in ***spring wheat***, ***winter wheat***
- Yellow rust in ***spring barley***, ***spring wheat***, ***winter barley***, ***winter wheat***

Approval information

- Approval expiry 31 Aug 2008 [1]

Efficacy guidance

- Apply at any time from the tillering stage up to the latest times indicated for each crop
- Best results achieved from applications before disease becomes established
- If disease pressure persists a second treatment may be required to prevent late season attacks
- Best control of ear diseases in wheat obtained from application at or after the ear completely emerged stage
- Tetraconazole is a DMI fungicide. Resistance to some DMI fungicides has been identified in Septoria leaf blotch which may seriously affect performance of some products. For further advice contact a specialist advisor and visit the Fungicide Resistance Action Group (FRAG)-UK website

Restrictions

- Maximum total dose equivalent to three full dose applications on wheat and two full dose applications on other cereals

Crop-specific information

- Latest use: before end of ear emergence (GS 59) for oats; before end of flowering (GS 69) for barley; before grain watery ripe (GS 71) for wheat

Environmental safety

- Dangerous for the environment
- Harmful to fish or other aquatic life. Do not contaminate surface waters or ditches with chemical or used container
- LERAP Category B

Hazard classification and safety precautions

Hazard H03, H11
Risk phrases R20, R22a, R38, R41, R53a
Operator protection A, C, H; U05a, U11, U20a
Environmental protection E13c, E15a, E16a, E16b, E34
Storage and disposal D01, D02, D05, D09a, D10b
Medical advice M03

SECTION 2

419 tetramethrin

A contact acting pyrethroid insecticide
IRAC mode of action code: 3

See also d-phenothrin + tetramethrin

Products

Killgerm Py-Kill W	Killgerm	10.2% w/v	EC	H4632

Uses
- Flies in ***agricultural premises***, ***livestock houses***

Efficacy guidance
- Dilute in accordance with directions and apply as space or surface spray

Restrictions
- For use only by professional operators
- Maximum number of treatments 1 per wk in agricultural premises and livestock houses
- Do not apply directly on food or livestock
- Remove exposed milk before application. Protect milk machinery and containers from contamination
- Do not use space sprays containing pyrethrins or pyrethroid more than once per wk in intensive or controlled environment animal houses in order to avoid development of resistance. If necessary, use a different control method or product

Environmental safety
- Dangerous for the environment
- Toxic to aquatic organisms

Hazard classification and safety precautions

Hazard H03, H08, H11
Risk phrases R22b, R41, R51, R53a
Operator protection A, H; U02b, U09a, U19a, U20b
Environmental protection E15a
Consumer protection C05, C06, C07, C08, C09, C10, C11
Storage and disposal D01, D09a, D12a
Medical advice M05b

420 thiabendazole

A systemic, curative and protectant benzimidazole (MBC) fungicide
FRAC mode of action code: 1

See also imazalil + thiabendazole

Products

1	Hykeep	Agrichem	2% w/w	DP	12704
2	Storite Clear Liquid	Frontier	220 g/l	SL	12706
3	Storite Excel	Frontier	500 g/l	SC	12705

Uses
- Dry rot in ***seed potatoes*** *(tuber treatment - post-harvest)* [3]; ***ware potatoes*** *(tuber treatment - post-harvest)* [1, 3]
- Fusarium basal neck rot in ***narcissi*** *(off-label - post-lifting or pre-planting dip)*, ***narcissi*** *(post-lifting spray)* [2]
- Gangrene in ***seed potatoes*** *(tuber treatment - post-harvest)* [3]; ***ware potatoes*** *(tuber treatment - post-harvest)* [1, 3]
- Silver scurf in ***seed potatoes*** *(tuber treatment - post-harvest)* [3]; ***ware potatoes*** *(tuber treatment - post-harvest)* [1, 3]
- Skin spot in ***seed potatoes*** *(tuber treatment - post-harvest)* [3]; ***ware potatoes*** *(tuber treatment - post-harvest)* [1, 3]

Approval information
- Thiabendazole included in Annex I under EC Directive 91/414

SEE SECTION 3 FOR PRODUCTS ALSO REGISTERED

- Use of thiabendazole products on ware potatoes requires that the discharge of thiabendazole to receiving water from washing plants is kept within emission limits set by the UK monitoring authority

Efficacy guidance

- For best results tuber treatments should be applied as soon as possible after lifting and always within 24 hr
- Dust treatments should be applied evenly over the whole tuber surface
- Thiabendazole should only be used on ware potatoes where there is a likely risk of disease during the storage period and in combination with good storage hygiene and maintenance [1, 3]
- Benzimidazole tolerant strains of silver scurf and skin spot are common in UK and tolerant strains of dry rot have been reported. To reduce the chance of such strains increasing benzimidazole based products should not be used more than once in the cropping cycle [1, 3]
- On narcissi treat as the crop goes into store and only if fungicidal treatment is essential. Adopt a resistance management strategy [2]

Restrictions

- Maximum number of treatments 1 per batch for ware or seed potato tuber treatments [1, 3]; 1 per yr for narcissi bulbs [2]
- Treated seed potatoes must not be used for food or feed [3]
- Do not remove treated potatoes from store for sale, processing or consumption for at least 21 d after application [1, 3]
- Do not mix with any other product
- Off-label use as post-lifting cold water dip or pre-planting hot water dip not to be used on narcissus bulbs grown in the Isles of Scilly [2]

Crop-specific information

- Latest use: 21 d before removal from store for sale, processing or consumption for ware potatoes [1, 3]
- Apply to potatoes as soon as possible after harvest using suitable equipment and always within 2 wk of lifting provided the skins are set. See label for details [1, 3]
- Potatoes should only be treated by systems that provide an accurate dose to tubers not carrying excessive quantities of soil [1, 3]
- Use as a post-lifting treatment to reduce basal and neck rot in narcissus bulbs. Ensure bulbs are clean [2]

Environmental safety

- Dangerous for the environment
- Toxic to aquatic organisms

Hazard classification and safety precautions

Hazard H04 [3]; H11 [1-3]
Risk phrases R43 [3]; R51 [2, 3]; R52 [1]; R53a [1-3]
Operator protection A, H [1-3]; C, D [2, 3]; U05a [2, 3]; U14 [3]; U19a, U20c [1-3]
Environmental protection E15b [1-3]; E38 [2, 3]
Storage and disposal D01, D09a, D12a [1-3]; D02, D05, D10c [2, 3]; D07, D11a [1]
Treated seed S04a [2]

421 thiabendazole + thiram

A fungicide seed dressing mixture for field and vegetable crops
FRAC mode of action code: 1 + M3

Products

Hy-TL	Agrichem	225:300 g/l	FS	06246

Uses

- Ascochyta in ***broad beans*** *(seed treatment)*, ***peas*** *(seed treatment)*, ***spring field beans*** *(seed treatment)*, ***winter field beans*** *(seed treatment)*
- Damping off in ***broad beans*** *(seed treatment)*, ***peas*** *(seed treatment)*, ***spring field beans*** *(seed treatment)*, ***winter field beans*** *(seed treatment)*
- Fusarium in ***bulb onions*** *(off-label - seed treatment)*
- Neck rot in ***bulb onions*** *(off-label - seed treatment)*

Specific Off-Label Approvals (SOLAs)

- ***bulb onions*** *(seed treatment) 20021299*

Approval information

- Thiabendazole and thiram included in Annex I under EC Directive 91/414

Efficacy guidance

- Dress seed as near to sowing as possible
- Dilution may be needed with particularly absorbent types of seed. If diluted material used, seed may require drying before storage

Restrictions

- Maximum number of treatments 1 per batch
- Seed to be treated should be of satisfactory quality and moisture content
- Do not use treated seed as food or feed

Crop-specific information

- Latest use: before drilling

Environmental safety

- Dangerous for the environment
- Treated seed harmful to game and wildlife

Hazard classification and safety precautions

Hazard H03, H11
Risk phrases R20, R22a, R36, R38, R43, R48, R50, R53a
Operator protection A, C, D, H, M; U05a, U08, U14, U15, U20a
Environmental protection E03, E34
Storage and disposal D01, D02, D05, D09a, D10c
Treated seed S01, S02, S03, S04a, S04b, S05, S06a
Medical advice M03, M04a

422 thiacloprid

A chloronicotinyl insecticide for use in agriculture and horticulture
IRAC mode of action code: 4A

Products

1	Biscaya	Bayer CropScience	240 g/l	OD	12471
2	Calypso	Bayer CropScience	480 g/l	SC	11257
3	Exemptor	Scotts	10% w/w	GR	13122
4	Standon Zero Tolerance	Standon	240 g/l	OD	13546

Uses

- Aphids in ***bedding plants***, ***hardy ornamental nursery stock***, ***ornamental plant production***, ***pot plants***, ***protected ornamentals*** [3]; ***cherries*** *(off-label - under temporary rain covers)*, ***mirabelles*** *(off-label - under temporary rain covers)*, ***protected chinese cabbage*** *(off-label)*, ***protected chives*** *(off-label)*, ***protected choi sum*** *(off-label)*, ***protected herbs (see appendix 6)*** *(off-label)*, ***protected leaf spinach*** *(off-label)*, ***protected lettuce*** *(off-label)*, ***protected pak choi*** *(off-label)*, ***protected parsley*** *(off-label)*, ***protected salad brassicas*** *(off-label - for baby leaf production)* [2]; ***seed potatoes***, ***ware potatoes*** [1, 4]
- Beetles in ***bedding plants***, ***hardy ornamental nursery stock***, ***ornamental plant production***, ***pot plants***, ***protected ornamentals*** [3]
- Bud mite in ***almonds*** *(off-label)*, ***chestnuts*** *(off-label)*, ***cob nuts*** *(off-label)*, ***hazel nuts*** *(off-label)*, ***walnuts*** *(off-label)* [2]
- Capsids in ***strawberries*** *(off-label)* [2]
- Common green capsid in ***blackberries*** *(off-label)*, ***protected blackberries*** *(off-label)*, ***protected raspberries*** *(off-label)*, ***raspberries*** *(off-label)*, ***rubus hybrids*** *(off-label)* [2]
- Damson-hop aphid in ***plums*** *(off-label)* [2]
- Palm thrips in ***protected aubergines*** *(off-label)*, ***protected courgettes*** *(off-label)*, ***protected cucumbers*** *(off-label)*, ***protected ornamentals*** *(off-label)*, ***protected peppers*** *(off-label)*, ***protected tomatoes*** *(off-label)* [2]
- Pollen beetle in ***mustard*** [1]; ***spring oilseed rape***, ***winter oilseed rape*** [1, 4]
- Raspberry beetle in ***protected blackberries*** *(off-label)*, ***protected raspberries*** *(off-label)* [2]
- Rosy apple aphid in ***apples*** [2]

- South American leaf miner in ***protected aubergines*** *(off-label)*, ***protected courgettes*** *(off-label)*, ***protected cucumbers*** *(off-label)*, ***protected ornamentals*** *(off-label)*, ***protected peppers*** *(off-label)*, ***protected tomatoes*** *(off-label)* [2]
- Tarnished plant bug in ***blackberries*** *(off-label)*, ***protected blackberries*** *(off-label)*, ***protected raspberries*** *(off-label)*, ***protected strawberries*** *(off-label)*, ***raspberries*** *(off-label)*, ***rubus hybrids*** *(off-label)* [2]
- Tobacco whitefly in ***protected aubergines*** *(off-label)*, ***protected courgettes*** *(off-label)*, ***protected cucumbers*** *(off-label)*, ***protected ornamentals*** *(off-label)*, ***protected peppers*** *(off-label)*, ***protected tomatoes*** *(off-label)* [2]
- Vine weevil in ***bedding plants***, ***hardy ornamental nursery stock***, ***ornamental plant production***, ***pot plants***, ***protected ornamentals*** [3]
- Western flower thrips in ***protected aubergines*** *(off-label)*, ***protected courgettes*** *(off-label)*, ***protected cucumbers*** *(off-label)*, ***protected ornamentals*** *(off-label)*, ***protected peppers*** *(off-label)*, ***protected tomatoes*** *(off-label)* [2]
- Wheat-blossom midge in ***spring wheat***, ***winter wheat*** [1]
- White rot in ***broccoli*** *(off-label)*, ***brussels sprouts*** *(off-label)*, ***cabbages*** *(off-label)*, ***calabrese*** *(off-label)*, ***cauliflowers*** *(off-label)* [1]
- Whitefly in ***bedding plants***, ***hardy ornamental nursery stock***, ***ornamental plant production***, ***pot plants***, ***protected ornamentals*** [3]
- Woolly currant scale in ***bilberries*** *(off-label)*, ***blackcurrants*** *(off-label)*, ***blueberries*** *(off-label)*, ***cranberries*** *(off-label)*, ***gooseberries*** *(off-label)*, ***redcurrants*** *(off-label)*, ***whitecurrants*** *(off-label)* [2]

Specific Off-Label Approvals (SOLAs)

- ***almonds*** *20060341* [2]
- ***bilberries*** *20060335* [2]
- ***blackberries*** *20060336* [2]
- ***blackcurrants*** *20060335* [2]
- ***blueberries*** *20060335* [2]
- ***broccoli*** *20062130* [1]
- ***brussels sprouts*** *20062130* [1]
- ***cabbages*** *20062130* [1]
- ***calabrese*** *20062130* [1]
- ***cauliflowers*** *20062130* [1]
- ***cherries*** *(under temporary rain covers) 20060338* [2]
- ***chestnuts*** *20060341* [2]
- ***cob nuts*** *20060341* [2]
- ***cranberries*** *20060335* [2]
- ***gooseberries*** *20060335* [2]
- ***hazel nuts*** *20060341* [2]
- ***mirabelles*** *(under temporary rain covers) 20060338* [2]
- ***plums*** *20060337* [2]
- ***protected aubergines*** *20063728* [2]
- ***protected blackberries*** *20070534* [2]
- ***protected chinese cabbage*** *20060592* [2]
- ***protected chives*** *20060453* [2]
- ***protected choi sum*** *20060592* [2]
- ***protected courgettes*** *20063728* [2]
- ***protected cucumbers*** *20063728* [2]
- ***protected herbs (see appendix 6)*** *20060453* [2]
- ***protected leaf spinach*** *20060592* [2]
- ***protected lettuce*** *20060453* [2]
- ***protected ornamentals*** *20063728* [2]
- ***protected pak choi*** *20060592* [2]
- ***protected parsley*** *20060340* [2]
- ***protected peppers*** *20063728* [2]
- ***protected raspberries*** *20070534* [2]
- ***protected salad brassicas*** *(for baby leaf production) 20060340* [2]
- ***protected strawberries*** *20060334* [2]
- ***protected tomatoes*** *20063728* [2]

- ***raspberries*** *20060336* [2]
- ***redcurrants*** *20060335* [2]
- ***rubus hybrids*** *20060336* [2]
- ***strawberries*** *20060333* [2]
- ***walnuts*** *20060341* [2]
- ***whitecurrants*** *20060335* [2]

Approval information

- Thiacloprid included in Annex I under EC Directive 91/414

Efficacy guidance

- Best results in apples obtained from a programme of sprays commencing pre-blossom at the first sign of aphids [2]
- For best pear midge control treat under warm conditions (mid-day) when the adults are flying (temperature higher than 12 °C). Applications on cool, damp days are unlikely to be very successful [2]
- Best results in field crops obtained from treatments when target pests reach threshold levels. For wheat blossom midge use pheromone or yellow sticky traps to monitor adult activity in crops at risk [1, 4]
- Treatment of ornamentals and protected ornamentals is by incorporation into peat-based growing media prior to sowing or planting, using suitable automated equipment [3]
- Ensure thorough mixing in the compost to achieve maximum control. Top dressing is ineffective [3]
- Unless transplants have been treated with a suitable pesticide prior to planting in treated compost full protection against target pests is not guaranteed [3]
- Minimise the possibility of development of resistance by alternating insecticides with different modes of action in the programme
- In dense canopies and on larger trees increase water volume to ensure full coverage

Restrictions

- Maximum number of treatments 2 per yr on apples, seed potatoes; 1 per yr on other field crops and for compost incorporation
- Consult processor before use on crops for processing [1, 4]
- Do not mix treated compost with any other bulky materials such as perlite or bark [3]
- Use treated compost as soon as possible, and within 4 wk of mixing [3]

Crop-specific information

- Latest use: up to and including flowering just complete (GS 69) for wheat; before sowing or planting for treated compost
- HI apples, potatoes 14 d;oilseed rape, mustard 30 d
- Test tolerance of ornamental species before large scale use [3]

Environmental safety

- Dangerous for the environment [1, 3, 4]
- Very toxic or toxic to aquatic organisms
- Risk to certain non-target insects or other arthropods. See directions for use [2]
- Broadcast air-assisted LERAP (30 m) [2]

Hazard classification and safety precautions

Hazard H03 [1-4]; H11 [1, 3, 4]
Risk phrases R20, R52 [2]; R22a [1, 2, 4]; R36, R38, R51 [1, 4]; R40, R53a [1-4]; R43 [2, 3]; R50 [3]
Operator protection A [1-4]; C [1, 4]; D [3]; H [2]; U05a [1-4]; U11, U19a [1, 4]; U14 [2]; U16a [1]; U20b [1, 2, 4]; U20c [3]
Environmental protection E13c, E22b [2]; E15a, E38 [3]; E15b [1, 4]; E17b [2] (30 m); E34 [1-4]
Storage and disposal D01, D02, D09a [1-4]; D10b [2]; D10c, D12a [1, 4]; D11a [3]
Treated seed S04a [3]
Medical advice M03 [1-4]; M05b [2]

423 thiamethoxam

A neonicotinoid insecticide for beet crops
IRAC mode of action code: 4A

Products

Cruiser SB	Syngenta	600 g/l	FS	12958

Uses

- Beet leaf miner in ***fodder beet*** *(seed treatment),* ***sugar beet*** *(seed treatment)*
- Flea beetle in ***fodder beet*** *(seed treatment),* ***sugar beet*** *(seed treatment)*
- Millipedes in ***fodder beet*** *(seed treatment),* ***sugar beet*** *(seed treatment)*
- Pygmy beetle in ***fodder beet*** *(seed treatment),* ***sugar beet*** *(seed treatment)*
- Springtails in ***fodder beet*** *(seed treatment),* ***sugar beet*** *(seed treatment)*
- Symphylids in ***fodder beet*** *(seed treatment),* ***sugar beet*** *(seed treatment)*
- Wireworm in ***fodder beet*** *(seed treatment - reduction),* ***sugar beet*** *(seed treatment - reduction)*

Approval information

- Thiamethoxam included in Annex I under EC Directive 91/414

Efficacy guidance

- Apply to sugar beet or fodder beet seed as part of the normal commercial pelleting process using special treatment machinery
- Seed drills must be suitable for use with polymer-coated seeds. Standard drill settings should not need to be changed
- Drill treated seed into a firm even seedbed. Poor seedbed quality or seedbed conditions may results in delayed emergence and poor establishment
- Where very high populations of soil pests are present protection may be inadequate to achieve an optimum plant stand
- Use in line with latest IRAG guidelines

Restrictions

- Maximum number of treatments 1 per seed batch
- Avoid deep or shallow drilling which may adversely affect establishment and reduce the level of pest control
- Do not use herbicides containing lenacil pre-emergence on treated crops

Crop-specific information

- Latest use: before drilling for fodder beet, sugar beet

Environmental safety

- Dangerous for the environment
- Very toxic to aquatic organisms

Hazard classification and safety precautions

Hazard H11
Risk phrases R50, R53a
Operator protection A, D, H; U05a, U07
Environmental protection E03, E15b, E34, E36a, E38
Storage and disposal D01, D02, D05, D09a, D11a, D12a, D14
Treated seed S02, S04b, S05, S06a, S06b, S08

424 thifensulfuron-methyl

A translocated sulfonylurea herbicide
HRAC mode of action code: B

See also *flupyrsulfuron-methyl + thifensulfuron-methyl*
fluroxypyr + thifensulfuron-methyl + tribenuron-methyl
metsulfuron-methyl + thifensulfuron-methyl

Products

Pinnacle	Headland	50% w/w	SG	12285

Uses

- Docks in ***permanent grassland*, *rotational grass***
- Green cover in ***land temporarily removed from production***

Approval information

- Thifensulfuron-methyl included in Annex I under EC Directive 91/414
- Accepted by BBPA for use on malting barley

Efficacy guidance

- Best results achieved from application to small emerged weeds when growing actively. Broad-leaved docks are susceptible during the rosette stage up to onset of stem extension
- Ensure good spray coverage and apply to dry foliage
- Susceptible weeds stop growing almost immediately but symptoms may not be visible for about 2 wk
- Product sold in twin pack with mecoprop-P to provide option for improved control of cleavers
- Only broad-leaved docks (*Rumex obtusifolius*) are controlled; curled docks (*Rumex crispus*) are resistant
- Docks with developing or mature seed heads should be topped and the regrowth treated later
- Established docks with large tap roots may require follow-up treatment
- High populations of docks in grassland will require further treatment in following yr
- Thifensulfuron-methyl is a member of the ALS-inhibitor group of herbicides and products should be used in a planned Resistance Management strategy. See Section 5 for more information

Restrictions

- Maximum number of treatments 1 per crop for cereals or one per yr for grassland and cereals. Must only be applied from 1 Feb in yr of harvest
- Do not treat new leys in year of sowing
- Do not treat where nutrient imbalances, drought, waterlogging, low temperatures, lime deficiency, pest or disease attack have reduced crop or sward vigour
- Do not roll or harrow within 7 d of spraying
- Do not graze grass crops within 7 d of spraying
- Specific restrictions apply to use in sequence or tank mixture with other sulfonylurea or ALS-inhibiting herbicides. See label for details
- Only one application of a sulfonylurea product may be applied per calendar yr to grassland and green cover on land temporarily removed from production

Crop-specific information

- Latest use: before 1 Aug on grass; flag leaf fully emerged stage (GS 39) on cereals
- On grass apply 7-10 d before grazing and do not graze for 7 d afterwards
- Product may cause a check to both sward and clover which is usually outgrown

Following crops guidance

- Only grass or cereals may be sown within 4 wk of application to grassland or setaside, or in the event of failure of any treated crop
- No restrictions apply after normal harvest of a treated cereal crop

Environmental safety

- Dangerous for the environment
- Very toxic to aquatic organisms
- Keep livestock out of treated areas for at least 7 d following treatment
- Take extreme care to avoid drift onto broad-leaved plants outside the target area or onto surface waters or ditches, or land intended for cropping
- Spraying equipment should not be drained or flushed onto land planted, or to be planted, with trees or crops other than cereals and should be thoroughly cleansed after use - see label for instructions

Hazard classification and safety precautions

Hazard H11
Risk phrases R50, R53a
Operator protection U19a, U20b
Environmental protection E06a (7 d); E15a, E38
Storage and disposal D09a, D11a, D12a

SEE SECTION 3 FOR PRODUCTS ALSO REGISTERED

425 thifensulfuron-methyl + tribenuron-methyl

A mixture of two sulfonylurea herbicides for cereals
HRAC mode of action code: B + B

Products

1	Calibre SX	DuPont	33.3:16.7% w/w	SG	12241
2	Inka SX	DuPont	25:25% w/w	SG	13601
3	Ratio SX	DuPont	40:10% w/w	SG	12601

Uses

- Annual dicotyledons in ***spring barley***, ***spring wheat***, ***winter barley***, ***winter wheat*** [1-3]
- Charlock in ***spring barley***, ***spring wheat***, ***winter barley***, ***winter wheat*** [1-3]
- Chickweed in ***spring barley***, ***spring wheat***, ***winter barley***, ***winter wheat*** [1-3]
- Docks in ***spring barley***, ***spring wheat***, ***winter barley***, ***winter wheat*** [2]
- Mayweeds in ***spring barley***, ***spring wheat***, ***winter barley***, ***winter wheat*** [1-3]

Approval information

- Thifensulfuron-methyl and tribenuron-methyl included in Annex I under EC Directive 91/414
- Accepted by BBPA for use on malting barley

Efficacy guidance

- Apply when weeds are small and actively growing
- Apply after 1 Feb
- Ensure good spray cover of the weeds
- Susceptible weeds cease growth almost immediately after application and symptoms become evident 2 wk later
- Effectiveness reduced by rain within 4 h of treatment and in very dry conditions
- Various tank mixtures recommended to broaden weed control spectrum [1]
- Thifensulfuron-methyl and tribenuron-methyl are members of the ALS-inhibitor group of herbicides and products should be used in a planned Resistance Management strategy. See Section 5 for more information

Restrictions

- Maximum number of treatments 1 per crop
- Do not apply to cereals undersown with grass, clover or other legumes, or any other broad-leaved crop
- Do not apply within 7 d of rolling
- Specific restrictions apply to use in sequence or tank mixture with other sulfonylurea or ALS-inhibiting herbicides. See label for details
- Do not apply to any crop suffering from stress
- Consult contract agents before use on crops grown for seed

Crop-specific information

- Latest use: before flag leaf ligule first visible (GS 39) for all crops

Following crops guidance

- Only cereals, field beans or oilseed rape may be sown in the same calendar year as harvest of a treated crop
- In the event of failure of a treated crop sow only a cereal crop within 3 mth of product application and after ploughing and cultivating to at least 15 cm. After 3 mth field beans or oilseed rape may also be sown

Environmental safety

- Dangerous for the environment
- Very toxic to aquatic organisms
- Spraying equipment should not be drained or flushed onto land planted, or to be planted, with trees or crops other than cereals and should be thoroughly cleansed after use - see label for instructions
- Take particular care to avoid damage by drift onto broad-leaved plants outside the target area or onto surface waters or ditches
- LERAP Category B [2, 3]

Hazard classification and safety precautions

Hazard H04 [1, 3]; H11 [1-3]

Risk phrases R43 [1, 3]; R50, R53a [1-3]
Operator protection A, H [1, 3]; U05a, U08, U20a [1, 3]; U14 [3]; U20c [2]
Environmental protection E15a [1, 3]; E15b [2]; E16a [2, 3]; E38 [1-3]
Storage and disposal D01, D02 [1, 3]; D09a, D11a, D12a [1-3]

426 thiodicarb

A carbamate insecticide with molluscicide uses
IRAC mode of action code: 1A

Products

1	Judge	Sipcam	4% w/w	RB	11847
2	Toro	Sipcam	4% w/w	RB	12138

Uses

- Slugs in ***brussels sprouts, durum wheat, potatoes, spring barley, spring oats, spring oilseed rape, spring wheat, triticale, winter barley, winter oats, winter oilseed rape, winter wheat***

Approval information

- Thiodicarb will not be included in Annex I of EC Directive 91/414. To comply with this EC decision all approvals for the storage and use of all products listed in this profile will expire on 25 November 2008
- In 2006 PSD required that all products containing this active ingredient should carry the following warning in the main area of the container label: "Thiodicarb is an anticholinesterase carbamate. Handle with care"
- Accepted by BBPA for use on malting barley
- Approval expiry 25 Nov 2008 [1, 2]

Efficacy guidance

- Apply bait as broadcast treatment or admixed with seed
- Additional broadcast treatments may be needed after drilling admixed seed
- Sow admixed seed as soon as possible after treatment

Restrictions

- Product contains an anticholinesterase carbamate compound. Do not use if under medical advice not to work with such compounds
- Maximum total dose per crop equivalent to one full dose treatment (admixture) or 3 full dose treatments (broadcast)
- Do not treat grain with more than 16% moisture content or allow treated seed to rise above this level
- May only be applied to oilseed rape and potatoes as broadcast application
- Must not be admixed with cereal seed already admixed with any other product containing thiodicarb
- Do not use treated seed as food or feed

Crop-specific information

- Latest use: before drilling when admixed with barley, durum wheat, oats, triticale, wheat; before second node detectable (GS 32) for barley, durum wheat, oats, triticale, wheat; before stem extension (GS 2,0) for oilseed rape
- HI potatoes, Brussels sprouts 3 wk

Environmental safety

- Dangerous for the environment
- Toxic to aquatic organisms
- Harmful to game, wild birds and animals

Hazard classification and safety precautions

Hazard H04, H11
Risk phrases R20, R22a, R36, R43, R51, R53a
Operator protection A, D, H, J; U05a, U13, U20b
Environmental protection E10b, E34
Storage and disposal D01, D02, D09a, D11a
Treated seed S04a
Medical advice M02, M05a

SEE SECTION 3 FOR PRODUCTS ALSO REGISTERED

427 thiophanate-methyl

A thiophanate fungicide with protectant and curative activity
FRAC mode of action code: 1

See also iprodione + thiophanate-methyl

Products

1	Mildothane Turf Liquid	Bayer Environ.	500 g/l	SC	09935
2	Snare	Headland Amenity	500 g/l	SC	11873

Uses

- Dollar spot in ***managed amenity turf***
- Fusarium patch in ***managed amenity turf***
- Red thread in ***managed amenity turf***
- Wormcast formation in ***managed amenity turf***

Approval information

- Thiophanate-methyl included in Annex I under EC Directive 91/414
- Following implementation of Directive 98/82/EC, approval for use of thiophanate-methyl on several crops was revoked in 1999
- Approval expiry 31 Aug 2008 [1, 2]

Efficacy guidance

- Treat turf at the first sign of disease and repeat at monthly intervals as necessary
- For earthworm cast suppression apply to moist turf at first sign of casting activity and repeat as necessary
- Apply to turf during period of active growth and do not mow for 48 h
- Control of Red Thread and Dollar Spot will be helped by application of a nitrogenous fertilizer 7 d before fungicide treatment

Restrictions

- On mown turf apply after cutting and at least 48 hr before the next cut
- Do not use when earthworms are inactive during periods of drought or when ground is frozen

Environmental safety

- Dangerous for the environment
- Very toxic to aquatic organisms

Hazard classification and safety precautions

Hazard H03, H11
Risk phrases R43, R50, R53a, R68
Operator protection A, C, H, M; U05a, U14, U20c
Environmental protection E15a, E34, E38
Storage and disposal D01, D02, D05, D09a, D10a, D12a
Medical advice M03

428 thiram

A protectant dithiocarbamate fungicide
FRAC mode of action code: M3

See also carboxin + thiram
prochloraz + thiram
thiabendazole + thiram

Products

1	Agrichem Flowable Thiram	Agrichem	600 g/l	FS	10784
2	Thiraflo	Chemtura	480 g/l	FS	13338
3	Thyram Plus	Agrichem	600 g/l	FS	10785
4	Unicrop Thianosan DG	Unicrop	80% w/w	WG	05454

Uses

- Botrytis in ***chrysanthemums***, ***freesias***, ***lettuce*** *(outdoor crops)*, ***ornamental plant production*** *(except Hydrangea)*, ***protected lettuce***, ***protected tomatoes***, ***raspberries***, ***strawberries***, ***tomatoes (outdoor)*** [4]

- Botrytis fruit rot in ***apples*** [4]
- Cane spot in ***raspberries*** [4]
- Damping off in ***broad beans*** *(seed treatment)*, ***combining peas*** *(seed treatment)*, ***dwarf beans*** *(seed treatment)*, ***forage maize*** *(seed treatment)*, ***grass seed*** *(seed treatment)*, ***runner beans*** *(seed treatment)*, ***spring field beans*** *(seed treatment)*, ***spring oilseed rape*** *(seed treatment)*, ***vining peas*** *(seed treatment)*, ***winter field beans*** *(seed treatment)*, ***winter oilseed rape*** *(seed treatment)* [1-3]; ***bulb onions*** *(seed treatment)*, ***cabbages*** *(seed treatment)*, ***cauliflowers*** *(seed treatment)*, ***edible podded peas*** *(seed treatment)*, ***leeks*** *(seed treatment)*, ***radishes*** *(seed treatment)*, ***salad onions*** *(seed treatment)*, ***turnips*** *(seed treatment)* [1, 3]; ***cress*** *(off-label - seed treatment)*, ***endives*** *(off-label - seed treatment)*, ***frise*** *(off-label - seed treatment)*, ***garlic*** *(off-label - seed treatment)*, ***lamb's lettuce*** *(off-label - seed treatment)*, ***lettuce*** *(off-label - seed treatment)*, ***lupins*** *(off-label - seed treatment)*, ***poppies for morphine production*** *(off-label - seed treatment)*, ***radicchio*** *(off-label - seed treatment)*, ***rhubarb*** *(off-label - seed treatment)*, ***scarole*** *(off-label - seed treatment)*, ***shallots*** *(off-label - seed treatment)*, ***soya beans*** *(seed treatment - qualified minor use))*, ***swedes*** *(off-label - seed treatment)* [1]; ***soya beans*** *(seed treatment - qualified minor use)* [3]
- Downy mildew in ***protected lettuce*** [4]
- Fire in ***tulips*** [4]
- Foliar disease control in ***chives (off-label)***, ***frise*** *(off-label)*, ***herbs (see appendix 6)*** *(off-label)*, ***lamb's lettuce*** *(off-label)*, ***leaf spinach*** *(off-label)*, ***parsley*** *(off-label)*, ***protected chives*** *(off-label)*, ***protected herbs (see appendix 6)*** *(off-label)*, ***protected leaf spinach*** *(off-label)*, ***protected parsley*** *(off-label)*, ***protected salad brassicas*** *(off-label - for baby leaf production)*, ***radicchio*** *(off-label)*, ***salad brassicas*** *(off-label - for baby leaf production)*, ***scarole*** *(off-label)* [4]
- Gloeosporium in ***apples*** [4]
- Pythium in ***poppies for morphine production*** *(off-label - seed treatment)* [1]
- Rust in ***blackcurrants***, ***carnations***, ***chrysanthemums*** [4]
- Scab in ***apples***, ***pears*** [4]
- Seed-borne diseases in ***carrots*** *(seed soak)*, ***celery (outdoor)*** *(seed soak)*, ***fodder beet*** *(seed soak)*, ***mangels*** *(seed soak)*, ***parsley*** *(seed soak)*, ***red beet*** *(seed soak)*, ***sugar beet*** *(seed soak)* [1]
- Spur blight in ***raspberries*** [4]

Specific Off-Label Approvals (SOLAs)

- ***chives*** *20064161* [4]
- ***cress*** *(seed treatment) 20070792* [1]
- ***endives*** *(seed treatment) 20070792* [1]
- ***frise*** *(seed treatment) 20070792* [1], *20064161* [4]
- ***garlic*** *(seed treatment) 20052395* [1]
- ***herbs (see appendix 6)*** *20064161* [4]
- ***lamb's lettuce*** *(seed treatment) 20070792* [1], *20064161* [4]
- ***leaf spinach*** *20064161* [4]
- ***lettuce*** *(seed treatment) 20070792* [1]
- ***lupins*** *(seed treatment) 20052394* [1]
- ***parsley*** *20064161* [4]
- ***poppies for morphine production*** *(seed treatment) 20040681* [1]
- ***protected chives*** *20064161* [4]
- ***protected herbs (see appendix 6)*** *20064161* [4]
- ***protected leaf spinach*** *20064161* [4]
- ***protected parsley*** *20064161* [4]
- ***protected salad brassicas*** *(for baby leaf production) 20064161* [4]
- ***radicchio*** *(seed treatment) 20070792* [1], *20064161* [4]
- ***rhubarb*** *(seed treatment) 20052396* [1]
- ***salad brassicas*** *(for baby leaf production) 20064161* [4]
- ***scarole*** *(seed treatment) 20070792* [1], *20064161* [4]
- ***shallots*** *(seed treatment) 20052395* [1]
- ***swedes*** *(seed treatment) 20052397* [1]

Approval information

- Thiram included in Annex I under EC Directive 91/414
- Approval expiry 31 Dec 2008 [4]

SEE SECTION 3 FOR PRODUCTS ALSO REGISTERED

Efficacy guidance

- Spray before onset of disease and repeat every 7-14 d. Spray interval varies with crop and disease. See label for details [4]
- Seed treatments may be applied through most types of seed treatment machinery, from automated continuous flow machines to smaller batch treating apparatus [1-3]
- Co-application of 175 ml water per 100 kg of seed likely to improve evenness of seed coverage [1-3]

Restrictions

- Maximum number of treatments 3 per crop for protected winter lettuce (thiram based products only; 2 per crop if sequence of thiram and other EBDC fungicides used) [4]; 2 per crop for protected summer lettuce [4]; 1 per batch of seed for seed treatments [1-3]
- Do not apply to hydrangeas [4]
- Notify processor before dusting or spraying crops for processing [4]
- Do not dip roots of forestry transplants [4]
- Do not spray when rain imminent [4]
- Do not treat seed of tomatoes, peppers or aubergines [1-3]
- Soya bean seed treatments restricted to crops that are to be harvested as a mature pulse crop only [1, 3]
- Treated seed should not be stored from one season to the next [1-3]

Crop-specific information

- Latest use: pre-drilling for seed treatments and seed soaks [1-3]; 21 d after planting out or 21 d before harvest, whichever is earlier, for protected winter lettuce; 14 d after planting out or 21 d before harvest, whichever is earlier, for spraying protected summer lettuce [4]
- HI 21 d for protected lettuce; 14 d for outdoor lettuce; 7 d for apples, pears, blackcurrants, raspberries, strawberries, tomatoes [4]
- Seed to be treated should be of satisfactory quality and moisture content [1-3]
- Follow label instructions for treating small quantities of seed [1-3]
- For use on tulips, chrysanthemums and carnations add non-ionic wetter [4]

Environmental safety

- Dangerous for the environment [1-3]
- Very toxic to aquatic organisms [1-3]
- Do not use treated seed as food or feed [1-3]
- Treated seed harmful to game and wildlife [1-3]
- A red dye is available from manufacturer to colour treated seed, but not recommended as part of the seed steep. See label for details [1]

Hazard classification and safety precautions

Hazard H03, H11 [1-3]; H04 [4]
Risk phrases R20, R43 [1, 3]; R22a, R48, R50, R53a [1-3]; R36, R38 [1, 3, 4]; R37 [4]
Operator protection A; U02a, U11 [1, 3]; U05a, U08, U20b [1-4]; U19a [4]
Environmental protection E13b [4]; E15a, E34 [1, 3]; E38 [2]
Storage and disposal D01, D02, D09a [1-4]; D05 [1-3]; D10a [2]; D10c, D12b [1, 3]; D11a [4]
Treated seed S01, S02, S03, S04b, S05, S06a, S07 [1-3]; S04a [1, 3]
Medical advice M03 [2]; M04a [1, 3]

429 tolclofos-methyl

A protectant nitroaniline fungicide for soil-borne diseases
FRAC mode of action code: 14

Products

1	Basilex	Scotts	50% w/w	WP	07494
2	Rizolex	Certis	10% w/w	DS	09673
3	Rizolex Flowable	Certis	500 g/l	FS	11399

Uses

- Black scurf and stem canker in ***potatoes*** *(tuber treatment)* [2, 3]
- Bottom rot in ***protected lettuce*** [1]
- Damping off in ***seedlings of ornamentals*** [1]

- Damping off and wirestem in ***brussels sprouts***, ***cabbages***, ***calabrese***, ***cauliflowers***, ***chinese cabbage*** [1]
- Foot rot in ***ornamental plant production***, ***seedlings of ornamentals*** [1]
- Rhizoctonia in ***potatoes*** *(off-label - chitted seed treatment)* [3]; ***protected celery*** *(off-label)*, ***protected radishes*** *(off-label)*, ***swedes*** *(with fleece or mesh covers)*, ***swedes*** *(without covers)*, ***turnips*** *(with fleece or mesh covers)*, ***turnips*** *(without covers)* [1]
- Root rot in ***ornamental plant production***, ***seedlings of ornamentals*** [1]

Specific Off-Label Approvals (SOLAs)

- ***potatoes*** *(chitted seed treatment) 20041323* [3]
- ***protected celery*** *20011055* [1]
- ***protected radishes*** *20011054* [1]

Approval information

- Tolclofos-methyl included in Annex I under EC Directive 91/414
- In 2006 PSD required that all products containing this active ingredient should carry the following warning in the main area of the container label: "Tolclofos-methyl is an anticholinesterase organophosphate. Handle with care"
- May be applied by misting equipment mounted over roller table. See label for details [2, 3]

Efficacy guidance

- Apply dust to seed potatoes during hopper loading [2]
- Apply flowable formulation to clean tubers with suitable misting equipment over a roller table. Spray as potatoes taken into store (first earlies) or as taken out of store (second earlies, maincrop, crops for seed) pre-chitting [3]
- Do not mix flowable formulation with any other product [3]
- To control Rhizoctonia in vegetables and ornamentals apply as drench before sowing, pricking out or planting [1]
- On established seedlings and pot plants apply as drench and rinse off foliage [1]

Restrictions

- Tolclofos-methyl is an atypical organophosphorus compound which has weak anticholinesterase activity. Do not use if under medical advice not to work with such compounds
- Maximum number of treatments 1 per batch for seed potatoes; 1 per crop for lettuce, brassicas, swedes, turnips; 1 at each stage of growth (ie sowing, pricking out, potting) to a maximum of 3, for ornamentals
- Only to be used with automatic planters [2]
- Not recommended for use on seed potatoes where hot water treatment used or to be used [2]
- Do not apply as overhead drench to vegetables or ornamentals when hot and sunny [1]
- Do not use on heathers [1]
- Must not be used via hand-held equipment
- Application to protected crops must only be made where the operator is outside the structure at the time of treatment

Crop-specific information

- Latest use: at planting of seed potatoes [2, 3]; before transplanting for lettuce, brassicas, ornamental specimens [1]; before 2 true lvs for swedes, turnips [1]

Environmental safety

- Dangerous for the environment [1-3]
- Very toxic to aquatic organisms
- LERAP Category B [1]
- Treated tubers to be used as seed only, not for food or feed [2]

Hazard classification and safety precautions

Hazard H04, H11

Risk phrases R36, R37, R38 [1-3]; R50 [2]; R51, R53a [1, 3]

Operator protection A, D, H [1-3]; C [3]; E [2, 3]; J, M [2]; U05a, U11 [2]; U16a [3]; U19a, U20b [1-3]

Environmental protection E15a [2, 3]; E16a [1]

Storage and disposal D01, D02, D12a [2]; D08 [2, 3]; D09a [1-3]; D10a [3]; D11a [1, 2]

Medical advice M01 [1]; M04a [2, 3]

SEE SECTION 3 FOR PRODUCTS ALSO REGISTERED

430 tralkoxydim

A foliar applied oxime herbicide for grass weed control in cereals.
HRAC mode of action code: A

Products

1	Alpha Tralkoxydim 250 SC	Makhteshim	250 g/l	SC	12884
2	Greencrop Gweedore	Greencrop	250 g/l	SC	09882
3	Landgold Tralkoxydim	Landgold	250 g/l	SC	08604
4	Landgold Tralkoxydim	Teliton	250 g/l	SC	12130
5	Standon Tralkoxydim	Standon	250 g/l	SC	09579
6	Strimma	Makhteshim	250 g/l	SC	12777

Uses

- Blackgrass in ***durum wheat***, ***spring barley***, ***spring wheat***, ***triticale***, ***winter barley***, ***winter rye***, ***winter wheat***
- Wild oats in ***durum wheat***, ***spring barley***, ***spring wheat***, ***triticale***, ***winter barley***, ***winter rye***, ***winter wheat***

Approval information

- Accepted by BBPA for use on malting barley

Efficacy guidance

- Product leaf-absorbed and translocated rapidly to growing points. Best results achieved after completion of emergence of grass weeds when growing actively in competitive crops under warm humid conditions with adequate soil moisture
- Activity not dependent on soil type or condition. Weeds germinating after application will not be controlled
- Best control of wild oats obtained from 2 leaf to 1st node detectable stage of weeds, and of blackgrass up to 3 tillers
- Tralkoxydim is an ACCase inhibitor herbicide. To avoid the build up of resistance do not apply products containing an ACCase inhibitor herbicide more than twice to any crop. In addition do not use any product containing tralkoxydim in mixture or sequence with any other product containing the same ingredient
- Use these products as part of a resistance management strategy that includes cultural methods of control and does not use ACCase inhibitors as the sole chemical method of grass weed control
- Applying a second product containing an ACCase inhibitor to a crop will increase the risk of resistance development; only use a second ACCase inhibitor to control different weeds at a different timing
- Always follow WRAG guidelines for preventing and managing herbicide resistant weeds. See Section 5

Restrictions

- Maximum number of treatments 1 per crop. See label
- Authorised adjuvant must always be added. See label
- Do not use on oats
- Do not spray undersown crops or crops to be undersown
- Do not spray when foliage wet or covered in ice or crop otherwise under stress
- Do not spray if a protracted period of cold weather forecast
- Do not spray crops under stress from chemical treatment, grazing, pest attack, mineral deficiency or low fertility. Treatment of stressed crops may be followed by transient foliar discolouration
- Do not roll or harrow within 1 wk of spraying
- Restrictions apply to mixtures or sequences with phenoxy hormone or sulfonylurea herbicides. See label

Crop-specific information

- Latest use: before booting (GS 41)
- Apply to winter cereals from 2 leaves unfolded. If necessary winter cereals may be sprayed twice: once in autumn and once in spring
- Apply to spring cereals from end of tillering

Environmental safety

- Dangerous for the environment [1, 6]

- Harmful to aquatic organisms [1-6]
- Take care to avoid drift onto neighbouring crops, especially oats

Hazard classification and safety precautions

Hazard H03, H11 [1, 6]; H04 [2-5]

Risk phrases R22a [1, 6]; R36, R43 [2-5]; R52, R53a [1-6]

Operator protection A, C, H; U02a, U04a, U08, U20b [1-6]; U05a [2-5]; U14 [1-4, 6]; U15 [1, 6]; U19a [2-4]

Environmental protection E15a [1-6]; E34 [5]; E38 [1, 6]

Storage and disposal D01, D02, D05, D09a [1-6]; D10b [2-5]; D10c, D12a [1, 6]

Medical advice M03 [5]

431 tri-allate

A soil-acting thiocarbamate herbicide for grass weed control
HRAC mode of action code: N

Products

Avadex Excel 15G	Gowan	15% w/w	GR	12109

Uses

- Blackgrass in ***broad beans*** *(off-label)*, ***canary seed*** *(off-label - for wild bird seed production)*, ***combining peas***, ***durum wheat***, ***fodder beet***, ***lucerne***, ***lupins*** *(off-label)*, ***mangels***, ***red beet***, ***red clover***, ***sainfoin***, ***spring barley***, ***spring field beans***, ***sugar beet***, ***triticale***, ***vetches***, ***vining peas***, ***white clover***, ***winter barley***, ***winter field beans***, ***winter wheat***
- Meadow grasses in ***broad beans*** *(off-label)*, ***canary seed*** *(off-label - for wild bird seed production)*, ***combining peas***, ***durum wheat***, ***fodder beet***, ***lucerne***, ***lupins*** *(off-label)*, ***mangels***, ***red beet***, ***red clover***, ***sainfoin***, ***spring barley***, ***spring field beans***, ***sugar beet***, ***triticale***, ***vetches***, ***vining peas***, ***white clover***, ***winter barley***, ***winter field beans***, ***winter wheat***
- Wild oats in ***broad beans*** *(off-label)*, ***canary seed*** *(off-label - for wild bird seed production)*, ***combining peas***, ***durum wheat***, ***fodder beet***, ***lucerne***, ***lupins*** *(off-label)*, ***mangels***, ***red beet***, ***red clover***, ***sainfoin***, ***spring barley***, ***spring field beans***, ***sugar beet***, ***triticale***, ***vetches***, ***vining peas***, ***white clover***, ***winter barley***, ***winter field beans***, ***winter wheat***

Specific Off-Label Approvals (SOLAs)

- ***broad beans*** *20063531*
- ***canary seed*** *(for wild bird seed production) 20070867*
- ***lupins*** *20063530*

Approval information

- Approved for aerial application on wheat, barley, rye, triticale, field beans, peas, fodder beet, sugar beet, red beet, lucerne, sainfoin, vetches, clover. See notes in Section 5
- Accepted by BBPA for use on malting barley

Efficacy guidance

- Incorporate or apply to surface pre-emergence (post-emergence application possible in winter cereals up to 2-leaf stage of wild oats)
- Do not use on soils with more than 10% organic matter
- Wild oats controlled up to 2-leaf stage
- If applied to dry soil, rainfall needed for full effectiveness, especially with granules on surface. Do not use if top 5-8 cm bone dry
- Do not apply with spinning disc granule applicator; see label for suitable types
- Do not apply to cloddy seedbeds
- Use sequential treatments to improve control of barren brome and annual dicotyledons (see label for details)

Restrictions

- Maximum number of treatments 1 per crop
- Consolidate loose, puffy seedbeds before drilling to avoid chemical contact with seed
- Drill cereals well below treated layer of soil (see label for safe drilling depths)
- Do not use on direct-drilled crops or undersow grasses into treated crops
- Do not sow oats or grasses within 1 yr of treatment

Crop-specific information

- Latest use: pre-drilling for beet crops; before crop emergence for field beans, spring barley, peas, forage legumes; before first node detectable stage (GS 31) for winter wheat, winter barley, durum wheat, triticale, winter rye

Environmental safety

- Irritating to eyes and skin
- May cause sensitization by skin contact
- Harmful to fish or other aquatic life. Do not contaminate surface waters or ditches with chemical or used container

Hazard classification and safety precautions

Hazard H04
Risk phrases R36, R38, R43
Operator protection A, C, H; U02a, U05a, U20a
Environmental protection E13c
Storage and disposal D01, D02, D09a, D11a

432 triazoxide

A benzotriazine fungicide available only in mixtures
FRAC mode of action code: 35

See also *clothianidin + prothioconazole + tebuconazole + triazoxide*
prothioconazole + tebuconazole + triazoxide

433 tribenuron-methyl

A foliar acting sulfonylurea herbicide with some root activity for use in cereals
HRAC mode of action code: B

See also *flupyrsulfuron-methyl + tribenuron-methyl*
fluroxypyr + thifensulfuron-methyl + tribenuron-methyl
metsulfuron-methyl + tribenuron-methyl
thifensulfuron-methyl + tribenuron-methyl

Products

1	Landgold Tribenuron 75	Teliton	75% w/w	WG	12131
2	Quantum SX	DuPont	50% w/w	SG	12239
3	Rockett	AgriGuard	75% w/w	WG	12344

Uses

- Annual dicotyledons in ***durum wheat*** [1, 3]; ***spring barley***, ***spring oats***, ***spring wheat***, ***triticale***, ***winter barley***, ***winter oats***, ***winter rye***, ***winter wheat*** [1-3]
- Charlock in ***durum wheat*** [1, 3]; ***spring barley***, ***spring oats***, ***spring wheat***, ***triticale***, ***winter barley***, ***winter oats***, ***winter rye***, ***winter wheat*** [1-3]
- Chickweed in ***durum wheat*** [1, 3]; ***spring barley***, ***spring oats***, ***spring wheat***, ***triticale***, ***winter barley***, ***winter oats***, ***winter rye***, ***winter wheat*** [1-3]
- Mayweeds in ***durum wheat*** [1, 3]; ***spring barley***, ***spring oats***, ***spring wheat***, ***triticale***, ***winter barley***, ***winter oats***, ***winter rye***, ***winter wheat*** [1-3]

Approval information

- Tribenuron-methyl included in Annex I under EC Directive 91/414
- Accepted by BBPA for use on malting barley
- Approval expiry 31 Aug 2008 [1, 3]

Efficacy guidance

- Best control achieved when weeds small and actively growing
- Good spray cover must be achieved since larger weeds often become less susceptible
- Susceptible weeds cease growth almost immediately after treatment and symptoms can be seen in about 2 wk
- Weed control may be reduced when conditions very dry

FOR FULL CONDITIONS OF USE ALWAYS READ THE PRODUCT LABEL

- Tribenuron-methyl is a member of the ALS-inhibitor group of herbicides and products should be used in a planned Resistance Management strategy. See Section 5 for more information

Restrictions

- Maximum number of treatments 1 per crop
- Specific restrictions apply to use in sequence or tank mixture with other sulfonylurea or ALS-inhibiting herbicides. See label for details
- Do not apply to crops undersown with grass, clover or other broad-leaved crops
- Do not apply to any crop suffering stress from any cause or not actively growing
- Do not apply within 7 d of rolling

Crop-specific information

- Latest use: up to and including flag leaf ligule/collar just visible (GS 39)
- Apply in autumn or in spring from 3 leaf stage of crop

Following crops guidance

- Only cereals, field beans or oilseed rape may be sown in the same calendar yr as harvest of a treated crop
- In the event of crop failure sow only a cereal within 3 mth of application. After 3 mth field beans or oilseed rape may also be sown

Environmental safety

- Dangerous for the environment
- Very toxic to aquatic organisms
- Take extreme care to avoid drift onto broad-leaved plants outside the target area or onto surface waters or ditches, or land intended for cropping
- Spraying equipment should not be drained or flushed onto land planted, or to be planted, with trees or crops other than cereals and should be thoroughly cleansed after use - see label for instructions

Hazard classification and safety precautions

Hazard H04, H11
Risk phrases R43, R50, R53a
Operator protection A [1-3]; H [2]; U05a, U08, U20b [1-3]; U19a [1, 3]
Environmental protection E15a [1-3]; E38 [2]
Storage and disposal D01, D02, D09a, D11a [1-3]; D12a [2]

434 triclopyr

A pyridinecarboxylic acid herbicide for perennial and woody weed control
HRAC mode of action code: O

See also *2,4-D + dicamba + triclopyr*
aminopyralid + triclopyr
clopyralid + fluroxypyr + triclopyr
clopyralid + triclopyr
fluroxypyr + triclopyr

Products

1	Garlon 4	Dow	480 g/l	EC	05090
2	Nomix Garlon 4	Nomix Enviro	480 g/l	EC	12081
3	Timbrel	Dow	480 g/l	EC	05815

Uses

- Brambles in ***forest***, ***land not intended to bear vegetation*** [1-3]
- Broom in ***forest***, ***land not intended to bear vegetation*** [1-3]
- Brush clearance in ***industrial sites*** [1-3]
- Docks in ***forest***, ***land not intended to bear vegetation*** [1-3]
- Gorse in ***forest***, ***land not intended to bear vegetation*** [1-3]
- Perennial dicotyledons in ***asparagus*** *(off-label - directed treatment)* [3]; ***forest***, ***industrial sites***, ***land not intended to bear vegetation*** [1-3]
- Perennial grasses in ***asparagus*** *(off-label - directed treatment)* [3]
- Rhododendrons in ***forest***, ***land not intended to bear vegetation*** [1-3]
- Scrub clearance in ***industrial sites*** [1-3]

SEE SECTION 3 FOR PRODUCTS ALSO REGISTERED

- Stinging nettle in ***forest*, *land not intended to bear vegetation*** [1-3]
- Woody weeds in ***forest*, *industrial sites*, *land not intended to bear vegetation*** [1-3]

Specific Off-Label Approvals (SOLAs)

- ***asparagus*** *(directed treatment) 20050530* [3]

Approval information

- Triclopyr included in Annex I under EC Directive 91/414

Efficacy guidance

- Apply in grassland as spot treatment or overall foliage spray when weeds in active growth in spring or summer. Details of dose and timing vary with species. See label
- Apply to woody weeds as summer foliage, winter shoot, basal bark, cut stump or tree injection treatment
- Apply foliage spray in water when leaves fully expanded but not senescent
- Do not spray in drought, in very hot or cold conditions
- Control may be reduced if rain falls within 2 h of application
- Control of rhododendron can be variable. If higher than 1.8 m cut stump treatment recommended. A follow-up shoot treatment may be required
- See label for maximum concentrations when applying in oil, water or via watering can

Restrictions

- Maximum number of treatments 2 per yr
- Do not apply to grass leys less than 1 yr old
- Do not drill kale, swedes, turnips, grass or mixtures containing clover within 6 wk of treatment. Allow at least 6 wk before planting trees
- Not to be used on food crops
- Do not apply through hand held rotary atomisers
- Not to be applied in or near water

Crop-specific information

- Latest use: 6 wk before replanting; 7 d before grazing
- Uses on land not intended for cropping include grassland of no agricultural interest such as roadside verges, railway and motorway embankments
- Apply winter shoot, basal bark or cut stump sprays in paraffin or diesel oil. Dose and timing vary with species. See label for details
- Inject undiluted or 1:1 dilution into cuts spaced every 7.5 cm round trunk
- Clover will be killed or severely checked by application in grassland

Environmental safety

- Dangerous for the environment
- Very toxic to aquatic organisms
- LERAP Category B
- Do not allow spray to drift onto agricultural or horticultural crops, amenity plantings, gardens, ponds, lakes or water courses. Vapour drift may occur under hot conditions
- Keep livestock out of treated areas for at least 7 d and until foliage of any poisonous weeds such as buttercups or ragwort has died and become unpalatable

Hazard classification and safety precautions

Hazard H03, H11
Risk phrases R22a, R22b, R38, R43, R50, R53a
Operator protection A, C, H, M; U02a, U05a, U08, U14, U20b
Environmental protection E07a, E15a, E16a, E16b, E23, E34, E38
Consumer protection C01
Storage and disposal D01, D02, D09a, D10b, D12a
Medical advice M05b

435 trifloxystrobin

A protectant strobilurin fungicide for cereals and managed amenity turf
FRAC mode of action code: 11

See also *cyproconazole + trifloxystrobin*
fluoxastrobin + prothioconazole + trifloxystrobin
prothioconazole + trifloxystrobin
tebuconazole + trifloxystrobin

Products

1	Micro-Turf	AgriGuard	50% w/w	WG	12988
2	Pan Aquarius	Pan Agriculture	50% w/w	WG	13018
3	Pan Tees	Pan Agriculture	50% w/w	WG	12894
4	Scorpio	Bayer Environ.	50% w/w	WG	12293
5	Swift SC	Bayer CropScience	500 g/l	SC	11227

Uses

- Brown rust in ***spring barley***, ***winter barley***, ***winter wheat*** [5]
- Fusarium patch in ***amenity grassland***, ***managed amenity turf*** [1-4]
- Glume blotch in ***winter wheat*** [5]
- Net blotch in ***spring barley***, ***winter barley*** [5]
- Red thread in ***amenity grassland***, ***managed amenity turf*** [1-4]
- Rhynchosporium in ***spring barley***, ***winter barley*** [5]
- Septoria leaf blotch in ***winter wheat*** [5]

Approval information

- Trifloxystrobin included in Annex I under EC Directive 91/414
- Accepted by BBPA for use on malting barley

Efficacy guidance

- Should be used protectively before disease is established in crop. Further treatment may be necessary if disease attack prolonged
- Treat grass after cutting and do not mow for at least 48 h afterwards to allow adequate systemic movement [1, 2, 4]
- Trifloxystrobin is a member of the QoI cross resistance group. Product should be used preventatively and not relied on for its curative potential
- Use product as part of an Integrated Crop Management strategy incorporating other methods of control, including where appropriate other fungicides with a different mode of action. Do not apply more than two foliar applications of QoI containing products to any cereal crop
- There is a significant risk of widespread resistance occurring in *Septoria tritici* populations in UK. Failure to follow resistance management action may result in reduced levels of disease control
- On cereal crops product must always be used in mixture with another product, recommended for control of the same target disease, that contains a fungicide from a different cross resistance group and is applied at a dose that will give robust control
- Strains of barley powdery mildew resistant to QoIs are common in the UK

Restrictions

- Maximum number of treatments 2 per crop per yr
- Do not apply to turf during dought conditions or to frozen turf [1, 2, 4]

Crop-specific information

- HI barley, wheat 35 d [5]

Environmental safety

- Dangerous for the environment
- Very toxic to aquatic organisms
- LERAP category B [1-4]

Hazard classification and safety precautions

Hazard H04 [1-4]; H11 [1-5]
Risk phrases R43 [1-4]; R50, R53a [1-5]
Operator protection A, H; U02a, U09a, U19a [5]; U05a, U20b [1-5]; U11, U13, U14, U15 [1-4]
Environmental protection E13b [5]; E15b, E16a, E16b, E34 [1-4]; E38 [1-5]
Consumer protection C02a [5] (35 d)

SEE SECTION 3 FOR PRODUCTS ALSO REGISTERED

Storage and disposal D01, D02, D09a, D10c, D12a
Medical advice M03 [1-4]

436 trifluralin

A soil-incorporated dinitroaniline herbicide for use in various crops
HRAC mode of action code: K1

See also *clodinafop-propargyl + trifluralin*
isoxaben + trifluralin
linuron + trifluralin
pendimethalin + trifluralin

Products

1	Alpha Trifluralin 48 EC	Makhteshim	480 g/l	EC	13433
2	Tandril 48	DAPT	480 g/l	EC	10643
3	Treflan	Dow	480 g/l	EC	05817
4	Triflurex 48 EC	Makhteshim	480 g/l	EC	13445

Uses

- Annual dicotyledons in ***broad beans, broccoli, brussels sprouts, cabbages, carrots, cauliflowers, french beans, kale, lettuce, parsnips, raspberries, runner beans, spring oilseed rape, strawberries, swedes, turnips, winter barley, winter oilseed rape, winter wheat*** [1-4]; ***calabrese*** [1, 3, 4]; ***linseed*** [1]; ***mustard*** [1-3]; ***navy beans, parsley, sugar beet*** [3]; ***spring field beans, winter field beans*** [1, 2]; ***spring linseed, winter linseed*** [3, 4]
- Annual grasses in ***broad beans, broccoli, brussels sprouts, cabbages, carrots, cauliflowers, french beans, kale, lettuce, parsnips, raspberries, runner beans, spring oilseed rape, strawberries, swedes, turnips, winter barley, winter oilseed rape, winter wheat*** [1-4]; ***calabrese*** [1, 3, 4]; ***linseed*** [1]; ***mustard*** [1-3]; ***navy beans, parsley, sugar beet*** [3]; ***spring field beans, winter field beans*** [1, 2]; ***spring linseed, winter linseed*** [3, 4]

Approval information

- Trifluralin will not be included in Annex I of EC Directive 91/414. All approvals for the storage and use of all products listed in this profile will expire on 30 March 2009 unless an earlier date has been set
- Accepted by BBPA for use on malting barley
- Approval expiry 31 Aug 2008 [3]

Efficacy guidance

- Acts on germinating weeds and requires soil incorporation to 5 cm (10 cm for crops to be grown on ridges) within 30 min of spraying. See label for details of suitable application equipment
- Apply and incorporate at any time during 2 wk before sowing or planting
- Best results achieved by application to fine, firm seedbed, free of clods, crop residues and established weeds
- In winter cereals normally applied as surface treatment without incorporation in tank-mixture with other herbicides to increase spectrum of control. See label for details
- Follow-up herbicide treatment recommended with some crops. See label for details

Restrictions

- Maximum number of treatments 1 per crop or per yr
- Do not apply to brassica plant raising beds
- Minimum interval between application and drilling or planting may be up to 12 mth. See label for details
- Do not use on sand, fen soil or soils with more than 10% organic matter

Crop-specific information

- Latest use: varies between products; check labels. Normally before 4 leaves unfolded (GS 13) for cereals; up to 6, 8 or 10 leaves for sugar beet; pre-sowing/planting for other crops
- Transplants should be hardened off prior to transplanting
- Apply in sugar beet after plants 10 cm high with 4-8 leaves and harrow into soil
- Apply in cereals after drilling up to and including 3 leaf stage (GS 13)

Environmental safety

- Dangerous for the environment

- Very toxic to aquatic organisms
- Harmful to fish or other aquatic life. Do not contaminate surface waters or ditches with chemical or used container
- Keep livestock out of treated areas for at least two weeks following treatment
- Flammable

Hazard classification and safety precautions

Hazard H03, H08 [1-4]; H04 [2]; H11 [1, 3, 4]
Risk phrases R22b [1-4]; R36 [1, 2, 4]; R37, R43 [3]; R38 [2]; R50, R53a, R67 [1, 3, 4]
Operator protection A, C [1, 2, 4]; U05a [1, 2, 4]; U08, U13, U20b [1-4]; U11 [1, 4]; U14, U19a [3]
Environmental protection E13c [1, 2, 4]; E15a, E34 [1, 4]; E38 [1, 3, 4]
Storage and disposal D01, D02, D06a [1, 2, 4]; D05, D07 [2]; D06b [3]; D08 [1, 4]; D09a, D10b [1-4]; D12a [1, 3, 4]
Medical advice M05b

437 triflusulfuron-methyl

A sulfonyl urea herbicide for beet crops
HRAC mode of action code: B

See also *lenacil + triflusulfuron-methyl*

Products

Debut	DuPont	50% w/w	WG	07804

Uses

- Annual dicotyledons in ***chicory*** *(off-label)*, ***fodder beet***, ***sugar beet***
- Charlock in ***red beet*** *(off-label)*
- Cleavers in ***red beet*** *(off-label)*
- Flixweed in ***red beet*** *(off-label)*
- Fool's parsley in ***red beet*** *(off-label)*
- Nipplewort in ***red beet*** *(off-label)*
- Ox-eye daisy in ***red beet*** *(off-label)*

Specific Off-Label Approvals (SOLAs)

- ***chicory*** *20050937*
- ***red beet*** *20023629*

Efficacy guidance

- Product should be used with a recommended adjuvant or a suitable herbicide tank-mix partner - see label for details
- Product acts by foliar action. Best results obtained from good spray cover of small actively growing weeds
- Susceptible weeds cease growth immediately and symptoms can be seen 5-10 d later
- Best results achieved from a programme of up to 4 treatments starting when first weeds have emerged with subsequent applications every 5-14 d when new weed flushes at cotyledon stage
- Weed spectrum can be broadened by tank mixture with other herbicides. See label for details
- Product may be applied overall or via band sprayer
- Triflusulfuron-methyl is a member of the ALS-inhibitor group of herbicides

Restrictions

- Maximum number of treatments 4 per crop
- Do not apply to any crop stressed by drought, water-logging, low temperatures, pest or disease attack, nutrient or lime deficiency

Crop-specific information

- Latest use: before crop leaves meet between rows
- HI 4 wk for red beet
- All varieties of sugar beet and fodder beet may be treated from early cotyledon stage until the leaves begin to meet between the rows

Following crops guidance

- Only winter cereals should follow a treated crop in the same calendar yr. Any crop may be sown in the next calendar yr

SEE SECTION 3 FOR PRODUCTS ALSO REGISTERED

- After failure of a treated crop, sow only spring barley, linseed or sugar beet within 4 mth of spraying unless prohibited by tank-mix partner

Environmental safety

- Dangerous for the environment
- Very toxic to aquatic organisms
- Extremely dangerous to fish or other aquatic life. Do not contaminate surface waters or ditches with chemical or used container
- LERAP Category B
- Take extreme care to avoid drift onto broad-leaved plants outside the target area or onto surface waters or ditches, or land intended for cropping
- Spraying equipment should not be drained or flushed onto land planted, or to be planted, with trees or crops other than sugar beet and should be thoroughly cleansed after use - see label for instructions

Hazard classification and safety precautions

Hazard H11
Risk phrases R50, R53a
Operator protection A; U05a, U08, U19a, U20a
Environmental protection E13a, E15a, E16a, E16b, E38
Storage and disposal D01, D02, D05, D09a, D11a, D12a

438 trinexapac-ethyl

A novel cyclohexanecarboxylate plant growth regulator for cereals, turf and amenity grassland

Products

1	Clayton Truss	Clayton	250 g/l	EC	13129
2	Groslow	Sherriff Amenity	25% w/w	WB	13694
3	Landgold Tacet	Teliton	250 g/l	EC	13036
4	Moddus	Syngenta	250 g/l	EC	08801
5	Pan Tepee	Pan Agriculture	250 g/l	EC	12279
6	Primo Maxx	Scotts	121 g/l	SL	11878
7	Standon Pygmy	Standon	250 g/l	EC	13001
8	Stunt	AgriGuard	250 g/l	EC	12859

Uses

- Growth retardation in ***amenity grassland***, ***managed amenity turf*** [2, 6]
- Lodging control in ***canary seed*** *(off-label - for wild bird seed production)*, ***spring wheat*** *(off-label - cv A C Barrie only)* [4]; ***durum wheat***, ***ryegrass seed crops***, ***spring barley***, ***spring oats***, ***spring rye***, ***triticale***, ***winter barley***, ***winter oats***, ***winter rye***, ***winter wheat*** [1, 3-5, 7, 8]

Specific Off-Label Approvals (SOLAs)

- ***canary seed*** *(for wild bird seed production) 20070869* [4]
- ***spring wheat*** *(cv A C Barrie only) 20042300* [4]

Approval information

- Trinexapac-ethyl included in Annex I under EC Directive 91/414
- Accepted by BBPA for use on malting barley
- Approval expiry 31 Aug 2008 [6]

Efficacy guidance

- Best results on cereals and ryegrass seed crops obtained from treatment from the leaf sheath erect stage [1, 3-5, 7, 8]
- Best results on turf achieved from application to actively growing weed free turf grass that is adequately fertilized and watered and is not under stress. Adequate soil moisture is essential [2, 6]
- Turf should be dry and weed free before application [2, 6]
- Environmental conditions, management and cultural practices that affect turf growth and vigour will influence effectiveness of treatment [2, 6]
- Repeat treatments on turf up to the maximum approved dose may be made as soon as growth resumes [2, 6]

Restrictions

- Maximum total dose equivalent to one full dose on cereals, ryegrass seed crops [1, 3-5, 7, 8]
- Maximum number of treatments on turf equivalent to five full dose treatments [2, 6]
- Do not apply if rain or frost expected or if crop wet. Products are rainfast after 12 h

- Only use on crops at risk of lodging [1, 3-5, 7, 8]
- Do not apply within 12 h of mowing turf [2, 6]
- Not recommended for closely mown fine turf [2]
- Do not treat newly sown turf [2, 6]
- Not to be used on food crops [2, 6]
- Do not compost or mulch grass clippings [2, 6]

Crop-specific information

- Latest use: before 2nd node detectable (GS 32) for oats, ryegrass seed crops; before 3rd node detectable (GS 33) for durum wheat, spring barley, triticale, rye; before flag leaf sheath extending (GS 41) for winter barley, winter wheat
- On wheat apply as single treatment between leaf sheath erect stage (GS 30) and flag leaf fully emerged (GS 39) [1, 3-5, 7, 8]
- On barley, rye, triticale and durum wheat apply as single treatment between leaf sheath erect stage (GS 30) and second node detectable (GS 32), or on winter barley at higher dose between flag leaf just visible (GS 37) and flag leaf fully emerged (GS 39) [1, 3-5, 7, 8]
- On oats and ryegrass seed crops apply between leaf sheath erect stage (GS 30) and first node detectable stage (GS 32) [1, 3-5, 7, 8]
- Treatment may cause ears of cereals to remain erect through to harvest [1, 3-5, 7, 8]
- Turf under stress when treated may show signs of damage [2, 6]
- Any weed control in turf must be carried out before application of the growth regulator [2, 6]

Environmental safety

- Dangerous for the environment [1, 3-5, 7, 8]
- Toxic to aquatic organisms [1, 3-5, 7, 8]

Hazard classification and safety precautions

Hazard H04, H11 [1, 3-5, 7, 8]
Risk phrases R43, R51, R53a [1, 3-5, 7, 8]
Operator protection A [1-8]; C [1, 3-8]; H [1, 3, 5-8]; K [6]; U05a [1, 3-8]; U15, U20c [1, 3-5, 7, 8]; U20b [2, 6]
Environmental protection E13c [2]; E15a [4, 5, 8]; E15b [1, 3, 7]; E34 [8]; E38 [3-5, 7]
Consumer protection C01 [2, 6]
Storage and disposal D01, D02 [1, 3-8]; D05 [1-8]; D09a [1-5, 7, 8]; D10a [2]; D10b [8]; D10c [1, 3-7]; D12a [1, 3-5, 7]

439 triticonazole

A triazole fungicide available only in mixtures
FRAC mode of action code: 3

See also prochloraz + triticonazole

440 urea (commodity substance)

A fungicide for use in forestry

Products

urea	various	37% w/v	SL

Uses

- Fungus diseases in ***cut stumps***

Approval information

- Approval for the use of urea as a commodity substance was granted on 25 July 2003 by Ministers under regulation 5 of the Control of Pesticides Regulations 1986
- Urea is being supported for review in the fourth stage of the EC Review Programme under Directive 91/414. Whether or not it is included in Annex I, the existing commodity chemical approval will have to be revoked because substances listed in Annex I must be approved for marketing and use under the EC regime. If urea is included in Annex I, products containing it will need to gain approval in the normal way if they carry label claims for pesticidal activity
- Approval expiry 31 Dec 2008 [1]

SEE SECTION 3 FOR PRODUCTS ALSO REGISTERED

Efficacy guidance

- For the treatment of cut stumps of trees, the following dyes may be used at 0.04% w/v concentration when mixed with solution of urea: Kenacid Turquoise V5898 (CI Acid Blue 42045), Denacid Turquoise AN 200 (CI Acid Blue 9 42090), Duasyn Acid Blue AE-20 (CI Acid Blue 9 42090)

Restrictions

- Maximum number of treatments: 1 per stump per yr
- Operators should wear rubber or other chemical-proof gloves, a nuisance dust mask and a cotton/terylene overall when handling dye powder

Hazard classification and safety precautions

Operator protection A

441 Verticillium lecanii

A fungal parasite of aphids and whitefly

Products

1	Mycotal	Koppert	16.1% w/w	WP	04782
2	Vertalec	Koppert	2.5% w/w	WP	04781

Uses

- Aphids in ***aubergines***, ***protected beans***, ***protected chrysanthemums***, ***protected cucumbers***, ***protected lettuce***, ***protected ornamentals***, ***protected peppers***, ***protected roses***, ***protected tomatoes*** [2]
- Whitefly in ***aubergines***, ***protected beans***, ***protected cayenne peppers*** *(off-label)*, ***protected chives*** *(off-label)*, ***protected cucumbers***, ***protected edible podded peas*** *(off-label)*, ***protected herbs (see appendix 6)*** *(off-label)*, ***protected lettuce***, ***protected ornamentals***, ***protected parsley*** *(off-label)*, ***protected peppers***, ***protected salad brassicas*** *(off-label)*, ***protected tomatoes*** [1]

Specific Off-Label Approvals (SOLAs)

- ***protected cayenne peppers*** *20070862* [1]
- ***protected chives*** *20070861* [1]
- ***protected edible podded peas*** *20070863* [1]
- ***protected herbs (see appendix 6)*** *20070861* [1]
- ***protected parsley*** *20070861* [1]
- ***protected salad brassicas*** *20070861* [1]

Efficacy guidance

- *Verticillium lecanii* is a pathogenic fungus that infects the target pests and destroys them
- Apply spore powder as spray as part of biological control programme keeping the spray liquid well agitated
- Pre-soak the product for 2-4 h before application to rehydrate the spores and assist in dispersion
- Treat before infestations build to high levels and repeat as directed on the label
- Spray during late afternoon and early evening directing spray onto underside of leaves and to growing points
- Best results require minimum 80% relative humidity and 18 °C within the crop canopy
- Product highly infective to many aphid species except the chrysanthemum aphid. Follow specific label directions for this pest [2]

Restrictions

- Never use in tank mixture
- Do not use a fungicide within 3 d of treatment. Pesticides containing captan, chlorothalonil, fenarimol, dichlofluanid, imazalil, maneb, prochloraz, quinomethionate, thiram or tolylfluanid may not be used on the same crop
- Keep in a refrigerated store at 2-6 °C but do not freeze

Environmental safety

- Products have negligible effects on commercially available natural predators or parasites but consult manufacturer before using with a particular biological control agent for the first time

Hazard classification and safety precautions

Operator protection U19a, U20b

Environmental protection E15a
Storage and disposal D09a, D11a

442 warfarin

A coumarin anti-coagulant rodenticide

Products

1	Grey Squirrel Bait	Killgerm	0.02% w/w	GB	13020
2	Sakarat Ready-to-Use (Cut Wheat Base)	Killgerm	0.025% w/w	RB	H6807
3	Sakarat Ready-to-Use (Whole Wheat)	Killgerm	0.025% w/w	RB	H6808
4	Sakarat X	Killgerm	0.05% w/w	RB	H6809
5	Sewarin Extra	Killgerm	0.05% w/w	RB	H6810
6	Sewarin P	Killgerm	0.025% w/w	RB	H6811
7	Sewercide Cut Wheat Rat Bait	Killgerm	0.05% w/w	RB	H6805
8	Sewercide Whole Wheat Rat Bait	Killgerm	0.05% w/w	RB	H6806
9	Warfarin 0.5% Concentrate	B H & B	0.5% w/w	CB	H6815
10	Warfarin Ready Mixed Bait	B H & B	0.025% w/w	RB	H6816

Uses
- Grey squirrels in ***agricultural premises***, ***forest***, ***trees*** *(nuts)* [1]
- Rats in ***agricultural premises*** [2-10]

Approval information
- Warfarin included in Annex I under EC Directive 91/414

Efficacy guidance
- For rodent control place ready-to-use or prepared baits at many points wherever rats active. Out of doors shelter bait from weather [2-10]
- Inspect baits frequently and replace or top up as long as evidence of feeding. Do not underbait [2-10]
- Use grey squirrel baits when bark stripping is evident in farm forestry or forestry or where there is risk of damage to trees grown for their nuts [1]
- Use squirrel bait in specially constructed hoppers and inspect every 2-3 d. Place bait hoppers so as to prevent rainwater entry and replace as necessary [1]

Restrictions
- For use only by local authorities, professional operators providing a pest control service and persons occupying industrial, agricultural or horticultural premises
- For use only between 15 Mar and 15 Aug for tree protection [1]
- Squirrel bait hoppers must be clearly labelled "Poison" or carry the instruction "Tree protection, do not disturb" [1]
- Warfarin baits for grey squirrel control must only be used outdoors in specially constructed hoppers which comply with the Grey Squirrel Order 1973, in Scotland and specified counties of England and Wales. See label [1]
- The use of warfarin to control grey squirrels is illegal unless the provisions of the Grey Squirrels Order 1973 are observed [1]

Environmental safety
- Harmful to wildlife
- Prevent access to baits by children and animals, especially cats, dogs and pigs
- Rodent bodies must be searched for and burned or buried, not placed in refuse bins or rubbish tips. Remains of bait and containers must be removed after treatment and burned or buried
- Bait must not be used where food, feed or water could become contaminated
- If squirrel bait hoppers have obviously been disturbed by badgers or other animals, change the site or lift onto tables or platforms [1]
- Warfarin baits must nor be used outdoors where pine martens are know to occur naturally [1]
- Under the terms of the Wildlife and Countryside Act 1981 warfarin squirrel baits must not be used outdoors in England, Scotland or Wales where red squrrels are know to occur [1]

Hazard classification and safety precautions

Operator protection A [1-8]; C, D, E, H [2-8]; U13 [1-10]; U20a [2-6]; U20b [1, 7-10]

Environmental protection E10b, E15a [1]
Storage and disposal D09a [1-10]; D10a [9, 10]; D11a [1-8]
Vertebrate/rodent control products V01a, V03a [4, 7-10]; V01b [1-3, 5, 6]; V02 [1-10]; V03b, V04b [2, 3, 5, 6]; V04a [1, 4, 7-10]
Medical advice M03 [2, 3, 5, 6]

443 zeta-cypermethrin

A contact and stomach acting pyrethroid insecticide
IRAC mode of action code: 3

Products

1	Fury 10 EW	Belchim	100 g/l	EW	12248
2	Minuet EW	Belchim	100 g/l	EW	12304
3	Synphony	AgriGuard	100 g/l	EW	13021

Uses

- Aphids in ***spring barley, spring oats, spring wheat, winter barley, winter oats, winter wheat***
- Barley yellow dwarf virus vectors in ***spring barley, spring wheat, winter barley, winter wheat***
- Cabbage seed weevil in ***spring oilseed rape, winter oilseed rape***
- Cabbage stem flea beetle in ***spring oilseed rape, winter oilseed rape***
- Cutworms in ***potatoes, sugar beet***
- Flax flea beetle in ***linseed***
- Flea beetle in ***spring oilseed rape, winter oilseed rape***
- Large flax flea beetle in ***linseed***
- Pea and bean weevil in ***combining peas, spring field beans, vining peas, winter field beans***
- Pea aphid in ***combining peas, vining peas***
- Pea moth in ***combining peas, vining peas***
- Pod midge in ***spring oilseed rape, winter oilseed rape***
- Pollen beetle in ***spring oilseed rape, winter oilseed rape***
- Rape winter stem weevil in ***spring oilseed rape, winter oilseed rape***

Approval information

- Zeta-cypermethrin included in Annex I under EC Directive 91/414
- Accepted by BBPA for use on malting barley

Efficacy guidance

- On winter cereals spray when aphids first found in the autumn for BYDV control. A second spray may be required on late drilled crops or in mild conditions
- For summer aphids on cereals spray when treatment threshold reached
- For listed pests in other crops spray when feeding damage first seen or when treatment threshold reached. Under high infestation pressure a second treatment may be necessary
- Best results for pod midge and seed weevil control in oilseed rape obtained from treatment after pod set but before 80% petal fall
- Pea moth treatments should be applied according to ADAS/PGRO warnings or when economic thresholds reached as indicated by pheromone traps
- Treatments for cutworms should be made at egg hatch and repeated no sooner than 10 d later

Restrictions

- Maximum number of treatments 2 per crop [3]
- Maximum total dose equivalent to two full dose treatments on all crops [1, 2]
- Consult processors before use on crops for processing

Crop-specific information

- Latest use: before 4 true leaves for linseed, before end of flowering for oilseed rape, cereals
- HI potatoes, field beans, peas 14 d; sugar beet 60 d

Environmental safety

- Dangerous for the environment
- Very toxic to aquatic organisms
- High risk to non-target insects or other arthropods. Do not spray within 6 m of the field boundary
- LERAP Category A

Hazard classification and safety precautions

Hazard H03, H11

Risk phrases R20, R22a, R43, R50, R53a
Operator protection A, C, H; U05a, U08, U14, U15, U19a, U20b
Environmental protection E15b, E16c, E16d, E22a, E34 [1-3]; E38 [1, 2]
Storage and disposal D01, D02, D09a, D10b [1-3]; D05 [3]; D12a [1, 2]
Medical advice M05a [1, 2]

444 ziram

A dithiocarbamate bird and animal repellent

Products

AAprotect	Unicrop	32% w/w	PA	03784

Uses

- Birds in ***all top fruit, field crops, forest, ornamental plant production***
- Deer in ***all top fruit, field crops, forest, ornamental plant production***
- Hares in ***all top fruit, field crops, forest, ornamental plant production***
- Rabbits in ***all top fruit, field crops, forest, ornamental plant production***

Approval information

- Ziram included in Annex I under EC Directive 91/414

Efficacy guidance

- Apply undiluted to main stems up to knee height to protect against browsing animals at any time of yr or spray 1:1 dilution on stems and branches in dormant season
- Use dilute spray on fully dormant fruit buds to protect against bullfinches
- Only apply to dry stems, branches or buds
- Use of diluted spray can give limited protection to field crops in areas of high risk during establishment period

Restrictions

- Maximum number of treatments 3 per season as spray for all top fruit, field crops, forest and ornamental specimens; 1 per season as paste treatment on trees and ornamental specimens
- Do not spray elongating shoots or buds about to open
- Do not apply concentrated spray to foliage, fruit buds or field crops

Crop-specific information

- HI edible crops 8 wk

Environmental safety

- Harmful to fish or other aquatic life. Do not contaminate surface waters or ditches with chemical or used container

Hazard classification and safety precautions

Hazard H04
Risk phrases R36, R37, R38
Operator protection A, C; U05a, U08, U19a, U20b
Environmental protection E13c
Consumer protection C02a (8 wk)
Storage and disposal D01, D02, D09a, D10c

445 zoxamide

A substituted benzamide fungicide available only in mixtures
FRAC mode of action code: 22

See also *mancozeb + zoxamide*

SECTION 3
PRODUCTS ALSO REGISTERED

Products also Registered

Products listed in the table below have not been notified for inclusion in Section 2 of this edition of the *Guide*. However most have extant approval until 31 December 2013 unless an earlier expiry date is shown. These products may legally be stored and used in accordance with their label until their approval expires, but they may not still be available for purchase.

Product	Approval holder	MAPP No.	Expiry Date
2,4-D + triclopyr			
Genoxone ZX EC	Agriphar	13131	
abamectin			
Dynamec	Syngenta	12539	31-07-08
RouteOne Abamectin 18	Albaugh UK	13560	
acetamiprid			
Gazelle SG	Certis	13725	10-09-11
acetic acid			
Wyevale Weed Killer	Wyevale	12687	
acibenzolar-S-methyl			
Bion	Syngenta	09803	31-10-11
alpha-cypermethrin			
Alert	BASF	13632	
Alert	BASF	10272	31-08-08
Alpha C 6 ED	Techneat	13611	28-02-15
Alpha C 6 ED	Techneat	11838	
Cleancrop Scope	United Agri	10273	31-08-08
Fastac	BASF	13604	
Fastac	BASF	10220	31-08-08
Fastac Dry	BASF	10221	
aluminium ammonium sulphate			
Curb S	Sphere	13564	
Guardsman B	Chiltern	05494	
Rezist	Barrettine	08576	
aluminium phosphide			
Degesch Fumigation Pellets	Rentokil	11436	
Detia Gas Ex-P	Igrox	09802	
amidosulfuron			
Barclay Cleave	Barclay	11340	
Pursuit	Bayer CropScience	07333	
Squire Ultra	Bayer CropScience	13125	

SECTION 3

Product	Approval holder	MAPP No.	Expiry Date
aminopyralid			
Pro-Banish	Dow	13767	23-10-11
aminopyralid + fluroxypyr			
Halcyon	Dow	12749	29-01-10
amisulbrom			
Shinkon	Procam	13722	27-09-11
amitrole			
Aminotriazole Technical	Nufarm UK	11137	
Weedazol Pro	Nufarm UK	11995	
amitrole + 2,4-D + diuron			
Trik	Nufarm UK	11441	13-12-08
ammonium sulphamate			
Amcide	B H & B	04246	22-05-08
asulam			
Agrotech Asulam	Agrotech-Trading	12243	
Asulox	Bayer CropScience	09969	30-09-08
Cleancrop Asulam	United Agri	10465	
Formule 1	Agrichem	13255	
I T Asulam	I T Agro	10186	
Milentus Asulam	Milentus	12245	
Noble Asulam	Barclay	12939	
Spitfire	AgriGuard	12002	30-04-08
azoxystrobin			
5504	Syngenta	12351	31-12-11
Aubrac	AgChem Access	13483	31-12-11
Barclay ZX	Barclay	11336	31-12-11
Clayton Stobik	Clayton	09440	31-12-11
Cleancrop Celeb	United Agri	13294	31-12-11
Heritage	Syngenta	11383	31-08-09
Ortiva	Syngenta	10542	31-12-11
Priori	Syngenta	10543	31-12-11
RouteOne Roxybin 25	Albaugh UK	13653	31-12-11
Roxybin 25	RouteOne	13268	31-12-11
azoxystrobin + chlorothalonil			
BTP	Syngenta	13400	
azoxystrobin + fenpropimorph			
RouteOne Roxypro FP	Albaugh UK	13634	
benalaxyl + mancozeb			
Galben M	Sipcam	07220	

Product	Approval holder	MAPP No.	Expiry Date
bentazone			
Basagran	BASF	00188	31-07-11
IT Bentazone	I T Agro	13132	31-07-11
RouteOne Benta 48	Albaugh UK	13662	31-07-11
RouteOne Bentazone 48	Albaugh UK	13664	31-07-11
Troy 480	Agrichem	12341	31-07-11
bentazone + MCPB			
Pulsar	BASF	04002	
bentazone + pendimethalin			
Impuls	BASF	13372	31-07-11
Impuls	BASF	11780	31-08-08
benzoic acid			
Menno Florades	Brinkman	12472	31-05-14
beta-cyfluthrin			
Gandalf	Makhteshim	12865	
beta-cyfluthrin + imidacloprid			
Chinook	Bayer CropScience	13696	
Chinook	Bayer CropScience	11421	
bifenox + isoproturon			
Banco	Makhteshim	11191	31-08-08
RP 4169	Makhteshim	13027	30-06-09
RP 4169	Makhteshim	11190	31-08-08
bifenox + MCPA + mecoprop-P			
Quickfire	Headland Amenity	13053	
Sirocco	Makhteshim	13051	
bitertanol + fuberidazole			
Sibutol	Bayer CropScience	11305	
Sibutol New Formula	Bayer CropScience	11307	
UK 743	Bayer CropScience	11315	
bitertanol + fuberidazole + imidacloprid			
Sibutol Secur	Bayer CropScience	11308	
Bordeaux mixture			
Wetcol 3	Law Fertilisers	02360	
boscalid			
Boscalid 50	RouteOne	13450	30-09-13
RouteOne Boscalid 50	Albaugh UK	13660	30-09-13
boscalid + epoxiconazole			
RouteOne Rocker	Albaugh UK	13642	

Product	Approval holder	MAPP No.	Expiry Date
bromoxynil			
Alpha Bromolin 225 EC	Makhteshim	08255	31-08-08
Alpha Bromotril P	Makhteshim	07099	31-08-08
Flagon 400 EC	Makhteshim	08875	31-08-08
bromoxynil + diflufenican + ioxynil			
Capture	Bayer CropScience	09982	
bromoxynil + fluroxypyr			
Tomahawk Plus	Makhteshim	09836	31-08-08
bromoxynil + ioxynil			
Alpha Briotril	Makhteshim	04876	31-08-08
Alpha Briotril Plus 19/19	Makhteshim	04740	31-08-08
Deloxil	Bayer CropScience	09987	
bromoxynil + ioxynil + triasulfuron			
Teal	Nufarm UK	11940	
bromoxynil + prosulfuron			
Jester	Syngenta	08681	30-11-08
bromoxynil + terbuthylazine			
Alpha Bromotril PT	Makhteshim	09435	
Cleancrop Amaize	United Agri	11990	
bromuconazole			
Granit	Bayer CropScience	09995	31-08-08
Tote	Interfarm	13780	
bupirimate			
Nimrod	Makhteshim	10563	31-08-08
Candida Oleophila Strain 0			
NEXY 1	BioNext	13609	22-07-11
captan			
Alpha Captan 50 WP	Makhteshim	04797	30-09-10
PP Captan 80 WG	Calliope SAS	12435	
PP Captan 83	Calliope SAS	12330	
captan + penconazole			
Topas C 50 WP	Novartis	08459	
carbendazim			
Barclay Shelter	Barclay	09000	30-06-08
BASF Turf Systemic Fungicide	BASF	05774	30-06-08
Bavistin	BASF	00217	30-06-08
Bavistin DF	BASF	03848	30-06-08
Bavistin FL	BASF	00218	30-06-08

EXPIRY DATE IS 31/12/2013 UNLESS OTHERWISE SHOWN

Product	Approval holder	MAPP No.	Expiry Date
Clayton Am-Carb	Clayton	11906	
Clayton Chizm FL	Clayton	11050	30-06-08
Cleancrop Curve	United Agri	11774	
Goldazim 500 SC	Chimac-Agriphar	10675	30-06-08
Goldazim 500 SC	Agriphar	13361	30-06-08
Headland Addstem	Headland	06755	30-06-08
Headland Addstem DF	Headland	08904	30-06-08
I.T. Carbendazim	I T Agro	12155	30-06-08
Mascot Systemic	Scotts	09132	
Mascot Systemic	Rigby Taylor	13423	
Nuturf Carbendazim	Nufarm UK	11469	
Occidor 500 SC	Chimac-Agriphar	10937	30-06-08
Occidor 500 SC	Agriphar	13364	30-06-08
Ringer	Barclay	13289	
S.M.I Carbendazim	Sub-Micron	12040	31-07-08
Tripart Defensor FL	Tripart	02752	30-06-08
Turf Systemic Fungicide	Barclay	13039	31-08-08
Turf Systemic Fungicide	Barclay	13291	
Turf Systemic Fungicide	PBI	09349	30-06-08
Turf Systemic Fungicide	SumiAgro	10821	31-08-08
Turfclear	Scotts	07506	
carbendazim + chlorothalonil			
Bravocarb	Syngenta	10520	30-06-08
Greenshield	Scotts	07988	30-06-08
carbendazim + epoxiconazole			
Langin	Agro Trade	12608	30-06-08
carbendazim + flusilazole			
Contrast	DuPont	06150	30-06-08
Harvesan	DuPont	12238	30-06-08
Punch C	DuPont	06801	30-06-08
Standon Flusilazole Plus	Standon	07403	31-08-08
carbendazim + prochloraz			
Novak	BASF	11693	30-06-08
Sportak Alpha HF	BASF	11698	30-06-08
carbetamide			
Carbetamex	Makhteshim	11150	31-08-08
Riot	Makhteshim	13380	
Scrum	Makhteshim	13386	
carbosulfan			
Marshal 10G	Belchim	11682	13-12-08

SECTION 3

EXPIRY DATE IS 31/12/2013 UNLESS OTHERWISE SHOWN

Product	Approval holder	MAPP No.	Expiry Date
carfentrazone-ethyl			
Aurora 40 WG	Belchim	11614	30-09-13
Headlite	AgChem Access	13624	24-08-09
Platform	Belchim	11615	30-09-13
RouteOne Fetraz 60	Albaugh UK	13626	24-08-09
carfentrazone-ethyl + mecoprop-P			
Platform S	FMC	10726	
chloridazon			
Burex 430 SC	Interfarm	09494	31-05-08
Cleancrop Akazon	United Agri	11903	
Gladiator DF	Tripart	06342	
Luxan Chloridazon	Luxan	06304	
Portman Weedmaster	Agform	06018	
Pyramin FL	BASF	11628	
Sculptor	Sipcam	08836	
Starter Flowable	Truchem	03421	
Tripart Gladiator 2	Tripart	06618	31-08-08
Weedmaster SC	Agform	08793	
chloridazon + ethofumesate			
Gremlin	Sipcam	09468	
chloridazon + metamitron			
Volcan Combi FL	Sipcam	11442	
chlormequat			
Adjust	Mandops	05589	31-08-08
Agriguard 5C Chlormequat 460	AgriGuard	09851	
Agriguard Chlormequat 760	AgriGuard	10290	
Alpha Chlormequat 460	Makhteshim	04804	
Alpha Pentagan	Makhteshim	04794	
Alpha Pentagan Extra	Makhteshim	04796	
Atlas 5C Quintacel	Nufarm UK	11130	
Barclay Holdup	Barclay	06799	
Barclay Holdup	Barclay	11365	
Barclay Holdup 600	Barclay	08794	
Barclay Holdup 600	Barclay	11373	
Barclay Holdup 640	Barclay	11374	
Barclay Holdup 640	Barclay	08795	
Barclay Liffey	Barclay	09856	
Barclay Liffey	Barclay	11366	
Barclay Lucan	Barclay	09855	
Barclay Lucan	Barclay	11367	
Barclay Take 5	Barclay	08524	
Barleyquat B	Mandops	06001	31-08-08

Product	Approval holder	MAPP No.	Expiry Date
Barleyquat B	Mandops	07051	31-08-08
BASF 3C Chlormequat 600	BASF	04077	
BASF 3C Chlormequat 750	BASF	06878	
Belcocel	Taminco	11881	
Bettaquat B	Mandops	07050	31-08-08
Chlormequat 46	Nufarm UK	11504	
Ciba Chlormequat 460	Ciba Specialty	09525	
Ciba Chlormequat 5C 460:320	Ciba Specialty	09527	
Ciba Chlormequat 730	Ciba Specialty	09526	
Clayton CCC 750	Clayton	07952	
Clayton Manquat	Clayton	09916	
Cleancrop Chlormequat 700	United Agri	10143	
Cropsafe 5C Chlormequat	Certis	07897	
Cropsafe 5C Chlormequat	Certis	11179	
Hyquat 70	Agrichem	03364	
K2	Mandops	10370	31-08-08
Larke	Nufarm UK	11453	31-08-08
Manipulator	Mandops	05871	31-08-08
Midget	Nufarm UK	12283	
MSS Mirquat	Mirfield	08166	
Portman Chlormequat 400	Agform	01523	
Portman Chlormequat 460	Agform	02549	
Portman Chlormequat 700	Agform	03465	
Portman Supaquat	Agform	03466	
Stabilan 750	Nufarm UK	09303	
Supaquat 720	Agform	09381	
Tricol	Nufarm UK	12682	
Tripart 5C	Tripart	04726	
Tripart Brevis	Tripart	03754	
Tripart Brevis 2	Tripart	06612	
Tripart Chlormequat 460	Tripart	03685	
Uplift	United Phosphorus	07527	
chlormequat + 2-chloroethylphosphonic acid			
Barclay Banshee XL	Barclay	11339	
Sypex	BASF	04650	
Terpal C	BASF	07062	
chlormequat + 2-chloroethylphosphonic acid + imazaquin			
Satellite	BASF	10395	
chlormequat + imazaquin			
Matrix	BASF	13216	29-02-08
2-chloroethylphosphonic acid			
Agrotech Ethephon	Agrotech-Trading	12170	
Barclay Coolmore	Barclay	11349	

Product	Approval holder	MAPP No.	Expiry Date
Dithane NT Dry Flowable	United Agri	09868	
Ethro 48	RouteOne	13333	31-07-12
Milentus Ethephon	Milentus	12185	
RouteOne Ethro 48	Albaugh UK	13639	31-07-12
2-chloroethylphosphonic acid + mepiquat chloride			
CleanCrop Fonic M	United Agri	09553	
Mepiquat Plus	RouteOne	13297	
RouteOne Mepiquat Plus	Albaugh UK	13636	
chloropicrin			
Custo-Fume	Custodian	13272	
chlorothalonil			
Bravo 720	Syngenta	10519	29-02-08
Busa	AgChem Access	13452	28-02-11
CleanCrop Rio	United Agri	13332	
Cropguard 2	Nufarm UK	13193	
Jupital	Syngenta	10528	
Mainstay	Quadrangle	05625	
Repulse	Certis	07641	30-09-11
RouteOne Ronil 50	Albaugh UK	13637	28-02-11
chlorothalonil + cyproconazole			
Bravo Xtra	Syngenta	11824	
Cleancrop Cyprothal	United Agri	09580	
Octolan	Syngenta	11675	
SAN 703	Syngenta	11676	
chlorothalonil + cyproconazole + propiconazole			
UK 282	Syngenta	13204	
chlorothalonil + flutriafol			
Argon	Headland	12312	
chlorothalonil + propamocarb hydrochloride			
Pan Wizard	Pan Agriculture	11953	
chlorothalonil + tebuconazole			
Pentangle	Nufarm UK	13746	
chlorotoluron			
Chute	Makhteshim	13520	
Rotor	Makhteshim	13511	
chlorotoluron + diflufenican			
Buckler	Nufarm UK	13476	
chlorotoluron + isoproturon			
Uppercut	Nufarm UK	12591	30-06-09

Product	Approval holder	MAPP No.	Expiry Date
chlorpropham			
Atlas CIPC 40	Atlas	07710	
Comrade	United Phosphorus	13566	31-01-10
Gro-Stop Innovator	Certis	13743	
Luxan Gro Stop 100	Luxan	12052	
Luxan Gro-Stop Innovator	Luxan	12340	
MSS CIPC 30M	Nufarm UK	10064	
Sprout Nip	Aceto	11786	
chlorpropham + metamitron			
Golmet	Interfarm	12858	31-08-08
chlorpyrifos			
Agriguard Chlorpyrifos	AgriGuard	13298	30-06-11
Alpha Chlorpyrifos 48 EC	Makhteshim	04821	31-08-08
Barclay Clinch II	Barclay	11346	31-08-08
Chlobber	AgChem Access	13723	
Govern	United Agri	12870	31-05-08
Lorsban 480	Dow	08076	31-08-08
Lorsban WG	Dow	10139	
Lorsban WG	Landseer	11962	
Pirisect	AgriGuard	13371	30-06-11
Pyrinex 48 EC	Makhteshim	13534	
Pyrinex 48EC	Makhteshim	08644	31-08-08
Spannit	SumiAgro Amenity	08744	31-08-08
suSCon Indigo	Fargro	09902	
chlorthal-dimethyl			
Dacthal W-75	AMVAC	10289	
RouteOne Chlorthal-D75	Albaugh UK	13649	
cinidon-ethyl			
Lotus	Nufarm UK	13744	30-09-12
Lotus	BASF	09231	31-10-11
clodinafop-propargyl			
RouteOne Roclod 24	RouteOne	13638	31-01-12
Topik	Syngenta	08461	31-08-08
Tuli	AgChem Access	13302	31-01-12
Viscount	Syngenta	12642	
Viscount	Syngenta	12309	31-08-08
clodinafop-propargyl + diflufenican			
Amazon	Bayer CropScience	10266	
Amazon	Syngenta	08128	

SECTION 3

Product	Approval holder	MAPP No.	Expiry Date
clodinafop-propargyl + trifluralin			
Hawk	Syngenta	08417	31-08-08
Reserve	Syngenta	12635	20-03-09
clomazone			
RouteOne Clomaz 36	Albaugh UK	13697	
clopyralid			
Agrotech-Clopyralid 200 sl	Agrotech-Trading	12445	
Barclay Karaoke	Barclay	11357	30-04-09
Cliophar	Agriphar	13360	
Cliophar	Chimac-Agriphar	09430	30-04-09
Loncid	I T Agro	10832	30-04-09
Milentus Clopyralid	Milentus	12448	
Vivendi 200	Agrichem	12782	
clopyralid + 2,4-D + MCPA			
Lonpar	Dow	08686	
clopyralid + florasulam + fluroxypyr			
Bofix FFC	Dow	13152	31-10-12
clopyralid + picloram			
Galaxy	AgriGuard	12228	30-04-08
Prevail	Dow	13205	
Prevail	United Agri	13033	31-05-08
clopyralid + propyzamide			
Matrikerb	Dow	10806	
clopyralid + triclopyr			
Blaster	Headland Amenity	13267	
Grazon 90	Dow	13117	
clothianidin			
Bayer UK 978	Bayer CropScience	11201	31-01-09
Poncho 250	Bayer CropScience	12287	13-02-09
copper hydroxide			
Spin Out	DuPont	12069	
copper silicate			
Socusil Slug Spray	Doff Portland	11817	
copper sulphate			
Cuproxat FL	Nufarm UK	13241	
cyazofamid + polyalkyleneoxide modheptamethylsiloxane			
Roazafod	RouteOne	13530	30-06-13
RouteOne Roazafod	Albaugh UK	13666	30-06-13

Product	Approval holder	MAPP No.	Expiry Date
cycloxydim			
Stratos	BASF	13012	
Stratos	BASF	06891	31-08-08
cyfluthrin			
Baythroid	Makhteshim	11663	31-08-08
cymoxanil			
Cymoxanil 45% WG	Globachem	13727	
cymoxanil + mancozeb			
Cleancrop Cyman	United Agri	12232	
Cleancrop Xanilite	United Agri	10050	
Curzate M68	DuPont	08072	
Curzate M68 WSB	DuPont	08073	
Matilda WG	Nufarm UK	13732	
Solace WG	Nufarm UK	13729	
Standon Cymoxanil Extra	Standon	09442	
Zetanil	Sipcam	11993	
cypermethrin			
Afrisect 10	Agriphar	13159	
Afrisect 10	Chimac-Agriphar	12027	31-08-08
Cleancrop Pyrimet	United Agri	10619	31-08-08
Cyperkill 10	Agriphar	13157	31-12-12
Cyperkill 10	Chimac-Agriphar	12028	31-08-08
Cyperkill 25	Agriphar	13160	
Cyperkill 25	Chimac-Agriphar	12029	31-08-08
Cyperkill 5	Agriphar	13162	
Cyperkill 5	Chimac-Agriphar	12026	31-08-08
I T Cyper	I T Agro	10557	31-08-08
MCC 25 EC	Agriphar	13161	
MCC 25 EC	Chimac-Agriphar	12030	31-08-08
Permasect C	Nufarm UK	11121	31-08-08
Sherpa 100 EC	SBM	13570	28-02-11
Toppel 10	United Phosphorus	08772	31-08-08
cyproconazole			
Alto 240 EC	Syngenta	12613	31-08-08
Cabaret	Bayer CropScience	11915	31-08-08
Caddy 240 EC	Bayer CropScience	11256	31-08-08
cyproconazole + propiconazole			
Alto Xtra	Syngenta	11839	
Menara	Syngenta	09321	
cyproconazole + quinoxyfen			
DOE 1762	Dow	08962	

Product	Approval holder	MAPP No.	Expiry Date
cyprodinil			
Barclay Amtrak	Barclay	11338	
Cleancrop Cyprodinil	United Agri	09668	
Standon Cyprodinil	Standon	09345	
2,4-D			
Damine	Agriphar	13366	19-09-08
Damine	Chimac-Agriphar	12567	19-09-08
Dioweed 50	United Phosphorus	13197	30-09-12
Growell 2,4-D Amine	GroWell	13146	30-09-12
Herboxone 60	Marks	13145	30-09-12
Maton	Headland	13234	30-11-10
2,4-D + dicamba			
Lawn Builder Plus Weed Control	Scotts	08499	
Lawn Spot Weeder	Bayer CropScience	13116	
New Estermone	Vitax	13792	
New Estermone	Vitax	06336	31-08-08
2,4-D + dicamba + fluroxypyr			
Holster XL	Barclay	13596	
2,4-D + dicamba + MCPA + mecoprop-P			
Dicophar	Agriphar	13256	
2,4-D + dicamba + triclopyr			
Cleancrop Broadshot	United Agri	11664	
Nu-Shot	Nufarm UK	11148	31-05-08
2,4-D + MCPA			
Agroxone Combi	Marks	11025	
2,4-D + mecoprop-P			
Sydex	Vitax	06412	
2,4-D + picloram			
Atladox HI	Nomix Enviro	05559	
daminozide			
B-NINE SG	Chemtura	12698	
Dazide	Fine	02691	31-08-08
dazomet			
Assassin	AgriGuard	13737	
Basamid	Kanesho	12895	
2,4-DB			
Butoxone DB	Marks	13680	
Butoxone DB	Marks	10482	31-12-08
DB Straight	United Phosphorus	13736	

Product	Approval holder	MAPP No.	Expiry Date
2,4-DB + MCPA			
Butoxone DB Extra	Marks	12152	
MSS 2,4-DB + MCPA	Mirfield	01392	
deltamethrin			
Agrotech Deltamethrin	Agrotech-Trading	12165	01-11-08
Cleancrop Decathlon	United Agri	12834	01-11-08
Deleet	Rentokil	09312	01-11-08
Delta-M 2.5 EC	MAC	12604	01-11-08
K-Obiol EC 25	Bayer Environ.	13573	14-06-11
K-Obiol ULV 6	Bayer Environ.	13572	14-06-11
Milentus Deltamethrin	Milentus	12219	01-11-08
Routeone Deltam 10	Albaugh UK	13615	01-11-08
desmedipham + ethofumesate + metamitron + phenmedipham			
Betanal Quattro	Bayer CropScience	12779	
desmedipham + ethofumesate + phenmedipham			
Agrotech-Desmedipham Plus EC	Agrotech-Trading	12381	
dicamba			
Cadence	Syngenta	08796	
Cadence	Barclay	09578	
I T Dicamba	I T Agro	10976	
dicamba + dichlorprop-P + MCPA			
Cleanrun 3	Scotts	11410	31-08-08
dicamba + MCPA + mecoprop-P			
ALS Premier Selective Plus	Amenity Land	08940	
Mascot Super Selective Plus	Barclay	12839	
Mircam Super	Nufarm UK	11836	
Premier Amenity Selective	Amenity Land	10098	
Trireme	Nufarm UK	11524	
dicamba + mecoprop-P			
Foundation	Syngenta	08475	
Headland Swift	Headland	11945	
Optica Forte	Marks	11913	
dichlobenil			
Casoron G	Chemtura	09022	
Casoron G	Miracle	07926	
Casoron G4	Chemtura	09215	
Casoron G4	Miracle	07927	
Casoron G-SR	Miracle	07925	
Dichlobenil Granules	Certis	11871	
Dicore	Agrichem	12542	
Embargo G	Barclay	12690	

Product	Approval holder	MAPP No.	Expiry Date
Luxan Dichlobenil Granules	Luxan	09250	30-09-11
Sierraron G	Scotts	09263	
Sierraron G	Scotts	09675	
Sierraron G4	Scotts	10491	
Standon Dichlobenil 6 G	Standon	08874	
Viking Granules	Nomix Enviro	11859	
dichlorprop-P			
Optica DP	Marks	07818	
Optica DP	Marks	11067	
dichlorprop-P + ferrous sulphate + MCPA			
Vitagrow Granular Feed, Weed & Mosskiller	Sinclair	10971	
dichlorprop-P + ioxynil			
Duet	DuPont	11553	
Mextrol DP	Nufarm UK	11529	
dichlorprop-P + MCPA			
Optica Duo	Marks	10343	
dichlorprop-P + MCPA + mecoprop-P			
Optica Trio	Marks	09747	
diclofop-methyl + fenoxaprop-P-ethyl			
Corniche	Bayer CropScience	08947	
Tigress Ultra	Bayer CropScience	08946	
difenoconazole			
Landgold Difenoconazole	Landgold	09964	
diflubenzuron			
Dimilin 25-WP	Chemtura	08902	
diflufenican			
Alliance Pro	Nufarm UK	12434	
Alpha DFF 500 SC	Makhteshim	12187	31-05-08
Alpha Diflufenican 500 SC	Makhteshim	13182	
CleanCrop Precinct	United Agri	13472	
Crater	Makhteshim	13084	
Dazzler	AgriGuard	13620	
DFF 500 WG	Goldengrass	13569	
Dican	Agform	13685	
Dina 50	Globachem	13627	
IT Diflufenican	I T Agro	13044	
Overlord	Makhteshim	13521	
Sempra	Agrichem	13525	
Viola	Bayer CropScience	13321	

Product	Approval holder	MAPP No.	Expiry Date
diflufenican + flurtamone + isoproturon			
Ingot	Bayer CropScience	09997	30-06-09
diflufenican + isoproturon			
Flarepath	Headland	13156	30-06-09
Javelin	Bayer CropScience	12367	30-06-09
Javelin Gold	Bayer CropScience	12536	30-06-09
Panther	Bayer CropScience	10745	30-06-09
Panther WDG	Bayer CropScience	10650	30-06-09
Wombat	Bayer CropScience	13028	30-06-09
diflufenican + trifluralin			
Ardent	Makhteshim	09968	31-08-08
Magdelin	Nufarm UK	13184	20-03-09
dimethoate			
Danadim	Headland	11550	
Rogor L40	Interfarm	07611	
Sector	Cheminova	10492	
dimethomorph + mancozeb			
Saracen	BASF	12005	
dinocap			
GF-1478	Dow	13399	07-03-11
diphenylamine			
No-Scald DPA	Cerexagri	08312	31-08-08
diquat			
A1412A2	Syngenta	13440	31-12-11
Agriguard diquat	AgriGuard	13763	05-03-11
Cleancrop Diquat	United Agri	13348	31-12-11
I.T. Diquat	I T Agro	13557	31-12-11
Mission	Agrichem	13411	31-12-11
Quad S	Q-Chem	13578	28-06-11
diquat + paraquat			
Speedway 2	Scotts	10790	31-12-08
dithianon			
Barclay Cluster	Barclay	11347	
diuron			
Barclay Diuron 500 SC	Barclay	13003	13-12-08
Diurex 90 WDG	Makhteshim	12767	13-12-08
Dorac 80 WP	Lanxess	13133	13-12-08
Freeway	Bayer Environ.	13247	13-12-08
I.T. Diuron	I T Agro	12240	13-12-08

Product	Approval holder	MAPP No.	Expiry Date
Karmex	DuPont	12068	13-12-08
Karmex Flowable	DuPont	12284	13-12-08
MSS Diuron 500 FL	Nufarm UK	11493	13-12-08
MSS Diuron 80 WP	Nufarm UK	11503	13-12-08
n2n Diuron Flowable	Nomix Enviro	11820	13-12-08
n2n Diuron Flowable	Nomix Enviro	11822	13-12-08
Nomix Diuron 80	Nomix Enviro	12079	13-12-08
Rescind	RP Amenity	08036	13-12-08
diuron + glyphosate			
CleanGuard	Amenity Land	12055	13-12-08
NX 2075	Nomix Enviro	11104	13-12-08
NX 3083	Nomix Enviro	08712	13-12-08
NX 3083	Nomix Enviro	11100	13-12-08
Total	Nomix Enviro	07932	13-12-08
Total	Nomix Enviro	11099	13-12-08
Touche	Nomix Enviro	11097	13-12-08
Trymark	Nomix Enviro	11867	13-12-08
dodeca-8,10-dienyl acetate			
Exosex CM	Exosect	12103	
dodemorph			
F238	BASF	00206	31-08-08
dodine			
Barclay Dodex	Barclay	11351	
Syllit 400 SC	Agriphar	13363	
Syllit 400 SC	Chimac-Agriphar	11079	
epoxiconazole			
Agrotech-Epoxiconazole 125 SC	Agrotech-Trading	12382	
Cleancrop EPX	United Agri	12135	31-08-08
Epro 125	RouteOne	13375	
Milentus Epoxiconazole	Milentus	12363	
Oropa	AgChem Access	13435	
RouteOne Epro 125	Albaugh UK	13640	
epoxiconazole + fenpropimorph			
Barclay Riverdance	Barclay	11341	
Landgold Epoxiconazole FM	Landgold	08806	
Optiorb	AgriGuard	13426	
Tango Super	BASF	13283	
epoxiconazole + fenpropimorph + kresoxim-methyl			
Allegro Plus	BASF	12218	
Asana	BASF	11934	
BAS 493F	BASF	11748	

Product	Approval holder	MAPP No.	Expiry Date
Bullseye	AgriGuard	12278	
Cleancrop Chant	United Agri	11746	
Mastiff	BASF	11747	
epoxiconazole + fenpropimorph + pyraclostrobin			
Diamant	BASF	11557	
RouteOne Super 3	Albaugh UK	13721	
epoxiconazole + kresoxim-methyl			
Allegro	BASF	12220	
Barclay Avalon	Barclay	11337	
Cleancrop Kresoxazole	United Agri	09698	
Landgold Strobilurin KE	Landgold	09908	
Serial Duo	AgriGuard	12252	
ethanol			
Ethy-Gen II	Banana-Rite	13412	
ethofumesate			
Agriguard Ethofumesate 200	AgriGuard	10698	29-02-08
Alpha Ethofumesate 50SC	Makhteshim	12697	31-03-08
Barclay Keeper 500 Flow	Barclay	11843	29-02-08
Barclay Keeper 500 SC	Barclay	13430	28-02-13
Barclay Keeper EC	Barclay	11887	29-02-08
Catalyst	AgriGuard	12592	29-02-08
Cleancrop EC 200	Barclay	11886	29-02-08
Cleancrop SC 500	Barclay	11844	29-02-08
Ethofumesate 2000	Barclay	12406	29-02-08
Ethosat 500	Makhteshim	11501	29-02-08
I T Ethofumesate	I T Agro	11108	29-02-08
Kubist 2	Nufarm UK	11126	29-02-08
Kubist Flo	Nufarm UK	11167	29-02-08
Landgold Ethofumesate 200	Landgold	08980	29-02-08
Linesman 200 EC	Barclay	11885	29-02-08
Linesman 500 SC	Barclay	11845	29-02-08
Nortron Flo	Bayer CropScience	08154	29-02-08
Oblix 500	Agrichem	12349	28-02-13
Salute	United Phosphorus	07660	29-02-08
ethofumesate + metamitron			
Goltix Uno	Makhteshim	12602	
ethofumesate + metamitron + phenmedipham			
MAUK 540	Makhteshim	11545	

Product	Approval holder	MAPP No.	Expiry Date
ethofumesate + phenmedipham			
Fenlander 2	Makhteshim	12282	31-08-08
Magic Tandem	Bayer CropScience	11814	
Twin	Makhteshim	11172	31-08-08
etridiazole			
Aaterra WP	Zeneca	06625	31-08-08
Terrazole 35 WP	Chemtura	09800	31-08-08
Terrazole 35 WP	Scotts	10468	31-08-08
fatty acids			
Finalsan	Certis	13102	
Finalsan Moss Control for Lawns Concentrate	Vitax	13141	
Finalsan Weed Killer RTU	Vitax	13143	
NEU1170 H	Growing Success	12971	
Organic Moss Control for Lawns Concentrate	Growing Success	13140	
Organic Weed Killer RTU	Growing Success	13142	
Safers Insecticidal Soap	Woodstream	07197	
fatty acids + maleic hydrazide			
Ultima	Neudorff	13347	
fenamidone + mancozeb			
Sonata	Bayer CropScience	11570	
fenarimol			
Pan Fenarimol	Pan Agriculture	13377	30-06-08
fenazaquin			
Matador 200 SC	Margarita	11058	
fenbuconazole			
Indar 5 EW	Whelehan	09644	
fenbuconazole + propiconazole			
Graphic	Interfarm	10987	
fenbutatin oxide			
Torq	Fargro	08370	31-08-08
Torque	BASF	10399	31-08-08
fenhexamid			
Agrovista Fenamid	Agrovista	13733	31-05-11
RouteOne Fenhex 50	Albaugh UK	13665	31-05-11
fenhexamid + tolylfluanid			
Talat	Certis	11311	

Product	Approval holder	MAPP No.	Expiry Date
fenoxaprop-P-ethyl			
Cheetah Energy	Bayer CropScience	13599	
fenpropidin			
Cleancrop Fulmar	United Agri	12033	
Landgold Fenpropidin 750	Landgold	08973	
Mallard	Syngenta	08662	
Patrol	Syngenta	10531	
fenpropidin + prochloraz			
SL 552A	Syngenta	08673	
Sponsor	BASF	11902	
fenpropidin + propiconazole			
Prophet	Syngenta	08433	
Sheen	Syngenta	08442	31-08-08
Zulu	Syngenta	08464	31-08-08
fenpropidin + tebuconazole			
SL 556 500 EC	Syngenta	08449	
fenpropimorph			
Cleancrop Fenpro	United Agri	09885	
Cleancrop Fenpropimorph	United Agri	09445	
Keetak	BASF	06950	
Raven	AgriGuard	13188	
fenpropimorph + flusilazole			
BAS 48500F	BASF	06784	30-06-08
fenpropimorph + kresoxim-methyl			
Cleancrop Duster	United Agri	12306	
Landgold Strobilurin KF	Landgold	09196	
fenpropimorph + pyraclostrobin			
BAS 528 00f	BASF	11444	
Jemker	BASF	12225	
fenpropimorph + quinoxyfen			
Jackdraw	BASF	13769	
fenpyroximate			
NNI-850 5 SC	Certis	12658	
ferric phosphate			
Growing Success Advanced Slug Killer	Growing Success	12378	31-10-11
ferrous sulphate			
Aitken's Lawn Sand	Aitken	05253	
Greenmaster Mosskiller	Scotts	12197	

SECTION 3

Product	Approval holder	MAPP No.	Expiry Date
Maxicrop Moss Killer & Conditioner	Maxicrop	04635	
Moss Control Plus Lawn Fertilizer	Miracle	07912	
No More Moss Lawn Feed	Wolf	10754	
Pentagon Prestige Lawn Sand	Sinclair	10456	
Phostrogen Mosskiller and Lawn Tonic	Bayer CropScience	13112	
Vitagrow Lawn Sand	Vitagrow	05097	
fipronil			
Regent 1 GR	BASF	11743	31-12-08
Regent 1GR	BASF	13090	30-09-09
flazasulfuron			
Chikara	Nomix Enviro	13775	31-05-14
flonicamid			
Mainman	Belchim	13123	04-05-09
florasulam			
Primus 25 SC	Dow	10175	30-09-12
RouteOne Florasul 50	Albaugh UK	13668	30-09-12
florasulam + fluroxypyr			
Image	Dow	13771	31-12-11
Slalom	Dow	13772	31-12-11
florasulam + triclopyr			
Gf-660	Dow	12417	
fluazifop-P-butyl			
Clayton Crowe	Clayton	12572	
Clean Crop Clifford	United Agri	13597	
Fluazifop +	RouteOne	13382	
Fusilade 250 EW	Syngenta	10525	
PP 007	Zeneca	06533	
PP 007	Syngenta	10533	
RouteOne Fluazifop +	Albaugh UK	13641	
Wizzard	Zeneca	06521	
Wizzard	Syngenta	10539	
fluazinam			
Alpha Fluazinam 50SC	Makhteshim	13622	
Barclay Cobbler	Barclay	11348	
Legacy	Syngenta	13735	
Legacy	ISK Biosciences	09966	31-10-11
Ohayo	Syngenta	13734	
Shirlan Programme	Syngenta	10574	
Standon Fluazinam 500	Standon	08670	

Product	Approval holder	MAPP No.	Expiry Date
fluazinam + metalaxyl-M			
Fluazmet M	RouteOne	13519	
RouteOne Fluazmet M	Albaugh UK	13650	
fludioxonil			
Beret Gold	BASF	12656	
fludioxonil + tefluthrin			
Austral Plus	Bayer CropScience	13376	
flufenacet + pendimethalin			
Cleancrop Hector	United Agri	12163	31-12-08
RouteOne Tenac 360	Albaugh UK	13778	31-12-08
fluoxastrobin			
Bayer UK 831	Bayer CropScience	12091	30-04-13
fluoxastrobin + prothioconazole			
Kurdi	AgChem Access	13531	30-04-13
fluoxastrobin + prothioconazole + tebuconazole			
Scenic	Bayer CropScience	12289	
flupyrsulfuron-methyl			
Lexus 50 DF	DuPont	09026	29-02-08
fluquinconazole			
Diablo	BASF	11688	
Jockey F	BASF	11690	
Triplex	BASF	12443	
fluquinconazole + prochloraz			
Baron	BASF	11686	
fluroxypyr			
Agrotech Fluroxypyr	Agrotech-Trading	12294	31-12-11
Alpha Fluroxypyr 20EC	Makhteshim	13352	31-12-11
Barclay Hudson	Barclay	13606	03-04-11
Barclay Hurler	Barclay	12425	31-12-11
Cerastar	Cera Chem	13183	31-12-11
Clean Crop Gallifrey	Inter-Trade	13040	08-05-10
Floxy	Agriphar	13367	31-12-11
Floxy	Chimac-Agriphar	12984	31-12-11
Hurler	Barclay	13458	31-12-11
IT Fluroxypyr	Inter-Trade	12914	08-05-10
Milentus Fluroxypyr	Milentus	12266	31-12-11
Minstrel	United Phosphorus	13745	08-05-10
Scorane	AgChem Access	13301	30-09-08

Product	Approval holder	MAPP No.	Expiry Date
Skora	AgChem Access	13422	31-12-11
Snapper	Nufarm UK	13460	30-04-11
fluroxypyr + triclopyr			
Evade	Dow	08071	31-08-08
flusilazole			
Genie 25	DuPont	10285	29-02-08
Lyric	DuPont	08252	28-02-08
Sanction 25	DuPont	10284	29-02-08
flutolanil			
NNF-136	Nihon Nohyaku	13120	
NNF-136	Nihon Nohyaku	11585	31-08-08
flutriafol			
Impact	Headland	12776	
fosetyl-aluminium			
Clayton Vite	Clayton	12727	31-08-08
Clean Crop Chicane	United Agri	12607	
Fosal	AgriGuard	13384	
I T Fosetyl-AL	I T Agro	11717	
Routeone Fosetyl 80	Albaugh UK	13625	
fosetyl-aluminium + propamocarb hydrochloride			
Previcur Energy	Bayer CropScience	13342	
fuberidazole + imidacloprid + triadimenol			
Baytan Secur	Bayer CropScience	11253	
fuberidazole + triadimenol			
Baytan Flowable	Makhteshim	11714	
gibberellins			
Berelex	Nufarm UK	08903	
Regulex	Sumitomo	10147	
Regulex 10 SG	Sumitomo	13323	
glufosinate-ammonium			
Challenge	Bayer CropScience	07306	
Nomix Touchweed	Nomix Enviro	09596	
glyphosate			
Accelerate	Headland	13390	30-06-12
Acrion	Bayer Environ.	12677	30-06-12
Barclay Barbarian	Barclay	12714	30-06-12
Barclay Gallup 360	Barclay	12659	30-06-12
Barclay Gallup Amenity	Barclay	13250	30-06-12
Barclay Gallup Biograde 360	Barclay	12660	30-06-12

Product	Approval holder	MAPP No.	Expiry Date
Barclay Gallup Biograde Amenity	Barclay	12716	30-06-12
Barclay Gallup Hi-Aktiv	Barclay	12661	30-06-12
Barclay Garryowen	Barclay	12715	30-06-12
Cleancrop Egret	United Agri	12667	30-06-12
Cleancrop Hoedown	United Agri	12913	30-06-12
Cleancrop Tungsten	United Agri	13049	30-06-12
Clinic	Nufarm UK	12678	30-06-12
Clipper	Makhteshim	13217	30-06-12
Deep Root Path & Patio Weedkiller 2	Growing Success	13150	06-06-10
Discman Biograde	Barclay	12856	01-03-10
First Line	Linemark	12759	02-02-09
Gallup Hi- Aktiv Amenity	Barclay	12898	30-06-12
Glister	Sinon EU	12990	30-06-12
Glydate	Nufarm UK	12679	30-06-12
Glymark	Nomix Enviro	12994	30-06-12
Glyper	Interfarm	13105	30-06-12
Glyphosate 360	SumiAgro	12680	30-06-12
Glyphosate 360	Monsanto	11726	30-06-08
Glyphosate 360	Cardel	12670	30-06-12
Initial Line	Linemark	12758	02-02-09
KN 540	Monsanto	12009	30-06-12
MON 79351	Monsanto	12860	07-03-09
MON 79376	Monsanto	12664	30-11-08
MON 79545	Monsanto	12663	30-11-08
MON 79632	Monsanto	12582	30-06-12
Nomix Conqueror	Nomix Enviro	12370	10-04-08
Nomix G	Nomix Enviro	12921	30-06-12
Nomix Nova	Nomix Enviro	12920	30-06-12
Nomix Revenge	Nomix Enviro	12993	30-06-12
Nufosate	Nufarm UK	12699	30-06-12
Pontil 360	Mastra	12733	30-06-12
Preline	Linemark	12756	02-02-09
Preline Plus	Linemark	12757	02-02-09
Proliance Quattro	Syngenta	13670	30-06-12
Reliance	Cardel	12676	30-06-12
Romany	Greencrop	12681	30-06-12
Roundup	Monsanto	12645	30-06-12
Roundup Amenity	Monsanto	12672	30-06-12
Roundup Amenity	Monsanto	10319	30-06-08
Roundup Biactive 3G	Monsanto	13409	20-12-10
Roundup Biactive Dry	Monsanto	12646	30-06-12
Roundup Biactive K	Monsanto	13248	20-12-10
Roundup Express	Monsanto	12526	30-06-12
Roundup Gold	Monsanto	10975	30-06-12
Roundup Greenscape	Monsanto	12731	30-06-12

Product	Approval holder	MAPP No.	Expiry Date
Roundup Pro-Green	Monsanto	11907	30-06-12
Roundup ProVide	Monsanto	11721	30-06-08
Roundup Provide	Monsanto	12953	30-06-12
Roundup Rail	Monsanto	12671	30-06-12
Roundup Ultimate	Monsanto	12774	30-11-08
Roundup Ultra ST	Monsanto	10199	30-06-08
Roundup Ultra ST	Monsanto	12732	30-06-12
RouteOne Rosate 360	Albaugh UK	13659	30-06-12
RouteOne Rosate 360	Albaugh UK	13661	30-06-12
Scorpion	Cardel	11656	
Stacato	Sipcam	12675	30-06-12
Statis	AgriGuard	13079	30-06-12
Tesco Complete Weedkiller Spray	Tesco	13151	06-06-10
Trustee Amenity	Barclay	12897	30-06-12
Trustee Elite	Barclay	12662	30-06-12
Tumbleweed Pro-Active	AgChem Access	12729	30-06-12
Weedbuster Super Concentrate	Syngenta	13316	25-11-08
Weedbuster Super Concentrate	Westland Horticulture	13319	25-11-08
imazalil			
Imaz 100 SL	Agrichem	13322	14-02-11
Imaz 200 EC	Agrichem	12991	31-12-08
IT Imazalil 100 SL	I T Agro	12308	01-02-08
Neozil 10 SL	Laboratorios Agrochem	13034	31-12-08
Sphinx	BASF	11764	31-12-11
imazalil + triticonazole			
Robust	BASF	11807	
imidacloprid			
Gaucho FS	Bayer CropScience	11282	
Imidasect	Biologic HC	12899	
Imidasect 70 WS	Biologic HC	13191	
Levington Professional Plus Intercept	Scotts	08569	31-03-08
Neptune	Makhteshim	12992	
Warrant 700 WG	Headland	13229	
imidacloprid + tebuconazole + triazoxide			
Raxil Secur	Bayer CropScience	11298	
4-indol-3-ylbutyric acid			
Seradix 1	Certis	10422	
Seradix 2	Certis	10423	
Seradix 3	Certis	10424	
iodosulfuron-methyl-sodium + mesosulfuron-methyl			
RouteOne Seafarer	Albaugh UK	13654	

Product	Approval holder	MAPP No.	Expiry Date
ioxynil			
Totril	Bayer CropScience	10026	31-08-08
iprodione			
Agrotech-Iprodione 50 WP	Agrotech-Trading	12400	31-08-08
Noble Iprodione	Barclay	12927	31-12-08
Rovral WP	BASF	11694	31-08-08
Standon Iprodione 50 WP	Standon	13781	31-12-08
iprodione + thiophanate-methyl			
Snooker	BASF	11744	31-08-08
isoproturon			
Aligran	Nufarm UK	11761	30-06-09
Aligran 83 WDG	Agform	13698	30-06-09
Alpha Isoproturon 650	Makhteshim	07034	30-06-09
Arelon 2	Nufarm UK	11670	30-06-09
Arelon 500	Nufarm UK	11639	30-06-09
Arelon 700	Nufarm UK	11542	30-06-09
Bison	Gharda	08699	30-06-09
Bison 83 WG	Gharda	10062	30-06-09
Bison 83 WG	Nufarm UK	11580	30-06-09
Clayton Impasse	Clayton	12520	30-06-09
Clayton Siptu 50 FL	Clayton	08751	31-08-08
Dexter	AgChem Access	12551	30-06-09
Fieldgard	Nufarm UK	11770	30-06-09
Fieldgard	Nufarm UK	11541	30-06-09
Genfarm IPU 500	Genfarm	12977	30-06-09
Isoguard	Nufarm UK	10829	30-06-09
Isoguard 83 WG	Gharda	10061	30-06-09
Luxan Isoproturon 500 Flowable	Luxan	12426	30-06-09
Tolkan Liquid	Bayer CropScience	11994	30-06-09
isoproturon + pendimethalin			
Encore	BASF	13506	31-12-08
Trump	BASF	13507	30-06-09
isoproturon + trifluralin			
Autumn Kite	Nufarm UK	11549	20-03-09
Snitch	Nufarm UK	12466	20-03-09
isoxaben			
Flexidor	Dow	05121	
Flexidor 125	Dow	05104	
Gallery 125	Rigby Taylor	06889	
Knot Out	Vitax	05163	

EXPIRY DATE IS 31/12/2013 UNLESS OTHERWISE SHOWN

Product	Approval holder	MAPP No.	Expiry Date
lambda-cyhalothrin			
Barbarossa	Clayton	13510	03-04-11
Clayton Zen	Clayton	13512	03-04-11
Cleancrop Silo	United Agri	13351	13-11-09
Dalda 5	Globachem	13688	28-06-11
IT Lambda	I T Agro	13451	03-04-11
Lambda C	RouteOne	13471	13-11-09
Markate 50	Agrovista	13529	28-06-11
RouteOne Lambda C	RouteOne	13663	13-11-09
Stealth	Syngenta	11514	31-12-11
lambda-cyhalothrin + pirimicarb			
Dovetail	Syngenta	10479	31-08-08
laminarin			
Vacciplant	Goemar	13260	31-03-15
lenacil			
Cleancrop Lenflow	United Agri	10059	
Venzar 80 WP	DuPont	09981	
linuron			
Alpha Linuron 50 WP	Makhteshim	04870	31-12-08
Linurex 50 SC	Makhteshim	07950	31-12-08
maleic hydrazide			
Cleancrop Malahide	United Agri	13629	22-07-11
Cleancrop Malahide	United Agri	11066	31-12-08
Fazor	Dow	13679	
Fazor	Chemtura	13617	
Fazor	Chemtura	05461	31-12-08
Mazide 25	Vitax	02067	31-12-08
Rouge	Nufarm UK	11090	31-12-08
Royal MH 180	Chemtura	07043	31-08-08
Royal MH 180	Certis	11062	31-08-08
Source II	Chiltern	13618	22-07-11
mancozeb			
Agrizeb	Agriphar	13365	
Agrizeb	Chimac-Agriphar	10980	
Beacon	AgriGuard	13368	30-06-11
Cleancrop Mancozeb	United Agri	11193	
Cleancrop Mandrake	United Agri	12500	
Dithane Dry Flowable Newtec	Dow	12564	
Laminator DG	Interfarm	11073	
Laminator FL	Interfarm	11072	
Laminator WP	Interfarm	11071	
Malvi	AgChem Access	12862	

Product	Approval holder	MAPP No.	Expiry Date
Manfil 80 WP	Indofil	12941	
Manfill 75 WG	Indofil	12419	
Mewati	AgChem Access	13029	
Penncozeb WDG	Nufarm UK	11622	
Trimanzone	Intracrop	09584	
Zebra	Headland	13155	30-06-11
mancozeb + metalaxyl-M			
MetMan 680	RouteOne	13553	30-06-11
RouteOne MetMan 680	Albaugh UK	13652	30-06-11
mancozeb + zoxamide			
Unikat 75 WG	Dow	11018	
maneb			
Trimangol 80	Cerexagri	06871	
Trimangol WDG	Cerexagri	06992	
MCPA			
Agrichem MCPA-50	Agrichem	04097	
Agricorn 500 II	Nufarm UK	09155	
Agritox Dry	Nufarm UK	10554	
Agroxone 40	Marks	10264	
Agroxone 75	Marks	09208	
Circium II	Nufarm UK	11801	
Growell MCPA DMA 500	GroWell	12989	
MCPA 25%	Nufarm UK	07998	
MCPA 500	Nufarm UK	08655	
Nufarm MCPA 750	Nufarm UK	11768	30-04-11
Nufarm MCPA Amine 50	Nufarm UK	12046	
MCPA + MCPB			
Butoxone Plus	Marks	11080	
Trifolex-Tra	BASF	10396	
MCPA + mecoprop-P			
No More Weeds Lawn Feed	Wolf	10747	
Optica Combi	Marks	10118	
mecoprop-P			
Duplosan 500	Nufarm UK	12108	
Duplosan KV 500	Nufarm UK	12107	
Duplosan New System CMPP	Nufarm UK	12106	
Dupont Mecoprop-p	DuPont	11739	
Headland Charge	Headland	10981	
Landgold Mecoprop-p	Teliton	12121	
Landgold Mecoprop-p	Landgold	06052	

Product	Approval holder	MAPP No.	Expiry Date
mecoprop-P + metribuzin			
Optica Plus	Marks	09325	
mecoprop-P + metsulfuron-methyl			
Headland Neptune	Headland	10230	
mepiquat chloride + prohexadione-calcium			
BAS 122 06W	BASF	12831	31-03-08
Medax Top	BASF	13282	
mesosulfuron-methyl			
AEF 6012-33H	Bayer CropScience	11218	14-07-09
mesotrione			
RouteOne Trione 10	Albaugh UK	13765	17-02-09
mesotrione + terbuthylazine			
RouteOne Mesot	Albaugh UK	13576	
metalaxyl-M			
Fongarid Gold	Syngenta	12547	30-09-12
metaldehyde			
Allure	Chiltern	12651	
Antares	De Sangosse	12998	
Appeal	Chiltern	12713	
Aristo	De Sangosse	11797	
Aristo 5	Certis	13495	
Aristo 6	Certis	13494	
Aristo M	De Sangosse	13278	
Aristo M	De Sangosse	12394	31-07-08
Attract	Chiltern	12712	
Barclay Metaldehyde Dry	Barclay	12446	
Blue Prince	Doff Portland	13061	
Brits	Doff Portland	11792	
Brits M	De Sangosse	12956	
Carakol Plus	Kollant	13004	
Certis 9363	Certis	13486	
Certis 9363 Blue	Certis	13487	
Certis Metaldehyde	Certis	13489	
Chiltern Blues	Chiltern	12702	
Chiltern Hardy	Chiltern	12647	
Chiltern Hundreds	Chiltern	12710	
Clartex	De Sangosse	12935	
Clean Crop Hyde	United Agri	12025	
Clean Crop Hyde	United Agri	12707	
Clean Crop Jekyll D	United Agri	12490	
Clean Crop Jekyll D 5	Doff Portland	13060	

Product	Approval holder	MAPP No.	Expiry Date
Cleancrop Hyde 2	United Agri	13199	
Cleancrop Jekyll	United Agri	11724	
Cleancrop Jekyll	Chiltern	12703	
Dixie 6 M	De Sangosse	13277	30-11-08
Dixie 6 M	De Sangosse	12398	31-07-08
Dixie 6 M	Clayton	13478	
Duracoat Slug Killer	Doff Portland	13134	
Entice	De Sangosse	11096	30-09-08
Escar-Go 6	Chiltern	12649	
ESP	De Sangosse	10938	30-09-08
Helimax	De Sangosse	11720	31-03-08
Helimax	De Sangosse	13108	
Helimax S	De Sangosse	12964	31-10-08
Keel Over	De Sangosse	13110	
Kimble	Chiltern	12907	
Limagri GR	SBM	13475	
Luxan 9363	Luxan	07359	
Luxan 9363 Blue	Luxan	12391	
Luxan Deal	Luxan	12934	
Luxan Metaldehyde	Luxan	06564	
Luxan Metaldehyde 5	Luxan	12401	
Luxan Red 5	Luxan	12390	
Luxan Trigger	Luxan	10419	
Luxan Trigger 5	Luxan	12389	
Lynx	De Sangosse	11767	
Lynx M	De Sangosse	11598	
Lynx M	De Sangosse	13280	
Lynx S	Lonza	12397	
Metarex Amba	De Sangosse	13245	
Metarex Amba	De Sangosse	13068	31-12-08
Metarex Green	De Sangosse	13069	
Metarex RG	De Sangosse	13070	
Metasec M	De Sangosse	13279	
Metasec M	De Sangosse	12396	
Mifaslug	FCC	10292	
Mifaslug 5	Luxan	12392	
Molotov	Chiltern	12650	
Monza	Makhteshim	13600	
Optimol XL	De Sangosse	12937	
Pastel M	Clayton	13479	
Pastel M	De Sangosse	13274	30-11-08
Pastel M	De Sangosse	12395	
Pesta	De Sangosse	11796	
Pesta M	De Sangosse	12399	31-07-08
Regel	De Sangosse	13115	

SECTION 3

EXPIRY DATE IS 31/12/2013 UNLESS OTHERWISE SHOWN

Product	Approval holder	MAPP No.	Expiry Date
Regel Star	De Sangosse	12441	29-02-08
Regel Star	De Sangosse	13071	
Rowent	Chiltern	12709	
Sloggy	SBM	13373	
Slug Pellets	De Sangosse	12936	
Slugdown	Doff Portland	13063	
Slug-Lentils	Frunol Delicia	13613	
Steadfast	Doff Portland	13062	
Super Six	Doff Portland	11794	
Super-Flor 6% Metaldehyde Slug Killer Mini Pellets	CMI	11789	
Trigger	Certis	13492	
Unicrop 6% Mini Slug Pellets	Unicrop	11795	
metamitron			
Barclay Seismic	Barclay	11378	
Bettix 70 WG	United Phosphorus	11019	
Celmitron 70WDG	SD Agchem	13630	
Cobra	Interfarm	10578	31-08-08
Fernpath Haptol Flo	AgriGuard	13549	
Homer	Makhteshim	11544	
Inter Metamitron WG	I T Agro	12388	
IT Metamitron SC	I T Agro	13059	
IT Metamitron WG	I T Agro	12467	
Lektan	Makhteshim	11543	
Maymat 70 WDG	Global Crop Care	12477	
MM 70	Nufarm UK	11582	
MM 70 Flo	Nufarm UK	11197	31-08-08
MM70	Gharda	09490	
Standon Metamitron	Standon	07885	
Target SC	Unicrop	11070	31-08-08
Target SC	Agrichem	13307	
metam-sodium			
Fumetham	Chemical Nutrition	10047	
Metam 510	Taminco	09796	
Metham Sodium 400	United Phosphorus	08051	
Sistan	Unicrop	01957	
Sistan 38	Unicrop	08646	
Vapam	Willmot Pertwee	09194	
metazachlor			
Agrotech Metazachlor 500 SC	Agrotech-Trading	12454	
Barclay Metaza	Barclay	11359	
Butisan S	BASF	00357	
Clean Crop MTZ 500	United Agri	09222	
Fuego	Makhteshim	11177	

Product	Approval holder	MAPP No.	Expiry Date
Fuego 50	Makhteshim	11473	
Gharda Bonanza	Nufarm UK	11571	31-08-08
Gharda Bonanza 500 SC	Nufarm UK	13628	
Luxan Taurus	Luxan	12442	
Mashona	AgChem Access	13006	
Metarap 500 SC	Global Crop Care	12540	
Metaz 50	RouteOne	13467	
Metazachlore GL 500	Globachem	12528	
Mezzanine	AgriGuard	13719	
Noble Metazachlor	Barclay	12892	
RouteOne Metaz 50	Albaugh UK	13648	
metconazole			
Conzole	AgriGuard	13444	
Pan Metconazole	Pan Agriculture	10473	
methiocarb			
Barclay Poacher	Barclay	09031	
Decoy	Bayer CropScience	11264	
Draza Wetex	Bayer CropScience	11271	
Lupus	Bayer CropScience	11291	
New Draza	Bayer CropScience	11294	
methyl bromide with chloropicrin			
Methyl Bromide 98	Bromine & Chem.	01335	
1-methylcyclopropene			
SmartFresh SmartTabs	Landseer	12684	24-09-10
metiram			
Polyram DF	BASF	08234	
metribuzin			
Citation 70	United Phosphorus	09370	
Clean Crop Solmet B	United Agri	12273	
Cleancrop Metribuzin	United Agri	09621	
Cleancrop Solmet	United Agri	10865	
Lanxess Metribuzin 70 WG	Lanxess	12487	
Lexone 2	DuPont	04991	
Metriphar	Chimac-Agriphar	12354	
Milentus Metribuzin 70 WG	Milentus	12224	
Promor	Bayer CropScience	12502	
Python	Makhteshim	13802	
Sarabi	AgChem Access	13453	30-09-12
Shotput	Makhteshim	13788	
Tuberon	Agrichem	12683	

Product	Approval holder	MAPP No.	Expiry Date
metsulfuron-methyl			
Accurate	Headland	13224	28-11-10
Agriguard Metsulfuron	AgriGuard	13324	29-01-10
Alias	DuPont	13397	30-06-11
Alias SX	DuPont	13398	30-06-11
Ally SX	DuPont	12059	30-06-11
Anvil XT	AgriGuard	12849	31-05-08
Cimarron	DuPont	13408	15-03-11
Cleancrop Mondial	United Agri	13353	30-06-11
Finy	Agrichem	12855	30-06-11
Forge	Interfarm	12848	30-06-11
Goldron-M	Goldengrass	13466	30-06-11
Lorate	DuPont	12743	30-06-11
Metro 20	AgriGuard	13496	25-04-11
Pike PX	Nufarm UK	13249	13-12-10
Revenge 20 SG	Agform	13633	30-06-11
RouteOne Romet 20	Albaugh UK	13657	30-06-11
metsulfuron-methyl + thifensulfuron-methyl			
Finish SX	DuPont	12259	30-06-11
Presite SX	DuPont	12291	30-06-11
metsulfuron-methyl + tribenuron-methyl			
DP 911 WSB	DuPont	09867	
myclobutanil			
Robut 20	RouteOne	13437	
RouteOne Robut 20	Albaugh UK	13646	
Systhane 20 EW	Whelehan	09397	
Systhane 6 Flo	Whelehan	06551	31-08-08
Systhane 6 Flo	Landseer	07334	31-08-08
Systhane 6 W	Dow	10808	
1-naphthylacetic acid			
Tipoff	Hatrim	12292	
2-naphthyloxyacetic acid			
Betapal Concentrate	Vitax	00234	31-08-08
napropamide			
Banweed	United Phosphorus	09376	
Devrinol	United Phosphorus	09375	
nicosulfuron			
Agrotech-Nicosulfron 40 SC	Agrotech-Trading	12938	
Maizon	AgriGuard	13222	
Maizon	AgriGuard	12833	31-08-08

Product	Approval holder	MAPP No.	Expiry Date
Samson	Syngenta	10433	31-08-08
Samson Extra 6%	Syngenta	13436	
nicotine			
No-FID	Certis	07959	
XL-All Insecticide	Vitax	02369	31-08-08
oxadiazon			
Noble Oxadiazon	Barclay	12889	
paclobutrazol			
Bonzi	Syngenta Bioline	13623	
Bonzi	Zeneca	06640	
Pirouette	Fine	13137	
paraquat			
Clayton Paragon	Clayton	11210	31-08-08
CleanCrop Parachute	United Agri	11876	31-08-08
Dextrone X	Nomix Enviro	00687	31-08-08
Fernpath Graminite	AgriGuard	11646	31-08-08
Gramoxone 100	Syngenta	10526	31-08-08
penconazole			
RouteOne Pencon 10	Albaugh UK	13651	
pencycuron			
Curon SC	Globachem	12449	31-12-08
Luxan Pencycuron	Luxan	12144	
Luxan Pencycuron 250 SC	Luxan	12142	
Monceren Flowable	Bayer CropScience	11425	
Pan Pencycuron P	Pan Agriculture	10494	
Tubercare 12.5 DS	Agrichem	12194	
pendimethalin			
Agriguard Pendimethalin	AgriGuard	13355	31-12-08
Agrotech Pendimethalin 400 SC	Agrotech-Trading	12665	31-12-08
Claymore	BASF	11779	31-12-08
CleanCrop Stomp	United Agri	13443	
Cleancrop Stomp	United Agri	11778	31-12-08
PDM 330 EC	BASF	11084	31-08-08
Routeone Penthal 330	Albaugh UK	13768	
Sovereign	BASF	11781	31-12-08
Sovereign	BASF	13442	
Standon Pendimethalin 400 SC	Standon	10729	31-08-08
Stomp 400 SC	BASF	11777	31-12-08
pendimethalin + picolinafen			
PicoMax	BASF	10739	31-12-08
Picona	BASF	10715	31-12-08

Product	Approval holder	MAPP No.	Expiry Date
PicoPro	BASF	10740	31-12-08
PicoStomp	BASF	11987	31-12-08
Stomp Pico	BASF	10738	31-12-08
pendimethalin + trifluralin			
Sword	Makhteshim	13015	20-03-09
Peniophora gigantea			
PG Suspension	Forest Research	11772	
phenmedipham			
Beetup Flo 160	United Phosphorus	10382	
Pump	Makhteshim	12830	
Shrapnel	Makhteshim	13593	
picloram			
Tordon 22K	Dow	05083	
picolinafen			
AC 900001	BASF	10714	30-09-12
picoxystrobin			
Acanto	DuPont	13043	30-06-09
pinoxaden			
Aoraki	AgChem Access	13528	19-09-09
Roaxe 100	RouteOne	13304	19-09-09
RouteOne Roaxe 100	Albaugh UK	13655	19-09-09
Symmetry	AgriGuard	13577	19-09-09
pirimicarb			
Agrotech-Pirimicarb 50 WG	Agrotech-Trading	12269	
Barclay Pirimisect	Barclay	11360	
Clayton Pirimicarb 50 SG	Clayton	09221	
Cleancrop Miricide	United Agri	11776	
Milentus Pirimicarb	Milentus	12268	
Noble Pirimicarb	Barclay	12877	
Pirimate	AgChem Access	13047	
RouteOne Primro 50 WG	Albaugh UK	13644	31-01-12
prochloraz			
Mirage 45 SC	Makhteshim	10894	
Octave	BASF	11741	
Prelude 20 LF	BASF	11904	
Sporgon 50 WP	Sylvan	08802	
Sportak 45 EW	BASF	11696	
Sportak 45 HF	BASF	11697	
prochloraz + propiconazole			
Mambo	Headland Amenity	13387	

Product	Approval holder	MAPP No.	Expiry Date
prochloraz + tebuconazole			
Orius P	Makhteshim	12880	
prochloraz + thiram			
Hy-Pro Duet	Agrichem	12334	
propachlor			
Alpha Propachlor 65 WP	Makhteshim	04807	
Brasson	Monsanto	10560	
Ramrod 20 Granular	Monsanto	10313	
Tripart Sentinel	Tripart	03250	
Tripart Sentinel 2	Tripart	05140	
propamocarb hydrochloride			
Previcur N	Bayer CropScience	08575	
Propeller	Agrichem	12192	
Propeller	PG - Crop Protection	11811	
propaquizafop			
Agil	Makhteshim	11048	
Alpha Propaquizafop 100 EC	Makhteshim	13689	
Barclay Rebel	Barclay	11648	
Longhorn	AgChem Access	12512	
Noble Propaquizafop	Barclay	12890	
Osprey	Makhteshim	13731	
RouteOne Quizro 10	Albaugh UK	13643	
propiconazole			
Anode	Makhteshim	12943	
propineb			
Antracol	Bayer CropScience	11237	31-03-09
propoxycarbazone-sodium			
Ethos	Bayer CropScience	12740	17-01-09
propyzamide			
Agrotech Propyzamide 400 SC	Agrotech Trading	13179	31-03-09
Bulwark Flo	Dow	10755	31-03-09
Compose	Spectrum	12505	31-03-09
Conform	AgChem Access	13584	31-03-09
Conform	Spectrum	12508	31-03-09
Flomide	Alfa	11103	31-03-09
Kerb 80 EDF	Dow	10761	31-03-09
Kerb Granules	Dow	10803	31-08-08
Mithras 80 EDF	Dow	10767	31-03-09
Propose	Interfarm	12504	31-03-09
Propose	Interfarm	11195	31-08-08
Redeem Flo	Dow	10764	31-03-09

Product	Approval holder	MAPP No.	Expiry Date
RouteOne Zamide Flo	Albaugh UK	13610	31-03-09
RouteOne Zamide WP	Albaugh UK	13589	31-03-09
Wage	Dow	13153	31-03-09
prosulfocarb			
Arcade	Syngenta	12641	
Dian	AgChem Access	13567	
RouteOne Rosulfocarb 80	Albaugh UK	13635	
prothioconazole			
Banguy De	PSI	13575	30-04-13
Prothro 25	RouteOne	13392	30-04-13
RouteOne Prothro 25	Albaugh UK	13658	16-08-08
prothioconazole + spiroxamine			
Prothioxamine	RouteOne	13357	03-01-09
RouteOne Prothioxamine	Albaugh UK	13656	03-01-09
prothioconazole + tebuconazole			
Standon Mastana	Standon	13795	
prothioconazole + trifloxystrobin			
Mobius	Bayer CropScience	13395	30-04-13
Zephyr	Bayer CropScience	13174	24-10-10
pymetrozine			
Chess WG	Syngenta	10651	31-07-08
pyraclostrobin			
BAS 500 06F	BASF	12338	31-05-14
BASF Insignia	BASF	11900	31-05-14
Comet	BASF	10875	31-05-14
Platoon	BASF	12325	31-05-14
pyraflufen-ethyl			
OS159	Ceres	12723	31-10-11
pyrethrins			
Pyrethrum 5 EC	Gooddeed	12638	
SuperSwift Natural Insecticide	Omex	13007	31-10-08
pyrimethanil			
Pyrus 400 SC	Agriphar	13286	31-05-12
RouteOne Pyrimet 40	Albaugh UK	13574	
Standon Pyrimethanil	Standon	10576	
quinoxyfen			
Clean Crop QFN	United Agri	11966	01-09-14

Product	Approval holder	MAPP No.	Expiry Date
quizalofop-P-ethyl			
CoPilot	Interfarm	08042	31-08-08
Sceptre	Bayer CropScience	08043	
rimsulfuron			
Agrotech-Rimsulfuron 25 WG	Agrotech-Trading	12427	
Landgold Rimsulfuron	Landgold	08959	
rotenone			
Devcol Liquid Derris	Nehra	06063	
Liquid Derris	Law Fertilisers	01213	
sodium chlorate			
Aztec Complete Weedkiller	Aztec Chemicals	12485	
Blagden Professional Sodium Chlorate Weedkiller	Blagden	12624	
Blitzweed	Micromix	12393	
Cooke's Professional Sodium Chlorate Weedkiller with Fire Depressant	Cooke's Chemicals	06796	
Deadfast Sodium Chlorate Weed Killer - Professional Use	Growing Success	12510	
Deosan Chlorate Weedkiller (Fire Suppressed)	JohnsonDiversey	08521	
Gem Sodium Chlorate Weedkiller	Joseph Metcalf	04276	
Growing Success Sodium Chlorate Weedkiller	Growing Success	10787	
Sodium Chlorate	Marlow Chemical	06294	
Sodium Chlorate Total Weedkiller	Vitax	13111	
Strathclyde Sodium Chlorate Weedclear	Dremm	12015	
t.w.k	Apex	12453	
spiroxamine			
Zenon	Bayer CropScience	11232	31-12-11
spiroxamine + tebuconazole			
Draco	Bayer CropScience	11267	
sulfuryl fluoride			
Profume	Dow	12035	30-04-13
sulphur			
Headland Venus	Headland	10611	
Microthiol Special	Cerexagri	06268	
Tripart Imber	Tripart	04050	31-08-08
tau-fluvalinate			
Alpha Tau-Fluvalinate 240 EW	Makhteshim	13605	
Revolt	Makhteshim	13383	

EXPIRY DATE IS 31/12/2013 UNLESS OTHERWISE SHOWN

Product	Approval holder	MAPP No.	Expiry Date
tebuconazole			
Agrotech-Tebuconazole 250 EW	Agrotech-Trading	12523	
Aurigen	Syngenta	12043	
Bayer UK226	Bayer CropScience	12412	
Chani	AgChem Access	13325	
Fezan	Sipcam	13681	
Flail	Headland	12728	
I.T. Tebu	I T Agro	12882	
Noble Tebuconazole	Barclay	12874	
Odin	Rotam Agrochemical	13468	
Raxil	Bayer CropScience	11295	
RouteOne Tebbro 25	RouteOne	13647	
Tebucur	Globachem	11890	31-08-08
tebuconazole + triazoxide			
Raxil S	Bayer CropScience	11296	
tebuconazole + trifloxystrobin			
Dedicate	Bayer Environ.	13612	
tebufenpyrad			
Masai G	BASF	10224	
tefluthrin			
Evict	Syngenta	11735	31-08-08
Evict	Bayer CropScience	11673	31-08-08
Force ST	Bayer CropScience	11671	
tepraloxydim			
Omara	Chem-Wise	13682	31-05-15
Omarra	BASF	13693	31-05-15
RouteOne Tepralox 50	Albaugh UK	13667	31-05-15
tetraconazole			
Domark	Isagro	13048	
Domark	Isagro	10452	31-08-08
Emerald	Isagro	10449	31-08-08
Eminent	Monsanto	10504	31-08-08
Juggler	Sipcam	13077	
Juggler	Sipcam	09391	31-08-08
Omen	Sipcam	09413	31-08-08
Tetra 5634	Monsanto	10507	31-08-08
thiabendazole + thiram			
Hy-Vic	Agrichem	06247	
sHYlin	Agrichem	10030	
thiacloprid			
Agrovista Reggae	Agrovista	13706	31-12-14

Product	Approval holder	MAPP No.	Expiry Date
thiamethoxam			
Actara	Syngenta	13728	12-09-11
thifensulfuron-methyl			
Harmony SX	DuPont	12181	30-06-12
Prospect SX	DuPont	12212	30-06-12
thifensulfuron-methyl + tribenuron-methyl			
Calibre	DuPont	07795	
thiodicarb			
Genesis	Sipcam	11846	25-11-08
Genesis ST	Sipcam	11853	25-11-08
thiram			
Thianosan DG	Unicrop	13404	31-07-09
Thiraflo	Chemtura	11526	31-08-08
tolclofos-methyl			
Rizolex 50 WP	Sumitomo	13592	
tolylfluanid			
Elvaron Multi	Bayer CropScience	11422	
tralkoxydim			
Clayton Lansdown	Clayton	12327	31-08-08
Grasp	Syngenta	10441	31-08-08
Grasp 400 SC	Nufarm UK	13594	
Grasp 400 SC	Syngenta	12280	
triadimenol			
Bayfidan	Bayer CropScience	11417	
Hi-Shot	BASF	10397	
Spinnaker	BASF	10398	
tri-allate			
Avadex BW Granular	Gowan	12104	
tribenuron-methyl			
Quantum	DuPont	06270	
Quantum PX	DuPont	10843	
Triad	Headland	12751	
triclopyr			
Agriguard Triclopyr	AgriGuard	10679	
Altix 240 EC	Agriphar	13362	
Altix 240 EC	Chimac-Agriphar	10648	
Cleancrop Triptic 48 EC	United Agri	11568	
Cleancrop Unival	United Agri	11388	
Garlon 2	Dow	05682	31-08-08

Product	Approval holder	MAPP No.	Expiry Date
Garlon 2	Syngenta	10672	31-08-08
n2n Garlon 4	Nomix Enviro	11806	
Tricle	Agrichem	12918	
trifloxystrobin			
Flint	Bayer CropScience	11259	30-09-13
Twist	Bayer CropScience	11230	30-09-13
Twist 500 SC	Bayer CropScience	11231	30-09-13
Zest SC	Bayer CropScience	11233	30-09-13
trifluralin			
Alpha Trifluralin 48 EC	Makhteshim	07406	31-08-08
Das-320	Dow	12375	30-03-09
Digermin	Sipcam	07221	30-03-09
Ipifluor	I Pi Ci	04692	30-03-09
Portman Trifluralin	Agform	05751	31-08-08
Treflan	Dow	13232	20-03-09
Triflurex 48 EC	Makhteshim	07947	31-08-08
Trilogy	United Phosphorus	08996	30-03-09
Trimaran	Nufarm UK	11400	31-08-08
Triplen	Sipcam	05897	30-03-09
triflusulfuron-methyl			
Debut WSB	DuPont	07809	
Landgold TFS 50	Landgold	08941	
Standon Triflusulfuron	Standon	09487	
trinexapac-ethyl			
Cleancrop Alatrin	United Agri	11805	
Primo Maxx	Syngenta	13374	30-04-12
RouteOne Rotrinex 25	RouteOne	13645	
Shortcut	Syngenta	13320	
Shortcut	Scotts	09254	31-07-08
Shrink	AgChem Access	13762	
zeta-cypermethrin			
Angri	AgChem Access	13730	
RouteOne Zeta 10	Albaugh UK	13608	
zucchini yellow mosaic virus weak strain			
Curbit	Bio-Oz	12745	22-01-10

SECTION 4
ADJUVANTS

Adjuvants

Adjuvants are not themselves classed as pesticides and there is considerable misunderstanding over the extent to which they are legally controlled under the Food and Environment Protection Act. An adjuvant is a substance other than water which enhances the effectiveness of a pesticide with which it is mixed. Consent C(i)5 under the Control of Pesticides Regulations allows that an adjuvant can be used with a pesticide only if that adjuvant is authorised and on a list published from time to time in *The Pesticides Register*. An authorised adjuvant has an *adjuvant number* and may have specific requirements about the circumstances in which it may be used.

Adjuvant product labels must be consulted for full details of authorised use, but the table below provides a summary of the label information to indicate the area of use of the adjuvant. Label precautions refer to the keys given in Appendix 4, and may include warnings about products harmful or dangerous to fish. The table includes all adjuvants notified by suppliers as available in 2008.

Product	**Supplier**	**Adj. No.**	**Type**
Abacus	De Sangosse	A0434	adjuvant/wetter
Contains	505.2 g/l esterified rapeseed oil, 300 g/l alkylphenyl hydroxypolyoxyethylene and 60 g/l natural fatty acids		
Use with	All approved pesticides on edible and non-edible crops when used at half their approved dose or less. Also with approved pesticides on specified crops, up to specified growth stages, at up to their full approved dose		
Protective clothing	A, C, H		
Precautions	R36, R58, U05a, U11, U14, U19a, U20b, E15a, E19b, E34, D01, D02, D05, D10a, D12a, H04		
Abacus	De Sangosse	A0543	vegetable oil
Contains	53.43% esterified rapeseed oil, 20% w/w alcohol ethoxylate and 9.0% w/w natural fatty acids		
Use with	All approved pesticides on all edible crops when used at half their recommended dose or less, and on all non-edible crops up their full recommended dose. Alsoat a maximum concentration of 0.1% with approved pesticides on listed crops up to specified growth stages		
Protective clothing	A, C, H		
Precautions	R36, U05a, U11, U14, U19a, U20b, E15a, E19b, E34, D01, D02, D05, D10a, D12a, H04		
Activator 90	De Sangosse	A0337	non-ionic surfactant/spreader/wetter
Contains	750 g/l alkylphenyl hydroxypolyoxyethylene and 150 g/l natural fatty acids		
Use with	Any spray for which additional wetter is approved and recommended		
Protective clothing	A, C, H		
Precautions	R36, R58, U02a, U04a, U05a, U10, U11, U20b, E15a, E19b, D01, D02, D05, D10a, H04		

SECTION 4

Product	Supplier	Adj. No.	Type
Activator 90	De Sangosse	A0547	non-ionic surfactant/wetter
Contains	750 g/l alcohol ethoxylates and 150 g/l natural fatty acids		
Use with	All approved pesticides on all edible crops when used at half their recommended dose or less, and on all non-edible crops up their full recommended dose. Alsoat a maximum concentration of 0.1% with approved pesticides on listed crops up to specified growth stages		
Protective clothing	A, C, H		
Precautions	R36, R38, R53a, U02a, U04a, U05a, U10, U11, U20b, E15a, E19b, D01, D02, D05, D10a, H04		
Adigor	Syngenta	A0522	wetter
Contains	47% w/w methylated rapeseed oil		
Use with	Topik, Axial, Hawk, Amazon and Viscount on cereals in accordance with recommendations on the respective herbicide labels		
Protective clothing	A, C, H		
Precautions	R43, R51, R53a, U02a, U05a, U09a, U20b, E15b, E38, D01, D02, D05, D09a, D10c, D12a, H04, H11		
Admix-P	De Sangosse	A0301	wetter
Contains	80% w/w polyalkylene oxide modified heptamethyltrisiloxane and a maximum of 20 % w/w allyloxypolyethylene glycol methyl ether		
Use with	A wide range of pesticides applied as corm, tuber, onion and other bulb treatments in seed production and in seed potato treatment		
Protective clothing	A, C, H		
Precautions	R20, R21, R22a, R36, R43, R48, R51, R58, U11, U15, U19a, E15a, E19b, D01, D02, D05, D10a, D12a, H03, H11		
Agropen	Intracrop	A0227	vegetable oil
Contains	95% refined rapeseed oil		
Use with	Recommended rates of approved pesticides up to the growth stage indicated for specified crops, and with half or less than the recommended rate on these crops after the stated growth stages. Also for use with pesticides on grassland at half their recommended rate		
Protective clothing	A, C		
Precautions	U08, U20b, E13c, E37, D09a, D10b		
Amber	Interagro	A0367	vegetable oil
Contains	95% w/w methylated rapeseed oil		
Use with	Sugar beet herbicides, oilseed rape herbicides, cereal graminicides and a wide range of other pesticides that have a label recommendation for use with authorised adjuvant oils on specified crops. See label for details		
Precautions	U05a, U20b, E15a, D01, D02, D05, D09a, D10a		
Arma	Interagro	A0306	penetrant
Contains	500 g/l alkoxylated fatty amine + 500 g/l polyoxyethylene monolaurate		
Use with	Cereal growth regulators, cereal herbicides, cereal fungicides, oilseed rape fungicides and a wide range of other pesticides on specified crops		
Precautions	R51, R58, U05a, E15a, E34, E37, D01, D02, D05, D09a, D10a, H11		

Product	Supplier	Adj. No.	Type
BackRow	Interagro	A0472	mineral oil
Contains	60% w/w refined paraffinic petroleum oil		
Use with	Pre-emergence herbicides. Refer to label or contact supplier for further details		
Protective clothing	A		
Precautions	R22b, R38, U02a, U05a, U08, U20b, E15a, D01, D02, D05, D09a, D10a, M05b, H03		
Ballista	Interagro	A0524	spreader/wetter
Contains	10% w/w alkoxylated triglycerides		
Use with	All approved pesticides and plant growth regulators for use in winter and spring cereals applied at full rate up to and including GS 52, and at half the approved rate thereafter. May be applied with approved pesticides for other specified crops at full rate up to specified growth stages and at half rate thereafter		
Precautions	U05a, U20b, E15a, E34, D01, D02, D09a		
Bandrift Plus	Ciba Specialty	-	drift retardant
Contains	A non-ionic polyamide dispersed in oil		
Use with	A wide range of pesticides. See label for restrictions on use with wettable powders, suspension concentrates and water dispersible granules, and other usage limitations		
Protective clothing	A, C, M		
Precautions	U05a, U08, U20a, E13c, E34, D01, D02, D09a		
Banka	Interagro	A0245	spreader/wetter
Contains	29.2% w/w alkyl pyrrolidones		
Use with	Potato fungicides and a wide range of other pesticides on specified crops		
Protective clothing	A, C		
Precautions	R38, R41, R52, R58, U02a, U05a, U11, U19a, U20b, E15a, E34, E37, D01, D02, D05, D09a, H04		
Barclay Clinger	Barclay	A0198	sticker/wetter
Contains	96% w/v poly-1-p-menthene		
Use with	Barclay Gallup, Barclay Gallup Amenity and a wide range of approved fungicides and insecticides. See label for details		
Precautions	U19a, U20b, E13c, D09a, D10a		
Barramundi	Interagro	A0376	mineral oil
Contains	95% w/w mineral oil		
Use with	Sugar beet herbicides, oilseed rape herbicides, cereal graminicides and a wide range of other pesticides that have a label recommendation for use with authorised adjuvant oils on specified crops. Refer to label or contact supplier for further details		
Protective clothing	A		
Precautions	R22b, R38, U02a, U05a, U08, U20b, E15a, D01, D02, D05, D09a, D10a, M05b, H03		

SECTION 4

Product	Supplier	Adj. No.	Type
Binder	Nufarm UK	A0598	spreader/wetter
Contains	30% alkyl ethoxylate		
Use with	Recommended rates of approved pesticides up to the growth stage indicated for specified crops, and with half or less than the recommended rate on these crops after the stated growth stages		
Precautions	R41, R52, R53a, U02a, U05a, U08, U11, U13, U15, U20b, E34, E37, D01, D09a, M03, H04		
Bio Syl	Intracrop	A0385	spreader/sticker/wetter
Contains	1% w/w polyoxyethylen-alpha-methyl-omega-[3-(1,3,3,3-tetramethyl-3-trimethylsiloxy)-disiloxanyl] propylether and 32.67% w/w ethylene oxide condensate		
Use with	Recommended rates of approved pesticides on a range of specified fruit and vegetable crops and on managed amenity turf. See label for details		
Protective clothing	A, C		
Precautions	R36, R38, U05a, U08, U20c, E15a, D01, D02, D10a, H04		
Bioduo	Intracrop	A0606	wetter
Contains	85% alkyl polyglycol ether and fatty acids		
Use with	A wide range of pesticides used in grassland, agriculture and horticulture and with pesticides used in non-crop situations		
Protective clothing	A, C		
Precautions	R22a, R36, R38, U05a, U08, U20c, E13c, E34, D01, D02, D09a, D10a, M03, H03, H04, H08		
Biofilm	Intracrop	A0634	sticker/wetter
Contains	96% poly-1-p-menthene		
Use with	All approved fungicides and insecticides on edible crops and all approved formulations of glyphosate used pre-harvest on wheat, barley, oilseed rape, stubble, and in non-crop situations and grassland destruction		
Precautions	U19a, U20c, E13c, D09a, D11a		
BioPower	Bayer CropScience	A0617	wetter
Contains	6.7% w/w 3,6-dioxaeicosylsulphate sodium salt and 20.2% w/w 3,6-dioxaoctadecylsulphate sodium salt		
Use with	Atlantis and all other approved cereal herbicides		
Protective clothing	A, C		
Precautions	R36, R38, U02a, U05a, U08, U13, U19a, U20b, E13c, E34, D01, D02, D05, D09a, D10a, H04		
Biothene	Intracrop	A0633	sticker/wetter
Contains	96% poly-1-p-menthene		
Use with	All approved fungicides and insecticides on edible crops up to 30 d before harvest, and all approved formulations of glyphosate used pre-harvest on wheat, barley, oilseed rape, stubble, and in non-crop situations and grassland destruction. Must not be used in mixture with adjuvant oils or surfactants		
Precautions	U19a, U20c, E13c, D09a, D11a		

Product	Supplier	Adj. No.	Type
Bond	De Sangosse	A0184	drift retardant/extender/sticker/wetter
Contains	450 g/l synthetic latex and 100 g/l alkylphenylhydroxy-polyoxyethylene		
Use with	Blight sprays on potatoes and other pesticides on a wide range of crops. See label or contact supplier for details		
Protective clothing	A, C, H		
Precautions	R36, R38, U11, U14, U19a, E15a, E19b, D01, D02, D05, D10a, D12a, H04		
Bond	De Sangosse	A0556	extender/sticker/wetter
Contains	900 g/l synthetic latex solution and 101 g/l alcohol alkoxylate		
Use with	All approved potato blight fungicides. Also with all approved pesticides on all edible crops when used at half their recommended dose or less, and on all non-edible crops up their full recommended dose. Also at a maximum concentration of 0.14% with approved pesticides on listed crops up to specified growth stages		
Protective clothing	A, C, H		
Precautions	R36, R38, U11, U14, U16b, U19a, E15a, E19b, D01, D02, D05, D10a, D12a, H04		
Bonser	Nufarm UK	A0601	spreader/wetter
Contains	258 g/l ammonium sulphate, 15.0 g/l modified soya lecithin, 15.0 g/l propionic acid and 4.0 g/l polyoxyethylene (5-7EO) C10-C15 primary alcohol		
Use with	Credit (MAPP 13032) on specified crops, and with all approved pesticides on non-edible crops at up to their full dose, and with all approved pesticides on edible crops at up to half their approved dose		
Break-Thru S 240	PP Products	A0431	wetter
Contains	85% w/w polyalkylene oxide modified heptamethyl trisiloxane		
Use with	All approved pesticides which have a recommendation for use with a wetting agent applied to cereals, permanent grassland and rotational grass. Also with all approved pesticides at full rate on non-edible crops and on listed edible crops up to specified growth stages and on all edible crops when used at half or less of their approved full rate		
Protective clothing	A, C		
Precautions	R20, R21, R38, R41, R51, R53a, U11, U16b, U19a, E19b, D01, D02, D09a, D10a, M05a, H03, H11		
Byo-Flex	Greenaway	A0545	sticker/wetter
Contains	95 g/l vegetable oil		
Use with	GLY 490 (MAPP 12718)		
Protective clothing	A, C		
Precautions	U05a, U08, U20b, C11, D05, D10a		

Product	Supplier	Adj. No.	Type
CA 237	Syngenta	A0579	spreader/wetter
Contains	655 g/l polyalkyleneoxide modified heptamethyltrisiloxane		
Use with	Cherokee (MAPP 12768). Also with all approved fungicides on winter and spring cereals applied at full dose, and all approved herbicides, plant growth regulators and insecticides on winter and spring cereals applied at full dose up to GS 52, or at half dose when applied thereafter		
Protective clothing	A, C, H		
Precautions	R20, R41, R48, R51, R53a, U02a, U05a, U08, U11, U14, U15, U19a, U20a, E15b, E34, E38, D01, D02, D09a, D10a, D12a, H03, H11		
C-Cure	Interagro	A0467	mineral oil
Contains	60% w/w refined mineral oil		
Use with	Pre-emergence herbicides		
Protective clothing	A		
Precautions	R22b, R38, U02a, U05a, U08, U20b, E15a, D01, D02, D05, D09a, D10a, M05b, H03		
Ceres Platinum	Interagro	A0445	penetrant/spreader
Contains	500 g/l alkoxylated fatty amine + 500 g/l polyoxyethylene monolaurate		
Use with	Cereal growth regulators, cereal herbicides, cereal fungicides, oilseed rape fungicides and a wide range of other pesticides on specified crops		
Precautions	R51, R58, U05a, E15a, E34, D01, D02, D05, D09a, D10a, H11		
Cerround	Helena	A0510	adjuvant/vegetable oil
Contains	80% w/w methylated soybean oil and 15% w/w polyalkylene oxide modified heptamethyl trisiloxane		
Use with	Approved pesticides on non-edible crops, cereals and stubbles of all edible crops, grassland (destruction), and pesticides approved for use on growing edible crops when used at half recommended dose or less. On specified crops product may be used at a maximum spray concentration of 0.5% with approved pesticides at their full approved rate up to the growth stages shown in the label		
Protective clothing	A, H		
Precautions	R38, R41, U02a, U05a, U08, U11, U19a, U20b, E13c, E34, E37, D01, D02, D05, D09a, M03, H04		
Codacide Oil	Microcide	A0011	vegetable oil
Contains	95% emulsified vegetable oil		
Use with	All approved pesticides and tank mixes. See label for details		
Protective clothing	A, C		
Precautions	U20b, D09a, D10b		
Companion PCT12	Ciba Specialty	A0482	spreader/sticker/wetter
Contains	25% w/w polyacryalmide		
Use with	Approved herbicides on cereals and any pesticide on non-food crops. See label for restrictions on use with wettable powders, suspension concentrates and water dispersible granules, and other usage limitations		
Precautions	U05a, U08, E13c, D01, D02, D09a		

Product	Supplier	Adj. No.	Type
Compliment	United Agri	A0578	spreader/vegetable oil/wetter
Contains	75% mixed fatty acid esters of rapeseed oil		
Use with	All approved pesticides in non-edible crops, all approved pesticides on edible crops when used at half or less their recommended rate, morpholine or triazine fungicides on cereals, and with all pesticides on listed crops up to specified growth stages		
Protective clothing	A, C		
Precautions	R43, R52, R53a, U14, E38, D01, D05, D07, D08, H04		
Contact Plus	Interagro	A0418	mineral oil
Contains	95% w/w mineral oil		
Use with	Sugar beet herbicides, oilseed rape herbicides, cereal graminicides and a wide range of other pesticides that have a label recommendation for use with authorised adjuvant oils on specified crops. Refer to label or contact supplier for further details		
Protective clothing	A, C		
Precautions	R22b, R38, U02a, U05a, U08, U20b, E15a, D01, D02, D09a, D10a, M05b, H03		
Cropspray 11-E	De Sangosse	A0282	adjuvant/mineral oil
Contains	99% highly refined paraffinic oil		
Use with	All approved pesticides on all edible crops when used at half their recommended dose or less, and on all non-edible crops up to their full recommended dose. Also with listed herbicides on a range of specified crops and with all pesticides on a range of specified crops up to specified growth stages.		
Protective clothing	A, C		
Precautions	R22a, U10, U16b, U19a, E15a, E19b, E34, E37, D01, D02, D05, D10a, D12a, M05b, H03		
Designer	De Sangosse	A0322	drift retardant/extender/sticker/wetter
Contains	8.44% w/w organosilicone wetter and 50% w/w synthetic latex		
Use with	A wide range of fungicides, insecticides and trace elements for cereals and specified agricultural and horticultural crops		
Protective clothing	A, C, H		
Precautions	R36, R38, R52, U02a, U11, U14, U15, U19a, E15a, E37, D01, D02, D05, D09a, D10a, H04		
Desikote Max	Mandops	A0551	extender/sticker/wetter
Contains	400 g/l di-1-p-menthene		
Use with	Approved pesticides on all edible crops (except herbicides on peas) and as an anti-transpirant on vegetables before transplanting, evergreens, deciduous trees, shrubs, bushes and on turf		
Protective clothing	A		
Precautions	R38, R50, R53a, U05a, U14, U20b, E15a, E34, E37, D01, D09a, D10b, D12b, M03, H04, H11		

Product	Supplier	Adj. No.	Type
Drill	De Sangosse	A0409	adjuvant/wetter
Contains	589.4 g/l esterified rapeseed oil, 225 g/l alkylphenyl hydroxypolyoxyethylene and 45 g/l natural fatty acids		
Use with	All approved pesticides on edible and non-edible crops when used at half their approved dose or less. Also with approved pesticides on specified crops, up to specified growth stages, at up to their full approved dose		
Protective clothing	A, C, H		
Precautions	R36, R58, U05a, U11, U14, U19a, U20b, E15a, E19b, E34, D01, D02, D05, D10a, D12a, H04		
Drill	De Sangosse	A0544	adjuvant
Contains	63.34% w/w esterified rapeseed oil, 15% w/w alcohol ethoxylate and 7.5% w/w natural fatty acids		
Use with	All approved pesticides on all edible crops when used at half their recommended dose or less, and on all non-edible crops up their full recommended dose. Alsoat a maximum concentration of 0.1% with approved pesticides on listed crops up to specified growth stages		
Protective clothing	A, C, H		
Precautions	R36, U05a, U11, U14, U19a, U20b, E15a, E19b, E34, D01, D02, D05, D10a, D12a, H04		
Elan	Intracrop	A0392	spreader/sticker/wetter
Contains	82% w/w polyoxyethylen-alpha-methyl-omega-disiloxanyl] propylether		
Use with	Recommended rates of approved pesticides up to the growth stage indicated for specified crops, and with half or less than the recommended rate on these crops after the stated growth stages. Also for use with recommended rates of herbicides on managed amenity turf		
Protective clothing	A, C, H		
Precautions	R36, R38, U05a, U08, U20c, E13c, E15a, E37, D01, D02, D09a, D10a, H04		
Emerald	Intracrop	A0636	anti-transpirant/extender
Contains	96% di-1-p-menthene		
Use with	Recommended rates of approved pesticides up to the growth stage indicated for specified crops, and with half or less than the recommended rate on these crops after the stated growth stages. Also for use alone on transplants, turf, fruit crops, glasshouse crops and Christmas trees. Must not be used in mixture with adjuvant oils or surfactants		
Precautions	U19a, U20c, E13c, E37, D09a, D11a		
Esterol	Interagro	A0330	vegetable oil
Contains	95% w/w methylated rapeseed oil		
Use with	Plant protection products that have a label recommendation for use with adjuvant oils. Refer to label or contact supplier for further details		
Precautions	U05a, U20b, E15a, D01, D02, D09a, D10a		
Felix	Intracrop	A0178	spreader/wetter
Contains	60% ethylene oxide condensate		
Use with	Mecoprop, 2,4-D in cereals and amenity turf, and a range of grass weedkillers in agriculture. See label for details		
Protective clothing	A, C		
Precautions	R36, R38, U05a, U08, U20a, E15a, D01, D02, D09a, D10a, H04, H08		

Product	Supplier	Adj. No.	Type
Frigate	Unicrop	A0325	sticker/wetter
Contains	800 g/l tallow amine ethoxylate		
Use with	Roundup formulations		
Protective clothing	A, C		
Precautions	R22a, R23, R41, R51, R58, R67, U05a, U11, U19a, E38, D01, D02, D12a, M04a, H02, H08, H11		
Gateway	United Agri	A0587	extender/sticker/wetter
Contains	73% w/v synthetic latex solution and 8.5% w/v polyether modified trisiloxane		
Use with	All approved pesticides in crops not destined for human or animal consumption, all approved pesticides on edible crops when used at half or less their recommended rate, and with all pesticides on listed crops up to specified growth stages		
Protective clothing	A, C, H		
Precautions	R41, R52, R53a, U11, U15, E37, E38, D01, D05, D07, D08, H04		
Gladiator	De Sangosse	A0542	penetrant/wetter
Contains	210 g/kg alkoxylated tallow amine, 380 g/kg alcohol ethoxylates, 75 g/l natural fatty acids and 210 g/kg polyalkylene glycol		
Use with	All approved pesticides on all edible crops when used at half their recommended dose or less, and on all non-edible crops up their full recommended dose. Alsoat a maximum concentration of 0.1% with approved pesticides on listed crops up to specified growth stages		
Protective clothing	A, C, H		
Precautions	R36, R38, R41, R52, R58, U05a, U10, U11, U19a, E15a, E19b, E34, E37, D01, D02, D05, D10a, D12a, M03, H04		
Gly-Flex	Greenaway	A0588	spreader/wetter
Contains	140 g/l refined mineral oil		
Use with	GLY-490 (MAPP 12718) or any approved formulations of 490 g/l glyphosate		
Protective clothing	A, C		
Precautions	R36, U05a, U08, U11, U12, U15, U20b, E13c, E34, E37, E38, C11, D01, D10a, D10b, H04		
Green Gold	Intracrop	A0250	spreader/wetter
Contains	95% refined rapeseed oil		
Use with	All pesticides which have a recommendation for the addition of a wetter/spreader		
Precautions	U08, U20b, E13c, E34, D09a, D10b		
Greencrop Astra	Greencrop	A0417	adjuvant/penetrant/spreader
Contains	500 g/l alkoxylated fatty amine and 500 g/l polyoxylene monolaurate		
Use with	Approved pesticides on a range of specified arable crops, and in forestry, managed amenity turf, stubbles and non-crop areas and situations		
Precautions	U05a, E13c, E34, D01, D02, D05, D09a, D10b		

SECTION 4

Product	Supplier	Adj. No.	Type
Greencrop Dolmen	Greencrop	A0419	adjuvant/spreader
Contains	64% w/w polyalkeneoxide modified heptamethyltrisiloxane		
Use with	Approved pesticides on a range of specified arable crops, and in forestry, managed amenity turf, stubbles and non-crop areas and situations		
Protective clothing	A, C, H		
Precautions	R21, R22a, R38, R41, R43, U02a, U05a, U08, U11, E13c, E34, D01, D02, D09a, D10b, M03, H03, H04		
Greencrop Fenit	Greencrop	A0602	wetter
Contains	47% methylated rapeseed oil		
Use with	Greencrop Helvick		
Precautions	R43, R50, R53a, U02a, U05a, U09a, U20b, E15b, D01, D02, D05, D09a, D10c, D12a, H04, H11		
Greencrop Rookie	Greencrop	A0515	wetter
Contains	20.2% w/w 3,6-dioxoctadecylsulphate sodium salt and 6.7% w/w 3,6-dioxaeicosylsulphate sodium salt		
Use with	All approved herbicides in cereals		
Protective clothing	A, C		
Precautions	R36, R38, R52, U02a, U05a, U08, U11, U13, U19a, E15a, D01, D02, D05, D09a, D10b, H04		
Grip	De Sangosse	A0558	extender/sticker/wetter
Contains	900 g/l synthetic latex solution and 101 g/l alcohol alkoxylate		
Use with	All approved potato blight fungicides. Also with all approved pesticides on all edible crops when used at half their recommended dose or less, and on all non-edible crops up their full recommended dose. Also at a maximum concentration of 0.14% with approved pesticides on listed crops up to specified growth stages		
Protective clothing	A, C, H		
Precautions	R36, R38, U11, U14, U16b, U19a, E15a, E19b, D01, D02, D05, D10a, D12a, H04		
Grip	De Sangosse	A0211	drift retardant/extender/sticker/wetter
Contains	450 g/l synthetic latex and 100 g/l alkylphenyl hydroxypolyoxyethylene		
Use with	Blight sprays on potatoes and other pesticides on a wide range of crops. See label or contact supplier for details		
Protective clothing	A, C, H		
Precautions	R36, R38, U11, U14, U19a, U20b, E15a, E19b, D01, D02, D05, D10a, D12a, M03, H04		
Grounded	Helena	A0456	mineral oil
Contains	732 g/l refined paraffinic petroleum oil		
Use with	All approved pesticides on edible and non-edible crops when used at half their approved dose or less. Also with approved pesticides on specified crops, up to specified growth stages, at up to their full approved dose		
Protective clothing	A, C		
Precautions	R53a, U02a, U05a, U08, U20b, E15a, E34, E37, D01, D02, D05, D09a, D10c, H11		

Product	Supplier	Adj. No.	Type
Guide	De Sangosse	A0208	acidifier/drift retardant/penetrant
Contains	350 g/l modified soya lecithin, 100 g/l alkylphenyl hydroxypolyoxyethylene and 350 g/l propionic acid		
Use with	All plant growth regulators for cereals and oilseed rape, dimethoate and chlorpyrifos for wheat bulb fly control. For other uses see label or contact supplier for details		
Protective clothing	A, C, H		
Precautions	R36, R38, U11, U14, U15, U19a, E15a, E19b, D01, D02, D05, D10a, D12a, H04		
Guide	De Sangosse	A0528	acidifier/drift retardant/penetrant
Contains	35% w/w lecithin, 35% w/w propionic acid and 9.39%w/w alcohol ethoxylate		
Use with	All approved pesticides on all edible crops when used at half their recommended dose or less, and on all non-edible crops up their full recommended dose. Also with morpholine fungicides on cereals and with pirimicarb on legumes when used at full dose; with iprodione on brassicas when used at 75% dose, and at a maximum concentration of 0.5% with all pesticides on listed crops at full dose up to specified growth stages		
Protective clothing	A, C, H		
Precautions	R36, R38, U11, U14, U15, U19a, E15a, E19b, D01, D02, D05, D10a, D12a, H04		
Headland Fortune	Headland	A0277	penetrant/spreader/vegetable oil/wetter
Contains	75% w/w mixed methylated fatty acid esters of seed oil and N-butanol		
Use with	Herbicides and fungicides in a wide range of crops. See label for details		
Protective clothing	A, C		
Precautions	R43, U02a, U05a, U14, U20a, E15a, E34, E38, D01, D02, D05, D09a, D10b, H04		
Headland Guard 2000	Headland	A0369	extender/sticker/wetter
Contains	52% w/w synthetic latex solution		
Use with	All approved pesticides that recommend the use of a sticking, wetting or extending agent; also with all approved pesticides for use on crops not destined for human or animal consumption; also with all approved pesticides on beans, peas, edible podded peas, oilseed rape, linseed, sugar beet, cereals, maize, Brussels sprouts, potatoes, cauliflowers		
Precautions	U05a, U08, U19a, U20b, E13b, E34, E37, D01, D02, D05, D10b		
Headland Guard Pro	Headland Amenity	A0423	extender/sticker
Contains	52% synthetic latex solution		
Use with	All approved pesticides, micronutrients and sea-weed based plant growth stimulants in amenity grass, managed amenity turf and amenity vegetation		
Precautions	U05a, U08, U19a, U20b, E13b, E34, E37, D01, D02, D05, D10b		

Product	**Supplier**	**Adj. No.**	**Type**
Headland Inflo XL	Headland Amenity	A0329	wetter
Contains	85% w/w polyalkylene oxide modified heptamethyl siloxane		
Use with	Any pesticide for use on amenity grassland or managed amenity turf (except any product applied in or near water) where the use of a wetting, spreading and penetrating surfactant is recommended to improve foliar coverage, and with any approved pesticide on crops not intended for human or animal consumption		
Protective clothing	A, C		
Precautions	R21, R22a, R38, R41, R43, R51, R58, U02a, U05a, U08, U14, U19a, U20b, E15a, E34, E38, D01, D02, D05, D09a, D10b, M03, H03, H11		
Headland Intake	Headland	A0074	penetrant
Contains	450 g/l propionic acid		
Use with	All approved pesticides on any crop not intended for human or animal consumption, and with all approved pesticides on beans, peas, edible podded peas, oilseed rape, linseed, sugar beet, cereals (except triazole fungicides), maize, Brussels sprouts, potatoes, cauliflowers. See label for detailed advice on timing on these crops		
Protective clothing	A, C		
Precautions	R34, U02a, U05a, U10, U11, U14, U15, U19a, E34, D01, D02, D05, D09b, D10b, M04a, H05		
Headland Rhino	Headland	A0328	wetter
Contains	85% w/w polyalkylene oxide modified heptamethyl siloxane		
Use with	Any herbicide, systemic fungicide, systemic insecticide or plant growth regulator (except any product applied in or near water) where the use of a wetting, spreading and penetrating surfactant is recommended to improve foliar coverage		
Protective clothing	A, C		
Precautions	R21, R22a, R38, R41, R43, R51, R58, U02a, U05a, U08, U11, U14, U19a, U20b, E13c, E34, D01, D02, D05, D09a, D10b, M03, H03, H11		
Intracrop BLA	Intracrop	A0453	extender/sticker
Contains	41.5% w/w styrene butadiene co-polymer solution		
Use with	All potato blight fungicides. Also with recommended rates of approved pesticides in certain non-crop situations and up to the growth stage indicated for specified crops, and with half or less than the recommended rate on these crops after the stated growth stages. Also with pesticides in grassland at half or less their recommended rate		
Precautions	U08, U20b, E13c, E34, E37, D01, D05, D09a, D10a		
Intracrop Green Oil	Intracrop	A0262	wetter
Contains	95% refined rapeseed oil		
Use with	Recommended rates of approved pesticides up to the growth stage indicated for specified crops, and with half or less than the recommended rate on these crops after the stated growth stages. Also for use with pesticides on grassland at half their recommended rate		
Precautions	U08, U20b, E13c, E37, D09a, D10b		

Product	Supplier	Adj. No.	Type
Intracrop Neotex	Intracrop	A0460	extender/sticker
Contains	42.5% w/w styrene butadiene co-polymer solution		
Use with	Recommended rates of approved pesticides up to specified growth stages of a wide range of agricultural arable and horticultural crops, and for non-crop uses. Use on edible crops beyond specified growth stages, and in grass, should only be with half recommended rates of the pesticide or less. See label for details of growth stage restrictions. In addition may be used with all potato blight fungicides at their recommended rates of use up to the latest recommended timing of the fungicide		
Precautions	U08, U20b, E13c, E34, E37, D01, D05, D09a, D10a		
Intracrop Novatex	Intracrop	A0566	extender/sticker
Contains	42.5% w/w styrene butadiene co-polymer solution		
Use with	Recommended rates of approved pesticides up to the growth stage indicated for specified crops, and with half or less than the recommended rate on these crops after the stated growth stages. Also for use with recommended rates of pesticides on grassland and specified non-crop situations		
Precautions	U08, U20b, E13c, E34, E37, D01, D09a, D10a		
Intracrop Predict	Intracrop	A0503	adjuvant/vegetable oil
Contains	91% w/w methylated rapeseed oil		
Use with	All approved pesticides on non-edible crops, and all pesticides approved for use on growing edible crops when used at half recommended dose or less. On specified crops product may be used at a maximum spray concentration of 1% with approved pesticides at their full approved rate up to the growth stages shown in the label		
Precautions	U19a, U20b, D05, D09a, D10b		
Intracrop Questor	Intracrop	A0495	activator/non-ionic surfactant/spreader
Contains	75% polyoxyethylene polypropoxypropanol		
Use with	All approved pesticides on non-edible crops and pesticides used in non-crop production, and all pesticides approved for use on growing edible crops when used at half recommended dose or less. On specified crops product may be used at a maximum spray concentration of 0.3% with approved pesticides at their full approved rate up to the growth stages shown in the label		
Protective clothing	A, C		
Precautions	R36, R38, U04a, U05a, U08, U19a, E13b, D01, D02, D09a, D10b, M03, M05a, H04		
Intracrop Retainer	Intracrop	A0508	adjuvant/vegetable oil
Contains	60% w/w methylated rapeseed oil		
Use with	All approved pesticides on non-edible crops, and all pesticides approved for use on growing edible crops when used at half recommended dose or less. On specified crops product may be used at a maximum spray concentration of 1% with approved pesticides at their full approved rate up to the growth stages shown in the label		
Precautions	U19a, U20b, D05, D09a, D10b		

SECTION 4

Product	Supplier	Adj. No.	Type
Intracrop Rigger	Intracrop	A0572	adjuvant/vegetable oil
Contains	91.7% w/w methylated rapeseed oil		
Use with	All approved pesticides on non-edible crops, and all pesticides approved for use on growing edible crops when used at half recommended dose or less. On specified crops product may be used at a maximum spray concentration of 1.78% with approved pesticides at their full approved rate up to the growth stages shown in the label		
Precautions	U19a, U20b, D05, D09a, D10b		
Intracrop Saturn	Intracrop	A0494	activator/non-ionic surfactant/spreader
Contains	75% polyoxyethylene polypropoxypropanol		
Use with	All approved pesticides on non-edible crops and pesticides used in non-crop production, and all pesticides approved for use on growing edible crops when used at half recommended dose or less. On specified crops product may be used at a maximum spray concentration of 0.3% with approved pesticides at their full approved rate up to the growth stages shown in the label		
Protective clothing	A, C		
Precautions	R36, R38, U04a, U05a, U08, U19a, E13b, D01, D02, D09a, D10b, M03, M05a, H04		
Intracrop Sprinter	Intracrop	A0513	spreader/wetter
Contains	192 g/l primary alcohol ethoxylate		
Use with	Recommended rates of approved pesticides up to the growth stage indicated for specified crops, and with half or less than the recommended rate on these crops after the stated growth stages. Also with herbicides on managed amenity turf at recommended rates, and on grassland at half or less than recommended rates		
Protective clothing	A, C		
Precautions	R36, R38, R41, U05a, U08, U11, U14, U15, U20c, E15a, D01, D02, D09a, D10b, M03, H04		
Intracrop Status	Intracrop	A0506	adjuvant/vegetable oil
Contains	91% w/w methylated rapeseed oil		
Use with	All approved pesticides on non-edible crops, and all pesticides approved for use on growing edible crops when used at half recommended dose or less. On specified crops product may be used at a maximum spray concentration of 1.0% with approved pesticides at their full approved rate up to the growth stages shown in the label		
Precautions	U19a, U20b, D05, D09a, D10b		
Intracrop Stay-Put	Intracrop	A0507	vegetable oil
Contains	91% w/w methylated rapeseed oil		
Use with	Recommended rates of approved pesticides for non-crop uses and with recommended rates of approved pesticides up to the growth stage indicated for specified crops, and with half or less than the recommended rate on these crops after the stated growth stages		
Precautions	U19a, U20b, D05, D09a, D10b		

Product	Supplier	Adj. No.	Type
Intracrop Super Rapeze MSO	Intracrop	A0491	adjuvant/vegetable oil
Contains	91.7% w/w methylated rapeseed oil		
Use with	All approved pesticides on non-edible crops, and all pesticides approved for use on growing edible crops when used at half recommended dose or less. On specified crops product may be used at a maximum spray concentration of 1.78% with approved pesticides at their full approved rate up to the growth stages shown in the label		
Precautions	U19a, U20b, D05, D09a, D10b		
Intracrop Warrier	Intracrop	0514	spreader/wetter
Contains	192 g/l primary alcohol ethylene		
Use with	Recommended rates of approved pesticides up to the growth stage indicated for specified crops, and with half or less than the recommended rate on these crops after the stated growth stages. Also with herbicides on managed amenity turf at recommended rates, and on grassland at half or less than recommended rates		
Protective clothing	A, C		
Precautions	R36, R38, R41, U05a, U08, U11, U14, U15, U20c, E15a, D01, D02, D09a, D10b, M03, H04		
Kantor	Interagro	A0623	spreader/wetter
Contains	790 g/l alkoxylated triglycerides		
Use with	All approved pesticides		
Precautions	U05a, U20b, E15b, E19b, E34, D01, D02, D09a, D10a, D12a		
Katalyst	Interagro	A0450	penetrant/water conditioner
Contains	90% w/w alkoxylated fatty amine		
Use with	Glyphosate and a wide range of other pesticides on specified crops. Refer to label or contact supplier for further details		
Protective clothing	A, C		
Precautions	R22a, R36, R38, R50, R58, U02a, U05a, U08, U19a, U20b, E15a, E34, E37, D01, D02, D05, D09a, D10a, M03, H03, H11		
Kinetic	Helena	A0252	spreader/wetter
Contains	80% w/w polyoxypropylene-polyoxyethylene glycol and 20% w/w polyalkylene oxide modified heptamethyl trisiloxane		
Use with	Approved pesticides on non-edible crops and cereals and stubbles of all edible crops when used at full recommended dose, and with pesticides approved for use on growing edible crops when used at half recommended dose or less. On specified crops product may be used at a maximum spray concentration of 0.2% with approved pesticides at their full approved rate up to the growth stages shown in the label		
Protective clothing	A, H		
Precautions	R38, R41, U02a, U05a, U08, U19a, U20b, E13c, E34, E37, D01, D02, D05, D09a, D10c, M03, H04		

Product	Supplier	Adj. No.	Type
Klipper	Amega	A0260	spreader/wetter
Contains	600 g/l ethylene oxide condensate		
Use with	Approved salt formulations of mecoprop alone or in mixtures with 2,4-D on managed amenity turf		
Protective clothing	A, C		
Precautions	R36, R38, U05a, U08, U20c, E15a, D01, D02, D09a, D10a, H04, H08		
Leaf-Koat	Helena	A0511	spreader/wetter
Contains	80% w/w polyoxypropylene-polyoxyethylene glycol and 20% w/w polyalkylene oxide modified heptamethyl trisiloxane		
Use with	Approved pesticides on non-edible crops and cereals and stubbles of all edible crops when used at full recommended dose, and with pesticides approved for use on growing edible crops when used at half recommended dose or less. On specified crops product may be used at a maximum spray concentration of 0.2% with approved pesticides at their full approved rate up to the growth stages shown in the label		
Protective clothing	A, H		
Precautions	R38, R41, U02a, U05a, U08, U19a, U20b, E13c, E34, E37, D01, D02, D05, D09a, D10c, M03, H04		
Level	United Agri	A0580	extender/sticker
Contains	52% synthetic latex solution		
Use with	All approved pesticides in non-edible crops, all approved pesticides on edible crops when used at half or less their recommended rate, potato blight fungicides and with all pesticides on listed crops up to specified growth stages		
Precautions	E37, D01, D02, D05		
Li-700	De Sangosse	A0529	acidifier/drift retardant/penetrant
Contains	35% w/w lecithin, 35% w/w propionic acid and 9.39%w/w alcohol ethoxylate		
Use with	All approved pesticides on all edible crops when used at half their recommended dose or less, and on all non-edible crops up their full recommended dose. Also with morpholine fungicides on cereals and with pirimicarb on legumes when used at full dose; with iprodione on brassicas when used at 75% dose, and at a maximum concentration of 0.5% with all pesticides on listed crops at full dose up to specified growth stages		
Protective clothing	A, C, H		
Precautions	R36, R38, U11, U14, U15, U19a, E15a, E19b, D01, D02, D05, D10a, D12a, H04		
LI-700	De Sangosse	A0176	acidifier/drift retardant/penetrant
Contains	350 g/l modified soya lecithin, 100 g/l alkylphenyl hydroxypolyoxyethylene and 350 g/l propionic acid		
Use with	All plant growth regulators for cereals and oilseed rape, dimethoate and chlorpyrifos for wheat bulb fly control. For other uses see label or contact supplier for details		
Protective clothing	A, C, H		
Precautions	R36, R38, U11, U14, U15, U19a, E15a, E19b, D01, D02, D05, D10a, D12a, H04		

Product	Supplier	Adj. No.	Type
Libsorb	Ciba Specialty	A0438	spreader/wetter
Contains	alkyl alcohol ethoxylate		
Use with	Any spray for which additional wetter is approved and recommended		
Protective clothing	A, C		
Precautions	R36, R38, U05a, U08, U20a, E13c, D01, D02, D09a, D10a, H04		
Logic	Microcide	A0288	vegetable oil
Contains	95% emulsified vegetable oil		
Use with	All approved pesticides on edible and non-edible crops for ground or aerial application		
Protective clothing	A, C		
Precautions	U19a, U20b, D09a, D10b		
Logic Oil	Microcide	A0293	vegetable oil
Contains	95% emulsified vegetable oil		
Use with	All approved pesticides on edible and non-edible crops for ground or aerial application		
Protective clothing	A, C		
Precautions	U19a, U20b, D09a, D10b		
Low Down	Helena	A0459	mineral oil
Contains	732 g/l refined paraffinic petroleum oil		
Use with	All approved pesticides on edible and non-edible crops when used at half their approved dose or less. Also with approved pesticides on specified crops, up to specified growth stages, at up to their full approved dose		
Protective clothing	A, C		
Precautions	R53a, U02a, U05a, U08, U20b, E15a, E34, E37, D01, D02, D05, D09a, D10c, H11		
Mangard	Mandops	A0132	adjuvant/anti-transpirant
Contains	96% w/w di-1-p-menthene		
Use with	Glyphosate and a wide range of other herbicides, fungicides and insecticides. See label for details		
Precautions	U20c, E15a, D09a, D10a		
Mediator Sun	Nufarm UK	A0512	wetter
Contains	655 g/l terpenic alcohols		
Use with	All approved pesticides and plant growth regulators on a wide range of specified crops at full dose up to specified growth stages, and at half dose at later growth stages		
Precautions	R36, R52, R53a, U11, U15, D01, M05a, H04		

SECTION 4

Product	Supplier	Adj. No.	Type
Mixture B NF	Amega	A0570	non-ionic surfactant/spreader/wetter
Contains	42.5% polyoxyethylene (3EO) C12-C15 primary alcohol and 38.25 % w/w polyoxyethylene (7EO) C12-C15 primary alcohol		
Use with	All approved pesticides on non-edible crops at their full recommended rate. Also with Timbrel (MAPP 05815) and all approved formulations of glyphosate in non-crop situations. Also with a range of specified herbicides when used at less than half their recommended rate on forest and grassland, and with all approved pesticides at full rate on a wide range of specified edible crops up to specified growth stages		
Protective clothing	A, C		
Precautions	R21, R22a, R36, R38, U05a, U08, U20b, E13c, E34, E37, D01, D02, D09a, D10a, M03, H03, H04		
Nettle	Interagro	A0466	mineral oil
Contains	60% w/w mineral oil		
Use with	Cereal graminicides and a wide range of other pesticides that have a label recommendation for use with authorised adjuvant oils on specified crops. Refer to label or contact supplier for further details		
Protective clothing	A		
Precautions	R22b, R38, U02a, U05a, U08, U20b, E15a, D01, D02, D05, D09a, D10a, M05b, H03		
Newman Cropspray 11E	De Sangosse	A0195	mineral oil
Contains	99% highly refined mineral oil		
Use with	Approved herbicides for which the addition of an adjuvant oil is recommended		
Protective clothing	A, C		
Precautions	R22a, U10, U19a, U20a, E15a, E19b, E34, E37, D01, D02, D05, D10a, M05b, H03		
Newman's T-80	De Sangosse	A0192	spreader/wetter
Contains	78% w/w polyoxyethylene tallow amine		
Use with	Glyphosate		
Protective clothing	A, C, H		
Precautions	R22a, R37, R38, R41, R50, R58, R67, U05a, U11, U14, U19a, U20a, E15a, E19b, E34, E37, D01, D02, D05, D10a, D12a, M03, M04a, H03, H08, H11		
Nion	Amega	A0415	spreader/wetter
Contains	90% (900 g/l) ethylene oxide condensate		
Use with	Recommended rates of approved pesticides up to specified growth stages of a wide range of agricultural arable and horticultural crops. Use on edible crops beyond specified growth stages should only be with half recommended rates of the pesticide or less. See label for details of growth stage restrictions.		
Protective clothing	A, C		
Precautions	R36, R38, U05a, U08, U19a, U20b, E13c, E37, D01, D02, D09a, D10a, H04		

Product	Supplier	Adj. No.	Type
Nu Film P	Intracrop	A0635	sticker/wetter
Contains	96% poly-1-p-menthene		
Use with	Glyphosate and many other pesticides and growth regulators for which a protectant is recommended. Do not use in mixture with adjuvant oils or surfactants		
Precautions	U19a, U20b, E13c, D09a, D10a		
Nufarm Cropoil	Nufarm UK	A0447	mineral oil
Contains	99% highly refined mineral oil		
Use with	Approved pesticides for use on certain specified crops (cereals, combinable break crops, field and dwarf beans, peas, oilseed rape, brassicas, potatoes, carrots, parsnips, sugar beet, fodder beet, mangels, red beet, maize, sweetcorn, onions, leeks, horticultural crops, forestry, amenity, grassland)		
Protective clothing	A, C		
Precautions	R43, U19a, U20b, E13c, D05, D09a, D10b, H04		
Nufarm Cropoil Gold	Nufarm UK	A0446	adjuvant/vegetable oil
Contains	97.5% w/w methylated rapeseed oil		
Use with	A wide range of pesticides used in agriculture, horticulture, forestry and amenity situations. See label for details of crops and dose rates.		
Protective clothing	A, H		
Precautions	R43, U05a, U15, U19a, U20b, E13c, D01, D02, D05, D09a, D10b, H04		
Output	Syngenta	A0429	mineral oil/surfactant
Contains	60% mineral oil and 40% surfactants		
Use with	Grasp		
Protective clothing	A, C		
Precautions	R38, U05a, U08, U19a, U20a, E13c, D01, D02, D09a, D10b, H04		
Pan Oasis	Pan Agriculture	A0411	vegetable oil
Contains	95% w/w methylated rapeseed oil		
Use with	A wide range of pesticides that have a label recommendation for use with authorised adjuvant oils. Contact distributor for further details		
Protective clothing	A		
Precautions	R36, U02a, U05a, U08, U20b, E15a, D01, D02, D05, D09a, D10a, H04		
Pan Panorama	Pan Agriculture	A0412	mineral oil
Contains	95% w/w mineral oil		
Use with	A wide range of pesticides that have a label recommendation for use with adjuvant oils. Contact distributor for details		
Protective clothing	A		
Precautions	R22b, R38, U02a, U05a, U08, U20b, E13c, D01, D02, D05, D09a, D10a, M05b, H03		

Product	Supplier	Adj. No.	Type
Paramount	United Agri	0585	wetter
Contains	85% polyalkylene oxide modified heptamethyl siloxane		
Use with	Approved pesticides on cereals and grassland for which a wetter is recommended, all approved pesticides in non-edible crops, all approved pesticides on edible crops when used at half or less their recommended rate, and with all pesticides on listed crops up to specified growth stages		
Protective clothing	A, C		
Precautions	R21, R22a, R38, R41, R51, R53a, U11, U14, U16b, E37, E38, D01, D05, D07, D08, H03, H11		
Phase II	De Sangosse	A0331	vegetable oil
Contains	95% w/w esterified rapeseed oil		
Use with	Pesticides approved for use in sugar beet, oilseed rape, cereals (for grass weed control), and other specified agricultural and horticultural crops		
Protective clothing	A, C, H		
Precautions	U19a, U20b, E15a, E19b, E34, D01, D02, D05, D10a, D12a		
Pin-o-Film	Intracrop	A0637	sticker/wetter
Contains	96% di-1-p-menthene		
Use with	Recommended rates of approved pesticides up to the growth stage indicated for specified crops, and with half or less than the recommended rate on these crops after the stated growth stages. Also for use with pesticides on grassland at half or less than their recommended rates. Must not be used in mixture with adjuvant oils or surfactants		
Precautions	U19a, U20c, E13c, C02a, D09a, D11a		
Planet	Intracrop	A0605	non-ionic surfactant/spreader/wetter
Contains	85% alkyl polyglycol ether and fatty acid		
Use with	Any spray for which additional wetter is recommended		
Precautions	R22a, R36, U05a, U08, U19a, U20b, E13c, D01, D09a, D10a, D11a, H03, H04, H08		
Prima	De Sangosse	A0310	mineral oil
Contains	99% highly refined mineral oil		
Use with	All pesticides on all crops when used up to 50% of maximum approved dose for that use (mixtures with Roundup must only be used for treatment of stubbles); all approved pesticides at full dose on listed crops (see label). For other uses see label or contact supplier for details		
Protective clothing	A, C		
Precautions	R22a, U10, U19a, E15a, E19b, E34, E37, D01, D02, D05, D10a, D12a, M05b, H03		
Prima	De Sangosse	A0531	mineral oil
Contains	99% highly refined paraffinic oil		
Use with	All pesticides on all crops when used up to 50% of maximum approved dose for that use (mixtures with Roundup must only be used for treatment of stubbles). Also with all approved pesticides at full dose on listed crops (see label). Also with listed herbicides on specified crops		
Protective clothing	A, C		
Precautions	R22a, U10, U16b, U19a, E15a, E19b, E34, E37, D01, D02, D05, D10a, D12a, M05b, H03		

Product	Supplier	Adj. No.	Type
Profit Oil	Microcide	A0294	extender/sticker/wetter
Contains	95% rape/soya oil		
Use with	All approved pesticides		
Protective clothing	A, C		
Precautions	U19a, U20b, D09a, D10b		
Ranger	De Sangosse	A0296	acidifier/drift retardant/penetrant
Contains	350 g/l modified soya lecithin, 100 g/l alkylphenyl hydroxypolyoxyethylene and 350 g/l propionic acid		
Use with	All plant growth regulators for cereals and oilseed rape, dimethoate and chlorpyrifos for wheat bulb fly control. For other uses see label or contact supplier for details		
Protective clothing	A, C, H		
Precautions	R36, R38, U11, U14, U15, U19a, E15a, E19b, D01, D02, D05, D10a, D12a, H04		
Ranger	De Sangosse	A0532	acidifier/drift retardant/penetrant
Contains	35% w/w lecithin, 35% w/w propionic acid and 9.39%w/w alcohol ethoxylate		
Use with	All approved pesticides on all edible crops when used at half their recommended dose or less, and on all non-edible crops up their full recommended dose. Also with morpholine fungicides on cereals and with pirimicarb on legumes when used at full dose; with iprodione on brassicas when used at 75% dose, and at a maximum concentration of 0.5% with all pesticides on listed crops at full dose up to specified growth stages		
Protective clothing	A, C, H		
Precautions	R36, R38, U11, U14, U15, U19a, E15a, E19b, D01, D02, D05, D10a, D12a, H04		
Rapide	Intracrop	A0563	penetrant/surfactant
Contains	40% propionic acid		
Use with	A wide range of pesticides and growth regulators, especially chlormequat		
Protective clothing	A, C		
Precautions	R34, R36, R38, U02a, U05a, U08, U19a, U20a, D01, D02, D09a, D10a, H04, H05		
Reward Oil	Microcide	A0295	extender/sticker/wetter
Contains	95% rape/soya oil		
Use with	All approved pesticides		
Protective clothing	A, C		
Precautions	U19a, U20b, D09a, D10b		
Ryda	Interagro	A0168	cationic surfactant/wetter
Contains	800 g/l polyethoxylated tallow amine		
Use with	Glyphosate		
Protective clothing	A, C		
Precautions	R22a, R36, R38, R41, R51, R58, U02a, U05a, U08, U13, U19a, U20a, E15a, E34, E37, D01, D02, D05, D09a, D10b, M03, H03, H08, H11		

Product	Supplier	Adj. No.	Type
Saracen	Interagro	A0368	vegetable oil
Contains	95% w/w methylated vegetable oil		
Use with	Cereal graminicides and a wide range of other pesticides that have a recommendation for use with authorised adjuvant oils on specified crops. Refer to label or contact supplier for further details		
Precautions	U05a, U20b, E15a, D01, D02, D09a, D10a		
SAS 90	Intracrop	A0311	spreader/wetter
Contains	100% polyoxyethylen-alpha-methyl-omega-[3-(1,3,3,3-tetramethyl-3-trimethylsiloxy)-disiloxanyl] propylether		
Use with	A wide range of pesticides on specified crops in agriculture and horticulture, and on amenity vegetation and land not intended for cropping		
Precautions	U02a, U08, U19a, U20b, E13c, E34, D09a, D10a		
Scout	De Sangosse	A0207	acidifier/drift retardant/non-ionic surfactant/penetrant
Contains	350 g/l modified soya lecithin, 100 g/l alkylphenyl hydroxypolyoxyethylene and 350 g/l propionic acid		
Use with	All plant growth regulators for cereals and oilseed rape, dimethoate and chlorpyrifos for wheat bulb fly control. For other uses see label or contact supplier for details		
Protective clothing	A, C, H		
Precautions	R36, R38, U11, U14, U15, U19a, E15a, E19b, D01, D02, D05, D10a, D12a, H04		
Scout	De Sangosse	A0533	acidifier/drift retardant/penetrant
Contains	35% w/w lecithin, 35% w/w propionic acid and 9.39%w/w alcohol ethoxylate		
Use with	All approved pesticides on all edible crops when used at half their recommended dose or less, and on all non-edible crops up their full recommended dose. Also with morpholine fungicides on cereals and with pirimicarb on legumes when used at full dose; with iprodione on brassicas when used at 75% dose, and at a maximum concentration of 0.5% with all pesticides on listed crops at full dose up to specified growth stages		
Protective clothing	A, C, H		
Precautions	R36, R38, U11, U14, U15, U19a, E15a, E19b, D01, D02, D05, D10a, D12a, H04		
Signal	Intracrop	A0461	extender/sticker
Contains	450 g/l synthetic latex		
Use with	All approved pesticides where addition of a sticker/extender is required and recommended		
Protective clothing	A, C		
Precautions	R36, R38, U02a, U05a, U08, U19a, U20b, E13c, D01, D02, D09a, D10a, M03, H04		

Product	Supplier	Adj. No.	Type
Siltex	Intracrop	A0398	spreader/sticker/wetter
Contains	27% w/w polyoxyethylen-alpha-methyl-omega-disiloxanyl] propylether		
Use with	Approved pesticides used at full dose on a range of specified fruit and vegetable crops, and on managed amenity turf		
Protective clothing	A, C		
Precautions	U05a, U08, U20c, E13c, E37, D01, D02, D09a, D10a		
Silwet L-77	De Sangosse	A0193	drift retardant/spreader/wetter
Contains	80% w/w polyalkylene oxide modified heptamethyltrisiloxane and a maximum of 20 % w/w allyloxypolyethylene glycol methyl ether		
Use with	All approved fungicides on winter and spring sown cereals; all approved pesticides applied at 50% or less of their full approved dose. A wide range of other uses. See label or contact supplier for details		
Protective clothing	A, C, H		
Precautions	R20, R21, R22a, R36, R43, R48, R51, R58, U11, U15, U19a, E15a, E19b, D01, D02, D05, D10a, D12a, H03, H11		
Slippa	Interagro	A0206	spreader/wetter
Contains	655 g/l polyalkyleneoxide modified heptamethyltrisiloxane		
Use with	Cereal fungicides and a wide range of other pesticides and trace elements on specified crops		
Protective clothing	A, C, H		
Precautions	R20, R38, R41, R43, R48, R51, R58, U02a, U05a, U08, U11, U19a, U20a, E15a, E34, E37, D01, D02, D05, D09a, D10a, M03, H03, H11		
SM-99	De Sangosse	A0134	mineral oil
Contains	99% w/w highly refined paraffinic oil		
Use with	All pesticides on all crops when used up to 50% of maximum approved dose for that use (mixtures with Roundup must only be used for treatment of stubbles). Also with all approved pesticides at full dose on listed crops (see label). Also with listed herbicides on specified crops		
Protective clothing	A, C		
Precautions	R22a, U10, U19a, E15a, E19b, E34, E37, D01, D02, D05, D10a, D12a, M05b, H03		
SM-99	De Sangosse	A0534	mineral oil
Contains	99% highly refined paraffinic oil		
Use with	All pesticides on all crops when used up to 50% of maximum approved dose for that use (mixtures with Roundup must only be used for treatment of stubbles). Also with all approved pesticides at full dose on listed crops (see label). Also with listed herbicides on specified crops		
Protective clothing	A, C		
Precautions	R22a, U10, U16b, U19a, E15a, E19b, E34, E37, D01, D02, D05, D10a, D12a, M05b, H03		

Product	Supplier	Adj. No.	Type
Solar	Intracrop	A0225	activator/non-ionic surfactant/spreader
Contains	75% polypropoxypropanol		
Use with	Foliar applied plant growth regulators		
Protective clothing	A, C		
Precautions	R36, R38, U04a, U05a, U08, U20a, E13c, E34, D01, D02, D09a, D10b, D11a, M03, H04		
Spartan	Interagro	A0375	penetrant/water conditioner
Contains	500 g/l alkoxylated fatty amine + 400 g/l polyoxyethylene monolaurate		
Use with	Cereal growth regulators, cereal herbicides, oilseed rape fungicides and a wide range of other pesticides on specified crops. Refer to label or contact supplier for further details		
Protective clothing	A, C		
Precautions	R22a, R36, R38, R51, R58, U02a, U05a, U19a, U20b, E15a, E34, E37, D01, D02, D05, D09a, D10a, M03, H03, H11		
Spray-fix	De Sangosse	A0559	extender/sticker/wetter
Contains	900 g/l synthetic latex solution and 101 g/l alcohol alkoxylate		
Use with	All approved potato blight fungicides. Also with all approved pesticides on all edible crops when used at half their recommended dose or less, and on all non-edible crops up their full recommended dose. Also at a maximum concentration of 0.14% with approved pesticides on listed crops up to specified growth stages		
Protective clothing	A, C, H		
Precautions	R36, R38, U11, U14, U16b, U19a, E15a, E19b, D01, D02, D05, D10a, D12a, H04		
Spray-fix	De Sangosse	A0297	drift retardant/extender/sticker/wetter
Contains	450 g/l synthetic latex and 100 g/l alkylphenyl hydroxypolyoxyethylene		
Use with	Blight sprays on potatoes and other pesticides on a wide range of crops. See label or contact supplier for details		
Protective clothing	A, C, H		
Precautions	R36, R38, U11, U14, U19a, E15a, E19b, D01, D02, D05, D09a, D10a, D12a, H04		
Spraygard	Mandops	A0552	extender/sticker/wetter
Contains	400 g/l di-1-p-menthene		
Use with	Approved pesticides on all edible crops (except herbicides on peas) and as an anti-transpirant on vegetables before transplanting, evergreens, deciduous trees, shrubs, bushes and on turf		
Protective clothing	A		
Precautions	R38, R50, R53a, U05a, U14, U20b, E15a, E34, E37, D01, D09a, D10b, D12b, M03, H04, H11		

Product	Supplier	Adj. No.	Type
Spraymac	De Sangosse	A0549	acidifier/non-ionic surfactant
Contains	350 g/l propionic acid and 100 g/l alcohol ethoxylate		
Use with	All approved pesticides on all edible crops when used at half their recommended dose or less, and on all non-edible crops up their full recommended dose. Alsoat a maximum concentration of 0.5% with approved pesticides on listed crops up to specified growth stages		
Protective clothing	A, C, H		
Precautions	R34, U04a, U10, U11, U14, U19a, U20b, E15a, E19b, E34, D01, D02, D05, D10a, D12a, M04a, H05		
Spraymac	De Sangosse	A0298	acidifier/non-ionic surfactant
Contains	350 g/l propionic acid and 100 g/l alkylphenyl hydroxypolyoxyethylene		
Use with	All plant growth regulators for cereals and oilseed rape and any spray for which the addition of a wetter/spreader is recommended. See label or contact supplier for details		
Protective clothing	A, C, H		
Precautions	R34, U04a, U10, U11, U14, U19a, U20b, E15a, E19b, E34, D01, D02, D05, D10a, D12a, M04a, H05		
Spread and Seal	Mandops	A0449	extender/sticker/wetter
Contains	334 g/l d-1-p-menthene		
Use with	All pesticides on all edible crops when used at up to 50% of their approved dose, and all approved forulations of glyphosate when used on cereals. Must not be used in combination with herbicides applied to pea crops		
Precautions	U20c, E13c, E15a, D09a, D10a		
Stamina	Interagro	A0202	penetrant
Contains	100% w/w alkoxylated fatty amine		
Use with	Glyphosate and a wide range of other pesticides on specified crops		
Protective clothing	A, C		
Precautions	R22a, R38, R50, R58, U02a, U05a, U20a, E15a, E34, D01, D02, D05, D09a, D10a, M03, H03, H11		
Standon Shiva	Standon	A0519	wetter
Contains	20.2 % w/w sodium salt of 3,6-dioxaoctadecylsulphate and 6.7% w/w sodium salt of 3,6-dioxaeicosylsulphate		
Use with	Approved cereal herbicides		
Protective clothing	A, C		
Precautions	R36, R38, R52, U02a, U05a, U08, U13, U19a, E15a, D01, D02, D05, D09a, D10b, H04		

Product	Supplier	Adj. No.	Type
Stika	De Sangosse	A0557	extender/sticker/wetter
Contains	450 g/l synthetic latex solution and 101 g/l alcohol alkoxylate		
Use with	All approved potato blight fungicides. Also with all approved pesticides on all edible crops when used at half their recommended dose or less, and on all non-edible crops up their full recommended dose. Also at a maximum concentration of 0.14% with approved pesticides on listed crops up to specified growth stages		
Protective clothing	A, C, H		
Precautions	R36, R38, U11, U14, U16b, U19a, E15a, E19b, D01, D02, D05, D10a, D12a, H04		
Stika	De Sangosse	A0452	extender/sticker/wetter
Contains	225 g/l synthetic latex and 100 g/l alkylphenyl hydroxypolyoxyethylene		
Use with	Approved pesticides on non-edible crops where the use of a sticking, wetting or extending agent is recommended, and pesticides approved for use on growing edible crops when used at half recommended dose or less. On specified crops product may be used at a maximum spray concentration of 0.14% with approved pesticides at their full approved rate up to the growth stages shown on the label		
Protective clothing	A, C, H		
Precautions	U11, U14, U19a, E15a, E19b, D01, D02, D05, D10a, D12a		
Super Nova	Interagro	A0364	penetrant
Contains	500 g/l alkoxylated fatty amine + 500 g/l polyoxyethylene monolaurate		
Use with	Cereal growth regulators, cereal herbicides, cereal fungicides, oilseed rape fungicides and a wide range of other pesticides on specified crops		
Precautions	R51, R58, U05a, E15a, E34, D01, D02, D05, D09a, D10a, H11		
Sward	Amega	A0215	spreader/wetter
Contains	15.2% w/w polyalkylene oxide modified heptomethyltrisiloxane		
Use with	Specified pesticides in amenity situations		
Protective clothing	A, C		
Precautions	R41, R43, U05a, U08, U20c, E13c, D01, D02, D09a, D10a, H04		
Talzene	Intracrop	A0233	wetter
Contains	800 g/l polyoxyethylene tallow amine		
Use with	Herbicides and dessicants up to their latest recommended time of application in agriculture, horticulture, amenity and forestry		
Protective clothing	A, C		
Precautions	R22a, R36, R38, R41, R48, U05a, U08, U11, U13, U19a, U20b, E13c, E34, D01, D02, D09a, D10a, M03, H03		
Teliton Ocean Power	Teliton	A0615	wetter
Contains	20.2% w/w 3,6-dioxoctadecylsulphate sodium salt and 6.7% w/w 3,6-dioxaeicosylsulphate sodium salt		
Use with	All approved herbicides for use on cereals		
Protective clothing	A, C		
Precautions	R36, R38, R52, U02a, U05a, U08, U13, U19a, E15a, D01, D02, D09a, D10a, H04		

Product	Supplier	Adj. No.	Type
Toil	Interagro	A0248	vegetable oil
Contains	95% w/w methylated rapeseed oil		
Use with	Sugar beet herbicides, oilseed rape herbicides, cereal graminicides and a wide range of other pesticides that have a label recommendation for use with authorised adjuvant oils on specified crops. Refer to label or contact supplier for further details		
Precautions	U05a, U20b, E15a, D01, D02, D05, D09a, D10a		
Torpedo II	De Sangosse	A0393	penetrant/wetter
Contains	215.5 g/litre alkoxylated tallow amine, 375 g/litre alkylphenyl hydroxypolyoxyethylene and 75 g/litre natural fatty acids		
Use with	All pesticides on edible and non-edible crops when used at half or less than the full approved dose and with products up to their full approved dose on a wide range of specified fruit and vegetable crops.		
Protective clothing	A, C, H		
Precautions	R38, R41, R52, R58, U05a, U10, U11, U19a, E15a, E19b, E34, E37, D01, D02, D05, D10a, D12a, M03, H04, H08, H11		
Torpedo-II	De Sangosse	A0541	penetrant/wetter
Contains	210 g/kg alkoxylated tallow amine, 380 g/kg alcohol ethoxylates, 75 g/l natural fatty acids and 210 g/kg polyalkylene glycol		
Use with	All approved pesticides on all edible crops when used at half their recommended dose or less, and on all non-edible crops up their full recommended dose. Alsoat a maximum concentration of 0.1% with approved pesticides on listed crops up to specified growth stages		
Protective clothing	A, C, H		
Precautions	R36, R38, R41, R52, R58, U05a, U10, U11, U19a, E15a, E19b, E34, E37, D01, D02, D05, D10a, D12a, M03, H04		
Transact	United Agri	A0584	penetrant
Contains	40% w/w propionic acid		
Use with	Plant growth regulators in cereals, all approved pesticides in non-edible crops, all approved pesticides on edible crops when used at half or less their recommended rate, and with all pesticides on listed crops up to specified growth stages		
Protective clothing	A, C		
Precautions	R34, U10, U11, U14, U15, U19a, D01, D05, D09b, M04a, H05		
Transcend	Helena	A0333	adjuvant/vegetable oil
Contains	80% w/w methylated soybean oil and 15% w/w polyalkylene oxide modified heptamethyl trisiloxane		
Use with	Approved pesticides on non-edible crops, cereals and stubbles of all edible crops, grassland (destruction), and pesticides approved for use on growing edible crops when used at half recommended dose or less. On specified crops product may be used at a maximum spray concentration of 0.5% with approved pesticides at their full approved rate up to the growth stages shown in the label		
Protective clothing	A, H		
Precautions	R38, R41, U02a, U05a, U08, U11, U19a, U20b, E13c, E34, E37, D01, D02, D05, D09a, M03, H04		

SECTION 4

Product	Supplier	Adj. No.	Type
Try-Flex	Greenaway	A0489	sticker/wetter
Contains	95% refined rapeseed oil		
Use with	Gly-480 (MAPP 12718) and Freeway (MAPP 11129) on natural surfaces not intended to bear vegetation, permeable surfaces overlying soil, hard surfaces overlying soil and forest		
Protective clothing	A, C, D, H, M		
Precautions	U08, U20b, E37, C11, D05, D10a		
Wetcit	Plant Solutions	A0586	penetrant/wetter
Contains	8.15% alcohol ethoxylate		
Use with	All approved pesticides on all edible crops when used at half their recommended dose or less, and on all non-edible crops up their full recommended dose. Alsoat a maximum concentration of 0.25% with approved pesticides on listed crops up to specified growth stages at full rate, and at half rate thereafter		
Precautions	R22a, R38, R41, U08, U11, U19a, U20b, E13c, E15a, E34, E37, D05, D08, D09a, D10b, H03		
X-Wet	De Sangosse	A0548	non-ionic surfactant/wetter
Contains	750 g/l alcohol ethoxylates and 150 g/l natural fatty acids		
Use with	All approved pesticides on all edible crops when used at half their recommended dose or less, and on all non-edible crops up their full recommended dose. Alsoat a maximum concentration of 0.1% with approved pesticides on listed crops up to specified growth stages		
Protective clothing	A, C, H		
Precautions	R36, R38, R53a, U02a, U04a, U05a, U10, U11, U20b, E15a, E19b, D01, D02, D05, D10a, H04		
Zarado	De Sangosse	A0516	vegetable oil
Contains	70% w/w esterified rapeseed oil		
Use with	All approved pesticides on edible and non-edible crops when used at half their approved dose or less. Also with approved pesticides on specified crops, up to specified growth stages, at up to their full approved dose		
Protective clothing	A, C, H		
Precautions	R43, R52, U05a, U13, U19a, U20b, E15a, E19b, E34, D01, D02, D05, D10a, M05a, H04		
Zinzan	Certis	A0600	extender/spreader/wetter
Contains	70% 1,2 bis (2-ethylhexyloxycarbonyl) ethanesulponate		
Use with	All approved pesticides on non-edible crops, and all pesticides approved for use on growing edible crops when used at half recommended dose or less. On specified edible crops product may be used at a maximum spray concentration of 0.075% with approved pesticides at their full approved rate up to the growth stages shown in the label		
Protective clothing	A, C		
Precautions	R38, R41, U11, U14, U20c, D01, D09a, D10a, H04		

SECTION 5
USEFUL INFORMATION

Pesticide Legislation

Anyone who advertises, sells, supplies, stores or uses a pesticide is bound by legislation, including those who use pesticides in their own homes, gardens or allotments. There are numerous UK statutory controls, but the major legal instruments are outlined below.

The Food and Environment Protection Act 1985 (FEPA) and Control of Pesticides Regulations 1986 (COPR)

FEPA introduced statutory powers to control pesticides, with the aims of protecting human beings, creatures and plants, safeguarding the environment, ensuring safe, effective and humane methods of controlling pests, and making pesticide information available to the public.

Control of pesticides is achieved by COPR. These Regulations lay down the Approvals required before any pesticide may be sold, stored, supplied, advertised or used, subject to conditions which are contained in a series of Schedules. Schedule 1 relates to advertisement; Schedule 2 to sale, supply and storage; Schedule 3 to use; and Schedule 4 to aerial application of pesticides. The Schedules may be changed at any time, following Parliamentary approval. Details are given on the websites of the Pesticides Safety Directorate (PSD, www.pesticides.gov.uk) and the Health and Safety Executive (HSE, www.hse.gov.uk).

The controls currently in force include the following.

- Only approved products may be sold, supplied, stored, advertised or used.
- No advertisement may contain any claim for safety beyond that which is permitted in the approved label text.
- Only products specifically approved for the purpose may be applied from the air.
- A recognised Storeman's Certificate of Competence is required by anyone who stores for sale or supply pesticides approved for agricultural use.
- A recognised Certificate of Competence is required by anyone who gives advice when selling or supplying pesticides approved for agricultural use.
- Users of pesticides must comply with the Conditions of Approval relating to use.
- A recognised Certificate of Competence is required for all contractors and for persons born after 31 December 1964 applying pesticides approved for agricultural use (unless working under direct supervision of a certificate holder). Proposals are now in place that every user of agricultural pesticides will have to hold a Certificate of Competence regardless of age or supervision. A suitable transition period will allow this requirement to be enacted.
- Only those adjuvants authorised by PSD may be used.
- Regarding tank-mixes, 'no person shall combine or mix for use two or more pesticides which are anti-cholinesterase compounds unless such a mixture is expressly permitted by the conditions of approval given in relation to at least one of those pesticides', and 'no person shall combine or mix for use two or more pesticides unless all the conditions of Approval given in relation to each of those pesticides can be complied with'.

The Plant Protection Products Directive

European Council Directive 91/414/EEC is intended to harmonise national arrangements for the authorisation of plant protection products within the European Community. It became effective in the UK on 25 July 1993. Under the provisions of the Directive, individual Member States are responsible for authorisation within their own territory of products containing active substances that appear in a list agreed at Community level. This list, known as Annex I, is being created over a period of time by review of existing active ingredients (to ensure they meet present safety standards) and authorisation of new ones. Over 120 active substances had achieved Annex I listing by the time of printing this edition.

Individual Member States are required to amend their national arrangements and legislation in order to meet the requirements of Directive 91/414/EEC. In the UK this has been achieved by a series of Plant Protection Products Regulations (PPPR) under which, over a period of time, all agricultural and horticultural pesticides will come to be regulated. Meanwhile, existing product approvals are being maintained under COPR, and new ones are granted for products containing active ingredients that were already on the market by 25 July 1993. Products containing new active substances that were not on the market at this date are being granted provisional approval in advance of Annex I listing of their active ingredients. As active ingredients are placed on Annex I of the Directive, products containing only those active ingredients will be regulated solely under PPPR. Active ingredients in this *Guide* that have been included in Annex I are identified in the 'Approval' section of the profile. The Directive also provides for a system of 'mutual recognition' of products registered in other Member States. Annex I listing, and a relevant approval in the Member State on which the mutual recognition is to be based, are essential prerequisites. Authorisation may be subject to conditions set to take account of differences such as climate and agricultural practice.

The Dangerous Preparations Directive (1999/45/EC)

The Dangerous Preparations Directive (DPD) came into force in UK for pesticide and biocidal products on 30 July 2004. Its aim is to achieve a uniform approach, across all Member States, to the classification, packaging and labelling of most dangerous preparations, including crop protection products. The Directive is implemented in UK under the Chemicals (Hazard Information and Packaging Supply) Regulations 2002, often referred to under the acronym CHIP3. In most cases, the Regulations have led to additional hazard symbols and associated risk and safety phrases relating to environmental and health hazards appearing on the label. All products affected by the DPD entering the supply chain from the implementation date above must be so labelled.

Under the DPD, the labels of all plant protection products now state 'To avoid risks to man and the environment, comply with the instructions for use.' As this instruction applies to all products, it has not been repeated in each entry of this *Guide*.

The Review Programme

The process of reviewing existing active ingredients, as required by the Plant Protection Products Directive 91/414, is taking considerably longer than originally anticipated. The programme is designed to ensure that all available plant protection products are supported by up-to-date information on safety and efficacy. Approximately 850 substances on the EU market are being handled in four stages but, because of the cost of providing this information, many are not being supported and their approvals are being revoked. Over half the original list has been lost because of this, and others are being revoked following scientific risk assessments.

At the start of the programme, it was envisaged that all active substances that were on the market on 25 July 1993 would have been reviewed by 25 July 2003. The process was rescheduled in 2006 for completion by 31 December 2008 but, although stages 1 and 2 have been completed, stages 3 and 4 are unlikely to be completed by this date. To speed up the process of review, the Commission has adopted a Regulation to amend some of the procedures allowing some active substances to follow a 'short-cut' route to inclusion ('green route') or non-inclusion ('red route') on Annex I within 6 months of receipt by the Standing Committee of the Draft Assessment Report.

Derogations for 'Essential Uses'

These uses, permitted since 2003 for specified active substances and products, and published in earlier editions of this *Guide*, terminated in December 2007 and may no longer be used.

The Control of Substances Hazardous to Health Regulations 1988 (COSHH)

The COSHH regulations, which came into force on 1 October 1989, were made under the Health and Safety at Work Act 1974, and are also important as a means of regulating the use of pesticides. The regulations cover virtually all substances hazardous to health, including those pesticides classed as Very toxic, Toxic, Harmful, Irritant or Corrosive; other chemicals used in farming or industry; and substances with occupational exposure limits. They also cover harmful micro-organisms, dusts and any other material, mixture, or compound used at work which can harm people's health.

The original Regulations, together with all subsequent amendments, have been consolidated into a single set of regulations: The Control of Substances Hazardous to Health Regulations 1994 (COSHH 1994).

The basic principle underlying the COSHH regulations is that the risks associated with the use of any substance hazardous to health must be assessed before it is used, and the appropriate measures taken to manage the risk. The emphasis is changed from that pertaining under the Poisonous Substances in Agriculture Regulations 1984 (now repealed), whereby the principal method of ensuring safety was the use of protective clothing, to preventing or controlling exposure to hazardous substances by a combination of measures. In order of preference, the measures should be:

1 substitution with a less hazardous chemical or product
2 technical or engineering controls (e.g. the use of closed handling systems, etc.)
3 operational controls (e.g. operators located in cabs fitted with air-filtration systems, etc.)
4 use of personal protective equipment, which includes protective clothing.

Consideration must be given to whether it is necessary to use a pesticide at all in a given situation, and if so, the product posing the least risk to humans, animals and the environment must be selected. Where other measures do not provide adequate control of exposure and the use of personal protective equipment is necessary, the items stipulated on the product label must be used as a minimum. It is essential that equipment is properly maintained and the correct procedures adopted. Where necessary, the exposure of workers must be monitored, health checks carried out, and employees must be instructed and trained in precautionary techniques. Adequate records of all operations involving pesticide application must be made and retained for at least 3 years.

Certificates of Competence – the roles of BASIS and the NPTC

COPR, COSHH and other legislation places certain obligations on those who handle and use pesticides. Minimum standards are laid down for the transport, storage and use of pesticides, and the law requires those who act as storekeepers, sellers and advisors to hold recognised Certificates of Competence.

BASIS is an independent registration scheme for the pesticide industry. It is responsible for organising training courses and examinations to enable staff to obtain a Certificate of Competence. In addition, BASIS undertakes annual assessments of pesticide supply stores, enabling distributors, contractors and seedsmen to meet their obligations under the Code of Practice for Suppliers of Pesticides. Further information can be obtained from BASIS (www.basis-reg.co.uk; see Appendix 2).

Certain spray operators also require Certificates of Competence under the Control of Pesticides Regulations. These certificates are awarded by NPTC to candidates who pass an assessment carried by an approved NPTC or Scottish Skills Testing Service Assessor in one or more of a series of competence modules. Holders are required to produce their certificate on demand for inspection to any authorised person. Further information can be obtained from NPTC (www.nptc.org.uk; see Appendix 2).

In addition to the above, most farm assurance schemes in England now require sprayer operators to be members of the National Register of Sprayer Operators (NRoSO). The register, which is administered by NPTC, was established in 2003 under the Voluntary Initiative using continuing professional development (CPD) as a means of ensuring ongoing training. Members are required to collect 30 CPD points in each 3-year period to retain their membership of the Register.

Maximum Residue Levels

A small number of pesticides are liable to leave residues in foodstuffs, even when used correctly. Where residues can occur, statutory limits, known as Maximum Residue Levels (MRLs), have been established. MRLs provide a check that products have been used as directed; they are not safety limits. However, they do take account of consumer safety because they are set at levels that ensure normal dietary intake of residues presents no risk to health. Wide safety margins are built in, and eating food containing residues above the MRL does not automatically imply a risk to health. Nevertheless, it is an offence to put into circulation any produce where the MRL is exceeded.

The surrounding legislation is complex. MRLs may be specified by several different bodies. The UK has set statutory MRLs since 1988. The European Union intends eventually to introduce MRLs for all pesticide/commodity combinations. These are being introduced initially by a series of priority lists, but will subsequently be covered by the review programme under Directive 91/414. However, in cases where no information is available, EU Directive 2000/42/EC requires many MRLs to be set at the limit of determination (LOD). As a result of this, certain approvals are being withdrawn where such use would leave residues above the MRL set in the Directive.

MRLs apply to imported as well as to home-produced foodstuffs. Details of those that have been set have been published in *The Pesticides (Maximum Residue Levels in Crops, Food and Feeding Stuffs) Regulations 1994*, and successive amendments to these Regulations. These Statutory Instruments are available from The Stationery Office.

MRLs are set for many chemicals not currently marketed in Britain. Because of this, and the ever-changing information on MRLs, no quantitative information is given in this *Guide*. Instead, users of the online version of the *Guide* (www.plantprotection.co.uk) may gain direct weblink access to the MRL databases on the Pesticides Safety Directorate website (https://secure.pesticides.gov.uk/MRLs). This online database sets out in table form those levels specified by UK Regulations, EC Directives, and the *Codex Alimentarius* for each commodity.

Approval (On-label and Off-label)

Only officially approved pesticides may be marketed and used in the UK. Approvals are granted by UK Government Ministers in response to applications that are supported by satisfactory data on safety, efficacy and, where relevant, humaneness. The Pesticides Safety Directorate (PSD, www.pesticides.gov.uk) is the UK Government Agency for regulating agricultural pesticides and plant protection products. The Health and Safety Executive (HSE, www.hse.gov.uk) fulfils the same role for other pesticides. The main focus of the regulatory process is the protection of human health and the environment.

Approvals are normally granted only in relation to individual products and for specified uses. It is an offence to use non-approved products or to use approved products in a manner that does not comply with the statutory conditions of use, except where the crop or situation is the subject of an off-label extension of use (see below).

Statutory Conditions of Use

Statutory conditions are laid down for the use of individual products and may include:

- field of use (e.g. agriculture, horticulture etc.)
- crop or situations for which treatment is permitted
- maximum individual dose
- maximum number of treatments or maximum total dose
- maximum area or quantity which may be treated
- latest time of application or harvest interval
- operator protection or training requirements
- environmental protection
- any other specific restrictions relating to particular pesticides.

Products must display these statutory conditions in a boxed area on the label entitled 'Important Information', or words to that effect. At the bottom of the boxed area must be shown a bold text statement: **'Read the label before use. Using this product in a manner that is inconsistent with the label may be an offence. Follow the Code of Practice for Using Plant Protection Products'**. This requirement came into effect in October 2006, and replaced the previous 'Statutory Box'.

Types of Approval

There are three categories of approval that may be granted, under the Control of Pesticides Regulations, to products containing active ingredients that were already on the market by 25 July 1993:

- **Full** (granted for an unstipulated period)
- **Provisional** (granted for a stipulated period)
- **Experimental Permit** (granted for the purposes of testing and developing new products, formulations or uses). Products with only an Experimental Permit may not be advertised or sold and do not appear in this *Guide*.

Products containing new active substances, and those containing older ingredients once they have been listed in Annex I to Directive 91/414 (see previous section), are granted approval under the Plant Protection Products Regulations. Again, there are three categories, similar to those listed above:

- **Standard approval** (granted for a period not exceeding 10 years). Only applicable to products whose active substances are listed in Annex I.
- **Provisional approval** (granted for a period not exceeding three years, but renewable). Applicable where Annex I listing of the active substance is awaited.
- **Approval for research and development** (to enable field experiments or tests to be carried out with active substances not otherwise approved).

The official lists of approved products, including all the above categories except those for experimental purposes, are shown on the websites of PSD (www.pesticides.gov.uk) and HSE (www.hse.gov.uk).
Details of new approvals, amendments to existing approvals and off-label approvals (see below) are published regularly on the above websites.

Withdrawal of Approval

Product approvals may be reviewed, amended, suspended or revoked at any time. Revocation may occur for various reasons, such as commercial withdrawal, or failure by the approval holder to meet data requirements.

From September 2007, where an approval is being revoked for purely administrative, 'housekeeping' reasons, the existing approval will be revoked and in its place will be issued:

- an approval for advertisement, sale and supply by any person for 24 months, and
- an approval for storage and use by any person for 48 months.

Where an approval is replaced by a newer approval, such that there are no safety concerns with the previous approval, but the newer updated approval is more appropriate in terms of reflecting the latest regulatory standard, the existing approval will be revoked and in its place will be issued:

- an approval for advertisement, sale and supply by any person for 12 months, and
- an approval for storage and use by any person for 24 months.

Where there is a need for tighter control of the withdrawal of the product from the supply chain, for example failure to meet data submission deadlines, the current timelines will be retained:

- immediate revocation for advertisement, sale or supply for the approval holder,
- approval for 6 months for advertisement, sale and supply by 'others', and
- approval for storage and use by any person for 18 months.

Immediate revocation and product withdrawal remain an option where serious concerns are identified, with immediate revocation of all approvals and approval for storage only by anyone for 3 months, to allow for disposal of product.

The expiry dates shown in this *Guide* are the final date of legal use of the product.

Approval of Commodity Substances

Commodity substances are compounds that have a variety of alternative and often widespread non-pesticidal uses, but also have potential for use as a pesticide. For a commodity substance to be used as a pesticide, it requires approval under the Control of Pesticides Regulations. Approval is given only for the *use* of the substance; approval is not given for sale, storage, supply or advertisement. There is no approval holder, and no pesticide product label, but the approval and associated conditions of use are published in the Guide to Approved Pesticides on the PSD website. Twenty such substances have been approved for certain specified uses as laid down in the Guide to Approved Pesticides. Of these, ethylene, carbon dioxide, formaldehyde, paraffin oil, peroxyacetic acid, potassium bicarbonate, sodium chloride, sodium hypochlorite, strychnine hydrochloride, sulphuric acid, and urea are approved for agricultural or horticultural use, and are included in this *Guide*.

Off-label Extension of Use

Products may legally be used in a manner not covered by the printed label in several ways:

- in accordance with the 'Long-Term Arrangements for Extension of Use' on non-edible crops (see below)
- in accordance with a specific off-label approval (SOLA). SOLAs are uses for which approval has been sought by individuals or organisations other than the manufacturer. The Notices of Approval are published by Defra and are widely available from ADAS or National Farmers'

Union offices. Users of SOLAs must first obtain a copy of the relevant Notice of Approval and comply strictly with the conditions laid down therein
- in tank mixture with other approved pesticides in accordance with Consent C(i) made under FEPA. Full details of Consent C(i) are given in Annex A of the Guide to Approved Pesticides on the PSD website (currently www.pesticides.gov.uk/publications.asp?id=499), but there are two essential requirements for tank mixes. First, all the conditions of approval of all the components of a mixture must be complied with. Second, no person may mix or combine pesticides which are cholinesterase compounds unless allowed by the label of at least one of the pesticides in the mixture
- in conjunction with authorised adjuvants
- in reduced spray volume under certain conditions
- the use of certain herbicides on specified set-aside areas subject to restrictions which differ between Scotland and the rest of the UK
- by mutual recognition of a use fully approved in another Member State of the European Union and authorised by PSD.

Although approved, off-label uses are not endorsed by manufacturers and such treatments are made entirely at the risk of the user.

Long-Term Arrangements for Extension of Use

Since 1 January 1990, arrangements have been in place allowing many approved products to be used for additional minor uses, without the need for Specific Off-Label Approvals (SOLAs). The arrangements – the Long-Term Arrangements for Extension of Use (LTAEU) – permitted growers to use pesticides on certain named minor crops provided an approval existed for use on specified major crops. Growers of minor crops were thus provided with access to a large number of important crop protection products.

The LTAEU have no parallel in European Community (EC) legislation and are not compatible with the Plant Protection Products Directive and procedures for setting maximum residue levels. After consultation, the Pesticides Safety Directorate concluded in November 2004 that the best and most cost-effective way of retaining these uses was by converting them to SOLAs. This exercise related only to uses on edible crops and is now complete.

For uses on non-edible crops, PSD decided in 2007 that, rather than remove the arrangements completely, they would be allowed to continue for products approved under UK legislation (Control of Pesticides Regulations) until the European Pesticide Review process (see Pesticide Legislation) is complete and the products have been re-registered under the Plant Protection Products Regulations (PPPR). For products already approved under PPPR, these uses will be replaced by SOLAs. It will be a two-stage process.

The first stage will be to issue SOLAs for 'PPPR' products. One SOLA will be issued for each product detailing all the uses requested.

The second stage will involve products currently approved under COPR. As they become re-registered, PSD will issue appropriate SOLAs for uses requested by growers. The arrangements will continue until the review process has been completed and all UK products are approved under PPPR.

Specific restrictions for extension of use

The uses and products that may be requested will be based on the current arrangements and restrictions, under the following rules:

1. Hardy ornamental nursery stock, Ornamental plants, Ornamental bulbs and flowers, Forest nursery crops prior to final planting out.

These will all come under the terms 'Ornamental plant production' and 'Forest nursery'. To negate any possible risk to workers, the following additional points must be considered:

- the crop from which extrapolation is to be made must be a hand-harvested crop, and
- use on protected crops must be extrapolated from a protected crop. Note: This also applies to all other LTAEU uses.

2. Non-ornamental and ornamental crops grown for seed

This use will be changed to 'All edible crops (seed crop)' and 'All non-edible crops (seed crop)'. This extrapolation currently applies to any product approved for use on any growing edible crop, but does not apply to crops of potatoes, cereals, oilseeds, peas, beans and other pulses grown for seed.

3. Hemp grown for fibre

Only products approved for use on oilseed rape will be considered.

4. *Miscanthus* spp. (elephant grass)

Only products approved for use on cereal, grass and forage maize will be considered.

5. Woad (*Isatis tinctoria, Isatis indigotica*)

Only products approved for use on oilseed rape will be considered.

6. Sunflower, quinoa, fodder radish and sweet clover (including white melilot, bokhara clover and *Melilotus alba*) grown as game cover

Only herbicides approved for use on oilseed rape will be considered.

7. Canary grass, tanka millet, white millet and sorghum grown as game cover

Only herbicides approved for use on cereal and forage maize will be considered.

8. Farm forestry and Coppices (short term, rotational, intensive wood production e.g. poplar or willow biofuel production)

For use in the first 5 years of establishment, only herbicides approved for use on cereals will be considered.

For use after coppicing, only herbicides approved for use on cereal, oilseed rape, sugar beet, potatoes, peas and beans will be considered.

As this use may involve application via hand-held equipment, which may not have been assessed for the extrapolated use, a restriction will be added to exclude use via hand-held equipment unless specifically permitted on the product label.

9. Nursery fruit trees, Nursery grape vines prior to final planting out, Bush fruit, Cane fruit, Non-fruiting strawberry plants

These uses will be changed to 'Top fruit' and 'Soft fruit'.

Only products approved for use on an edible crop will be considered.

10. Mature stock or mother hop plants which are kept specifically for the supply of propagation material, Propagation of hop planting material – propagules prior to final planting out, Nursery hops. First year plants not taken to harvest that year, in their final planting out position

These uses will be changed to 'Hops'.

This extrapolation applies to any product.

As this use may involve application via hand-held and broadcast air-assisted equipment, which may not have been assessed for the extrapolated use, a restriction will be added to exclude use via hand-held and broadcast air-assisted equipment, unless specifically permitted on the product label.

Use of Herbicides In or Near Water

Products in this *Guide* approved for use in or near water are listed in Table 5.2. Before use of any product in or near water, the appropriate water regulatory body (Environment Agency/Local Rivers Purification Authority; or in Scotland the Scottish Environment Protection Agency) must be consulted. Guidance and definitions of the situation covered by approved labels are given in the Defra publication *Guidelines for the Use of Herbicides on Weeds in or near Watercourses and Lakes*. Always read the label before use.

Table 5.2 Products approved for use in or near water

Chemical	Product
2,4-D	Depitox
dichlobenil	Casoron G, Certis Dichlobenil Granules, Midstream GSR, Osorno, Scotts Dichlo G Macro, Scotts Dichlo G Micro
glyphosate	Asteroid, Buggy XTG, Dow Agrosciences Glyphosate 360, Envision, Glyfos Dakar, Glyfos Dakar Pro, Glyfos Gold, Glyfos Proactive, Greenaway Gly-490, Kernel, Manifest, Roundup Biactive, Roundup Pro Biactive, Roundup ProBiactive 450, Tangent
maleic hydrazide	Regulox K

Use of Pesticides in Forestry

Table 5.3 Products in this *Guide* approved for use in forestry

Chemical	Product	Use
2,4-D + dicamba + triclopyr	Broadsword, Nu-Shot	Herbicide
4-indol-3-ylbutyric acid + 1-naphthylacetic acid	Synergol	Growth regulator/stimulator/sprout suppressant
aluminium ammonium sulphate	Curb Crop Spray Powder, Liquid Curb Crop Spray	Animal deterrent/repellent
ammonium sulphamate	Root-Out	Herbicide, Tree killer
asulam	Asulox, Greencrop Frond	Herbicide
carbosulfan	Marshal Soil Insecticide suSCon CR granules	Insecticide
chlorpyrifos	Clayton Pontoon, Dursban WG, Equity, Govern, Greencrop Pontoon, Parapet	Acaricide, Insecticide
cycloxydim	Greencrop Pomeroy, Greencrop Valentia, Laser, Standon Cycloxydim	Herbicide
cypermethrin	Forester	Insecticide
diflubenzuron	Dimilin Flo	Insecticide
diquat + paraquat	Clayton Paradigm, PDQ	Desiccant, Herbicide
fluazifop-P-butyl	Clayton Maximus, Fusilade Max, Greencrop Bantry	Herbicide
glufosinate-ammonium	Harvest, Kaspar	Herbicide
glyphosate	Amega Duo, Amenity Glyphosate 360, Buggy XTG, Clinic Ace, Credit, Dow Agrosciences Glyphosate 360, Envision, Glyfos, Glyfos Dakar Pro, Glyfos Gold, Glyfos Proactive, Glyfos Supreme, Glyphogan, Greenaway Gly-490, Hilite, Kernel, Manifest, Roundup Pro Biactive, Roundup ProBiactive 450, Stirrup, Tangent, Tumbleweed Pro-Active, Typhoon 360	Herbicide
isoxaben	Agriguard Isoxaben, Flexidor 125	Herbicide

Chemical	Product	Use
metazachlor	Alpha Metazachlor 50 SC, Clayton Buzz, Clayton Metazachlor, Clayton Metazachlor 50 SC, Greencrop Monogram, Landgold Metazachlor 50 SC, Marksman, Rapsan 500 SC, Standon Metazachlor 500, Sultan 50 SC	Herbicide
paraquat	Agriguard Paraquat	Herbicide
propaquizafop	Bulldog, Cleancrop GYR, Falcon, Greencrop Satchmo, Landgold PQF 100, Raptor, Shogun, Standon Propaquizafop, Zealot	Herbicide
propyzamide	Agriguard Propyzamide 50 WP, Bulwark Flo, Clayton Propel 80 WG, Engage, Kerb 50 W, Kerb Pro Flo, Menace 80 EDF, Setanta 50 WP	Herbicide
pyrethrins	Spruzit	Insecticide
triclopyr	Garlon 4, Nomix Garlon 4, Timbrel	Herbicide
urea (commodity substance)	urea	Fungicide
ziram	AAprotect	Animal deterrent/repellent

Pesticides Used as Seed Treatments

Information on the target pests for these products can be found in the relevant pesticide profile in Section 2.

Table 5.4 Products used as seed treatments (including treatments on seed potatoes)

Chemical	Product	Formulation	Crop(s)
aluminium ammonium sulphate	Guardsman STP Seed Dressing Powder	DS	Corms, Flower bulbs, Seeds
beta-cyfluthrin + clothianidin	Poncho Beta	FS	Fodder beet, Sugar beet
beta-cyfluthrin + imidacloprid	Chinook Blue	LS	Spring oilseed rape, Winter oilseed rape
	Chinook Colourless	LS	Winter oilseed rape
carboxin + thiram	Anchor	FS	Spring barley, Spring oats, Spring rye, Spring wheat, Triticale, Winter barley, Winter oats, Winter rye, Winter wheat
clothianidin	Deter	FS	Durum wheat, Triticale, Winter barley, Winter oats, Winter rye, Winter wheat
clothianidin + prothioconazole	Redigo Deter	FS	Durum wheat, Triticale, Winter barley, Winter oats, Winter rye, Winter wheat
clothianidin + prothioconazole + tebuconazole + triazoxide	Raxil Deter	FS	Winter barley
cymoxanil + fludioxonil + metalaxyl-M	Wakil XL	WS	Broad beans, Carrots, Chard, Chives, Combining peas, Herbs (see appendix 6), Leaf spinach, Lupins, Parsley, Parsnips, Poppies for morphine production, Red beet, Salad brassicas, Spinach beet, Spring field beans, Vining peas, Winter field beans
fludioxonil	Beret Gold	FS	Spring barley, Spring oats, Spring wheat, Winter barley, Winter oats, Winter wheat
fludioxonil + flutriafol	Beret Multi	FS	Spring barley, Spring oats, Spring wheat, Winter barley, Winter oats, Winter wheat

Chemical	Product	Formulation	Crop(s)
fludioxonil + tefluthrin	Austral Plus	FS	Spring barley, Spring oats, Spring wheat, Winter barley, Winter oats, Winter wheat
fluoxastrobin + prothioconazole	Redigo Twin	FS	Durum wheat, Spring rye, Triticale, Winter oats, Winter rye, Winter wheat
fluquinconazole	Galmano	FS	Winter wheat
fluquinconazole + prochloraz	Epona	FS	Winter wheat
	Galmano Plus	FS	Winter wheat
	Jockey	FS	Winter barley, Winter wheat
flutolanil	Rhino	FS	Potatoes
	Rhino DS	DS	Potatoes
fuberidazole + imidacloprid + triadimenol	Tripod Plus	LS	Winter barley, Winter oats, Winter wheat
fuberidazole + triadimenol	Tripod	FS	Spring barley, Spring oats, Spring rye, Spring wheat, Triticale, Winter barley, Winter oats, Winter rye, Winter wheat
guazatine	Panoctine	LS	Spring barley, Spring oats, Spring wheat, Winter barley, Winter oats, Winter wheat
	Ravine	LS	Spring barley, Spring oats, Spring wheat, Winter barley, Winter oats, Winter wheat
guazatine + imazalil	Panoctine Plus	LS	Spring barley, Spring oats, Winter barley, Winter oats
	Ravine Plus	LS	Spring barley, Spring oats, Winter barley, Winter oats
imazalil	Fungazil 100 SL	LS	Seed potatoes, Ware potatoes
	IMA 100 SL	LS	Seed potatoes, Ware potatoes
imazalil + pencycuron	Monceren IM	DS	Potatoes
imazalil + thiabendazole	Storite Super	FS	Seed potatoes

Chemical	Product	Formulation	Crop(s)
imidacloprid	Gaucho	WS	Broccoli, Brussels sprouts, Cabbages, Calabrese, Cauliflowers, Collards, Kale, Sugar beet
iprodione	Rovral Liquid FS	FS	Flax, Seed potatoes, Spring oilseed rape, Winter oilseed rape
pencycuron	Agriguard Pencycuron	DS	Potatoes
	Monceren DS	DS	Potatoes
	Pency Flowable	FS	Potatoes
	Standon Pencycuron DP	DS	Potatoes
prochloraz	Prelude 20LF	LS	Flax, Hemp for oilseed production, Linseed
prochloraz + thiram	Agrichem Hy-Pro Duet	FS	Winter oilseed rape
prochloraz + triticonazole	Kinto	FS	Durum wheat, Triticale, Winter barley, Winter oats, Winter rye, Winter wheat
prothioconazole	Redigo	FS	Durum wheat, Spring oats, Spring rye, Spring wheat, Triticale, Winter barley, Winter oats, Winter rye, Winter wheat
prothioconazole + tebuconazole + triazoxide	Raxil Pro	FS	Spring barley, Winter barley
silthiofam	Latitude	FS	Spring wheat, Winter barley, Winter wheat
thiabendazole + thiram	Hy-TL	FS	Broad beans, Bulb onions, Peas, Spring field beans, Winter field beans
thiamethoxam	Cruiser SB	FS	Fodder beet, Sugar beet
thiram	Agrichem Flowable Thiram	FS	Broad beans, Bulb onions, Cabbages, Carrots, Cauliflowers, Celery (outdoor), Combining peas, Cress, Dwarf beans, Edible podded peas, Endives, Fodder beet, Forage maize, Frise, Garlic, Grass seed, Lamb's lettuce, Leeks, Lettuce, Lupins, Mangels,

Chemical	Product	Formulation	Crop(s)
			Parsley, Poppies for morphine production, Radicchio, Radishes, Red beet, Rhubarb, Runner beans, Salad onions, Scarole, Shallots, Soya beans, Spring field beans, Spring oilseed rape, Sugar beet, Swedes, Turnips, Vining peas, Winter field beans, Winter oilseed rape
	Thiraflo	FS	Broad beans, Combining peas, Dwarf beans, Forage maize, Grass seed, Runner beans, Spring field beans, Spring oilseed rape, Vining peas, Winter field beans, Winter oilseed rape
	Thyram Plus	FS	Broad beans, Bulb onions, Cabbages, Cauliflowers, Combining peas, Dwarf beans, Edible podded peas, Forage maize, Grass seed, Leeks, Radishes, Runner beans, Salad onions, Soya beans, Spring field beans, Spring oilseed rape, Turnips, Vining peas, Winter field beans, Winter oilseed rape
tolclofos-methyl	Rizolex	DS	Potatoes
	Rizolex Flowable	FS	Potatoes

Aerial Application of Pesticides

Only those products specifically approved for aerial application may be so applied, and they may only be applied to specific crops or for specified uses. A complete list of products approved for application from the air can be accessed in Guide to Pesticides on the website of the Pesticides Safety Directorate (PSD, www.pesticides.gov.uk). The list in Table 5.5 is taken mainly from this source, with updating from issues of *The Pesticides Monitor* and product labels.

It is emphasised that the list is for guidance only – reference must be made to the product labels for detailed conditions of use, which must be complied with. The list does not include those products which have been granted restricted aerial application approval limiting the area which may be treated.

Detailed rules are imposed on aerial application regarding prior notification of the Nature Conservancy, water authorities, bee keepers, Environmental Health Officers, neighbours, hospitals, schools, etc. and the conditions under which application may be made. The full conditions are available from Defra and must be consulted before any aerial application is made.

Table 5.5 Products approved for aerial application

Chemical	Product	Crops
2-chloroethylphosphonic acid	Agriguard Cerusite	Winter barley
	Cerone	Winter barley
	Pan Stiffen	Winter barley
asulam	Asulox	Amenity grassland, Forest, Permanent grassland, Rotational grass
	Brack-N	Amenity grassland, Forest, Permanent grassland
	Greencrop Frond	Amenity grassland, Forest, Permanent grassland
benalaxyl + mancozeb	Intro Plus	Potatoes
	Tairel	Potatoes
chlormequat	Agriguard Chlormequat 700	Spring oats, Spring wheat, Winter oats, Winter wheat
	Agriguard Chlormequat 720	Spring oats, Spring wheat, Winter oats, Winter wheat
	Barclay Take 5	Spring oats, Spring wheat, Triticale, Winter barley, Winter oats, Winter rye, Winter wheat
	BASF 3C Chlormequat 720	Spring oats, Spring rye, Spring wheat, Triticale, Winter barley, Winter oats, Winter rye, Winter wheat
	Greencrop Carna	Spring oats, Spring wheat, Winter barley, Winter oats, Winter wheat

Chemical	Product	Crops
	Hive	Winter wheat
	Mandops Chlormequat 700	Spring oats, Spring wheat, Winter oats, Winter wheat
	Mirquat	Spring oats, Spring wheat, Winter oats, Winter wheat
	New 5C Cycocel	Spring oats, Spring rye, Spring wheat, Triticale, Winter barley, Winter oats, Winter rye, Winter wheat
	New 5C Quintacel	Spring oats, Spring rye, Spring wheat, Triticale, Winter barley, Winter oats, Winter rye, Winter wheat
	Sigma PCT	Spring wheat, Winter barley, Winter wheat
	Terbine	Spring oats, Spring wheat, Winter oats, Winter wheat
chlorothalonil	Agriguard Chlorothalonil	Potatoes
	Bravo 500	Potatoes
	Greencrop Orchid 2	Potatoes
	Greencrop Orchid B	Potatoes
	Landgold Chlorothalonil 50	Potatoes
	Repulse	Potatoes
	Sanspor	Potatoes
	Sonar	Potatoes
chlorotoluron	Lentipur CL 500	Durum wheat, Triticale, Winter barley, Winter wheat
copper oxychloride	Cuprokylt	Potatoes
diflubenzuron	Dimilin Flo	Forest
dimethoate	BASF Dimethoate 40	Spring rye, Spring wheat, Sugar beet, Sugar beet seed crops, Triticale, Winter rye, Winter wheat
	Danadim Progress	Spring rye, Spring wheat, Sugar beet, Triticale, Winter rye, Winter wheat

Chemical	Product	Crops
mancozeb	Dithane 945	Potatoes
	Dithane NT Dry Flowable	Potatoes
	Micene 80	Potatoes
	Micene DF	Potatoes
	Penncozeb WDG	Potatoes
	Quell Flo	Potatoes
metaldehyde	Bristol Blues	All edible crops (outdoor), All non-edible crops (outdoor), Natural surfaces not intended to bear vegetation
	CDP Minis	All edible crops (outdoor), All non-edible crops (outdoor), Natural surfaces not intended to bear vegetation
	Condor	All edible crops (outdoor), All non-edible crops (outdoor), Natural surfaces not intended to bear vegetation
	Condor 5	All edible crops (outdoor), All non-edible crops (outdoor), Natural surfaces not intended to bear vegetation
	Dixie 6	All edible crops (outdoor), All non-edible crops (outdoor), Natural surfaces not intended to bear vegetation
	Doff Horticultural Slug Killer Blue Mini Pellets	All edible crops (outdoor), All non-edible crops (outdoor), Cultivated land/soil
	Entice	All edible crops (outdoor), All non-edible crops (outdoor), Natural surfaces not intended to bear vegetation
	ESP	All edible crops (outdoor), All non-edible crops (outdoor)
	Lincoln VI	All edible crops (outdoor), All non-edible crops (outdoor), Natural surfaces not intended to bear vegetation

Chemical	Product	Crops
	Lynx S	All edible crops (outdoor), All non-edible crops (outdoor)
	Optimol	All edible crops (outdoor), All non-edible crops (outdoor), Natural surfaces not intended to bear vegetation
	Pesta M	All edible crops (outdoor), All non-edible crops (outdoor)
	Polymet 5	All edible crops (outdoor), All non-edible crops (outdoor), Natural surfaces not intended to bear vegetation
	Super 5	All edible crops (outdoor), All non-edible crops (outdoor), Natural surfaces not intended to bear vegetation
pirimicarb	Agriguard Pirimicarb	Spring barley, Spring oats, Spring wheat, Winter barley, Winter oats, Winter wheat
	Aphox	Durum wheat, Spring barley, Spring oats, Spring rye, Spring wheat, Triticale, Winter barley, Winter oats, Winter rye, Winter wheat
	Greencrop Glenroe	Durum wheat, Spring barley, Spring oats, Spring rye, Spring wheat, Triticale, Winter barley, Winter oats, Winter rye, Winter wheat
	Phantom	Durum wheat, Spring barley, Spring oats, Spring rye, Spring wheat, Triticale, Winter barley, Winter oats, Winter rye, Winter wheat
	Standon Pirimicarb 50	Durum wheat, Spring barley, Spring oats, Spring rye, Spring wheat, Triticale, Winter barley, Winter oats, Winter rye, Winter wheat
tri-allate	Avadex Excel 15G	Combining peas, Fodder beet, Lucerne, Red beet, Red clover, Sainfoin, Spring barley, Spring field beans, Sugar beet, Triticale, Vetches, Vining peas, White clover, Winter barley, Winter field beans, Winter wheat

Resistance Management

Pest species are, by definition, adaptable organisms. The development of resistance to some crop protection chemicals is just one example of this adaptability. Repeated use of products with the same mode of action will clearly favour those individuals in the pest population able to tolerate the treatment. This leads to a situation where the tolerant (or resistant) individuals can dominate the population and the product becomes ineffective. In general, the more rapidly the pest species reproduces and the more mobile it is, the faster is the emergence of resistant populations, although some weeds seem able to evolve resistance more quickly than would be expected.

In the UK, key independent research organisations, chemical manufacturers and other organisations have collaborated to share knowledge and expertise on resistance issues through three action groups. Participants include ADAS, the Pesticides Safety Directorate, universities, colleges and the Home Grown Cereals Authority. The groups have a common aim of monitoring resistance in UK and devising and publishing management strategies designed to combat it where it occurs. The groups are:

The Weed Resistance Action Group (WRAG), formed in 1989 (Secretary: Dr Stephen Moss, Rothamsted Research, Harpenden, Herts AL5 2JQ *Tel: 01582 7631330 ext. 2521*)

The Fungicide Resistance Action Group (FRAG), formed in 1995 (Secretary: Mr Oliver Macdonald, Pesticides Safety Directorate, Mallard House, King's Pool, 3 Peasholme Green, York YO1 2PX *Tel: 01904 455864*)

The Insecticide Resistance Action Group (IRAG), formed in 1997 (Secretary: Dr Stephen Foster, Rothamsted Research, Harpenden, Herts AL5 2JQ *Tel: 01582 763133 ext. 2324*)

The above groups publish detailed advice on resistance management relevant for each sector and, in some cases, specific to a pest problem. This information, together with further details about the function of each group, can be obtained from the Pesticides Safety Directorate web site: www.pesticides.gov.uk/rags.asp

The speed of appearance of resistance depends on the mode of action of the crop protection chemicals, as well as the manner in which they are used. Resistance among insects and fungal diseases has been evident for much longer than weed resistance to herbicides, but examples in all three categories are now widespread and increasing. This has created a need for agreement on the advice given for the use of crop protection chemicals in order to reduce the likelihood of the development of resistance and to avoid the loss of potentially valuable products in the chemical armoury.

Mixing or alternating modes of action is one of the guiding principles of resistance management. To assist appropriate product choices, mode of action codes, published by the international Resistance Action Committees (see below), are shown in the respective active substance profiles of this *Guide*. The product label and/or a professional advisor should always be consulted before making decisions.

The general guidelines for resistance management are similar for all three problem areas.

Preparation in advance

- Be aware of the factors that favour the development of resistance, such as repeated annual use of the same product, and assess the risk.
- Plan ahead and aim to integrate all possible means of control.
- Use cultural measures such as rotations, stubble hygiene, variety selection and, for fungicides, removal of primary inoculum sources, to reduce reliance on chemical control.
- Monitor crops regularly.
- Keep aware of local resistance problems.
- Monitor effectiveness of actions taken and take professional advice, especially in cases of unexplained poor control.

Using crop protection products

- Optimise product efficacy by using it as directed, at the right time, in good conditions.
- Treat pest problems early.
- Mix or alternate chemicals with different modes of action.
- Avoid repeated applications of very low doses.
- Keep accurate field records.

Label guidance depends on the appropriate strategy for the product. Most frequently it consists of a warning of the possibility of poor performance due to resistance, and a restriction on the number of treatments that should be applied in order to minimise the development of resistance. This information is summarised in the profiles in this *Guide*, but detailed guidance must always be obtained by reading the label itself before use.

International action committees

Resistance to crop protection products is an international problem. Agrochemical industry collaboration on a global scale is via three action committees whose aims are to support a coordinated industry approach to the management of resistance worldwide. In particular they produce lists of crop protection chemicals classified according to their mode of action. These lists and other information can be obtained from the respective websites:

Herbicide Resistance Action Committee (HRAC): www.plantprotection.org/hrac

Fungicide Resistance Action Committee (FRAC): www.frac.info

Insecticide Resistance Action Committee (IRAC): www.irac-online.org

Poisons and Poisoning

Chemicals Subject to the Poisons Law

Certain products in this book are subject to the provisions of the Poisons Act 1972, the Poisons List order 1982 and the Poisons Rules 1982 (copies of all these are obtainable from The Stationery Office). These rules include general and specific provisions for the storage, sale and supply of listed non-medicine poisons. Full details can be accessed in Annex C of the Guide to Approved Pesticides on the PSD website (currently www.pesticides.gov.uk/publications.asp?id=499). The nature of the formulation and the concentration of the active ingredient allow some products to be exempted from the rules (see below).

The chemicals approved for use in the UK and included in this book are specified under Parts I and II of the Poisons List as follows.

Part I Poisons

Sale restricted to registered retail pharmacists and to registered non-pharmacy businesses provided sales do not take place on retail premises:

aluminium phosphide
chloropicrin
magnesium phosphide
strychnine

Part II Poisons

Sale restricted to registered retail pharmacists and listed sellers registered with a local authority:

formaldehyde
nicotine (b)
oxamyl (a)
paraquat (c)
sulphuric acid

(a) Granular formulations containing up to 12% w/w of this, or a combination of similarly flagged poisons, are exempt.
(b) Formulations containing not more than 7.5% of nicotine are exempt.
(c) Pellets containing not more than 5% paraquat ion are exempt.

Note:

The European Commission Ozone Depleting Substances Regulation (EC Regulation 2037/2000) banned the use of methyl bromide in the UK, with the exception of a few 'Critical Use Exemptions', which will be reviewed annually by the EC. Suppliers of products containing methyl bromide which had previously been listed in this *Guide* withdrew them in 2004.

Occupational Exposure Limits

A fundamental requirement of the COSHH Regulations is that exposure of employees to substances hazardous to health should be prevented or adequately controlled. Exposure by inhalation is usually the main hazard, and in order to measure the adequacy of control of exposure by this route, various substances have been assigned occupational exposure limits.

There are two types of occupational exposure limits defined under COSHH: Occupational Exposure Standards (OES) and Maximum Exposure Limits (MEL). The key difference is that an OES is set at a level at which there is no indication of risk to health; for an MEL a residual risk may exist and the level takes socio-economic factors into account. In practice, MELs have most often been allocated to carcinogens and other substances for which no threshold of effect can be identified, and for which there is no doubt about the seriousness of the effects of exposure.

OESs and MELs are set on the recommendation of the Advisory Committee on Toxic Substances. Full details are published by the Health and Safety Executive (HSE) in *Occupational Exposure Limits* 1995, EH 40/95 (ISBN 0 7176 0876 X).

As far as pesticides are concerned, OESs and MELs have been set for relatively few active ingredients. This is because pesticide products usually contain other substances in their formulation, including solvents, which may have their own OES/MEL. In practice, inhalation of solvent may be at least, or more, important than that of the active ingredient. These factors are taken into account in the approval process of the pesticide product under the Regulations. This indicates one of the reasons why a change of pesticide formulation usually necessitates a new approval assessment.

First Aid Measures

If pesticides are handled in accordance with the required safety precautions, as given on the container label, poisoning should not occur. It is difficult, however, to guard completely against the occasional accidental exposure. Thus if a person handling, or exposed to, pesticides becomes ill, it is a wise precaution to apply first aid measures appropriate to pesticide poisoning even though the cause of illness may eventually prove to have been quite different. An employer has a legal duty to make adequate first aid provision for employees. Regular pesticide users should consider appointing a trained first aider even if numbers of employees are not large, as there is a specific hazard.

The first essential in a case of suspected poisoning is for the person involved to stop work, to be moved away from any area of possible contamination, and for a doctor to be called at once. If no doctor is available the patient should be taken to hospital as quickly as possible. In either event it is most important that the name of the chemical being used should be recorded, and preferably the whole product label or leaflet should be shown to the doctor or hospital concerned.

Some pesticides which are unlikely to cause poisoning in normal use are extremely toxic if swallowed accidentally or deliberately. In such cases get the patient to hospital as quickly as possible, with all the information you have.

General measures

Measures appropriate in all cases of suspected poisoning include:

- remove any protective or other contaminated clothing (taking care to avoid personal contamination)
- wash any contaminated areas carefully with water or with soap and water if available
- in cases of eye contamination, flush with plenty of clean water for at least 15 minutes
- lay the patient down, keep at rest and under shelter; cover with one clean blanket or coat, and avoid overheating
- monitor level of consciousness, breathing and pulse rate
- if consciousness is lost, place the casualty in the recovery position (on his or her side with head down and tongue forward to prevent inhalation of vomit)

Reporting pesticide poisoning

Any cases of poisoning by pesticides must be reported without delay to an HM Agricultural Inspector of the Health and Safety Executive. In addition, any cases of poisoning by substances named in Schedule 2 of The Reporting of Injuries, Diseases and Dangerous Occurrences Regulations 1985 must also be reported to HM Agricultural Inspectorate (this includes organophosphorus chemicals, mercury and some fumigants).

Cases of pesticide poisoning should also be reported to the manufacturer concerned.

Additional information

General advice on the safe use of pesticides is given in a range of Health and Safety Executive leaflets available from HSE Books (see Appendix 2).

A useful booklet, *Guidelines for Emergency Measures in Cases of Pesticide Poisoning*, is available from CropLife International (see Appendix 2).

The major agrochemical companies are able to provide authoritative medical advice about their own pesticide products. Detailed advice is also available to doctors from the National Poisons Information Service (NPIS).

New arrangements for the provision of information to doctors about poisons and the management of poisonings have been introduced as part of the modernisation of the NPIS. They provide a year-round, 24-hour-a-day service for healthcare staff, on the diagnosis, treatment and management of patients who may have been poisoned. The new arrangements are aimed at moving away from the telephone as the first point of contact for poisons information, to the use by doctors of an online database, supported by a second-tier, consultant-led information service for more complex clinical advice.

As a result of these changes the contact addresses and telephone numbers listed in previous editions of this *Guide* are no longer applicable. **Anyone suspecting poisoning by pesticides should seek professional medical help immediately via their GP or NHS Direct/NHS 24.**

Environmental Protection

Protection of Bees

Honey bees

Honey bees are a source of income for their owners, and important to farmers and growers as pollinators of their crops. It is irresponsible and unnecessary to use pesticides in such a way that may endanger them. Pesticides vary in their toxicity to bees, but those that present a special hazard carry a specific warning in the precautions section of the label. These are indicated in this *Guide* in the **hazard classification and safety precautions** section of the pesticide profile.

Product labels indicate the necessary environmental precautions to take, but where use of an insecticide on a flowering crop is contemplated, the British Beekeepers Association (see Appendix 2) has produced the following guidelines for growers:

- target insect pests with the most appropriate product
- choose a product that will cause minimal harm to beneficial species
- follow the manufacturer's instructions carefully
- inspect and monitor crops regularly
- avoid spraying crops in flower or where bees are actively foraging
- keep down flowering weeds
- spray late in the day, in still conditions
- avoid excessive spray volume and run-off
- adjust sprayer pressure to reduce production of fine droplets and drift
- give local beekeepers as much warning of your intention as possible.

Wild bees

Wild bees also play an important role. Bumblebees are useful pollinators of spring-flowering crops and fruit trees because they forage in cool, dull weather when honey bees are inactive. They play a particularly important part in pollinating field beans, red and white clover, lucerne and borage. Bumblebees nest and overwinter in field margins and woodland edges. Avoidance of direct or indirect spray contamination of these areas, in addition to the creation of hedgerows and field margins, and late cutting or grazing of meadows, all helps the survival of these valuable insects.

The Campaign Against Illegal Poisoning of Wildlife

The Campaign Against Illegal Poisoning of Wildlife, aimed at protecting some of Britain's rarest birds of prey and wildlife while also safeguarding domestic animals, was launched in March 1991 by the (then) Ministry of Agriculture, Fisheries and Food and the Department of the Environment, Transport and the Regions. The main objective is to deter those who may be considering using pesticides illegally. The Campaign is supported by a range of organisations associated with animal welfare, nature preservation, field sports and game keeping including the Royal Society for the Protection of Birds, English Nature, the Countryside Alliance and the Game Conservancy Trust.

The three objectives are to:

- advise farmers, gamekeepers and other land managers on legal ways of controlling pests
- advise the public on how to report illegal poisoning incidents and to respect the need for legal alternatives
- investigate incidents and prosecute offenders.

A freephone number (0800 321 600) is available to make it easier for the public to report incidents, and numerous leaflets, posters, postcards, coasters and stickers have been created to publicise the existence of the Campaign.

The Campaign arose from the results of the Wildlife Incident Investigation Scheme for the investigation of possible cases of illegal poisoning. Under this scheme, all reported incidents are considered and thoroughly investigated where appropriate. Enforcement action is taken wherever sufficient evidence of an offence can be obtained, and numerous prosecutions have been made since the start of the Campaign.

Further information about the Campaign is available from:

Pesticides Safety Directorate
Room 317, Mallard House
3 Peasholme Green
York YO1 7PX
e-mail: wiis@psd.defra.gsi.gov.uk
website: www.pesticides.gov.uk/environment.asp?id=504

Water Quality

The Food and Environment Protection Act 1985 (FEPA) places a special obligation on users of pesticides to 'safeguard the environment and in particular avoid the pollution of water'. Under the Water Resources Act 1991 it is an offence to pollute any controlled waters (watercourses or groundwater), either deliberately or accidentally. Protection of controlled waters from pollution is the responsibility of the Environment Agency. Users of pesticides therefore have a duty to adopt responsible working practices and, unless they are applying herbicides in or near water, to prevent them getting into water. Guidance on how to achieve this is given in the Defra *Code of Good Agricultural Practice for the Protection of Water* (available from The Stationery Office; see Appendix 2). The duty of care covers not only the way in which a pesticide is sprayed, but also its storage, preparation and disposal of surplus, sprayer washings and the container. Products in this *Guide* that are a major hazard to fish, other aquatic life or aquatic higher plants carry one of several specific label precautions in their profile, depending on the assessed hazard level. Advice on any water pollution problems is available from the Environment Agency and the Crop Protection Association (see Appendix 2).

Protecting surface waters

Surface waters are particularly vulnerable to contamination. One of the best ways of preventing those pesticides that carry the greatest risk to aquatic wildlife from reaching surface waters is to prohibit their application within a boundary adjacent to the water. Such areas are known as no-spray, or buffer, zones. Certain products in this *Guide* are restricted in this way, and have a legally binding label precaution to make sure the potential exposure of aquatic organisms to pesticides that might harm them is minimised. Before 1999 the protected zones were measured from the edge of the water. The distances were 2 m for hand-held or knapsack sprayers, 6 m for ground crop sprayers, and a variable distance (but often 18 m) for broadcast air-assisted applications, such as in orchards. The introduction of LERAPs (see below) has changed the method of measuring buffer zones.

'Surface water' includes lakes, ponds, reservoirs, streams, rivers and watercourses (natural or artificial). It also includes temporarily or seasonally dry ditches, which have the potential to carry water at different times of the year. Buffer zone restrictions do not necessarily apply to all products containing the same active ingredient. Those in formulations that are not likely to contaminate surface water through spray drift do not pose the same risk to aquatic life and are not subject to the restrictions.

Local Environmental Risk Assessments for Pesticides

Local Environment Risk Assessments for Pesticides (LERAPs) were introduced in March 1999, and revised guidelines were issued in January 2002. They give users of most products currently subject to a buffer zone restriction the option of continuing to comply with the existing buffer zone restriction (using the new method of measurement), or carrying out a LERAP and possibly reducing the size of the buffer zone as a result. In either case, there is a new legal obligation for users to record his decision, including the results of the LERAP. The scheme has changed the method of measuring the buffer zone. Previously the zone was measured from the edge of the water, but it is now the distance from the top of the bank of the watercourse to the edge of the spray area.

The LERAP provides a mechanism for taking into account other factors that may reduce the risk, such as dose reduction, the use of low drift nozzles, and whether the watercourse is dry or flowing. The previous arrangements applied the same restriction regardless of whether there was actually water present. Now there is a standard zone of 1 m from the top of a dry ditch bank.

Other factors to include in a LERAP that may allow a reduction in the buffer zone are:

- size of the watercourse: the wider it is, the greater the dilution factor and the lower the risk of serious pollution
- dose applied: the lower the dose, the less is the risk
- application equipment: sprayers and nozzles are star-rated according to their ability to reduce spray drift fallout - equipment offering the greatest reductions achieves the highest rating of three stars. The scheme was initially restricted to ground crop sprayers; new, more flexible rules introduced in February 2002 included broadcast air-assisted orchard and hop sprayers.

Other changes introduced in 2002 allow the reduction of a buffer zone if there is an appropriate living windbreak between the sprayed area and a watercourse.

Not all products that had a label buffer zone restriction are included in the LERAP scheme. The option to reduce the buffer zone does not apply to organophosphorus or synthetic pyrethroid insecticides. These groups are classified as Category A products. All other products that had a label buffer zone restriction are classified as Category B. The wording of the buffer zone label precautions has been amended for all products to take account of the new method of measurement, and whether or not the particular product qualifies for inclusion in the LERAP scheme.

Products in this *Guide* that are in Category A or B are identified in the **Environmental Safety** section of the profile. Updates to the list are published regularly by PSD and details can be obtained from www.pesticides.gov.uk/psd_databases.asp?id=325

The introduction of LERAPs is an important step forward because it demonstrates a willingness to reduce the impact of regulation on users of pesticides by allowing flexibility where local conditions make it safe to do so. This places a legal responsibility on users to ensure the risk assessment is done either personally, or by the spray operator or a professional consultant or advisor. It is compulsory to record the LERAP and make it available for inspection by enforcement authorities.

Groundwater regulations

New Groundwater Regulations were introduced in 1999 to complete the implementation in the UK of the EU Groundwater Directive (Protection of Groundwater Against Pollution Caused by Certain Dangerous Substances - 80/68/EEC). These Regulations help prevent the pollution of groundwater by controlling discharges or disposal of certain substances, including all pesticides.

Groundwater is defined under the Regulations as any water contained in the ground below the water table. Pesticides must not enter groundwater unless it is deemed by the appropriate Agency to be permanently unsuitable for other uses. The Regulations make it a criminal offence to dispose of pesticides onto land without official authorisation from the Environment Agency (in England and Wales) or the Scottish Environmental Protection Agency. Normal use of a pesticide in accordance with product approval does not require authorisation. This includes spraying the washings and rinsings back on the crop provided that, in so doing, the maximum approved dose for that product on that crop is not exceeded.

The Agencies will review all authorisations within a 4-year interval. They may also grant authorisations for a limited period. The Agencies can serve notice at any time to modify the conditions of an authorisation where necessary to prevent pollution of groundwater. In practice, the best advice to farmers and growers is to plan to use all diluted spray within the crop and to dispose of all washings via the same route, making sure they stay within the conditions of approval of the product.

SECTION 6
APPENDICES

Appendix 1
Suppliers of Pesticides and Adjuvants

Agrichem: Agrichem (International) Ltd
Industrial Estate
Station Road
Whittlesey
Cambs.
PE7 2EY
Tel: (01733) 204019
Fax: (01733) 204162
Email: info@agrichem.co.uk
Web: www.agrichem.co.uk

AgriGuard: AgriGuard Ltd
Unit 1
Broomfield Business Park
Malahide
Co. Dublin
Ireland
Tel: (+353) 1 846 2044
Fax: (+353) 1 846 2489
Email: info@agriguard.ie
Web: www.agriguard.ie

Agropharm: Agropharm Limited
Buckingham Place
Church Road
Penn
High Wycombe
Bucks.
HP10 8LN
Tel: (01494) 816575
Fax: (01494) 816578
Web: www.agropharm.co.uk

Agrovista: Agrovista UK Ltd
Cambridge House
Nottingham Road
Stapleford
Nottingham
NG9 8AB
Tel: (0115) 939 0202
Fax: (0115) 921 8498
Web: www.agrivista.co.uk

Amega: Amega Sciences
Lanchester Way
Royal Oak Industrial Estate
Daventry
Northants.
NN11 5PH
Tel: (01327) 704444
Fax: (01327) 71154
Email: admin@amega-sciences.com
Web: www.amega-sciences.com

Antec Biosentry: Antec Biosentry
DuPont Animal Health Solutions
Windham Road
Chilton Industrial Estate
Sudbury
Suffolk
CO10 2XD
Tel: (01787) 377305
Fax: (01787) 310846
Email: biosecurity@antecint.com
Web: www.antecint.com

B H & B: Battle Hayward & Bower Ltd
Victoria Chemical Works
Crofton Drive
Allenby Road Industrial Estate
Lincoln
LN3 4NP
Tel: (01522) 529206
Fax: (01522) 538960

Barclay: Barclay Chemicals Manufacturing Ltd
Damastown Way
Damastown Industrial Park
Mulhuddart
Dublin 15
Ireland
Tel: (+353) 1 822 4555
Fax: (+353) 1 822 4678
Email: info@barclay.ie
Web: www.barclay.ie

Barrier: Barrier BioTech Ltd
36/37 Haverscroft Industrial Estate
New Road
Attleborough
Norfolk
NR17 1YE
Tel: (01953) 456363
Fax: (01953) 455594
Email: sales@barrier-biotech.com
Web: www.barrier-biotech.com

BASF: BASF plc
Agricultural Divison
PO Box 4, Earl Road
Cheadle Hulme
Cheshire
SK8 6QG
Tel: (0845) 602 2553
Fax: (0161) 485 2229
Web: www.agricentre.co.uk

Bayer CropScience:
Bayer CropScience Limited
230 Cambridge Science Park
Milton Road
Cambridge
CB4 0WB
Tel: (01223) 226500
Fax: (01223) 426240
Web: www.bayercropscience.co.uk

Bayer Environ.: Bayer Environmental Science
230 Cambridge Science Park
Milton Road
Cambridge
CB4 0WB
Tel: (01223) 226680
Fax: (01223) 226635
Web: www.bayer-escience.co.uk

Belchim: Belchim Crop Protection Ltd
Unit 1b, Fenice Court
Eaton Socon
St Neots
Cambs.
PE19 8EP
Tel: (01480) 403333
Fax: (01480) 403444
Email: info@belchim.com
Web: www.belchim.com

Cardel: Cardel Agro SPRL
Avenue de Tervuren 270-272
B-1150 Brussels
Belgium
Tel: (+32) 2776 7652
Fax: (+32) 2776 7659

Certis: Certis
1b Mills Way
Boscombe Down Business Park
Amesbury
Wilts.
SP4 7RX
Tel: (01980) 676500
Fax: (01980) 626555
Email: certis@certiseurope.co.uk
Web: www.certiseurope.co.uk

Chemtura: Chemtura Europe Ltd
Kennet House
4 Langley Quay
Slough
Berks.
SL3 6EH
Tel: (01753) 603000
Fax: (01753) 603077
Web: www.chemtura.com

Chiltern: Chiltern Farm Chemicals Ltd
East Mellwaters
Stainmore
Bowes
Barnard Castle
Co. Durham
DL12 9RH
Tel: (01833) 628282
Fax: (01833) 628020
Web: www.chilternfarm.com

Ciba Specialty: Ciba Specialty Chemicals
Water Treatments Limited
P O Box 38
Low Moor
Bradford
W. Yorks.
BD12 0JZ
Tel: (01274) 417549
Fax: (01274) 417305
Email: soil.additives@cibasc.com
Web: www.cibasc.com

Clayton: Clayton Plant Protection Ltd
Bracetown Business Park
Clonee
Co. Meath
Ireland
Tel: (+353) 1 821 0127
Fax: (+353) 81 841 1084
Email: info@cppltd.eu
Web: www.cppltd.eu

DAPT: DAPT Agrochemicals Ltd
14 Monks Walk
Southfleet
Gravesend
Kent
DA13 9NZ
Tel: (01474) 834448
Fax: (01474) 834449
Email: rkjltd@supanet.com

Dax: Dax Products Ltd
18 Marlborough Road
Woodthorpe
Nottingham
NG5 4FG
Tel: (0115) 926 9996
Fax: (0115) 966 1173
Email: info@daxproducts.co.uk
Web: www.daxproducts.co.uk

De Sangosse: De Sangosse Ltd
Hillside Mill
Quarry Lane
Swaffham Bulbeck
Cambridge
CB25 0LU
Tel: (01223) 811215
Fax: (01223) 810020
Email: info@desangosse.co.uk
Web: www.desangosse.co.uk

Dewco-Lloyd: Dewco-Lloyd Ltd
Cyder House
Ixworth
Suffolk
IP31 2HT
Tel: (01359) 230555
Fax: (01359) 232553

Doff Portland: Doff Portland Ltd
Aerial Way
Hucknall
Nottingham
NG15 6DW
Tel: (0115) 963 2842
Fax: (0115) 963 8657
Email: info@doff.co.uk
Web: www.doff.co.uk

Dow: Dow AgroSciences Ltd
Latchmore Court
Brand Street
Hitchin
Herts.
SG5 1NH
Tel: (01462) 457272
Fax: (01462) 426605
Email: fhihotl@dow.com
Web: www.dowagro.com/uk

DuPont: DuPont (UK) Ltd
Crop Protection Products Department
Wedgwood Way
Stevenage
Herts.
SG1 4QN
Tel: (01438) 734450
Fax: (01438) 734452
Web: www.gbr.ag.dupont.com

Elliott: Thomas Elliott Fertilisers
Bencewell Granary
Oakley Road
Bromley
Kent
BR2 8HG
Tel: (0208) 462 6622
Fax: (0208) 462 5599

Fargro: Fargro Ltd
Toddington Lane
Littlehampton
Sussex
BN17 7PP
Tel: (01903) 721591
Fax: (01903) 730737
Email: promos@fargro.co.uk
Web: www.fargro.co.uk

Fine: Fine Agrochemicals Ltd
Hill End House
Whittington
Worcester
WR5 2RQ
Tel: (01905) 361800
Fax: (01905) 361810
Email: enquire@fine.eu
Web: www.fine.eu

Frontier: Frontier Agriculture Ltd
Fleet Road Industrial Estate
Holbeach
Spalding
Lincs.
PE12 8LY
Tel: (01406) 421405
Email: les_sykes@bankscargill.co.uk
Web: www.bankscargill.co.uk

Globachem: Globachem NV
Leeuwerweg 138
BE-3803 Sint Truiden
Belgium
Tel: (+32) 11 785 717
Fax: (+32) 11 681 565

Gowan: Gowan International
51 Harlow Moor Drive
Harrogate
N. Yorkshire
HG2 0LE
Tel: (01432) 524500
Fax: (01432) 524500

Greenaway: Greenaway Amenity Ltd
7 Browntoft Lane
Donington
Spalding
Lincs.
PE11 4TQ
Tel: (01775) 821031
Fax: (01775) 821034
Email: greenawayamenity@aol.com
Web: www.greenawaycda.com

Greencrop: Greencrop Technology Ltd
See Clayton Plant Protection Ltd

Headland: Headland Agrochemicals Ltd
Norfolk House
Gt. Chesterford Court
Gt. Chesterford
Saffron Walden
Essex
CB10 1PF
Tel: (01799) 530146
Fax: (01799) 530229
Email: enquiry@headlandgroup.com
Web: www.headland-ag.co.uk

Headland Amenity:
Headland Amenity Limited
1010 Cambourne Business Park
Cambourne
Cambs.
CB3 6DP
Tel: (01223) 597834
Fax: (01223) 598052
Email: info@headlandamenity.com
Web: www.headlandamenity.com

Helena: Helena Chemical Company
Cambridge House
Nottingham Road
Stapleford
Nottingham
NG9 8AB
Tel: (0115) 939 0202
Fax: (0115) 939 8031

Hermoo: Hermoo (UK)
21 Victoria Road
Wargrave
Berks.
RG10 8AD
Tel: (0118) 940 4264
Fax: (0118) 940 4264
Email: johnhudson23@aol.com
Web: www.hermoo.com

Interagro: Interagro (UK) Ltd
Sworders Barn
Sworders Yard
North Street
Bishop's Stortford
Herts.
CM23 2LD
Tel: (01279) 501995
Fax: (01279) 501996
Email: info@interagro.co.uk
Web: www.interagro.co.uk

Interfarm: Interfarm (UK) Ltd
Kinghams's Place
36 Newgate Street
Doddington
Cambs.
PE15 0SR
Tel: (01354) 741414
Fax: (01354) 741004
Email: technical@interfarm.co.uk
Web: www.interfarm.co.uk

Intracrop: Intracrop
Little Hay
Broadwell
Lechlade
Glos.
GL7 3QS
Tel: (01367) 860255
Fax: (01926) 634798

Irish Drugs: Irish Drugs Ltd
Burnfoot
Lifford
Co. Donegal
Ireland
Tel: (+353) 74 9368104
Fax: (+353) 74 9368311
Email: idl@eircom.net

K & S Fumigation:
K & S Fumigation Services Ltd
Asparagus Farm
Court Lodge Road
Appledore
Ashford
Kent
TN26 2DH
Tel: (01233) 758252
Fax: (01233) 758343
Email: david@kstreatments.co.uk

Killgerm: Killgerm Chemicals Ltd
115 Wakefield Road
Flushdyke
Ossett
W. Yorks.
WF5 9AR
Tel: (01924) 268400
Fax: (01924) 264757
Email: info@killgerm.com
Web: www.killgerm.com

Koppert: Koppert (UK) Ltd
Unit 8
53 Hollands Road
Haverhill
Suffolk
CB9 8PJ
Tel: (01440) 704488
Fax: (01440) 704487
Email: info@koppert.co.uk
Web: www.koppert.com

Landgold: Landgold & Co. Ltd
P O Box 374
Richmond House
St Julian's Avenue
St. Peter Port
Guernsey
GY1 3YS
Tel: (01481) 713667

Landseer: Landseer Ltd
Lodge Farm
Goat Hall Lane
Galleywood
Chelmsford
Essex
CM2 8PH
Tel: (01245) 357109
Fax: (01245) 494165

Law Fertilisers: Law Fertilisers Ltd
Eastwood End
Wimblington
Cambs.
PE15 0QJ
Tel: (01354) 740740
Fax: (01354) 740720
Email: sales@lawfertilisers.co.uk
Web: www.lawfertilisers.co.uk

Luxan: Luxan (UK) Ltd
Crown Business Park
Old Dalby
Leics.
LE14 3NQ
Tel: (01664) 820052
Fax: (01664) 820216
Email: enquiry@luxan.co.uk
Web: www.luxan.co.uk

Makhteshim: Makhteshim-Agan (UK) Ltd
Unit 16
Thatcham Business Village
Colthrop Way
Thatcham
Berks.
RG19 4LW
Tel: (01635) 860555
Fax: (01635) 876622
Email: admin@mauk.co.uk
Web: www.mauk.co.uk

Mandops: Mandops (UK) Ltd
36 Leigh Road
Eastleigh
Hants.
SO50 9DT
Tel: (023) 8064 1826
Fax: (023) 8062 9106
Email: enquiries@mandops.co.uk
Web: www.mandops.co.uk

Microcide: Microcide Ltd
Shepherds Grove
Stanton
Bury St. Edmunds
Suffolk
IP31 2AR
Tel: (01359) 251077
Fax: (01359) 251545
Email: microcide@microcide co.uk
Web: www.microcide.co.uk

Monsanto: Monsanto (UK) Ltd
The Maris Centre
45 Hauxton Road
Trumpington
Cambridge
CB2 9LQ
Tel: (01223) 849540
Fax: (01223) 849414
Email:
technical.helpline.uk@monsanto.com
Web: www.monsanto-ag.co.uk

Nickerson: Nickerson UK Ltd
Rothwell
Market Rasen
Lincs.
LN7 6DT
Tel: (01472) 371471
Fax: (01472) 371602
Email: enquiries@Nickerson.co.uk
Web: www.nickerson.co.uk

Nissan: Nissan Chemical Europe Sarl
Parc d'Affaires de Crecy
2 Rue Claude Chappe
69371 St Didier au Mont d'Or
France
Tel: (+33) 4 376 44020
Fax: (+33) 4 376 46874

Nomix Enviro: Nomix Enviro Limited
Portland Building
Portland Street
Staple Hill
Bristol
BS16 4PS
Tel: (0117) 957 4574
Fax: (0117) 956 3461
Email: info@nomix.co.uk
Web: www.nomix.co.uk

Nufarm UK: Nufarm UK Ltd
Crabtree Manorway North
Belvedere
Kent
DA17 6BQ
Tel: (020) 8319 7222
Fax: (020) 8319 7280
Email: infouk@uk.nufarm.com
Web: www.nufarm.co.uk

Omex: Omex Agriculture Ltd
Bardney Airfield
Tupholme
Lincoln
LN3 5TP
Tel: (01526) 396000
Fax: (01526) 396001
Email: enquire@omex.com
Web: www.omex.co.uk

Pan Agriculture: Pan Agriculture Ltd
8 Cromwell Mews
Station Road
St Ives
Huntingdon
Cambs.
PE27 5HJ
Tel: (01480) 467790
Fax: (01480) 467041
Email: info@panagriculture.co.uk

Plant Solutions: Plant Solutions
Pyports
Downside Bridge Road
Cobham
Surrey
KT11 3EH
Tel: (01932) 576699
Fax: (01932) 868973
Email: sales@plantsolutionsltd.com
Web: www.plantsolutionsltd.com

PP Products: PP Products
Longmynd
Tunstead Road
Hoveton
Norwich
NR12 8QN
Tel: (01603) 784367
Fax: (01603) 784367
Email: mail@ppproducts.net

Q-Chem: Q-CHEM nv
Leeuwerweg 138
BE-3803
Belgium
Tel: (+32) 11 785 717
Fax: (+32) 11 681 565

Rentokil: Rentokil Initial plc
Felcourt
East Grinstead
W. Sussex
RH19 2JY
Tel: (01342) 833022
Fax: (01342) 836180
Web: www.ri-research.com

Rhizopon: Rhizopon UK Ltd
Croda Rosa
12 Bixley Road
Ipswich
Suffolk
IP3 8PL
Tel: (01473) 712666
Fax: (01473) 712666

Rigby Taylor: Rigby Taylor Ltd
Rigby Taylor House
Crown Lane
Horwich
Bolton
Lancs.
BL6 5HP
Tel: (01204) 677777
Fax: (01204) 677765
Email: info@rigbytaylor.com
Web: www.rigbytaylor.com

Scotts: The Scotts Company (UK) Ltd
Paper Mill Lane
Bramford
Ipswich
Suffolk
IP8 4BZ
Tel: (01473) 830492
Fax: (01473) 830814
Web: www.scottsprofessional.co.uk

Sherriff Amenity: Sherriff Amenity Services
The Pines
Fordham Road
Newmarket
Suffolk
CB8 7LG
Tel: (01638) 721888
Fax: (01638) 721815
Web: www.sherriffamenity.com

Sinclair: William Sinclair Horticulture Ltd
Firth Road
Lincoln
LN6 7AH
Tel: (01522) 537561
Fax: (01522) 513609
Web: www.william-sinclair.co.uk

Sipcam: Sipcam UK Ltd
3 The Barn
27 Kneesworth Street
Royston
Herts.
SG8 5AB
Tel: (01763) 212100
Fax: (01763) 212101

Sorex: Sorex Ltd
St Michael's Industrial Estate
Widnes
Cheshire
WA8 8TJ
Tel: (0151) 420 7151
Fax: (0151) 495 1163
Email: enquiries@sorex.com
Web: www.sorex.com

Sphere: Sphere Laboratories (London) Ltd
c/o Mainswood
Putley Common
Ledbury
Herefordshire
HR8 2RF
Tel: (07974) 732026
Fax: (01531) 670517
Email: homesfg@aol.com

Standon: Standon Chemicals Ltd
48 Grosvenor Square
London
W1K 2HT
Tel: (020) 7493 8648
Fax: (020) 7493 4219

SumiAgro: SumiAgro (UK) Ltd
See Barclay Chemicals Manufacturing Ltd

Summit Agro: Summit Agro Europe Limited
Vintners' Place
68 Upper Thames Street
London
EC4V 3BJ
Tel: (020) 7246 3688
Fax: (020) 7246 3799
Email: summit@summit-agro.com

Syngenta:
Syngenta Crop Protection UK Limited
CPC4
Capital Park
Fulbourn
Cambridge
CB21 5XE
Tel: (0800) 169 6058
Fax: (01223) 493700
Web: www.syngenta-crop.co.uk

Syngenta Bioline: Syngenta Bioline
Telstar Nursery
Holland Road
Little Clacton
Essex
CO16 9QG
Tel: (01255) 863200
Fax: (01255) 863206
Email: syngenta.bioline@syngenta.com
Web: www.syngenta-bioline.co.uk

Teliton: Teliton Ltd
Suite 1
Weston Homes Business Centre
Parsonage Road
Takeley
Essex
CM22 6PU
Tel: (01279) 874166
Fax: (01279) 874167
Email: teliton@btconnect.com

Truchem: Truchem Ltd
The Knoll
The Cross
Horsley
Stroud
Gloucestershire
GL6 0PR
Tel: (01453) 833293
Fax: (01453) 833293
Email: nichollstruchem@aol.com

Unicrop: Universal Crop Protection Ltd
Park House
Maidenhead Road
Cookham
Berks.
SL6 9DS
Tel: (01628) 526083
Fax: (01628) 810457
Email: enquiries@unicrop.com

United Agri: United Agri Products Ltd
The Crossways
Alconbury Hill
Huntingdon
PE28 4JH
Tel: (01480) 418000
Fax: (01480) 418010

United Phosphorus: United Phosphorus Ltd
Chadwick House
Birchwood Park
Warrington
Cheshire
WA3 6AE
Tel: (01925) 819999
Fax: (01925) 817425
Web: www.upleurope.com

Vitax: Vitax Ltd
Owen Street
Coalville
Leicester
LE67 3DE
Tel: (01530) 510060
Fax: (01530) 510299
Email: info@vitax.co.uk
Web: www.vitax.co.uk

Whyte Agrochemicals:
Whyte Agrochemicals Ltd
Marlborough House
298 Regents Park Road
Finchley
London
N3 2UA
Tel: (020) 8346 5946
Fax: (020) 8349 4589
Web: www.whytechemicals.co.uk

Appendix 2
Useful Contacts

Agricultural Industries Confederation Ltd (AIC)
Confederation House
East of England Showground
Peterborough PE2 6XE
Tel: (01733) 385236
Fax: (01733) 385270
Web: www.agindustries.org.uk

BASIS Ltd
Bank Chambers
34 St John Street
Ashbourne
Derbyshire DE6 1GH
Tel: (01335) 343945/346138
Fax: (01335) 346488
Web: www.basis-reg.co.uk

British Beekeepers' Association
National Agricultural Centre
Stoneleigh
Kenilworth
Warwickshire CV8 2LZ
Tel: (024) 7669 6679
Fax: (024) 7669 0682

British Beer & Pub Association
Market Towers
1 Nine Elms Lane
London SW8 5NQ
Tel: (0207) 627 9191
Fax: (0207) 627 9123

BCPC (British Crop Production Council)
7 Omni Business Centre
Omega Park
Alton
Hampshire GU34 2QD
Tel: (01420) 593200
Fax: (01420) 593209
Web: www.bcpc.org

BCPC Publications Sales
7 Omni Business Centre
Omega Park
Alton
Hampshire GU34 2QD
Tel: (01420) 593200
Fax: (01420) 593209
Web: www.bcpc.org/bookshop
e-mail: publications@bcpc.org

British Pest Control Association
1 Gleneagles House
Vernon Gate
South Street
Derby DE1 1UP
Tel: (01332) 294288
Fax: (01332) 295904
Web: www.bpca.org.uk

Crop Protection Association Ltd
18 & 20 Evans Business Centre
Cully Court
Bakewell Road
Orton Southgate
Peterborough PE2 6XS
Tel: (01733) 367213
Fax: (01733) 367212
Web: www.cropprotection.org.uk

CropLife International
(previously the Global Crop Protection Federation)
Avenue Louise 143
B-1050 Brussels
Belgium
Tel: (+32) 2 542 0410
Fax: (+32) 2 542 0419
Web: www.croplife.org

Department of Agriculture and Rural Development (Northern Ireland)
Pesticides Section
Dundonald House
Upper Newtownards Road
Belfast BT4 3SB
Tel: (028) 9052 4704
Fax: (028) 9052 4266

Department of Environment, Food and Rural Affairs (Defra)
Nobel House
17 Smith Square
London SW1P 3JR
Tel: (020) 7238 6000
Fax: (020) 7238 6591
Web: www.defra.gov.uk

Environment Agency
Rio House
Waterside Drive
Aztec West
Almondsbury
Bristol BS12 4UD
Tel: (01454) 624400
Fax: (01454) 624409
Web: www.environment-agency.gov.uk

European Crop Protection Association (ECPA)
Avenue E van Nieuwenhuyse 6
B-1160 Brussels
Belgium
Tel: (+32) 2 663 1550
Fax: (+32) 2 663 1560
Web: www.ecpa.be

Farmers' Union of Wales
Llys Amaeth
Queen's Square
Aberystwyth
Dyfed SY23 2EA
Tel: (01970) 612755
Fax: (01970) 624369

Forestry Commission
231 Corstorphine Road
Edinburgh EH12 7AT
Tel: (0131) 334 0303
Fax: (0131) 334 3047
Web: www.forestry.gov.uk

Health and Safety Executive
Information Services
Room 318, Daniel House
Stanley Precinct
Bootle
Merseyside L20 3TW
Tel: (0151) 951 3191
Fax: (0151) 951 3467

Health and Safety Executive
Biocides & Pesticides Assessment Unit
Room 123, Magdalen House
Bootle
Merseyside L20 3QZ
Tel: (0151) 951 3535
Fax: (0151) 951 3317

Health and Safety Executive – Books
PO Box 1999
Sudbury
Suffolk CO10 2WA
Tel: (01787) 881165
Fax: (01787) 313995

Horticultural Development Council
Bradbourne House
East Malling
Kent ME19 6DZ
Tel: (01732) 848383
Fax: (01732) 848498

Lantra
Lantra House
Stoneleigh Park
Nr Coventry
Warwickshire CV8 2LG
Tel: (024) 7669 6996
Fax: (024) 7669 6732
Web: www.lantra.co.uk

National Association of Agricultural Contractors (NAAC)
Samuelson House
Paxton Road
Orton Centre
Peterborough
Cambs. PE2 5LT
Tel: (01733) 362920
Fax: (01733) 362921

National Farmers' Union
Agriculture House
Stoneleigh Park
Stoneleigh
Warwickshire CV8 2TZ
Tel: (024) 7685 0662
Fax: (024) 7685 8501
Web: www.nfuonline.com

National Turfgrass Council
Hunter's Lodge
Dr Brown's Road
Minchinhampton
Gloucestershire GL6 9BT
Tel: (01453) 883588
Fax: (01453) 731449

NPTC
National Agricultural Centre
Stoneleigh
Kenilworth
Warwickshire CV8 2LG
Tel: (024) 7669 6553
Fax: (024) 7669 6128
Web: www.nptc.org.uk

Pesticides Safety Directorate
Mallard House
King's Pool
3 Peasholme Green
York YO1 2PX
Tel: (01904) 640500
Fax: (01904) 455733
Web: www.pesticides.gov.uk

Processors and Growers Research Organisation
The Research Station
Great North Road
Thornhaugh
Peterborough
Cambs. PE8 6HJ
Tel: (01780) 782585
Fax: (01780) 783993
Web: www.pgro.co.uk

Scottish Beekeepers' Association
North Trinity House
114 Trinity Road
Edinburgh EH5 3JZ
Tel: (0131) 552 5341

Scottish Environment Protection Agency (SEPA)
Erskine Court
The Castle Business Park
Stirling FK9 4TR
Tel: (01786) 457 700
Fax: (01786) 446 885
Web: www.sepa.org.uk

TSO (The Stationery Office)
Publications Centre
PO Box 276
London SW8 5DT
Tel: (020) 7873 9090 (orders)
(020) 7873 0011 (enquiries)
Fax: (020) 7873 8200
Web: www.thestationeryoffice.com

Ulster Beekeepers' Association
57 Liberty Road
Carrickfergus
Co. Antrim BT38 9DJ
Tel: (01960) 362998

Welsh Beekeepers' Association
Trem y Clawdd
Fron Isaf
Chirk
Wrexham
Clwyd LL14 5AH
Tel/Fax: (01691) 773300

SECTION 6

Appendix 3
Keys to Crop and Weed Growth Stages

Decimal Code for the Growth Stages of Cereals

Illustrations of these growth stages can be found in the reference indicated below and in some company product manuals.

0 Germination
- 00 Dryseed
- 03 Imbibition complete
- 05 Radicle emerged from caryopsis
- 07 Coleoptile emerged from caryopsis
- 09 Leaf at coleoptile tip

1 Seedling growth
- 10 First leaf through coleoptile
- 11 First leaf unfolded
- 12 2 leaves unfolded
- 13 3 leaves unfolded
- 14 4 leaves unfolded
- 15 5 leaves unfolded
- 16 6 leaves unfolded
- 17 7 leaves unfolded
- 18 8 leaves unfolded
- 19 9 or more leaves unfolded

2 Tillering
- 20 Main shoot only
- 21 Main shoot and 1 tiller
- 22 Main shoot and 2 tillers
- 23 Main shoot and 3 tillers
- 24 Main shoot and 4 tillers
- 25 Main shoot and 5 tillers
- 26 Main shoot and 6 tillers
- 27 Main shoot and 7 tillers
- 28 Main shoot and 8 tillers
- 29 Main shoot and 9 or more tillers

3 Stem elongation
- 30 ear at 1 cm
- 31 1st node detectable
- 32 2nd node detectable
- 33 3rd node detectable
- 34 4th node detectable
- 35 5th node detectable
- 36 6th node detectable
- 37 Flag leaf just visible
- 39 Flag leaf ligule/collar just visible

4 Booting
- 41 Flag leaf sheath extending
- 43 Boots just visibly swollen
- 45 Boots swollen
- 47 Flag leaf sheath opening
- 49 First awns visible

5 Inflorescence
- 51 First spikelet of inflorescence just visible
- 52 1/4 of inflorescence emerged
- 55 1/2 of inflorescence emerged
- 57 3/4 of inflorescence emerged
- 59 Emergence of inflorescence completed

6 Anthesis
- 60
- 61 } Beginning of anthesis
- 64
- 65 } Anthesis half way
- 68
- 69 } Anthesis complete

7 Milk development
- 71 Caryopsis watery ripe
- 73 Early milk
- 75 Medium milk
- 77 Late milk

8 Dough development
- 83 Early dough
- 85 Soft dough
- 87 Hard dough

9 Ripening
- 91 Caryopsis hard (difficult to divide by thumb-nail)
- 92 Caryopsis hard (can no longer be dented by thumb-nail)
- 93 Caryopsis loosening in daytime

(From Tottman, 1987. *Annals of Applied Biology*, **110**, 441–454)

Stages in Development of Oilseed Rape

Illustrations of these growth stages can be found in the reference indicated below and in some company product manuals.

0 Germination and emergence

1 Leaf production

- 1,0 Both cotyledons unfolded and green
- 1,1 First true leaf
- 1,2 Second true leaf
- 1,3 Third true leaf
- 1,4 Fourth true leaf
- 1,5 Fifth true leaf
- 1,10 About tenth true leaf
- 1,15 About fifteenth true leaf

2 Stem extension

- 2,0 No internodes ('rosette')
- 2,5 About five internodes

3 Flower bud development

- 3,0 Only leaf buds present
- 3,1 Flower buds present but enclosed by leaves
- 3,3 Flower buds visible from above ('green bud')
- 3,5 Flower buds raised above leaves
- 3,6 First flower stalks extending
- 3,7 First flower buds yellow ('yellow bud')

4 Flowering

- 4,0 First flower opened
- 4,1 10% all buds opened
- 4,3 30% all buds opened
- 4,5 50% all buds opened

5 Pod development

- 5,3 30% potential pods
- 5,5 50% potential pods
- 5,7 70% potential pods
- 5,9 All potential pods

6 Seed development

- 6,1 Seeds expanding
- 6,2 Most seeds translucent but full size
- 6,3 Most seeds green
- 6,4 Most seeds green-brown mottled
- 6,5 Most seeds brown
- 6,6 Most seeds dark brown
- 6,7 Most seeds black but soft
- 6,8 Most seeds black and hard
- 6,9 All seeds black and hard

7 Leaf senescence

8 Stem senescence

- 8,1 Most stem green
- 8,5 Half stem green
- 8,9 Little stem green

9 Pod senescence

- 9,1 Most pods green
- 9,5 Half pods green
- 9.9 Few pods green

(From Sylvester-Bradley, 1985. *Aspects of Applied Biology*, **10**, 395–400)

Stages in Development of Peas

Illustrations of these growth stages can be found in the reference indicated below and in some company product manuals.

0 Germination and emergence

000 Dry seed
001 Imbibed seed
002 Radicle apparent
003 Plumule and radicle apparent
004 Emergence

1 Vegetative stage

101 First node (leaf with one pair leaflets, no tendril)
102 Second node (leaf with one pair leaflets, simple tendril)
103 Third node (leaf with one pair leaflets, complex tendril)
•
•
l0x X nodes (leaf with more than one pair leaflets, complex tendril)
•
•
10n Last recorded node

2 Reproductive stage (main stem)

201 Enclosed buds
202 Visible buds
203 First open flower
204 Pod set (small immature pod)
205 Flat pod
206 Pod swell (seeds small, immature)
207 Podfill
208 Pod green, wrinkled
209 Pod yellow, wrinkled (seeds rubbery)
210 Dry seed

3 Senescence stage

301 Desiccant application stage. Lower pods dry and brown, middle yellow, upper green. Overall moisture content of seed less than 45%
302 Pre-harvest stage. Lower and middle pods dry and brown, upper yellow. Overall moisture content of seed less than 30%
303 Dry harvest stage. All pods dry and brown, seed dry

(From Knott, 1987. *Annals of Applied Biology*, **111**, 233–244)

Stages in Development of Faba Beans

Illustrations of these growth stages can be found in the reference indicated below and in some company product manuals.

0 Germination and emergence

000 Dry seed
001 Imbibed seed
002 Radicle apparent
003 Plumule and radicle apparent
004 Emergence
005 First leaf unfolding
006 First leaf unfolded

1 Vegetative stage

101 First node
102 Second node
103 Third node
•
•
l0x X nodes
•
•
10n N, last reoorded node

2 Reproductive stage (main stem)

201 Flower buds visible
203 First open flowers
204 First pod set
205 Pods fully formed, green
207 Pod fill, pods green
209 Seed rubbery, pods pliable, turning black
210 Seed dry and hard, pods dry and black

3 Pod senescence

301 10% pods dry and black
•
•
305 50% pods dry and black
•
•
308 80% pods dry and black, some upper pods green
309 90% pods dry and black, most seed dry. Desiccation stage.
310 All pods dry and black, seed hard. Pre-harvest (glyphosate application stage)

4 Stem senescence

401 10% stem brown/black
•
•
405 50% stem brown/black
•
•
409 90% stem brown/black
410 All stems brown/black. All pods dry and black, seed hard.

(From Knott, 1990. *Annals of Applied Biology*, **116**, 391–404)

SECTION 6

Stages in Development of Potato

Illustrations of these growth stages can be found in the reference indicated below and in some company product manuals.

0 Seed germination and seedling emergence

000 Dry seed
001 Imbibed seed
002 Radicle apparent
003 Elongation of hypocotyl
004 Seedling emergence
005 Cotyledons unfolded

1 Tuber dormancy

100 Innate dormancy (no sprout development under favourable conditions)
150 Enforced dormancy (sprout development inhibited by environmental conditions)

2 Tuber sprouting

200 Dormancy break, sprout development visible
21x Sprout with 1 node
22x Sprout with 2 nodes
•
•
29x Sprout with 9 nodes
21x(2) Second generation sprout with 1 node
22x(2) Second generation sprout with 2 nodes
•
•
29x(2) Second generation sprout with 9 nodes

Where x = 1, sprout <2 mm;
2, 2-5 mm; 3, 5-20 mm;
4, 20-30 mm; 5, 50-100 mm;
6, 100-150 mm long

3 Emergence and shoot expansion

300 Main stem emergence
301 Node 1
302 Node 2
•
•
319 Node 19
Second order branch
321 Node 1
•
•
Nth order branch
3N1 Node 1
•
•
3N9 Node 9

4 Flowering

Primary flower
400 No flowers
410 Appearance of flower bud
420 Flower unopen
430 Flower open
440 Flower closed
450 Berry swelling
460 Mature berry
Second order flowers
410(2) Appearance of flower bud
420(2) Flower unopen
430(2) Flower open
440(2) Flower closed
450(2) Berry swelling
460(2) Mature berry

5 Tuber development

500 No stolons
510 Stolon initials
520 Stolon elongation
530 Tuber initiation
540 Tuber bulking (>10 mm diam)
550 Skin set
560 Stolon development

6 Senescence

600 Onset of yellowing
650 Half leaves yellow
670 Yellowing of stems
690 Completely dead

(From Jefferies & Lawson, 1991. *Annals of Applied Biology*, **119**, 387–389)

Stages in Development of Linseed

Illustrations of these growth stages can be found in the reference indicated below and in some company product manuals.

0 Germination and emergence

- 00 Dry seed
- 01 Imbibed seed
- 02 Radicle apparent
- 04 Hypocotyl extending
- 05 Emergence
- 07 Cotyledon unfolding from seed case
- 09 Cotyledons unfolded and fully expanded

1 Vegetative stage (of main stem)

- 10 True leaves visible
- 12 First pair of true leaves fully expanded
- 13 Third pair of true leaves fully expanded
- 1*n* *n* leaf fully expanded

2 Basal branching

- 21 One branch
- 22 Two branches
- 23 Three branches
- 2*n* *n* branches

3 Flower bud development (on main stem)

- 31 Enclosed bud visible in leaf axils
- 33 Bud extending from axil
- 35 Corymb formed
- 37 Buds enclosed but petals visible
- 39 First flower open

4 Flowering (whole plant)

- 41 10% of flowers open
- 43 30% of flowers open
- 45 50% of flowers open
- 49 End of flowering

5 Capsule formation (whole plant)

- 51 10% of capsules formed
- 53 30% Of capsules formed
- 55 50% of capsules formed
- 59 End of capsule formation

6 Capsule senescence (on most advanced plant)

- 61 Capsules expanding
- 63 Capsules green and full size
- 65 Capsules turning yellow
- 67 Capsules all yellow brown but soft
- 69 Capsules brown, dry and senesced

7 Stem senescence (whole plant)

- 71 Stems mostly green below panicle
- 73 Most stems 30% brown
- 75 Most stems 50% brown
- 77 Stems 75% brown
- 79 Stems completely brown

8 Stems rotting (retting)

- 81 Outer tissue rotting
- 85 Vascular tissue easily removed
- 89 Stems completely collapsed

9 Seed development (whole plant)

- 91 Seeds expanding
- 92 Seeds white but full size
- 93 Most seeds turning ivory yellow
- 94 Most seeds turning brown
- 95 All seeds brown and hard
- 98 Some seeds shed from capsule
- 99 Most seeds shed from capsule

(From Freer, 1991. *Aspects of Applied Biology*, **28**, 33–40)

SECTION 6

Stages in Development of Annual Grass Weeds

Illustrations of these growth stages can be found in the reference indicated below and in some company product manuals.

0 Germination and emergence

- 00 Dry seed
- 01 Start of imbibition
- 03 Imbibition complete
- 05 Radicle emerged from caryopsis
- 07 Coleoptile emerged from caryopsis
- 09 Leaf just at coleoptile tip

1 Seedling growth

- 10 First leaf through coleoptile
- 11 First leaf unfolded
- 12 2 leaves unfolded
- 13 3 leaves unfolded
- 14 4 leaves unfolded
- 15 5 leaves unfolded
- 16 6 leaves unfolded
- 17 7 leaves unfolded
- 18 8 leaves unfolded
- 19 9 or more leaves unfolded

2 Tillering

- 20 Main shoot only
- 21 Main shoot and 1 tiller
- 22 Main shoot and 2 tillers
- 23 Main shoot and 3 tillers
- 24 Main shoot and 4 tillers
- 25 Main shoot and 5 tillers
- 26 Main shoot and 6 tillers
- 27 Main shoot and 7 tillers
- 28 Main shoot and 8 tillers
- 29 Main shoot and 9 or more tillers

3 Stem elongation

- 31 First node detectable
- 32 2nd node detectable
- 33 3rd node detectable
- 34 4th node detectable
- 35 5th node detectable
- 36 6th node detectable
- 37 Flag leaf just visible
- 39 Flag leaf ligule just visible

4 Booting

- 41 Flag leaf sheath extending
- 43 Boots just visibly swollen
- 45 Boots swollen
- 47 Flag leaf sheath opening
- 49 First awns visible

5 Inflorescence emergence

- 51 First spikelet of inflorescence just visible
- 53 1/4 of inflorescence emerged
- 55 1/2 of inflorescence emerged
- 57 3/4 of inflorescence emerged
- 59 Emergence of inflorescence completed

6 Anthesis

- 61 Beginning of anthesis
- 65 Anthesis half-way
- 69 Anthesis complete

(From Lawson & Read, 1992. *Annals of Applied Biology*, **12**, 211–214)

Growth Stages of Annual Broad-leaved Weeds

Preferred Descriptive Phrases

Illustrations of these growth stages can be found in the reference indicated below and in some company product manuals.

Pre-emergence
Early cotyledons
Expanded cotyledons
One expanded true leaf
Two expanded true leaves
Four expanded true leaves
Six expanded true leaves
Plants up to 25 mm across/high
Plants up to 50 mm across/high
Plants up to 100 mm across/high
Plants up to 150 mm across/high
Plants up to 250 mm across/high
Flower buds visible
Plant flowering
Plant senescent

(From Lutman & Tucker, 1987. *Annals of Applied Biology*, **110**, 683–687)

Appendix 4
Key to Hazard Classifications and Safety Precautions

Every product label contains information to warn users of the risks from using the product, together with precautions that must be followed in order to minimise the risks. A hazard classification (if any) and symbol is shown with associated risk phrases, followed by a series of safety precautions. These are represented in the pesticide profiles in Section 2 by code letters and numbers under the heading **Hazard classification and safety precautions**.

The codes are defined below, under the same sub-headings as they appear in the pesticide profiles.

Where a product label specifies the use of personal protective equipment (PPE), the requirements are listed under the sub-heading **Operator protection**, using letter codes to denote the protective items, according to the list below. Often PPE requirements are different for specified operations, e.g. handling the concentrate, cleaning equipment etc., but it is not possible to list them separately. The lists of PPE are therefore an indication of what the user may need to have available to use the product in different ways. **When making a COSHH assessment it is therefore essential that the product label is consulted for information on the particular use that is being assessed**.

Where the generalised wording includes a phrase such as '... for *xx* days', the specific requirement for each pesticide is shown in brackets after the code.

Hazard

H01 Very toxic
H02 Toxic
H03 Harmful
H04 Irritant
H05 Corrosive
H06 Extremely flammable
H07 Highly flammable
H08 Flammable
H09 Oxidising agent
H10 Explosive
H11 Dangerous for the environment

Risk phrases

R08 Contact with combustible material may cause fire
R09 Explosive when mixed with combustible material
R16 Explosive when mixed with oxidising substances
R20 Harmful by inhalation
R21 Harmful in contact with skin
R22a Harmful if swallowed
R22b May cause lung damage if swallowed
R23 Toxic by inhalation
R24 Toxic in contact with skin
R25 Toxic if swallowed
R26 Very toxic by inhalation
R27 Very toxic in contact with skin
R28 Very toxic if swallowed
R34 Causes burns
R35 Causes severe burns
R36 Irritating to eyes
R37 Irritating to respiratory system

R38 Irritating to skin
R39 Danger of very serious irreversible effects
R40 Limited evidence of a carcinogenic effect
R41 Risk of serious damage to eyes
R42 May cause sensitization by inhalation
R43 May cause sensitization by skin contact
R45 May cause cancer
R46 May cause heritable genetic damage
R48 Danger of serious damage to health by prolonged exposure
R50 Very toxic to aquatic organisms
R51 Toxic to aquatic organisms
R52 Harmful to aquatic organisms
R53a May cause long-term adverse effects in the aquatic environment
R53b Dangerous to aquatic organisms
R54 Toxic to flora
R55 Toxic to fauna
R56 Toxic to soil organisms
R57 Toxic to bees
R58 May cause long-term adverse effects in the environment
R60 May impair fertility
R61 May cause harm to the unborn child
R62 Possible risk of impaired fertility
R63 Possible risk of harm to the unborn child
R64 May cause harm to breast-fed babies
R66 Repeated exposure may cause skin dryness or cracking
R67 Vapours may cause drowsiness and dizziness
R68 Possible risk of irreversible effects

Operator protection

A Suitable protective gloves (the product label should be consulted for any specific requirements about the material of which the gloves should be made)
B Rubber gauntlet gloves
C Face-shield
D Approved respiratory protective equipment
E Goggles
F Dust mask
G Full face-piece respirator
H Coverall
J Hood
K Apron/Rubber apron
L Waterproof coat
M Rubber boots
N Waterproof jacket and trousers
P Suitable protective clothing
U01 To be used only by operators instructed or trained in the use of chemical/product/type of produce and familiar with the precautionary measures to be observed
U02a Wash all protective clothing thoroughly after use, especially the inside of gloves
U02b Avoid excessive contamination of coveralls and launder regularly
U02c Remove and wash contaminated gloves immediately
U03 Wash splashes off gloves immediately
U04a Take off immediately all contaminated clothing
U04b Take off immediately all contaminated clothing and wash underlying skin. Wash clothes before re-use
U04c Wash clothes before re-use
U05a When using do not eat, drink or smoke
U05b When using do not eat, drink, smoke or use naked lights
U06 Handle with care and mix only in a closed container
U07 Open the container only as directed (returnable containers only)
U08 Wash concentrate/dust from skin or eyes immediately
U09a Wash any contamination/splashes/dust/powder/concentrate from skin or eyes immediately
U09b Wash any contamination/splashes/dust/powder/concentrate from eyes immediately
U10 After contact with skin or eyes wash immediately with plenty of water

U11 In case of contact with eyes rinse immediately with plenty of water and seek medical advice
U12 In case of contact with skin rinse immediately with plenty of water and seek medical advice
U13 Avoid contact by mouth
U14 Avoid contact with skin
U15 Avoid contact with eyes
U16a Ensure adequate ventilation in confined spaces
U16b Use in a well ventilated area
U18 Extinguish all naked flames, including pilot lights, when applying the fumigant/dust/liquid/product
U19a Do not breathe dust/fog/fumes/gas/smoke/spray mist/vapour. Avoid working in spray mist
U19b Do not work in confined spaces or enter spaces in which high concentrations of vapour are present. Where this precaution cannot be observed distance breathing or self-contained breathing apparatus must be worn, and the work should be done by trained operators
U20a Wash hands and exposed skin before eating, drinking or smoking and after work
U20b Wash hands and exposed skin before eating and drinking and after work
U20c Wash hands before eating and drinking and after work
U20d Wash hands after use
U21 Before entering treated crops, cover exposed skin areas, particularly arms and legs
U22a Do not touch sachet with wet hands or gloves/Do not touch water soluble bag directly
U22b Protect sachets from rain or water
U23a Do not apply by knapsack sprayer/hand-held equipment
U23b Do not apply through hand held rotary atomisers
U23c Do not apply via tractor mounted horizontal boom sprayers
U24 Do not handle grain unnecessarily

Environmental protection

E02a Keep unprotected persons/animals out of treated/fumigation areas for at least xx hours/days
E02b Prevent access by livestock, pets and other non-target mammals and birds to buildings under fumigation and ventilation
E02c Vacate treatment areas before application
E02d Exclude all persons and animals during treatment
E03 Label treated seed with the appropriate precautions, using the printed sacks, labels or bag tags supplied
E05a Do not apply directly to livestock/poultry
E05b Keep poultry out of treated areas for at least xx days/weeks
E05c Do not apply directly to animals
E06a Keep livestock out of treated areas for at least xx days/weeks after treatment
E06b Dangerous to livestock. Keep all livestock out of treated areas/away from treated water for at least xx days/weeks. Bury or remove spillages
E06c Harmful to livestock. Keep all livestock out of treated areas/away from treated water for at least xx days/weeks. Bury or remove spillages
E06d Exclude livestock from treated fields. Livestock may not graze or be fed treated forage nor may it be used for hay silage or bedding
E07a Keep livestock out of treated areas for up to two weeks following treatment and until poisonous weeds, such as ragwort, have died down and become unpalatable
E07b Dangerous to livestock. Keep livestock out of treated areas/away from treated water for at least xx weeks and until foliage of any poisonous weeds, such as ragwort, has died and become unpalatable
E07c Harmful to livestock. Keep livestock out of treated areas/away from treated water for at least xx days/weeks and until foliage of any poisonous weeds such as ragwort has died and become unpalatable
E08a Do not feed treated straw or haulm to livestock within xx days/weeks of spraying
E08b Do not use on grassland if the crop is to be used as animal feed or bedding
E09 Do not use straw or haulm from treated crops as animal feed or bedding for at least xx days after last application
E10a Dangerous to game, wild birds and animals
E10b Harmful to game, wild birds and animals
E11 Paraquat can be harmful to hares; spray stubbles early in the day

E12a High risk to bees
E12b Extremely dangerous to bees
E12c Dangerous to bees
E12d Harmful to bees
E12e Do not apply to crops in flower or to those in which bees are actively foraging. Do not apply when flowering weeds are present
E12f Do not apply to crops in flower, or to those in which bees are actively foraging, except as directed on [crop]. Do not apply when flowering weeds are present
E12g Apply away from bees
E13a Extremely dangerous to fish or other aquatic life. Do not contaminate surface waters or ditches with chemical or used container
E13b Dangerous to fish or other aquatic life. Do not contaminate surface waters or ditches with chemical or used container
E13c Harmful to fish or other aquatic life. Do not contaminate surface waters or ditches with chemical or used container
E13d Apply away from fish
E14a Extremely dangerous to aquatic higher plants. Do not contaminate surface waters or ditches with chemical or used container
E14b Dangerous to aquatic higher plants. Do not contaminate surface waters or ditches with chemical or used container
E15a Do not contaminate surface waters or ditches with chemical or used container
E15b Do not contaminate water with product or its container. Do not clean application equipment near surface water. Avoid contamination via drains from farmyards or roads
E16a Do not allow direct spray from horizontal boom sprayers to fall within 5 m of the top of the bank of a static or flowing waterbody, unless a Local Environment Risk Assessment for Pesticides (LERAP) permits a narrower buffer zone, or within 1 m of the top of a ditch which is dry at the time of application. Aim spray away from water
E16b Do not allow direct spray from hand-held sprayers to fall within 1 m of the top of the bank of a static or flowing waterbody. Aim spray away from water
E16c Do not allow direct spray from horizontal boom sprayers to fall within 5 m of the top of the bank of a static or flowing waterbody, or within 1m of the top of a ditch which is dry at the time of application. Aim spray away from water. This product is not eligible for buffer zone reduction under the LERAP horizontal boom sprayers scheme.
E16d Do not allow direct spray from hand-held sprayers to fall within 1 m of the top of the bank of a static or flowing waterbody. Aim spray away from water. This product is not eligible for buffer zone reduction under the LERAP horizontal boom sprayers scheme scheme.
E16e Do not allow direct spray from horizontal boom sprayers to fall within 5 m of the top of the bank of a static or flowing water body or within 1 m from the top of any ditch which is dry at the time of application. Spray from hand held sprayers must not in any case be allowed to fall within 1 m of the top of the bank of a static or flowing water body. Always direct spray away from water. The LERAP scheme does not extend to adjuvants. This product is therefore not eligible for a reduced buffer zone under the LERAP scheme.
E16f Do not allow direct spray/granule applications from vehicle mounted/drawn hydraulic sprayers/applicators to fall within 6 m of surface waters or ditches/Do not allow direct spray/granule applications from hand-held sprayers/applicators to fall within 2 m of surface waters or ditches. Direct spray/applications away from water
E17a Do not allow direct spray from broadcast air-assisted sprayers to fall within xx m of surface waters or ditches. Direct spray away from water
E17b Do not allow direct spray from broadcast air-assisted sprayers to fall within xx m of the top of the bank of a static or flowing waterbody, unless a Local Environmental Risk Assessment for Pesticides (LERAP) permits a narrower buffer zone, or within 5 m of the top of a ditch which is dry at the time of application. Aim spray away from water
E18 Do not spray from the air within 250 m horizontal distance of surface waters or ditches
E19a Do not dump surplus herbicide in water or ditch bottoms
E19b Do not empty into drains
E20 Prevent any surface run-off from entering storm drains
E21 Do not use treated water for irrigation purposes within xx days/weeks of treatment
E22a High risk to non-target insects or other arthropods. Do not spray within 6 m of the field boundary
E22b Risk to certain non-target insects or other arthropods. For advice on risk management and use in Integrated Pest Management (IPM) see directions for use
E22c Risk to non-target insects or other arthropods

E23 Avoid damage by drift onto susceptible crops or water courses
E34 Do not re-use container for any purpose/Do not re-use container for any other purpose
E35 Do not burn this container
E36a Do not rinse out container (returnable containers only)
E36b Do not open or rinse out container (returnable containers only)
E37 Do not use with any pesticide which is to be applied in or near water
E38 Use appropriate containment to avoid environmental contamination

Consumer protection

C01 Do not use on food crops
C02a Do not harvest for human or animal consumption for at least xx days/weeks after last application
C02b Do not remove from store for sale or processing for at least 21 days after application
C04 Do not apply to surfaces on which food/feed is stored, prepared or eaten
C05 Remove/cover all foodstuffs before application
C06 Remove exposed milk before application
C07 Collect eggs before application
C08 Protect food preparing equipment and eating utensils from contamination during application
C09 Cover water storage tanks before application
C10 Protect exposed water/feed/milk machinery/milk containers from contamination
C11 Remove all pets/livestock/fish tanks before treatment/spraying
C12 Ventilate treated areas thoroughly when smoke has cleared/Ventilate treated rooms thoroughly before occupying

Storage and disposal

D01 Keep out of reach of children
D02 Keep away from food, drink and animal feeding-stuffs
D03 Store away from seeds, fertilizers, fungicides and insecticides
D04 Store well away from corms, bulbs, tubers and seeds
D05 Protect from frost
D06a Store away from heat
D06b Do not store near heat or open flame
D06c Do not store in direct sunlight
D06d Do not store above 30/35 C
D07 Store under cool, dry conditions
D08 Store in a safe, dry, frost-free place designated as an agrochemical store
D09a Keep in original container, tightly closed, in a safe place
D09b Keep in original container, tightly closed, in a safe place, under lock and key
D09c Store unused sachets in a safe place. Do not store half-used sachets
D10a Wash out container thoroughly and dispose of safely
D10b Wash out container thoroughly, empty washings into spray tank and dispose of safely
D10c Rinse container thoroughly by using an integrated pressure rinsing device or manually rinsing three times. Add washings to sprayer at time of filling and dispose of container safely
D11a Empty container completely and dispose of safely/Dispose of used generator safely
D11b Empty container completely and dispose of it in the specified manner
D12a This material (and its container) must be disposed of in a safe way
D12b This material and its container must be disposed of as hazardous waste
D13 Treat used container as if it contained pesticide
D14 Return empty container as instructed by supplier (returnable containers only)
D15 Store container in purpose built chemical store until returned to supplier for refilling (returnable containers only)

Treated Seed

S01 Do not handle treated seed unnecessarily
S02 Do not use treated seed as food or feed
S03 Keep treated seed secure from people, domestic stock/pets and wildlife at all times during storage and use
S04a Bury or remove spillages
S04b Harmful to birds/game and wildlife. Treated seed should not be left on the soil surface. Bury or remove spillages

S04c Dangerous to birds/game and wildlife. Treated seed should not be left on the soil surface. Bury or remove spillages

S04d To protect birds/wild animals, treated seed should not be left on the soil surface. Bury or remove spillages

S05 Do not reuse sacks or containers that have been used for treated seed for food or feed

S06a Wash hands and exposed skin before meals and after work

S06b Wash hands and exposed skin after cleaning and re-calibrating equipment

S07 Do not apply treated seed from the air

S08 Treated seed should not be broadcast

Vertebrate/Rodent control products

V01a Prevent access to baits/powder by children, birds and other animals, particularly cats, dogs, pigs and poultry

V01b Prevent access to bait/gel/dust by children, birds and non-target animals, particularly dogs, cats, pigs, poultry

V02 Do not prepare/use/lay baits/dust/spray where food/feed/water could become contaminated

V03a Remove all remains of bait, tracking powder or bait containers after use and burn or bury

V03b Remove all remains of bait and bait containers/exposed dust/after treatment (except where used in sewers) and dispose of safely (e.g. burn/bury). Do not dispose of in refuse sacks or on open rubbish tips.

V04a Search for and burn or bury all rodent bodies. Do not place in refuse bins or on rubbish tips

V04b Search for rodent bodies (except where used in sewers) and dispose of safely (e.g. burn/bury). Do not dispose of in refuse sacks or on open rubbish tips

V04c Dispose of safely any rodent bodies and remains of bait and bait containers that are recovered after treatment (e.g. burn/bury). Do not dispose of in refuse sacks or on open rubbish tips

V05 Use bait containers clearly marked POISON at all surface baiting points

Medical advice

M01 This product contains an anticholinesterase organophosphorus compound. DO NOT USE if under medical advice NOT to work with such compounds

M02 This product contains an anticholinesterase carbamate compound. DO NOT USE if under medical advice NOT to work with such compounds

M03 If you feel unwell, seek medical advice immediately (show the label where possible)

M04a In case of accident or if you feel unwell, seek medical advice immediately (show the label where possible)

M04b In case of accident by inhalation, remove casualty to fresh air and keep at rest

M05a If swallowed, seek medical advice immediately and show this container or label

M05b If swallowed, do not induce vomiting: seek medical advice immediately and show this container or label

M05c If swallowed induce vomiting if not already occurring and take patient to hospital immediately

M06 This product contains an anticholinesterase carbamoyl triazole compound. DO NOT USE if under medical advice NOT to work with such compounds

Appendix 5
Key to Abbreviations and Acronyms

The abbreviations of formulation types in the following list are used in Section 2 (Pesticide Profiles) and are derived from the *Catalogue of Pesticide Formulation Types and International Coding System* (CropLife International Technical Monograph 2, 5th edn, March 2002).

1 Formulation Types

AB	Grain bait
AE	Aerosol generator
AL	Other liquids to be applied undiluted
AP	Any other powder
BB	Block bait
BR	Briquette
CB	Bait concentrate
CF	Capsule suspension for seed treatment
CG	Encapsulated granule (controlled release)
CL	Contact liquid or gel (for direct application)
CP	Contact powder (for direct application)
CR	Crystals
CS	Capsule suspension
DC	Dispersible concentrate
DP	Dustable powder
DS	Powder for dry seed treatment
EC	Emulsifiable concentrate
EG	Emulsifiable granule
EO	Water in oil emulsion
ES	Emulsion for seed treatment
EW	Oil in water emulsion
FG	Fine granules
FP	Smoke cartridge
FS	Flowable concentrate for seed treatment
FT	Smoke tablet
FU	Smoke generator
FW	Smoke pellets
GA	Gas
GB	Granular bait
GE	Gas-generating product
GG	Macrogranules
GL	Emulsifiable gel
GP	Flo-dust (for pneumatic application)
GR	Granules
GS	Grease
GW	Water soluble gel
HN	Hot fogging concentrate
KK	Combi-pack (solid/liquid)
KL	Combi-pack (liquid/liquid)
KN	Cold-fogging concentrate
KP	Combi-pack (solid/solid)
LA	Lacquer
LI	Liquid, unspecified
LS	Solution for seed treatment
ME	Microemulsion
MG	Microgranules
OD	Oil dispersion
OL	Oil miscible liquid
PA	Paste
PC	Gel or paste concentrate
PS	Seed coated with a pesticide

PT	Pellet
RB	Ready-to-use bait
RH	Ready-to-use spray in hand-operated sprayer
SA	Sand
SC	Suspension concentrate (= flowable)
SE	Suspo-emulsion
SG	Water soluble granules
SL	Soluble concentrate
SP	Water soluble powder
SS	Water soluble powder for seed treatment
ST	Water soluble tablet
SU	Ultra low-volume suspension
TB	Tablets
TC	Technical material
TP	Tracking powder
UL	Ultra low-volume liquid
VP	Vapour releasing product
WB	Water soluble bags
WG	Water dispersible granules
WP	Wettable powder
WS	Water dispersible powder for slurry treatment of seed
WT	Water dispersible tablet
XX	Other formulations
ZZ	Not Applicable

2 Other Abbreviations and Acronyms

ACP	Advisory Committee on Pesticides
ACTS	Advisory Committee on Toxic Substances
ADAS	Agricultural Development and Advisory Service
a.i.	active ingredient
AIC	Agriculture Industries Confederation
BBPA	British Beer and Pub Association
CDA	Controlled droplet application
CPA	Crop Protection Association
cm	centimetre(s)
COPR	Control of Pesticides Regulations 1986
COSHH	Control of Substances Hazardous to Health Regulations
d	day(s)
Defra	Department for Environment, Food and Rural Affairs
EA	Environment Agency
EBDC	ethylene-bis-dithiocarbamate fungicide
FEPA	Food and Environment Protection Act 1985
g	gram(s)
GCPF	Global Crop Protection Federation (now CropLife International)
GS	growth stage (unless in formulation column)
h	hour(s)
ha	hectare(s)
HBN	hydroxybenzonitrile herbicide
HI	harvest interval
HSE	Health and Safety Executive
ICM	integrated crop management
IPM	integrated pest management
kg	kilogram(s)
l	litre(s)
LERAP	Local Environmental Risk Assessments for Pesticides
m	metre(s)
MAFF	Ministry of Agriculture, Fisheries and Food (now Defra)
MBC	methyl benzimidazole carbamate fungicide
MEL	maximum exposure limit
min	minute(s)
mm	millimetre(s)
MRL	maximum residue level

mth	month(s)
NA	Notice of Approval
NFU	National Farmers' Union
OES	Occupational Exposure Standard
OLA	off-label approval
PPE	personal protective equipment
PPPR	Plant Protection Products Regulations
PSD	Pesticides Safety Directorate
SOLA	specific off-label approval
ULV	ultra-low volume
VI	Voluntary Initiative
w/v	weight/volume
w/w	weight/weight
wk	week(s)
yr	year(s)

Appendix 6
Definitions

The descriptions used in this *Guide* for the crops or situations in which products are approved for use are those used on the approved product labels. These are now standardised in a Crop Hierarchy published by the Pesticides Safety Directorate in which definitions are given. To assist users of this *Guide* the definitions of some of the terminology where misunderstandings can occur are reproduced below.

Rotational grass: Short-term grass crops grown on land that is likely to be growing different crops in future years (*e.g. short-term intensively managed leys for one to three years that may include clover*)

Permanent grassland: Grazed areas that are intended to be permanent in nature (*e.g. permanent pasture and moorland that can be grazed*).

Ornamental Plant Production: All ornamental plants that are grown for sale or are produced for replanting into their final growing position (*e.g. flowers, house plants, nursery stock, bulbs grown in containers or in the ground*).

Managed Amenity Turf: Areas of frequently mown, intensively managed, turf that is not intended to flower and set seed. It includes areas that may be for intensive public use (*e.g. all types of sports turf*).

Amenity Grassland: Areas of semi-natural or planted grassland subject to minimal management. It includes areas that may be accessed by the public (*e.g. railway and motorway embankments, airfields, and grassland nature reserves*). These areas may be managed for their botanical interest, and the relevant authority should be contacted before using pesticides in such locations.

Amenity Vegetation: Areas of semi-natural or ornamental vegetation, including trees, or bare soil around ornamental plants, or soil intended for ornamental planting. It includes areas to which the public have access. It does NOT include hedgerows around arable fields.

Natural surfaces not intended to bear vegetation: Areas of soil or natural outcroppings of rock that are not intended to bear vegetation, including areas such as sterile strips around fields. It may include areas to which the public have access. It does not include the land between rows of crops.

Hard surfaces: Man-made impermeable surfaces that are not intended to bear vegetation (*e.g. pavements, tennis courts, industrial areas, railway ballast*).

Permeable surfaces overlying soil: Any man-made permeable surface (excluding railway ballast) such as gravel that overlies soil and is not intended to bear vegetation

Green Cover on Land Temporarily Removed from Production: Includes fields covered by natural regeneration or by a planted green cover crop that will not be harvested (*e.g. green cover on setaside*). It does NOT include industrial crops.

Forest Nursery: Areas where young trees are raised outside for subsequent forest planting.

Forest: Groups of trees being grown in their final positions. Covers all woodland grown for whatever objective, including commercial timber production, amenity and recreation, conservation and landscaping, ancient traditional coppice and farm forestry, and trees from natural regeneration, colonisation or coppicing. Also includes restocking of established woodlands and new planting on both improved and unimproved land.

Farm forestry: Groups of trees established on arable land or improved grassland including those planted for short rotation coppicing. It includes mature hedgerows around arable fields.

Indoors (for rodenticide use): Situations where the bait is placed within a building or other enclosed structure, and where the target is living or feeding predominantly within that building or structure.

Herbs: Reference to Herbs or Protected Herbs when used in Section 2 may include any or all of the following. The particular label or SOLA Notice will indicate which species are included in the approval.

Agastache spp.
Angelica
Applemint
Balm
Basil
Bay
Borage (except when grown for oilseed)
Camomile
Caraway
Catnip
Chervil
Clary
Clary sage
Coriander
Curry plant
Dill
Dragonhead
English chamomile
Fennel
Fenugreek
Feverfew
French lavender
Gingermint
Hyssop
Korean mint
Land cress
Lavandin
Lavender
Lemon balm
Lemon peppermint
Lemon thyme
Lemon verbena
Lovage
Marigold
Marjoram
Mint
Mother of thyme
Nasturtium
Nettle
Oregano
Origanum heracleoticum
Parsley root
Peppermint
Pineapplemint
Rocket
Rosemary
Rue
Sage
Salad burnet
Savory
Sorrel
Spearmint
Spike lavender
Tarragon
Thyme
Thymus camphoratus
Violet
Winter savory
Woodruff

Herbs for Medicinal Uses: Reference to Herbs for Medicinal Uses when used in Section 2 may include any or all of the following. The particular label or SOLA Notice will indicate which species are included in the approval

Black cohosh
Burdock
Dandelion
Echinacea
Ginseng
Goldenseal
Liquorice
Nettle
Valerian

Appendix 7
References

The information given in *The UK Pesticide Guide* provides some of the answers needed to assess health risks, including the hazard classification and the level of operator protection required. However, the *Guide* cannot provide all the details needed for a complete hazard assessment, which must be based on the product label itself and, where necessary, the Health and Safety Data Sheet and other official literature.

Detailed guidance on how to comply with the Regulations is available from several sources.

Pesticides: Code of Practice

Code of Practice for Using Plant Protection Products, 2006 (Defra publication PB 11090)

Known as the 'Green Code', this revised Code of Practice, which came into effect on 15 December 2005, replaced all previous editions of the *Code of Practice for the Safe Use of Pesticides on Farms and Holdings*. It also replaced the voluntary code of practice for the use of pesticides in amenity and industrial areas (the 'Orange Code'). The code explains how plant protection products can be used safely and so meet the legal requirements of the Control of Pesticides Regulations 1986 (COPR) (ISBN 0-110675-10-X), and the Control of Substances Hazardous to Health Regulations 2002 (COSHH) (ISBN 0-110429-19-2). The principal source of information for making a COSHH assessment is the approved product label. In most cases the label provides all the necessary information, but in certain circumstances other sources must be consulted, and these are listed in the Code of Practice.

The code, which applies to England and Wales only, is available in CD-ROM format and electronically from the PSD and Defra websites. A Welsh language version is available in printed and electronic formats. A Scottish version (approved by the Scottish Parliament) is being produced in printed and electronic formats. Northern Ireland will produce its own version of the code in due course.

Other Codes of Practice

Code of Practice for Suppliers of Pesticides to Agriculture, Horticulture and Forestry (the 'Yellow Code') (Defra Booklet PB 0091)

Code of Good Agricultural Practice for the Protection of Soil (Defra Booklet PB 0617)

Code of Good Agricultural Practice for the Protection of Water (Defra Booklet PB 0587)

Code of Good Agricultural Practice for the Protection of Air (Defra Booklet PB 0618)

Approved Code of Practice for the Control of Substances Hazardous to Health in Fumigation Operations. Health and Safety Commission (ISBN 0-717611-95-7)

Code of Best Practice: Safe use of Sulphuric Acid as an Agricultural Desiccant. National Association of Agricultural Contractors, 2002. (Also available at www.naac.co.uk/?Codes/acidcode.asp)

Safe Use of Pesticides for Non-agricultural Purposes. HSE Approved Code of Practice. HSE L21 (ISBN 0-717624-88-9)

Other Guidance and Practical Advice

HSE (by mail order from HSE Books – See Appendix 2)

COSHH – A brief guide to the Regulations 2003 (INDG136)

A Step by Step Guide to COSHH Assessment, 2004 (HSG97) (ISBN 0-717627-85-3)

Defra (from The Stationery Office – see Appendix 2)

Local Environment Risk Assessments for Pesticides (LERAP): Horizontal Boom Sprayers.

Local Environment Risk Assessments for Pesticides (LERAP): Broadcast Air-assisted Sprayers.

Crop Protection Association and the Voluntary Initiative (see Appendix 2)

Every Drop Counts: Keeping Water Clean

Best Practice Guides. A range of leaflets giving guidance on best practice when dealing with pesticides before, during and after application.

BCPC (British Crop Production Council – see Appendix 2)

The Pesticide Manual (14th edition) (ISBN 1-901396-14-2)

The e-Pesticide Manual PC CD-ROM (Version 4.0) (ISBN 978-1-901396-43-0); (Version 4.1 upgrade)

The Manual of Biocontrol Agents (3rd edition of *The BioPesticide Manual*) (ISBN 1-901396-35-5)

IdentiPest PC CD-ROM (ISBN 1-901396-05-3)

Garden Detective PC CD-ROM (ISBN 1-901396-32-0)

Small Scale Spraying (formerly *Hand-held & Amenity Sprayers Handbook*) (ISBN 1-901396-07-X)

Field Scale Spraying (formerly *Boom and Fruit Sprayers Handbook*) (ISBN 1-901396-08-8)

Using Pesticides (ISBN 1-901396-10-X)

Safety Equipment Handbook (ISBN 1-901396-06-1)

The Environment Agency (see Appendix 2)

Best Farming Practices: Profiting from a Good Environment

Use of Herbicides in or Near Water

SECTION 7
INDEX

Index of Proprietary Names of Products

The references are to entry numbers, not to pages. Adjuvant names are referred to as 'Adj' and are listed separately in Section 4

REFERENCES ARE TO ENTRY NUMBERS NOT PAGES

REFERENCES ARE TO ENTRY NUMBERS NOT PAGES

REFERENCES ARE TO ENTRY NUMBERS NOT PAGES

REFERENCES ARE TO ENTRY NUMBERS NOT PAGES

THE UK PESTICIDE GUIDE 2008

RE-ORDERS

☐ Please send me ____ more copies of *The UK Pesticide Guide 2008* at £39.50 each

Postage: single copy £3.50; orders up to £175, £7.00; orders over £175, no charge.

Name ____________________ Position ____________________

Institution ____________________ Department ____________________

Address ____

City ____________ Region ____________ Postcode ________ Country ____________

Tel ____________ Fax ____________ E-mail ____________

EU countries except UK – VAT No: ____________________

Payment (pre-payment is required):

☐ I enclose a cheque/draft for £________ payable to BCPE Ltd. Please send me a receipt.

☐ I wish to pay by credit card: ☐ Visa ☐ Mastercard ☐ Amex ☐ Switch

Please charge to my card £_______ and send me a receipt. Name of issuing bank ____________________

Card no. ☐☐☐☐ ☐☐☐☐ ☐☐☐☐ ☐☐☐☐

Expiry date ☐☐/☐☐ Security code ☐☐☐ Switch cards only: Start date ☐☐/☐☐ Issue no. ☐

Signature ________________________________ Date ____________

Name and address of cardholder if different from above: ____________________

____

BCPC

Please photocopy and return to:

BCPC Publications Sales, 7 Omni Business Centre, Omega Park, Alton, Hants GU34 2QD, UK

Tel: 01420 593 200, Fax: 01420 593 209, Email: publications@bcpc.org, Web: www.bcpc.org

FUTURE EDITIONS

☐ I wish to take out an annual order for ____ copies of each new edition of *The UK Pesticide Guide*

☐ Please send me advance price details for the 2009 edition of *The UK Pesticide Guide* when available

Order by phone: 01420 593 200 or online: www.bcpc.org/bookshop

Bookshop orders to:

www.cabi.org

Customer Services, CAB International, Nosworthy Way,
Wallingford, Oxfordshire OX10 8DE, UK
Tel: 01491 832111, Fax: 01491 829292,
Email: enquiries@cabi.org, Web: www.cabi.org

Bulk discount:

100+ copies	30%
50–99	25%
10–49	15%
List price £39.50	

Thank you for your order